IAVCEI Proceedings in Volcanology 3

P. Gasparini R. Scarpa K. Aki (Eds.)

Volcanic Seismology

With 234 Figures

Springer-Verlag
Berlin Heidelberg New York
London Paris Tokyo Hong Kong
Barcelona Budapest

Professor Paolo Gasparini
Dipartimento di Geofisica e Vulcanologia
Università di Napoli Federico II
Largo San Marcellino 10
I-80138 Napoli, Italy

Professor Roberto Scarpa
Dipartimento di Fisica
Università Degli Studi dell'Aquila
Via Vetoio
I-67010 Coppito (L'Aquila), Italy

Professor Keiiti Aki
Department of Geological Sciences
University of Southern California
3651 University Ave.
Los Angeles, CA 90089-1740, USA

ISBN-13: 978-3-642-77010-4 e-ISBN-13: 978-3-642-77008-1
DOI: 10.1007/978-3-642-77008-1

Library of Congress Cataloging-in-Publication Data. Volcanic seismology / P. Gasparini, R. Scarpa, K. Aki (eds.). p. cm. – (IAVCEI proceedings in volcanology; 3) Includes bibliographical references and index.
ISBN-13: 978-3-642-77010-4
1. Volcanism – Congresses.
2. Seismology – Congresses. I. Gasparini, Paolo. II. Scarpa, Roberto. III. Aki, Keiiti, 1930–. IV. Series.
QE521.V66 1992 551.2′–dc20 92-13939

Softcover reprint of the hardcover 1st edition 1992

Typesetting: Best-set Typesetters Ltd., Hong Kong
32/3145-5 4 3 2 1 0 – Printed on acid-free paper

Preface

For many centuries people living on volcanoes have known that the outset of seismic activity is often a forerunner of a volcanic eruption. This understanding allowed people living close to the sites of the Mt. Nuovo 1538 eruption at Campi Flegrei, Italy, and of the Mt. Usu 1663 eruption, in Hokkaido, Japan (to quote only two examples) to flee before the eruptions started. During the second half of the 19th century seismographs were installed on some volcanoes, and the link between seismic and eruptive activity started to be assessed on a firmer scientific basis. The first systematic observations of the correlations existing between seismic activity and volcanic eruptions were probably those carried out at Mt. Vesuvius by Luigi Palmieri in 1856. Palmieri was the Director of Osservatorio Vesuviano and built an electromagnetic seismograph with the aim of "making visible the smallest ground motions by recording them on paper and indicating direction, intensity and duration". He was able to show the relationship between earthquakes and the different phases of volcanic activity. He identified the harmonic tremor which he indicated was a precursor of volcanic activity: "the characteristic feature of the ground motions preceding eruption is its continuity ... (before the eruption of 1861) the electromagnetic seismograph began to show a continuous tremor".

The Palmieri seismograph was also utilized in Japan until 1883, when it was replaced by the new Gray-Milne seismographs, and, later, by the Omori instruments. Instrumental monitoring of seismic activity at the Hawaiian volcanoes started in the second decade of this century with the introduction of Omori and Imamura seismographs at Kilauea. In the following decades, the number of seismographs installed on volcanoes instruments increased slowly, but constantly.

By the mid-1970s seismographic arrays had already been deployed on many volcanoes. The consequent improvement of earthquake location precision and of sensitivity level, allowed us to observe seismic energy increases, over several weeks to hours before some large eruptions, and to acknowledge the intriguing pattern of volcanic tremors. During the last decade, advances in instrumentation (e.g., the use of broad-band and three-component digital seismographs), signal processing and theory of seismic wave radiation resulted in new ways of understanding complex volcanic earthquake sources. At the same time, the development of dense seismic networks and the development of powerful numerical methods for inverting travel-times and waveforms resulted in the use of tomographic techniques to investigate local 3-D structures, such as magmatic reservoirs. Use of these sophisticated techniques on volcanoes, emphasized their utmost complexity as they revealed the extreme heterogeneity existing at all scales.

Times were ripe, therefore, for volcano seismologists to meet and discuss new results and the consequent changes of perspective of the ever-lasting problems. The International Association of Volcanology and Chemistry of the Earth's Interior (IAVCEI) and the CNR National Group of Volcanology decided to sponsor a Workshop of Volcanic Seismology, which was held in Capri (Italy) in September, 1988. In view of what we have just written, the choice of this location, directly facing Mt. Vesuvius, had some historical merit.

This volume presents a collection of selected papers from the Capri Workshop. They are presented in four sections:

1. Seismic activity and other precursors of eruptions.
2. Modeling of volcanic earthquakes, tremors and long-period events.
3. Tomography and detection of magma chambers.
4. Seismic activity in resurgent calderas and geothermal fields.

The first section starts with a concise paper assessing how different precursors of eruptions can be treated statistically to evaluate the probability of eruptions. Earthquake and ground deformation data, on many volcanoes in different parts of the world, are then presented.

Each of the following sections begins with a paper giving a general view of our present knowledge on the subject, followed by more specific papers.

The high quality data collected on many volcanoes has stimulated theoretical modeling of volcanic earthquakes and of the various complex seismic sources on volcanoes. The discovery of earthquake families provided new clues to the study of inhomogeneous fracturing processes. It is now recognized that the dynamics of the gas phase and the magmatic fluids play an important role in the generation of seismic waves within volcanoes. The possible existence of non-double couple earthquakes is relevant for eruption prediction, as they can be generated by fluid injection in fractures. All of these problems, including recent observations on Kilauea, Izu-Oshima and Stromboli, are reviewed in the second section of the book.

The third section contains four papers dealing with the most recent advances of tomographic studies of volcanoes of Western United States, Kilauea and the geothermal area of Larderello. The structure of volcanoes is still very difficult to study, because of anisotropy, heterogeneity and strongly variable attenuation. The study of coda waves, therefore, as discussed in the last chapter of this section, gives an interesting contribution.

The last section deals mostly with the results of investigations carried out by Italian, North-American and Spanish groups in the Campi Flegrei resurgent caldera. They include both the results of seismic and ground deformation studies and theoretical models.

Unfortunately, in order to keep the size of the volume within acceptable limits, many articles concerning other methods of monitoring of active volcanoes (geochemical, electromagnetic, etc.), which were presented at the Workshop, have not been included.

The editors have greatly benefited from the excellent and painstaking work made by a team of referees, most of whom gave lively contributions to the discussions during the Capri Workshop. The editors wish to express their sincere appreciation to: Rolf Schick (Stuttgart Universität), Bernard Chouet, Dave P. Hill, H.M. Iyer, Jaert Kissling (U.S.G.S., Menlo Park, California), François Cornet, Jean Virieux, Aldo Zollo (Institute de Physique du Globe, Université Pierre et Marie Curie, Paris), John Dvorak (Hawaiian Volcano Observatory), Eduardo Del Pezzo (University of Catania), Giuseppe De Natale, Marcello Martini, Folco Pingue (Osservatorio Vesuviano, Napoli), Jack Foley (M.I.T., Boston), Enzo Faccioli (Politecnico di Milano), Lucia Civetta, Guiseppe Capaldi, Giovanni Orsi, Roberto Scandone, Filippo Vinale (Università di Napoli), Lian Sheng Gao (Seismological Brigade, Beijing), Hans U. Schmincke (Ruhr Universität, Bochum), M. Don Watts (Unocal, Santa Rosa).

We acknowledge all the sponsors of the Capri Workshop who allowed specialists from all over the world to convene and to discuss the topics presented in this book. Our grateful thanks to UNESCO, UNDRO, IAVCEI, CNR Group of Volcanology, University of Napoli, ENEA, ENEL, AGIP, AQUATER, SIGECOR.

Modern volcanoseismologists find a treasury of careful observation, clear classification and brightness of intuition in the papers which Takeshi Minakami published in the 1960s and 1970s. Those attending the Capri Workshop agreed to dedicate this book to his memory with their gratitude.

Paolo Gasparini
Roberto Scarpa
Keiiti Aki

Acknowledgements. We gratefully acknowledge the financial support of the Department of Physics of the University of Aquila, Italy, which subsidized the additional publication costs, thus making it possible to publish a volume longer than usual. 300 printed pages remain the standard for the series.

Contents

1 Introduction

2 Seismic Activity and Other Precursors of Eruptions

4 Tomography and Detection of Magma Chambers

5 Seismic Activity in Resurgent Calderas and Geothermal Fields

List of Contributors

Dr. R. AGUILERA
Escuela Superior de Ciencias
Universidad Autonoma
de Baja California
Apartado Postal 877
Ensenada Baja
California 22800, Mexico

Prof. KEIITI AKI
Dept. of Geological Sciences
University of Southern California
3651 University Ave.
Los Angeles, CA 90089-0740, USA

Dr. G.E. ALVARADO
Dept. de Geologia
Sección de Sismología e Ing.
Sísmica
Obs. Vulcanológico del Arenal
Instituto Costarricense
de Electricidad (ICE)
Apto. 10032,
San José, Costa Rica

Dr. W.B. AMBEH
Department of Earth Sciences
University of Leeds
Leeds, LS2 9JT, United Kingdom

Prof. R.C. ASTER
Institute of of Mining
and Technology
Department of Geoscience
New Mexico Tech
Socorro, NM 87801, USA

Dr. R.A. BARQUERO
Dept. de Geologia
Sección de Sismología e Ing.
Sísmica
Obs. Vulcanológico del Arenal
Instituto Costarricense
de Electricidad (ICE)
Apto. 10032
San José, Costa Rica

Dr. F. BATINI
ENEL – Ente Nazionale per
l'Energia Elettrica
56100 Pisa, Italy

Prof. P. BERNARD
Inst. de Physique du Globe
Laboratoire de Sismologie
4, Place Jussieu
75252 Paris Cedex 05, France

Prof. A. BOTTARI
Istituto Geofisico e Geodetico
Università di Messina
Via Osservatorio 4
98121 Messina, Italy

Dr. B. BRANDSDÓTTIR
Science Institut
Unversity of Iceland
Dunhaga 3
107 Reykjavik, Iceland

Dr. P. BRIOLE
Institut de Physique du Globe
Laboratoire de Sismologie
4, Place Jussieu
75252 Paris Cedex 05, France

Dr. D. CACCAMO
Istituto Geofisico e Geodetico
Università di Messina
Via Osservatorio 4
98121 Messina, Italy

Dr. B. CHOUET
U.S. Geological Survey
345 Middlefield Road
Mail Stop 977
California 94025, USA

Prof. Dr. M. COMO
Dipartimento di Ingegneria Civile
Edile
II Università Degli Studi di Roma
Via Orazio Raimondo
00173 Rome, Italy

Dr. A.M. CORREIG
DGDGP-GEOFISICA
Instituto d'Estudis Catalans
Facultad de Fisica, UB
Marti Franques 1
08028 Barcelona, Spain

Dr. F.H. CORNET
Institute de Physique du Globe
Paris
Laboratoire de Sismologie
4 Place Jussieu
75252 Paris Cedex 05, France

Dr. T. DAHM
Geophysik. Institut
Universität Karlsruhe
Hertzstr. 16
7500 Karlsruhe, FRG

Dr. P. B. DAWSON
U.S. Geological Survey
345 Middlefield Road
Mail Stop 977
Menlo Park, CA 94025, USA

Dr. G. DE NATALE
Osservatorio Vesuviano
80056 Ercolano (Napoli), Italy

Prof. E. DEL PEZZO
Dipartimento di Fisica
Università di Salerno
84081 Baronissi (Salerno), Italy

Dr. G. DEL RE
Dipartimento di Geofisica
e Vulcanología
Università di Napoli Federico II
Largo S. Marcellino 10
80138 Napoli, Italy

Dr. J.L. DIEZ
Museo Nacional de
Ciencias Naturales
CSIC
Dept. de Geología y Geofísica
José Gutiérrez Abascal 2
28006 Madrid, Spain

Dr. P. EINARSSON
Science Institut
University of Iceland
Dunhaga 3
107 Reykjavik, Iceland

Dr. J.M. ESPINDOLA
Instituto de Geofísica
Universidad Nacional Autónoma
de México
Apartado Postal 2681
Ensenada
Baja California 22800, Mexico

Dr. J.R. EVANS
U.S. Geological Survey
345 Middlefield Road MS 977
Mail Stop 977
Menlo Park, CA 94025, USA

Dr. A. FADELI
Physics Department
Gadjah Mada University
Yogyakarta, Indonesia

Prof J. D. FAIRHEAD
Department of Earth Sciences
University of Leeds
Leeds, LS2 9JT
United Kingdom

Dr. S. FALSAPERLA
Istituto Internazionale
di Vulcanologia
C.N.R.
Viale Regina Margherita 6
95123 Catania, Italy

Prof. Dr. S. A. FEDOTOV
Institute of Volcanology
Petropavlovsk-Kamchatsky
683006, USSR

Dr. V. FERRAZZINI
Dept. of Geological Sciences
University of Southern California
3651 University Ave.
Los Angeles, CA 90089-0740, USA

Dr. J. E. FOLEY
Earth Resources Laboratory
Dept. of Earth, Atmospheric
and Planetary Sciences
Massachusetts Institute
of Technology
Cambridge, MA 02139, USA

Dr. G. R. FOULGER
University of Durham
Dept. of Geological Sciences
Science Laboratories
South Road
Durham
DH1 3LE, United Kingdom

Dr. M. FURUMOTO
Faculty of Science
Dept. of Earth Sciences
Kanazawa University
1-1 Marunouchi
Kanazawa, Ishikawa 920, Japan

Dr. L. S. GAO
Department of Geological Sciences
University of Southern California
3651 University Ave.
Los Angeles, CA 90089-740, USA

Prof. P. GASPARINI
Dipartimento de Geofisica
e Vulcanologia
Università di Napoli Federico II
Largo S. Marcellino 10
80138 Napoli, Italy

Dr. R. GAULON
Institut de Physique du Globe
Laboratoire de Sismologie
4, Place Jussieu
75252 Paris Cedex 05, France

Dr. C. GODANO
Osservatorio Vesuviano
Via Manzoni 249
80129 Napoli, Italy

Dr. L. GONZALEZ
Instituto de Geofísica
Universidad National Autónoma
de México
Apartado Postal 2681
Ensenada
Baja California 22800, Mexico

Dr. V. I. GORELCHIK
Institute of Volcanology
Petropavlovsk-Kamchatsky
683006, USSR

Dr. A. GORINI
Osservatorio Vesuviano
Via Manzoni 249
80129 Napoli, Italy

Prof. G. GRANDORI
Politecnico di Milano
Piazza Leonardo da Vinci 32
20133 Milano, Italy

Dr. F. MARIA GUADAGNO
Dipartimento di Geofisica
e Vulcanologia
Università di Napoli Federico II
Largo S. Marcellino 10
80138 Napoli, Italy

Prof. E. GUAGENTI
Politecnico di Milano
Piazza Leonardo da Vinci 32
20133 Milano, Italy

Dr. C. GUTIERREZ
Instituto de Ingenieria
Universidad Nacional Autónoma
de Mexico
Cd. Universitaria
Mexico City 04510 D.F., Mexico

Dr. D.P. HILL
U.S. Geological Survey
345 Middlefield Road
Menlo Park, CA 94025, USA

Dr. G. IANNACCONE
Dipartimento di Geofisica
e Vulcanologia
Università di Napoli Federico II
Largo S. Marcellino 10
80138 Napoli, Italy

Dr. H. INOUE
Building Research Institute
Tsukuba
Ibaraki 305, Japan

Dr. H.M. IYER
U.S. Geological Survey
345 Middlefield Road MS 977
Menlo Park, CA 94025, USA

Dr. T. KUNITOMO
Dept. of Earth Sciences
Nagoya University
Chikusa
Nagoya 464, Japan

Dr. G. LA VOLPE
Dipartimento di Geofisica
e Vulcanologia
Università di Napoli Federico II
Largo S. Marcellino 10
80138 Napoli, Italy

Dr. M. LEMBO
Dipartimento di Ingegneria Civile
Edile
II Università Degli Studi di Roma
Via Orazio Raimondo
00173 Rome, Italy

Dr. R.E. LONG
University of Durham
Dept. of Geological Sciences
Science Laboratories
South Road
Durham
DH1 3LE, United Kingdom

Prof. R.P. MEYER
Department of Geology
and Geophysics
University of Wisconsin-Madison
Madison, WI, USA

Dr. M.L. MUÑOZ
E.U. Informatica
Universidad Politecnica
28006 Madrid, Spain

Dr. B. MARTINELLI
Swiss Seismological Service
Institute of Geophysics
Zürich, Switzerland

Dr. M. MARTINI
Osservatorio Vesuviano
Via Manzoni 249
80129 Napoli, Italy

Dr. F. Medina
Instituto de Geofísica
Universidad Nacional Autónoma de México
Apartado Postal 2681
Ensenada
Baja California 22800, Mexico

Dr. T. Matumoto
Geophysics Department
University of Texas
Austin, TX 77550, USA

Dr. A. Montalto
Istituto Internazionale di Vulcanologia
C.N.R.
Viale Regina Margherita 6
95123 Catania, Italy

Dr. G. Neri
Istituto Internazionale di Vulcanologia del C.N.R.
Viale Regina Margherita 6
95123 Catania, Italy

Dr. G. Nunnari
Istituto Internationale di Vulcanologia del C.N.R.
Piazza Roma 2
95123 Catania, Italy

Dr. C. Nunziata
Dipartimento di Geofisica e Vulcanologia
Università di Napoli Federico II
Largo S. Marcellino 10
80138 Napoli, Italy

Dr. R. Ortiz
Museo Nacional de Ciencias Naturales
CSIC
Dept. de Geología y Geofísica
José Gutiérrez Abascal 2
28006 Madrid, Spain

Dr. D. Patané
Istituto Internazionale di Vulcanologia del C.N.R.
Viale Regina Margherita 6
95123 Catania, Italy

Dr. F. Pingue
Osservatorio Vesuviano
Via Manzoni 249
80123 Napoli, Italy

Dr. E. Privitera
Istituto Internazionale di Vulcanologia del C.N.R.
Viale Regina Margherita 6
95123 Catania, Italy

Dr. G. Puglisi
Istituto Internazionale di Vulcanologia del C.N.R.
Piazza Roma 2
95123 Catania, Italy

Prof. A. Rapolla
Dipartimento di Geofisica e Vulcanologia
Università di Napoli Federico II
Largo S. Marcellino 10
80138 Napoli, Italy

Prof. G. R. Robson
Dept. of Geological Sciences
University of Durham
Durham
DH1 3LE, United Kingdom

Dr. J. C. Ruegg
Institut de Physique du Globe
Laboratoire de Sismologie
4, Place Jussieu
75252 Paris Cedex 05, France

Prof. R. Scarpa
Dipartimento di Fisica
Università Degli Studi dell'Aquila
Via Vetoio
67010 Coppito (L'Aquila), Italy

Prof. R. Schick
Institut f. Geophysik
Richard-Wagner-Str. 44
7000 Stuttgart 1, FRG

Dr. G.W. Stuart
Department of Earth Sciences
University of Leeds
Leeds, LS2 9JT, United Kingdom

Prof. M.N. Toksoz
Earth Resources Laboratory
Dept. of Earth, Atmospheric
and Planetary Sciences
Massachusetts Institute
of Technology
Cambridge, MA 02139, USA

Dr. K. Yamaoka
Regional Center of Earthquake
and Volcanoes
Nagoya University
Nagoya 464, Japan

Dr. N.A. Zharinov
Institute of Volcanology
Petropavlovsk-Kamchatsky
683006, USSR

Dr. A. Zollo
Dipartimento di Geofisica
e Vulcanologia
Università di Napoli Federico II
Largo S. Marcellino 10
80138 Napoli, Italy

1 Introduction

State of the Art in Volcanic Seismology

Keiiti Aki

Abstract

The ultimate goal of a volcanic seismologist is to predict a volcanic eruption using seismological observations.

Towards that goal, seismologists have worked on both conventional and unconventional approaches to the understanding of structure and processes inside a volcano. The structure inside a volcano is, probably, the most complex subject seismologists have ever encountered on Earth. It is extremely heterogeneous, anisotropic and absorptive with irregular interfaces and topography including cracks of all sizes and orientation. The source processes are also much more complicated than usual tectonic earthquakes, because they involve additional dynamics of gas and fluid magma in generating seismic signals, to be deciphered by seismologists.

In the present work, the author presents a personal view of various seismological methods currently used in studying structure and processes in volcanic areas, and recommends a unique design of experiment tailored to each volcanic area. With regard to the prediction of volcanic eruption, he advocates a combined monitoring of coda Q^{-1} and volcanic (harmonic) tremor. The study of volcanic tremor requires a new breed of seismologists who can deal with the dynamics of fluid and gas coupled to the seismic radiation. Non-linearity of dynamics, at the source of volcanic tremor, may be much stronger than that at an earthquake source. Older seismologists trained in linear solid dynamics, however, may still find a great deal of work to do in delineating the detailed internal structure of a volcano, and in accurately separating the source, path and site effects in the complicated seismograms observed in a volcanic area.

1 Introduction

The subject of volcanic seismology is not only the most beautiful and spectacular, but also the most difficult to study of all the subjects seismologists have encountered on Earth. This is because seismic sources in volcanoes involve dynamic motion of gas, fluid and solid, and propagation paths in volcanoes are usually extremely heterogeneous, anisotropic and absorptive, with irregular topographies and interfaces including cracks of all scales and orientation. Thus, volcanic seismology is the most challenging to seismologists requiring ingenuity in designing experiments and interpreting observations.

I have been involved in volcanic seismology since 1975 through participation in two major geothermal projects, namely, the Hot Dry Rock Geothermal Energy Development Project of the Los Alamos National Laboratory and the Magma Tap Project of Sandia National Laboratories. I worked with several colleagues in the collection, analysis and interpretation of seismic data from various geothermal areas including Fenton Hill Hot Dry Rock Project site,

Valles caldera, Kilauea Iki, Mt. St. Helens, Long Valley caldera, and most recently Puu Oo.

In the present paper, I shall discuss some of the basic problems encountered in practicing volcanic seismology, and try to identify key issues that need to be resolved for the future productive development of this most fascinating branch of seismology. As such, I am relying heavily on my own experiences, and apologize in advance, for omitting important contributions by others on the topics under discussion.

2 Conventional Versus Unconventional Approaches

Conventional approaches in seismology have been developed, primarily, for subsurface structures that may not deviate too far from a stack of homogeneous layers. From the beginning of my geothermal studies, I doubted that such approaches would be effective for delineating complex structures in geothermal areas, and my doubt has been intensified over the past decade or so.

For example, we do not very often find convincing cases for a successful reflection survey in a volcanic area, except in a few notable examples, such as, where the sedimentary cover acts as a cap for a vapor reservoir in Cerro Prieto, Baja California, a deep reflector in the Rio Grande rift, New Mexico (Brown et al. 1980), and possible magma chambers in the oceanic area (Morton and Sleep 1985).

Likewise, the standard refraction survey which is most effective in delineating the vertical structure of the Earth's crust does not seem to have produced results which strikingly reveal the internal structure of a volcano.

The tomographic method using teleseismic P time residuals, on the other hand, seems to have produced a much clearer picture of the interior of a volcano, probably because this method is specifically designed, to determine the lateral variation of seismic velocity, at various depths.

The use of teleseismic tomography has become almost a conventional approach in volcanic seismology, largely due to the effort of H.M. Iyer (1988) and his colleagues at the U.S. Geological Survey. This approach has contributed significantly to defining the shape and size of a low-velocity body or a high-velocity body (depending on the nature of surrounding rock) which may be related to the magma body.

This approach can be and should be improved by the inversion of full wave form data, using the solution of seismic wave scattering in 3-D structures, and a preliminary result on the Valles caldera by Roberts et al. (1987) appears to be promising.

Seismic tomography using local earthquake data should give a finer image of subsurface structures because of their shorter wavelengths than teleseismic P waves. A successful example is Thurber's (1984) detailed study of the structure of Kilauea, in which, the small low-velocity zone of the volcano's summit magma reservoir has been resolved within the otherwise high-velocity core of the volcano. Other examples can be found in a review paper by Thurber and Aki (1987).

A. Hirn and his colleagues at the University of Paris (Nercessian et al. 1984) improved the resolution of tomographic pictures by using Moho-reflected (PMP) waves generated by explosions which have higher frequencies than teleseismic P waves. The method worked well for a geothermal area in France, but not for the Long Valley caldera in California because of the complexity of PMP waves in the latter area. Thus, the PMP tomography has not yet succeeded in becoming "conventional". We must keep looking for a new "conventional" method that can apply to geothermnal areas in general.

In the absence of such a method, however, an alternative approach is to design a custom-made experiment tailored to a particular geothermal site. In our experiment with Kilauea Iki (Aki et al. 1978), a partially frozen lava lake in Hawaii, we found that a variety of unconventional methods was essential for determining its seismic properties. For example, the naturally occurring seismic events (caused by thermal stress due to cooling) above the magma reservoir, gave the most detailed information on the lateral extent of the reservoir. Furthermore, the dispersion of Love waves played an important role in defining the magma body, because we could not use high-frequency body waves which suffered strong attenuation and scattering in the severely cracked crust. Rayleigh waves also, could not exist because velocity was too low due to the presence of flat-layered magma. The SH waves from an earthquake outside Kilauea Iki provided important information on the S-wave velocity structure and the viscosity of magma body. The lesson we learned from this experience was that, both active and passive seismic experiments using many conventional and unconventional techniques, are needed for delineating a complex structure like a frozen lava lake.

There is another class of promising approaches in volcanic seismology which are unconventional and robust. For example, the frequency dependence of coda Q^{-1} measured from local earthquakes in geothermal areas such as Etna (Del Pezzo et al. 1987), Campi Fregrei (Castellano et al. 1984; De Natale et al. 1987), and Kilauea (Chouet 1976) showed higher and roughly constant values, for frequencies higher than a certain critical value, deviating from the trend for other tectonically active regions. Another example is the scaling law of seismic spectrum (Chouet et al. 1978; De Natale et al. 1987) which should reflect the unique character of faulting produced in freshly formed Earth's crust in volcanic regions.

Coda Q^{-1} was reported to change before and after a minor eruption at Mt. St. Helens by Fehler et al. (1988). Q^{-1} was higher before the eruption than after, probably because the increased pore pressure opened cracks before the eruption.

There is no doubt that the ultimate goal of a volcanic seismologist is to predict an eruption of a volcano using seismological observations. I believe that there are two promising approaches which should be combined for their maximum effect. One is the monitoring of coda Q^{-1} as mentioned above. The physical reality of temporal change in coda Q^{-1} in seismic regions and the understanding of its physical mechanism have made a significant advance recently (Jin and Aki 1989), and I feel that its promise for predicting volcanic eruption is even brighter. The other approach is monitoring the source process of volcanic (harmonic) tremor, which I now believe involves dynamics of both gas and fluid magma. In the following, I shall try to explain why I believe this is so, step by step.

3 Separation of Source, Path and Site Effects

When a seismologist looks at a seismogram, the first question he or she should ask is "Can the effects of source, propagation path and recording site on the seismogram be separated?" This is essentially an inverse problem, and can be non-unique if the available data are inadequate. When they are inadequate, the traditional approach has been to postulate a working hypothesis or doctrine by asserting that one of the effects dominates the others.

Let us consider volcanic (harmonic) tremor as the subject of our discussion. One of the earliest models of the tremor attributed its harmonic character to the propagation-path effect. According to Omer (1950), the shallow crust of Kilauea volcano is nothing but laminae composed of lava flows, ash beds and intrusives. The magma flow exerts a transverse force on one end of the layers and sets these partially free layers vibrating, in a way similar to bending a deck of playing cards clamped at one end. Thus, in Omer's model, the tremor is caused by a quasi-free vibration of solid layers forming the shallow part of the volcano.

Minakami (1974) also considered that the harmonic character of tremor is not due to the effect of a particular source mechanism but primarily due to the propagation path effect. According to him, the tremor is a swarm of b-type events, which show relatively low-frequency wave trains with a spindle-shaped envelope, because of the extreme shallowness of its focal depth.

A similar idea is shared by many seismologists, who know that seismograms similar to b-type earthquakes, are often observed for nonvolcanic sources, such as buried explosions or tectonic earthquakes. Here, the wave path is extremely heterogeneous and absorptive, as often found in a volcanic area.

Another school of seismologists, however, considered that the harmonic character of the tremor is due to the source effect, namely, the resonant vibration of a magma body. It is not easy to resolve this fundamental difference when the data are inadequate. For example, if only records at a single station are available, for tremors originating from the same source location, it is impossible to separate the source effect from the path effect.

Recently accumulated high-quality data, however, show unequivocal evidence for the source effect responsible for the harmonic character of tremor, at least for some cases. For example, the period of deep tremor observed in Hawaii is common to all the stations spread over a large area, (Koyanagi et al. 1987). Also, the period of tremor changes with time, during an episode, as in the case of an eruption of Sakurajima (Kamo et al. 1977). Ferrazzini et al. present another convincing case for the source effect, in the present volume.

The correct attitude to this issue seems to me that we need not argue which of the sources, path and site effects is responsible for the harmonic character of tremor, but to try to separate these effects in a given data set and determine the contribution of each effect case by case. We need an experiment designed to enable a unique separation of these effects. The theory of seismic inverse problems may be useful for designing an optimal experiment.

3.1 Modeling Tremor Sources

Ramberg's (1970) experimental study of magma transport using centrifuges showed that while a finger convection (blob) is the preferred mode of mass transport when the viscosity contrast is relatively low between magma and surrounding media, a flow through fractures is the prefered mode when the contrast is very high. For basaltic volcanoes, we often observe spectacular fissure eruptions obviously indicating magma transport through cracks. Even for andesitic volcanoes with a central crater in a cone-shaped body, Nakamura (1977) explained the distribution of sites of flank eruptions by hypothesizing that the vent beneath the central crator may be formed as an intersection of vertical fractures oriented in various directions. Further, sills and dikes are common geologic features of many volcanic regions. These are some of the reasons why I prefer a fluid-filled crack as the seismic model of tremor source.

A fluid-filled crack model was suggested first by Sassa (1935). He found that, the observed pattern of motion surrounding a line of craters of Aso volcano, during an active period in 1933, could be explained by an opening-closing mode of a crack, coincident with the crater alignment. Recent development in modeling tremor sources by a fluid-filled crack have been summarized by Chouet et al. (1987), and also discussed by Chouet, and Ferrazzini et al. in the present volume. One of their findings from the application of a fluid-filled crack model to observed spectra of tremor is that the narrowness of observed spectral peak requires a sharp impedance contrast between magma and surrounding solid, which is only made possible by the presence of bubbles in the magma.

The role of vesiculation or degassing in magmatic dynamics has been emphasized recently by various authors. Eaton et al. (1987) made an unusually complete study of magmatic dynamics during the 1959 eruption of Kilauea volcano, combining continuous tilt observation, by the Press-Ewing seismograph, with an accurate log of lava eruption into Kilauea Iki lava lake. The observed episodes of magma flow repeated many times into and draining back from Kilauea Iki, were explained in terms of large-scale vesiculation of lava in the conduit, effectively pumping lava from the reservoir up into the lake. During early episodes, volcanic tremor was observed in the eruption phase but not in the drain-back phase. In later episodes, however, tremor was observed also in the drain-back phase. In general, the amplitude of tremor correlated well with the eruption rate, and also with the tilt near the summit closely related to the amount of magma in the summit reservoir.

In addition to this large-scale cycling of magma between the summit reservoir and Kilauea Iki, Eaton et al. (1987) observed a periodic change in tilt, called "tilt storm" with a period of about 1 hour for several days before the beginning of the eruption. This could be an indication of a smaller-scale cycling involving degassing and magma flow. It is very interesting to note here that, a similar periodic tilt change accompanied by tremor activity, was observed during the recent eruption of Oshima (Fukuyama 1988).

As described by Ferrazzini et al. in this volume, we made a direct visual confirmation of the role of degassing in generating the "gas-piston events" named by Swanson et al. (1979). We found that, the gas-piston event associated with degassing to Puu Oo's crater, shows a very similar spectral signature as the long-period event and continuous tremor (which is essentially a swarm of long-period events) occurring under the summit. We feel, therefore, that degassing also plays a major role in generating long-period events and tremors.

Thus, degassing may be playing important role not only in sustaining the tremor by creating a high impedance contrast but also in generating it.

4 Conclusion

Volcanoes and geothermal areas are the most complex subject for seismologists encountered on Earth, and therefore conventional techniques developed for average earth's structure may not work with them. For each case, we need an ingeneous design of custom-made experiments for a particular volcano or geothermal site to gain a clear look at its internal structure and working.

With regard to the prediction of volcanic eruption based on seismological data, there are two promising approaches which should be combined for their maximum effect. One is the monitoring of coda Q^{-1}, and the other is the in-depth study of volcanic (harmonic) tremor. Recently accumulated evidence supports the view that the harmonic character of the tremor is due to the source effect, which probably involves the dynamics of both gas and fluid magma, key elements of volcanic eruption.

A complete study of volcanic tremor requires a new breed of seismologists who can deal with dynamics of gas and fluid coupled to the seismic radiation observable through a seismograph. Non-linearity of magmatic dynamics at a source of volcanic tremor is probably much stronger than that at an earthquake source.

Older seismologists trained in linear solid dynamics, however, may still find a great deal of work in delineating the detailed internal structure of a volcano, and in separating accurately the source, path and site effects in complex seismograms observed in a volcanic area.

Acknowledgement. I would like to dedicate this paper to Dr. George Kolstad of the Office of Basic Energy Sciences, Department of Energy, Washington, D.C., USA, who has consistently supported my research on volcanic seismology, from its beginning. Almost all of my ideas expressed in this work were born from the research works supported through his office.

This work was supported by the Department of Energy under grant DE-FG03-85ER13336.

References

Aki K, Chouet B, Fehler M, Zandt G, Koyanagi R, Colp J, Hay R (1978) Seismic properties of a shallow magma reservoir in Kilauea Iki by active and passive experiments. J Geophys Res 83:2273–2282

Brown LD, Chapin CE, Sanford AR, Kaufman S, Oliver J (1980) Deep structure of the Rio Grande rift from seismic reflection profiling. J Geophys Res 85:4773–4800
Castellano M, Del Pezzo E, de Natale G, Zollo A (1984) Seismic coda and turbidity coefficient at Campi Flegrei Volcanic area: Preliminary results. Bull Volcanol 47:219–224
Chouet B (1976) Source, scattering and attenuation effects on high frequency seismic waves. Thesis, Massachusetts Inst Tech, Massachusetts
Chouet B, Aki K, Tsujiura M (1978) Regional variation of the scaling law of earthquake source spectra. Bull Seismol Soc Am 68:49–80
Chouet B, Koyanagi RY, Aki K (1987) Origin of volcanic tremor in Hawaii, Part II. Theory and discussion. USGS Prof Pap 1350:1259–1280
De Natale G, Iannaccone G, Martini M, Zollo A (1987) Seismic sources and attenuation properties at the Campi Flegrei Volcanic area. In: Okal EA (ed) Advances in volcanic seismology. Pageoph 125, 6:883–918
Del Pezzo E, Gresta S, Patane G, Patane D, Scarcella G (1987) Attenuation of short period seismic waves at Etna as compared to other volcanic areas. In: Okal EA (ed) Advances in volcanic seismology. Pageoph 125, 6:1039–1050
Eaton JP, Richter DH, Krivoy HL (1987) Cycling of magma between the summit reservoir and Kilauea Iki lava lake during the 1959 eruption of Kilauea volcano. USGS Prof Pap 1350:1307–1336
Fehler M (1983) Observation of volcanic tremor at Mt. St. Helens volcano. J Geophys Res 88:3476–3484
Fehler M, Roberts P, Fairbanks T (1988) A temporal change in coda wave attenuation observed during an eruption of Mt. St. Helens. J Geophys Res 93:4367–4373
Fukuyama E (1988) Relationship between saw-teeth-shaped tilt change observed at Izu-Oshima volcano and its volcanic activities. Res Notes 77, Nat Res Cent Disaster Prev Sci Tech Agency JPN pp 1–14
Iyer HM (1988) Seismological detection and delineation of Magma chambers beneath intraplate volcanic centers in western USA. In: King CY, Scarpa R (eds) Modeling of volcanic processes. Vieweg, Braunschweig Wiesbaden, pp 1–56
Jin A, Aki K (1989) Spatial and temporal correlation between coda Q^{-1} and seismicity and its physical mechanism. J Geophys Res 94:14 041–14 099
Kamo K, Furuzawa T, Akamatsu J (1977) The nature of some volcanic micro-tremors at the Sakurajima volcano. Bull Volcanol Soc Jpn 22:41–58
Koyanagi RY, Chouet B, Aki K (1987) Origin of volcanic tremors in Hawaii, Part I. Data from the Hawaiian Volcano Observatory 1969–1985. USGS Prof Pap 1350:1221–1258
Minakami T (1974) Seismology of volcanoes in Japan. In: Civetta L, Gasparini P, Luongo G, Rapolla A (eds) Physical volcanology. Elsevier, Amsterdam, pp 1–27
Morton JL, Sleep NH (1985) Seismic reflections from a Lau Basin magma chamber in geology of offshore resources of Pacific island arcs – Tonga region. In: School DW, Vallier T (eds) Circum-Pacific Council for energy and mineral Resources. Earth Sci Ser 2. Am Assoc Pet Geol, p 441, Tulsa, OK
Nakamura K (1977) Volcanoes as possible indicators of tectonic stress orientation – principle and proposal. J Volcanol Geotherm Res 2:1–16
Nercessian A, Hirn A, Tarantola A (1984) Three-dimensional seismic transmission prospecting of the Mont Dore volcano, France. Geophys J R Astr Soc 17:307–315
Omer GC (1950) Volcanic tremor. Bull Seismol Soc Am 40:175–194
Ramberg H (1970) Model studies in relation to instrusion of plutonic bodies. In: Newall G, Rast N (eds) Mechanism of igneous intrusion. Gallery Press, Liverpool, pp 261–272
Roberts PM, Aki K, Fehler M (1987) A study of the 2-D seismic structure of the Valles Caldera, New Mexico using discrete wavenumber modelling of observed teleseismic P-waves. EOS Trans AGU 68:44
Sassa K (1935) Volcanic micro-tremors and eruption-earthquakes. Mem Coll Sci Univ Kyoto Ser A, 18:255–293
Swanson DA, Duffield W, Jackson D, Peterson D (1979) Chronological narrative of the 1969–71 Mauna Ulu eruption, Kilauea Volcano, Hawaii. USGS Prof Pap 1056:55

Thurber CH (1984) Seismic detection of the summit magma complex of Kilauea Volcano, Hawaii. Science 223:165–167
Thurber CH, Aki K (1987) Three-dimensional seismic imaging. Ann Rev Earth Planet Sci 15:115–139

2 Seismic Activity and Other Precursors of Eruptions

Short-Term Precursors of Natural Disastrous Events: A Statistical Assessment

G. GRANDORI and E. GUAGENTI

Abstract

This work refers to alarm systems based on the combination of different short-term precursors F_1, F_2, . . . Fn, related to the probability that a disastrous event E may occur soon. Combining two or more precursors means that alarm A is issued only if two or more, out of the monitored precursors, overlap.

The theory of independent precursors is first summarized. Then the influence of non-independence is discussed. For the sake of simplicity, the discussion is limited to a double-precursor system.

Non-independence of false alarms in the precursors and non-independence of successful alarms play quite different roles. Non-independence of false alarms has unfavourable influence on the effectiveness of the doubleprecursor system, while non-independence of successful alarms always leads to an increase in this effectiveness.

1 Introduction

Emergency measures, aimed at the prevention of some of the most dangerous effects of natural disastrous events (such as volcanic eruptions or severe earthquakes), can be taken, based on the observation of short-term precursors.

Peculiar characteristics of short-term precursors are: firstly, the short lifespan Δt of the warning and, secondly, a conditional probability of the disastrous event E (in the time interval Δt following the occurrence of the precursor) that is much larger than the background probability P(E).

To set up an alarm system based on short-term precursors, inevitably means that besides successful alarms, there will also be a certain number of false alarms. Moreover, (and this is obviously an important aspect of the problem), there will be a certain percentage of unpredicted events E. The following two quantities describe some characteristics of an alarm system that may prove useful in decision making.

1. The conditional probability P(E/A), that the event E will occur in the time interval Δt, following the alarm A. This probability is a measure of the hazard; its importance, therefore, needs not be stressed.
2. A quantity to provide a concise evaluation of the effectiveness of the alarm system, allowing for both false and missed alarms, and that could be defined as follows:

$$U = \frac{s}{f + m}, \tag{1}$$

where s is the expected number of successes (alarms followed by an event E within Δt) over a long period of time in which the system is actually operating; f and m are the expected numbers of false and missed alarms over the same period of time.

The effectiveness U, so defined, is substantially the ratio between the expected number of favourable cases and the expected number of unfavourable cases.

Expression (1) may be suitably modified whenever social, economic or psychological factors make it seem preferable to attribute different relative weights to the successes, the false alarms and the missed alarms.

The conditional probability P(E/A) has to be compared with the hazard threshold that is considered high enough to justify the proclamation of the alarm. The effectiveness U, describes the consequences, in the long range, of the systematic use of the considered alarm system.

For a single precursor F, we define "probability of false alarm" the quantity $p = P(\bar{E}/F)$ and "probability of missed alarm" the quantity $q = P(\bar{F}/E)$.

A precursor will be classified here as "weak" either if both p and q are very high, or if at least one of them is very high. Should a "strong" precursor be available (i.e., with both p and q sufficiently small) the prediction problem would be satisfactorily solved by this single precursor. However, in many cases, only weak precursors are available, in the sense that they show a high percentage of false alarms. Should a precursor of this type be used systematically as an alarm system, both the quantities P(E/A) and U would assume discouraging values, even if the gain P(E/A)/P(E) proves to be rather high.

As a possible means of reducing the expected number of false alarms, the combination of two or more precursors was suggested. Combining two or more precursors means that alarm A is issued only if two or more of the monitored precursors overlap.

The theory of multiprecursor systems has been developed quite extensively over the last few years in as much as concerns earthquakes precursors (Utsu 1979; Aki 1981; Grandori and Grandori 1983; G. Grandori et al. 1984; Jones 1985; Kagan and Knopoff 1987; Grandori and Grandori 1988; G. Grandori et al. 1988). In the present paper, some of the results obtained so far are summarized, with particular attention to aspects that may be of general interest. Moreover, a few new results are presented with regard to the fact that precursors may or may not be independent of each other and the relevant consequences.

For the sake of simplicity, the presentation is limited to the case of a double-precursor system in stationary conditions, i.e., in particular, with P(E), p and q constant in time.

Section 2 considers two independent precursors, F_1 and F_2. It should be made clear that the independence of the two *precursors* does not mean that the two *events*, F_1 and F_2, are statistically independent. This would be contradictory, to the fact that both are related to event E. We say that the two precursors are independent if events F_1 and F_2 are correlated to each other *only* through event

Table 1. Comparison between a single precursor and a double precursor system

	Single			Pair		
Type	p	q	U	$p^{(2)}$	$q^{(2)}$	$U^{(2)}$
a	0.96	0.96	0.02	0.125	0.998	0.002
b	0.4	0.96	0.04	0.0001	0.998	0.002
c	0.96	0.4	0.04	0.125	0.64	0.52

E, with no other direct correlation. In the general case of n precursors, such a conditional independence of events $F_1, F_2, \ldots,$ Fn exists if, and only if, (Utsu 1979: Grandori and Grandori 1983):

$$P\left(\bigcap_{i=1}^{h} F_i/E\right) = \prod_{i=1}^{h} P(F_i/E), \tag{2}$$

$$P\left(\bigcap_{i=1}^{h} F_i/E\right) = \prod_{i=1}^{h} P(F_i/E). \tag{3}$$

As will be seen, the effectiveness of an alarm system based on a pair of independent precursors does actually depend on a number of factors and may indeed be smaller than the effectiveness of a system based on the use of only one of the component precursors. The conditions under which a double-precursor system could be more effective than each component precursor will be analyzed.

Section 3 considers double precursor systems based on non-independent precursors. It will be shown that, if the non-independence refers to false precursors, the quality of the alarm system is lower than in the case of independence; on the contrary, a non-independence regarding successful precursors improves the quality of the alarm system.

Section 4 refers to the difficulties that are encountered in the collection of the statistical data necessary for a reliable definition of p and q of each precursor and for a judgement on the independence of two precursors. Some indications of possible ways of overcoming such difficulties are presented.

2 Is a Double Precursor System Always More Effective than Each Precursor?

The following discussion is derived from Grandori G. et al. (1988). The examples of Table 1 show what happens when changing over from a single precursor to a pair of independent precursors. Three types of weak precursors are considered (on the left-hand side of the Table): type *a* with both p and q very high, type *b* with q very high and p relatively low, and type *c* with q relatively low and p very high. The right-hand side of the Table shows the characteristics of alarm systems based on a pair of independent precursors, both of type *a*, or both of type *b*, or both of type *c*, respectively. As the probability of false alarm $p^2 = P(\bar{E}/F_1 \cap F_2)$ of the double precursor systems depends on P(E), a value P(E) = 0.00025 has been assumed. In stationary conditions, this value corresponds to a mean interoccurrence time of events E equal to 22 years, with an alarm lifespan $\Delta t = 2$ days.

The probability of false alarms in the double-precursor system is always extremely low compared to the single precursor. As an obvious counterbalance, the probability of missed alarm $q^2 = p(\bar{F}_1 \cap \bar{F}_2/E)$ is larger than that of the single precursor. In particular, in case *a* and *b* E would be predicted only twice in a 1000 events. The influence of this fact on the effectiveness of the double-precursor system is such that U^2 is 10 times smaller than U in case *a* and 20 times smaller in case *b*.

On the contrary, two precursors of type *c* provide a much more effective alarm system than the single precursor. In other words, an alarm system based on a pair of independent precursors, even if the probability of false alarm of each single precursor is very high, seems to have very interesting characteristics, provided that the probability of missed alarm of each precursor is relatively low.

3 Is Non-Independence Always Unfavourable?

In order to discuss the influence of non-independence, let us again consider, for the sake of simplicity, a pair of precursors with the same characteristics: $p_1 = p_2 = p$, $q_1 = q_2 = q$. In this case, Eqs. (2) and (3) can be rewritten as follows:

$$P(F_2/F_1 \cap E) = P(F_2/E), \tag{4}$$

$$P(F_2/F_1 \cap \bar{E}) = P(F_2/\bar{E}). \tag{5}$$

It is important to note that Eq. (4) examines successful precursors, while Eq. (5) examines false precursors. The quantity:

$$C_s = P(F_2/F_1 \cap E) \tag{6}$$

may be assumed as a "correlation index" between successful alarms of the two precursors, and the quantity:

$$C_f = P(F_2/F_1 \cap \bar{E}) \tag{7}$$

as a "correlation index" between false alarms. If the two precursors are not independent (and positively correlated) we have in general:

$$P(F_2/E) < C_s \leqslant 1; \quad P(F_2/\bar{E}) < C_f \leqslant 1.$$

Observe now that:

$$1 - p^2 = P(E/F_1 \cap F_2) = \frac{P(E \cap F_1 \cap F_2)}{P(F_1 \cap F_2)},$$

and hence, for the total probabilities theorem:

$$1 - p^2 = \left[1 + \frac{P(F_1 \cap F_2 \cap \bar{E})}{P(F_1 \cap F_2 \cap E)}\right]^{-1}.$$

Then, taking into account the definitions (6), (7), it is easy to derive the following expression of p:

$$p^2 = \left[1 + \frac{C_s}{C_f}\frac{1-p}{p}\right]^{-1}. \qquad (8)$$

It is also easy to prove that:

$$q^2 = 1 - C_s(1 - q). \qquad (9)$$

It can be seen that an increase of C_f leads to an increase of p^2 (unfavourable effect), while an increase of C_s leads to a decrease of both p^2 and q^2 (favourable effect). In other words: if the non-independence regards false alarms, the quality of the system is worse than in the case of independence; however, a non-independence regarding successful alarms improve the quality of the system.

Figure 1 shows the quantitative influence of C_s and C_f on p^2 and on U^2 for two cases. As can be seen, it is crucial that the correlation between false alarms is very small: even a slight degree of correlation leads to large values of p^2 and to small values of U^2.

On the contrary, large values of C_s are always welcome. The favourable influence of C_s on p^2 is actually not large; however, in the field of interest (C_f very small) the favourable influence of C_s on U^2 may be appreciable.

In the general case in which $p_1 \neq p_2$, $q_1 \neq q_2$, the correlation indexes:

$$C_{s,\,21} = P(F_2/F_1 \cap E) \text{ and } C_{s,\,12} = P(F_1/F_2 \cap \bar{E})$$

are different from each other. The same happens for:

$$C_{f,\,21} = P(F_2/F_1 \cap \bar{E}) \text{ and } C_{f,\,12} = P(F_1/F_2 \cap \bar{E}).$$

The analysis can be carried out in an analogous way. In particular, Eq. (8) becomes:

$$p^2 = \left[1 + \frac{C_{s,\,21}}{C_{p,\,21}}\frac{1-p_1}{p_1}\right]^{-1} = \left[1 + \frac{C_{s,\,12}}{C_{f,\,12}}\frac{1-p_2}{p_2}\right]^{-1}.$$

4 Gathering Statistical Data

As we have seen, only weak precursors of type *c* (p very high, but q relatively low) can lead to a good double precursor system. Moreover, it is of crucial importance that the correlation index C_f be very small.

Observe now that (again for the example $p_1 = p_2$) the statistical definition of C_f is:

$$C_f = \frac{K_{12}}{K_1},$$

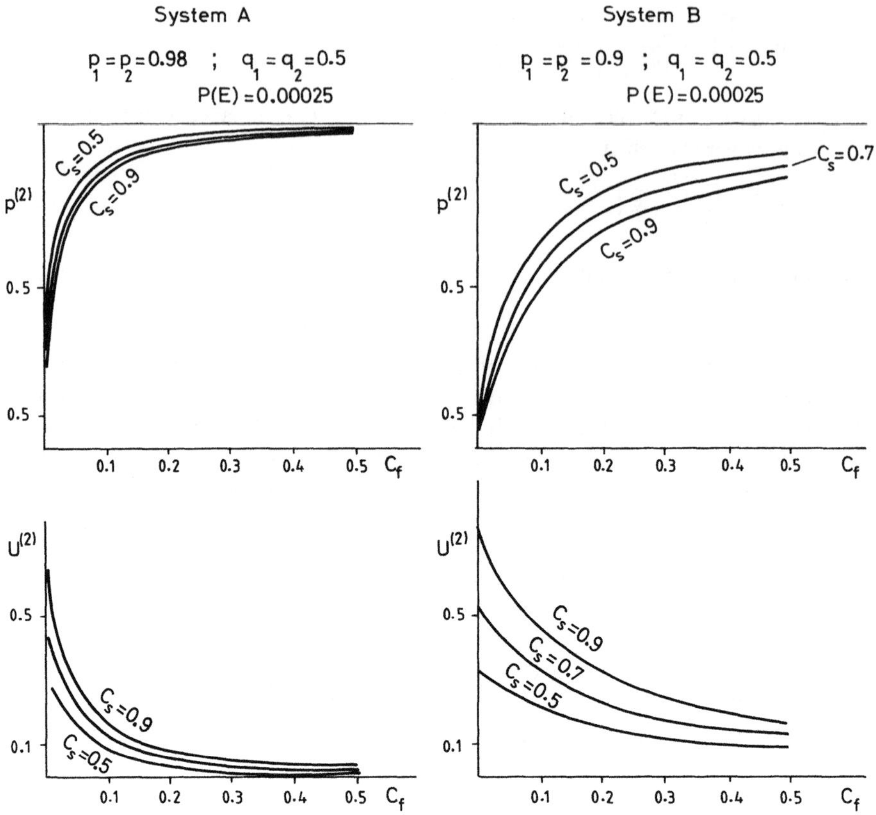

Fig. 1. Influence of correlation indices C_f and C_s on p^2 and on U^2 for two different double precursor systems

where K_1 is the number of false F_1 over a given time period, and K_{12} is the number of overlapping false F_1 and false F_2 over the same time period. In the case of precursors of type *c*, the number of false precursors in the gap between two events E is, by definition, rather large. As a consequence, a period of observation lasting one or two interarrival times of events E may be sufficient for a reliable evaluation of C_f.

A statistical evaluation of C_s would require a much longer time period. In fact, successful precursors occur (when they occur) immediately before events E. As a consequence, a reliable statistical evaluation would involve monitoring the precursors during a considerable number of interarrival times of events E. However, as we have been able to note, fortune is on our side: any value of C_s (even if unknown) larger than its minimum, would not greatly influence the conditional probability $P(E/F_1 \cap F_2)$ and would simply improve the effectiveness of the system, compared with the same quantities evaluated in the hypothesis of independence between successful precursors.

In stands that a reliable statistic evaluation of p and q of each precursor would entail very long observation periods. A gleam of hope is offered by the fact that a precursory phenomenon may have the same characteristics p and q in many regions or sites [this has been ascertained in the seismic precursor consisting of potential foreshocks, see Grandori G. et al. (1988)]. In this case, a shorter time window could be compensated for by a large space window.

References

Aki K (1981) A probabilistic synthesis of precursory phenomena. In: Simpson DN, Richard PG (eds) Earthquake prediction, an international review. Am Geophys Union, Washington, DC, pp 566–574

Grandori E, Grandori G (1983) Some probabilistic aspects of earthquake prediction. Meccanica J Ital Assoc Theor Appl Mech 18:246–253

Grandori E, Grandori G (1988) Multiprecursor alarm systems. Int Symp Earthq Countermeasures, Beijing, China

Grandori G, Guagenti E, Perotti F (1984) Some observations on the probabilistic interpretation of short-term earthquake precursors. Earthq Eng Struct Dyn 12:749–760

Grandori G, Grandori E, Perotti F (1988) Alarm systems based on a pair of short-term earthquake precursors. Bull Seismol Soc Am 78:1538–1549

Jones L (1985) Foreshocks and time-dependent earthquake hazard assessment in southern California. Bull Seismol Soc Am 75:1669–1679

Kagan YY, Knopoff L (1987) Statistical short-term earthquake prediction. Science 236:1563–1567

Utsu T (1979) Calculation of the probability of success of an earthquake prediction (in the case of Izu-Oshima-Kinkai earthquake of 1979). Rep Coordinating Comm Earthq Prediction 21:1964–1966 (in Japanese)

Deformation, Earthquakes and Mechanism of Activity of Klyuchevskoy Volcano

S.A. Fedotov, V.I. Gorelchik, and N.A. Zharinov

Abstract

Klyuchevskoy is a gigantic, approximately 7000 year old basaltic volcano 4750 m in height. In this paper, a model of its activity, feeding channels and magma chambers is presented; it was developed from observations of summit and flank eruptions, magma flow rate, vertical ground deformation and location and migration of earthquake hypocentres and theoretical calculations.

Variations in the depth, (Z_o) of the effective source of magmatic pressure and of the centre of seismic energy release, ($\bar{H}_E$) are traced. Relations have been established between Z_o, vertical deformation (U_z) and expected magma accumulation in the feeding system (ω). Volume increments in the feeding system of the volcano δV have been calculated. Correlations were found between Z_o and $\bar{H}_E$ (correlation coefficient $r = 0.63$), on one hand, U_z and $\omega (r = 0.66)$, and ω and δV ($r = 0.62$) on the other, which support the validity of the model.

From 1979 to 1987 the volume increment δV changed by a factor of 3–5, the position of its centre migrated in depth from 3 km to 15 km below sea level, whereas the lava level remained constant in the summit crater and, consequently, pressure in the feeding channel was constant as well. The maximum increase of δV occurred at a depth of 15 km before the large flank eruption of 1983.

This model predicts that the basaltic magma rises into the crust from a large system of magma chambers located in the crust-mantle transition layer. It ascends through the crust at depths of 30–0 km in a vertical channel. Here, the feeding conduit is surrounded by a plastic shell about 3 km in diameter, around which numerous earthquakes occur. Most of them occur within the column-like zone 5–7 km in diameter. Dikes and sills are common here and frequently intrude from the vertical channel, the longest dikes reaching 20 km. No large peripheral or crustal magma chambers of the volcano were detected.

1 Introduction

Klyuchevskoy is one of the most active and powerful basaltic volcanoes in the Kuril-Kamchatka arc and belongs to the Klyuchevskoy volcanic group where the Bezymianny and Plosky Tolbachik active volcanoes are located.

Klyuchevskoy is 7000–8000 years old, about 4750 m in height with an average magma flow rate of 60×10^6 t/year. The depth of the Wadati-Zavaritsky-Benioff zone beneath the volcano is 160 km; erupted products are calc-alkaline alumina basalts; magnesian basalts and basaltic andesites are also present.

Detailed seismological studies over almost the last two decades (1971–1987) have allowed location of earthquake foci in the crust beneath Klyuchevskoy and correlation of their temporal-spatial relations to eruptions (Gorelchik 1985; Gorelchik et al. 1987; Gorelchik et al. 1988; Fedotov et al. 1988).

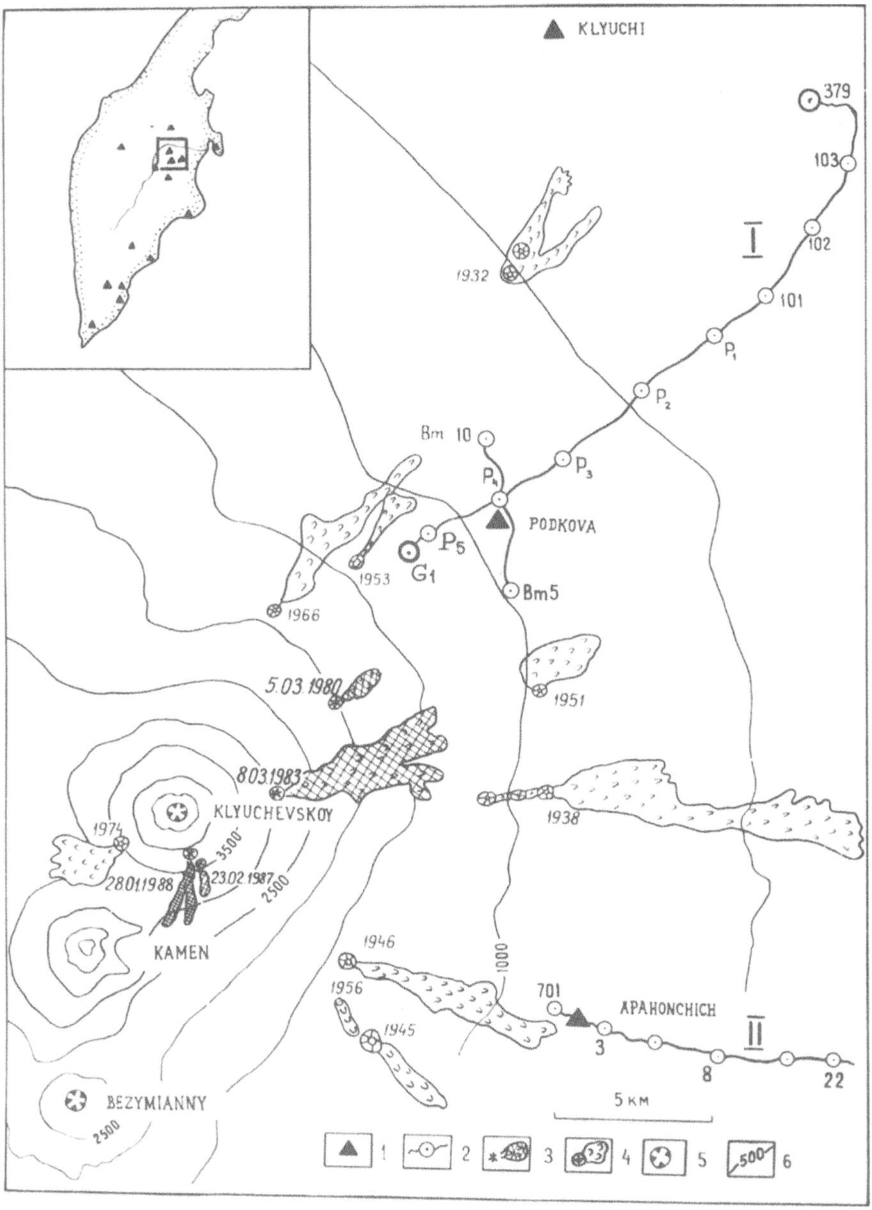

Fig. 1. Sketch map of the Klyuchevskoy volcanic area. *1* Seismic stations; *2* bench marks of levelling profiles; *3* lava flows from 1980–1988 flank eruptions; *4* lava flows from 1932–1974 flank eruptions; *5* summit craters; *6* contour lines

Data from geodetic measurements, which have been conducted systematically since 1978, allow us to obtain quantitative characteristics of relative vertical and horizontal displacements in the vicinity of the volcano and thus to study their relations to the activity of Klyuchevskoy (Zharinov and Eremeev 1985).

Figure 1 shows the region of Klyuchevskoy together with locations of levelling profiles and permanent seismic stations.

The present study discusses the data on eruptions, deformation and seismicity of Klyuchevskoy from 1978 to 1987.

2 Review of the Activity of Klyuchevskoy from 1978 to 1988

Recent activity is characterized by alternation of summit eruptions of different duration and intensity with flank eruptions. Thirteen flank eruptions occurred from 1932 to 1988. In the 20th century, paroxysmal summit eruptions occurred in 1926 and 1945 (Piyp 1956). The volume of erupted loose products was about 1 km^3. Slight and moderate eruptions at the summit crater (interparoxysmal) are more frequent. Detailed analysis of Klyuchevskoy activity from 1932 to 1986 is given in Fedotov et al. (1987).

The Klyuchevskoy volcano was in an active explosive-effusive state for more than 30% of the time over the period 1978–1987. Within a 10 year span two explosive-effusive summit eruptions occurred (in 1978 and 1984–1987) during which lava poured from the summit crater down the slopes of the volcano. In addition, four flank eruptions took place (March 1980, March–June 1983, February–March 1987 and January–September 1988).

The power of summit eruptions in 1984–1987 exceeded all previous eruptions since 1972. The highest explosive-effusive activities were observed during June–July 1984, November 1984, November–December 1985, December 1986–March 1987. During November–early December 1985 a great lava outpouring occurred from the summit crater and was accompanied by ash-gas outbursts to a height of 2.5–4 km above the crater. As a result of interaction of magma with glaciers covering the flanks of the volcano, violent phreatic explosions and lahars originated. Thermal power of the eruptive column reached 100×10^6 kW, and about 100 tons/s of ash was ejected.

During December–March 1987, the ash column occasionally rose to a height of 10 km above the crater, thermal power reached 400×10^6 kW, and ejection of pyroclastics increased to as much as 400 t/s. Thus, thermal power of the eruptive column and ejection of pyroclastics increased with time during summit eruptions of 1984–1987.

From 1978 to 1987 the summit crater gradually filled with lava and pyroclastics. From aerial survey data, the crater floor rose by more than 100 m over this particular period. This process was slowed down only temporarily by subsidence of the cinder cone in the summit crater in 1982. A new large cinder cone was formed at this subsidence area during April 1984–early 1987. The summit crater was buried with lava and scoria completely, for the first time, during the last half of the century.

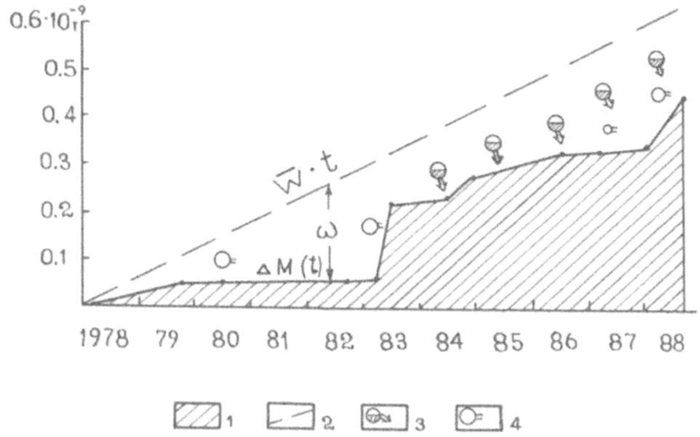

Fig. 2. Mass of products erupted at Klyuchevskoy during 1978–1987 *1* Mass of actually erupted materials, ΔM(t) (tons), excluding ashes; *2* erupted material accumulation at the average rate $\overline{W}$; *3* summit eruptions with outpouring of lava downslope; *4* flank eruptions

Information on the depth and structure of the Klyuchevskoy summit crater has been available since 1936 (see e.g., Kirsanov and Markov 1979). A lava flow occurred from the Klyuchevskoy summit crater during the paroxysmal eruption in January 1945. The volume of ash ejected was close to 0.3 km^3. After this eruption, the Klyuchevskoy summit crater deepened to 500 m (Piyp 1956) and for several decades the crater kept its cup-shaped configuration, 600–700 m in diameter, the height of the walls 100–200 m. From 1964–1968, the summit crater deepened owing to explosions. In 1968, the floor of the summit crater was located 300 m lower than the base of the cinder cone which was observed in the crater later in December 1984.

Three flank eruptions at Klyuchevskoy were effusive (in 1980, 1987 and 1988) and one was predominantly effusive (1983). Preparation and onset of flank eruptions in 1980, 1987 and 1988 proceeded simultaneously with the effusive-explosive activity of the central eruption when the level of the magma column was low and a 100 m deep pit-crater formed at the top. Despite individual differences, these four flank eruptions all showed a successive increase in absolute elevations from 500 to 4000 m (the elevation of the 1932–1988 flank eruptions) and a late January–early March onset (Fedotov et al. 1987).

1983 and 1988 flank eruptions and 1984–1987 summit eruptions were particularly large, the volumes of erupted products amounting to 0.066 km^3, 0.034 km^3 and 0.046 km^3, respectively. Nonetheless, during 1978–1988 a net accumulation of magma took place in the feeding system of the volcano. This is demonstrated in Fig. 2 with the cumulative plot of erupted products in 1987–1988 compared with the average magma flow rate at Klyuchevskoy. The average magma flow rate was determined from the volume and age of the volcano and also through the known magma flow rate during 1932–1987. These values turned out to be close: 60×10^6 and 51×10^6 t, respectively (Fedotov et al. 1987).

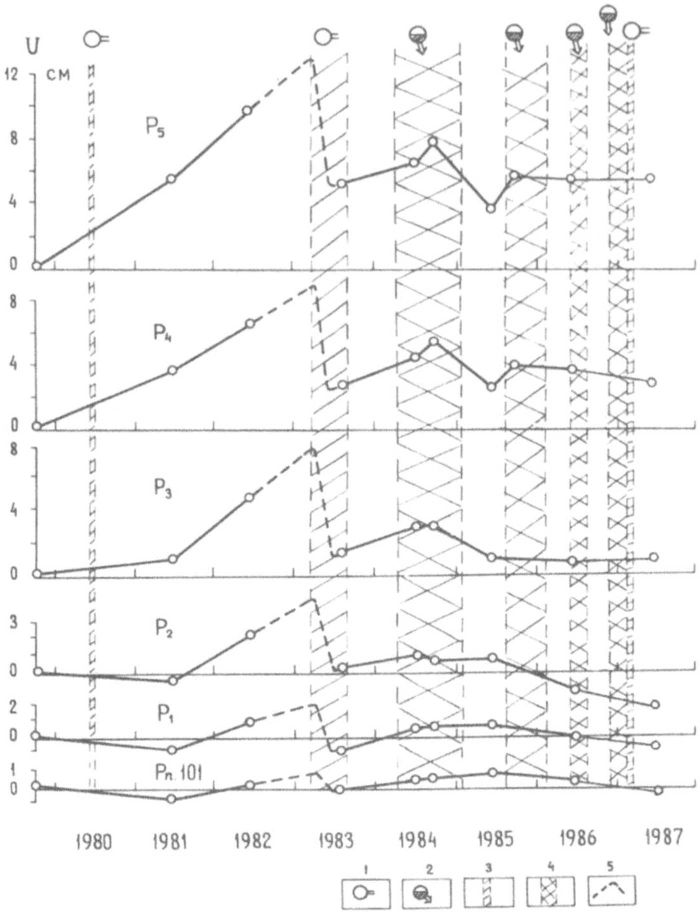

Fig. 3. Vertical displacements U_Z (cm) along the radial profile I at Klyuchevskoy during 1979–1987. *1* Flank eruptions; *2* summit eruptions with outpouring of lava downslope; *3* duration of flank eruptions; *4* duration of summit eruptions; *5* probable shape of graph $U_Z(t)$ (cm) in 1982–1983

2.1 Vertical Displacements During 1978–1987 and Their Interpretation

To study vertical displacements of the Earth's surface at Klyuchevskoy, two long levelling profiles were established on its north-eastern slope (Zharinov and Eremeev 1985; Fedotov et al. 1988). In a region 500 km^2 in area, a geodetic polygon was built with distance-angle and triangulation networks and tiltmeter stations. The present study discusses the results of repeated levelling along the radial profile I, on which 10 repeated levellings were carried out.

We would point out that establishment of levelling profiles and other geodetic surveys at Klyuchevskoy are complicated by complex topography, a forest at the northern foot of the volcano (up to 700–800 m above sea level), great differences in height and long winters when the thickness of snow reaches as much as 1.5–2 m.

Setting of bench marks on profile I was accomplished in 1978 along the radius to the volcano with azimuth of 45°. The closest bench mark is located at a distance of 12 km from the central crater, and the farthest reference bench mark is located at a distance of 31 km from the crater (see Fig. 1). Radial profile II was made in June 1985 on the south-eastern slope of the volcano. This profile has a length of 20 km and is sub-perpendicular to profile I. Levelling of profile I was carried out once a year in 1979 (initial measurements), 1981, 1982, 1983, 1986 and 1987 and twice a year in 1984 and 1985. Levelling of profile II was first made from June 25 to July 15, 1985. In September 1985 part of the profile was repeated and a third survey was made in 1986. In 1987 profile II was extended towards the volcano and initial levelling was made.

Geodetic measurements have revealed considerable variations in vertical displacements. After 1979 an uplift of the volcano slope occurred reaching a maximum some time within 6–9 months before the onset of flank eruption in March 1983 (Fig. 3). Levelling data obtained from 1978 to 1987 (see Fig. 3) were used to estimate the depth of the effective pressure source (or centre) under the volcano (Fedotov et al. 1988). This estimation is purely approximate since only relative surface displacements were determined. The results of levelling along profile II were not used because of their small number.[1] An expanding sphere was taken as a model for the pressure source (Mogi 1958; Yokoyama 1971; Dietrich and Decker 1975) and we assumed that the pressure source (sources) was located under the summit of the volcano. In this case, the depth and dimensions of the source can be estimated from vertical surface displacements. They can be expressed as (Efimov and Demin 1979):

$$U_z(r) = -2\Delta P R_o^3 \frac{1 - v^2}{C} \cdot \frac{Z_o}{(r^2 + Z_o^2)^{3/2}}, \qquad (1)$$

where U_z = the vertical displacements; C = Young's modulus; V = Poisson's ratio; ΔP = the excess pressure in the source (chamber); R_o = the radius of the pressure source (spherical space); Z_o = the depth of the effective pressure source (the centre of the sphere).

The axial-symmetric pressure source is the simplest model. Vertical deformations triggered by this source should diminish monotonously with increasing distance from the crater. The real situation, however, may be more complex as different magma chambers may exist at different depths and the feeding channel may offshoot. In a favourable case, the volume of these additional pressure centres is small, in comparison with the principal effective pressure source, and the vertical displacements correspond satisfactorily to Eq. (1).

Graphs of relative displacements illustrate the character of deformation diminishing with increasing distance from the volcano (Fig. 4). Ratios of varia-

[1] Vertical displacements were considered only along one profile on a flank of the volcano. This produced uncertainties in the location of the source. But later data, provided by the additional triangulation net, have shown that deformations related to a central source prevailed.

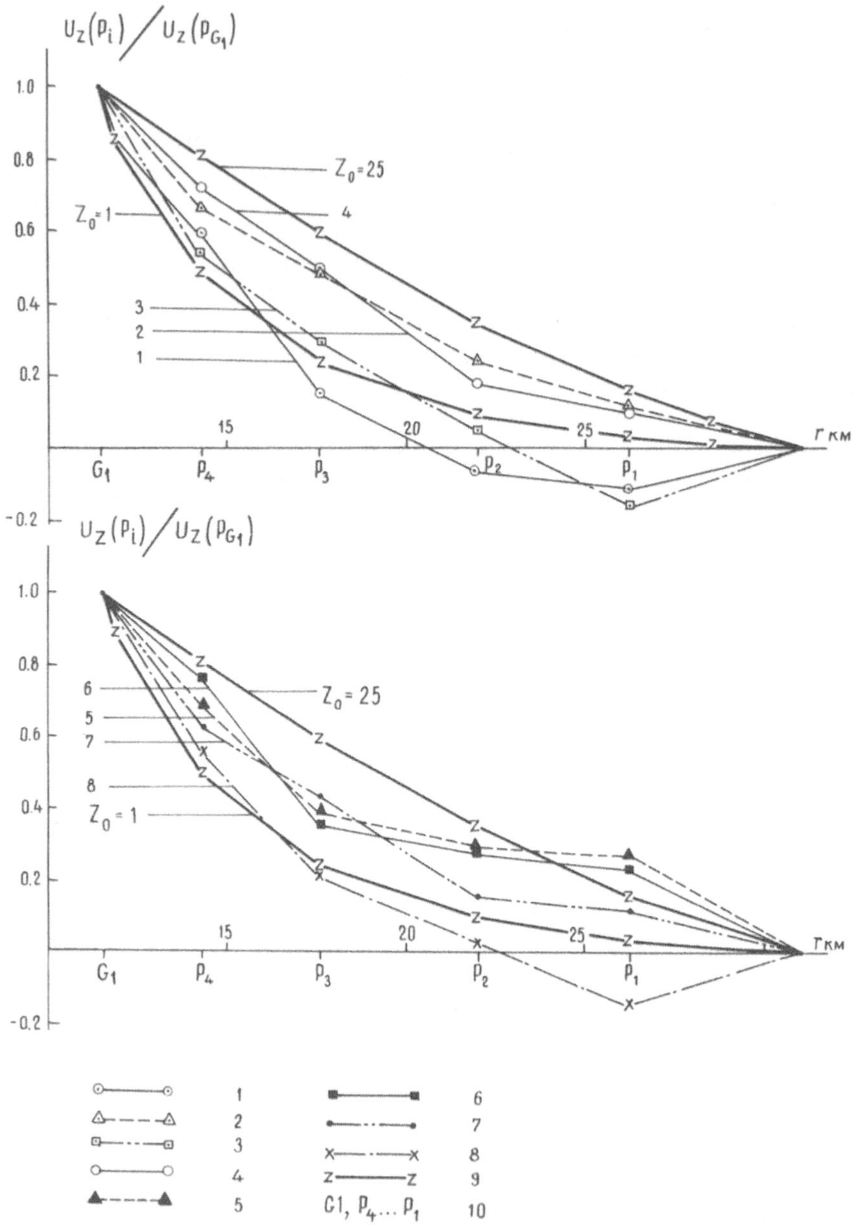

Fig. 4. Relative vertical displacements of bench marks $U_Z(P_i)/U_Z(G_1)$ along profile I over the period 1981–1987. Levelling of 1979 was taken to be the initial date, *1* 1981; *2* 1982; *3* Oct 1983; *4* Jun 1984; *5* Oct 1984; *6* Jun 1985; *7* Jul 1986; *8* Jun 1987; *9* calculated relative displacements from a point spherical source at depths of Z_o = 1 km and Z_o = 25 km; *10* deep-seated bench mark G and ground bench marks $P_4 \ldots P_1$

tion in elevation difference at bench mark P_i (i = 1, 2, 3, ...) (P_4, P_3, P_2, etc. in Fig. 1) to that bench mark G_1, closest to the crater, are put on the vertical axis opposite the corresponding bench marks (see Fig. 4). Displacement of the reference bench mark (379 in Fig. 1), the absolute height of which is taken to be invariable, brings about distortions in the plots of relative excesses in Fig. 4. The relative error is smaller for bench marks having larger displacement. These bench marks are located upslope of the volcano: G_1, P_5, P_4, P_3, P_2.

Thus, for the levelling line approximately 30 km long, taking into account curves and route tilt, the error in determining the mark of the remote bench mark may reach 11–16 mm. The bench marks closest to the crater (P_3, P_4, P_5 and G_1) show the maximum displacements. Displacements of bench marks P_1 and P_2 lie within the range of measurement errors (see Fig. 1).

As one can see in Fig. 4, a noticeable deviation from the monotonous U_z (r) decline was observed in 3 out of 9 cases and in particular during 1981, March 1983 and 1987 levellings. In the second case, the values of the elevation differences were obtained by interpolation and are therefore less reliable. In 1981 and 1987, the pressure source (or sources) differed more significantly from the accepted model than in other cases. The uplift with respect to the initial levelling made in 1979 is characteristic of all the levelling made. Furthermore, Fig. 4 shows the calculated theoretical curves for pressure sources at depths of 1 and 25 km. For the bench marks located close to the crater (P_3, P_4, P_5, and G_1), the relative displacements of which were determined confidently, U_z $(P_i)/U_z$ (P_{G1}) ratios lie between the theoretical curves for pressure sources at depths of 1 and 25 km.

In all cases where relative uplifts increase steadily approaching the volcano, one may assume that the extension centre was located beneath the volcano.

The depth of the pressure centre Z_o (Eq. 1) was obtained using two methods.

Taking a logarithm of Eq. (1) we obtain:

$$\log U_z = \log A + \log B \qquad (2)$$

where

$$A = -2\Delta P R_o^3 \frac{1 - \nu^2}{C}; \quad B = \frac{Z_o}{(r^2 + Z_o^2)^{3/2}}. \qquad (3)$$

In this case, for the effective pressure source acting for a short period of time during measurements, the values of ΔP, R_o, ν, C and log A are constant and the shape of plot log U_z (Z_o, r) depends entirely on the depth of the effective pressure source $Z_o \cdot Z_o$ was estimated by comparing the observed curves log U_z (Z_o, r) with the theoretical ones which were computed for different values of Z_o (Fedotov et al. 1988).

The Z_o values were determined using the second method from the equation:

$$\frac{U_z(r_i)}{U_z(r_{i+1})} = K, \qquad (4)$$

where r is the horizontal distance of bench mark i from the central crater, U_z the relative uplifts of bench marks i and i + 1 with respect to the reference bench

Table 1. Depths of the centers of seismic energy release $\bar{H}_E$ and of the point pressure source Z_o determined from geodetic data[a]

No.	Date	$\bar{H}_E$ (km)	Z_o (km)			δA	δV
			I	IIa	IIb		
1.	June 1981	4.5	3.0	(6.3)	(9.3)	1.0	1.0
2.	July 1982	11.5	10.0	9.9 ± 1.1	9.0 ± 1.1	1.8	3.5
3.	March 1983*	7.5	15.0	14.1 ± 1.5	13.8 ± 1.7	2.7	5.3
4.	June–July 7, 1983	3.3	3.0	(2.9)	(2.9)	1.6	1.2
5.	June 21–July 19, 1984	6.2	5.0	9.7 ± 1.9	9.3 ± 2.4	1.5	2.9
6.	October 1984	6.1	4.5	8.2 ± 1.8	9.2	1.3	2.5
7.	May 29–July 2, 1985	12.6	10.0	12.4 ± 1.4	13.6 ± 2.0	0.6	1.2
8.	June 2–July 4, 1986	12.0	10.0	9.0 ± 1.3	8.8 ± 1.6	1.2	2.3
9.	June 2–July 20, 1987	5.0	3.5	(4.9)	(4.7)	0.9	0.9

[a] Z_0 was determined using method I, taking a logarithm of equation for the point source and in method II, by introducing factor $K = U_Z(P_i)/U_Z(G_1)$ – IIa variant and $K = U_Z(P_i)/U_Z(P_5)$ – IIb variant. In parentheses are Z_o values obtained from three upper bench marks without accuracy evaluation; *marks the less reliable value of Z_o for March 1983 eruption; in this case the increment has been obtained by interpolation. Values of δA-relative variation in excess pressure and volume and δV-relative variation in volume of the feeding system were calculated from the differences between excesses of deep-seated bench mark G_1 and adjacent upper bench marks of the profile P_5, P_4, P_3, and P_2.

mark which is located at a distance where surface displacements are small. In our case this distance is 20–30 km.

It follows from Eqs. (1) and (4) that

$$Z_o = \sqrt{\frac{K^{2/3}\, r_i^2 - r_{i+1}^2}{1 - K^{2/3}}}. \tag{5}$$

Values of Z_o were determined for the pairs of bench marks located upslope where variations in U_z are greater than the leveling errors.

In determining the coefficient "K" using the second method, in one case (IIa) G_1 and in another case (IIb) P were taken as the reference bench marks.

The values of Z_o estimated using the two methods are reported in Table 1. They change from 3 to 15 km below sea level. Z_o estimated with method IIa proved to be 1.5 ± 2.0 km higher for all 9 cases. A standard error of Z_o calculated by means of method IIa was within the ranges of ±(1–2) km. Judging from these values, the depth of the effective centre of the pressure source was estimated with an accuracy of 2–3 km. Estimates obtained through the second method are, on average, somewhat more accurate than those obtained using the first method. Variations of Z_o during 1981–1987 estimated by means of the second method (IIa) are shown in Fig. 5.

After determining Z_o one can firstly estimate the absolute values of A from Eqs. (1) and (3) and then R_o.

The magnitude of uplift is proportional to A. Variations in factor A through time are shown in Fig. 5. In constructing the graph, the values were normalized

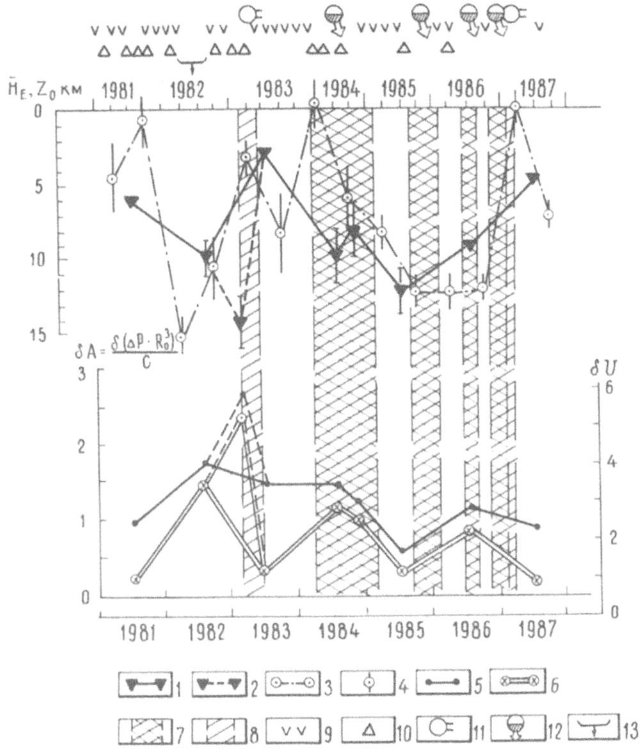

Fig. 5. Variations in depth of the magmatic pressure source Z_o beneath Klyuchevskoy from geodetic data and of the seismic energy release centre ($\bar{H}_E$); relative variations in $\delta_A = \frac{\delta P(R_o^3)}{C}$ and relative volume increments δV of the source of magma pressure during 1981–1987. *1* Variations in Z_o; *2* interpolated segments; *3* $\bar{H}_E$ variation for half-year intervals of time; *4* error bar of $\bar{H}_E$; *5* variation in the relative value of $\delta A = A_i/A_{1981}$; *6* variation in the relative value of $\delta V = \Delta V_i/\Delta V_{1981}$; *7* duration of summit eruptions; *8* duration of flank eruptions; *9* fumarolic activity at the summit crater; *10* explosive activity at the summit crater; *11* flank eruptions; *12* summit eruptions with downslope outpouring of lava; *13* subsidence in the middle part of the summit crater

to value A of 1981. One can see that the relative value δA = (t)/A (1981) increased by a factor of 2 or 3 prior to the large flank eruption of 1983.

Factor A includes the excess pressure ΔP, geometrical dimensions of pressure source R_o as well as Poisson's coefficient ν and Young's modulus C which characterize the elastic properties of the surrounding medium. V, C and ΔP can be determined from seismic wave velocities and densities of the medium and magma. Poisson's coefficient was taken to be, as usual, ν = 0.25.

Young's modulus C for various depths was determined from relations:

$$C = 2\rho V_s^2(1 + \nu),\ V_s = \frac{V_P}{K}, \tag{6}$$

where ρ is the density, V_p and V_s are P- and S-wave velocities. Values of V_p, K and ρ in the edifice of Klyuchevskoy and in the crust beneath this volcano were determined from deep seismic sounding data and gravimetric investigations (Gorelchik and Stepanov 1976).

The excess pressure of magma in the feeding channel may be expressed as:

$$\Delta P(H) = (\rho - \rho_m)g \cdot \Delta H, \tag{7}$$

where ρ_m is the density of magma, g = the gravitational acceleration, ΔH = the increase in depth.

Volume increments of the effective pressure source which took place during 1981–1987 were estimated. The volume increment, ΔV, during 1979–1981 was taken to be 1. Then the ratio $\delta V = \Delta V(t)/\Delta V$ (1981) was determined.

Utilizing the value A, which was obtained from geodetic observations, $\nu = 0.25$ and the calculated C and $\Delta P(H)$ in Eq. (3), one can estimate the variation R_o^3.

During 1979–1987, the level of the lava head in the Klyuchevskoy summit crater increased by around 100 m (see Sect. 1). Simultaneously, the excess pressure of magma ΔP in the feeding system in the volcano should have increased by 20–25 bar which constitutes only a few percent of the value of ΔP in the crust at depths down to 20 km. ΔP at depths of 3–15 km varies within the ranges of ±15% of the average 740 bar. Thus, at first approximation, ΔP may be assumed to be time-independent. In this case, R_o^3/C is proportional to A for the period under consideration. The graph of $\delta A = A(t)/A(1981)$ is given in Fig. 5.

More accurate estimations of volume increment δV of the effective source of pressure have been obtained by using ΔP and C values for various layers. From these, we calculated the average value of Young's modulus in the layer located above the effective source, which lies at depths of 3, 5, 10 and 15 km. Then, the value of δV was calculated. In calculating δV we have taken the values of Z_o determined by using the second method. A graph of relative variations in $\delta V(t)$ is given in Fig. 5. As can be seen, from 1981 to 1987, δV increased by a factor of 3–5 as compared with its value during 1979–1981. The largest increment δV was observed prior to their large eruption in 1983.

On the whole, the relative values of δA and δV were close to each other and varied in a similar fashion during 1981–1987. The excess volume of the effective source of pressure increased prior to the large flank eruption of 1983, diminishing afterwards and in 1987, the dimensions of the effective source of pressure were the same as in 1981.

The observed vertical displacements on the slope of the volcano, U_z, correlate well with the supposed magma accumulation ω in the feeding channels and chambers (correlation factor r = 0.7) (see Figs. 2 and 3). A good agreement (correlation factor r = 0.6) is also observed between the proposed magma accumulation ω and computed volume increment of the magma pressure source δV (Figs. 2 and 5). This indicates that ground deformations reflected processes occurring inside the volcano and that the application of Eq. 1 was admissible.

In addition to Z_o, δA and δV graphs, Fig. 5 also shows the $\bar{H}_E$ graph which demonstrates the migration in depth of the centre of the seismic energy release (see details in section 3). The $\bar{H}_E$ values were determined for 6- and 3-month period. Figure 5 shows $\bar{H}_E$ for 6-month periods.

As one can see in Fig. 5, depths $\bar{H}_E$ and Z_o determined independently, were close in value and changed in a similar fashion through time. This correspondence was the best for 6 more reliable Z_o estimates made from 1982 to 1986.

Correlation factor r between $\bar{H}_E$ and Z_o, which was determined using the first method from the log B graphs, equals 0.7. For $\bar{H}_E$ and eight more reliable Z_o estimates obtained through method IIa from relations $U_Z(r_i)/U_E(r_{i+1})$, correlation factor $r = 0.83$ and the difference $H - Z_o = 0.27 \pm 1.8$ km. Similarity of Z_o (t) and $\bar{H}_E$ (t) graphs obtained independently, implies that the depth variations in the effective source of excess magma pressure were real and that Z_o was estimated with an accuracy of a few km.

Table 1, Fig. 5 and the correlation factors given indicate that the observed vertical deformations U_Z correspond in approximately 2/3 of the cases to the accepted model for a centre of expansion located in the crust under the summit of Klyuchevskoy. In 1/4 of the cases, the pressure source could involve multiple centres of expansion. Comparison of geodetic and seismological data is given in section 4.

3 Seismic Monitoring

General characteristics of seismicity in the region of Klyuchevskoy and the results of seismological studies during the 1978–1987 eruptions are discussed in the present section. Information on seismic stations, processing procedure and detailed description of seismic phenomena related to the eruptions of 1974, 1978, 1980, 1983 and 1984–1987 have been given earlier (Gorelchik 1985; Gorelchik et al. 1987, 1988; Fedotov et al. 1988). Here we shall note that the accuracy of determining the coordinates of the foci in the region of Klyuchevskoy with the existing network of seismic stations (see Fig. 1) and accepted 6-layered velocity model of the medium (Gorelchik and Stepanov 1976) was generally ±2–3 km for epicentres and ±5 km for depth. Theoretical estimations of the accuracy were confirmed in 1986, during a deep seismic sounding survey, in the region of the Klyuchevskoy volcanic group. Located earthquakes were recorded at a minimum of 3–4 seismic stations. Energy class K_S of earthquakes was determined from S waves according to Fedotov's nomogram. $K_S = 4.6 + 1.5M$, where M is the magnitude determined from the surface waves (Fedotov 1972).

For crustal earthquakes beneath the Klyuchevskoy volcano, the detection was complete or at least consistent for the events of $K \geqslant 6.0$ ($M \geqslant 1$), except during the most intense swarm or strong volcanic tremor periods. Sometimes, under favourable conditions, especially after network improvements over the last decade, foci of many smaller earthquakes ($K_S \geqslant 5.0$) were determined. The seismic characteristics discussed in the present paper were evaluated from events of $K_S \geqslant 5.5$ ($M \geqslant 0.6$) and/or $K_S \geqslant 6.0$ ($M \geqslant 1$).

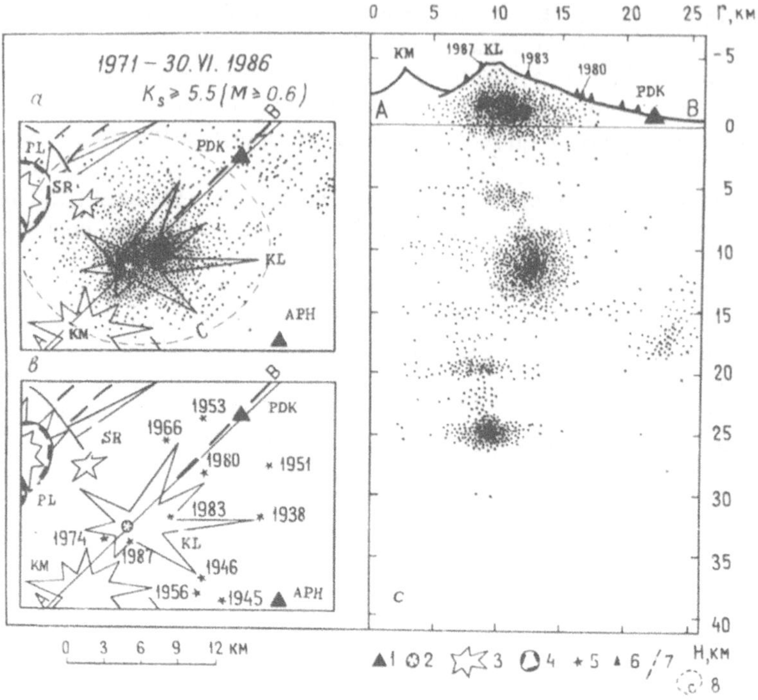

Fig. 6. a Epicentres of earthquakes of $K_S \geq 5.5$ ($M \geq 0.6$); **b** location of flank eruptions; **c** distribution of earthquake foci in a vertical plane crossing the line AB in the region of Klyuchevskoy during 1971–June 30, 1986. *1* Seismic stations; *2* summit crater of Klyuchevskoy; *3* volcanic built-up structure; *4* caldera of Plosky volcano; *5* flank eruptions; *6* flank eruptions in section; *7* main assumed tectonic features; *8* boundaries of the central seismic zone. Active volcanoes: *KL* Klyuchevskoy; extinct volcanoes: *KM* Kamen; *PL* Plosky; *SR* Sredny

3.1 General Characteristics of Seismicity at Klyuchevskoy

Many years of detailed seismological studies (Gorelchik 1985; Gorelchik et al. 1988) have shown that, the principal seismicity in the crust under the Kluychevskoy volcanic group, is concentrated in a few seismic zones. These zones are located within the region under which a large system of intermediate magma chambers is thought to exist in lower crustal layers or in the crust-mantle transition layer (Fedotov et al. 1980). These magma chambers feed the long-lived Klyuchevskoy volcanic centre.

The seismic zone related to summit and flank eruptions of Klyuchevskoy is located around the central crater (Central zone). Nearly all historic flank eruption centres are located inside this zone (contour C in Fig. 6a), excluding the 1932 eruptions, which occurred at a distance of 20 km from the summit crater (Fig. 1; 6a,b). Data given in Fig. 6 imply that seismicity in the crust under Klyuchevskoy is distributed unevenly in depth between 0–30 km.

No earthquakes, at least of energy class $K_S \geq 7.5$ ($M \geq 2$), occur under

Klyuchevskoy between the crust mantle transition layer (30 km) and the Zavaritsky-Benioff zone (~160 km).

Considering the vertical cross-section in Fig. 6c, one should consider that hypocentre concentrations at depths of 5, 10, 15, 20 and 25 km and their rare occurrence, between these depths, may largely be due to depth determination accuracy of ±5 km, and therefore all the depths mentioned are given as multiples of 5. Hence, the decrease in number of hypocentres at depths of 1–5 km under Klyuchevskoy (Fig. 6c) does not prove the presence of a large shallow magma chamber.

Figure 7 shows the density of earthquakes at different depths with a step of 5 km. Analysis of these maps shows that, not only the number of seismic events changes with depth, but also dimensions and structure of the seismic zone under Klyuchevskoy. The slice between H = 0 km and above sea level has the largest area, while the one within the depth interval of 15–20 km is the smallest. The highest density of epicentres directly around the summit crater is observed at depths shallower than 5 km. Small aseismic "patches" arise in the layers between 5–10, 10–15 and 15–20 km around the vertical projection of the summit crater. Configuration and dimensions of the "patches" vary at different depths, their diameter being no more than 3 km. Judging from the data, at depths between 5 and 20 km under Klyuchevskoy, either a hot plastic zone exists, the vertical feeding channel, or crustal magma chambers barely exceeding 3 km in diameter are located along a vertical line.

3.2 Seismic Activity Related to Klyuchevskoy Eruptions of 1978–1987

Here we analyze the distribution of earthquakes in the crust beneath Klyuchevskoy in space and time during 1978–1987, depending on the state of the volcano. More detailed reports on the volcano seismicity during this period are contained in a number of papers (Gorelchik 1985; Fedotov et al. 1988).

Seismicity in crustal layers at different depths under Klyuchevskoy is related differently to volcanic activity.

Earthquakes with a focal depth of less than 5 km generally occur before (3–5 days) and during flank eruptions and also during central eruptions of the volcano. In mid- and lower crustal layers (10–30 km) earthquakes can originate at any state of the volcano. However, they are predominantly observed in the periods between eruptions, a few weeks or a few months before the increase in activity of the volcano, preceeding either the summit crater eruption or a flank eruption, or both.

From detailed seismological studies in 1971–1987, some quantitative characteristics of the Klyuchevskoy seismic activity were determined over six-month periods: the average depth of the centre of seismic energy released $\bar{H}_E$

$$\bar{H}_E = \frac{\sum_{i=1}^{N} H_i E_i}{\sum_{i=1}^{N} E_i},$$

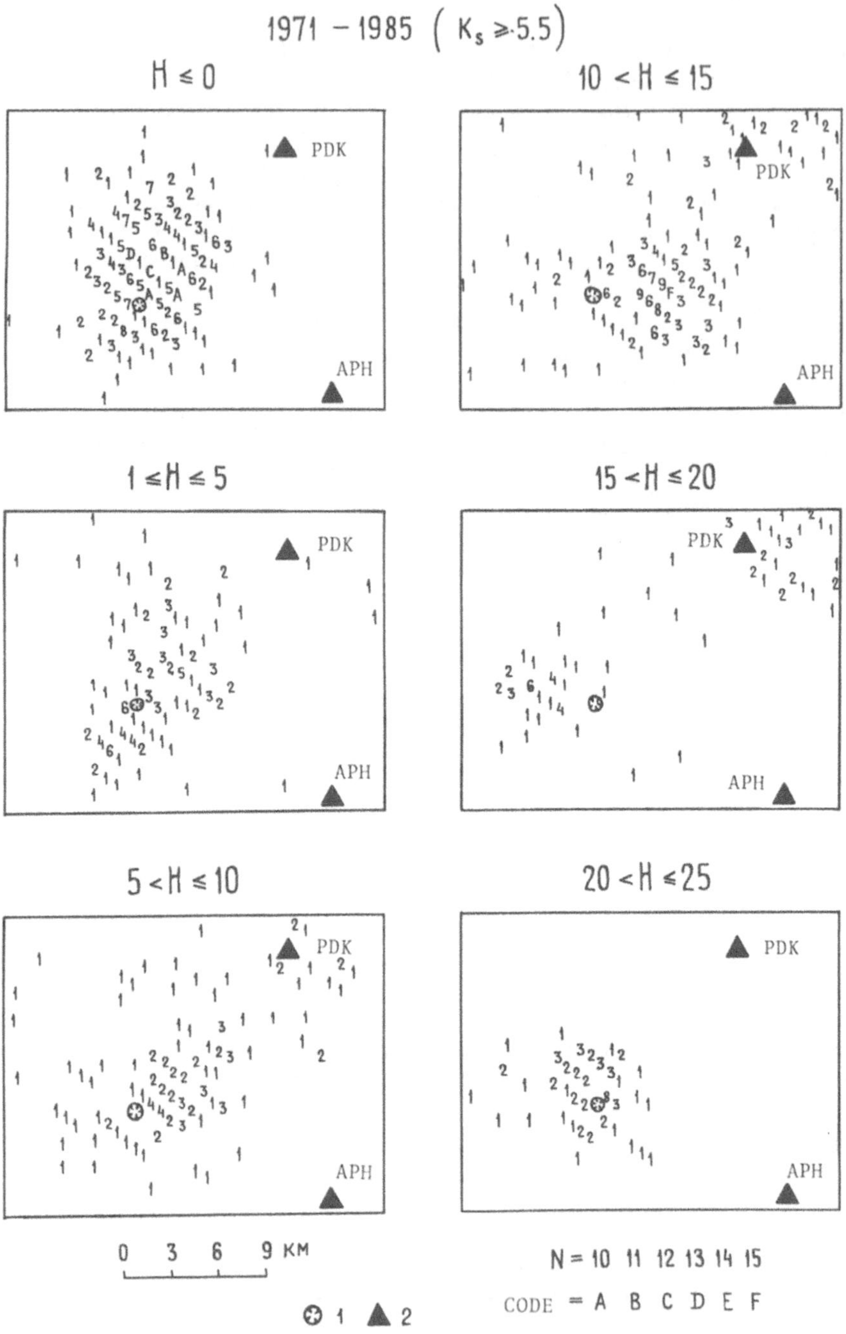

Fig. 7. Depth slices from 0 to 25 km showing foci of Klyuchevskoy earthquakes, 1971–1985. *Digits* show the number (N) of epicentres at definite sites in each depth inteval. Values of $N \geqslant 10$ are coded by *letters* as shown on the *lower right-hand side*. *1* Summit crater of Klyuchevskoy; *2* seismic stations

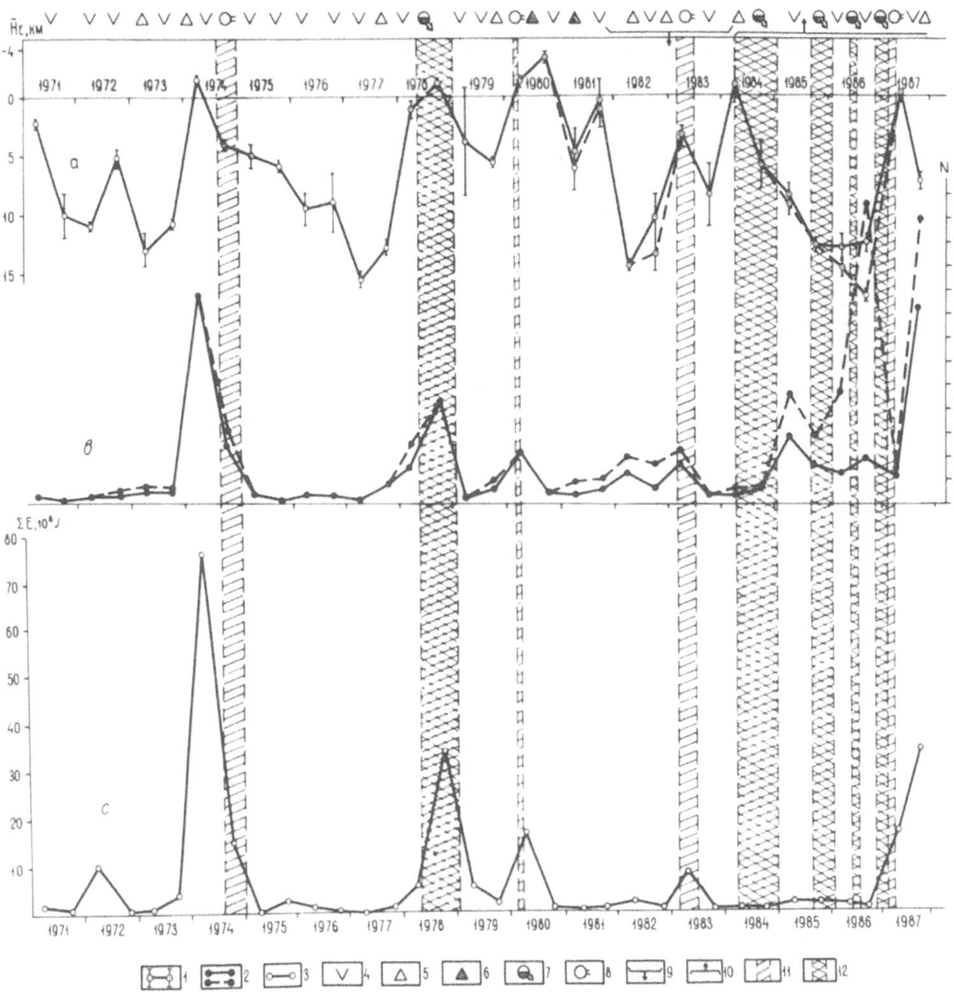

Fig. 8a–c. Variations with time of the half-year values of: **a** the depth of the centre of seismic energy release ($\bar{H}_E$); **b** the number N of earthquakes of $K_S \geqslant 5.5$ ($M \geqslant 0.6$) – (*dashed line*); and $K_S \geqslant 6.0$ ($M \geqslant 1$) – (*solid line*); **c** total seismic energy release ΣE in the Central zone under Klyuchevskoy during 1971–1987. *1* $\bar{H}_E$ vs. time; *2* N vs. time; *3* ΣE vs. time; *4* fumarolic activity; *5* explosive activity at the summit crater; *6* summit eruptions with lava outpouring within the crater; *7* summit eruptions with lava outpouring downslope; *8* flank eruptions; *9* collapse of the summit crater; *10* growth of cinder cone within the summit crater; *11* duration of flank eruptions; *12* duration of summit eruptions with lava outpourings downslope

where E is the energy of a single earthquake with a focal depth H; the number of earthquakes N of $K_S \geqslant 5.5$ ($M \geqslant 0.6$) and $K_S \geqslant 6.0$ ($M \geqslant 1$) at a depth between 30 km to the crater; the total energy of earthquakes

$$\Sigma E = \sum_{i=1}^{N} E_i J.$$

Figure 9 a–c shows variations in these parameters through time.

From the $\bar{H}_E$ (t) graph (see Fig. 8) one may note that the shallowest depth (above sea level) of the centre of seismic energy released $\bar{H}_E$ was observed during the summit eruptions of 1974, 1978 and 1984 and flank eruptions of 1980 and 1987. During the flank eruptions of 1974 and 1983 the depth of the centre $\bar{H}_E$ was 3–5 km and during comparatively calm periods between eruptions the depth of the centre $\bar{H}_E$ increased up to 12–15 km. On the whole, the dependence $\bar{H}_E$ (t) in Fig. 8a has a quasi-periodical nature. Alternations of depth increase and decrease are traced systematically, throughout the overall period of observations, from 1971–1987 and this is obviously an important feature in the mechanism of Klyuchevskoy activity.

The N(t) graph in Fig. 8 shows that, except for 1984, all the essential variations in the state of Klyuchevskoy (summit eruptions, flank eruptions, subsidence in the crater) were characterized by an increase in the number of earthquakes within the volcanic edifice and in the crust beneath it.

In all, from 1971 to 1987, 907 earthquakes of $K_S \geqslant 6.0$ ($M \geqslant 1$) were recorded in the crust under Klyuchevskoy, with an average of one earthquake of $K_S \geqslant 6.0$ every 6.7 ± 5.5 days. Analysis of the distribution in monthly numbers (N_m) of earthquakes of $K_S \geqslant 6.0$ ($M \geqslant 1$) for 17 years has demonstrated that in cases when $N_m \leqslant 15$, seismic events at a significance level of 0.05 correspond to Poisson's distribution with $\lambda = 2.5$. That is, in 50% of the cases 0, 1 or 2 earthquakes per month occurred, whilst for the rest of the time 3 to 14 events per month. If $N_m > 15$ events, groups of earthquakes occurred that did not correspond to Poisson's distribution. The total number of grouping events for 17 years within 13-month intervals was 539. All these intervals fall in the years during which summit and (or) flank eruptions were observed, that is, the origin of groups of earthquakes in the crust under Klyuchevskoy is non-random.

The character of variations in $\Sigma E(t)$ in Fig. 8c is, on the whole, similar to that of the curve N(t), but at the end of the period under study divergences arise. The number of earthquakes of $K_S \geqslant 5.5$ and $K_S \geqslant 6$ beneath Klyuchevskoy during the last 6 months of 1986 was greater than during the first 6 months of 1987. At the same time, the value of seismic energy released during these earthquakes (see Fig. 8c) during the first 6 months of 1987 (namely, during the earthquake swarm which preceeded the flank eruption in February 1987) was by an order of magnitude greater than during the last 6 months of 1986. Concurrently with energy ΣE increase, during the first 6 months of 1987, a sharp decrease of the average depth $\bar{H}_E$ of the effective seismic source occurred, i.e., during the transition from the state of "calm" in the middle of 1986 to the summit and then to the flank eruption in late December 1986–February 1987 (see section 1) a migration of seismic activity under Klyuchevskoy occurred from the lower and middle layers of the crust to higher layers.

Thus, from Fig. 8, in evaluating the state of Klyuchevskoy from seismological data, seismic parameters, such as $\bar{H}_E$, N and ΣE, are of importance.

4 Discussion of Results. Mechanism of Klyuchevskoy Activity

Comparison of information given in sections 1–3 allows us to design the feeding scheme of the volcano, and its activity and to construct an appropriate model (Fedotov et al. 1988). Principal data already known concerning the mechanism of Klyuchevskoy activity and a number of propositions can be used as reference. Various data on magmatic processes which take place beneath the Klyuchevskoy volcano in a depth range from Wadati-Zavaritsky-Benioff zone (or focal layer) up to the lower part of the lithosphere are published in many papers (Tokarev 1970; Fedotov 1976a,b,c,d; Shirokov 1978; Fedotov et al. 1980; Balesta 1981; Fedotov 1984; Fedotov et al. 1984a,b; 1985; Balesta et al. 1984). A review of these results is omitted in the present paper.

The mechanism of magma rise in the asthenosphere is the gravitational convection of viscous liquids. It is probably responsible for magma migration from the asthenosphere up to the lower crust.

The mechanism of basaltic magma-rise from the intermediate magma chambers through the lithosphere and solid crust to the surface is, however, different. Buoyancy remains a main driving force. But here, the motion of magma, occurs by virtue of excess pressure in the feeding magma chambers and channels through short-lived dikes or stable, roundish-cylindrical channels.

In the present study, we consider the position, structure and activity of the part of the Klyuchevskoy feeding system within the crust. Its essential features were known before this study.

Feeding channels and magma chambers of summit and flank eruptions are located in the cone of the volcano about 4750 m high and in the crust at a radius of 20–25 km around the central crater (20 km is the distance to the farthest, historical flank crater) at depths between 0 and 30 km.

Klyuchevskoy volcano has existed for about 7000 years and is still in the initial stage of development. This basaltic volcano is permanently active, actively erupting about half the time. Its magma flow rate is about 60×10^6 t/year. According to theoretical calculations, the summit crater of such a volcano should be fed by a stable vertical channel, the diameter of which can reach 1 km (Fedotov and Goritsky 1980; Fedotov 1981), while the flank eruptions should be fed by the fissures (dikes), which are formed prior to such eruptions and which die soon after (Fedotov 1976a,b,c,d; Fedotov 1984).

The age of the volcano is not old enough for large crustal and subsidiary magma chambers to be formed by its feeding channel (Fedotov 1980; Fedotov 1982). Theoretical conclusions about the crustal and subsidiary magma chambers of Klyuchevskoy agree with data on the structure of the volcano and its "roots". There are no signs of the presence of a caldera which could point to the existence of a large shallow magma chamber beneath the volcano.

The new data discussed in Sections 1, 2 and 3 permit us to develop ideas about the structure and mechanism of magma chambers feeding Klyuchevskoy that lie in the crust at depths of 0–30 km. Thanks to seismic data, the determination of the shape and position of the feeding magma system of the volcano, was improved.

Earthquakes originate in an elastic medium when its strain field changes and associated stresses exceed the local brittle strength of the rocks. This occurs commonly around the magma channels and chambers during sill or dike intrusion or when the pressure in the existing magma channels or chambers changes. Seismological evidence has illustrated that, such a magmatic activity is concentrated beneath Klyuchevskoy, in the vertical area extending from the summit crater to depths of 25 km, 30 km at the most.[2] The highest level of seismicity has a diameter of about 5–7 km (Fig. 7).

The walls of the feeding channel have a temperature of 1100–1200 °C which gradually decreases to the temperature of the enclosing rock with increasing distance from the channel. In the area around the channel where the temperature is 1100–700 °C, rocks should be plastic, and earthquakes either decrease strongly in number or are lacking. Therefore an aseismic column-like body may exist and was expected, in the middle part of the focal area of earthquakes surrounding the feeding channel (or channels), of the volcano. Indeed, inside this column-like zone at depths between 5 and 15–20 km, there is a central aseismic part with cross dimensions of less than 3 km (see Fig. 7) which may be a trace of the plastic zone which should exist around the main feeding channel of the volcano.

No aseismic isometrical volumes with dimensions of 5 km and more, which could be identified with large magma chambers, were detected. As the errors in determing the focal depth are as much as 5 km, one cannot infer, using seismological data, the existence of magma chambers of smaller dimensions (see Figs. 6, 7).

Earthquake swarms taking place outside the vertical seismic column under Klyuchevskoy are observed during the preparation and development of flank eruptions occurring at a distance of more than 10 km from the crater.

Judging from the variation in the number of earthquakes with depth, seismic activity around the feeding system occurs with approximately the same intensity at depths of 0–25 km and increases twofold within the edifice (above sea level). Such increase may be caused by the lower strength of the rocks of the volcanic edifice compared with the rocks of the crustal layers.

Earthquakes of $K_S \geqslant 6$ ($M \geqslant 1$) and, consequently, variations in stresses around the feeding system of the volcano occur with breaks of no more than two or three months (see section 3). This is probably because of the large excess pressure in magma channels, which, according to calculations, may reach 1000 bar at depths of 0–30 km under basaltic volcanoes (Fedotov 1976b,d). The excess magma pressure ΔP at depths between 0 and 20 km under Klyuchevskoy, according to approximate estimates, may exceed 700 bar (see section 2). Viscous liquid at this pressure breaks rocks and intrudes into them. The shell of highly

[2] Calculations of temperature under the Klyuchevskoy volcanic group indicate that, the temperature rises above 1000 °C at depths of more than 28–29 km. This explains the lack of earthquakes at depths of more than 30 km.

heated plastic rocks surrounding the magma chambers and channels, mentioned above, must have existed for a long time, and should resist this intrusion. However, intrusion occurs and the feeding magmatic system frequently leaks into the crust at depths shallower than 25 km. A study of the mechanism of such magma intrusions and conditions of the channel existence, is beyond the scope of the present paper. A discussion on the essential role of the plastic shell of magma chambers can be found in (Yoder 1972).

Geodetic surveys carried out from 1981 to 1987 (Sect. 2, Figs. 3–5) show that the centre of effective pressure under Klyuchevskoy migrated between the depths of 3 and 15 km. The centre of released seismic energy migrated, along with the centre of effective pressure (see Fig. 5). The correlation factor is 0.6–0.7. The agreement between the results of independent geodetic and seismological observations indicates the reality of this process.

It has been established, although less reliably, that the change of the effective volume of magma channels and chambers exerting pressure upon the ambient elastic medium occurred simultaneously. The increase in the effective volume changed in 1981 through 1987 two- or threefold, the maximum increase being observed before the large 1983 flank eruption and after this eruption this value was twice less (see Fig. 5). This seemed to be one of the principal phenomena in the activity of the Klyuchevskoy feeding system. A possible explanation is the following.

In 1981 through 1987 lava flowed out from the vents in the summit crater; the level of lava and magma column in the vent and feeding channel was approximately stable, and consequently, the excess pressure in the channel did not vary significantly. A large variation in effective pressure on the ambient medium would be exerted by rapid changes in the surface and by the shape of channels and chambers filled up with magma. Since a rapid formation of large isometric magma chambers in the crust is impossible (Fedotov 1980, 1982), the rapid change in volume can be produced in two ways, either by formation of dikes and sills, or by the joining of previously isolated bodies of melt (magma chambers, sills) with the feeding system.

The temporal increase in the effective pressure volume during the 1983 flank eruption (see Fig. 5) seems to be due to the formation of a large feeding dike and excess magma pressure within. Perhaps, the source volume has taken the same dimensions as in 1981 when the eruption stopped and the excess magma vanished in the new dike.

Dikes, offshooting from the main feeding channel, originate before typical flank eruptions. The joining of isolated magma bodies with the feeding medium is still a theory. However, similar phenomena are observed in the eastern rift of Kilauea (Wright and Helz 1987).

Previous and new data can be used to reconstruct a scheme of magma feeding and activity of Klyuchevskoy. It is briefly described below.

Depths of 70–40 km (Asthenosphere and Lower Layers of the Lithosphere). Here, a constant ascent of magma occurs from the area of its fusion and intrudes diapirically through the asthenospheric magma column.

Depths of 40–25 km (Crust-Mantle Transition and Lower Crust). Here, an area of large intermediate magma chambers is located, in which basic and ultrabasic magma accumulates and differentiates. This is a possible source of eccentric eruptions.

Depths of 25–0 km (Earth's crust). Here there is a constant vertical channel (or a few close and connecting channels) through which the magma is supplied from the intermediate chambers to the volcano. The diameter of the channel is less than 1 km.

The constant crustal channel (or channels) is surrounded by heated plastic rocks. The overall diameter of the channels and plastic shell enclosing them is less than 3 km. Because of high excess pressure dikes and sills offshoot frequently from the constant channels. When dikes reach the surface, flank eruptions, of the effusive-explosive style, commence.

The radial fissure-dikes occur at a distance of about 20 km from the feeding channel.

No large crustal magma chambers were detected. They have probably had no time to form, since the volcano is in the initial stage of development.

The feeding magma channels in the crust and the areas radiating a few kilometers around them, are in a state of constant change and faulting caused by magma flow in the channels with a rate of 60×10^6 t/year and by the excess pressure of up to 700 bar and more. The excess pressure decreases with distance away from the channel but is still high around the plastic shell of the channel; the medium contains rapidly cooling intrusions (dikes and thin sills) and the stress field is composite and variable. These conditions facilitate the origin of a great number of faultings and earthquakes here.

A redistribution of magma in the feeding system causes shifts of the magma pressure source. The centre of seismic energy release and the centre of the effective magmatic pressure and tend to migrate together at a depth interval of 3–15 km beneath the volcano. Their magnitude and movement indicate the development of processes in the feeding system and may be used to forecast the volcano activity.

Depths of 0–4.5 km Above Sea Level (Edifice of the Volcano). Here, the feeding system terminates. The central channel comes up to the summit crater which has a diameter of 700 m. The tension fissures emerge. Magma degassing and volcanic explosions occur. During the flank eruptions, their vents act frequently as the lava boccas, whereas the major part of gases escape through the summit crater. Subterminal eruptions are predominantly effusive.

Klyuchevskoy's activity has much in common with another giant basaltic volcano, Etna. In 1985, a scheme of the inner structure, feeding channels and chambers of Etna was developed. This scheme was constructed on the basis of 300-year monitoring of the volcano and recent geological, geophysical and geochemical investigations (Chester et al. 1985). This scheme and the above-described model for Klyuchevskoy coincide and support each other as far as some main features are concerned: the accumulation of basalts in a large intermediate chamber in the crust-mantle transitional layer; ascent of basalts to the volcano through the crust in a narrow vertical channel; the occurrence of flank eruptions when dikes intrude from this channel; decrease in gas emission and calm effusive character of flank eruptions when the dike diverges from the upper part of the feeding channel, etc. A comparative study of the mechanism of Klyuchevskoy's and Etna's activity would be very interesting.

5 Application of the Model to Forecast the Klyuchevskoy Eruptions

The data discussed in Section 3 and a model for magma feeding (Sect. 4) enable estimation of the potential hazards of the Klyuchevskoy eruptions and to conduct more purposeful observation on its state and activity.

No signs of the existence of a large subsidiary chamber under Klyuchevskoy were detected. The volcano is in an active state and discharges are almost constant. This means that the probability of large eruptions accompanied by formation of a subsidence caldera under the cone or ejection of some km^3 of juvenile basalts, at present, is very low.

A good correlation (coefficient $r = 0.7$) between the value of bulging up of Klyuchevskoy's slopes and quantity of accumulated magma (assuming its even supply) is worthy of note (Figs. 2, 3). Figure 2 shows that the excess quantity of magma accumulated in the feeding system during 1978–1987 is estimated as 200×10^6 t in spite of the summit eruptions of 1984–1987. Taking into account that a comparable quantity of magma could accumulate over previous years (Fedotov et al. 1987) we may conclude that the weight of juvenile basalts of a large eruption, if it occurs in nearest years, would reach 200×10^6 t.

It should be noted that, during the whole period 1981–1987, the slope of the volcano was higher than in 1979. The bulging up of the volcano slope in 1986 was moderate (Fig. 3) and the volume increase in the feeding magma system (effective source of pressure) is smaller than before the large flank eruption in 1983 (Fig. 5). Judging from this, no preparation for a new but similar eruption occurred in 1986.

During 1985–1987, the bench marks of levelling profile I, close to the crater (P_5, P_4 and P_3), stayed at one level while bench marks away from the crater were 3 lowered (P_2, P_1, P_{101}) (see Fig. 3). According to the accepted model, the increase in the profile's steepness can be explained by a rise in the effective centre of pressure at that time from 1.3 to 3.7–7.7 km (see Fig. 5): the reality of this migration is confirmed by similar migration of the centre of seismic energy released $\bar{H}_E$ which occurred in 1986–1987 (see Fig. 5). The maximum difference between depths $\bar{H}_E$ and Z_o falls in the second half of 1986 according to Fig. 5. This is not extremely significant, since no measurements were conducted at that time, on the levelling profile. At the same time, no remarkable increase in the volume δV of the upper feeding magma system occurred (see Fig. 5). In all probability, the simultaneous rise of the pressure centre and centre of seismic energy released from depths of 10–13 km to 3–4 km, at constant volume of the feeding system, testifies to magma flowing upwards at these depths in 1987. This would mean that the possibility of eruption increased. Its power, however, would not be very great for there is no evidence of large excess volume accumulation of magma, in the upper part of the Klyuchevskoy feeding system.

During 1932–1983, the absolute elevations of flank eruptions increased steadily with time. Based on this tendency, in 1985 we correctly estimated the position of the ensuing flank eruption which occurred on February 23–March 17, 1987 (Fedotov et al. 1987) (Sect. 1). The top of the radial fissure of this eruption was located 700 m lower than the lava head in the summit crater. The excess pressure of magma or lava in its feeding fissure was low for Klyuchevskoy, from less than 100 to a maximum 150 bar. It was an effusive eruption with a small volume of lava. At the end of 1987, it was proposed that, if the tendency to increasing elevations of flank eruptions was maintained then the new flank eruptions will be mostly small, subterminal ones.

These expectations have come true to a great extent. On January 28, 1988 a system of radial fissures originated in the upper part of the Klyuchevskoy volcano at an altitude of 4000–4300 m. In its southern part, at an altitude of 4000 m, a lava vent emerged and a subterminal lava eruption commenced. A continuous issue of lava continued up to September 14, 1988. Lava flow-rate was small and did not exceed $3.3\,m^3/s$, being equal on average to $1.2\,m^3/s$. However, because of the long duration of eruption, its volume proved to be equal to $30 \times 10^6\,m^3$ (N.F. Dobrynin, Pers. Comm.). Noteworthy is the fact that the lava flow rate $1.2\,m^3/s$ is very close in value to the average runoff of basaltic magma for Klyuchevskoy (1.9 t/s). Such a flank eruption could continue for an indefinite period of time.

The subterminal position of low lava flow-rate, during the January–September 1988 eruption, corresponded to the prediction, but the duration of the eruption and the final volume of lava proved to be larger than expected.

In December 1988, two more small flank eruptions occurred in the upper part of the cone of Klyuchevskoy. The related phenomena will be considered in the next paper.

As observations progress and experience accumulates, the evaluations of the state of Klyuchevskoy will be more definite and valid.

Acknowledgements. The authors are grateful to their colleagues at the Institute of Volcanology: Eremeev N.N. (deceased), V.T. Garbuzova, A.P. Ivanov, V.I. Levina, K.S. Kirishev, V.P. Khanzutin, V.G. Ushakov, A.G. Zharinova and others who selflessly conducted observations of the Klyuchevskoy activity and helped in processing the data during 1978–1987.

References

Balesta ST (1981) Earth's crust and magma chambers in the regions of recent volcanism. Nauka, Moscow, p 133 (in Russian)

Balesta ST, Zubin MI, Kargopoltsev AA, Fedorchenko IA (1984) Deep structure of the eruption region. In: Fedotov SA (ed) The large Tolbachik fissure eruption. Nauka, Moscow, 15:514–537 (in Russian)

Chester DK, Duncan AM, Guest IE, Kilburn CRJ (1985) Mount Etna. The anatomy of a volcano. Lond, p 404

Dietrich H, Derker W (1975) Finite element modeling of surface deformation associated with volcanism. J Geophys Res 80, 29:4094–4102

Efimov AB, Demin SS (1979) Studies of the strain-deformational state close to the magma chamber. Volcanol Seismol 1:16–27 (in Russian)

Fedotov SA (1972) Energy classification of the Kuril-Kamchatka earthquakes and the problem of magnitudes. Nauka, Moscow, p 116 (in Russian)
Fedotov SA (1976a) Geophysical data on the deep magma activity beneath Kamchatka and estimation of forces which cause magma rise to the volcanoes. Izv Akad Nauk SSSR Ser Geol 4:5–16 (in Russian)
Fedotov SA (1976b) Geophysical data on the deep magma activity beneath island arc volcanoes and similar structures. Izv Akad Nauk SSSR Ser Geol 5:25–37 (in Russian)
Fedotov SA (1976c) On basic magma rise in the earth's crust and mechanism of fissure basaltic eruptions. Izv Akad Nauk SSSR Ser Geol 10:5–23 (in Russian)
Fedotov SA (1976d) Mechanism of magma ascent and deep feeding channels of island arc volcanoes. Bull Volcanol 39, 2:241–254
Fedotov SA (1980) On temperatures of entering magmas, formation, dimensions and evolution of volcanic magma chambers. Volcanol Seismol 4:3–29 (in Russian)
Fedotov SA (1981) Magma rates in feeding conduits of different volcanic centers. J Volcanol Geotherm Res 9:379–394
Fedotov SA (1982) Temperatures of entering magma, formation and dimensions of magma chambers of volcanoes. Bull Volcanol 45, 4:333–348
Fedotov SA (1984) Large fissure eruptions: their theory. Mechanism of the Tolbachik eruption. In: Fedotov SA (ed) The large Tolbachik fissure eruption. Nauka, Moscow, 16:537–575 (in Russian)
Fedotov SA, Goritsky YA (1980) Thermal computation of the cylindrical feeding channels and magma discharge for central type volcanoes. Volcanol Seismol 1:3–15 (in Russian)
Fedotov SA, Gorelchik VI, Stepanov VV (1980) Seismological studies on the mechanism of the large Tolbachik fissure eruption, 1975–1976. Bull Volcanol 43:73–84
Fedotov SA, Gorelchik VI, Zobin VM, Stepanov VV, Chubarova OS, Shirokov VA (1984a) Seismological data on mechanism and development of eruption. In: Fedotov SA (ed) The large Tolbachik fissure eruption. Nauka, Moscow, 12:389–448 (in Russian)
Fedotov SA, Enman VB, Nikitenko YP, Maguskin MA, Levin VE, Zharinov NA, Enman SV (1984b) Geodetic studies. In: Fedotov SA (ed) The large Tolbachik fissure eruption. Nauka, Moscow, 14:472–513 (in Russian)
Fedotov SA, Gusev AA, Chernysheva GA, Shumilina LS (1985) Kamchatka seismofocal zone (geometry, location of earthquake foci and connection with volcanism). Volcanol Seismol 4:91–107 (in Russian)
Fedotov SA, Khrenov AP, Zharinov NA (1987) Klyuchevskoy volcano, its activity in 1932–1986 and probable development. Volcanol Seismol 4:3–17 (in Russian)
Fedotov SA, Zharinov NA, Gorelchik VI (1988) Deformations and earthquakes of the Klyuchevskoy volcano, a model of its activity. Volcanol Seismol 2:3–42 (in Russian)
Gorelchik VI (1985) Seismic activity of Klyuchevskoy during preparation and development of the Predskazanny breakthrough in March–July 1983. Volcanol Seismol 1:71–87 (in Russian)
Gorelchik VI, Stepanov VV (1976) Seismicity in the region of the northern volcanic group in Kamchatka in 1971–1972. In: Ivanov BV, Balesta ST (eds) Deep structure, seismicity and recent activity of the Klyuchevskoy volcanic group. Nauka, Vladivostok, pp 198–218 (in Russian)
Gorelchik VI, Zobin VM, Tokarev PI (1987) Seismicity of volcanoes. Volcanol Seismol 6:61–77 (in Russian)
Gorelchik VI, Chubarova OS, Garbuzova VT (1988) Seismicity in the region of the northern volcanic group in Kamchatka, 1971–1983. Volcanol Seismol 1:90–100 (in Russian)
Kirsanov IT, Markov IT (1979) Evolution of basalts in the process of formation of Klyuchevskoy volcano. In: Sobolev VS (ed) The problems of deep magmatism. Nauka, Moscow, pp 80–97 (in Russian)
Mogi K (1958) Relation between the eruptions of various volcanoes and the deformations of the ground surface around them. Bull Earth Res Inst 36:99–134
Piyp BI (1956) Mount Klyuchevskoy and its eruptions in 1944–1945 and in the past. Proc Lab Volcanol 11:308 (in Russian)
Shirokov VA (1978) Relation betwee volcanic eruptions in Kamchatka and earthquakes in the upper mantle. Bull Volcanol Stations 54:3–8 (in Russian)

Tokarev PI (1970) On focal layer, seismicity and volcanism in the Kuril-Kamchatka zone. Izv Akad Nauk SSSR Ser Phys Earth 3:15–30 (in Russian)
Wright TL, Helz RJ (1987) Recent advances in Hawaiian petrology and geochemistry. In: Decker RW, Wright TL, Stauffer PH (eds) Volcanism in Hawaii. US Geol Surv Prof Pap 1350, VII:625–640, Washtington
Yoder H (1972) Formation of basaltic magma. Mir, Moscow, p 238 (in Russian)
Yokoyama I (1971) A model for the crustal deformation around volcanoes. J Phys Earth 19:199–207
Zharinov NA, Eremeev NN (1985) Deformations of the earth's surface at Klyuchevskoy which preceded and accompanied eruption Predskazannoe in 1983. Volcanol Seismol 1:88–97 (in Russian)

Seismotectonics of the Mount Cameroon Volcanic Region, West Africa

W.B. Ambeh, J.D. Fairhead, and G.W. Stuart

Abstract

Mount Cameroon is an active volcano located on the continental margin midway along a 1600 km chain of volcanic centres called the Cameroon line. Detailed microearthquake surveys conducted in the Mount Cameroon region, during three consecutive field seasons from 1984 to 1987, reveal an overall seismicity pattern characterised by, the predominant clustering of earthquakes at a few distinct localities, with no apparent clear lineations of epicentres. The earthquakes occur both as discrete events and in swarms at a rate of about two events per 3 days. Duration magnitudes ranged from 0.5 to 4, although most of the events are smaller than $M_D = 3$. Hypocentre depths extend from near the surface to about 55 km, indicating the presence of both crustal and subcrustal activity. Calculated b values are low compared to those reported for most volcanic regions. Focal mechanisms are highly variable and show mostly normal and strike slip solutions. The subcrustal earthquakes have mechanisms which are consistent with one another and these show nearly pure strike slip faulting, with NE oriented P axes. The focal mechanisms of the subcrustal events, coupled with the general SW-NE alignment of the volcanic cones on the mountain, suggest an approximate SW-NE orientation of the maximum compressional axis of the regional stress field in the region. Apart from Kilauea volcano, Hawaii, Mount Cameroon is the only other intraplate volcanic centre having such deep seismic activity. The cause of such subcrustal activity is unknown, but probably relates to the presence of a zone of weakness, which both concentrates regional stress and fails more easily than its surroundings.

1 Introduction

Mount Cameroon is an active volcanic centre located on the continental margin midway along a 1600-km-long volcanic chain called the Cameroon line (Fig. 1). Mount Cameroon is an impressive southwest-northeast trending, almost elliptical edifice, 4095 m high, measuring about 50 by 35 km at its base and covering an area of some 700 km^2. Its simple geometry is interrupted to the SSW by Mount Etinde, a sharply pointed peak which rises to an elevation of 1715 m and whose base is partially buried by the apron of lava flows from the main massif. From LANDSAT satellite imagery, Mount Cameroon appears as a stepped horst, limited to the west and east by subsident sedimentary basins, in which up to 7000 m of sediments have accumulated, since the end of the Cretaceous (Déruelle et al. 1987).

Mount Cameroon is the only historically active volcano in the whole of West Africa and has erupted five times this century in 1909, 1922, 1954, 1959 and 1982. The morphology of the volcanic vents reveals that most of the eruptions

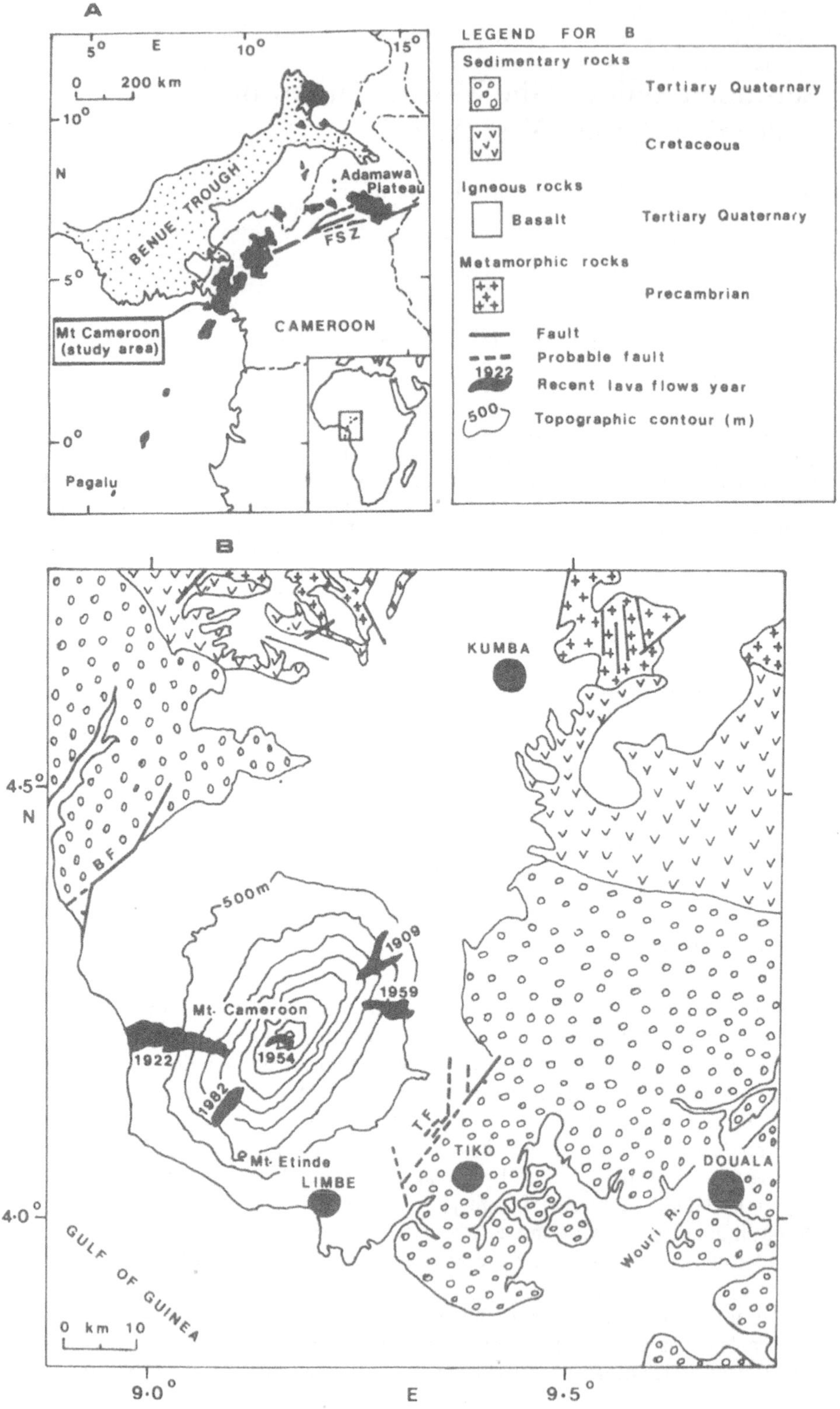

A
5°
E
10°
15°
0
200 km
10°
N
5°
0°
BENUE TROUGH
Adamawa Plateau
FSZ
CAMEROON
Mt Cameroon (study area)
Pagalu
LEGEND FOR B
Sedimentary rocks
Tertiary Quaternary
Cretaceous
Igneous rocks
Basalt
Tertiary Quaternary
Metamorphic rocks
Precambrian
Fault
Probable fault
1922
Recent lava flows year
500
Topographic contour (m)
B
KUMBA
4·5°
N
BF
500m
1909
1959
Mt· Cameroon
1922
1954
1982
TF
TIKO
DOUALA
Mt· Etinde
LIMBE
4·0°
GULF OF GUINEA
Wouri R.
0
km
10
9·0°
E
9·5°

appear to have been of Strombolian type, though Hawaiian and Vulcanian types are also present (Gèze 1953; De Swardt 1956). Déruelle (1982) has evaluated the volcanic hazards in the immediate vicinity of Mount Cameroon and concluded that risks from explosive eruptions are minimal. The only serious risks seem to be from lava flows which, in the 1922 eruption, destroyed oil palm and rubber plantations. Lava flows from the 1982 eruption, the most voluminous this century, also threatened plantations and a small village. The erupted lavas are moderately alkaline basic lavas.

Detailed microearthquake surveys in the Mount Cameroon region were conducted by the University of Leeds (United Kingdom) and the Institut de Recherches Géologiques et Minières (IRGM-Cameroon) during three field seasons, from late 1984 to 1987, following the 1982 eruption, in the hope of monitoring its long term seismo-volcanic character. Preliminary results, on hypocentral locations and the nature of the waveforms of the different event types, recorded during the period from December 1985 to March 1986, were reported by Ambeh et al. (1989). In this chapter, we present an expanded data set which better defines the seismicity patterns, and with focal mechanism solutions, enables us to investigate the tectonic significance of these results.

2 Local Geology and Regional Tectonics

The Cameroon Line is a unique feature and represents a 1600 km long chain of volcanic centres that extends from the Atlantic island of Pagalu (2° S) into mainland Africa (Fig. 1). The volcanic rocks of this line are dominantly alkaline, ranging from transitional basalt to nephelinite and alkali rhyolite to phonolite (Fitton 1987).

Mount Cameroon is composed of alkali basalt to basanite lava flows interbedded with small amounts of pyroclastic material. These lie on Cretaceous to Plio-Quaternary sediments, consisting mostly of sandstones, which are, in turn, underlain by a Precambrian metamorphic basement of schists and gneisses (Dumort 1968). Structural information available on the Mount Cameroon region is sparse, due largely to the thick accumulations of volcanics, which mask the underlying bedrock structure, the difficulty of carrying out fieldwork in the area and the unavailability of exposures because of weathering and the dense vegetation. Only two significant faults, the Tiko fault (TF) and the Boa fault (BF), have been mapped close to the volcano (Fig. 1). According to Hedberg (1968), these faults are normal faults and appear to have been active during the Quaternary. Marine seismic reflection work, in the Gulf of Guinea, reveals a

Fig. 1A,B. Study area. **A** The Cameroon volcanic line (*black areas* are volcanics; *stippled area* represents the Cretaceous sedimentary rocks of the Benue trough, Nigeria; *FSZ* Foumban shear zone). **B** Simplified geological and structural map of the Mount Cameroon region (*TF* Tiko fault; *BF* Boa fault; topographic *contours* at 500-m intervals) broad asymmetric SW-NE trending uplift between Bioko and the mainland, with the northwest limb dipping gradually toward the Niger basin, while the southeast flank is marked by high angle faulting downthrown to the Douala basin (Hedberg 1968).

Age dating of volcanic rocks from the Cameroon line (e.g., Jacquemin et al. 1982), indicates that the earliest igneous activity, consisting mostly of granite and syenite intrusive ring complexes, have ages ranging from 66 to 30 Ma. The oldest dated extrusive rocks, trachyte lavas and ignimbrite flows from a ring complex at Kirawa on the Nigeria-Cameroon border, give Rb-Sr whole-rock ages of 51.2 and 45.5 Ma respectively (e.g., Dunlop 1983). The younger extrusive centres give K-Ar ages ranging from 35 Ma (Mandara Mts.) to present (Mt. Cameroon) (e.g., Fitton and Dunlop 1985). The greatest outpouring has occurred within the last 10 Ma forming the volcanic line seen today. Morphologically recent cinder cones are present on most of the volcanic centres, thereby implying that, volcanic activity has occurred over the whole length of the line in the last few million years (Fitton 1987). There is no evidence of a consistent migration of volcanic activity with time along the line, but of the whole volcanic line being an active unit within the last 10 Ma.

The origin of the Cameroon line is still the subject of much speculation. Morgan (1983) and Van Houten (1983) postulated that it is an expression of the movement of the African plate over a deep mantle plume. This hotspot origin is ruled out, for at least the last 10 Ma of its development, by the fact that there is no evidence of a systematic age progression of volcanic activity along the line during this time. Such volcanism could have utilized a hotspot trace which is now buried beneath the recent lavas.

The suggestion of Freeth (1979) that the Cameroon line is an extensional feature, resulting from membrane stresses generated by the northward movement of the African plate away from the equator, is consistent with the regional uplift and alkaline magmatism, but there is a lack of direct field evidence for extension. Furthermore, the region should have been under compression at the time of emplacement of the intrusive complexes (Freeth 1979).

For the greater part of its length, the continental segment of the Cameroon line is closely associated with a major Precambrian shear zone, the Foumban shear zone (FSZ), which is considered to be the continuation of the Pernambuco lineament in Brazil prior to continental separation (Burke et al. 1971; Gorini and Bryan 1976; Sykes 1978; Browne and Fairhead 1983). Although the Foumban shear zone cannot be traced beneath the southwest section of the volcanic line, it can be traced as a gravitational feature to Mount Cameroon (University of Leeds 1987, unpubl. data). These workers have suggested that the Cameroon line volcanism, is the result of reactivation of this ancient lineament, during Cretaceous and Tertiary times. Such a reactivation has been clearly identified along the Foumban shear zone (further to the NE where it crosses the Adamawa plateau) (Le Maréchal and Vincent 1971).

The remarkable similarity in shape and size of the Benue Trough and the Cameroon line has led Fitton (1980) to hypothesize that, the Cameroon line is due to the sudden displacement, at about 85 to 60 Ma, of an asthenospheric swell from beneath the Benue Trough to its present position, as a result of the major reorganization of the African continental plate boundaries. It is, however, difficult to reconcile this suggestion with the fact that no major volcanic event has been found within the Benue Trough, and there is an absence of seamounts of Cretaceous age aligning with the Benue Trough.

Whatever the cause of the Cameroon line volcanism, since basalts erupted in the continental and oceanic sectors of the line are chemically and isotopically indistinguishable, there can be no doubt that magmas erupted on both its continental and oceanic sectors share a common origin, most probably from the upper mantle (Fitton and Dunlop 1985; Halliday et al. 1988). Presently, we do not believe that the available geological and geophysical data are sufficient to distinguish between the proposed models for the origin of the Cameroon line described above.

3 Data Collection and Analysis

The University of Leeds (UK) and IRGM (Cameroon) have conducted detailed seismic monitoring in the Mount Cameroon area, during three consecutive field seasons from October 1984 to April 1987 (Fig. 2). The first survey, from November 1984 to March 1985, involved the installation of five permanent seismic stations, each equipped with a single vertical component seismometer, belonging to IRGM. During the second field season, November 1985 to April 1986, one permanent 3-component station and up to eight temporary stations were set up, increasing the total number of seismic stations around the volcano to fourteen. The temporary seismic stations consisted of a mixture of single- and three-component seismometers. The third field season took place from December 1986 to April 1987 and also involved the operation of eight temporary stations, in addition to the six permanent ones, mostly at the same sites as during the second field season. Instrumentation throughout consists of Willmore Mk III seismometers with high gain amplifier modulators recording on FM magnetic tape recorders. The recording speed was 15/320 inch/s, resulting in a sharp falloff in the frequency response above 16 Hz. Data from the permanent stations was also written on paper charts by ink and pen drum recorders. A common and absolute time base was achieved using electronic clocks, synchronized by VLF radio signals, from the Omega navigation transmitter in Liberia. The results of the first two field seasons were described briefly by Ambeh et al. (1989). Here, we are now able to report on the overall seismotectonic patterns derived from the detailed monitoring of the region during the second and third field seasons.

The hypocentral parameters, azimuth to the stations and the angles of emergence at the source for the local earthquakes were calculated using the computer program HYPO71 (Lee and Lahr 1975). The accuracy of the arrival time picks generally depended on the nature of the phase onset and the background noise. Subjective estimates of the picking errors, backed by repicks of a representative sample of events after a few months, indicate errors between ± 0.05 and ± 0.15 s for impulsive and clear emergent P-wave onsets. S-wave

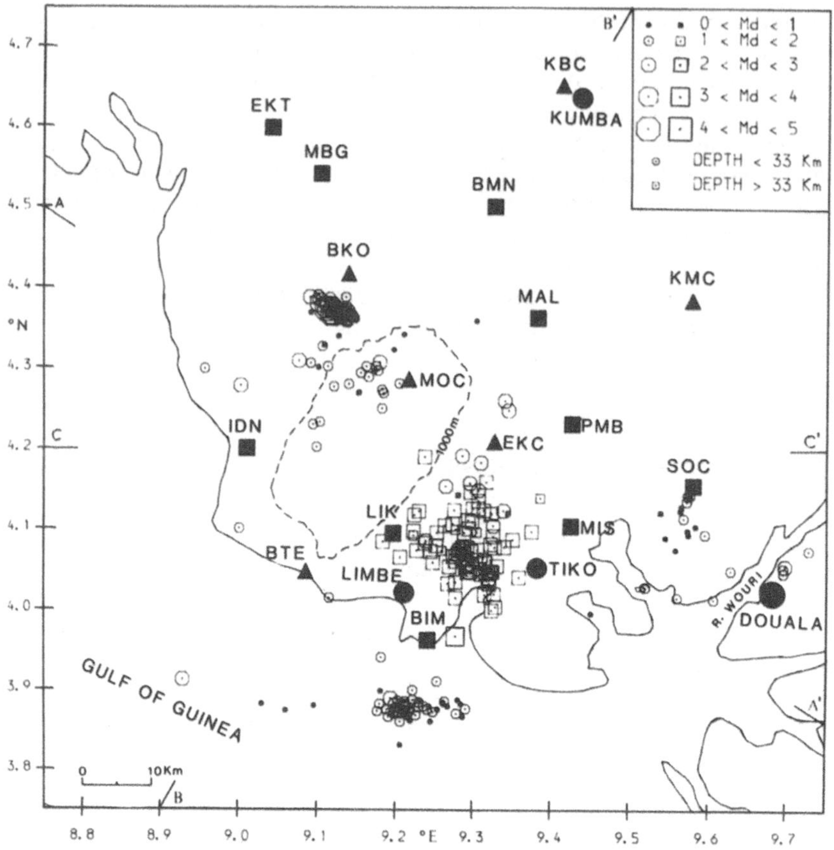

Fig. 2. Seismicity of the Mount Cameroon region (December 1985 to March 1986; January to April 1987). *Triangles* are permanent stations while *squares* are temporary stations. *Solid circles* are towns

arrival times were always included where they were available. They have been weighted by a factor of 0.5 times the weight for the impulsive P wave arrivals. Arrival times of emergent P phases are also downweighted according to the approximate reading error.

The velocity model (Table 1), for use in the hypocentre locations, was determined by simultaneous inversion of local earthquake arrival time data, using the inversion scheme and computer program of Roecker (1981). Layer boundaries were fixed to those determined by Stuart et al. (1985) for the northern section of the Cameroon line (the Adamawa plateau) using the explosion seismic refraction method. This was because attempts to determine the layer boundaries, by varying each in turn and then testing for minimum mean travel time residuals for a selected set of events, did not result in a well defined minimum. We do recognize that such a simple structure cannot totally reflect the

Table 1. Velocity model

Layer thickness (km)	P wave velocity (km/s)
15	5.95
18	6.60
	8.00

geological complexity of the area. For determining the S-wave structure from the P-wave model, 121 selected earthquakes, each consisting of at least four pairs of P and S phases, were used to evaluate V_p/V_s, the ratio of P- to S-wave velocity, using the Wadati method. A weighted mean V_p/V_s ratio of 1.75 ± 0.01 was calculated.

In general, only earthquakes with at least six phases of which at least one was an S phase were located. This resulted in a total of 208 earthquakes being located during the 1985/86 and 84 earthquakes during the 1986/87 field seasons. Hypocentres presented in this paper have rms time residuals of less than 0.5 s.

The absolute accuracy of the hypocentral locations depends on the correctness of the velocity model, the accuracy of the arrival time picks, the distribution and number of stations recording the event, and the number of P and S phases used to constrain the location. The best way, to get true estimates of the location accuracies, is to set off explosions at known locations during the recording period. No explosions were available during this study and so to evaluate the uncertainties in the locations, a series of tests using both synthetic and real data were made. The uncertainty in identifying the onsets of the phases results in the largest error in measuring the arrival times. Since all the picks in this study were made by one person (WBA), we expect this error to be largely systematic. Systematic errors will result in a shift in the origin time of the earthquake but not in the hypocentre. However, the influence of random error in the arrival times on the calculated location was examined by calculating travel times for hypothetical events in different regions, adding random errors of 0.1 s for P and 0.3 s for S and relocating the events. We also examined the variations in the calculated locations of 31 selected earthquakes due to perturbations in the assumed velocity model, V_p/V_s ratio, and group of stations and number of phases used in locating the event. The results of these tests suggest that location accuracies, for events within or just outside the network, are better than about ± 3 km and ± 5 km for the epicentral coordinates and the focal depths respectively. For events outside the network, the location accuracies will progressively deteriorate with increasing distance.

Magnitudes are assigned using a signal duration formula developed for this area following, for example, Lee et al. (1972). Local magnitudes for 21 earthquakes and a distance correction function were first obtained by linear regression on equivalent Wood-Anderson amplitudes (e.g., Hutton and Boore 1987). The local magnitudes were then used to calibrate the signal duration magnitude formula, giving the final expression $M_D = -1.55 + 2.31 \log\tau + 0.0008\Delta$, where τ is signal duration in seconds and Δ is epicentral distance in km.

3.1 Hypocentral Distribution

Figure 2 shows the detailed seismicity of the Mount Cameroon region plotted according to magnitude ranges while Fig. 3 shows depth cross-sections along NW-SE, SW-NE and E-W profiles. These plots show that most of the seismicity is spatially associated with the volcano though few events occur directly beneath it. The overall seismicity pattern shows a predominant clustering of earthquakes at a few distinct localities, with no apparent clear lineations of epicentres. The frequency of earthquake occurrence as a whole was about two events per 3 days and magnitudes ranged from 0.5 to 4, although most of the events were smaller than $M_D = 3$ Focal depths range from near surface to about 55 km, indicating the presence of both crustal and subcrustal activity.

The most seismically active area is the zone to the immediate southeast of the mountain, encompassed by the seismic stations BIM, LIK, EKC and MIS. Events in this region, which is termed the southeast seismic zone, usually occur as single earthquakes about once every 3 days. They exhibit similar waveforms, emergent onsets and constant magnitudes of 2.8 ± 0.1. The similarity in waveform for these events probably suggests that their source locations and mechanisms are nearly identical. A few of these events are sometimes accompanied by one or two very small ($M_D < 1$) aftershocks. They occur at depths between 30 and 55 km. This zone shows a poorly defined north-south trend of epicentres and seems to show a vertical alignment on the E-W profile. Though this zone of activity coincides with the Tiko fault, the fact that these events are relatively deep makes it unlikely that they are generated as a result of movements of this fault.

The northwest seismic zone is the area of seismicity to the northwest of the mountain between the stations IDN, BKO and MOC. Earthquakes in this area occur both as discrete events and in swarms. The concentration of epicentres very close to station BKO marks the site of the Bokosso swarm, a short intense swarm for which about 60 events were recorded during a period of 12 h. Five of the earthquakes for this swarm were felt at the village of Bokosso (near BKO) with a maximum Modified Mercalli (MM) intensity of approximately V. Focal depths for events in the northwest seismic zone are usually less than 16 km, with the Bokosso earthquakes concentrating at depths of 9 to 12 km. Relocation of the Bokosso swarm, earthquakes, using a master event technique, shows them to be clustered at depths between 8 and 10 km.

All the seismicity in the Gulf of Guinea, just offshore of station BIM, also occurred as a single short swarm during the 1985/86 field season. This area was not active at all during the 1986/87 field season. These events have focal depths in the range 13 to 22 km, though relative relocations show them to be clustered between 15 and 19 km. The relative relocations also indicate that the E-W trend in these epicentres is most probably an artifact, possibly due to the distribution of seismic stations.

Small concentrations of earthquakes also occur close to station SOC and the Wouri estuary, far from the mountain massif. Except for a few isolated events, there is no significant seismicity in the NE and SW areas of the mountain. No earthquakes were located on the Boa fault or in the area of the 1982 eruption.

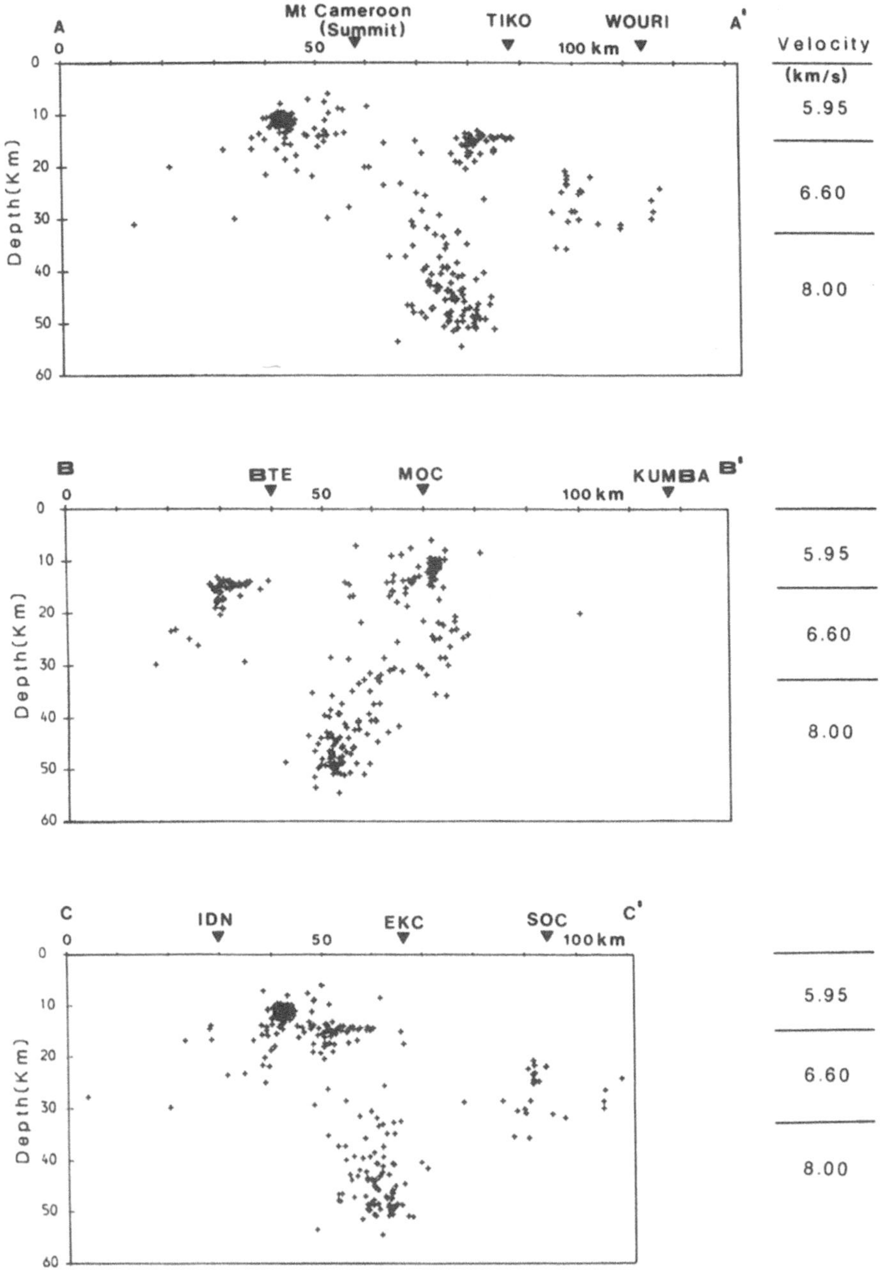

Fig. 3A–C. Earthquake hypocentres projected onto vertical planes

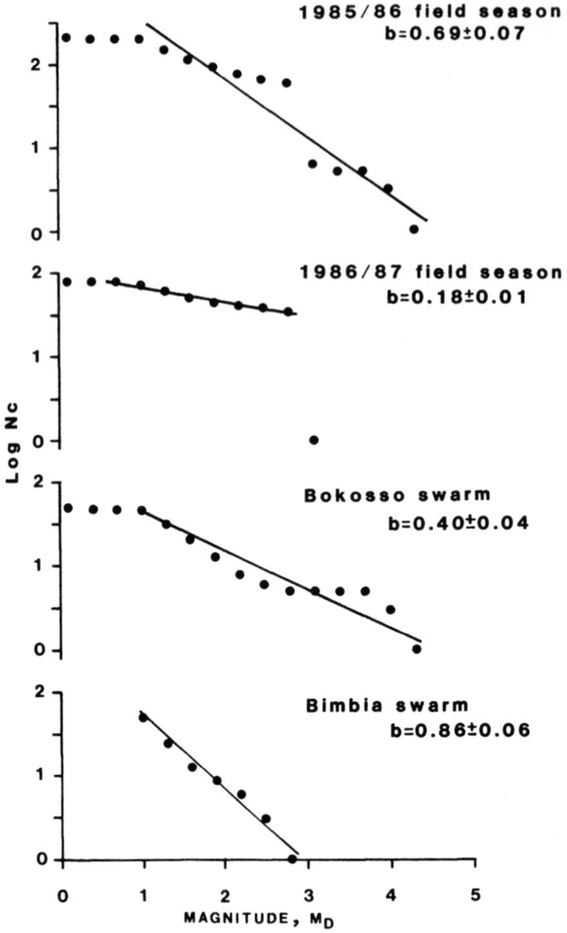

Fig. 4. Magnitude-frequency relations (magnitude interval is 0.3)

3.2 Magnitude-Frequency Relations

The frequency of occurrence of earthquakes as a function of magnitude is commonly expressed as logNc = a − bM, a slightly modified form of the Gutenberg-Richter relation (Gutenberg and Richter 1954), where Nc is the cumulative number of earthquakes of magnitude greater than or equal to M and a and b are constants. The constant a depends on the sample size while b, commonly called the b value, is a measure of the relative numbers of large and small events in the sample. Frequently a large b value has been taken as an indication of greater heterogeneity of the structure (Mogi 1963). However, Scholz (1968), from the results of laboratory experiments on microfracturing in rocks, concluded that the b value is more an indicator of the state of stress in the source region, rather than the heterogeneity of the material, with high b values being associated with low stress conditions and vice versa. Some studies have suggested that the b values in volcanic areas are higher than those in tectonic environments (e.g. Ward and Matumoto 1967; Minakami 1974). Utsu (1971) has also concluded that b values are usually higher for swarms than for earthquakes in general.

Figure 4 reveals b values of 0.69 ± 0.07 for the 1985/86 field season (208 events), 0.18 ± 0.01 for the 1986/87 field season (81 events), 0.40 ± 0.04 for the Bokosso swarm (49 events) and 0.86 ± 0.06 for the Bimbia swarm (49 events) computed by standard linear regression. These b values are less than most of those reported in the literature for volcanic regions and individual swarms. Particularly striking is the extremely low b value for the 1986/87 data set, which is probably a more representative value for the Mount Cameroon region, since it does not include significant swarms, unlike the 1985/86 data. It shows the preponderance of relatively large events (mainly from the southeast seismic zone) compared to smaller magnitude events.

According to Schwarz and Coppersmith (1984), data from paleoseismology suggest that a specific fault segment may generate characteristic earthquakes having a relatively narrow range of magnitudes. The main subcrustal southeast seismic zone events may be considered as characteristic earthquakes for this region. A direct implication of the characteristic earthquake model is that, the occurrence of earthquakes on individual faults or fault segments, often does not follow the form of the Gutenberg-Richter frequency-magnitude relationship (Schwarz and Coppersmith 1984). Since the southeast seismic zone is the most active area in the Mount Cameroon region, its effect on the frequency-magnitude curve is greatest, and this may explain the overall low b value observed for the region, especially during the 1986/87 field season. However, care must be taken in basing inferences on these simple statistics because of the relatively small sample sizes and short time periods considered.

3.3 Focal Mechanisms

Mapping the contemporary stress field, in the Mount Cameroon region, could provide some information on why Mount Cameroon is currently the only active volcano on the Cameroon line. Thirteen single and two composite focal mechanism solutions determined from P wave first motion data are shown in Fig. 5. Equal area projections of the lower focal hemisphere are used to plot all the first motion data; for waves leaving the source through the upper hemisphere, first motions are shown at the equivalent lower hemisphere position. In almost all the solutions presented, the rays from the earthquakes to the stations are for the most part upgoing and are not very much affected by the uncertainties in the structure.

The quality of the focal mechanism solutions varies considerably; for example solutions L and N are poorly constrained while the rest are moderately to well constrained. This is due to the fact that most of the seismicity occurs towards the edges of the seismic network, making it difficult to achieve complete coverage of the focal sphere.

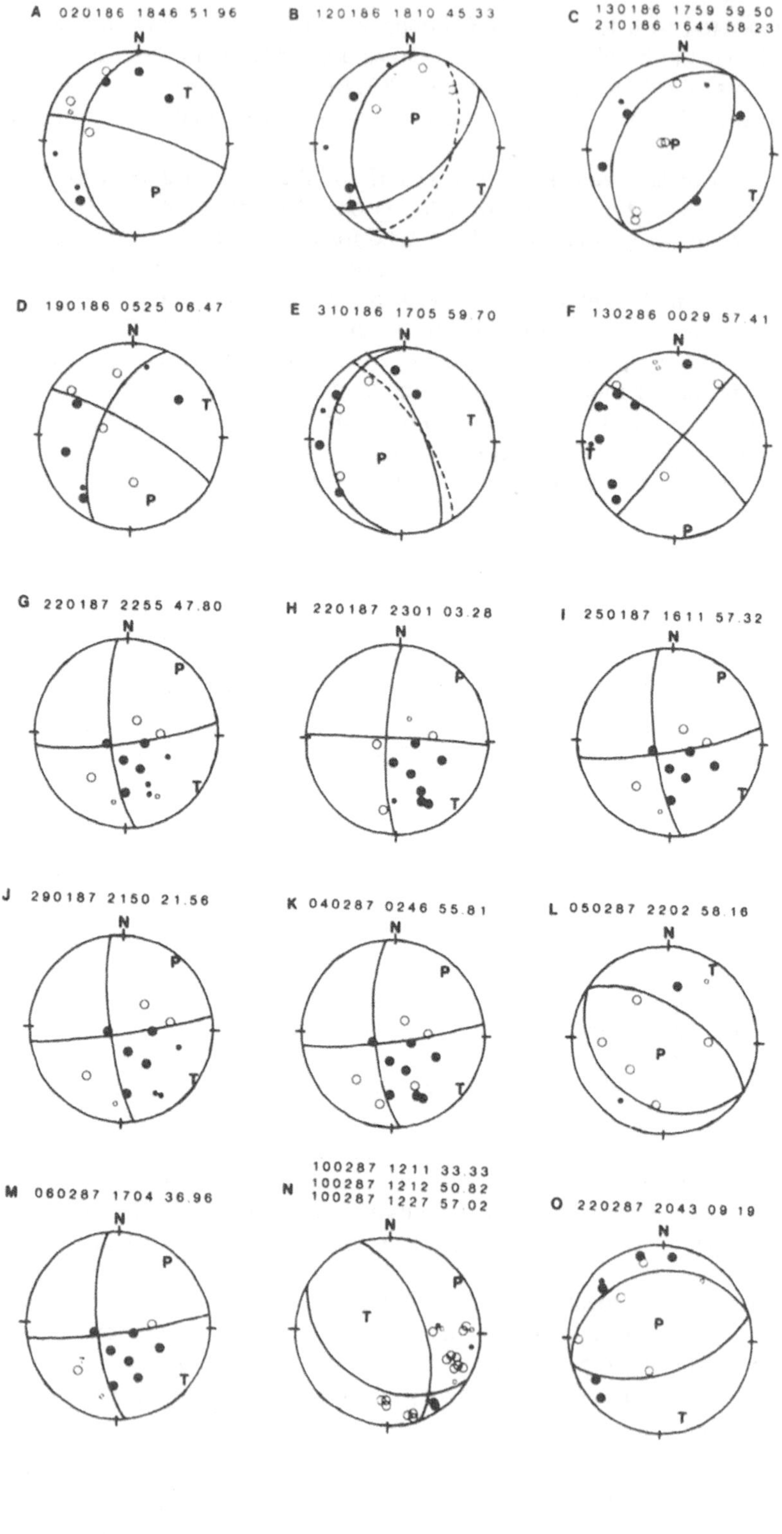
A 020186 1846 51.96
B 120186 1810 45.33
C 130186 1759 59.50
210186 1644 58.23
D 190186 0525 06.47
E 310186 1705 59.70
F 130286 0029 57.41
G 220187 2255 47.80
H 220187 2301 03.28
I 250187 1611 57.32
J 290187 2150 21.56
K 040287 0246 55.81
L 050287 2202 58.16
M 060287 1704 36.96
N 100287 1211 33.33
100287 1212 50.82
100287 1227 57.02
O 220287 2043 09.19
N
P
T

Figure 6 shows the location of epicentres of events whose mechanisms have been determined. Focal mechanisms for the southeast seismic zone, are by far the most consistent and show nearly pure strike slip faulting on steeply dipping nodal planes, striking approximately north-south and east-west. The corresponding P axis is oriented NE with the T axis being SE. Focal mechanisms for the northwest seismic zone show both normal and strike slip solutions, with normal solutions the more numerous. Some of the normal mechanisms possess some strike slip component. There is, however, no consistent orientation of any of the nodal planes for this northwest region. A composite focal mechanism solution for three earthquakes close to SOC is of thrust type, though as has already been mentioned, this mechanism is poorly constrained. A compilation of all the P and T axes (Fig. 6) reveals that, except for the subcrustal events from the southeast seismic zone, both are highly scattered, indicating no consistent orientation for the region as a whole which may reflect the complexity of the local stress field.

4 Discussion and Conclusions

The seismicity of the Mount Cameroon region is almost unique because of the presence of a regular zone of subcrustal activity. The distribution of earthquake epicentres does not clearly define any linear patterns which could represent buried faults. No events were located on the Boa fault and though the southeast seismic zone coincides with the Tiko fault, these events are probably too deep to be attributable to it. Since these faults were found to have been active during the past 1 Ma (Hedberg 1968), deformation on them may be occurring aseismically.

The occurrence of earthquakes at depths from near surface to up to 55 km, though there seems to be an ill-defined paucity between 20 and 30 km, may imply brittle behaviour throughout most of the crust and uppermost mantle. Except for Hawaii, Chen and Molnar (1983) in their detailed investigation of the distributions of focal depths for earthquakes in intracontinental and intraplate regions, did not report of any other areas with a focal depth distribution similar to the Mount Cameroon region. The subcrustal earthquakes of Kilauea volcano, Hawaii, are associated mostly with the magma conduit. In this case, the conduit is interpreted as a low-rigidity passive intruder in the lithosphere, a zone weakened by magma which both concentrates regional stress and fails more easily than its surrounding (Klein et al. 1987).

The variability of the focal mechanisms to the northwest of the Cameroon volcano, for these predominantly crustal earthquakes, may be a reflection of the perturbation of the regional stress field by local effects such as load

Fig. 5A–O. Focal mechanisms in equal area projection of the lower hemisphere. *Solid circles* represent compressional first motions and *open circles* dilatations. *Smaller circles* are uncertain readings

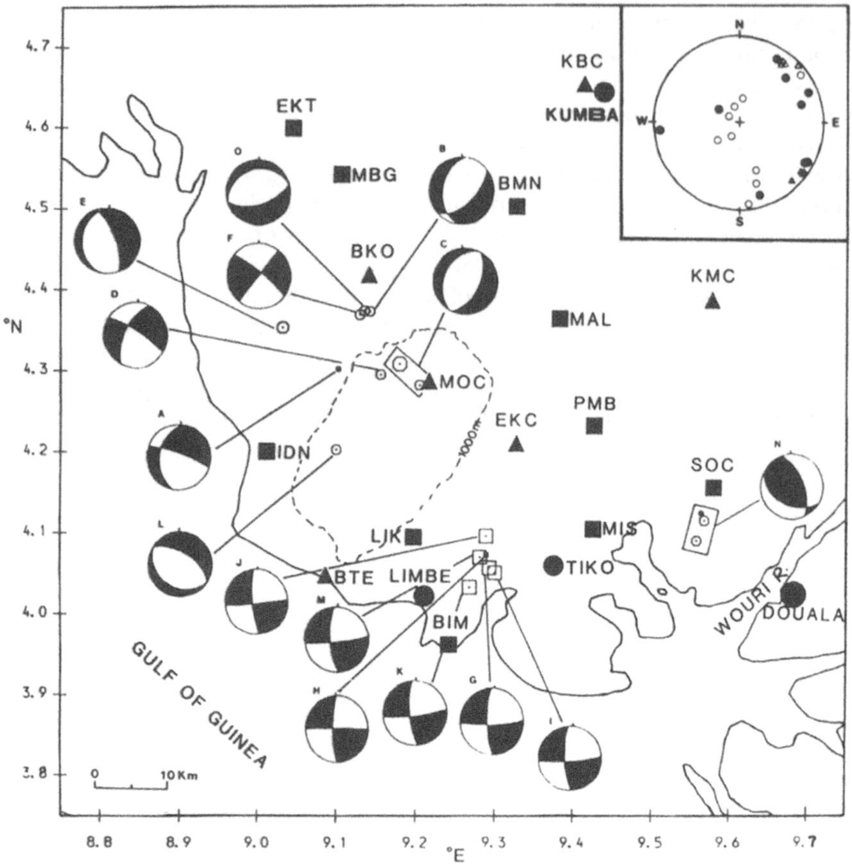

Fig. 6. Focal mechanisms in lower hemisphere abbreviated beach ball format. *Black areas* are quadrants with compressional first motions. *Inset* Lower hemisphere stereographic projection of P and T axes of the focal mechanisms. *Open* and *solid circles* are respectively the P and T axes of crustal events, while *open* and *solid triangles* are the P and T axes of subcrustal events

inhomogeneities and magmatic pressure. Furthermore, it is well known that preexisting zones of weakness, such as old faults, are often the preferential sites for earthquakes and that these local structures may, in fact, distort the ambient stress field (e.g., Mckenzie 1969). Because of the significant occurrence of earthquake swarms in the Mount Cameroon region, the possible direct relation of Mount Cameroon seismicity to magmatic activity must be considered. The Bokosso and Bimbia swarms may have been caused by the intrusion of small magma bodies. At the moment, however, nothing but the common association of swarms with volcanoes suggests magmatic activity as being responsible for these swarms. The b values for the Bokosso and Bimbia swarms are low compared to the high values recorded for some confirmed volcanic swarms (e.g., Einarsson and Brandsdottir 1980; McNutt and Beavan 1984).

The SW-NE trend of the P axes of focal mechanisms for the subcrustal earthquakes is similar to the trend of the Cameroon line. According to Nakamura (1977) and Nakamura et al. (1977), the orientation of feeder dyke patterns and the alignment of volcanic cones on polygenetic volcanoes reflects the effect of the regional non-magmatic stress superimposed on the magmatic pressure, with the trends of volcanic cones and dykes concentrating mostly in a direction of either the maximum compressional or the intermediate stress axis of the regional stress field. Thus, the orientation of the maximum compressional axis of the regional stress field, in the Mount Cameroon area, may be approximately SW-NE. Similarly to Klein et al. (1987), we suggest that these subcrustal events may be due to the presence of a zone of weakness, in which earthquakes seem to be ultimately driven by a SW-NE oriented regional lithospheric stress field, possibly due to the forces which cause plate motions. It may be a magmatic conduit as in Hawaii or an expression of the reactivation of the Foumban shear zone.

Acknowledgements. We would like to thank IRGM (Cameroon), NERC (United Kingdom) and Leeds University (United Kingdom) for providing funds and logistic support which made it possible to carry out this project. Dave Francis and Janusz Bak (University of Leeds), Joe Nnange, Festus Akah, Charles Kome, Simon Epie and many others (IRGM) assisted with the fieldwork. We wish to extend our thanks to the CDC, IRZ-Batoke, Socapalm-Tillo, and the many chiefs and people who cooperated with us during this work. Thanks to Dr. S.W. Roecker of Rensselaer Polytechnic Institute, Troy, New York, for supplying me with a copy of his simultaneous inversion program. Financial support for W.B. Ambeh to attend the IWVS in Capri was provided by the Organizing Committee of the IWVS and Leeds University. W.B. Ambeh also acknowledges tenure of a Cameroon government scholarship. Helpful criticisms and suggestions by an anonymous reviewer are gratefully acknowledged.

References

Ambeh WB, Fairhead JD, Francis DJ, Nnange JM, Soba D (1989) Seismicity of the Mount Cameroon region, West Africa. J Afr Earth Sci 9:1–7

Browne SE, Fairhead JD (1983) Gravity study of the Central African rift system: A model of continental disruption. 1: the Ngaoundere and Abu Gabra rifts. Tectonophysics 94:187–203

Burke KC, Dessauvagie TFJ, Whiteman AJ (1971) Opening of the Gulf of Guinea and the geological history of the Benue depression and Niger delta. Nature (Lond) 233:51–55

Chen WP, Molnar P (1983) Focal depths of intracontinental and intraplate earthquakes and their implications for the thermal and mechanical properties of the lithosphere. J Geophys Res 88:4183–4214

De Swardt AMJ (1956) The 1954 eruption of Cameroon mountain. Rec Geol Surv Nigeria 1954:35–40

Déruelle B (1982) Risques volcaniques au Mont Cameroun. Rev Géogr Cameroun 3:33–40

Déruelle B, N'ni J, Kambou R (1987) Mount Cameroon: an active volcano of the Cameroon line. J Afr Earth Sci 6:197–214

Dumort JC (1968) Notice explicative sur la feuille Douala-Quest. Direction Mines Géol, Yaounde-Cameroon

Dunlop H (1983) Strontium isotope geochemistry and potassium-argon studies on volcanic rocks from the Cameroon line, West Africa. PhD thesis, Edinburgh Univ, Edinburgh

Einarsson P, Brandsdottir B (1980) Seismic evidence for lateral magma intrusion during the July 1978 deflation of the Krafla volcano in NE Iceland. J Geophys 47:160–165

Fitton JG (1980) The Benue trough and the Cameroon line – a migrating rift system in West Africa. Earth Planet Sci Lett 51:132–138

Fitton JG (1987) The Cameroon line, West Africa: a comparison between oceanic and continental alkaline volcanism. In: Fitton JG, Upton BGJ (eds) Alkaline igneous rocks. Geol Soc Lond, Spec Publ 30:273–291

Fitton JG, Dunlop HM (1985) The Cameroon line, West Africa and its bearing on the origin of oceanic and continental alkali basalt. Earth Planet Sci Lett 72:23–38

Freeth SJ (1979) Deformation of the African plate as a consequence of membrane stress domains generated by post-Jurassic drift. Earth Planet Sci Lett 45:93–104

Gèze B (1953) Les volcans du Cameroun Occidental. Bull Volcanol 13:63–92

Gorini MA, Bryan GM (1976) The tectonic fabric of the Equatorial Atlantic and adjoining continental margins: Gulf of Guinea to northeastern Brazil. An Acad Bras Cienc 48:101–119

Gutenberg B, Richter CF (1954) Seismicity of the Earth and associated phenomena. Princeton Univ Press, Princeton

Halliday AN, Dickin AP, Fallick AE, Fitton JG (1988) Mantle dynamics: a Nd, Sr, Pb and O isotopic study of the Cameroon line volcanic chain. J Petrol 29:181–211

Hedberg JD (1968) A geological analysis of the Cameroon trend. PhD Thesis, Princeton Univ, Princeton

Hutton LK, Boore DM (1987) The M_L scale in southern California. Bull Seismol Soc Am 77:2074–2094

Jacquemin H, Sheppard SMF, Vidal P (1982) Isotopic geochemistry (O, Sr, Pb) of the Golda Zuelva and Mboutou anorogenic complexes, North Cameroon: mantle origin with evidence for crustal contamination. Earth Planet Sci Lett 61:97–111

Klein FW, Koyanagi RY, Nakata JS, Tanigawa WR (1987) The seismicity of Kilauea's magma system. In: Decker RW, Wright TL, Stauffer PH (eds) Volcanism in Hawaii. US Geol Surv Prof Pap 1350:1019–1185

Lee WHK, Lahr JC (1975) HYYPO71 (revised): A computer program for determining hypocentre, magnitude, and first motion pattern of local earthquakes. US Geol Surv Open File Rep 75/311, 116 pp

Lee WHK, Bennett RE, Meagher KL (1972) A method of estimating magnitude of local earthquakes from signal duration. US Geol Surv Open File Rep, 28 pp

Le Maréchal A, Vincent P (1971) Le fossé Cretace du sud-Adamaous (Cameroun). Cah ORSTOM Ser Géol 3:60–67

Mckenzie DP (1969) The relation between fault plane solutions for earthquakes and the directions of the principal stresses. Bull Seismol Soc Am 59:591–601

McNutt SR, Beavan RJ (1984) Patterns of earthquakes and the effect of the solid earth and ocean load tides at Mount St. Helens prior to the May 18, 1980 eruption. J Geophys Res 89:3075–3086

Minakami T (1974) Seismology of volcanoes in Japan. In: Civetta L, Gasparini P, Luongo G, Rapolla A (eds) Physical volcanolgy. Elsevier, Amsterdam, pp 1–27

Mogi K (1963) Some discussions on aftershocks, foreshocks, and earthquake swarms- The fracture of a semi-infinite body caused by an inner stress origin and its relation to the earthquake phenomena. 3. Bull Earthq Res Inst Tokyo Univ 41:615–658

Morgan WJ (1983) Hotspot tracks and the early rifting of the Atlantic. Tectonophysics 94:123–139

Nakamura K (1977) Volcanoes as possible indicators of tectonic stress orientation. J Volcanol Geotherm Res 2:1–16

Nakamura K, Jacob KH, Davies JN (1977) Volcanoes as possible indicators of tectonic stress orientation- Aleutians and Alaska. Pure Appl Geophys 115:87–112

Roecker SW (1981) Seismicity and tectonics of the Pamir-Hindu Kush region of Central Asia. PhD Thesis, Massachusetts Inst Tech, Cambridge

Scholz CH (1968) The frequency-magnitude relation of microfracturing in rock and its relation to earthquakes. Bull Seismol Soc Am 58:399–415

Schwarz DP, Coppersmith KJ (1984) Fault behaviour and characteristic earthquakes: examples from the Wasatch and San Andreas fault zones. J Geophys Res 89:5681–5698

Stuart GW, Fairhead JD, Dorbath L, Dorbath C (1985) A seismic refraction study of the crustal structure associated with the Adamawa plateau and Garoua rift, Cameroon, West Africa. Geophys JR Astr Soc 81:1–12

Sykes LR (1978) Intraplate seismicity, reactivation of pre-existing zones of weakness, alkaline magmatism and other tectonism post-dating continental fragmentation. Rev Geophys Space Phys 16:621–688

Utsu T (1971) Aftershocks and earthquake statistics III. J Fac Sci Hokkaido Univ Ser VII 3:379–441

Van Houten FB (1983) Sirte basin, north-central Libya: Cretaceous rifting above a fixed mantle hotspot? Geology 11:115–118

Ward PL, Matumoto T (1967) A summary of volcanic and seismic activity in Katmai National Monument, Alaska. Bull Volcanol 31:107–129

Volcanic Earthquakes at Merapi (Central Java) During the Lava Dome Building Beginning in October 1986

A. Fadeli

Abstract

This study presents results of seismic measurements on Merapi volcano from August 1986 to July 1987. This time period includes the volume increase of the magma dome by about 2 million m^3 in October 1986. The seismic data were recorded by three stations which are located within 5 km distance from the summit, (KLT: 1.5 km from the summit. north-west; LBH and KKN: 2.2 km and 5 km, south).

With the beginning of the lava doming, the following changes in the seismicity were observed: the number of shocks per hour increased, with an increase in the ratio of recorded shocks between the stations LBH to KLT and KKN to KLT. Before the building of the lava dome, a significant periodicity at 24 and 12 h in the occurrence of the shocks was noted. This phenomenon was greatly reduced with the accumulation of outcoming magma. The diurnal and semidiurnal occurrence possibly follows barometric pressure fluctuations acting on the volcano.

Furthermore, the dominant frequencies in the seismic shocks shifted to lower values. As the first onset of the shocks can not be accurately determined (Minakami B-type), epicentres and source depths were estimated using the maximum amplitude ratios found at the different stations. The locations of the shocks with respect to the lava dome building are discussed.

1 Introduction

Merapi volcano, which is located in central Java (7° 32′ 30″ S, 100° 26′ 30″ E), at a height of 2911 m above sea level, is one of the 26 most active volcanoes in Indonesia. The historical record of eruptions, which started in the year 1006 (Van Bemmelen 1950), shows that, on the average, small eruptions occur at about 2–3 year intervals and larger eruptions at intervals of 9–16 years. Merapi volcano is also well-known for lava dome building processes, which have been observed since 1883 (Data Dasar Gunung Api, Direktorat Volkanologi 1979).

The seismic network at Merapi volcano which was installed by the Physics Department, Gadjah Mada University, Yogyakarta (GMU) in 1984 (Fig. 1) consists of three seismic stations which are located within 5 km from the summit (KLT, LBH and KNN). A fourth station (MTK), which is located 50 km south of the summit, can be used to distinguish qualitatively between volcanic and non-volcanic signals. Each station consists of a 1 Hz vertical seismometer, and the seismic data are radio telemetered to the Geophysical Laboratory in Yogyakarta. The seismicity of Merapi is characterized by a continuous occurrence of volcanic shocks and sporadic tremor (Fadeli et al. 1987). The

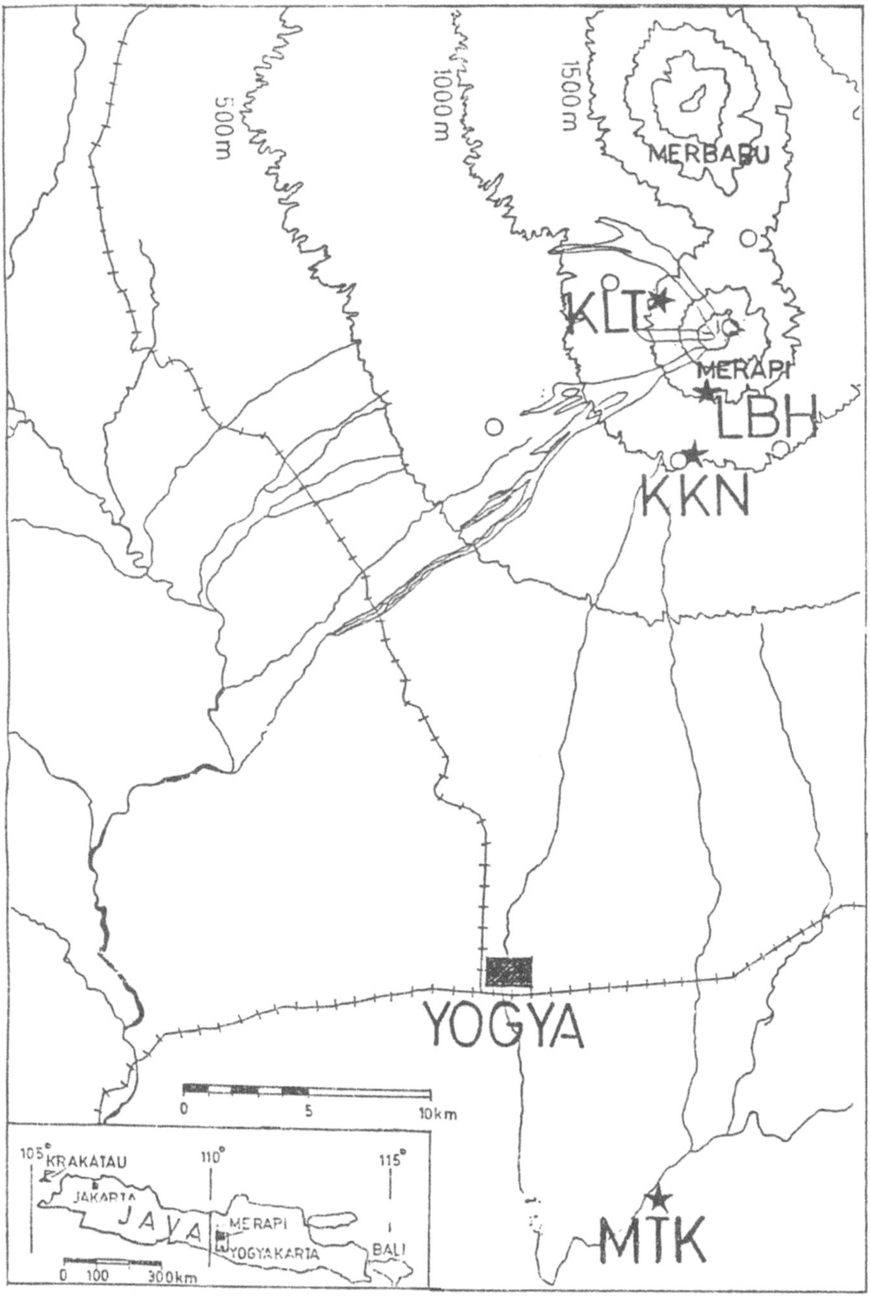

Fig. 1. A sketch map showing the locations of the GMU stations (★) and the VSI stations (○)

seismograms of the volcanic shocks show an emergent onset (Fig. 2). This fact makes it difficult to compute hypocentres using arrival time data.

At the beginning of October 1986, a dome-building process started, with an increase in volume of about 2 million m^3. The change in volume of the lava dome was measured by the Volcanological Survey of Indonesia (VSI) using a photographic method (Irwan Bahar, pers. commun.)

In this chapter, the seismic data recorded at KLT, LBH and KKN during the lava dome-building period will be analyzed and discussed. The data will be studied particularly in regard to the hourly and daily changes in the recorded number of volcanic shocks. As the arrival times of the shocks cannot be accurately determined, the source locations, with respect to the lava doming process, will be estimated, using the maximum amplitude ratios of seismograms recorded at the three stations.

2 Data Analysis

2.1 Number of Volcanic Shocks

This section mainly concerns the following points:

1. Change in the number of volcanic shocks in time and space;
2. Diurnal and semidiurnal periodicities in the occurrence of the volcanic shocks.

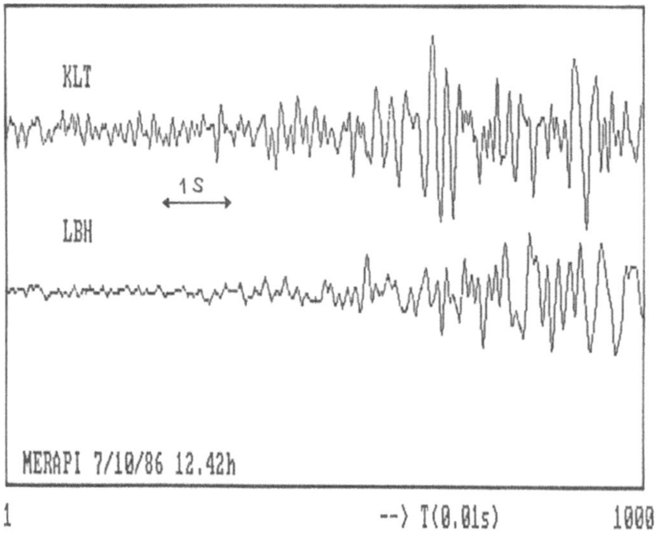

Fig. 2. Sample recording of volcanic earthquakes recorded at *KLT* and *LBH* stations, showing unclear time onset

Figure 3 shows the hourly average of the number of shocks recorded at the three stations within 1 year, starting on August 1, 1986. The average values were calculated with a 48-h window and then they were smoothed by a five-point running mean. As we can see in Fig. 3 during the "normal" state of volcanic activity, KLT, which is the closest station to the active dome, recorded an average of nine events per hour, i.e., about five times as many as those recorded at LBH. The most distant station from the summit (KKN), recorded 50% fewer shocks than LBH did. This indicates that the location of the sources of the volcanic shocks were at a shallow region close to the summit. On October 9, 1986 a new swarm of volcanic shocks started. The ratio of the recorded number of shocks at the three stations changed significantly, indicating deeper source depths. A first maximum in the number of volcanic shocks occurred on October 17, when KLT and LBH recorded more than 30 events per hour, KKN more than 20 events per hour. The lava dome-building process took about 6 months.

The onset of the lava dome-building episode can be clearly seen in the cumulative number of daily shocks as a function of time (Fig. 4). It is interesting to see that the data can be approximated by piecewise linear functions where each segment shows a duration of more than 1 month. With 95% confidence level, a linear regression analysis shows that the linearity of the segment is better

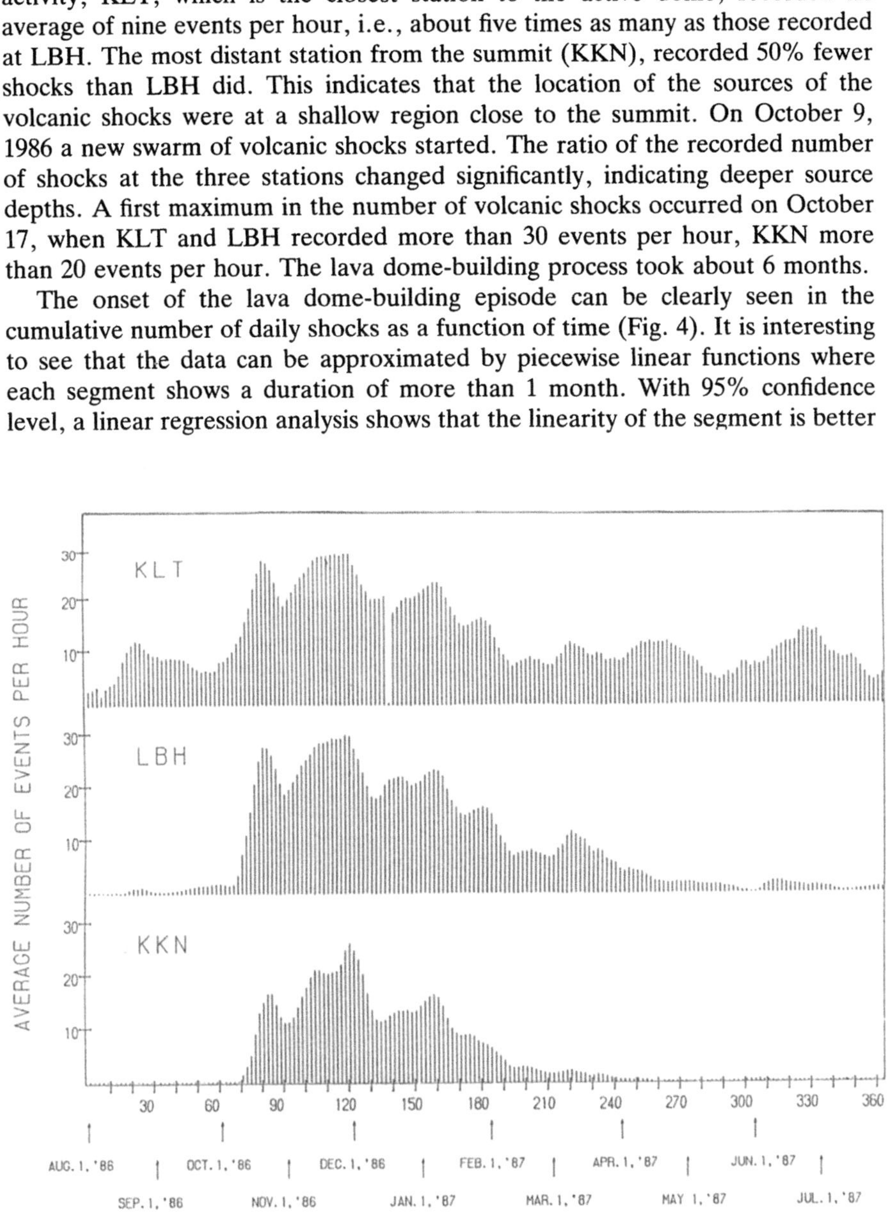

Fig. 3. Time series showing the average number of volcanic shocks per hour in 1-year period started on August 1, 1986

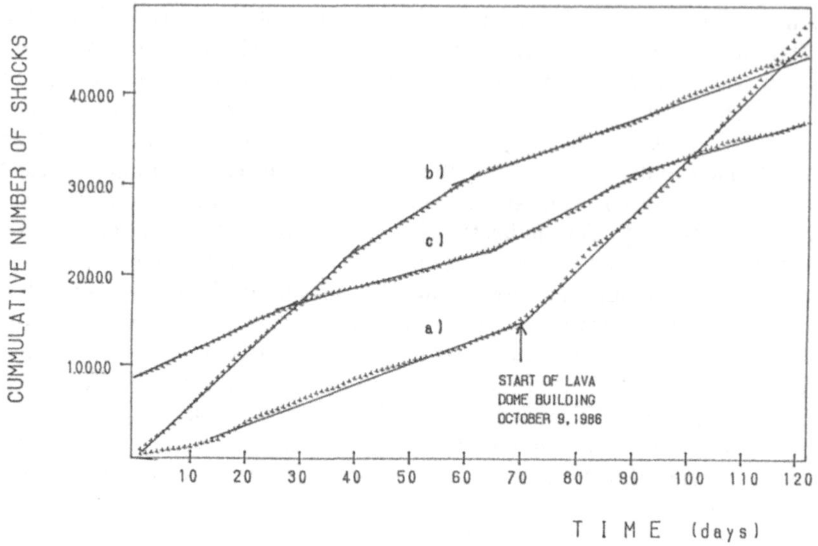

Fig. 4. Cumulative number of shocks recorded at KLT in three periods of 4 months each. *a* Start: August 1, 1986; *b* start: January 1, 1987; *c* start: April 1, 1987. The plots are in 1-day intervals

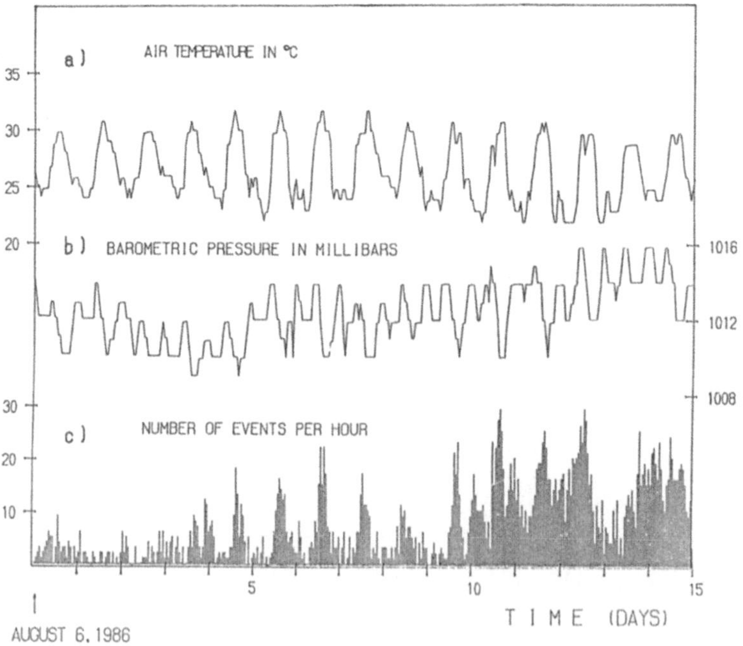

Fig. 5. a Time series showing air temperature and **b** the hourly barometric pressure, recorded in Semarang (50 km north of Merapi volcano). **c** Hourly number of shocks recorded at station KLT in August 1986 showing the diurnal periodicity and the occurrence of the maximum values at 13–16 h

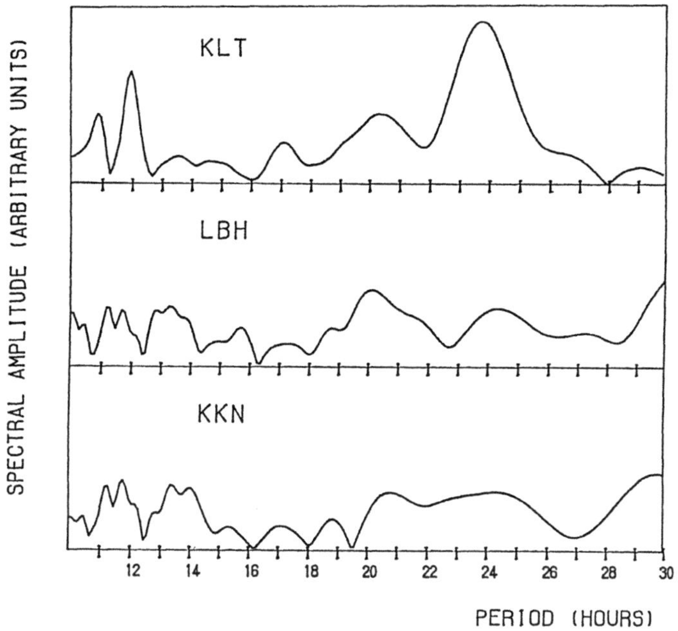

Fig. 6. The DFT (Goertzel algorithm) of the shock time series recorded at *KLT* (*above*), *LBH* (*middle*) and *KKN* (*below*) before the lava dome building episode in August 1986

than 5%. A similar phenomenon with the cumulative number of eruptions was also observed at Stromboli (Schick and Müller 1988) and at Piton de la Fournaise volcano (Lenat and Bachelery 1988).

The time series of the number of volcanic shocks per hour recorded at KLT during a normal state of activity, for example August 1986, shows a clear diurnal periodicity (Fig. 5c). The maximum of the hourly number of shocks seems to be relatively stable in the daytime from 13 to 16 h. A discrete Fourier transform (DFT) analysis using the Görtzel algorithm (Oppenheim and Schafer 1975) shows that the 24 and 12 h spectral peaks are less significant with increasing station distances from the active dome (Fig. 6). A similar spectral analysis for the station KLT shows that during the lava dome-building episode the 12 and 24 h spectral amplitude peaks became less significant (Fig. 7).

Many studies have been reported on the possibility that earthquakes and volcanic eruptions may be triggered by tidal forces (e.g., Knopoff 1964; Mauk and Johnston 1973). Emter et al. (1986) find that the correlation between the hourly number of shocks and the corresponding theoretical tidal strain was positive for Etna but was not so for Stromboli. According to Kirbani (1987), no phase correlation exists between hourly average tremor amplitudes at Merapi and gravitational tides. Furthermore, an investigation of tidal correlation of the occurrence of volcanic shocks at Merapi gave partial significance (Fadeli 1987). As the occurrence of the maximum number of shocks at Merapi is clearly related to solar phases (S1, S2) tidal influence can be neglected, and we must look for triggering forces with a corresponding S1 and S2 variation. Both air

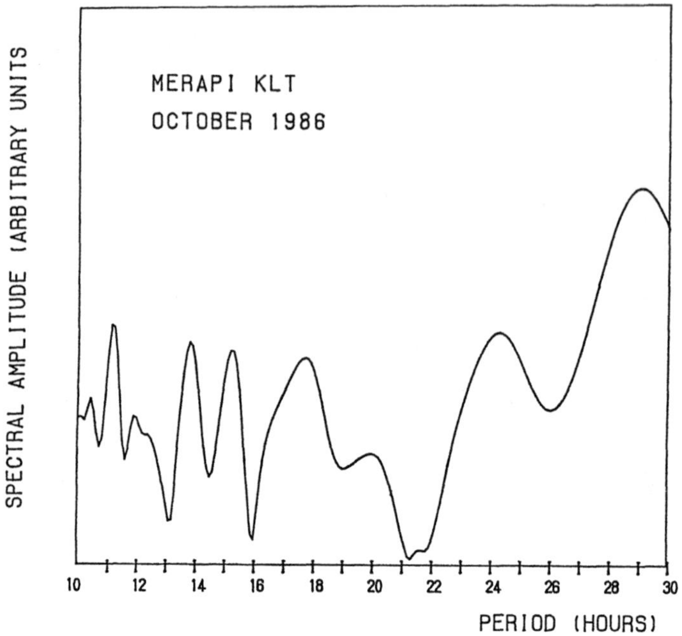

Fig. 7. The DFT (Goertzel algorithm) of the shock time series recorded at station *KLT* during the lava dome building episode in October 1986

temperature (Fig. 5a) and barometric pressure (Fig. 5b) variations have a strong S1 and S2 periodicity. Figure 8 shows the spectral amplitude of (a) the hourly air temperature, (b) the hourly barometric pressure and (c) the hourly number of shocks. There are clearly evident peaks at 12 and 24 h in each spectrum. The spectral energy is more dominant for the 24 h period than for the 12 h period for both the number of shocks and air temperature, but this trend is reversed for the barometric pressure. To investigate statistically the correlation between the number of volcanic shocks and the barometric pressure or air temperature fluctuations, the correlation coefficient as a function of time lag τ was calculated using the formula:

$$r_{xy}(\tau) = \frac{\Sigma(x_i - \bar{x})(y_i + \tau - \bar{y})}{\{\Sigma(x_i - \bar{x})^2 \Sigma(y_i - \bar{y})^2\}^{1/2}}, \qquad (1)$$

where x_i is the barometric pressure or air temperature data and y_i is the number of shocks per hour. For a time series of 744 h duration recorded in August 1986, the results show that the first maximum value of the correlation coefficient is 0.47 occurring with a time lag of 6 h for barometric pressure, and 0.35 occurring with a time lag of 1 h for air temperature. These results mean that with a confidence level of 99.9% we can statistically accept that there is a correlation between the hourly number of shocks and barometric pressure or air temperature. It was clearly observed that rainfall can also act as an external triggering force for volcanic activity. Schick (1988) suggests a possible hydrostatic loading effect. The influence of barometric pressure and rainfall to the volcanic activity was also observed at Aso volcano (Sassa 1936).

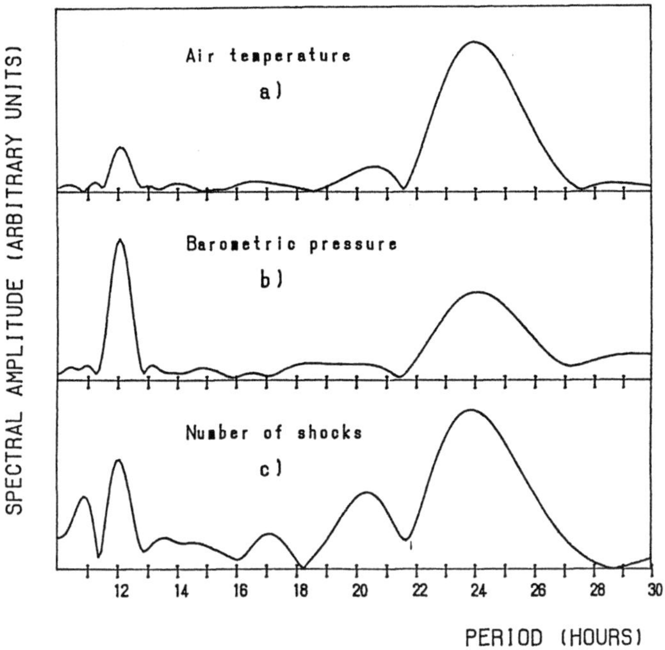

Fig. 8. Spectral amplitude of **a** air temperature; **b** corresponding barometric pressure; **c** hourly number of volcanic shocks recorded at station KLT

2.2 Maximum Amplitude of Seismograms

To estimate the locations of the sources of volcanic shocks using the ratios of maximum amplitudes on the seismograms, the medium is assumed to be homogeneous and isotropic. The loss of the seismic wave energy is mainly caused by internal friction and geometrical spreading. The attenuation of the body wave amplitude with distance is then:

$$A(r) = \frac{A_o}{r} \exp.(-\alpha r) \tag{2}$$

and the amplitude ratio of seismogram recorded at two stations is:

$$\frac{A(r1)}{A(r2)} = \frac{r2}{r1} \exp.(r2 - r1)\alpha, \tag{3}$$

where A (r) is the amplitude of the body wave at a distance r from the origin, A_o is the amplitude of the body wave at the source, and α is the attenuation coefficient.

Here $r^2 = (x - x_o)^2 + (y - y_o)^2 + (z - z_o)^2$, where (x, y, z) and (x_o, y_o, z_o) are the coordinates of the station and the source, respectively. With the three stations we have on Merapi, for one event there are only two independent equations of amplitude ratios in the form of Eq. (3). This is not enough to compute analytically the value of x_o, y_o, and z_o. To find the location of the source, a trial and error method was used. For a given value of x_o, y_o, and z_o the values of amplitude ratios LBH-KLT and KNN-KLT are calculated. By tabulating values of these amplitude ratios for the whole values of x_o, y_o, and z_o in the body of the volcano with a space interval 100 m, we can then approximate the location of the source from the amplitude ratios. For the calculation of the source locations the medium was treated as a rock with an attenuation coefficient between 0.2 and 0.35 km^{-1}. With a wave velocity 3 km/s these parameters are comparable to Q-values ranging from 30 to 50. To estimate the migration of the sources during the lava doming process, the change of the amplitude ratios was studied. Only seismic data recorded by at least two stations were used. Typical scatter plots of the maximum amplitude of seismograms recorded at KKN-KLT and LBH-KLT before the lava dome building episode are shown in Fig. 9a,b, respectively. The difference in the value of N in Fig. 9c,d means that a number of seismic signals was not recorded at station KKN. Out of 500 volcanic shocks which were recorded from August 1 to August 21, 331 events (60%) were not detected by station KKN. For 75% of them the amplitude ratios LBH/KLT and KKN/KLT are less than 0.4 and 0.1, respectively (hypocenters shallower than 1 km). Out of 169 volcanic shocks which were recorded by station KKN, 50% show amplitude ratios at KKN/KLT and LBH/KLT exceeding 0.7 and thus indicating a source deeper than 5 km. After the lava

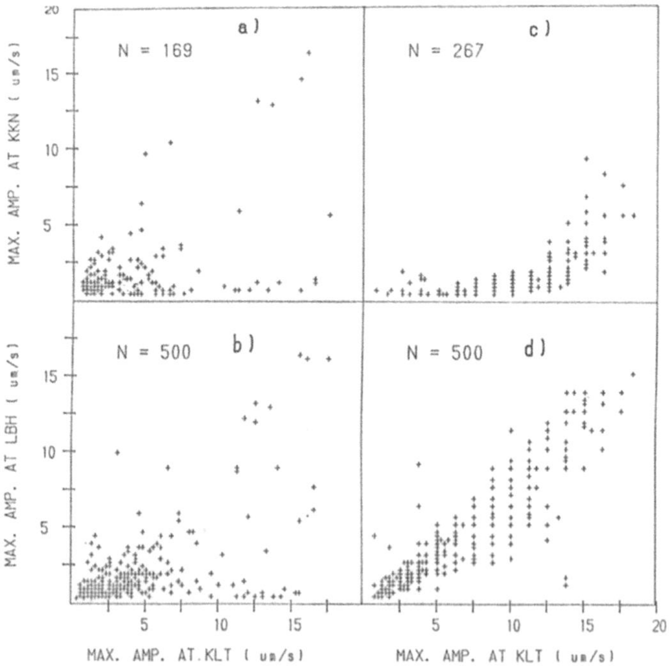

Fig. 9. **a** Scatter plot relating the maximum amplitude of the seismograms recorded at KKN and KLT stations before the lava dome building period. **b** Corresponds to **a**, but for LBH and KLT stations. **c** Similar to **a**, **d** similar to **b**, but during the lava dome building process

dome building episode started on October 9, 1986, the amplitude ratio at LBH/KLT increased gradually and station KKN recorded more shocks. Figure 9c,d shows the corresponding plots within 2 days on the 15th and 16th of October, when the amplitude ratio became roughly constant in time. From the figures one can see that the LBH/KLT amplitude ratios were in the range 0.7–0.9, and most shocks, which were also recorded by KKN, have KKN/KLT amplitude ratios of about 0.2. The amplitude ratios of these values show source locations deeper than 1800 m.

2.3 Dominant Frequencies in the Seismograms

The frequency components of the seismic shocks from Merapi range from several tenths of Hz up to 30 Hz. During the lava dome-building episode, when the shocks were also recorded by the more distant stations, the dominating part of the spectrum of the shocks recorded at KLT shifted to lower frequencies. For the shocks which were only recorded at KLT, the frequency components were in a higher frequency range (Fig. 10a). The dominant frequencies shifted to a lower range for the shocks which were detected by two stations, KLT and LBH (Fig. 10b). The frequency range of the shocks detected by all three stations was lowered further (Fig. 10c).

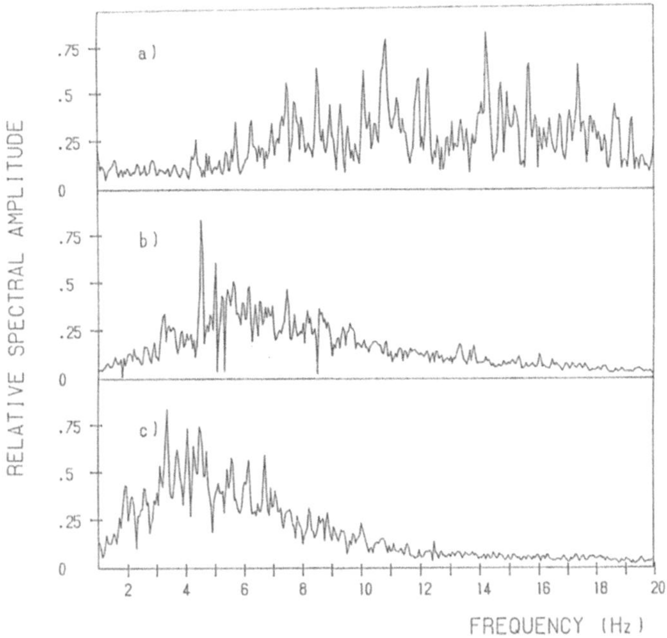

Fig. 10. a Typical spectrum of volcanic shocks recorded at KLT station, which were not detected by LBH and KKN stations. **b** Similar to **a**; spectrum of the shocks which were also detected by LBH station but not by KKN station. **c** Similar to **a**; spectrum of the shocks which were also detected by LBH and KKN stations

3 Discussion and Conclusions

In summary, a seismic data analysis at Merapi from August 1986 to July 1987 leads to the following results:

1. The lava dome building episode at Merapi, which started at the beginning of October 1986, was associated with shallow volcanic earthquakes. The amplitude ratio calculation showed that during the lava dome-building period the sources of the majority of the volcanic shocks were moving down from about 400 to 1800 m depth from the surface. The moving down of the volcanic foci producing shocks, is also in agreement with the change of the dominant frequencies of the shocks, if, we assume that the attenuation of amplitudes is proportional to frequency. This result leads to a conclusion that the lava dome-building episode at Merapi beginning in October 1986 was probably triggered by external forces such as rainfall.
2. Although it still needs further observation, it is suggestive that the diurnal and semidiurnal periodicities in the hourly number of volcanic shocks recorded at Merapi are controlled by the air pressure fluctuations at the Earth's surface on the volcano. The air pressure changes could be caused by the simultaneous effect of barometric pressure and air temperature fluctuations. The response of the volcano to external pressure disturbances possibly indicates the instability of the volcano.

Although the focal mechanism of volcanic earthquakes is still in discussion, it is generally clear that the generation of volcanic shocks is associated with internal magma movements. Thus, frequency of occurrence of volcanic shocks can be used to indicate the state of volcanic activity. The increasing number of volcanic earthquakes at Merapi, during the lava dome-building episode, was possibly a consequence of the upward movement of magma. The piecewise linear plot of the cumulative number of shocks as a function of time, shown in Fig. 4, could mean that the Merapi volcano system has a number of steady states of activity. The change in the steady states of the volcanic activity from one to another is probably caused by the change in the magma supply rate. In the case of an instability condition, the system is sensitive to disturbances. One open question is: what parameters control the stability of the volcano system? The ability of magma to move upwards through fissures depends on several factors, among which are the pressure gradient and the viscosity of the magma fluid (Wilson and Head 1981). Following this direction, atmospheric pressure triggering volcanic earthquakes at Merapi, may be more understandable.

References

Emter D, Zuern W, Schick R, Lombardo G (1986) Search for tidal effects on volcanic activities at Mt. Etna and Stromboli. Proc Int Symp Earth Tides, Madrid, pp 765–774

Fadeli A (1987) An Analysis of seismic data from Merapi volcano (central Java). 1987 Contrib Merapi Volcano Res Group. Gadjah Mada Univ, Sekip Unit III, PO Box Bls 21, Yogyakarta, Indonesia

Fadeli A, Kirbani SB, Mugiono R, Schick R (1987) First Results on a new seismic station network on Merapi volcano (central Java, Indonesia). 1987 Contrib Merapi Res Group. Gadjah Mada Univ, Sekip Unit III, PO Box Bls 21, Yogyakarta, Indonesia

Heaton TH (1975) Tidal triggering of earthquakes. Geophys J R Astr Soc 43:307–326

Kirbani SB (1987) Seismicity (Tremors) of Merapi volcano (central Java, Indonesia). 1987 Contrib Merapi Volcano Res Group. Gadjah Mada Univ, Sekip Unit III, PO Box Bls 21, Yogyakarta, Indonesia

Klein FW (1976) Earthquakes swarm and semidiurnal solid earth tide. Geophys J R Astr Soc 45:245–295

Knopoff L (1964) Earth tides as triggering mechanism for earthquakes. Bull Seismol Soc Am 54:1865–1870

Lenat JF, Bachelery P (1988) Dynamics of magma transfer at Piton de la Fournaise volcano (Reunion island, Indian ocean). In: Yu King C, Scarpa R (eds) Modeling of volcanic processes. Vieweg, Braunschweig, pp 57–72

Mauk SJ, Johnston MJS (1973) On the triggering of volcanic eruptions by earth tides. J Geophys Res 78:3356–3362

Nicolis G (1987) Bifurcation and stochastic Analysis of nonlinear systems: an introduction. In: Nicolis C, Nicolis G (eds) Irreversible phenomena and dynamical systems analysis in geosciences. NATO ASI Ser C: Mathematical and physical sciences, vol 192, pp 3–29

Oppenheim AV, Schafer RW (1975) Digital signal processing. Prentice Hall, Englewood Cliffs, New Jersey

Rydelek PA, Davis PM, Koyanagi RY (1988) Tidal triggering of earthquakes swarms at Kilauea volcano, Hawai. J Geophys Res 93:4401–4411

Sassa K (1936) Micro-seismometric study on eruptions of the volcano Aso. Mem Coll Sci, Kyoto Imp Univ, Ser A 19:11–56

Schick R (1988) Volcanic tremor-source mechanism and correlation with eruptive activity. Natural Hazard 1:125–144

Schick R, Mueller W (1988) Volcanic activity and eruption sequences at Stromboli during 1983–1984. In: Yu King C, Scarpa R (eds) Modelling of volcanic processes. Vieweg, Braunschweig, pp 120–139

Ulbrich U, Ahorner L, Ebel A (1987) Statistical investigations on diurnal and annual periodicity and on tidal triggering of local earthquakes in central Europe. J Geophys Res 61:150–157

Van Bemmelen (1950) The geology of Indonesia, vol 1A. General Geology Government Printing office, The Hague

Wilson L, Head JW (1981) Ascent and eruption of basaltic magma on earth and moon. J Geophys Res 81:2971–3001

Young D, Zuern W (1979) Tidal triggering of earthquakes in the Swabian Jura? J Geophys 45:171–182

Recent Seismicity at Vulcano Island (Southern Italy) and Adjacent Regions: Time Patterns and Periodicities

A. Bottari, D. Caccamo, A. Montalto, G. Neri, and E. Privitera

Abstract

Data from the seismic network of the Aeolian Islands (January 1978–June 1988) show that both regional faulting earthquakes and shocks related to fluid dynamics at Vulcano volcano, are characterized by higher levels of activity at the start and end of the time interval investigated. In spite of this analogy, the hypothesis of a strict relationship between the two classes of events is not, at the present stage of the investigation, supported by shorter-term analyses, such as comparisons of the most relevant peaks of activity in the frame of the two different processes. Possible cause-effect relationships could, however, be masked by intrinsic complexities of the same phenomena and by the large number of factors and parameters involved in their occurrence. On the other hand, both tectonic and volcanic earthquakes display a remarkable periodicity of 420 days, which can tentatively be explained as an effect of cyclic stress variations produced on seismogenetic structures by the Earth's free nutation process (Chandler wobble).

1 Introduction

On the basis of the waveform features, earthquakes in the Vulcano area can be essentially distinguished into two main categories:

1. typical high-frequency shocks which show impulsive onsets and clear S-phases (an example is given in Fig. 1a);
2. fairly monochromatic low-frequency events (Fig. 1b) which are generally located at very shallow depth in correspondence with the main fumarolic fields (Fig. 2).

The first category includes earthquakes related to faulting mechanisms and involves a broad range of magnitudes (up to $M \approx 6$).

Shocks of the second type are characterized by lower magnitudes ($M < 2.5$) and have been interpreted (Blot 1971; Latter 1971; Del Pezzo and Martini 1981) as due to gas pressure variation phenomena occurring in the upper structures of the volcano. The monochromatic and prolonged character of these events is attributed (Blot 1971; Latter 1971) to resonance effects inside the gas-driving conduits.

Seismic data from the short-period local networks operating in northeastern Sicily, southern Calabria and the Aeolian Islands (Fig. 2) are used in order to investigate earthquakes of type-1 with magnitude $M > 2.5$, located within a radius of 30 km around Vulcano (focal depth $\leqslant$30 km). Considering the network

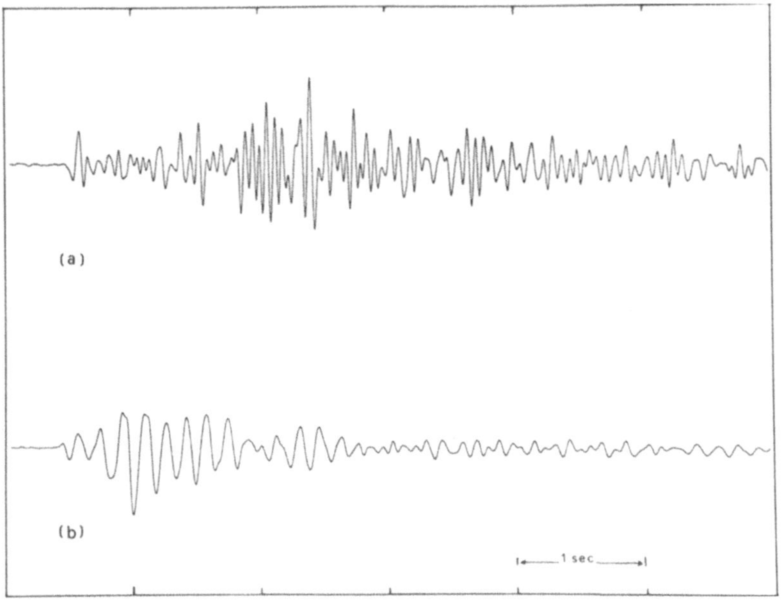

Fig. 1a,b. Morphotypes of high-frequency (**a**) and low-frequency (**b**) seismic events at Vulcano

features, no "loss" of earthquakes occurring in this sector is expected above M ≈ 2 (Di Prima et al. 1985). In some cases, data are integrated by information from the National Network catalogue (Istituto Nazionale di Geofisica, Rome).

Type-2 events, due to their low magnitude values, are only recorded by stations located in the Lipari-Vulcano area (Fig. 2). In this case, a magnitude threshold equal to 1 has been adopted, together with an hypocentral distance upper boundary of about 10 km.

An analysis of medium-term (10 years) time evolution of both these classes is one of the purposes of the present study, along with a comparison of their occurrence pattern. Such an approach is also suggested by recent episodes of temporal coincidence between noticeable short-term variations in the two earthquake classes. Finally, the autocorrelation technique is applied for analyzing possible periodicities in earthquake time occurrence, with particular reference to lower energy events.

2 Regional Faulting Earthquakes

Type-1 earthquakes located within a radius of 30 km around Vulcano at depth ≤30 km have been considered in the present study. Earthquake locations have been performed using the Equivalent Velocity Method (Caccamo and Neri 1984) and the Earth model as given from Jeffreys and Bullen (1967). Magnitudes have been estimated by magnitude-duration empirical relationships (Di

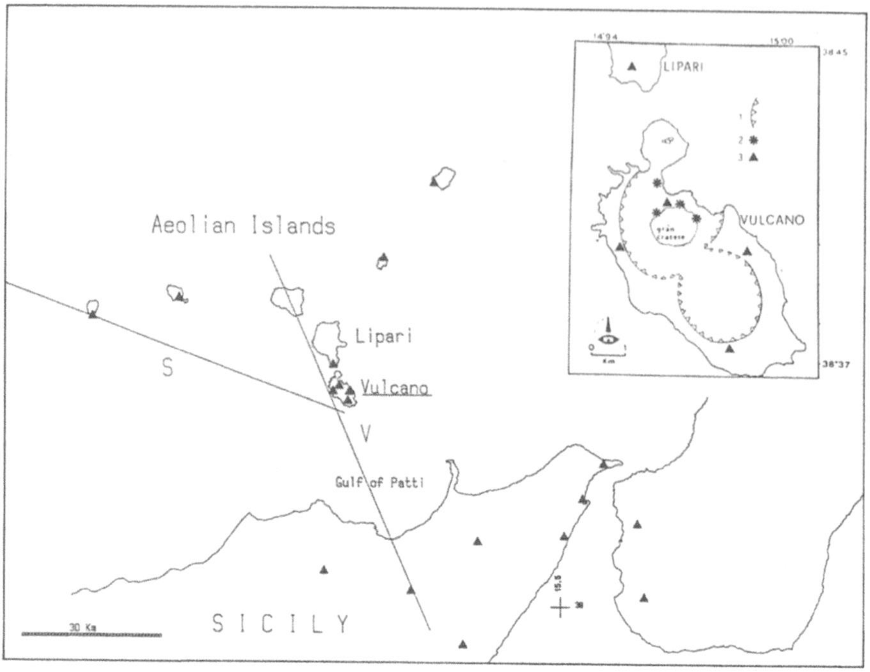

Fig. 2. Map of the Aeolian Island region with location of both seismic stations (*triangles*) and regional faults (*S* Sisifo, *V* Vulcano). A more detailed map of Vulcano island is given in the *inset* (*1* caldera rims; *2* main fumarolic fields; *3* seismic stations)

Prima et al. 1985) which are valid for stations of the Aeolian Island network. Two different magnitude thresholds (3.5 and 2.5) have been chosen for data collected in the Jan 78–Jun 88 and Jul 84–Jun 88 time intervals. Hereinafter, the two data sets will be respectively denoted as M35 and M25.

Figure 3 shows the cumulative strain release CSR ($\Sigma i \sqrt{Ei}$) for sample M35; the relation $\log(E) = 9.15 + 2.15 * M$, proposed by Di Filippo and Marcelli (1950) for Italian earthquakes has been used for converting magnitude into energy values. The main phase of activity occurred at the beginning of the sample (April 78–June 79) and is associated with a mainshock-aftershock sequence (Del Pezzo et al. 1983; Neri 1985a; Bottari et al. 1986) in the Gulf of Patti, south of Vulcano (Fig. 3, inset). The magnitude of the mainshock was 5.5, which is also the highest value in the whole 10-year period under study. The seismic strain release was at fairly low levels during the July 79–Dec. 84 time interval. A new, but less remarkable, increase is observed in the last part of the sample. The total amount of seismic energy released after the Gulf of Patti sequence (2.8 E + 18 erg) is compatible with the storage-release model proposed by Neri (1985b) for this region and implies a quite modest (<10%) release of the accumulated energy.

Sectors mainly affected by seismicity are located to the south and west of Vulcano (Fig. 3, inset). According to Bottari et al. (1986), most of the seismic

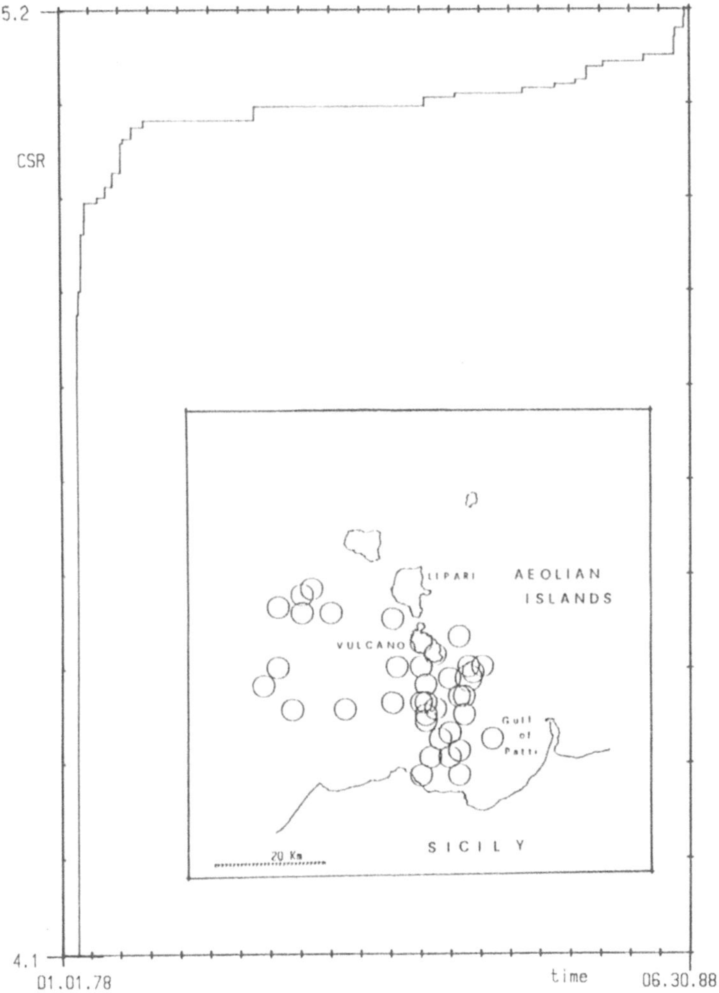

Fig. 3. Cumulative strain release *CSR* associated with earthquakes of magnitude ⩾3.5 which occured within a radius of 30 km around Vulcano (focal depth <30 km) during the January 1978–June 1988 period. The *CSR* unit is 10 ** $10\sqrt{\text{erg}}$. Epicenters of the same earthquakes are given in the *inset*

energy release can be related to the activity of two regional tectonic systems referred by some authors (see for example Finetti and Del Ben 1986) as the "Vulcano" and "Sisifo" faults (Fig. 2). In particular, the "Vulcano" fault (also known as Eolie-Tindari-Letoianni tectonic line) is a NNW-SSE lithospheric structure crossing northeastern Sicily and the Lipari-Vulcano eruptive complex (Cristofolini et al. 1985). This tectonic system has been considered as responsible for the 1978 Gulf of Patti sequence (Bottari and Neri 1980; Del Pezzo et al. 1983; Neri 1985a).

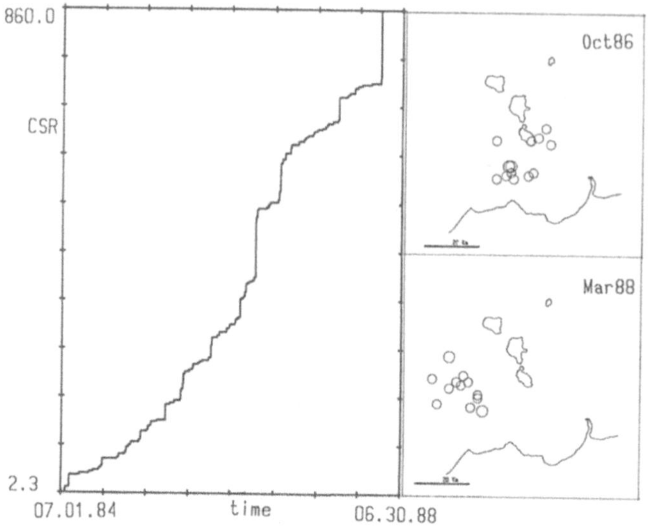

Fig. 4. Cumulative strain release (*CSR*) associated with earthquakes of magnitude >2.5 which occurred within a radius of 30 km around Vulcano (focal depth ≤30 km) during the July 1984–June 1988 period. The *CSR* unit is 10 ** 7$\sqrt{\text{erg}}$. Epicenter locations relative to the main phases of seismic strain release are also given

Figure 4 shows the cumulative strain release for sample M25. A comparison with Fig. 3 indicates that the general pattern of the seismic energy release in the Vulcano area is strongly governed by earthquakes of magnitude ≥3.5. According to Fig. 4, the most remarkable phases occurred in October 1986 and March 1988 and were located in the sectors defined as the most active ones on the basis of investigations on sample M35.

The autocorrelation technique has been applied to the temporal pattern (number of events per 10 days) of the M25 earthquakes. Both autocorrelation and significance computations have been carried out using standard procedures reported in Davis (1973). Results are shown in Fig. 5. Only two peaks (LAG = 10 and 42 decades, respectively) exceed the significance threshold 0.95. The first of these peaks (10 decades) is well "constrained" by the time extension of the M25 sample (138 decades). On the other hand, the period relative to the second peak (42 decades) is fairly close to the limit generally adopted in this kind of study: according to Davis (1973), the reliability of the results can be assumed if the condition SD ≥ 4P is satisfied, where SD is the sample duration and P the cycle length.

2.1 Fluid-Dynamics Earthquakes

Since the end of the last eruption (1890), volcanic activity at Vulcano has been mainly fumarolic. The main fumarolic fields are located in the northern part of the island, corresponding to the centers (Fig. 2) of the most recent eruptive

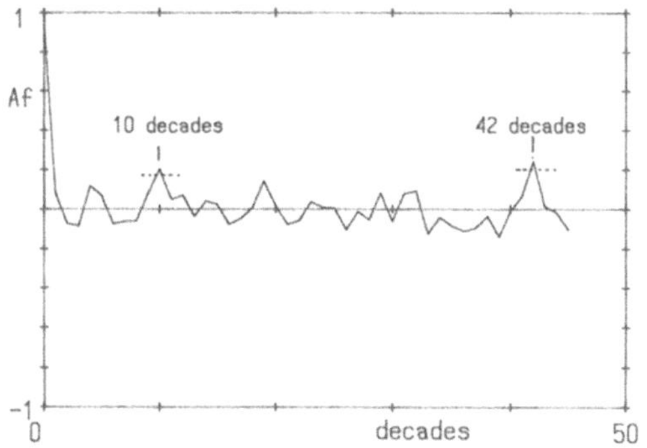

Fig. 5. Autocorrelation function (*Af*) of the type-1 earthquake temporal pattern (sample M25) expressed as number of events per 10 days. Peaks exceeding the significance threshold 0.95 (*dashed line*) are indicated

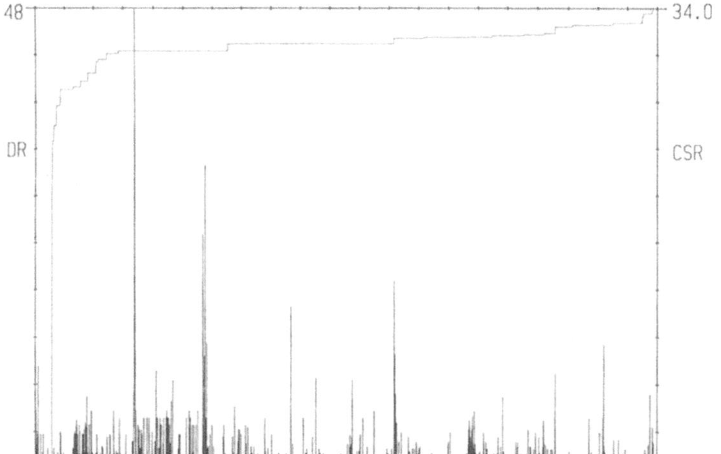

Fig. 6. Daily rate (*DR*) of type-2 earthquakes during the January 1978–June 1988 period and cumulative strain release (*CSR*) associated with M35 earthquakes. In this figure, the strain value of each event has been corrected by introducing a weighting factor 1/r, where r is the hypocentral distance from the Gran Cratere area. The CSR unit is 10 ** $8\sqrt{erg}$/km

activity. Fluid dynamics inside the volcano have been hypothesized (Blot 1971; Latter 1971; Del Pezzo and Martini 1981) as responsible for most of the local seismic events recorded on the island. These events (defined as type-2 earthquakes in the Introduction) are characterized by fairly monochromatic waveforms, low magnitude values (<2.5) and are mostly located at shallow depth (<1 km) beneath the Gran Cratere area (Fig. 2).

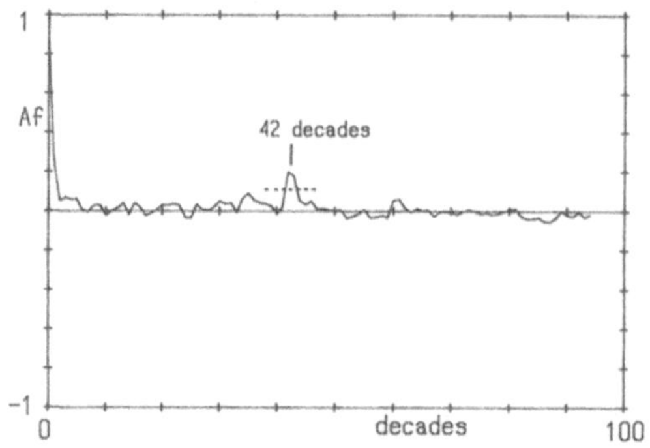

Fig. 7. Autocorrelation function (Af) of the type-2 earthquake temporal pattern expressed as number of events per 10 days. Basic data refer to the January 1978–June 1988 time interval. The only peak exceeding the significance threshold 0.95 (*dashed line*) is indicated

In this section, the seismic events with magnitude M > 1, located in a radius of about 10 km around the Gran Cratere area, have been analyzed. The sample so-defined is very close to coincidence with the type-2 earthquake set. The daily rate temporal pattern of this seismic class during the Jan. 78–June 88 time interval is reported in Fig. 6 and shows that the activity was high until early 1981, low in the 1981–1983 time interval, and attained fairly high levels again in the remaining part of the sample. This reveals some analogy with the temporal pattern of the regional faulting earthquakes, given in Fig. 3 in terms of cumulative strain release. Moreover, Fig. 6 evidences a remarkable tendency of type-2 shocks to cluster in the time domain; detailed analyses of the main clusters reveal their swarm character, because of the absence of mainshocks.

The temporal pattern (number of events per 10 days) of type-2 earthquakes has been analyzed by the autocorrelation approach. Results are reported in Fig. 7 and reveal that only one peak (LAG = 42 decades) exceeds the significance threshold 0.95. The ratio between sample duration (SD) and cycle length (P) is quite high (≈9). This periodicity is also confirmed by using different sampling criteria (1 and 30 days, respectively).

Figure 6 also allows us a comparison between time evolution of type-2 events and cumulative strain release associated with faulting earthquakes (M35). In this case, conversely with Fig. 3, the strain value for each earthquake has been "corrected" by introducing a weighting factor 1/r, where r is the hypocentral distance from the sector (Gran Cratere) where the most of type-2 events occurred. This procedure has been applied in order to better investigate possible causal relationships between the two seismic sequences. However, no clear evidence of such a phenomenon appears on the basis of the adopted criteria. This result is confirmed by using different values of the weighting factor ($1/r^2$, 1 /r and "no correction") and a lower magnitude threshold for faulting earthquakes (sample M25).

3 Discussion

A first consideration which arises from the results of this study is that fluid-dynamics quakes and faulting seismic events in the Vulcano area show some similarity regarding time evolution during the 1978–1988 period. In particular, both classes of events display higher levels of activity at the beginning of the sample, a clear decrease in its central part and a new, but less remarkable, increase in the last years investigated. Approximately, the low-activity period corresponds to 1980–1984 and 1981–1983 for faulting and fluid-dynamics earthquakes, respectively.

In spite of the above background analogy, no strict temporal relationship is, however, observed between the most relevant episodes of seismic energy release by faulting processes and peaks of volcanic earthquake recurrence (Fig. 6). The only case of evident correspondence is observed in early 1984 when a faulting seismic phase was followed in few days by a remarkable increase in the rate of fluid-dynamics' quakes. A similar episode was observed out of the period of the present investigation (August 1988). In this case, however, faulting earthquakes showed fairly low magnitude values ($M \leq 2.5$), but they took place very close to Gran Cratere area (Neri et al. 1989).

On the whole, available information does not give any clear evidence of a direct and mutual influence between the two seismic classes investigated. It is possible that triggering phenomena (if they exist) are masked by factors acting as noise sources, such as the ones produced by earthquake selecting criteria. In order to minimize these problems, more detailed analyses are needed. For instance, the above episode of August 1988 underlines the opportunity of both lowering magnitude thresholds and refining methods for earthquake selection, according to hypocentral location with respect to volcanic centers. Focal mechanisms should also be computed for better identifying faulting earthquakes able to perturb volcanic structures with consequent effects on fluid-dynamics' quakes. Of course, more information about the physico-chemical evolution of the volcanic system, and about its "sensitivity" to induced stresses, is also required in this context.

Another finding of this investigation is that both seismic classes show a marked periodicity close to 420 days. Such a finding recalls the "Chandler wobble", the motion of the earth's instantaneous rotation axis superimposed with "procession" and "forced nutation" phenomena. The Chandler wobble, also known as Earth's free nutation process, produces cyclic variations of geographic coordinates over the whole Earth's surface (Poma and Proverbio 1981; Vanicek and Krakiwsky 1982; Poma et al. 1987). Endogenous causes have been suggested (O'Connell and Dziewonski 1976; Dragoni et al. 1983) for explaining the origin of this phenomenon which is also widely considered as linked to seismic occurrence (Mansinha and Smylie 1967; Dahlen 1971, 1973; Press and Briggs 1975; O'Connell and Dziewonski 1976). Our hypothesis is that processes associated with the Chandler wobble, in the frame of a medium strongly affected by structural heterogeneities, could produce periodic modifications in the local stress field, with consequent effects on both classes of seismic phenomena investigated in the present work. Starting from this hypothesis, and in order to explain observed similarities in medium-term time patterns, it seems reasonable (1) to plan further detailed investigations on temporal variations of the Chandler wobble amplitude and (2) to attempt an evaluation of their local stress effects able to influence seismic processes.

References

Blot C (1971) Etudes sismologiques de Vulcano. Cah ORSTOM Sér Geophys 11:1–21

Bottari A, Neri G (1980) On locating local earthquakes at the Messina Straits network. Ann Geof XXXIII:281–296

Bottari A, Caccamo D, Cefali F, Lo Giudice E, Neri G (1986) Recent shallow seismicity in the southern Calabro-Peloritan Arc region. Ann Geophys 4B 1:91–98

Caccamo D, Neri G (1984) A new analytic procedure to determine hypocentral parameters of local seismic events. Bull Seismol Soc Am 74:665–677

Cristofolini R, Ghisetti F, Scarpa R, Vezzani L (1985) Characters of the stress field in the Calabrian Arc and southern Apennines (Italy) as deduced by geological, seismological and volcanological information. Tectonophysics 117:39–58

Dahlen FA (1971) The excitation of the Chandler wobble by earthquakes. Geophys J R Astr Soc 25:157–206

Dahlen FA (1973) A correction to the excitation of the Chandler wobble by earthquakes. Geophys J R Astr Soc 32:203–217

Davis JC (1973) Statistics and data analysis in Geology. Wiley, New York

Del Pezzo E, Martini M (1981) Seismic events under Vulcano, Aeolian Islands, Italy. Bull Volcanol 44 3:521–525

Del Pezzo E, Maresca R, Martini M, Scarpa R (1983) Seismicity of the Aeolian Islands, southern Italy. Ann Geophys 2 2:173–180

Di Filippo D, Marcelli L (1950) Magnitudo ed energia dei terremoti in Italia. Ann Geofis III:337–348

Di Prima S, Falsaperla S, Neri G (1985) Determinazione della magnitudo nelle stazioni della rete sismica delle isole Eolie mediante l'impiego della durata. Considerazioni critiche sulla applicazione di un siffatto criterio ai terremoti di origine vulcanica. Rend Oss Geofis Reggino XXIX:91–98

Dragoni M, Yüen DA, Boschi E (1983) Global postseismic deformation in a stratified earth: effects on Chandler wobble excitation. JGR 88:2240–2250

Finetti I, Del Ben A (1986) Geophysical study of the Tyrrhenian opening. Boll Geof Teor Appl XXVIII 110:75–155

Jeffreys H, Bullen KE (1967) Seismological tables. Br Assoc, Gray-Milne Trust

Latter JH (1971) Near surface seismicity of Vulcano, Aeolian Islands, Sicily. Bull Volcanol 351:117–126

Mansinha L, Smylie DE (1967) Effect of earthquakes on the Chandler wobble and the secular polar shift. J Geophys Res 72:4731–4743

Neri G (1985a) Distribuzione spazio-temporale della sismicità delle Eolie. Atti IV Convegno Gruppo Nazionale Geofis Terra Solida CNR: 825–837

Neri G (1985b) Considerazioni metodologiche sull'analisi del rilascio di energia sismica. Atti IV Convegno Gruppo Nazionale Geofis Terra Solida CNR: 135–144

Neri G, Patanè D, Privitera E, Montalto A, Spampinato S, Velardita S (1989) Intervento straordinario per la sorveglianza vulcanica di Vulcano, Estate 1988: il contributo dell'U.O. "Sismologia Eolie" dell'Istituto Internaznonale di Vulcanologia di Catania. Open-file Rep IIV-CNR 32/89

O'Connell RJ, Dziewonski AM (1976) Excitation of the Chandler wobble by large earthquakes. Nature (Lond) 262:259–262

Poma A, Proverbio E (1981) Random and long periodic variations in the Earth's motion. J Interdiscip Cycle Res 12 3:237–246
Poma A, Proverbio E, Uras S (1987) Long-term variations in the earth's motion and crustal movements. J Geodyn 8:245–261
Press F, Briggs P (1975) Chandler wobble, earthquakes, rotation, and geomagnetic changes. Nature (Lond) 256:270–273
Vanicek P, Krakiwsky E (1982) Geodesy: the concepts. North-Holland Publ Co, Amsterdam, pp 63–69

Arenal Volcano (Costa Rica) Premonitory Seismicity

R. Barquero, G.E. Alvarado, and T. Matumoto

Abstract

Arenal volcano has shown almost continuous seismic activity since 1968. A swarm of A-type earthquakes was recorded before and during the opening Pelean explosive phase of July 1968. A-type events were recorded again 5 months before the explosive Merapi-Soufriere-type eruptive phase of June 1975 which produced hot avalanches. There is no clear relation between local tectonic seismicity and the strong Strombolian explosive phases of 1984 and 1987. B-type earthquakes were recorded mainly during the first decade of activity. Explosion quakes with or without an air shock wave phase were recorded particularly during 1968, 1980 and from 1984 to the present day. The foci of the explosions were at least 100 m below the summit crater. Of all the seismic signals recorded at Arenal, the most relevant type for forecasting eruptive activity is tremor, since it appears to be related to crater degassing, block lava avalanches and resonance and oscillation in the conduit system. Weeks before a major explosive or effusive phase occurs, the A-type earthquakes and the tremors increase in frequency and amplitude and then gradually decrease, after which volcanic earthquakes begin to increase.

1 Introduction

Arenal volcano is a small stratovolcano in northwestern Costa Rica. The volcano is closely paired with an inactive cone, Chato, lying immediately southeast of the Arenal edifice (Fig. 1).

The Arenal-Chato system is located at the offset between two linear volcanic ranges, the Guanacaste Range to the north and the Cordillera Volcánica Central to the south. The limit between the two ranges has been postulated to represent a sharp break between different segments of the Caribbean plate overriding the active Central American subduction zone (Matumoto and Latham 1976; Carr and Stoiber 1977; Umaña et al. 1979; Burbach et al. 1984). Alternatively, Guendel (1986) suggests that the change in trend of the volcanic arc is due to gradual southward shallowing of the dip angle of the subduction zone. Volumetric variation (Carr and Stoiber 1977) and chemical differences along the length of the volcanic arc (Kussmaul et al. 1982; Carr 1984) may be due to differences in dip of the subduction zone.

The volcano began its formation about 3000 years ago (Borgia et al. 1988) at an intersection of N-S, NNE-SSW and WNW-ESE regional structural lineaments. An interesting and possibly significant feature is the progressive migration of the eruptive centres from SSE to NNW.

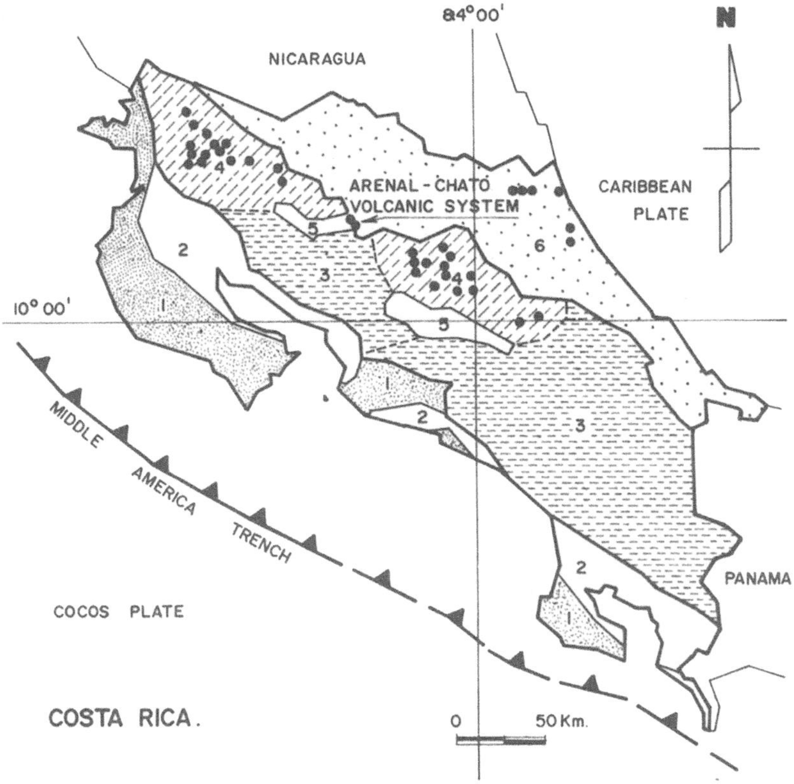

Fig. 1. Map of the geology of Costa Rica, showing structural trends and location of Arenal and Chato Volcanoes (see text for details). Geology of *Costa Rica*: *1*. Cretaceous to Middle Tertiary ophiolitic suite. *2*. Tertiary basin. *3*. Tertiary volcanic range: Cordillera de Tilarán and Montes del Aguacate to the northwest, Cordillera de Talamanca to the southeast. *4*. Active Quaternary volcanic range: Cordillera de Guanacaste to the northwest, Cordillera Central to the southeast; dots indicate major volcanoes. *5*. Intraarc basin: Arenal graben to the northwest, Valle Central to the southeast. *6*. Caribbean coastal plain

The Arenal volcano-tectonic depression and the nearby Tilaran igneous horst have axes trending NW-SE, parallel to the Nicaraguan volcano-tectonic depression; (Alvarado 1984; Fig. 1). The Tilaran range also shows a sub-radial pattern of normal and/or strike slip faults with vertex at Cerro Poco Sol, just a few kilometers from Arenal. On April 13, 1973, an earthquake of $M_s = 6.5$ and IMM = VIII (Plafker 1973) took place at the strike-slip Chiripa fault (Matumoto and Latham 1976).

In 1965, the Departamento de Geologia of Instituto Costarricense de Electricidad (ICE) began geological studies in the Arenal volcanic region for the Arenal Hydroelectrical Project. This project is the second largest in Central America and the Caribbean region, and provides 55% of the electrical energy for Costa Rica.

Feasibility studies for the Arenal Hydroelectrical Project were carried out from 1965 to 1968. Design activities were underway when the Arenal volcano erupted in July 1968 after approximately 300 years repose. Construction of the Sangregado dam was initiated in 1974 and completed in 1978. The dam is only 6.5 km away from the summit crater of Arenal. The intense and continuous activity of this volcano represents a potential hazard to the dam; thus, the Oficina de Sismología e Ing. Sísmica of the Geology Department of ICE has been continuously monitoring the volcano since 1974. A permanent observatory, the Observatorio Vulcanologico del Arenal (OVA), was created in 1988.

Matumoto (1968, 1976) and Minakami et al. (1969) have carried out seismic observation programs at the volcano since 1968. Since 1974, ICE has carried out numerous studies and collected much of the seismic information documented and summarized in this work.

2 Instrumentation

A permanent seismographic network was established to monitor the northwestern Costa Rican region around Arenal lake and Arenal volcano between 1974 and 1978. When Arenal volcano began another explosive phase in 1975, two more stations (AR-9 and AR-10) were installed closer to the volcano (Fig. 2).

The radio telemetered seismic stations were equipped with 2 Hz vertical seismometers (1 Hz for stations AR-6, AR-7 and AR-8). The signals from these stations were sent to the recording centre, the Observatorio Sismologico de Chiripa in Tilaran by VHF radio. The signals were then connected to a strip chart recorder through a 20-s digital memory delay channel and triggering circuit. In addition to this recording system, the seismic signal from station AR-1, located on the western flank of Arenal, was continuously monitored on a drum recorder.

At the present time, two seismographic stations are continuously monitoring the volcano: AR-6, located 22 km west of Arenal, and FOR, located 4 km east of the summit crater. These stations have short period vertical seismometers (1 Hz) and are telemetered by a telephone line to a recording centre at ICE in San José. These two stations are part of the joint ICE-Universidad de Costa Rica National Network (RSN: ICE-UCR).

2.1 Volcanic Activity

Beginning in 1965, various premonitory manifestations of volcanic activity were detected (Fig. 2): (1) affected animals and vegetation; (2) colorless gas was

Fig. 2. Detailed map of Arenal Volcano showing location of seismic stations. Inset shows Sangregado dam (see text for details)

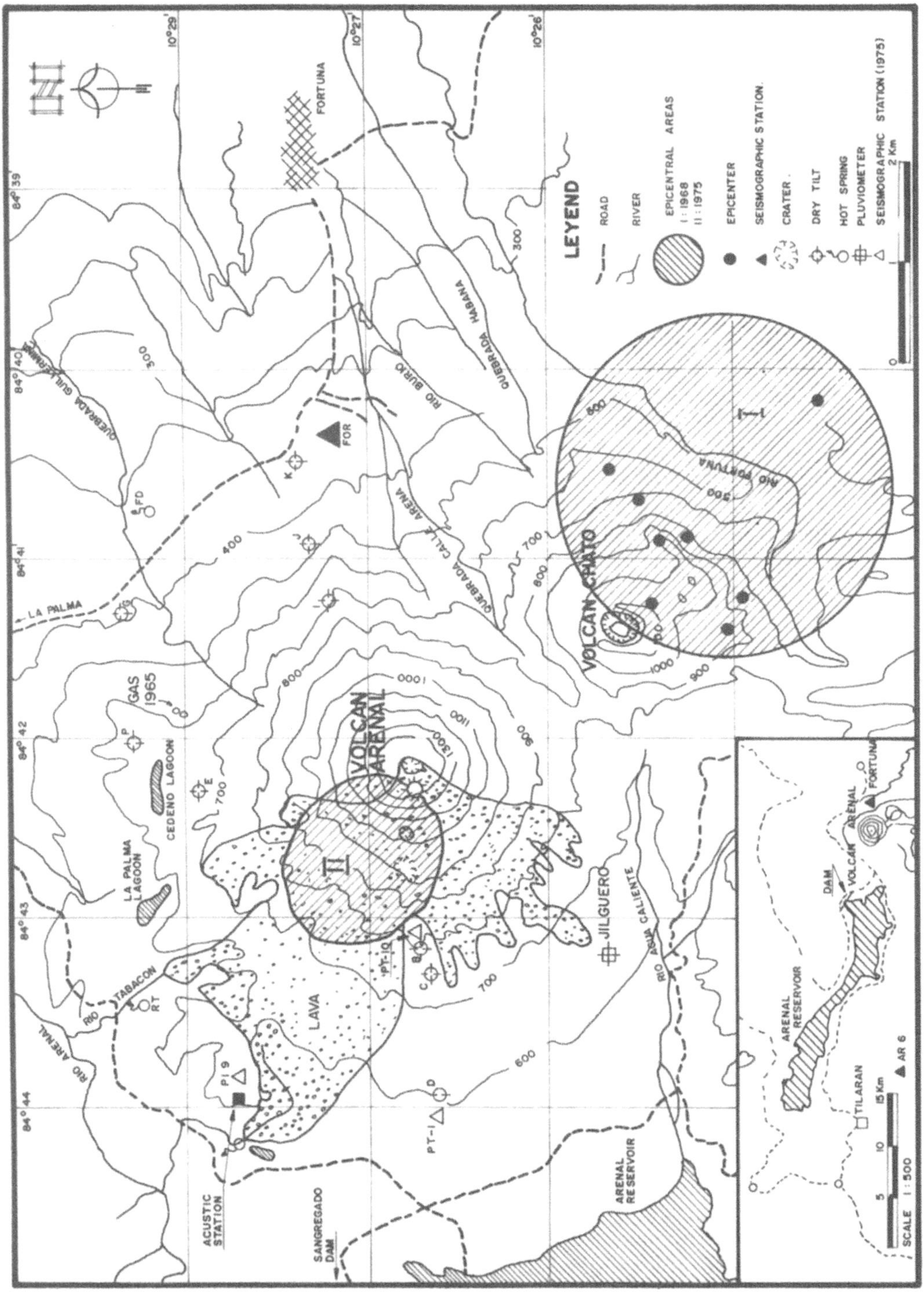
LEYEND
ROAD
RIVER
EPICENTRAL AREAS
I : 1968
II : 1975
EPICENTER
SEISMOGRAPHIC STATION
CRATER
DRY TILT
HOT SPRING
PLUVIOMETER
SEISMOGRAPHIC STATION (1975)
2 Km
VOLCAN ARENAL
VOLCAN CHATO
FORTUNA
LA PALMA
LA PALMA LAGOON
CEDENO LAGOON
GAS 1965
RIO BURIO
QUEBRADA HABANA
RIO FORTUNA
QUEBRADA CALLE ARENA
RIO TABACON
RIO ARENAL
RIO AGUA CALIENTE
JILGUERO
LAVA
ACUSTIC STATION
SANGREGADO DAM
ARENAL RESERVOIR
FOR
PT-I
PT-10
P I 9
ARENAL RESERVOIR
DAM
TILARAN
AR 6
SCALE 1 : 500

reported on the NE flank of the volcano; (3) the Cedeño lagoon (on the north flank of Arenal) dried out and the fish died; (4) a new hot spring (Fuente Doris) appeared; and (5) the Tabacón river became hot (Avila 1978; Alvarado and Barquero 1987).

On the morning of July 1968, an explosive phase started at Arenal volcano. During the first 3 days this phase destroyed an area of 15 km^2, killing 78 people, and affecting an area of 230 km^2. Three new craters A, B and C (Fig. 2) were formed during the initial explosions (Melson and Saenz 1968; Alvarado et al. 1988).

On September 19, 1968, lava started to flow from the lower crater (A) and continued until 1973. In 1974, the lateral activity at crater A migrated up the flank to crater C (Melson and Saenz 1968; Bennet and Raccichini 1977). Rheologic aspects of the dynamics and structure of lava flows are treated by Borgia et al. (1988) and Borgia (1988).

Between June 17 and 21, 1975 several glowing avalanches were generated from crater C. This phase was preceeded by a seismic swarm at the beginning of 1975 and two months before by an increase in volcanic earthquakes and tremor (Matumoto 1976; Matumoto and Umaña 1976; Van der Bilt et al. 1976).

During the period 1975 through 1983, the effusive activity continued intermittently. In 1984 Arenal entered a new Strombolian phase characterized by an increment in the explosive activity, changing to Strombolian-effusive from 1985 to the present (Alvarado and Barquero 1987).

2.2 Volcanic Seismicity

Earthquake activity in volcanic areas is defined as "volcanic seismicity" even though it is often difficult to differentiate between events of volcanic (particularly volcano-tectonic) origin and those of tectonic origin. An earthquake is considered "volcanic" when it is directly related to magmatic processes, such as magma injection or gas expansion, but this relationship is usually hard to demonstrate. Table 1 shows the classifications proposed by Minakami (1969) and Latter (1979), modified for Arenal using previous work by Matumoto (1976), Alvarado and Barquero (1987), and Morales et al. (1988).

The spatial and temporal relation between recorded seismic signals type and eruptive phase at Arenal are described as follows:

A-Type. The volcano-tectonic earthquakes were distributed toward the SE part of Chato volcano (Fig. 2) during a short period of seismic observations from August 26 to September 29, 1968. A total of nine earthquakes were located with hypocentres between 2 and 8 km in depth (Minakami et al. 1969; Minakami 1974). A-type events were again recorded between January and Febraury, 1975 (Fig. 2).

B-Type. These events were also recorded more frequently during the 1968 observations, and were distributed around the new craters, especially at the lowest crater, A (Minakami et al. 1969). B-type earthquakes were recorded

Table 1. Classification of volcanic earthquakes at Arenal volcano

Type	Name	Frequency (Hz)	Characteristics
A	Volcano-tectonic	⩾3	Well-defined P-S wave phases, event generated by displacement in fractures.
B	Volcanic	<3	Low frequency and low magnitude shallow events.
E	Explosive E_H (E + T_H coda) E_I (E + T_I coda) E_L (E + T_L coda)	<3	Directly related to Strombolian explosions (E_c), sometimes showing a high frequency air shock wave phase.
T_H	Tremor (high f.)	3–15	Continuous seismic noise associated with avalanches and rock fall from the lava front or strong degassing.
T_I	Tremor (Intermed. f.)	1–3	Vibration of short to long duration originated from diverse sources: degassing, gas fluctuation, resonance in conduits.
T_L	Tremor (low f.)	0.3–0.6	Oscillation of magmatic conduits.

again during another period of observation in 1975. Their source was the active crater C (Matumoto 1976) (Fig. 2).

E-Type. Two different kinds of explosive earthquakes have been recognized at Arenal:

E_A. Events with a very clear air-shock wave phase emergent arrivals. These seismic signals are related to stronger explosions and are recorded mainly in periods of strong activity. Nevertheless, atmospheric conditions (wind direction, cloud cover, rain) as well as airborne ash or dust may interfere with the arrival time and intensity of the sound wave.

E_C. Explosive events with impulsive arrivals and intense amplitudes. Several wave phases of 1 and 2 Hz and another of high frequency (3 to 5 Hz) can be distinguished. All are related apparently to the rise of bubbles, and rupture and recuperation of the near surface conduit lava during and after an explosion. E_C-type are the most common events at Arenal, and an average of 3 to 5 are recorded daily at ordinary levels of activity. Usually these earhtquakes show tremor codas of high, medium and low frequencies (E_H, E_I, E_L, respectively).

Tremor. These signals have various frequencies and are the dominant type of seismic signal recorded. They can be classified as follows.

1. High frequency tremor (T_H): these signals are associated with block lava avalanches from the upper crater (C), collapse of lava flow fronts, or strong

degassing at the summit crater. They are a continuous noise with frequencies of more than 3 Hz and may last for several hours, usually appearing after explosions.

2. Medium frequency tremor (T_I): irregular vibrating patterns that last from minutes to hours, probably from a shallow source. At times, these events are so strong that they can be recorded at the Jicaral station located 80 km from the volcano. T_I-type tremor appears to correlate with vigorous degassing and oscillation of the conduit lava column. E-type earthquakes are sometimes accompanied by tremor codas of high (E_H), intermedium (E_I) and low (E_L) codas. In 1975, the volcanic activity was intermittent with durations of one to several days; the frequency of volcanic tremor T_I-type showed a strong correlation with the active phases. During the quiet phases the predominant frequency was 1.3 Hz and during pre-eruptive phases the frequency increased to 3.3 Hz. The change in the frequency of tremor T_I is apparently due to rise or subsidence of the magma column or changes in its rheology. If the fundamental mode corresponds to the resonance of the air column, it should be 120 m long during the quiet periods (1.3 Hz). When the activity begins, the magma rises and the air column shrinks to 50 m (3.3 Hz). As an example, after the explosions of June 1975, Rodrigo Saenz (in W. Melson, pers. comm. 1988) observed that the level of lava went down within the crater approximately 100 m. According to the model, the magma column goes down gradually as well as the resonance frequency. However, recent field observations indicate that there is no evidence relating subsidence of the magma column to decrease in frequency. This variation thus seems to be related to resonance in the conduits.
3. Low frequency tremor (T_L): tremor with a characteristic two-second period is associated with oscillation of the magma chamber or an organ-pipe effect in the conduit (Matumoto 1976; Alvarado et al. 1988).

3 Premonitory Seismicity at Arenal Volcano

Seismic phenomena previous to the paroxismic explosion of July 29, 1968 started as early as May, 1968 (Avila 1978). On July 28, beginning at approximately 23:00 LT (LT = GMT-6 h), the inhabitants of Tabacon and Pueblo Nuevo, two now extinct small towns at the base of the volcano, reported numerous earthquakes (Fig. 2). The small earthquakes gradually increased in magnitude and number. The intensity was great enough to awaken people who, fearing the possible collapse of their houses, fled outdoors. Survivors of the blast reported swinging lamps, rattling of dishes and shaking of furniture. The number of earthquakes felt ranged from 20 to over 100. This variation may be due to the observer's distance from the volcano. The swarm of earthquakes decreased from 30 to 120 min before the initial explosion. No earthquakes were felt after the first eruption. The maximum intensity of the ground movement is estimated to have been between IV and V (MM) at La Fortuna and La Palma, 4 and 6.5 km respectively from the volcano (Matumoto 1968, 1976).

The largest event in the swarm may have reached 4.5 magnitude, and the energy released may have been up to 10^{21} erg. (Matumoto 1976). During the period from August 6 to August 14, a short period seismograph in continuous operation recorded approximately 220 microearthquakes at an average rate of 1.4 per hour. The majority of the events were of shallow origin. The temporal distribution of microearthquakes was rather uneven, with some peak events of 5 to 30 times more than the average. No direct relation was observed between seismicity and eruptions, suggesting that magmatic activity decreased to such a degree that no external activity could be observed (Matumoto 1968, 1976).

A period of explosive volcanic activity occurred in June 1975. A-type earthquakes occurred between the end of January and February, before the reactivation of the volcano (Figs. 3 and 4). B-type events were present during the whole period and E-type were associated with the eruptions, sometimes showing a very clear air-shock wave phase, frequently followed by tremor codas of low (E_L) or high (E_H) frequency (Matumoto 1976; Alvarado et al. 1988).

A moderate increment in the explosive activity was recorded in August 1980 with eruptions of gas and ash reaching 500 m over the upper crater C. Recorded T seismic signals (2 Hz) suggest that the magma column may have been rising.

In June 1984, a new strombolian explosive phase was preceeded several weeks before by an increase in the frequency and amplitude of tremors from 0.8 to 3.3 Hz (Montero 1984), suggesting movement and resonance of the magma column. Most of the volcanic earthquakes recorded from this period up to the present time at stations FOR and AR6 correspond to explosion earthquakes (E-type and E_I types) and tremor (T_I to T_H). Montero (1984) noted six volcano-tectonic earthquakes between June 11–12, and one on August 28, 1984. But because of poor epicentral location, they could have been ordinary local tectonic events.

On March 7, 1987 a swarm of shallow (less than 10 km) tectonic earthquakes 15 km east of Arenal was recorded by the National Seismic Network stations. The maximum magnitude was 4.2 with an intensity of IV (MM) at La Fortuna, (6.5 km East of Arenal).

Approximately 12 days later, an increment in the eruptive activity took place, culminating with small nuees ardentes between March and September of that year (SEAN Bulletin 1987). The March 7 swarm was related to a local fault, but, as had apparently happened in the cases before, probably perturbed the eruptive system at depth, increasing the eruptive activity days or months afterwards.

Finally, the activity at Arenal from 1985 to 1992 has been very constant with Strombolian phases and lava flows (Fig. 5).

4 Conclusions

Studies of the seismicity at Arenal volcano during recent eruptive phases reveal several features. A-type earthquakes show systematic patterns associated with the strong explosive phases: (1) previous to and during the initial explosion of

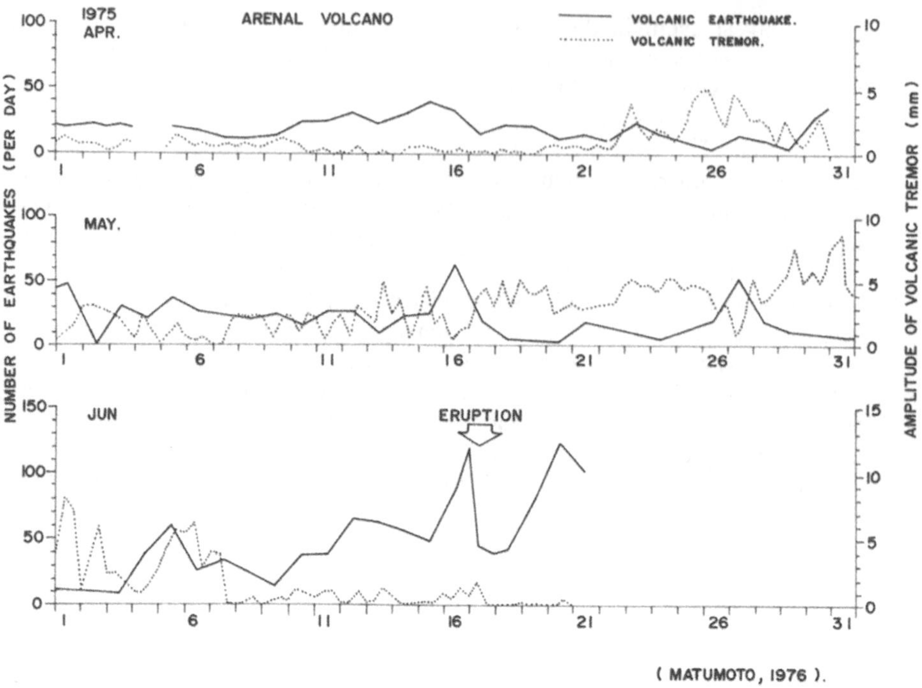

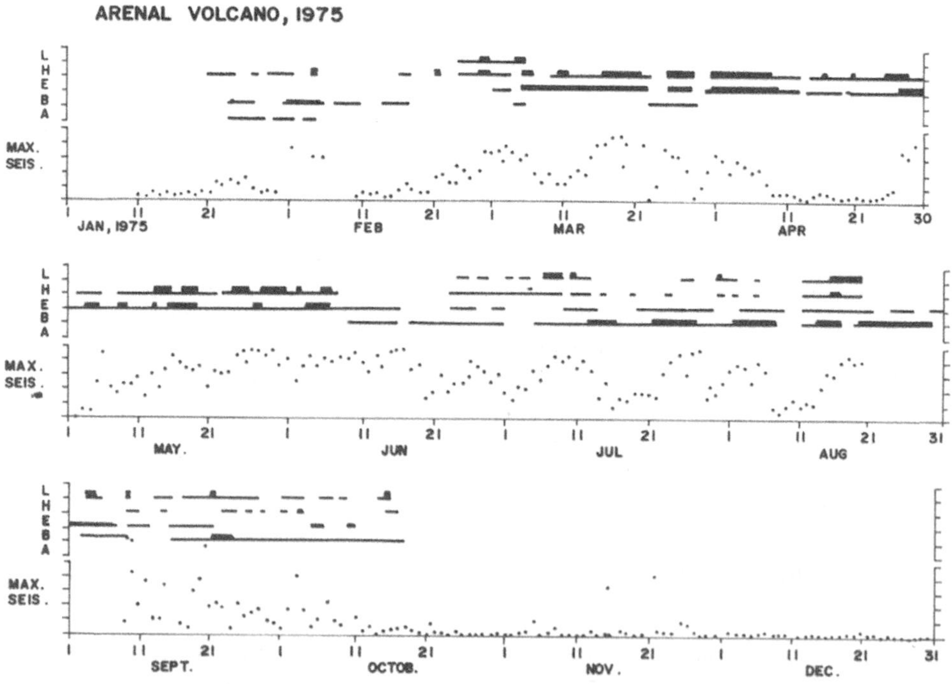

Figs. 3 and 4. Earthquake activity (1975) at Arenal (see text for details)

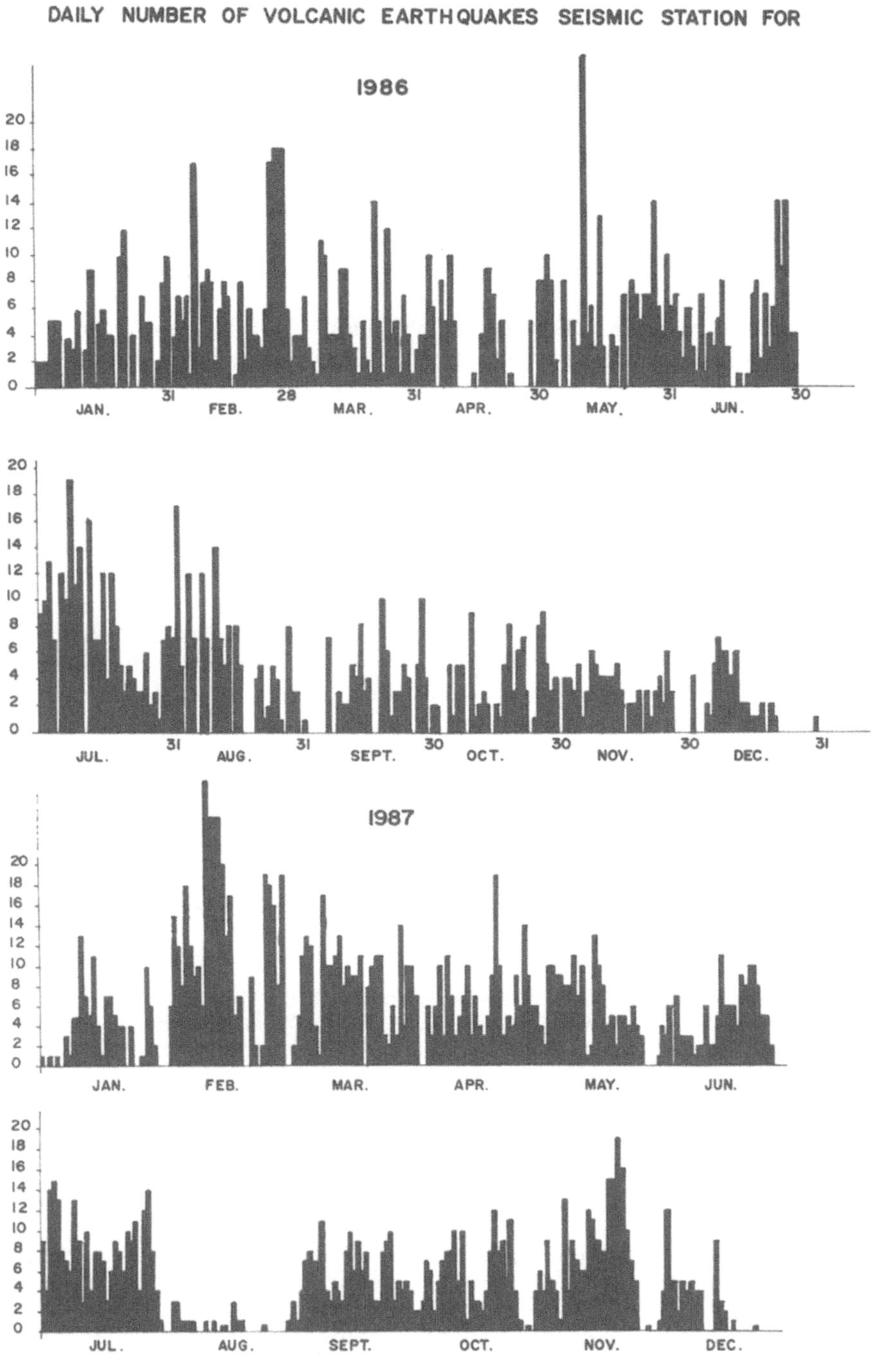

Fig. 5. Earthquake activity in 1986 and 1987 (see text)

1968 and (2) previous to the explosions and glowing avalanches emplacement of 1975. Tectonic-like earthquakes were recorded within a radius of 20 km from Arenal's crater during moderate increases in the eruptive activity in 1984 and 1987, but the temporal relationship is not clear. The infrequency of A-type events could be explained by an open and continuous conduit system under low lateral stress.

B-type events, originating near the active crater were recorded during the first decade of activity, but are not frequent thereafter. An irregular increase in the number of B-type earthquakes prior to main explosive phases has been observed.

Explosion earthquakes (E-type) are the most common events and have averaged three to five strong events per days during regular active periods. Two different kinds of E-type earthquakes have been recognized: E_A-type, events with a clear air-shock wave phase, and E_C events with impulsive arrivals and high amplitudes, varying in frequency.

Three main kinds of tremor have been recognized according to their frequency range: high frequency (T_H), related to block lava avalanches or vigorous degassing; intermediate frequency (T_I), associated with vigorous degassing and oscillation of the lava column, and low frequency (T_L), probably related to oscillation and resonance of the magmatic conduits.

An inverse relation between tremor and the number of volcanic earthquakes (B and E-type) has been observed at Arenal where an increment in duration and amplitude of tremor takes place several weeks before an explosive period. Several days before the eruption, there is a gradual decrease in tremor as the number of volcanic earthquakes increases.

The complexity of the eruptive conduit system of Arenal (including the variety of shapes of inactive and active pipes) as well as the changes in the rheology of an andesitic-basaltic magma column, are probable factors affecting the diversity and temporal stability of tremor.

Arenal has shown an extraordinary long period of activity lasting for 24 years up to the present time.

Acknowledgements. Our gratitude to the Departamento de Geologia of ICE, for its support to the authors over the past 20 years. To W. Melson of the Smithsonian Institution (USA), Andrea Borgia of the Jet Propulsion Laboratory (USA) and Deborah Jerez of Weston Consultants (USA) for discussions concerning this manuscript. To Milton Corrales of ICE for his technical assistance in the laboratory.

References

Alvarado GE (1984) Aspectos petrologicos-geologicos de los volcanes y unidades lavicas del Cenozoico Superior de Costa Rica. Thesis, Esc Centroam Geol, Univ Costa Rica, San Jose, xii + 183 pages

Alvarado GE, Barquero R (1987) Las señales sismicas del volcan Arenal (Costa Rica) y su relacion con las fases eruptivas (1968–1986). Cienc Tec II (1):19–35, San Jose

Alvarado GE, Matumoto T, Borgia A, Barquero R (1988) Sintesis geovulcanologica del Arenal (Costa Rica): 20 años de continua actividad eruptiva (1968–1988). Bol Obs Vulcanol Arenal I (1):1–55, San Jose

Avila G (1978) Investigacioń y vigilancia del volcan Arenal, Alajuela, Costa Rica. Inst. Costavricence Electricidad (ICE). Informe interno. San Jose, Costa Rica, 40 pp

Bennet FO, Racichini S (1977) Las erycione del volcań Arenal, Costa Rica. Rev Geogr. de Am. Central, no 5-6:7-35 Univ. Nacional, Heredia, Costa Rica.

Borgia A (1988) Physical aspects of eruptions at Arenal and Poas volcanoes, Costa Rica. Thesis, Princeton Univ, USA, 233 pp

Borgia A, Casertano L, Cigolini C (1983) Extructura y dinámica de los flujos de lova del Arenal. Bol. de Vulconol. Univ. Nacional Heredia 14:79-80

Borgia A, Poore C, Carr MJ, Melson WG, Alvarado GE (1988) Structural, stratigraphic, and petrologic aspects of the Arenal-Chato volcanic system, Costa Rica: Evolution of a young stratovolcanic complex. Bull Volcanol 50:86-105

Burbach G, Frohlich C, Pennigton W, Matumoto T (1984) Seismicity and tectonics of the subducted Cocos Plate. J Geophys Res 81:4885-4895

Carr MJ (1984) Symetrical and segmented variation of physical and geochemical characteristics of the Central American Volcanic Front. J Volcanol Geotherm Res 20:231-252

Carr MJ, Stoiber RE (1977) Geologic setting of some destructive earthquakes in Central America. Bull Geol Soc Am 5:387-410

Ferrick MG, Gamar A, Lawrence WF (1982) Source Mechanism of Volcanic Tremor. J Geoph Res 97 (B10):8675-8683

Guendel F (1978) On the relationship between earth tides and volcanic activity at Arenal Volcano. Thesis, Univ Texas, Costa Rica, MA, 32 pp

Guendel F (1986) Seismotectonics of Costa Rica, on Analytical view of the southern terminus of the Middle America French. P.H.D. Thesis, Univ. of California at Santa Criy, USA, 157 pp

Kussmaul S, Paniagua S, Gainza J (1982) Recopilacion, clasificacion e interpretacion petroquimica de las rocas igneas de Costa Rica. Inst Geogr Nac, Inf Sem July-Dec, 18:17-79, San Jose

Latter JH (1979) Volcanological observations at Tongariro National Park, 2. Types and classification of volcanic earthquakes, 1976-1978. Dep Sci Inst Res Geophys Div, New Zealand, Rep 150, ix + 60 pp

Malone SD, Endo ET, Weaver CS, Ramey JW (1981) Seismic monitoring for eruption prediction. US Geol Surv Prof Pap 1250:803-813

Matumoto T (1968) Seismological observations at Mt. Arenal and other volcanoes in Costa Rica. Preliminary Rep, 3 pp (unpublished)

Matumoto T (1976) Prediction of a volcanic eruption implied from seismic data Rev Geogr Inst Panam Geogr Historia 5:285-293, San Jose

Matumoto T, Latham GV (1976) Results from the Arenal Seismic Network Inst Costarricense Electricidad Dept Geol, 33 pp (unpublished)

Matumoto T, Umaña JC (1976) Informe sobre las erupciones del volcan Arenal ocurrida el 17 de junio de 1975. Rev Geogr Inst Panam Geogr Historia 5:299-235, San Jose

Melson WG, Saenz R (1968) The 1968 eruption of Volcan Arenal. Costa Rica: Preliminary summary of field and laboratory studies. Smith Inst Cent Short-lived phenom, Washington, 51 pp

Minakami T (1969) earthquakes originating from volcanoes. Atti xvii Conv Ass Geof Ital 1-4:1-38, Napoli

Minakami T (1974) Prediction of volcanic eruptions In: Civetta L, Gasparini P, Luongo G, Rapolla A (eds) Physical Volcanology. Elsevier Sci Publ Co, N.Y. pp 313-333

Minakami T, Utibori S, Hiraga S (1969) The 1968 eruption of volcano Arenal, Costa Rica. Bull Earth Res Inst 47:783-802

Montero W (1984) Informe sismologico del volcan Arenal, periodo de junio a agosto de 1984. 22 pags. + figs. Inst Costarricense Electricidad, Dept Geol, 22 pp, figures (unpublished)

Morales LD, Soley JF, Alvarado GE, Borgia A, Soto G (1988) Analisis espectral de algunas señales sismicas y su relacion con la actividad de los volcanes Arenal y Poas, Costa Rica. Bol Obs Vulcanol Arenal I (2):1-25 San Jose

Plafker G (1973) Field reconnaisance of the effects of the earthquake of April 13, 1973, near Laguna de Arenal, Costa Rica. Bull Seismol Soc Am 63:1847-1856

SEAN Bull (1987) vol 12, n°3, 4, 7 and 8. Washington, DC

Umaña J, Guendel F, Pautre A (1979) Contribution d'un barrage dans une zone sismiquement et volcanologiquement active Bull Int Assoc Eng Geol 20:117–120
Van der Bilt H, Paniagua S, Avila G (1976) Informe sobre la actividad del volcan Arenal iniciada el 17 de junio de 1975. Rev Geof Inst Panam Geogr Historia 5:295–298

Analysis of the Seismic Activity Related to the 1982 Eruption of the El Chichon Volcano, Mexico

F. MEDINA, L. GONZALEZ, G. GUITERREZ, R. AGUILERA, and J.M. ESPINDOLA

Abstract

Interesting seismic activity occurred before, during and after the 1982 paroxysmal explosions of the El Chichon volcano in southeastern Mexico. The first eruption was preceded by a large amount of seismic energy released (10^{19} ergs). Periodical increments (6–7 days) in the seismic energy released were observed, the same period separated the first from the second and third eruptions. The majority of events prior to eruptions were at intermediate to superficial depth.

b-Value variations were observed, maximum value was reached during the period between the first and the second third eruptions, this aspect was probably related to an intense shallow fracturing process.

During the 48 h before eruptions, seismic signals probably related to shallow explosions were detected. Location of the seismic events post eruptions revealed the probable location of the magmatic reservoir.

1 Introduction

Towards the end of March and the beginning of April, 1982, the El Chichon volcano (located in the State of Chiapas, Mexico) generated dense eruptive columns, pyroclastic flows, basal surges and the fragmentation of the upper parts of the volcanic edifice, developing a new 1 km-wide crater. This paroxysmal activity was generated after a 600 year repose period, and due to the direct action of the eruptive phases hundreds of people died and thousands were injured (Medina 1982).

Analysis of the preceded seismic activity due to volcanic eruptions is considered the most useful tool for hazard prediction. However, at the present time there is no specific indicator which allows us to estimate the possibility that a paroxysmal explosion will occur. Tokarev (1963) proposed, using the seismic energy released before eruptions, an empirical hyperbolic relation between the accumulated sum of the square root of the seismic energy of the precursory earthquake swarms and eruption time. Minakami (1974) pointed out the presence of shallow earthquakes, called type B with special characteristics, before eruptions. Gresta and Patané (1983) reported b value variations associated with eruptive activity. Yokoyama (1988) proposed that eruptions should occur when seismic energy release reaches a peak of activity and a cumulative level of 10^{17}–10^{18} ergs.

For the El Chichon volcano, important seismic activity occurred before, during and after the three main eruptive phases. In this study, an analysis of this seismic activity is presented.

2 Seismic Pattern

Seismic activity associated with the 1982 eruptions of the El Chichon volcano started in December, 1981. At this time, a seismic network had been in operation relatively close to the volcano, to study induced seismicity caused by impounding dams. This network was operated by the Federal Electricity Power Commission. One of this stations, located 60 km to the south of the volcano, with high gain and the only one in continuous operation during the period December-1981 to May-1982, provided the information for the premonitory seismic activity. During January and February, the background seismicity was scarce; during March, seismicity increased drastically from 20 events a day to 200 on 27–28 (that is, 36 h before the first eruption). Figure 1 shows the daily number of events during March and April. Most of the seismic events during March are characterized by an unclear definition in the P and S phases; the seismic signals appear as type B according to the Minakami (1974) description. However, Havskov et al. (1983a) classified these events as type 1 because sometimes the S phase is read with some difficulty; moreover, the events covered a wide range of magnitudes up to 4.1 inclusive.

Two seismic signals (~1 s in period) started to be seen 48 h before the first eruption (on March, 28 at 23:32 h, local time). These seismic signals are classified by Havskov et al. (1983a), as type 2 and type 3 events. In Fig. 2 the different events are shown. In general, type 2 events present an anomalous coda length compared to their maximum amplitude.

Background seismicity saturated the seismograms 6 h before the first eruption; therefore, the seismogram from March 28 (13 h local time) to March 29 (13 h local time) could not be read due to the probable tremor mark on it.

After this first eruption, a 300–400 m diameter crater was formed at the top of the old dome (Medina 1986). Little seismicity was observed in the following 24 h after the first eruption (12 to 15 events) and on March 30, seismicity increased saturating once again the seismogram. During April 1 and 2, little seismicity was observed, although background seismicity increased. During the morning of April 3, seismicity increased again; that day, at 19:35, the second large eruption took place followed by the third 11 h later (on April 4 at 5:33 h). During this eruptive activity, the crater was enlarged to one kilometer in diameter (Medina 1986).

Over the following 4 days, the background seismicity impeded reading of the seismograms. However, events of type A or typical tectonic events started to be seen. After April 7 (72 h after the third eruption) seismicity began to decay linearly with time, from more than 400 events/day to only 30 events. Only events of type A (or type 4 in the Havskov et al. 1983a, classification) were registered during the April 7 to April 30 time span.

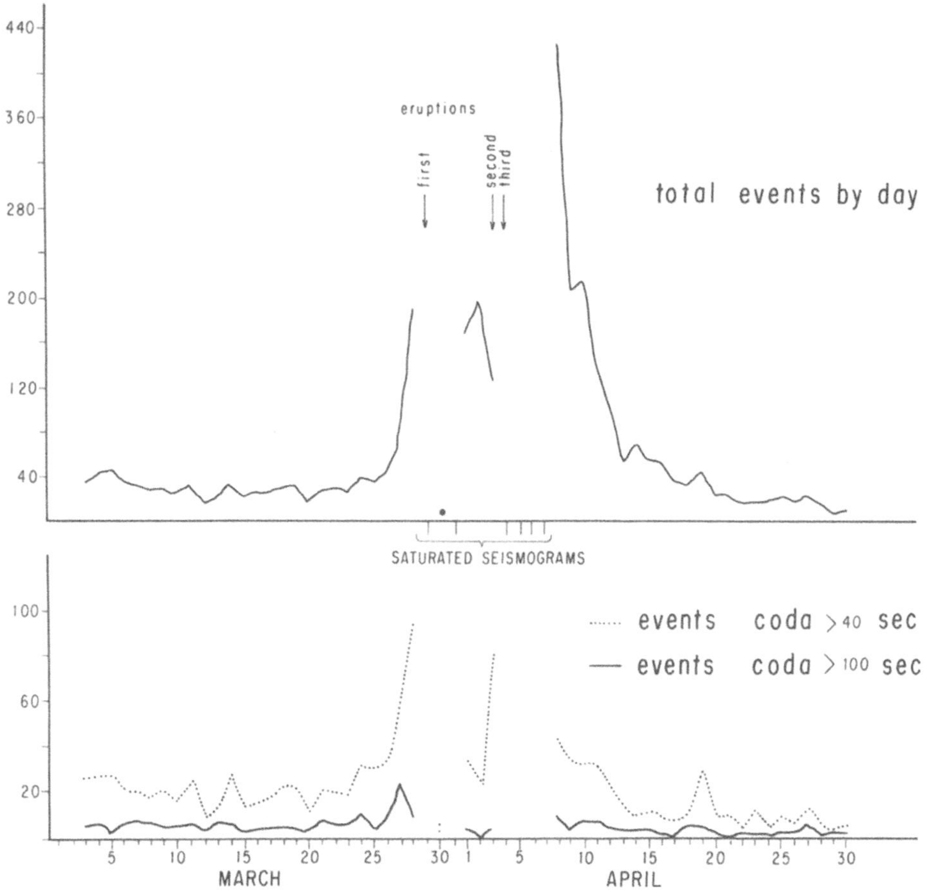

Fig. 1. Number of daily seismic events during March and April. *Below* Number of events greater than 40-s coda and 100-s coda

Frequent seismic events with coda magnitude greater than 3 (~110-s coda) were present during the period before eruptions on an average of 2–3/day. Events with coda magnitude greater than 4 (~450-s coda) were observed during the first week of March. Events with coda magnitude greater than 2.1 (~40-s coda) followed a similar distribution of daily seismicity; but 24 h before the eruptions, this seismicity increased to more than 80 events (Fig. 1).

2.1 b-Value Variations

The magnitude of the events was evaluated by coda length, according to the empirical formula reported by Gonzalez (1980):

$$Mc = -0.86 + 1.87 \log t,$$

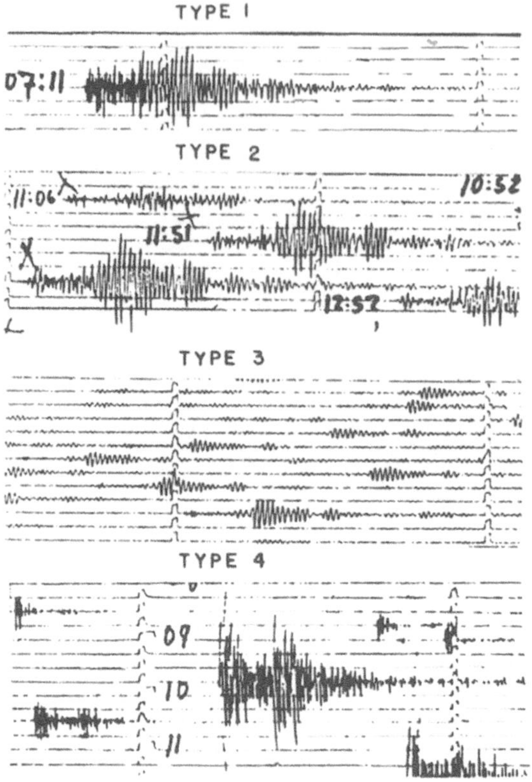

Fig. 2. Different types of seismic events from El Chichon volcano registered in the CH-3 station located ~60 km south of the volcano

where Mc is the coda magnitude and t corresponds to the signal duration time. Using this formula, the coda magnitude for ~3000 events that occurred between March 3 and April 30 is evaluated. During March 29 and 31 and April 4, 5, 6 and 7 the background seismicity saturated the seismogram and no readings were possible.

For the b value determination, the maximum likelihood method reported by Utsu (1965) was used:

$$b = 0.4343/\bar{M} - Mo,$$

where Mo is the lowest magnitude considered and $\bar{M}$ is the average magnitude for the analyzed events. The error in b value is evaluated (at 95% confidence limits for a normal distribution) using the formula reported by Aki (1965):

$$\Delta b = 1.96\,b/\sqrt{n}.$$

Even though some authors evaluate the b value based on increments of a specific number of events (Gresta and Patané 1983), we think that temporal

Table 1. b-Values for different time periods

	b	Δb	Events	Maximum coda MAG.
March 2–15	1.10	±0.18	400	4.1
March 16–28	1.39	±0.21	567	3.9
March 31–April 3	1.71	±0.25	483	3.3
April 7–15	1.21	±0.11	1091	3.8
April 16–30	0.93	±0.16	322	4.0

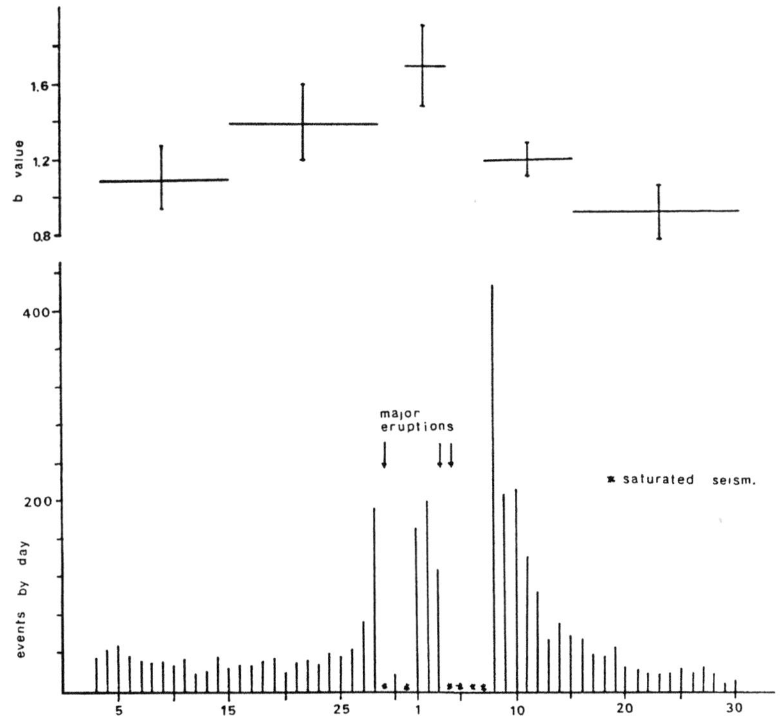

Fig. 3. Daily events recorded, and b-value variations for the different periods of March and April from El Chichon

basis is best for this analysis. This is due to the possible relation in the b value and the changes in the local stress field. In Table 1 five b-values are shown for different periods; before eruptions, b-values increased from 1.1 to 1.39. During the days between eruptions, b-values reached a maximum value of 1.71 and drastically decayed after eruptions to 1.21 and 0.93 at the end of April (Fig. 3).

The increase of b value during the period between eruptions can be correlated with high shallow fracturing; which was responsible for the sudden enlargement of the crater diameter from 400 meters to one kilometer during the second and third eruptions.

The most frequent event magnitude before eruptions was in the interval of 2.1–2.3 to 2.2–2.4 during March and decayed to 1.5–2 in the period between eruptions.

2.2 Energy Released

Even though coda magnitude represents a local scale, with this information the seismic energy released in the previous eruptions period can be estimated by the known formula:

$$\log E = 11.8 + 1.5\,M.$$

Figure 4 shows the cumulative energy released from March 3 to March 28; as may be observed, important increments in the seismic energy released were presented periodically every 6–7 days. The same period was observed between the first large eruption and the second/third eruptions. This periodical increment in the seismic energy released might be related to the internal magmatic phenomena.

The hyperbolic relationship for the accumulated sum of the square root of the seismic energy reported by Tokarev (1963) is not observed (Fig. 4b), this paramenter increased linearly with time.

2.3 Seismic Signals

Analysis of the four principal types of seismic events (see Fig. 2) observed during March and April, 1982, eruptions of the El Chichon volcano, can be explained as follows: continuous intermediate to superficial (~8–2 km in depth) activity initiated on the first few days of March, four weeks before eruptions. Type 1 signals correspond to this activity. The long coda length observed in most of this kind of event is the result of vibrations of the superficial layers of the area. A stratified sequence of shales, sandstones and conglomerates of Oligocene, Eocene and Paleocene age makes up the first 1 to 1.5 km layers above the Cretaceous basement in the area (Medina et al. 1990).

During the 48 h before eruptions and the period between eruptions, type 2 and 3 events appear as the consequence of shallow explosions; these explosions were probably originated by magma interaction with superficial aquifers or sudden gasification of anhidrite deposits.

After eruptions most of the events observed are of type 4, that is, as typical tectonic earthquakes. This seismicity is probably related to inverse fault activity produced by the evacuated area in the magma chamber. This activity is strongly concentrated in a cylindrical area (~8 km width and 15 km length) between 5 and 20 km in depth inclining down to E-SE of the volcano. Unfortunately no focal mechanism analysis was possible due to the actual dispersion (in four different institutions) of the original seismograms.

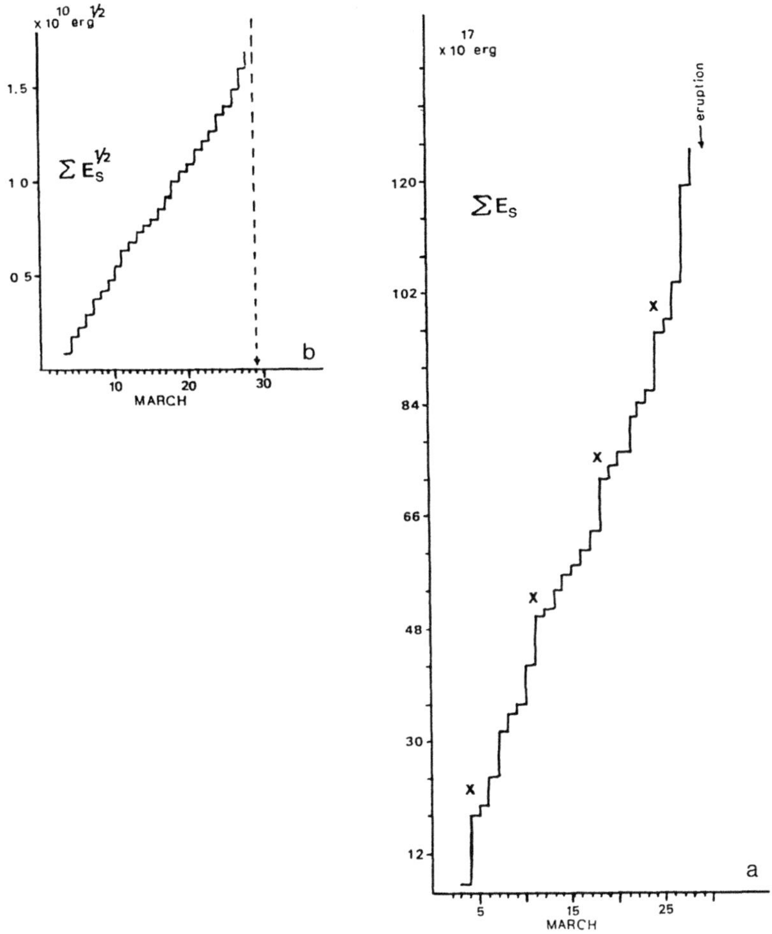

Fig. 4. a Cumulative seismic energy released from March 3 to March 28; **b** cumulative sum of the square root of the seismic energy released from El Chichon

2.4 Hypocentral Distribution

For hypocentral distribution analysis, 340 events were selected with well-defined P and S phases, collected during April and May, 1982, with a portable smoked paper network. The network consists basically of eight stations located around the volcano (Fig. 5a). Four stations were permanent during this period and the others operated only on a temporary basis. All the seismic events correspond to type 4, that is, as typical tectonic event signals.

Location of epicenters was made by the program HYPO 78 with rms errors in arrival times of less than 0.3 s. Horizontal and vertical standard errors of location were estimated to average 1.5 and 2 km, respectively. The velocity model employed is shown in Table 2.

Table 2. Crustal model

P wave velocity (km/s)	Layer thickness (km)
3.51	1.0
4.32	2.3
4.50	1.7
5.50	5.0
6.00	5.0
6.50	5.0
7.50	∞

Most of the seismic events are located to the SE of the volcano; this aspect is in contrast with the majority of location events before eruptions reported to the E of the volcano by Havskov et al. (1983a). Horizontal epicenter distribution shows that events of less than 10 km in depth form a cluster displaced to the S of the volcano (Fig. 5b), events between 10 and 20 km show the most intense activity and form a great cluster dispersed to the SE (Fig. 5c). Events with depth greater than 20 km are scarce (Fig. 5d).

All the post eruption seismicity appears limited to the north by the San Juan fault, apparently the most important in the area. This fault has the same orientation as the Holocene volcanic occurrences in the area (Medina et al. 1990).

Vertical distribution of events are shown in Fig. 6a (E-W profile) and 6b (N-S profile). The NS profile shows a cluster of events vertically elongated and inclined down to the S of the volcanic structure; the EW profile shows the cluster of events elongated and inclined eastwards. Most of the events (66%) are located between 10 and 20 km in depth. A roughly cylindrical zone 8 km wide and 15 km long, contains the majority of events, this cylindrical zone is inclined downwards E-SE of the volcano. The same area can be correlated with the maximum fault development zone (Canul et al. 1983; see Fig. 7) and with a deep gravity anomaly (Medina et al. 1990). These facts permit us to postulate the presence of magmatic emplacement in this area, which probably corresponds to the actual feeding magma chamber.

3 Discussion and Conclusions

Even though the El Chichon volcano is separated from the trench by 350 km, focal machanism analysis reveals that the state of stress in the area might be related with the compressive stress produced by subduction (Havskov et al. 1983b). A Benioff zone delineated for the area has a dip of 45° and strike of N 30° E (Havskov et al. 1983b). However, very scarce seismic activity is detected below the volcano, and local magma emplacements appear as the principal source of the magmatic activity. The possible presence of a great, partially melted zone below Chiapas crust has been reported by Yamamoto (1986).

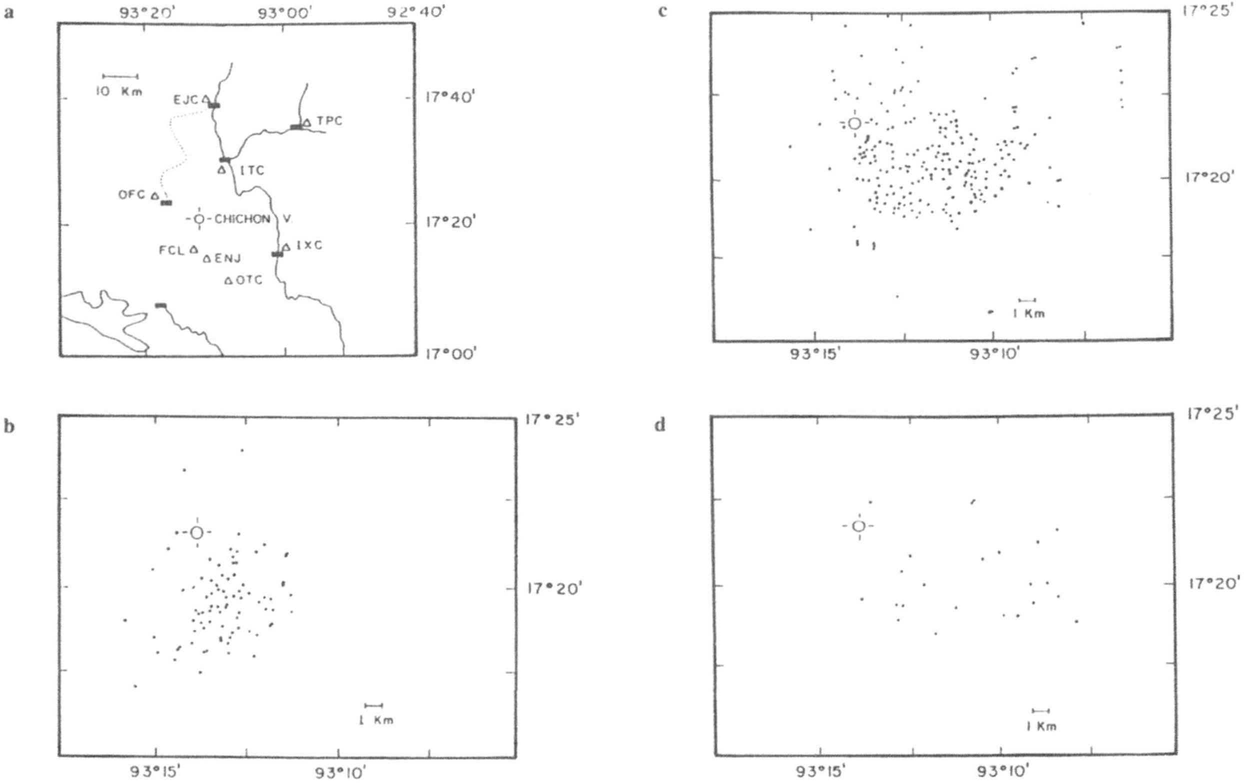

Fig. 5. a Location of the seismic network in operation after April 1 around the El Chichon volcano; **b** horizontal distribution of events with depth less than 10 km, **c** horizontal distribution of events with depth between 10 and 20 km; **d** horizontal distribution of events with depth greater than 20 km

a

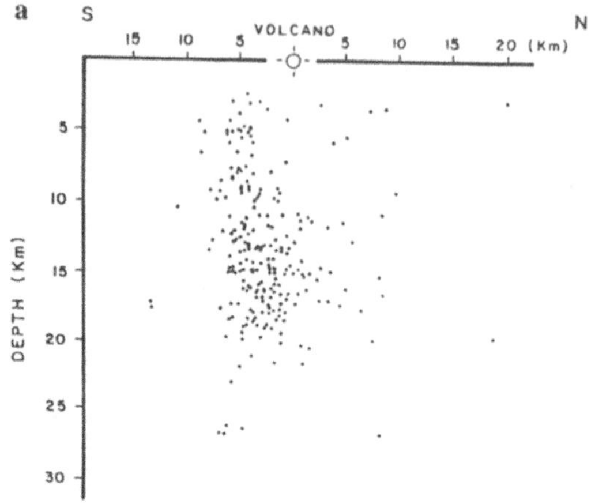

b

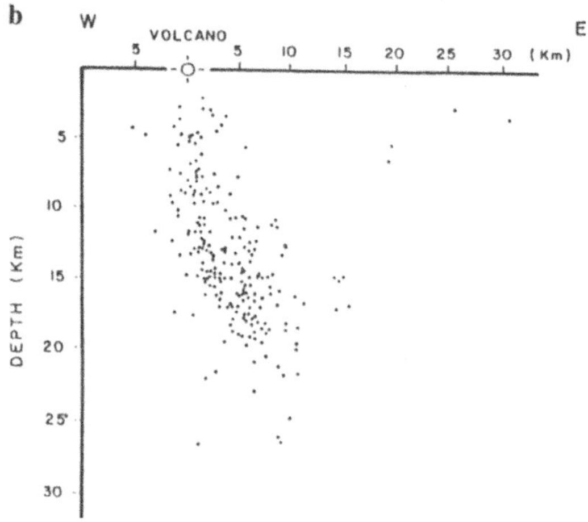

Fig. 6. a Vertical distribution of events, profile E-W; **b** vertical distribution of events, profile N-S

For the 1982 eruptions of the El Chichon volcano, the seismicity analysis of the different seismic signals observed permits us to infer that intermediate to superficial activity (8 to 2 km in depth) appears as the precursory activity followed by shallow explosions two days before eruptions. These shallow explosions were probably related to interaction between hot magma and aquifer deposits. The high increase in the b value observed during the period between the first and the second/third eruptions, can be correlated with a high shallow fracturing process developed during this period. This observation is in agreement with the sudden enlargement of the crater's diameter, observed during the second/third eruptions. Tephra deposits from this last activity contain a large amount of non juvenile material.

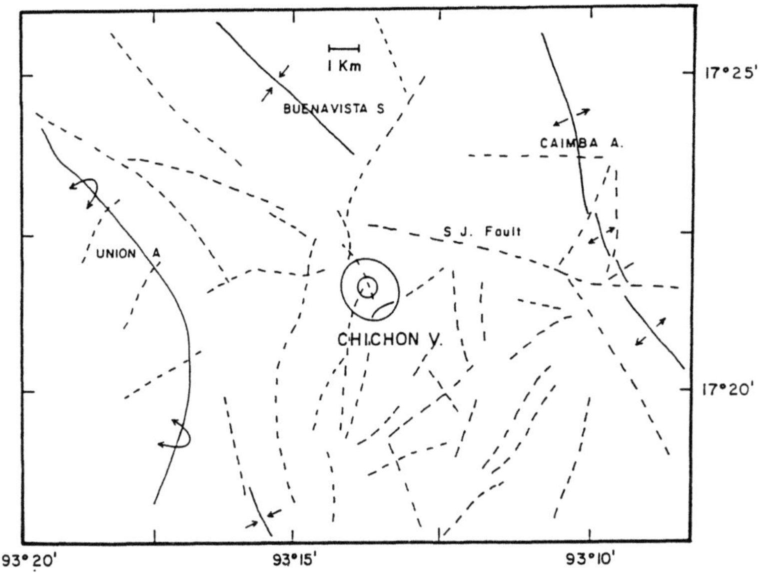

Fig. 7. Tectonic features of the El Chichon area reported by Canul et al. (1983) *Discontinuous line* faults; *convergent arrows* syncline structure; *divergent arrows* anticline structure

During the 24 h before the first and second/third eruptions, more than 80 events with coda magnitude greater than 2.1 were observed. Seismic energy released between March 3 and March 28 reached the 1.2×10^{19} ergs. Important periodical increments (6–7 days) in the seismic energy released were also observed; the same period separated the first and the second/third large activity.

Hypocentral distribution of the post-eruption seismic events reveals that the activity was strongly concentrated in a roughly cylindrical zone, 8 km in width and 15 km in length. Most of the events were located SE of the volcano; the same area can be correlated with the maximum fault development zone and with a deep gravity anomaly. The actual feeding magma chamber is probably located in this area.

Acknowledgements. We would like to thank G. Vilchis and F. Ruiz from IFUNAM for providing the figures.

References

Aki K (1965) Maximum likelihood estimate of b in the equation log N = a − bM and its confidence limits. Bull Earth Res Inst Tokyo Univ 43:237–239

Canul R, Razo A, y Rocha V (1983) Geologia e historia vulcanologica del volcan Chichonal. Inst Geol UNAM, Mexico, 1983:3–22

Gonzalez L (1980) Estudio de las replicas del temblor de Oaxaca del 29 de noviembre de 1978. BS Thesis Fac Cienc, UNAM, Mexico, 92 pp

Gresta S, Patane G (1983) Changes in b values before the Etnean eruption of March–August 1983. Pageoph 121:903–912

Havskov J, De la Cruz S, Singh S, Medina F, Gutierrez C (1983a) Seismic activity related to the March–April, 1982 eruptions of El Chichon volcano, Chiapas, Mexico. Geophys Res Lett 10:293–296

Havskov J, Singh S, Novelo D (1983b) Geometry of the Benioff zone in the Tehuantepec area in southern Mexico. Geofis Int 21:325–330

Medina F (1982) El volcan Chichon. GEOS Bull Mexican Geophys Union 2:4–19

Medina F (1986) Analisis de las columnas eruptivas del volcan Chichon, Marzo-Abril, 1982. Geofis Int 25:233–249

Medina F, Gonzalez T, Gonzalez L (1990) Gravity and seismicity analysis of El Chichon volcano area, Chiapas, Mexico. Pageoph 133:149–165

Minakami T (1974) Seismology of volcanoes of Japan. In: Civetta L, Gasparini P, Luongo G, Rapolla A (eds) Physical Volcanology. Elsevier, Amsterdam

Tokarev P (1963) On a possibility of forecasting of Bezymianny volcano eruptions according to seismic data. Bull Volcanol 26:386–397

Utsu T (1965) A method for determining the value of b in a formula $\log N = a - bM$ showing the magnitude – frequency for earthquakes. Geoph Bull Hokkaido Univ 13:99–103

Yamamoto J (1986) Evidences of the existence of an abnormal seismic signal attenuation in southern Mexico. Geofis Int 25:56–67

Yokoyama I (1988) Seismic energy releases fron volcanoes. Bull Volcanol 50:1–13

Coda Q Dependence on Time, Frequency and Coda Duration Interval at Mt. Etna, Sicily

Edoardo Del Pezzo and Domenico Patané

Abstract

A re-estimate of coda Q (Qc) was carried out on short period vertical seismograms of earthquakes located on Etna volcano, Sicily. Care was taken in selecting coda duration intervals in which the estimates are made. A clear dependence of Qc on frequency and coda duration interval has been evidenced. Qc is proportional to f^n with n equal to 1 for short coda duration intervals (25 s) and with n equal to or less than 0.8 for longer codas. A depth effect and/or the presence of multiple scattering can be the cause of the Qc dependence on coda duration.

Dependence of Qc on time was investigated in an 8-year time interval.

An increase of 1/Q seems to occur before the volcanic eruptions but we cannot place a statistical significance on the results.

1 Introduction

The term "coda" means the part of the seismogram which follows the arrival of all the body and surface waves. In small, local earthquakes the coda is composed of high frequency seismic waves, the most important properties being that: they are not regular plane waves coming from the source to the receiver; and the time decay of power spectra of coda waves is independent of the distance and the path between epicentre and station and of the magnitude (at least for $M < 6$). Moreover, coda length is determined by the magnitude.

These properties suggested that coda waves were generated by a simple mechanism which, in a first approximation, is not dependent on source radiation and on the path between hypocentre and the station. Aki (1969) suggested that these waves were single backscattered waves coming from the heterogeneities of the Earth medium and proposed a model based on the Born approximation and on the uniformity random medium assumption. Aki and Chouet (1975) revised this model extending it to the two extreme cases of single isotropic scattering and diffusion. Sato (1977) and Kopnichev (1977) developed the single isotropic scattering model. Aki (1980), basing his ideas on the similarity of the quality factors for S waves and coda waves, suggested that coda is mainly composed of S-S backscattered waves. In recent times, the importance of multiple scattering in the coda envelope generation was pointed out by many authors (see for example Gao et al. (1983); Kopnichev (1977); Frankel and Wennemberg (1987)) and some controversy about the validity of the single scattering model is raised. Despite this, the relative comparison among different values of single scattering coda Q, evaluated in different areas of the world, indicate that this last parameter is a tectonic indicator, because in some way it reflects the scattering and attenuative properties of the Earth's lithosphere. Moreover, the single scattering model seems to be applicable for short coda durations and for short hypocentral distances.

As the elastic properties of the Earth's volume presumably change during a seismic cycle, it is possible that the scattering properties of the coda waves could reflect these variations. Coda Q is, for this reason, a possible seismic precursor. Jin and Aki (1986), Gusev and Lemzikov (1985) among others, found changes in coda Q before large earthquakes, and Fehler et al. (1988) report a change in coda Q before the volcanic eruption of Mount St. Helens. Unfortunately, the meaning of the variations is not always the same, giving rise to difficulties in finding a physical explanation for the phenomenon.

The aim of this paper is to re-evaluate the coda Q at Mt. Etna, eastern Sicily, using care in the definition of coda duration interval upon which the values of the estimates seem to depend (Roecker et al. 1982) and to investigate the possible Q variations during the eruptive cycles of this active volcano.

2 Model and Method of Data Analysis

According to the single scattering model, the decay rate of the coda envelope can be expressed as:

$$A(f|t) = C(f)t^{-1}\exp(-\pi ft/Qc), \tag{1}$$

where $A(f|t)$ is the amplitude of the coda envelope band pass filtered in a frequency band centred at f: t is the time measured form origin along the coda envelope: C(f) includes the source and depends on the amount of scattering present in the Earth medium, t to minus 1 takes into account the geometrical spreading effect and the exponential term accounts for the attenuation.

Multiplying both terms of the above expression by t, and taking the natural logarithm of both sides we can write:

$$\mathrm{Ln}(A(f|t) * t) = \ln C(f) - \pi ft/Qc. \tag{2}$$

The relationship (2) can be least square fitted by the amplitudes of the coda envelope evaluated at different values of time t, giving an estimate of Qc.

Amplitudes of the coda envelope at different frequency bands can be easily estimated, filtering the seismograms with a band pass filter centred at the frequency bands we wish to investigate.

It is noteworthy that the relationship (1) is valid under the assumption that station and source are very close together and that the medium is random but uniform. In this hypothesis, a wavelet composing the coda at t seconds from the origin time, is generated at a scatterer located at a distance given by:

$$d = v * t/2.$$

At this time t, the contribution of all the scatterers located at the same distance d, can be summed because of the hypothesis of random medium (Aki and Chouet 1975). This means that for each value of the time t, the waves making up the coda are generated inside a spherical shell whose inner radius is d. If the medium is really uniform and the single scattering dominates, the factor Qc, which is estimated from the envelope, must be independent of the coda duration interval.

One of the aims of this paper is to investigate the possible dependence of Qc on the length of coda. This dependence could reflect a possible variation of Qc with depth if we strictly assume single scattering. Multiple scattering effects might also produce the same dependence, or both effects may be superimposed, but the possibility of fitting multiple scattering models will be the subject of future studies.

2.1 Data Analysis

Re-estimate of Qc as a Function of the Frequency. Data are local earthquakes recorded by the seismic network of Etna Volcano. A map of the station position and epicentres of the best located events are shown in Fig. 1. A detailed description of the earthquake location procedure and of the data is reported in Del Pezzo et al. (1985). Stations were equipped with vertical component seismograms and data were telemetered to a central recording site and analogically recorded on tape. 39 events were selected on the basis of signal to noise ratio in Del Pezzo et al. (1985), to estimate single scattering coda Q. These data were filtered using an analogical filter in the frequency band centred at 1,2,4,8,12 and 18 Hz with an attenuation rate of 48 dB/octave. The estimates of Qc obtained in Del Pezzo et al. (1987), were obtained using no constraints on the coda duration interval. The coda length in Del Pezzo et al. (1985) was determined for each event visually, evaluating the time point where the coda signal falls below the noise level, on the seismograms.

As already stated, the estimates of Qc can depend on coda duration interval and single scattering assumption strictly holds only for short coda durations. The importance that the definition of coda duration interval assumes, when estimates of Qc obtained in different regions have to be compared, was pointed out by Aki (1982). Moreover, the envelopes of different frequency bands may have different coda durations, and this effect could mask the real frequency dependence of Qc. These are the reasons why a re-estimate of Qc at Mt. Etna was considered necessary, prior to studying the possible variation of this parameter with the time.

Envelopes of the filtered traces were measured, joining the maximum amplitudes of the filtered traces and successively sampling this line at 1 sample/s.

Coda duration was estimated by plotting the quantity $\ln(A * t)$ versus t [see Eq. (2)]. An example is in Fig. 2, where vertical bars show the beginning and the end of the coda envelope. It is widely accepted that the end of the coda is the point where the amplitude of the signal falls below the noise level. As the noise is presumably stationary for short time intervals, the quantity $A * t$ begins

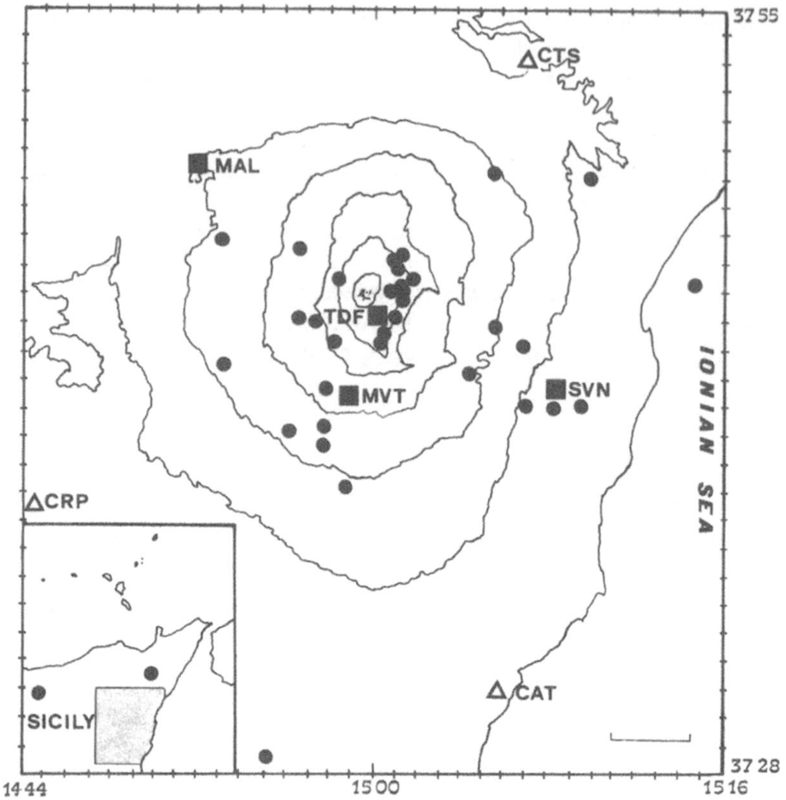

Fig. 1. Map of Etna Volcano. *Squares* show the location of the station used for the present study and *dots* the earthquake epicentres. *Triangles* are the stations of the network not used for coda analysis

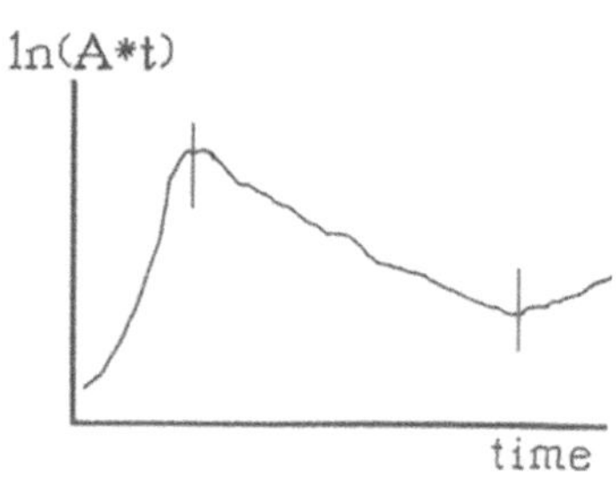

Fig. 2. Example of coda envelope. *Vertical* bars show the beginning and the end of the coda

to increase when the noise becomes predominant on the signal. The slope of the plot ln (A * t) versus t is proportional to Qc. The slope is obtained by fitting the plot with the least squares method. The fits with a correlation coefficient higher than 0.7 were selected for the analysis.

Figure 3 shows an example of the distribution of 1/Q for different frequency bands at MVT. The arithmetical average of the values roughly approximates to

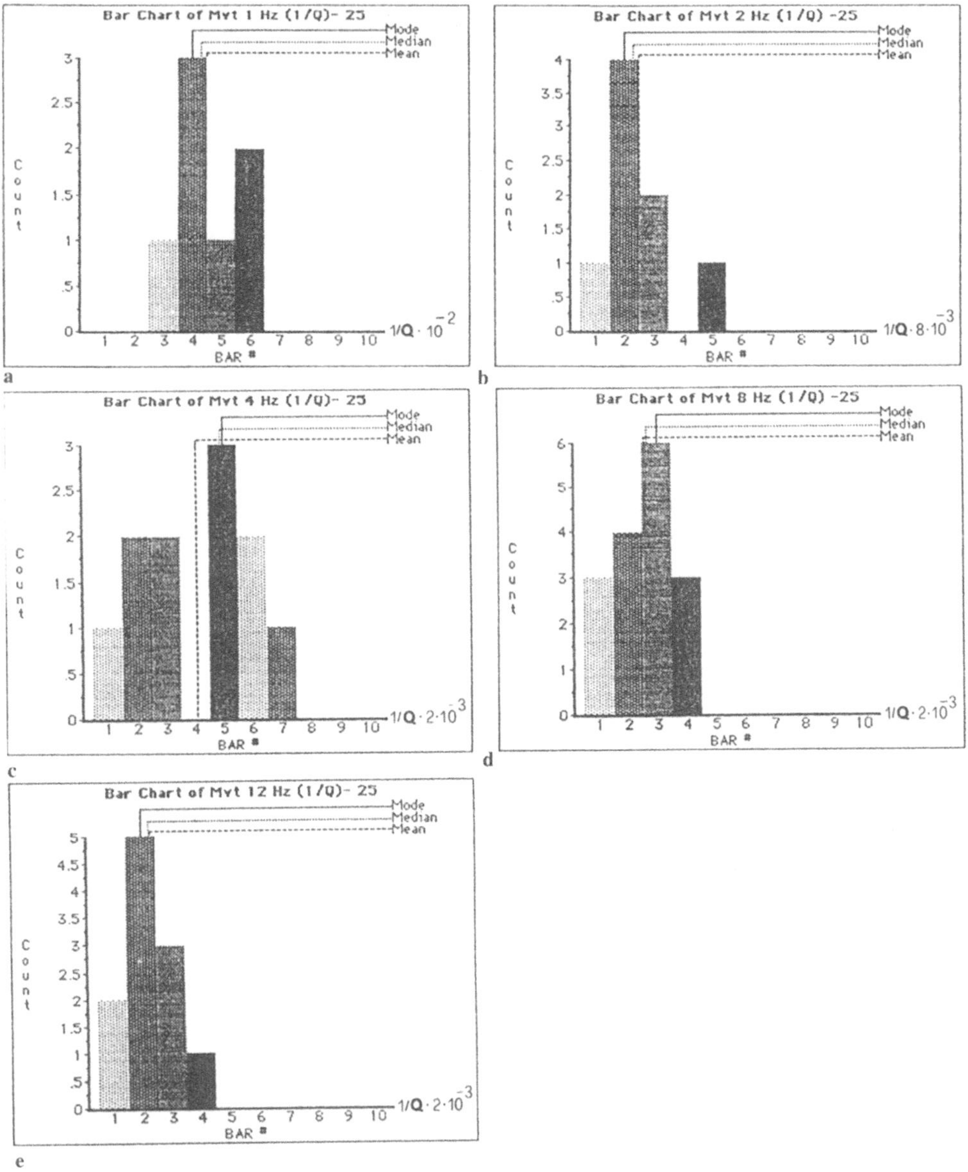

Fig. 3a–e. Example of the distribution of 1/Q estimates for different frequency bands

the maximum of the distribution. Nonpeaked distributions of the estimated 1/Q were not considered for this analysis.

Estimates of 1/Q as a function of the coda duration interval (estimated from the origin) were obtained by cutting off the envelope at 25, 40 and 60 s (the latter only for MVT stations, due to the better signal to noise ratio at long coda duration). Results are shown in Fig. 4. It seems that Q increases (better quality

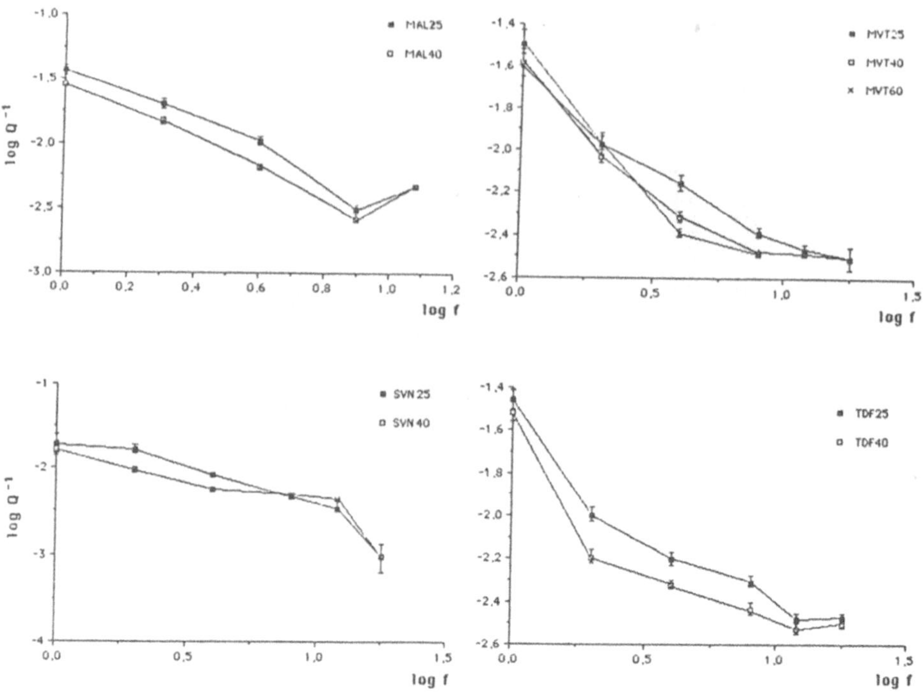

Fig. 4. Estimates of 1/Q as a function of frequency and coda duration interval. Coda duration interval is written near the station mark

of the medium) when the volume of Earth's crust, sampled by coda waves, increases. This effect can be explained with a depth variation of the quality factor.

In Fig. 4, a frequency variation of 1/Q as a function of frequency is shown. The relationship

$$Q = qf^n$$

fits data with "n" assuming different values when different coda lengths are used. In Fig. 5, for example, the result for MVT is reported. In Table 1 the results for all stations are summarized. Interestingly, n increases when coda duration decreases, approaching 1 at 25 s of coda duration interval. Values of n equal to 1 were found for tectonically active areas of Italy (Rovelli 1982) while a smaller value was found by Chouet (1976) for Hawaii: These differences found for an active volcano were interpreted by Aki (1980) as due to the presence of molten materials in the crust. The same interpretation was made for Etna in Del Pezzo et al. (1985), but it needs to be modified in the light of the present result.

Time Dependence of Qc. Data used for the study of time dependence of Qc were analogically recorded on tape from 1981 to 1988. During this period six eruptive episodes occurred in Etna volcano.

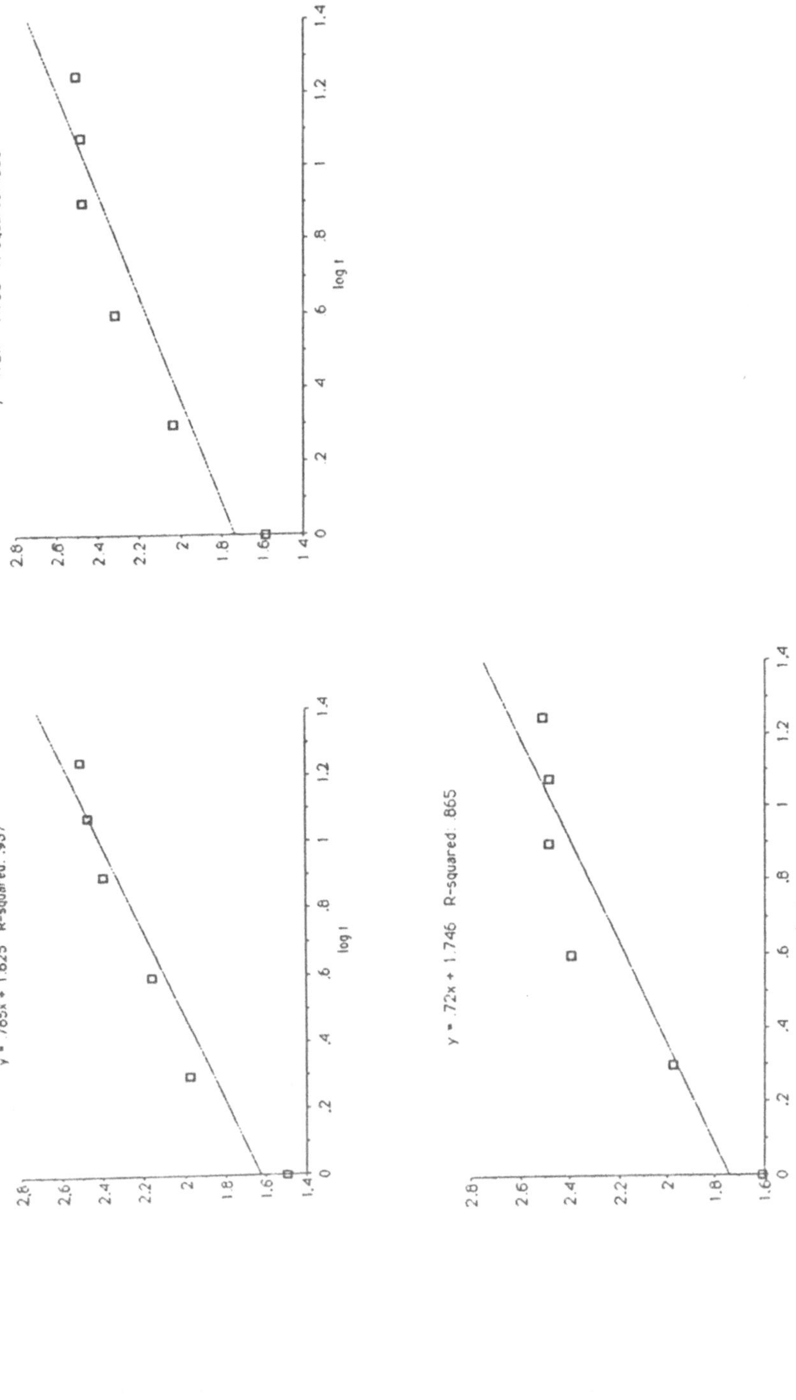

Fig. 5. Logarithm of Q versus Logarithm of frequency plotted for MVT station for different coda duration interval

Table 1. Summary of the results for all stations used and for different coda duration intervals

MVT – 25 Sec –	Q = (1.62 ± 0.08)	$* f^{0.78\pm0.10}$
MVT – 40 Sec –	Q = (1.74 ± 0.11)	$* f^{0.72\pm0.13}$
MVT – 60 Sec –	Q = (1.75 ± 0.12)	$* f^{0.72\pm0.14}$
MAL – 25 Sec –	Q = (1.43 ± 0.12)	$* f^{0.97\pm0.17}$
MAL – 40 Sec –	Q = (1.59 ± 0.14)	$* f^{0.87\pm0.20}$
SVN – 25 Sec –	Q = (1.64 ± 0.05)	$* f^{0.76\pm0.08}$
SVN – 40 Sec –	Q = (1.83 ± 0.06)	$* f^{0.56\pm0.09}$
TDF – 25 Sec –	Q = (1.62 ± 0.10)	$* f^{0.77\pm0.13}$
TDF – 40 Sec –	Q = (1.76 ± 0.15)	$* f^{0.70\pm0.18}$

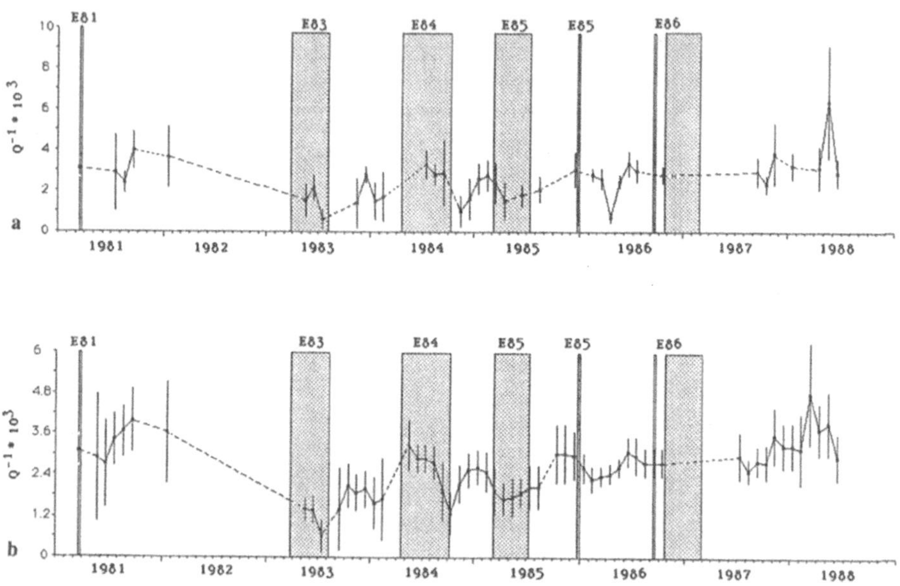

Fig. 6a,b. Time variation of 1/Q at Mt. Etna. **a** average of 1/Q over a three-month interval. **b** average over a three-month interval sliding of 1 month. Coda duration interval at 40 s

Data were band-pass filtered in the 3–7 Hz band.
Envelopes of the filtered traces were sampled as shown and the slope of log (A * t) versus t plot was estimated. Data with correlation coefficient lower than 0.7 were disregarded.

Coda length was truncated at 40 and 60 s. A running mean was carried out on data. The time window is 90 days long and slides 30 days each step. For each step the average 1/Q and its standard deviation are reported on the plot of Figs. 6 and 7. In the same plots, the average 1/Q over a 3-month period obtained without the sliding window is reported. Dotted bars show the time periods of the eruptions.

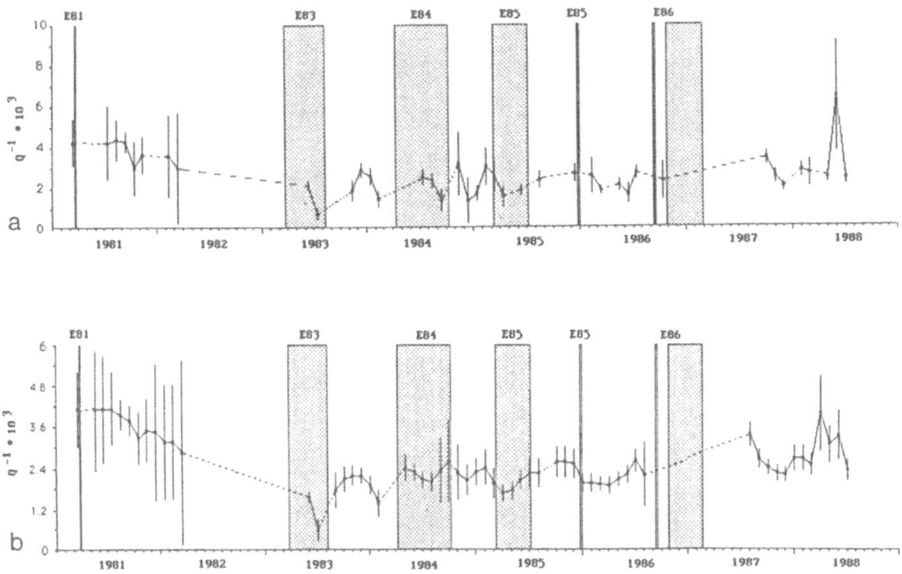

Fig. 7a,b. The same as Fig. 6 for coda duration interval of 60 s

Error bars are lower for data with coda truncated at 60 s than those at 40 s, but the trends are similar.

Some variations seem to occur before the volcanic eruptions of Mt. Etna. Looking at the trend of the plot, an increase of 1/Q before the volcanic eruptions seems to occur. Unfortunately, only a few points have associated error bars small enough to allow us to give a statistical test to these variations.

3 Conclusions

Single scattering coda Q, Qc, is sensitive to coda duration interval in which it is estimated. The relationship $Q = qf^n$ fits data well when the coda duration interval is truncated at 25 seconds. In this case, the parameter "n" is near unity and is similar to that found in most active tectonic zones. "n" decreases when coda length increases. A comparison with other volcanic areas in the world is shown in Fig. 8 where a maximum coda length of 60 s is used for Etna data. We have no information on the coda duration interval used in the other areas reported in Fig. 8, so this comparison is only qualitative.

The dependence of Qc on the coda duration interval could be an effect due to possible multiple scattering effects for long coda duration; whereas if we assume the single scattering model as valid, this effect has to be considered as a depth effect, because long codas sample greater portions of lithosphere which presumably have a higher Q than the crust. This last interpretation is reported in several papers on the same subject, as for example Roecker et al. (1982) and seem reasonable. Probably the dependence on coda duration is due to both

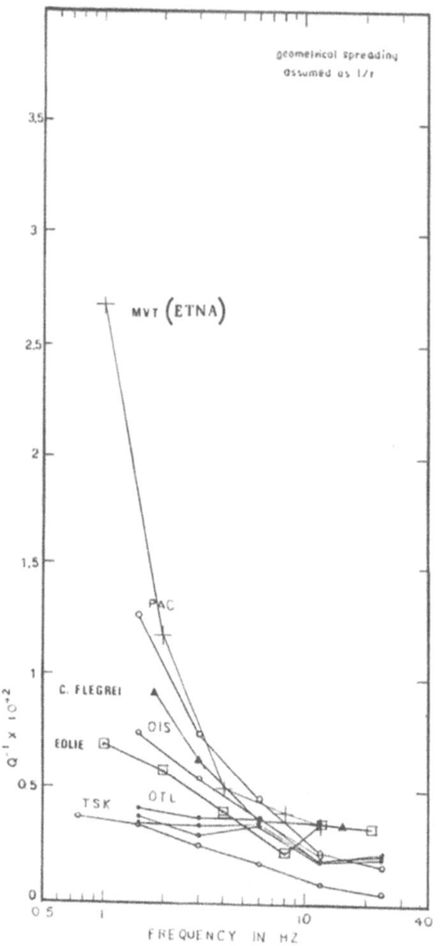

Fig. 8. Comparison of Etna with other areas around the world. Etna data are truncated at 60 s of duration interval

multiple scattering effects and also to an increase of Q with depth, but at present it is numerically difficult to separate them by estimating the Q using a multiple scattering model (Aki, pers. comm.). Improvements of the methods for fitting the non-linear multiple scattering model with coda envelopes are in progress and will be published elsewhere.

Despite the already vague physical meaning of the single scattering coda Q, without a doubt, this parameter is a sort of tectonic indicator of the state of the lithosphere. Hence, we can use this parameter to monitor the state of stress in the lithosphere during eruptive and seismic phenomena.

Some variation of Qc with time seem to occur at Etna during the eruptive periods. No statistical significance, however, can be placed because of poor quality data and high signal to noise ratio, but the method seems to be promising even if applied to very short time intervals.

References

Aki K (1969) Analysis of the seismic coda of local earthquakes as scattered waves. J Geophys Res 19, 74:615–631

Aki K (1980) Attenuation of shear waves in the lithosphere for frequencies from 0.05 to 25 Hz. Phys Earth Planet Int 21:50–60

Aki K (1982) Scattering and attenuation. Bull Seismol Soc Am 72:S319–S330

Aki K, Chouet B (1975) Origin of coda waves: source attenuation and scattering effects. J Geophys Res 80:3322–3342

Chouet B (1976) Source scattering and attenuation effects on high frequency seismic waves. PhD Thesis, MIT Cambridge, Massachussets

Del Pezzo E, Gresta S, Patane' G, Patane' D, Scarcella G (1985) Attenuation of short period seismic waves at Etna as compared to other volcanic areas. Pure Appl Geophys: 125:6

Fehler M, Roberts P, Fairbanks T (1988) A temporal change in coda wave attenuation observed during an eruption at Mt. St. Helens. J Geophys Res 93:4367–4373

Frankel A, Wennemberg L (1987) Energy-flux model of seismic coda: separation of Scattering and intrinsic attenuation. BSSA 77, 4:1223–1251

Gao LS, Biswas NN, Lee LC, Aki K (1983) Effects on multiple scattering on coda waves in three dimensional medium. PAGEOPH 121:3–15

Gusev AA, Lemzikov VK (1985) Properties of scattered elastic waves in lithosphere of Kamchatka parameters and temporal variations. Tectonophysics 112:137–153

Jin A, Aki K (1986) Temporal change in coda-Q before the Thangshan earthquake of 1976 and the Haichen earthquake of 1975. J Geophys Res 91:(B1)665–673

Jin A, Cao T, Aki T (1895) Regional change of coda Q in the ocean lithosphere. J Geophys Res 90:8651–8659

Kopnichev YF (1977) The role of multiple scattering in the formation of seismograms' tail. Izv Akad Nauk SSSR Fiz Zemli 13:394–398 (English translation)

Roecker SW, Tucker B, King J, Hatzfeld D (1982) Estimates of Q in central Asia as a function of frequency and depth using the coda of locally recorded earthquakes. BSSA 72:129–149

Rovelli A (1982) On the frequency dependence of Q in Friuli from short period digital records. BSSA 72:2369–2372

Sato H (1977) Energy propagation including scattering effects: single isotropic scattering approximation. J Phys Earth, 25:27–41

Measurements of Ground Movement on Mount Etna, Sicily: A Systematic Plan to Record Different Temporal and Spatial Components of Ground Movement Associated with Active Volcanism

P. Briole, R. Gaulon, G. Nunnari, G. Puglisi, and J.C. Ruegg

Abstract

Several techniques are being used on Mount Etna, to record ground movements. A network of automatic tiltmeters has been installed to record rapid ground movements associated both with the sudden changes in volcanic activity (that may occur over a time period of a few days) and the long-term deformation component. Tilt changes that occur over long time periods are recorded by repeated relevelling of short lines, using a standard levelling technique. Short aperture levelling lines have also been sited near some tiltmeters, so that we will be able to record both short-term and long-term changes in ground movements at several locations on Etna. Such information also allows us to validate the data recorded by automatic tiltmeters. Horizontal strain is recorded by making repeated trilateration surveys, using the global positioning system (GPS) to measure ground movements on Etna while electronic distance measurements (EDM) are performed whenever it was possible to compare the precision of the GPS and EDM techniques. Our first comparison shows that these two methods agree at about the 2 ppm level.

1 Introduction

Many investigators have performed ground deformation studies on Mount Etna, an active volcano located on the east coast of Sicily in southern Italy (Wadge et al. 1975; Wadge 1976; Murray et al. 1977; Henbest et al. 1978; Villari 1980; Murray 1982; Sanderson et al. 1983; Murray and Pullen 1984; Nunnari and Romano 1987; Bousquet et al. 1987; Nunnari and Puglisi 1987; Cosentino et al. 1987). These studies have shown that volcanic activity at Etna is often accompanied by measurable ground movement. The results of these studies (mostly concerning the 1974, 1981 and 1983 eruptions) have been used to develop models for the eruptive behaviour of the volcano. Most of these proposed models have considered only simple elastic models as possible sources for the measured ground movement. Two exceptions are the models proposed by Sanderson et al. and by Bousquet et al. who also considered micro-gravity measurements and structural features in deriving models for the 1981 and 1983 eruptions. The limitation in the spatial and temporal coverage of ground deformation measurements, which are needed to derive a more realistic model for Etna, is chiefly caused by the difficulties encountered when working in the severe environment around Etna.

Previous scientific teams have usually used only one technique to measure ground deformation at Etna. Scientists from the Osservatorio Vesuviano (OV)

have established a levelling line between Serra La Nave and Rifugio Citelli on the south, west and north flanks of the volcano at an average altitude of 1500 m above sea level. Measurements of this levelling line are performed at least every year. The scientific team from the Italian Istituto Internazionale di Vulcanologia (IIV) in Catania has established small trilateration networks on the south, northeast and west flanks of Etna. These trilateration surveys were begun in 1979 and are still in progress. J.B. Murray from the Open University of London has organized a team every year, since 1975, which undertakes a levelling traverse on the summit of the volcano between Piccolo Rifugio and Monte Pizzillo. A trilateration network has also been established by Murray. The locations of these levelling routes and trilateration networks are shown in Fig. 1.

Following the early work of Costantino Tomasi (1962), tiltmeters in operation on Etna have shown that the summit area rises before eruptions (Wadge 1976; Villari 1982; Bousquet et al. 1984; Nunnari and Romano 1987). Because very few tiltmeters have been simultaneously operated on Etna, only a qualitative interpretation can be given to data recorded by these instruments. Furthermore, the tilt patterns indicated by these instruments were usually not supported by the results of geodetic surveys.

To further advance the monitoring of surface deformation studies at Etna, a joint program was begun in 1986 by the Institut de Physique du Globe (IPG) of Paris and by the IIV to develop a systematic approach using tiltmeters, levelling and trilateration surveys, and space geodesy. Eight tiltmeters stations were installed in 1986–1987 at sites that surround the volcano at an average altitude of about 1700 m above sea level. The first GPS survey was made in May 1988 at nine stations. One purpose of this first survey was to compare distances determined by GPS and by the trilateration measurements and to connect the separate geodetic networks that already exist on Etna.

2 Tilt Measurements

The tiltmeter network shown in Fig. 2 was designed to provide coverage around the volcano at altitudes from 1000 to 2000 m above sea level. At this stage in our program, only one tiltmeter has been installed in the summit (TDL), near the Central Crater. Each tilt-station consists of two independent sensors capable of determining the amount of tilt in two perpendicular directions. The sensors have been installed so that one records tilt changes directed radially with respect to the summit crater, while the other one records tilt along the orthogonal direction (tangential). The basic design and operation of these instruments were developed by Blum (1963). The sensors have been modified so that the instrument can be used for the study of tilt changes associated with crustal movements (Saleh 1986): Each instrument has a resolution of about 0.15 µrad and a dynamic range of about 600 µrad. This range is sufficient to record the magnitude of tilt changes at Etna reported by earlier investigators (Wadge 1976; Villar 1982). Because of environmental noise caused by rainfall and temperature fluctuations, the actual resolution of these instruments for time periods longer

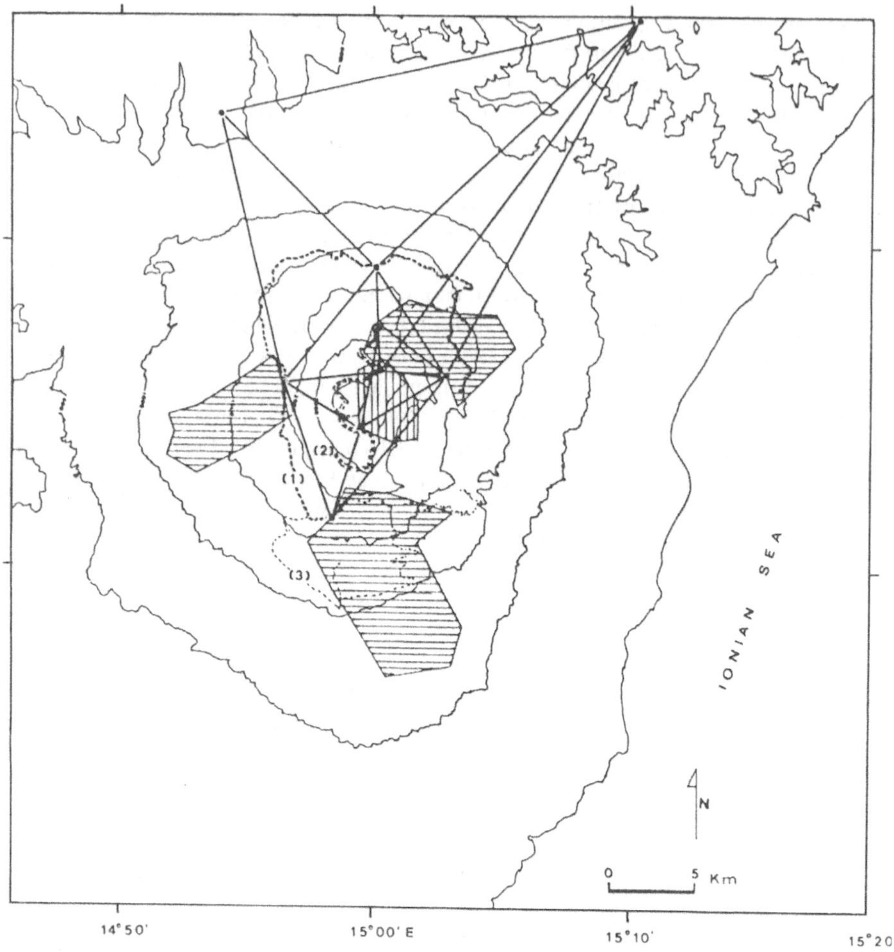

Fig. 1. Major trilateration networks and levelling routes on Etna. The three small trilateration networks *shaded by horizontal lines* were established by the Istituto Internazionale di Vulcanologia (IIV) of Catania; the small trilateration network shaded by *vertical lines* was established by J.B. Murray of the Open University of London. The large trilateration network, where *individual lines* are shown, was established by us; *dashed lines* represent the following levelling or gravity routes: *1* levelling route established by Osservatorio Vesuviano (OV) and part of the micro-gravity line established by OV and IIV; *2* levelling route established by J.B. Murray; and *3* the remainder of the OV-IIV micro-gravity line. 500-m Interval elevation contours around Etna are also shown

than a few days is lower than the design resolution. In order to minimize diurnal temperature fluctuations, wherever possible these instruments have been installed in lava tubes. Also, measurements of air temperature are made at each site so that the effect of temperature fluctuations on tilt change can be partially removed. Tilt and temperature are recorded at each site on a micro-computer system. A three hour sampling rate enables us to collect up to eight months of

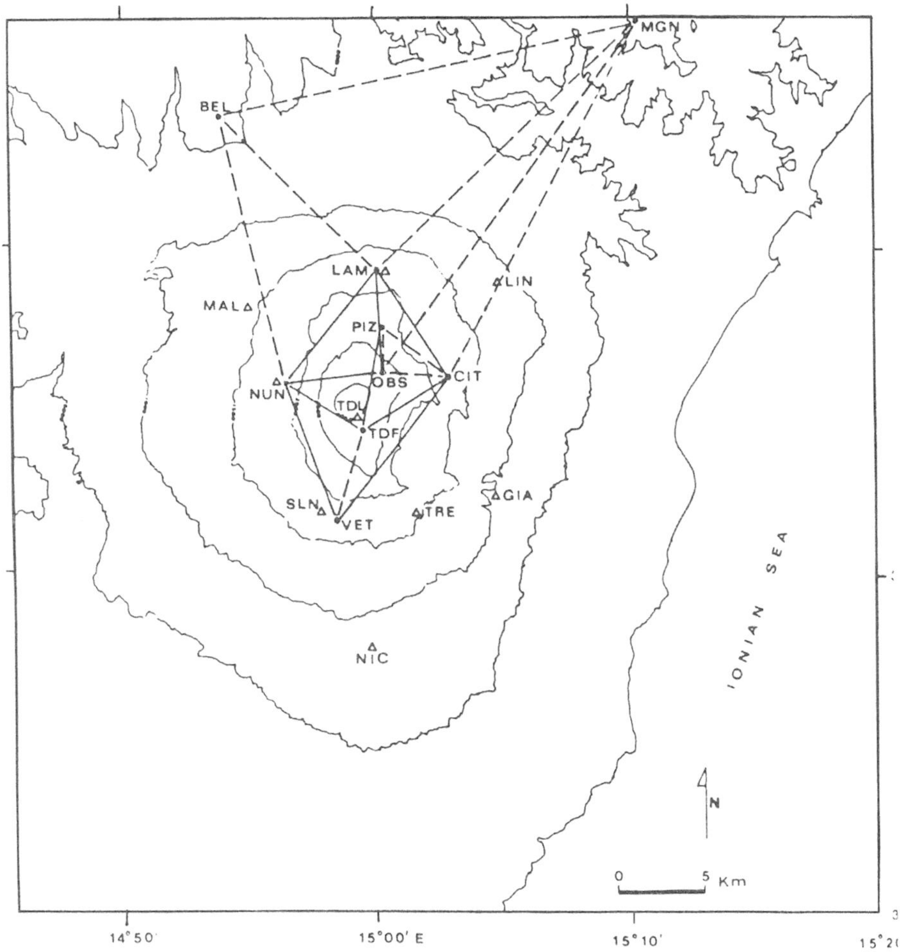

Fig. 2. Location of tiltmeter and GPS stations on Etna. Tiltmeter stations are indicated by *triangles*. *Dashed lines* show the network measured by both an electronic distance measuring instrument and by GPS receivers. Note that NUN and LAM benchmarks are located close to tiltmeter sites

data and to maintain the instruments. No failure of this system occurred during its first year of operation. Data collected at three stations (TDL, LIN and SLN) are telemetered to Toulouse, France, using the ARGOS satellite.

Cross-shaped arrays of benchmarks have been installed around TRE, LAM and NUN tilt-stations so that the long-term tilt changes at these sites can be measured by levelling of these benchmarks. Two different arrays of benchmarks have been used: one using a short-base of 50 m and the other a long-base of 2000 m. We estimate the accuracy of these tilt measurements to be 5 µrad for short-base arrays and 2 µrad for long-base arrays. This accuracy is in fact the a posteriori variance obtained by compensation of the data with an inversion

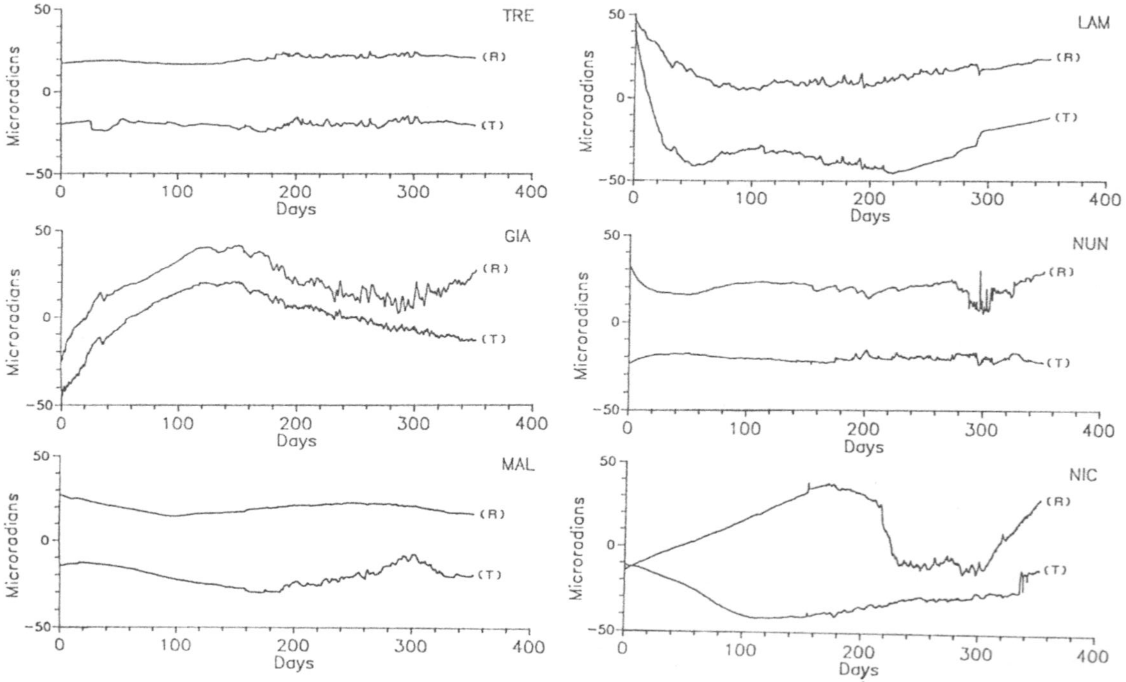

Fig. 3. Time series of tilt data for six stations on Etna. These data have been filtered and corrected for temperature fluctuations. Each diagram shows the tilt change recorded in the radial (R) and tangential (T) directions, with respect to the central summit crater of Etna

algorithm (Tarantola and Valette 1982). We have found that, despite the use of a precise levelling technique for the short-base arrays and a trigonometric levelling technique for long-base arrays, the long-base array is more accurate. This accuracy, which is close to intrinsic instrumental precision, can be reached only with a sufficient redundancy of data. In the particular analyzed time interval, we also assumed that there was no volcanic phenomena capable of influencing the levelling measurements.

Data recorded by six tilt instruments are shown in Fig. 3. These data have been filtered for reducing high-frequency noise and corrected for temperature fluctuations. The positive direction of tilt change is listed in Table 1 for each station. Temperature corrections were made by assuming a linear relationship between temperature change and tilt change for each tilt component for each station and by considering also the temporal shift between the two series of data, using a cross-correlation analysis.

We have analyzed the first year of tiltmeter data using a power spectra method (Mortenson and Hopkins 1988). From a third method we can determine the resolution of each station for various lengths of time. Assuming that none of the tiltmeters recorded a coherent pattern of ground movements caused by volcanic activity (that is, all tiltmeter recordings were caused by noise sources), we can compute the noise level for both tilt components for each station. This hypothesis is sustained by the volcanological observations made by Calvari et al. (1989), during the time interval considered. The results for 1- and 10-day time intervals are listed in Table 1.

The noise level of each tiltmeter increases as the length of time increases. For a time period of ten days, the resolution of our tiltmeters is from 5–10 μrad. This resolution is slightly larger than the resolution that can be attained by levelling our arrays of level benchmarks.

Table 1. The orientation of the sensor (tilt = signal increases) and the rms noise (in μrad) computed for two time intervals (1 and 10 days). The resolution for a 10-day period is of the same order as that obtained by using geodetic methods

Tilt station	Orientation up	RMS (μrad) 1 day	10 days
TRE radial	NW	1.8	4.2
TRE tangential	SW	2.2	5.3
LAM radial	N	2.6	6.3
LAM tangential	W	2.3	5.1
GIA radial	S	4.4	11.1
GIA tangential	E	2.7	5.3
NUN radial	NE	6.7	12.3
NUN tangential	SE	2.9	5.4
MAL radial	NW	0.6	1.1
MAL tangential	SW	2.0	4.4
NIC radial	S	5.1	10.1
NIC tangential	W	5.6	10.6

2.1 GPS Measurements

A GPS survey was performed on Etna in May 1988 at nine general locations: three in the summit area (TDF, OBS, PIZ), four on the flanks (VET, NUN, LAM, CIT) and two not on the volcanic cone (BEL, MGN), located in the Peloritan-Nebroden Mountains, about 25-km north of Etna. We located these GPS stations close to tiltmeter stations and pre-existing levelling and trilateration benchmarks so that we could interconnect all of the geodetic networks on Etna. The GPS stations TDF, LAM and NUN are each close to tiltmeter stations.

The benchmarks were occupied by GPS receivers over a period of ten days. Daily observations lasted about three hours and consisted of two observing sessions. Data were collected on three single-frequency, five channel receivers (two TR5S and one NR52) manufactured by SERCEL. Some individual lines of the GPS array were measured several (up to four) times in order to improve the precision in computing the baseline length.

The recorded GPS data have been processed using a "single-base analysis" software tool, developed at IPG, based on the well-known double difference method. After the computation of each individual baseline, a specific network compensation program, to obtain the final distance values, has been used. The network internal closure error has been estimated to be about 2 ppm (baseline residual less than 5 cm).

Our GPS measurements are not as accurate as similar measurements conducted in other areas and reported by other scientists. The principal reason concerns the correction used for the propagation delay. This point is particularly critical in mountain areas. We processed our data by using two different models: one used temperature, pressure and humidity collected on the ground during the survey, while the other one used a simple theoretical model representing the standard atmosphere. On the basis of the result, we observed that the components of the baseline vector were very sensitive to the particular model considered. On the other hand, the baseline length is not very sensitive to the model. We chose the standard atmosphere model because more complex models require accurate knowledge concerning the lower troposphere. However, we hope to improve the accuracy in the near future considering a more reliable model for the local troposphere.

There are two other reasons which limit accuracy:

The first reason is due to the fact that GPS receivers used for this survey used only L1 frequency so that correction for ionospheric propagation delay is not allowed and the accuracy of GPS measurements was limited to about 1 ppm.

The second reason is due to the fact that in computing the baseline vector, we used the ephemeris broadcast by each satellite during the survey. The errors when using broadcast ephemeris are in the range 0.1–1 ppm.

Trilateration measurements were made using a K&E Rangemaster III electronic distance measuring instrument. Whenever possible, the measurements were made by night, measuring the meteorological data at both ends of the line. The modulation frequency of the instrument has been controlled before and after the survey. The precision obtained in such conditions is about 1 ppm

(Ruegg and Kasser 1987). A comparison of baseline lengths (mark to mark) determined from trilateration measurements and from GPS measurements shows an agreement to within a few millimeters (Table 2). The results indicate a lower agreement for baselines up to 5 km with respect to shorter distances. This result, even if not obvious, since absolute errors on both GPS and EDM measurements increase with distance, can probably be justified considering that the baselines PIZ-OBS and PIZ-OBS1 refer both to benchmark PIZ, which has been positioned by just one GPS measurement. It is also important to observe that there is an implicit limit in comparing EDM and GPS measurements, concerning the different correction for propagation delay in the atmosphere. In fact the path of electromagnetic waves is essentially horizontal for EDM and quite vertical for the GPS measurements.

The first year of tiltmeter recordings shows that instruments at five stations (TRE, LAM, NUN, GIA and LIN) are suitable to record tilt changes that occur at Etna over time periods less than several days. However, installation of instruments in lava tubes has greatly reduced noise level. High noise level at two stations (LAM and NIC) indicates that these instruments should be installed at new sites. The array of level benchmarks will be reoccupied so that we can also obtain a long-term record of tilt changes at these same sites.

Our GPS measurements are not as accurate as similar measurements conducted in other areas and reported by other scientists. The reasons for this mismatch have been analyzed in the paper. Nevertheless, our ability to reproduce trilateration measurements to within a few parts per million, in a few days'

Table 2. Control of accuracy with direct EDM measurements. GPS distances are the distances obtained after compensation of the network. EDM distances are the measured distances. The benchmarks *CIT3*, *OBS1*, *TDF1*, *VET1*, *MGN1*, *LAM6* and *NUN1* are located close to the main corresponding benchmarks and connected together by trilateration and triangulation techniques

Baseline	Distance GPS (m)	Distance EDM (m)	Difference (m)	(ppm)
CIT-PIZ	4766.187	4766.190	0.003	0.6
CIT-OBS	4052.873	4052.880	0.007	1.7
CIT3-OBS1	4083.498	4083.500	0.002	0.5
PIZ-OBS	2326.949	2327.052	0.103	44.3
PIZ-OBS1	2334.639	2334.684	0.045	19.3
TDF1-VET1	5508.105	5508.104	−0.001	0.2
TDF-VET	5501.055	5500.975	−0.080	14.5
MGN1-LAM6	20713.550	20713.560	0.010	0.5
MGN1-CIT	22880.395	22880.401	0.006	0.3
MGN1-PIZ	22929.053	22929.086	0.033	1.4
MGN1-OBS	24725.186	24725.143	−0.043	1.7
MGN1-LAM	20628.183	20628.178	−0.005	0.2
MGN-LAM	20633.159	20633.124	−0.035	1.7
MGN-BEL	24681.894	24682.133	0.239	9.7
BEL-LAM6	13089.951	13089.955	0.004	0.3
BEL-NUN1	16203.699	16203.694	−0.005	0.3

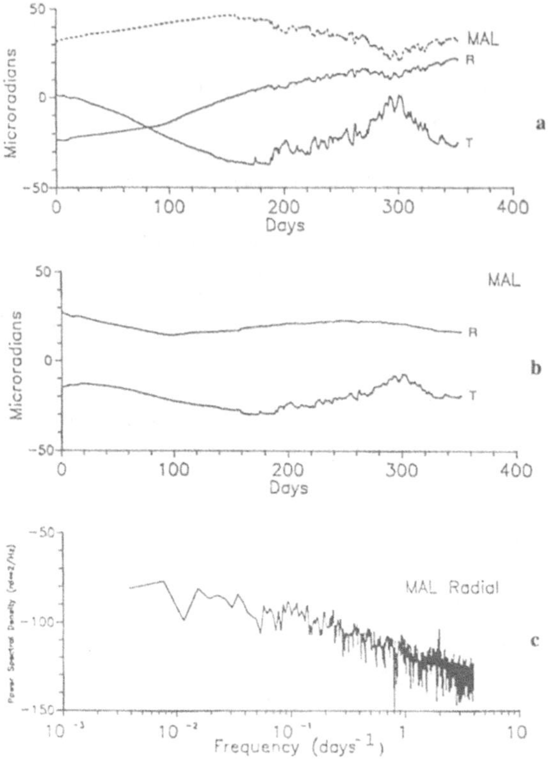

Fig. 4a–c. An example of power spectra analysis of tilt data (for the MAL station). **a** Tilt signals (*solid lines*) and temperature (*dashed lines*); **b** tilt signals after correction for temperature fluctuations; and **c** power spectrum for data shown in **b**

survey demonstrates that GPS is a useful technique to measure ground movements on active volcanoes.

Acknowledgements. This work has been supported by the French CNRS (PIRPSEV) and Italian CNR. P.A. Blum has been the initiator of this French-Italian cooperation in ground deformations monitoring. He also provided the tilt sensors. The authors are grateful to the anonymous reviewers for the critical reading of the manuscript and for many useful suggestions. Finally, all persons involved in the numerous fieldwork studies since 1986 are acknowledged.

References

Beutler G, Buersima I, Gurtner W, Rothacher M, Schilknecht T, Mader GL, Abell MD (1987) Evaluation of the 1984 GPS system campaign with the Bernese GPS software. J Geophys Res 87:1295–1303

Blum PA (1963) Contribution à l'étude des variations de la pesanteur en un lieu. Ann Geophys 19, 3:215–243

Bousquet JC, Lanzafame G, Villari L (1984) Les ruptures de surface liées à l'éruption du 28 mars 1983 de l'Etna (Sicile). Bull Volcanol 47:895–907

Bousquet JC, Gresta S, Lanzafame G, Paquin C (1987) Il campo degli sforzi quaternari e attuali nella regione etnea. Mem Soc Geol It 38:483–506

Calvari S, Coltelli M, Consoli O, Pompilio M, Scribano V (1989) L'attività eruttiva ai crateri sommitali. Rapp Attività Etna 1988, IIV-CNR, Catania

Cosentino M, Frazzetta G, Villari L (1987) Mt. Etna doings between 1983–1987. In: Volcanological research in Italy (1983–1987). CNR-GNV, August 1987, pp 38–64

Costantino Tomasi G (1962) L'applicazione della "tiltmetria" per la sorveglianza dell'attivita' dell'Etna. Boll Acc Gioenia Sci Nat, Catania, 1962–1963, pp 35–46

Henbest SN, Mills AA, Ottey P (1978) Two tiltmeters and one integrating seismometer for the monitoring of volcanic activity, and results of some trials on Mount Etna. J Volcanol Geotherm Res 4:133–149

Mortenson CE, Hopkins DG (1988) Tiltmeter measurement in Long Valley caldera, California. J Geophys Res 92:13767–13776

Murray JB (1982) Les déformations de l'Etna à la suite de l'éruption de Mars 1981. Bull Pirpsev 57:1–30

Murray JB, Pullen AD (1984) Three-dimensional model of the feeder conduit of the 1983 eruption of Mount Etna volcano from ground deformation measurements. Bull Volcanol 47.4(2):1145–1163

Murray JB, Guest JE, Butterworth PS (1977) Large ground deformation on Mount Etna volcano. Nature (Lond) 266:338–340

Nunnari G, Puglisi G (1987) Lo studio delle deformazioni del suolo sull'Etna mediante misure distanziometriche dal 1979 al 1985. Boll GNV-CNR, pp 505–520

Nunnari G, Romano R (1987) Ground tilt related to volcanological observations to analyse the December 1985 Mt. Etna eruption. Per Mineral 55:177–186

Ruegg JC, Kasser M (1987) Deformation across the Asal-Ghoubbet rift, Djibouti, uplift and crustal extension 1979–1986. Geophys Res Lett 14:745–748

Saleh B (1986) Developpement d'une nouvelle instrumentation pour les mesures de déformations. Application au génie civil. Thèse de doctorat, Univ Paris VI

Sanderson TJO, Berrino G, Corrado G, Grimaldi M (1983) Ground deformation and gravity changes accompanying the March 1981 eruption of Mount Etna. J Volcanol Geotherm Res 16:299–315

Tarantola A, Valette B (1982) Generalized nonlinear inverse problems solved using the least square criterion. Rev Geophys Space Phys 20(2):219–232

Tralli DM, Dixon TH, Stephens SA (1988) Effect of wet tropospheric path delay on estimation of geodetic baseline in the Gulf of California. J Geophys Res 93:6545–6557

Villari L (1980) Comment on "Two tiltmeters and one integrating seismometer for the monitoring of volcanic activity, and results of some trials on Mount Etna" by SN Henbest, AA Mills and P Ottey. J Volcanol Geotherm Res 8:343–348

Villari L (1982) L'Etna. Le Scienze 168:80–91

Wadge G (1976) Deformation of Mount Etna, 1971–1974. J Volcanol Geotherm Res 1:237–263

Wadge G, Horsfall JAC, Brunder JL (1975) Tilt and strain monitoring of the 1974 eruption of Mount Etna. Nature (Lond) 254:21–23

Walsh JB, Decker RW (1971) Surface deformation associated with volcanism. J Geophys Res 76:3291–3302

3 Modeling of Volcanic Earthquakes, Tremors and Long-Term Events

A Seismic Model for the Source of Long-Period Events and Harmonic Tremor

BERNARD CHOUET

Abstract

The sharply peaked spectra of long-period events and harmonic tremor can be explained by the resonance of a three-dimensional, fluid-driven crack induced by a pressure transient applied over a small area ΔS of the crack wall. The crack resonance is sustained by a very slow dispersive wave called the crack wave, the phase velocity of which decreases rapidly with increasing wavelength and with increasing values of the crack stiffness, $C = (b/\mu)(L/d)$, a dimensionless parameter where b is the bulk modulus of the fluid, μ is the rigidity of the solid, L is the crack length and d the crack thickness. The excitation of modes depends on the position of the pressure transient, the extent of crack surface affected by the transient, the time history of the transient, and the boundary conditions in effect at the crack perimeter. The far-field spectrum depends critically on the crack stiffness and on the impedance contrast between the fluid and solid, $Z = (\rho_s\alpha)/(\rho_f a)$, in which ρ_s, ρ_f are the densities of the solid and fluid, and α and a represent the velocity of the compressional wave in the solid and acoustic wavespeed of the liquid, respectively. Lowering the ratio b/μ, or increasing the ratios α/a and ρ_s/ρ_f increases Z, which increases the signal duration. The duration of resonance is measured by the quality factor Q describing the damping of oscillations of the dominant mode observed in the far field. Values of $Q \simeq 10 - 20$ that are consistent with those observed for shallow tremor and long-period events can be obtained with $Z \simeq 3 - 40$, $b/\mu \simeq 0.5 - 0.01$, $\alpha/a \simeq 2 - 8$, and $\rho_s/\rho_f \simeq 1.5 - 5$. A high impedance contrast Z and sustained oscillatory source within the fluid are required to explain $Q > 20$ observed in some events. The low ratio b/μ associated with the large values of Z can be achieved with void fractions of gas in the liquid ranging up to a few percent. The presence of these bubbles drastically reduces the acoustic wavespeed of the fluid so that resonance at a long period is possible in a crack of small dimensions. The far-field signature is strongly dependent on the spatio-temporal characteristics of the transient driving the crack. These features of the model suggest that the dynamics of the gas phase may play a critical role in the excitation mechanism of both long-period events and tremor.

1 Introduction

An understanding of the dynamics of fluid-driven cracks is critical for the study of the sources of long-period events and harmonic tremor and for the quantitative assessment of eruptive hazards posed by active volcanoes. My purpose in this paper is to examine the far-field elastic wave field radiated by a fluid-driven crack using a model recently developed by Chouet (1986, 1988). After briefly reviewing the observed properties of long-period events and tremor and summarizing the basic features of Chouet's model, I analyze the dependence of the far-field radiation on the physical properties of the fluid and solid, and demonstrate the sensitivity of the radiated waveform on the spatio-temporal characteristics of the pressure transient that triggers the crack excitation. I conclude with a discussion of the implication of the model for the fluid properties and for the driving mechanism of long-period events and tremor.

2 Characteristics of Long-Period Events and Tremor

Among the many manifestations of seismicity associated with active volcanism, two types of signals, the long-period event and tremor, are ubiquitous. Tremor is identified by its continuous signal of sustained amplitude at frequencies within a limited bandwidth. Long-period events are discrete signals having an amplitude envelope resembling those of normal earthquakes but spectral features resembling those of tremor. Examples of cyclic tremor, long-period events, and short-period earthquakes observed at Nevado del Ruiz, Colombia, are shown in Fig. 1. The cyclic tremor appears as four bands of sustained signal at a frequency of 4 Hz spaced at regular intervals in the record. Tremor of this type is commonly associated with the activity of shallow hydrothermal systems on volcanoes

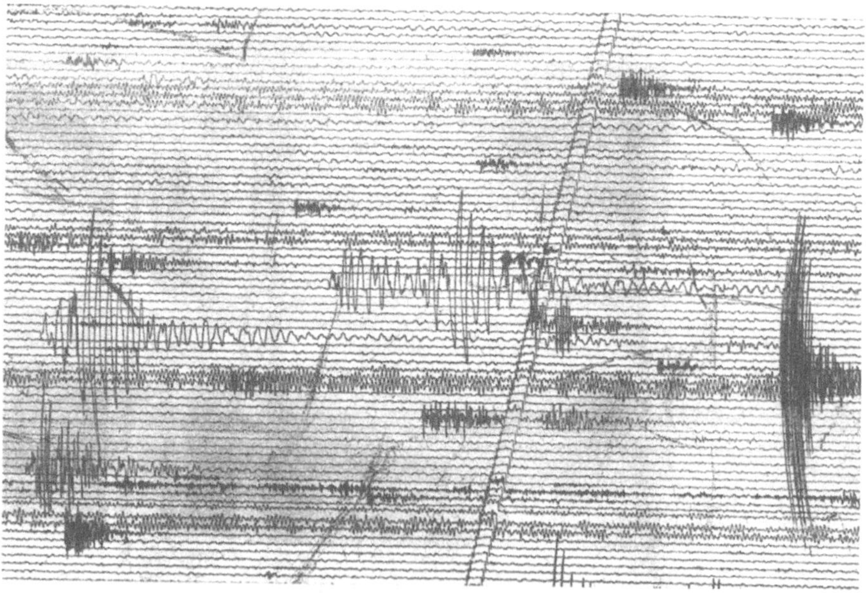

Fig. 1. Smoked-paper record depicting the vertical component of ground velocity obtained at a distance of 4 km from the active crater of Nevado del Ruiz volcano, Colombia, on 9 September, 1985. The record shows two long-period events (*center* and *left-center*), cyclic tremor episodes (the *four bands* of sustained signal at small amplitudes spaced at regular intervals in the record), and high-frequency volcano-tectonic earthquakes interspersed throughout the window (see, for example, large event at the *right*). The width of the tick marks is 1 s (Reproduced with permission of Bruno Martinelli, Swiss Seismological Service)

(McKee et al. 1981; Gil Cruz et al. 1987) and geysers (Kieffer 1984). The two long-period events displayed near the center of the figure show a quasi monochromatic waveform with a dominant frequency near 2 Hz lasting approximately 20 s. Long-period events often display high frequencies superimposed on their waveforms near the onset of the signal (Fehler and Chouet 1982; Bame and Fehler 1986). These events are quite distinct from short-period earthquakes, which typically have broadband signatures (see large event at right in Fig. 1). Other examples of long-period events are found in the seismic activity associated with some man-made fractures. Studies of microseismic events induced during hydraulic fracturing experiments carried out in low permeability rock at the Los Alamos National Laboratory, New Mexico, show that long-period events are the first events to occur at the beginning of an experiment (Bame and Fehler 1986). The long-period events observed in the Los Alamos experiments have a signature similar to those of long-period events observed on active volcanoes, but they are dominated by frequencies of 100–300 Hz and have durations of the order of 150 ms (Bame and Fehler 1986).

Long-period events and tremor are characterized by the following properties: (1) their spectra are sharply peaked, showing a dominant and several subdominant peaks (Fehler and Chouet 1982; Fehler 1983; McNutt 1986; Koyanagi et al. 1987). (2) Spectral analysis of the signals for both types of events shows that there is little variation of the dominant frequency with station location (e.g., Aki and Koyanagi 1981). (3) Temporal changes in the tremor spectrum are observed to correlate with temporal changes in volcanic activity, a behavior that cannot be ascribed to path effects (Kamo et al. 1977; Aki and Koyanagi 1981; Koyanagi et al. 1987). (4) During steady activity, the wide variation in amplitude of the spectral peak of tremor and long-period events and corresponding small variation in the frequency at which the peak occurs suggest that the amplitude of the signal is determined by the force generating the signal rather than the size of the source region (Fehler and Chouet 1982; Koyanagi et al. 1987). (5) The frequent occurrence of long-period events together with tremor, their common spectral features, and their common locations suggest a common source (Koyanagi et al. 1987).

2.1 The Fluid-Driven Crack Model

The relative consistency of the spectrum of long-period events and tremor observed at various locations means that the unique waveform of these events is due to a source effect rather than a path effect (Aki and Koyanagi 1981; Fehler and Chouet 1982; Fehler 1983; Aki 1984; Koyanagi et al. 1987). In other words, these signals can be interpreted as representing events in which the fluid participates actively in the source process. Included in this family of sources are various types of fluid-filled cavities (cracks, pipes and other volcanic conduits) in which a transient or sustained pressure disturbance of any origin (such as unsteady flow, degassing, rock failure, etc.) drives the resonance of the cavity. Observations at Kilauea, Nevado del Ruiz, and other volcanoes indicate that long-period events and tremor are related to both magmatic and hydrothermal activities (McKee et al. 1981; Kieffer 1984; Gil Cruz et al. 1987; Koyanagi et al. 1987).

Several models have been proposed to account for the observations of long-period events and tremor (see Chouet et al. 1987, for a review). Most attempts at a quantitative description of this type of source have considered the long-period event as some sort of impulse response of the tremor-generating system (Aki et al. 1977; Aki and Koyanagi 1981; Chouet 1985; Crosson and Bame 1985; Chouet 1988). Recently, Chouet and Julian (1985) and Chouet (1986) have developed fluid-driven crack models in which the coupling between the fluid and solid is fully analyzed, leading to quantitative results that allow a direct comparison with seismic observations. In these models, the equations for mass and momentum transfer in the fluid are solved simultaneously with the equations of elastodynamics in the solid. A solution for the motion of the crack wall is obtained under appropriate boundary conditions for stresses on the crack surface and boundary conditions for fluid flow at the crack perimeter with an initial condition provided by a pressure transient applied over an area ΔS of the crack wall.

An example of numerical solution obtained for the crack excitation using the model of Chouet (1986) is given in Fig. 2. Depicted in the figure are snapshots of the normal component of particle velocity on the crack wall taken at four different times near the onset of crack resonance. Time is represented by the number of time steps, N, completed in Chouet's finite difference calculation. The first snapshot at the upper left provides the crack orientation in the Cartesian coordinates system x,y,z; the crack is rectangular and lies in the plane $z = 0$ with width W in the x direction and length L in the y direction. The plot shows the propagation of the elastic disturbance on one face of the crack shortly after the onset of the transient, which is represented by a step of amplitude ΔP in excess pressure applied at the origin. The second snapshot represents the motion of the crack wall after 156 time steps and depicts the first reflections from the front, left, and right edges of the crack. The third snapshot, taken after 280 time steps, shows the motion of the wall immediately preceding the reflection from the back edge of the crack, and the last snapshot, taken after 600 time steps, depicts the shape of the lateral and longitudinal modes of crack resonance, which by then are well established.

The frequencies of the wave modes of the crack are fixed by the crack geometry and physical properties of the fluid and solid, and the excitation of these modes depends on the position of the pressure transient, the area ΔS over which the transient is applied, the time history of the transient, and the boundary conditions for fluid flow at the crack perimeter. In Chouet's model there is no mass transfer in or out of the crack so that the x and y components of flow (see Fig. 2) vanish at the crack perimeter. This precludes the existence of modes with wavelengths $2L$ and $2W$ for the normal component of wall displacement since these particular modes require a net inflow and outflow of fluid to satisfy the conservation of mass in the crack. The wavelengths of the modes that are allowed in this model are thus given by $2L/n$, $2W/n$; $n = 2,3,4,\ldots$.

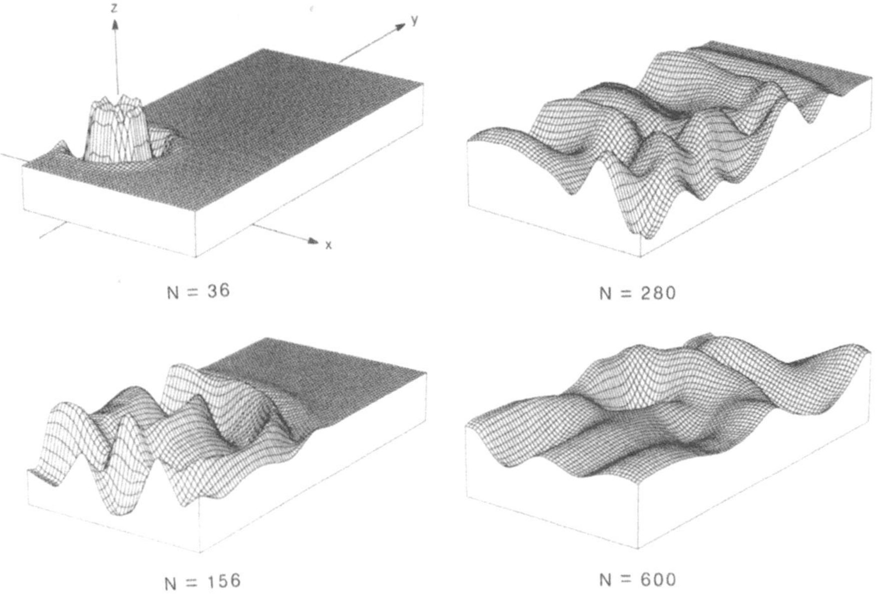

Fig. 2. Snapshots of the normal component of particle velocity on the surface of a fluid-filled crack excited by a step in pressure applied to a small area ΔS of the crack wall. The transient is applied at the origin of the coordinate system depicted at the *upper left*, which is located at a quarter of the crack width from the *left edge* of the crack and a quarter of the crack length from the *front edge* of the crack. The snapshots show the propagation of the disturbance on the crack wall at four different times in the history of crack oscillation specified by the number of time steps, N, in the finite difference solution. The time step has size $\Delta t = 0.00625L/\alpha$ (see text for explanation of symbols). Depicted are the shape of the initial disturbance at N = 36, the first reflections from the *front*, *left* and *right edges* of the crack at $N = 156$, and the shape of the lateral and longitudinal modes of resonance of the crack at $N = 280$ and $N = 600$

Chouet (1986) found that the resonant period of the fluid-filled crack is much longer than that expected from acoustic resonance because of the presence of a slow wave, which he called the crack wave. This trapped wave is always slower than the acoustic speed of the fluid and is inversely dispersive, showing a strong decrease in wavespeed for increasing wavelengths. The phase velocity of the crack wave also decreases rapidly with increasing values of a dimensionless parameter called the crack stiffness and defined as (Aki et al. 1977; Chouet 1986)

$$C = \frac{bL}{\mu d} \quad , \tag{1}$$

where b is the bulk modulus of the fluid, μ is the rigidity of the solid, L is the crack length and d the crack thickness.

The dispersion characteristics of the crack wave are reproduced from Chouet (1986) and Ferrazzini et al. (1990) in Fig. 3. These curves express the dimension-

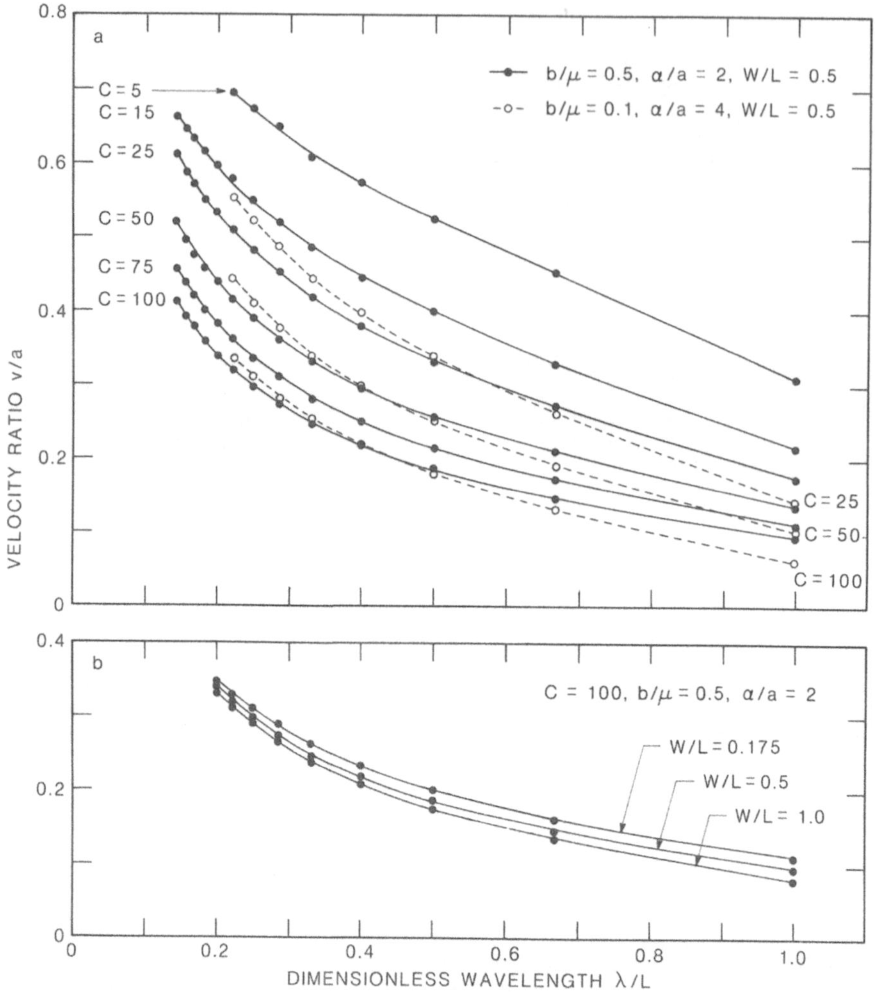

Fig. 3a,b. Ratio v/a of the phase velocity of the crack wave to acoustic velocity in the fluid plotted as a function of dimensionless wavelength λ/L. **a** Results obtained for various values of the crack stiffness factor ranging between 5 and 100 and for two selections of the ratios b/μ and α/a; $b/\mu = 0.5$, $\alpha/a = 2$ (solid curves); $b/\mu = 0.1$, $\alpha/a = 4$ (dashed curves) assuming a fixed ratio $W/L = 0.5$. **b** Results obtained for several ratios W/L with $C = 100$, $b/\mu = 0.5$, $\alpha/a = 2$ (from Chouet 1986, and Ferrazzini et al. 1990)

less ratio v/a of the phase velocity of the crack wave to the acoustic wavespeed of the fluid as a function of dimensionless wavelength λ/L for several values of the crack stiffness ranging between 5 and 100 and for selected values of the ratios b/μ, α/a and W/L, where α is the wavespeed of the compressional wave in the solid. Figure 3a demonstrates the dependence of the crack wave velocity on wavelength, crack stiffness, and ratios b/μ and α/a. Lowering b/μ at constant C decreases the phase velocity of the crack wave at long wavelengths and increases it at short wavelengths. As shown in Fig. 3b, the phase velocity also decreases with increasing values of the ratio W/L. At short wavelengths the phase velocity converges to a constant, the value of which approaches the acoustic speed of the fluid for increasing values of α/a and decreasing values of b/μ. The existence of the crack wave was demonstrated analytically by Ferrazzini and Aki (1987) for the limiting case of a fluid layer sandwiched between two elastic half spaces. An experimental confirmation of the numerical and analytical predictions concerning the behavior of this slow fracture wave was also obtained recently by Tang and Cheng (1988). This wave has important implications for the estimation of crack size based on the observations of long-period events as it shows that resonance at a long-period is possible in a crack of small dimensions.

Chouet (1988) investigated the near-field and far-field radiation properties of the fluid-driven crack model as a function of crack stiffness, fluid viscosity, trigger position, and trigger area, for fixed ratio $b/\mu = 0.5$. The far-field spectra he calculated are marked by the presence of many narrow-band peaks resulting from the interaction of the longitudinal and lateral modes of crack resonance. The dominant spectral peak radiated by the crack usually represents the mixing of lateral and longitudinal overtones of the crack rather than a fundamental mode. Chouet found this dominant peak to be insensitive to the radiation direction, in harmony with observations.

3 Dependence of the Far-Field Compressional Wave on Source Properties

The elastic wavefield radiated by the fluid-driven crack is a function of the crack geometry, the physical properties of the fluid and solid, and the spatio-temporal characteristics of the trigger that excites the crack. For the rectangular crack model, the geometrical parameters are expressed by the dimensionless ratios L/d and W/L. Similarly, the properties of the fluid and solid are fixed by the dimensionless ratios α/a, ρ_f/ρ_s and b/μ. These ratios define the impedance contrast between the fluid and solid through the relation (Aki et al. 1977)

$$Z = \frac{\rho_s \alpha}{\rho_f a} \quad , \tag{2}$$

where $\alpha = [(\lambda + 2\mu)/\rho_s]^{1/2}$ and $a = (b/\rho_f)^{1/2}$, in which λ and μ are the Lamé elastic constants. Assuming $\lambda = \mu$ in the solid (2) becomes:

$$Z = \sqrt{\frac{3\rho_s \mu}{\rho_f b}} \quad , \tag{3}$$

showing that the impedance contrast is inversely proportional to the square root of the ratio b/μ. This ratio, coupled with the ratio L/d, defines the crack stiffness parameter C seen in Eq. 1. Finally, the trigger parameters are the location of the pressure transient on the crack surface given by l and w (see Fig. 4), the area ΔS over which the transient is applied, and the source-time function of the transient characterized by the rise time τ.

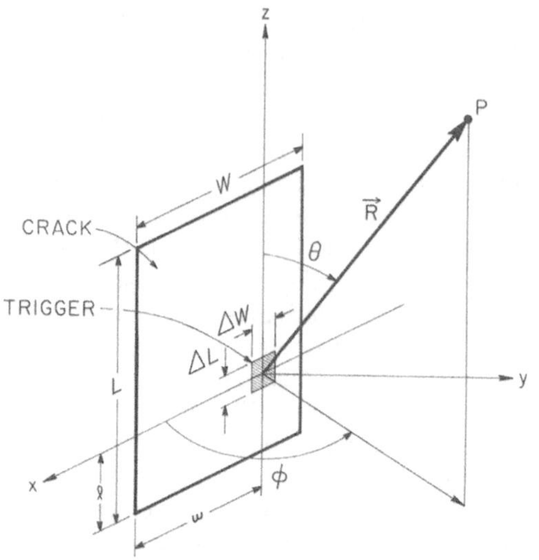

Fig. 4. Crack geometry and coordinate system used in the calculation of the far-field radiation. The pressure transient triggering the crack resonance is applied over an area $\Delta S = \Delta L \cdot \Delta W$ of the crack surface located at a distance w from the left edge of the crack and a distance l from the bottom edge

Table 1. Crack parameters

Figure	C	b/μ	α/a	ρ_f/ρ_s	Z	L/d	W/L	l	w	ΔW	ΔL	ΔS	τ'
5a,b	100	0.5	2	0.67	3	200	0.5	$L/2$	$W/2$	$W/10$	$L/40$	$L^2/800$	0
5c,d	20	0.1	4	0.53	7.5	200	0.5	$L/2$	$W/2$	$W/10$	$L/40$	$L^2/800$	0
5e,f	2	0.01	8	0.21	37.5	200	0.5	$L/2$	$W/2$	$W/10$	$L/40$	$L^2/800$	0
6a,b	100	0.1	4	0.53	7.5	1000	0.5	$L/2$	$W/2$	$W/10$	$L/40$	$L^2/800$	0
6c,d	100	0.03	6	0.37	16.2	3247	0.5	$L/2$	$W/2$	$W/10$	$L/40$	$L^2/800$	0
6e,f	100	0.01	8	0.21	37.5	10000	0.5	$L/2$	$W/2$	$W/10$	$L/40$	$L^2/800$	0
7a,b	100	0.5	2	0.67	3	200	0.5	$L/4$	$W/4$	$W/20$	$L/40$	$L^2/1600$	0
7c,d	100	0.5	2	0.67	3	200	0.5	$L/4$	$W/4$	$W/20$	$L/40$	$L^2/1600$	0.9434
8a	5	0.5	2	0.67	3	10	0.5	0	$W/2$	$W/10$	$L/40$	$L^2/800$	1.2987
8b	50	0.5	2	0.67	3	100	0.5	0	$W/2$	$W/10$	$L/40$	$L^2/800$	1.2987
8c	100	0.5	2	0.67	3	200	0.5	0	$W/2$	$W/10$	$L/40$	$L^2/800$	1.2987
8d	100	0.5	2	0.67	3	200	0.5	$L/4$	$W/2$	$W/10$	$L/40$	$L^2/800$	1.2987
8e	100	0.5	2	0.67	3	200	0.5	$L/4$	$W/2$	W	$L/40$	$L^2/80$	1.2987
8f	100	0.5	2	0.67	3	200	0.5	$L/2$	$W/2$	$W/10$	$L/40$	$L^2/800$	1.2987

My purpose here is to investigate the dependence of the far-field compressional wave radiated by the crack on impedance contrast for constant aspect ratio L/d and for constant stiffness C, and to demonstrate the sensitivity of the far-field synthetics on the mechanism driving the crack excitation using simple kinematic models of the transient source-space-time function. The relevant crack parameters are listed in Table 1. All the models assume an inviscid fluid and $\lambda = \mu$ in the solid.

The aspect ratios L/d considered for the models span the range 10^1–10^4, which is consistent with the ratios measured on dikes (MacDonald 1972; Delaney and Pollard 1981), and the ratios inferred for man-made hydrofractures (McFarland and Murphy 1976; Ferrazzini et al. 1990). The physical parameters of the fluid are more difficult to ascertain. For example, the values selected in Fig. 6c,d are compatible with the densities and wave velocities for a water-filled crack in granite, namely $\rho_f = 1\,\mathrm{g/cm^3}$, $a = 1.09\,\mathrm{km/s}$ for water at 200 °C, and $\rho_s = 2.69\,\mathrm{g/cm^3}$, $\alpha = 6\,\mathrm{km/s}$ for the rock. Such parameters are appropriate for the hydrofractures observed in the experiments conducted at Fenton Hill, New Mexico (Fehler 1979, Table 1). The parameters selected in the remaining figures cover the ranges $5 \times 10^{-1} - 10^{-2}$ for b/μ, $2 - 8$ for α/a, and $0.21 - 0.67$ for ρ_f/ρ_s. These are compatible with magmatic and hydrothermal fluids containing void fractions of gas ranging up to a few percent. These numbers can only be considered as rough guesses of the in situ properties of such fluids, as those properties are very poorly known (Kieffer 1977; Aki et al. 1978). These parameters represent a range of impedance contrasts from 3 to 37.5.

The geometry used for the computation of the far-field radiation is depicted in Fig. 4. The crack is rectangular with length L, width W, and thickness d, and lies in the plane $y = 0$ in the Cartesian coordinates x,y,z. The pressure transient is applied over an area ΔS of length ΔL and width ΔW located at a distance w from the left edge and l from the bottom edge of the crack. Observation of the crack is made in the far field at a distance R with orientation from the crack plane given by the angles θ and ϕ. The Fourier transform of the far-field radial displacement, $u^R(\vec{R}, \omega)$, in an infinite homogeneous elastic medium is given by (Chouet 1986)

$$u^R(\vec{R}, \omega) = \frac{1}{4\pi\rho_s\alpha^3 R}(\lambda + 2\mu \sin^2\theta \sin^2\phi) \cdot \int_{-l}^{L-l}\int_{w-W}^{w} \Delta\dot{v}(x, z, \omega)\exp\left[\frac{i\omega}{\alpha}(x \sin\theta \cos\phi + z \cos\theta)\right] dx\,dz, \qquad (4)$$

where $\omega = 2\pi\nu$ in which ν is the frequency, x and z are the coordinates of a point on the crack surface, and $\Delta\dot{v} = (\dot{v}^+ - \dot{v}^-)$ is the source space-time function in which $\dot{v}^+$, $\dot{v}^-$ represent the normal components of wall velocities on opposite faces of the crack in the positive and negative y directions.

Equation (4) is evaluated using the numerical solution for $\Delta\dot{v}$ obtained by the finite difference method of Chouet (1986). Examples of far-field spectra calculated with Eq. (4) and seismograms obtained by inverse Fourier transform of these spectra are shown in Figs. 5–8 for the selected models.

3.1 Dependence on Impedance Contrast for Constant Aspect Ratio

The crack resonance is strongly enhanced by increasing values of the impedance contrast. Figure 5 shows solutions obtained for a crack with fixed ratios $W/L = 0.5$, $L/d = 200$, and impedance contrast Z ranging from 3 to 37.5 (see Table 1).

The increase in the ratio α/a in Fig. 5a,c,e decreases the frequencies of the spectral peaks, but this effect is compensated in part by the associated decrease in the stiffness factor C, which increases the ratio v/a (see Fig. 3). These two competing factors explain why the dominant frequency remains the same in Fig. 5a,c but decreases in Fig. 5e. The opposite effects of α/a and C cancel each other out in the first two figures, while the effect of α/a dominates in Fig. 5e where it produces a marked shift in the spectral peaks with respect to those seen in Fig. 5c. The frequency shift due to the ratio α/a is associated with a decrease in the spacing between peaks and an overall increase in the amplitudes and decrease in the bandwidths of these peaks (cf. Fig. 5a,e). It is of interest to note also that the spectral envelope shows a marked increase in energy content at low frequencies for low values of the crack stiffness (Fig. 5e). The synthetic seismograms in Fig. 5b,d,f reflect those spectral features. They show that the signal amplitude and duration increase with the impedance contrast, and indicate that the shape of the coda envelope is dependent on the crack stiffness.

The crack resonance can be quantified through the use of the quality factor Q describing the damping of oscillations of a selected mode (Aki and Richards 1980)

$$Q = \frac{\nu}{\Delta\nu} \quad , \tag{5}$$

where ν is the frequency of the spectral peak due to this mode and $\Delta\nu$ is the bandwidth of the peak measured at one half of its amplitude. This formula was applied to the dominant frequencies seen in Fig. 5, yielding values of Q ranging from 7.8 to 16.4 over the range considered for Z (see Table 1). The factor Q does not increase linearly with the impedance contrast as both the frequency and bandwidth of the peak depend on the dispersion characteristics of the crack wave. For example, the quality factor in Fig. 5e is only marginally higher than that in Fig. 5c because the decrease in bandwidth of the dominant peak is associated with a decrease in the frequency of this peak.

3.2 Dependence on Impedance Contrast for Constant Stiffness

Figure 6 shows the dependence of the far-field solution on the impedance contrast for constant stiffness $C = 100$, constant ratio $W/L = 0.5$, and for various values of b/μ, α/a, and L/d. Of interest is the marked sensitivity of the dominant and subdominant spectral peaks on the latter three parameters. The increase in Z from 7.5 in Fig. 6a to 37.5 in Fig. 6e is accompanied by an increase in the excitation of the subdominant mode relative to the dominant mode to the extent that the former eventually dominates over the latter. This is quantified by the values of Q obtained for both peaks, which first show a decrease in the quality factor of the dominant peak from $Q = 13.4$ in Fig. 6a to $Q = 9.9$ in Fig. 6c, followed by a jump from values near 10 to values near 17 for crack

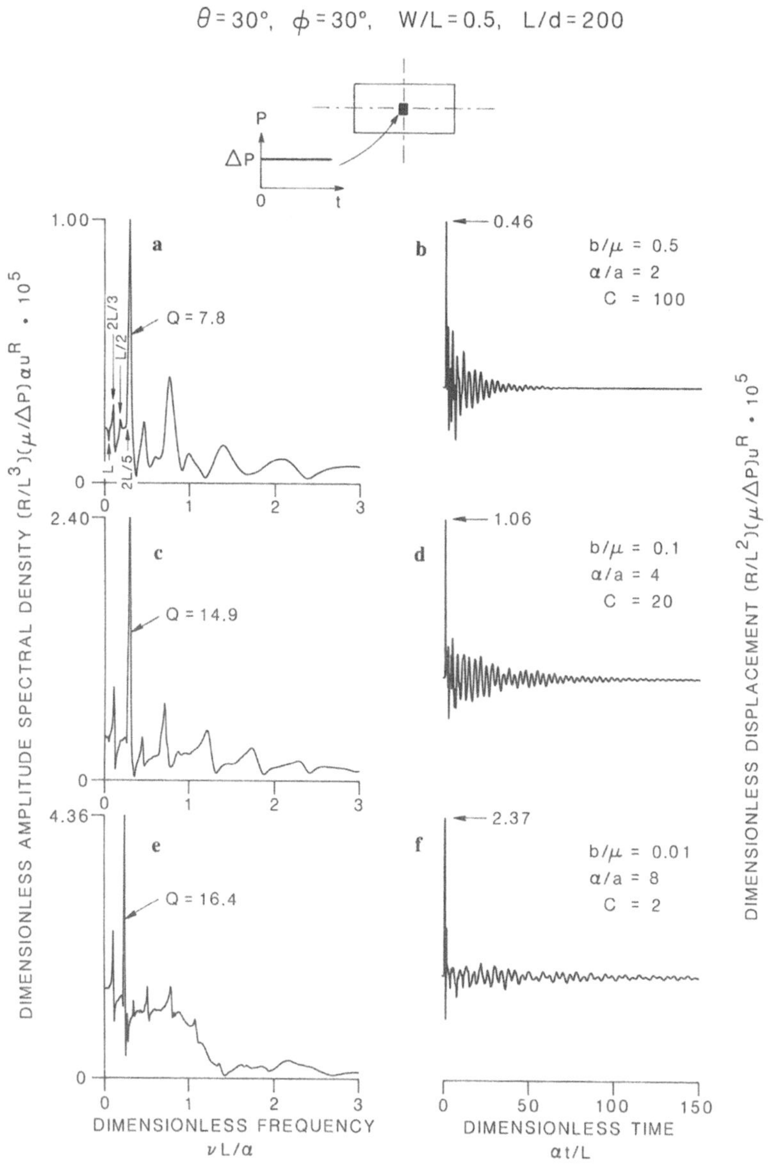

Fig. 5a–f. Far-field spectra and associated seismograms of the compressional wave train radiated by the excitation of a fluid-filled crack. The synthetics are obtained for $\theta = 30°$, $\phi = 30°$ (see Fig. 4) and are given in dimensionless form as a function of the crack and medium parameters. The ratios $W/L = 0.5$ and $L/d = 200$ are kept constant, and the ratios b/μ, α/a and stiffness C are varied: **a,b** $b/\mu = 0.5$, $\alpha/a = 2$, $C = 100$; **c,d** $b/\mu = 0.1$, $\alpha/a = 4$, $C = 20$; **e,f** $b/\mu = 0.01$, $\alpha/a = 8$, $C = 2$. The pressure transient is a step of amplitude ΔP applied at the center of the crack (see sketch at top of figure). Dimensionless peak amplitudes are indicated by *arrows* in the seismograms

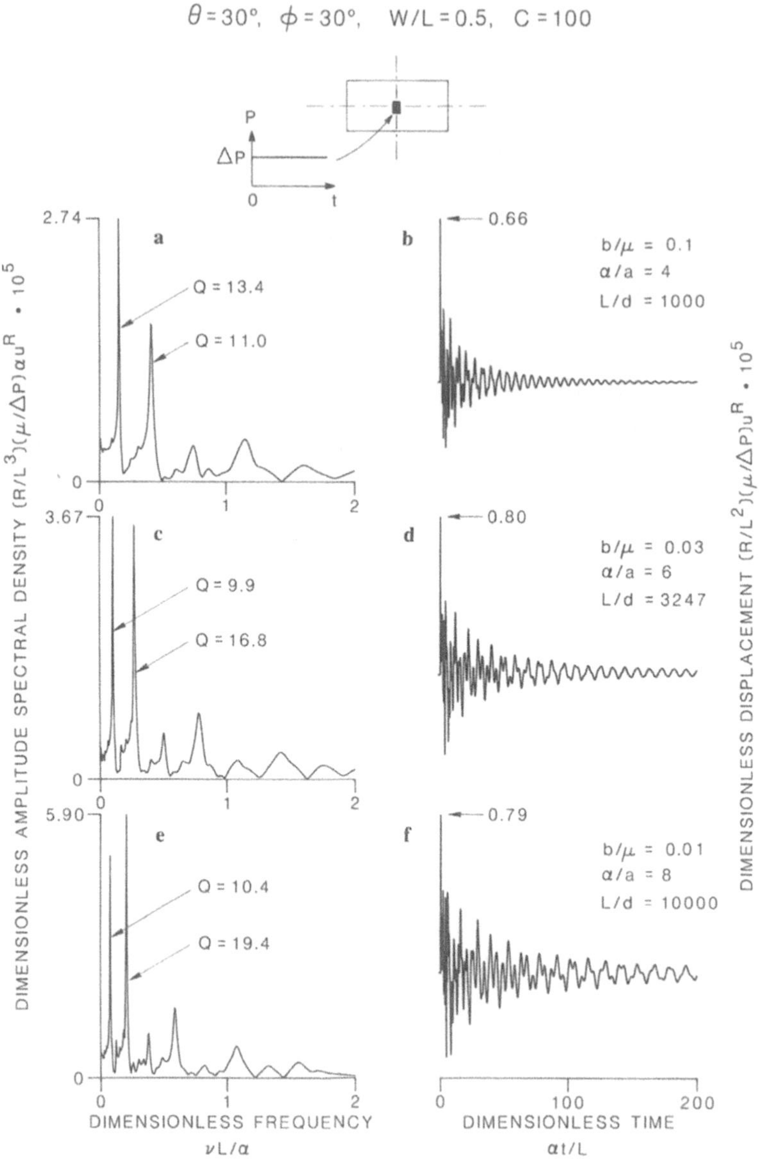

Fig. 6a–f. Same as Fig. 5 but here the ratio $W/L = 0.5$ and the stiffness $C = 100$ are fixed, and the ratios b/μ, α/a and L/d are varied: **a,b** $b/\mu = 0.1$, $\alpha/a = 4$, $L/d = 1000$; **c,d** $b/\mu = 0.03$, $\alpha/a = 6$, $L/d = 3247$; **e,f** $b/\mu = 0.01$, $\alpha/a = 8$, $L/d = 10000$

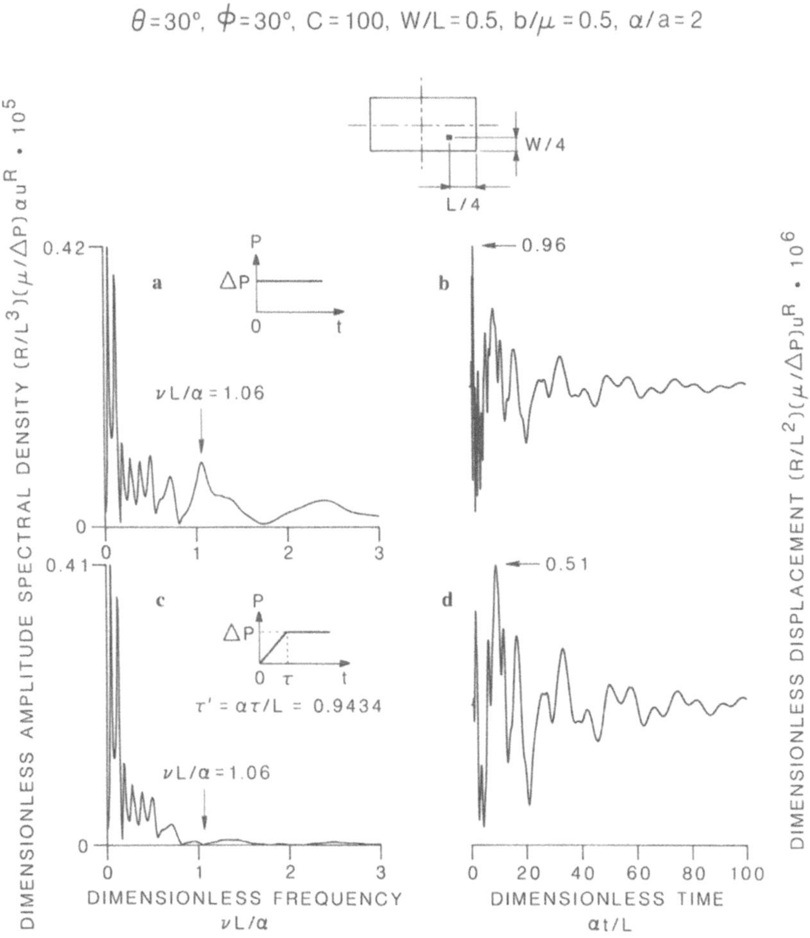

Fig. 7a–d. Same as Fig. 5 for two different transients applied over an area $\Delta S = L^2/1600$ positioned at $l = L/4$ and $w = W/4$. The stiffness C and the ratios W/L, L/d, b/μ and α/a are kept constant: $C = 100$, $W/L = 0.5$, $L/d = 200$, $b/\mu = 0.5$, $\alpha/a = 2$. **a,b** Results for a step transient of amplitude ΔP. **c,d** Results for a ramp step transient of amplitude ΔP and dimensionless rise time $\tau' = 0.9434$

parameters near those of Fig. 6c, followed by an increase to $Q = 19.4$ in Fig. 6e. The decrease in Q observed in Fig. 6a,c is related to the lowering of the dominant frequency associated with the increase in the ratio α/a as discussed above, while the sudden jump in that factor observed in Fig. 6c,e reflects the switching of the dominant frequency from a lower to a higher mode of resonance.

The position of the trigger naturally affects the relative excitation of the various modes of crack resonance. In Figs. 5 and 6, for example, the trigger is positioned at the center of the crack emphasizing the odd modes with wave-

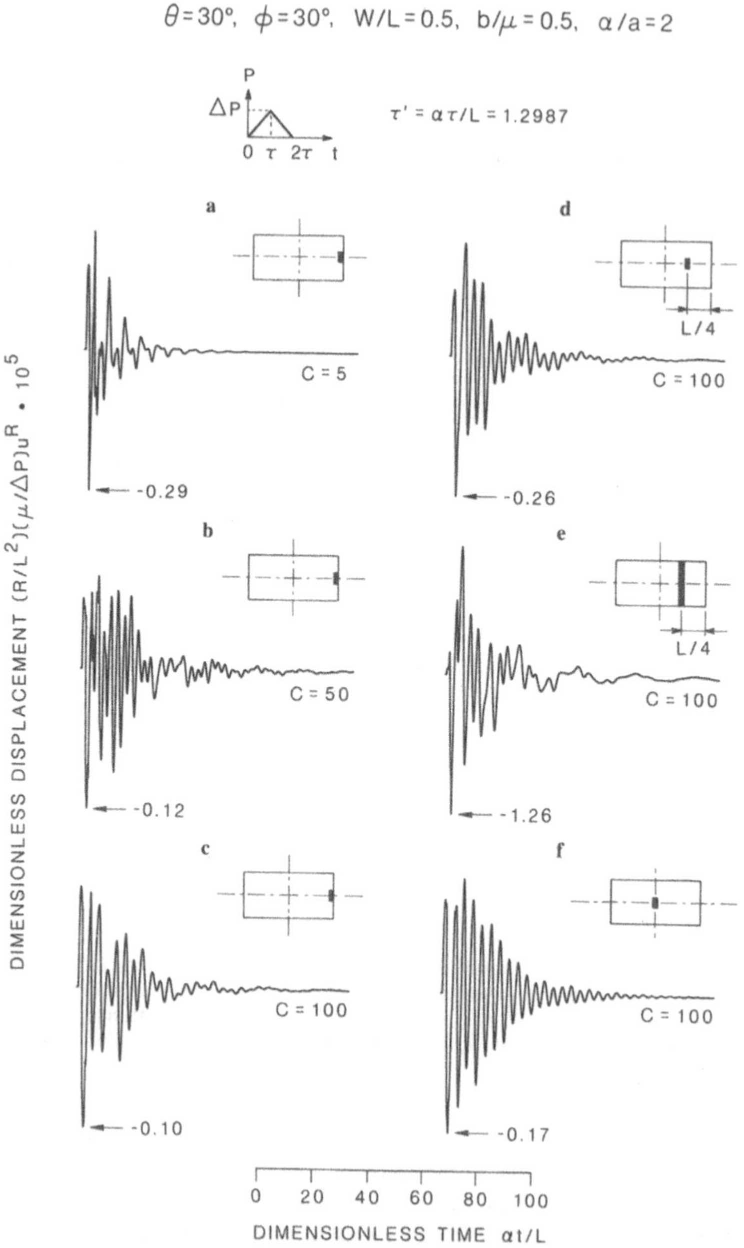

Fig. 8a–f. Far-field compressional wave trains produced by the excitation of a fluid-filled crack. The results are obtained for $\theta = 30°$, $\phi = 30°$ and are expressed in dimensionless units of the crack and medium parameters. Peak amplitudes are indicated by the *arrows*. The ratios W/L, b/μ and α/a are kept constant and the synthetics demonstrate the dependence of the far-field radial displacement on crack stiffness, trigger position, and trigger area for a triangular pressure transient with dimensionless rise time $\tau' = 1.2987$ and dimensionless duration $2\tau'$

lengths $2L/n$, $2W/n$; $n = 3, 5, 7, \ldots$. Notice that the even modes in these models are minimized but not totally removed because the area ΔS over which the transient is applied has a finite extent that provides limited excitation of such modes. The fundamental and first three longitudinal overtones of the crack obtained from Table 2 of Chouet (1986) are indicated by arrows in the far-field spectrum displayed in Fig. 5a. The lowest resonant frequency seen in this spectrum corresponds to the wavelength $2L/3$ and the dominant frequency occurs close to the frequency expected for the overtone with wavelength $2L/5$. The latter is interpreted as being due to the interaction of this longitudinal mode with one of the lateral modes of the crack. The spectral peaks at higher frequencies are produced by the interaction of higher modes. These peaks are not evenly spaced in frequency since they reflect the interference pattern between the longitudinal and lateral modes of the crack associated with the dispersion characteristics of the crack wave (Fig. 3), and they also depend on the crack orientation (Chouet 1988).

3.3 Dependence on Rise Time of Transient

Figure 7 illustrates the sensitivity of the radiated compressional wave to the rise time of the pressure transient that triggers the crack resonance. The results shown at the top are obtained for a step transient of amplitude ΔP with Fourier transform $\Delta P/(i\omega)$, and the bottom plots show the solutions for a step transient with rise time τ and Fourier transform $-(\Delta P/\tau)[1 - \exp(-i\omega\tau)]/\omega^2$. The crack and trigger geometry, and the physical parameters of the fluid and solid, are identical in both models. The amplitude spectrum of the first transient is expressed as $\Delta P/\omega$ and that of the latter is given by $(\Delta P/\tau)[2 - 2\cos(\omega\tau)]^{1/2}/\omega^2$. From these expressions one sees that the ramp transient contributes an extra factor ω^{-1} to the spectral roll-off at high frequencies and also introduces nodes at $\nu = n/\tau$, $n = 0, 1, 2, \ldots$ in the crack spectrum. In the examples shown in the figure, the value of τ was selected in such a way as to produce a node coinciding with the subdominant spectral peak positioned at the dimensionless frequency $\nu' = \nu L/\alpha = 1.06$ ($\tau' = 1/\nu' = 0.9434$), thereby eliminating this particular peak from the spectrum. Figure 7d shows the resulting seismogram in which the high-frequency content seen in Fig. 7b has been removed by this operation.

3.4 Dependence on Stiffness and on Position and Area of Transient

Figure 8 shows the radiated compressional wave trains for selected models assuming a transient shaped in the form of a triangular impulse of amplitude ΔP and width 2τ. The Fourier transform of this transient is given by $-(\Delta P/\tau)[1 - \exp(-i\omega\tau)]^2/\omega^2$, which has the amplitude spectrum $(\Delta P/\tau)[6 - 8\cos(\omega\tau) + 2\cos(2\omega\tau)]^{1/2}/\omega^2$. This function is similar to that obtained for the ramp transient, displaying spectral nodes at $\nu = n/\tau$; $n = 0, 1, 2, \ldots$ and a roll-off proportional to ω^{-2} at high frequencies. Six models are depicted showing the dependence of the radiated wave on the crack stiffness and trigger characteristics for sources with the common parameters $W/L = 0.5$, $b/\mu = 0.5$, and $\alpha/a = 2$. The three synthetics at the left show the results obtained for a transient applied over an area $\Delta S = L^2/800$ located at $l = 0$, $w = W/2$ on cracks with different stiffnesses ranging from $C = 5$ (top) to $C = 50$ (middle) and $C = 100$ (bottom). The synthetics at the right show the dependence of the solution on the position of the transient, and on the size and shape of the area ΔS over which the transient is applied for a crack with stiffness $C = 100$. The top trace (Fig. 8d) represents a trigger applied over an area $\Delta S = L^2/800$ centered on the main axis of the crack at $l = L/4$, the middle trace (Fig. 8e) represents a trigger applied over $\Delta S = L^2/80$ extending over the entire width of the crack at $l = L/4$, and the bottom trace (Fig. 8f) represents a trigger applied over $\Delta S = L^2/800$ positioned at the center of the crack.

The sensitivity of the far-field compressional wave on the characteristics of the transient is quite obvious in these signatures. The shape of the waveform reflects the excitation and interaction of the various modes of crack resonance, which depend on the position of the transient and size of ΔS. This is best observed in the excitation of the fundamental longitudinal mode, which is minimized in Fig. 8a–c,f by positioning the transient at a node, and maximized in Fig. 8d,e by positioning the transient at an antinode for this mode. The maximum excitation of this mode is seen in Fig. 8e, where the trigger is located at an antinode and ΔS is the largest. The lateral modes of the crack are not excited in this latter case because the transient is applied simultaneously over a band that covers the entire width of the crack. Of particular interest is the result presented in Fig. 8f, which shows a quasi monochromatic signal. This is due to the choice of τ, which coincides with the subdominant spectral peak at the dimensionless frequency $\nu' = \nu L/\alpha = 0.77$ in Fig. 5a ($\tau' = 1/\nu' = 1.2987$). The elimination of this peak leaves the spectrum with a single well-defined narrow-band peak, which is responsible for the harmonic signature seen in Fig. 8f.

4 The Importance of Gas Content

An important result of the crack model is the demonstrated sensitivity of the duration of radiated signal on the impedance contrast between fluid and solid (see Figs. 5 and 6). The far-field calculations presented in this study show that the quality factor Q of the resonating crack increases with the impedance contrast, reaching values near 20 for $Z \simeq 40$. These results are consistent with the quality factor $Q \simeq 5 - 10$ of tremor originally reported by Aki et al. (1977) but are below the estimates of $Q \leqslant 40$ reported by Ferrazzini and Aki (this Vol.) for tremor and gas-piston events observed at Kilauea, Hawaii. The high values of Q observed in the field imply a high impedance contrast Z or low ratio b/μ between magma and rock, which can be achieved through the presence of gas bubbles in the fluid. The bulk modulus of the liquid-gas mixture can be expressed as (Aki et al. 1978)

$$b = \frac{1}{[(1-\varepsilon)b_l^{-1} + \varepsilon b_g^{-1}]} \quad , \tag{6}$$

where ε is the volume fraction of gas, and b_l and b_g are the bulk moduli of the liquid and gas, respectively. For a bimolecular gas under ambient pressure P, the bulk modulus of gas is given by $b_g = 1.4P$ (Aki et al. 1978). At a depth of 100 m, $P \simeq 30$ bar yielding $b_g \simeq 4 \times 10^7$ dyne/cm^2. On the other hand, the experimental data for liquid magma of Murase and McBirney (1973) yields $b_l \simeq 10^{11}$ dyne/cm^2. Then, for $\varepsilon = 0.05$ one obtains $b \simeq 10^9$ dyne/cm^2, which yields $b/\mu \simeq 10^{-2}$ assuming $\mu \simeq 3 \times 10^{11}$ dyne/cm^2. The same calculation done for a depth of 5 km yields $b/\mu \simeq 10^{-1}$. Thus, the range of b/μ considered in our models (see Table 1) can be achieved with a few volume percent of bubbles in magmatic fluids at depths extending to a few kilometers. These depths are consistent with the shallow magmatic activity observed on active volcanoes.

The viscosity of the fluid, however, introduces an additional energy loss at the source which affects the above estimates of Q. The apparent quality factor of the tremor source represents the sum of the energy loss per cycle due to seismic radiation and the energy loss per cycle due to viscous damping in the resonating fluid. For small losses the observed Q can be written as (Aki 1984)

$$\frac{1}{Q} = \frac{1}{Q_r} + \frac{1}{Q_v} \quad , \tag{7}$$

where Q_r is the quality factor due to radiation and Q_v is that due to fluid viscosity.

Chouet (1988) showed that the viscosity of the fluid decreases the amplitudes of the spectral peaks observed in the far field and also produces some shift in the frequency of these peaks. A calculation he performed for a crack with the parameters $C = 100$, $\alpha/a = 2$, $b/\mu = 0.5$ shows that the Q of the dominant spectral peak decreased from 7.5 for an inviscid fluid to 4.7 for a fluid with viscosity of 2500 P (see Fig. 6c,d of Chouet 1988). The spectral broadening due to fluid viscosity suggests that the impedance contrast between fluid and solid required to explain the observed Q of tremor may need to be even larger than the values considered above, a requirement that may be met with a larger bubble content in the fluid. The presence of these bubbles will drastically reduce the sound speed of the fluid (Kieffer 1977) and hence the speed of the crack wave (see Fig. 3) so that resonance at a long period may be possible for a small crack.

One must, however, bear in mind other factors that complicate the simple picture presented so far. The first of these factors relates to the driving mechanism of tremor, which may involve the superposition of signals from many sources in the fluid and from many cracks. For example, it may be difficult to explain the high values of Q of up to 40 reported by Ferrazzini and Aki (this Vol.) solely through the expedient of adding more bubbles in the magmatic fluid. A more harmonious explanation is to consider a sustained oscillatory source in the fluid, the origin of which lies in the dynamics of the interactions between the gas and liquid phases. The second factor relates to the depth of the tremor source. Here we have to contend with the duration of the signals from deep (30–60 km) long-period events and tremor beneath Kilauea volcano (Aki and Koyanagi 1981; Koyanagi et al. 1987; Shaw and Chouet 1988), where confining pressure exceeds 1 GPa and effectively prevents the sustained presence of a significant gas phase. Based on those constraints it seems likely that the origin of deep tremor may be different from that of shallow tremor. I shall discuss this idea more extensively below.

5 The Driving Mechanism of Tremor

The first attempt at developing a seismic model of the source of tremor was undertaken by Aki et al. (1977), who considered the elastic radiation produced by the sudden extension of a fluid-filled crack in response to excess magma pressure. In their model, the extension of the crack tip or opening of a channel between adjacent cracks excite the vibration of the crack wall from which elastic waves are radiated into the solid. Tremor results from the superposition of many such events produced by sustained jerky extensions of the crack tip or by sustained jerky openings of channels connecting a chain of cracks. A similar model was used by Aki and Koyanagi (1981) to estimate the rate of magma transport under Kilauea volcano, Hawaii, from the observation of deep (30–60 km) tremor. This model was also adopted by Shaw and Chouet (1988, 1989) in studies of the intermittent structure of the same deep tremor data, in which individual tremor episodes were considered to be a composite of interacting long-period events, each of the latter representing an individual squirt of magma into a bifurcating set of magma-filled cracks within a small hypocentral volume in the volcanic edifice. The transport model of Shaw and Chouet (1988) is a model of the percolation of fluid in a random medium (Stanley and Meakin 1988), in which mass transfer is linked to tensile fracture in a dendritic network of cracks in response to a combination of factors involving magma pressure, magma concentration, fracture geometry, rock strength, and stress states. They treat magma transport and fracture propagation as a system of coupled non-linear oscillators in which the trigger mechanism punctuates the timing of coupled oscillations of fracture and flow in response to the relaxation frequency of the hypocentral plexus coupled with the characteristic frequencies of the fluid-filled cracks.

The common assumption of these studies is that tremor is composed of a succession of long-period events, the latter being interpreted as the crack response to a pressure transient, the origin of which lies in the brittle failure of the crack. Observed similarities in the source locations and spectral signatures of long-period events and tremor lend credence to the idea that the two types of events share a common source (Koyanagi et al. 1987), while the rough agreement between the observed characteristics of long-period events (e.g., Fig. 1) and the synthetics generated by the model (e.g., Fig. 8) lends additional support to the idea that the long-period event represents some sort of impulse response of the tremor generating system (Chouet 1985, 1988).

Although the above models provide a working hypothesis for the origin of deep tremor in Hawaii, the mechanism triggering the crack excitation is clearly not limited to the tensile failure postulated in those models. The brittle crack model, for example, cannot explain the occurrence of shallow tremor sustained over many months in the east rift of Kilauea volcano where magma flow has been observed to be quasi steady, suggesting that the dike system feeding an eruption can remain unobstructed over long periods of time (Dietel et al. 1989). There are many other instances also where tremor is obviously unrelated to the failure of rock. Examples are tremor associated with geyser activity (Kieffer 1984) and tremor generated by shallow hydrothermal activity (Gil Cruz et al. 1987), where exploding and imploding bubbles of steam may play a major role in the excitation mechanism of both long-period events and tremor (Kieffer 1984; Leet 1988).

As mentioned earlier, a fundamental implication of the model studied here, as well as models considered by other investigators (Aki et al. 1977; Chouet 1985, 1986, 1988; Crosson and Bame 1985; Ferrazzini and Aki this Vol.) is that the high Q observed for tremor requires the presence of bubbles in the fluid. This naturally suggests that the excitation mechanism of tremor may be closely related to the activity of these bubbles. Evidence in support of this idea comes from various observations and experiments. Photoballistic experiments conducted at Stromboli by Chouet et al. (1974) and observations of gas-piston activity at Kilauea volcano by Dietel et al. (1989) reveal a large volume ratio of gas to melt in the erupting magma and demonstrate that the degassing of the magma is an energetic phenomenon. Analytic studies of two-phase flow by Vergniolle and Jaupart (1986) and laboratory models of eruptions by Jaupart and Vergniolle (1988) also point to the critical role played by the gas phase.

What I am speculating here is that shallow magmatic tremor is driven by the acoustic emission from oscillating bubbles in response to pressure fluctuations in the fluid. Violent collapse is one possible behavior for a gas bubble in an oscillating pressure field that is associated with intense acoustic emission (Plesset and Prosperetti 1977). Theoretical calculations of the acoustic emission from collapsing bubbles show that for fixed relative gas pressure the maximum amplitude of the radiated pressure is proportional to the product of the collapse pressure and maximum bubble radius (Ross 1987). The radiation consists of a relatively low-frequency pressure oscillation associated with bubble growth, and a very sharp positive pressure pulse associated with bubble collapse. When the partial pressure of gas in the bubble is less than a few percent of the collapse pressure, the positive acoustic peak associated with this collapse can exceed the amplitude of the driving collapse pressure by a factor of several hundred [Ross 1987, Eq. (7.48)]. In other words, this is a mechanism by which small pressure fluctuations in the fluid can be greatly amplified. Since the effect of a single bubble is highly localized (Cramer 1980), one must consider the collective oscillations and implosions of clusters of bubbles containing a large number of bubbles to obtain a source of sound of significant spatial extent.

The acoustic emission from a field of collapsing bubbles has been studied under the assumption that individual collapses are random occurrences (Il'ichev and Lesunovskii 1963; Morozov 1968; Lyamshev 1969; Boguslavskii et al. 1970). These studies show that the acoustic power radiated by cavitation is proportional to the total volume of cavitation generated per unit time. The resultant spectrum covers a broad band of frequencies and is peaked at a frequency that is proportional to the square root of the collapse pressure and inversely proportional to the radius of the largest bubble in the field. The spectral amplitude increases with a slope bounded between ω and ω^2 at low frequencies and rolls off proportional to ω^{-1} at high frequencies (Ross 1987).

One possible source of pressure fluctuations driving the bubbles in magma may be the periodic shedding of vortices from alternate sides of an obstacle in the fluid stream. The role of viscosity, as measured by the Reynolds number *Re*, is crucial in the establishment of a von Kármán vortex street. This number is given by

$$Re = \frac{\bar{u}d}{\eta} \quad , \tag{8}$$

where $\bar{u}$ is the average flow speed, d is the cross-stream obstacle width, and η is the kinematic viscosity. For Reynolds numbers greater than 40, a regular street of vortices is shed with a shedding frequency ν characterized by the dimensionless Strouhal number $S = \nu d/\bar{u}$ (Cook and Harlow 1986). A Reynolds number of 40 can be achieved with a flow velocity of the oder of 1 m/s, width of obstacle of about 4 m, and kinematic viscosity of the fluid of $10^{-1}\,m^2/s$ (Chouet 1988). These figures are consistent with observations of fissure geometry and magma flow made by this author on the east rift of Kilauea volcano, Hawaii.

In this context it seems natural to view the long-period event as the impulse response of the tremor generating system, perhaps related to bubble collapses in a localized zone of the fluid, while tremor itself is viewed as the response of this system to sustained bubble oscillations in the fluid. In this dynamic view of tremor generation, the long-period event and tremor share the same source but have different driving mechanisms. The area ΔS (the shaded square in Figs. 4–8) then represents a patch of the crack surface over which the pressure disturbance due to a localized field of collapsing bubbles is applied, and the duration of the signal is controlled by a number of parameters that include the average flow speed, the kinematic viscosity of the fluid, the distribution of bubbles, and the pressure time history associated with the collapsing bubbles. We have seen that confining pressures restrict sustained magma degassing to sources that extend to depths of a few km only. This is a zone where magma-ground water interactions are important, so that the same model of tremor generation by acoustic emission from oscillating and collapsing bubbles may be applicable to tremor associated with hydrothermal sources such as geysers (Kieffer 1984), and to cyclic tremor episodes associated with phreatic eruptions (McKee et al. 1981; Gil Cruz et al. 1987), although the source of pressure fluctuations driving the bubbles may be different for each of those situations.

There are two advantages to this view of the origin of shallow tremor. First, it may explain the emergent onset of the signal that is observed on records of tremor, and second, it may provide an explanation for sustained tremor lasting sometimes over periods of many months in situations where mass transport is quasi steady or minimal (e.g., Gil Cruz et al. 1987). These considerations point to the necessity of including the dynamics of multiphase flow in future models of tremor generation. With the advent of large computers, such complex models may be within out grasp (Harlow and Amsden 1975; Cook and Harlow 1986;

Valentine and Wohletz 1989). However, the sensitivity of the radiated signal to the source space-time function of the trigger seen in the present study (see, for example, Figs. 7 and 8) suggests that unravelling the driving mechanism of tremor may be a difficult task indeed.

6 Conclusions

The three-dimensional fluid-driven crack model of Chouet (1986, 1988) was used to study the dependence of the radiated elastic wavefield on the physical parameters of the fluid and solid and on the spatio-temporal characteristics of the trigger. The spectral peaks of the far-field P-wave spectrum were found to be very sensitive to the impedance contrast between the fluid and solid. The quality factor Q of the dominant spectral peak shows values of 10 to 20 consistent with field observations of shallow tremor (Aki et al. 1977; Ferrazzini and Aki this Vol.) for values of the impedance contrast Z ranging between 3 and 40. The relationship between Q and Z is complex, however. Increasing Z decreases the bandwidth of the spectral peaks observed in the far field, but that increase is associated with an increase in the ratio α/a which shifts the peaks to lower frequencies, thus limiting the increase of Q associated with the narrowing of the spectral peaks. The dependence of the dispersion characteristics of the crack wave on Z through the ratio b/μ also changes the pattern of interaction between the lateral and longitudinal modes of crack resonance, so that the dominant spectral peak may switch from lower to higher modes, or vice versa, thereby producing a sudden jump in the estimates of Q.

The values $b/\mu \simeq 0.5 - 0.01$ required to produce the above ranges of Q and Z can be achieved with void fractions of gas in the liquid of a few percent. The presence of this gas suggests that the generation of shallow tremor may be intimately related to the degassing of magma. The similarities in the source locations and spectral signatures of long-period events and tremor observed in Hawaii (Koyanagi et al. 1987) suggest a common source. Accordingly, it is natural to view the long-period event as the impulse response of the tremor generating system, and tremor itself as the response of this system to sustained pressure oscillations, the origin of which may perhaps lie in the oscillations of gas bubbles in the fluid. Deep tremor under Hawaii almost certainly has a different origin as significant gas content is unlikely to be maintained at depths of 30–60 km where confining pressure exceeds 1 GPa (Chouet et al. 1987). Sustained tremor at these depths was interpreted by Aki and Koyanagi (1981) in terms of a fracture mechanism and viewed by Shaw and Chouet (1988, 1989) as representing nonlinear resonances in the coupled oscillations of fracture and flow in a local plexus of fluid-filled fractures.

The far-field signature of the model is in harmony with the characteristics of the seismograms observed for long-period events. The details of this signature are very sensitive to the spatio-temporal characteristics of the pressure transient triggering the crack oscillation, suggesting that the identification of the driving mechanism responsible for tremor may be an arduous task. Further comparisons between observations and analysis of crack models including multi-phase fluids are needed.

Acknowledgements. I thank Larry Baker whose optimization of my original finite difference code made possible the extensive computations required by the models presented in this paper. I am also grateful to Dave Hill and Randy White whose careful reviews helped improve the manuscript.

References

Aki K (1984) Evidence for magma intrusion during the Mammoth Lakes earthquakes of May 1980 and implications of the absence of volcanic (harmonic) tremor. J Geophys Res 89: 7689–7696

Aki K, Koyanagi R (1981) Deep volcanic tremor and magma ascent mechanism under Kilauea, Hawaii. J Geophys Res 86:7095–7109

Aki K, Richards PG (1980) Quantitative seismology. WH Freeman, San Francisco

Aki K, Fehler M, Das S (1977) Source mechanism of volcanic tremor: fluid-driven crack models and their application to the 1963 Kilauea eruption. J Volcanol Geotherm Res 2:259–287

Aki K, Chouet B, Fehler M, Zandt G, Koyanagi R, Colp J, Hay RG (1978) Seismic properties of a shallow magma reservoir in Kilauea Iki by active and passive experiments. J Geophys Res 83:2273–2282

Bame DA, Fehler MC (1986) Observations of long-period earthquakes accompanying hydraulic fracturing. Geophys Res Lett 13:149–152

Boguslavskii YY, Ioffe AI, Naugol'nykh KA (1970) Sound radiation by a cavitation zone. Sov Phys Acoustics 16:17–20

Chouet B (1985) Excitation of a buried magmatic pipe: a seismic source model for volcanic tremor. J Geophys Res 90:1881–1893

Chouet B (1986) Dynamics of a fluid-driven crack in three dimensions by the finite difference method. J Geophys Res 91:13 967–13 992

Chouet B (1988) Resonance of a fluid-driven crack: radiation properties and implications for the source of long-period events and harmonic tremor. J Geophys Res 93:4375–4400

Chouet B, Julian BR (1985) Dynamics of an expanding fluid filled crack. J Geophys Res 90:11 187–11 198

Chouet B, Hamisevicz N, McGetchin TR (1974) Photoballistics of volcanic jet activity at Stromboli, Italy. J Geophys Res 79:4961–4976

Chouet B, Koyanagi R, Aki K (1987) The origin of volcanic tremor in Hawaii. Part II: Theory and discussion. US Geol Surv Prof Pap 1350:1259–1280

Cook TL, Harlow FH (1986) Vortices in bubbly two-phase flow. Int J Multiphase Flow 12:35–61

Cramer E (1980) The dynamics and acoustic emission of bubbles driven by a sound field. In: Lauterborn W (ed) Cavitation and inhomogeneities in underwater acoustics. Proc 1st Int Conf Göttingen FRG, 9–11 July 1979. Springer, Berlin Heidelberg New York, p 54

Crosson RS, Bame DA (1985) A spherical source model for low frequency volcanic earthquakes. J Geophys Res 90:10 237–10 247

Delaney PT, Pollard DD (1981) Deformation of host rocks and flow of magma during growth of minette dikes and breccia-bearing intrusions near Ship Rock, New Mexico. US Geol Surv Prof Pap 1202:1–61

Dietel C, Chouet B, Aki K, Ferrazzini V, Roberts P, Koyanagi R (1989) Data summary for dense GEOS array observations of seismic activity associated with magma transport at Kilauea volcano, Hawaii. US Geol Surv Open File Rep 89-113:1–179

Fehler M (1979) Seismological investigation of the mechanical properties of a hot dry rock geothermal system. PhD Thesis, Massachusetts Inst Tech, Cambridge, Massachusetts
Fehler M (1983) Observations of volcanic tremor at Mount St. Helens volcano. J Geophys Res 88:3476–3484
Fehler M, Chouet B (1982) Operation of a digital seismic network on Mount St. Helens volcano and observations of long-period seismic events that originate under the volcano. Geophys Res Lett 9:1017–1020
Ferrazzini V, Aki K (1987) Slow waves trapped in a fluid-filled infinite crack: implication for volcanic tremor. J Geophys Res 92:9215–9223
Ferrazzini V, Chouet B, Fehler M, Aki K (1990) Quantitative analysis of long-period events recorded during hydrofracture experiments at Fenton Hill, New Mexico. J Geophys Res 95:21 871–21 884
Gil Cruz F, Meyer HJ, Chouet B, Harlow D (1987) Observations of long-period events and tremor at Nevado del Ruiz volcano 1985–1986. Abstr Hawaii Symp How volcanoes work, 19–25 Jan 1987. Hilo, Hawaii, p 90
Harlow FH, Amsden AA (1975) Numerical calculation of multiphase fluid flow. J Comput Phys 17:19–52
Il'ichev VI, Lesunovskii VP (1963) On the noise spectra associated with hydrodynamic cavitation. Sov Phys Acoustics 9:25–28
Jaupart C, Vergniolle S (1988) Laboratory models of Hawaiian and Strombolian eruptions. Nature (Lond) 331:58–60
Kamo K, Furuzawa T, Akamatsu J (1977) Some natures of volcanic microtremors at the Sakurajima volcano. Bull Volcanol Soc JPN 22:41–58
Kieffer SW (1977) Sound speed in liquid-gas mixtures: water-air and water-steam. J Geophys Res 82:2895–2904
Kieffer SW (1984) Seismicity of Old Faithful Geyser: an isolated source of geothermal noise and possible analogue of volcanic seismicity. J Volcanol Geotherm Res 22:59–95
Koyanagi RY, Chouet B, Aki K (1987) Origin of volcanic tremor in Hawaii. Part I: Compilation of seismic data from the Hawaiian Volcano Observatory, 1972 to 1985. US Geol Surv Prof Pap 1350:1221–1258
Leet RC (1988) Saturated and subcooled hydrothermal boiling in groundwater flow channels as a source of harmonic tremor. J Geophys Res 93:4835–4849
Lyamshev LM (1969) On the theory of hydrodynamic cavitation noise. Sov. Phys Acoustics 15:494–498
MacDonald GA (1972) Volcanoes. Prentice-Hall, Englewood Cliffs, New Jersey
McFarland RD, Murphy HD (1976) Extracting energy from hydraulically fractured geothermal reservoirs. Los Alamos Sci Lab Rep LA-UR-76-1324
McKee CO, Wallace DA, Almond RA, Talai B (1981) Fatal hydro-eruption of Karkar volcano in 1979: Development of a maar-like crater. In: Johnson RW (ed) Cooke-Ravian Volume of volcanological papers. Mem 10, Geol Surv Papua New Guinea, p 63
McNutt SR (1986) Observations and analysis of B-type earthquakes, explosions, and volcanic tremor at Pavlov volcano, Alaska. Bull Seismol Soc Am 76:153–175
Morozov VP (1968) Cavitation noise as a train of sound pulses generated at random times. Sov Phys Acoustics 14:361–365
Murase T, McBirney AR (1973) Properties of some common igneous rocks and their melts at high temperatures. Geol Soc Am Bull 84:3563–3592
Plesset MS, Prosperetti A (1977) Bubble dynamics and cavitation. Annu Rev Fluid Mech 9:145–185
Ross D (1987) Mechanics of underwater noise. Peninsula Publ, Los Altos, California
Shaw HR, Chouet B (1988) Application of nonlinear dynamics to the history of seismic tremor at Kilauea volcano, Hawaii. US Geol Surv Open File Rep 88-539:1–78
Shaw HR, Chouet B (1989) Singularity spectrum of intermittent seismic tremor at Kilauea volcano, Hawaii. Geophys Res Lett 16:195–198
Stanley HE, Meakin P (1988) Multifractal phenomena in physics and chemistry. Nature (Lond) 335:405–409

Tang XM, Cheng CH (1988) Wave propagation in a fluid-filled fracture – an experimental study. Geophys Res Lett 15:1463–1466
Valentine GA, Wohletz KH (1989) Numerical models of plinian eruption columns and pyroclastic flows. J Geophys Res 94:1867–1887
Vergniolle S, Jaupart C (1986) Separated two-phase flow and basaltic eruptions. J Geophys Res 91:12 842–12 860

Volcanic Tremor: Seismic Signals of (Almost) Unknown Origin[1]

ROLF SCHICK

"It is an old observation that like causes produce like effects, but it is just as well known that in nature very different causes can produce identical effects. It is indisputable that in these causes all observations are included which we normally sum up under the collective name 'earthquakes'..." (written by Karl Ludoph Griesbach, Austrian co-founder of seismotectonics, in the year 1869).

Abstract

There exists strong evidence that the origin of volcanic tremor is caused by pressure fluctuations which are directly associated with degassing processes of the magma. The tremor signals contain features that are also found in self-excited and self-sustained feed-back systems. Two-phase flow (liquid magma and gas/steam) phenomena offer good possibilities at modeling pressure sources which are in agreement with tremor observations. However, the observation of volcanic tremor alone does not allow us to draw specific conclusions on the source mechanism. Different physical seismic source processes associated with two-phase flow are in concordance with signal characteristics of volcanic tremor. This implies severe restrictions in the application of volcanic tremor in understanding the dynamics of volcanic activity. The problem is enhanced by the fact that, in general, only a limited frequency band of the flow-induced seismic signals is observed (ca. 1–10 Hz). The extension of, mainly, the lower cut-off frequency to values of 0.01 Hz or less might strongly improve the possibility of establishing less speculative source models for the origin of volcanic tremor than is possible at present.

1 Introduction

An active volcano represents a source of sound. The origin of the excited signals is manifold: mechanical instabilities within the solid parts of the volcano, non-stationary fluid-flows from magma, water or gas induced vibrations, and many kinds of fast occurring interactions between the solid, liquid and gaseous components of the volcano generate acoustic and seismic signals. The recording and interpretation of the signals is generally performed with instruments and techniques analogous to the study of earthquakes. This might not be the most appropriate approach, as only parts of the observed signals resemble seismograms having their origin in shear-fracture events. To allude to the uncertainty in the origin, the collective name "volcanic earthquake" is often used. Many types of seismic signals, which are recorded in active volcanic areas, clearly deviate in their waveform from earthquake signals. They are much longer in their duration

[1] Manuscript submitted end of 1988.

than the earthquake transients, and on several volcanoes, like Etna, Stromboli and Aso, they are, at least as far as we know from the time span of our records of these signals, always present. The source of the signals is clearly inside the volcano, and the high amplitudes yield, in many cases, a sufficient signal to noise ratio over the influence of meteorological or oceanic induced noise. Often the spectral power of these signals is concentrated in intensity peaks of small bandwidths, an observation which can be used to distinguish the signals from the broader spectra of more impulsive events. The correlation of the data in terms of activity phenomena observed on a volcano suggests an origin or at least a connection with internal fluid flow (for instance, Cosentino et al. 1989). This fact makes the signals interesting for investigation, as conclusions for internal magma flows might become possible. The complexity in the description of the mentioned signals, and even more, any assumption of their origin, is, however, very large. These problems can, for instance, easily be seen in the associated terminology. The nomenclature includes names like volcanic tremor (instead of tremor also the plural, tremors, is often used), harmonic tremor, microtremor from first to fourth kind or degree, long-period events, explosive and intrusive tremor, tremor-storms and others. Here, the use of a uniform notation would be highly desirable, but as different volcanoes in different times may generate similar signals from different source mechanisms, the introduction of clear definitions might, at the present state of our knowledge, also lead to unsatisfactory results. In the following text, we shall use the name "volcanic tremor".

2 Signal Characteristics of Volcanic Tremor

During the earlier days of power spectra calculations, the spectra of volcanic tremor measured at Etna and Stromboli, for instance, were described as "coloured Gaussian noise" (Schick and Riuscetti 1973). Later investigations, however, clearly showed that volcanic tremor may also contain important non-stochastic parts. Apparently there exist common features among tremor observations from such different volcanoes as Etna, Stromboli, Merapi (Central Java), Kliuchevskoi and Nevado del Ruiz (Seidl et al. 1981; Schick and Kirbani 1987; Gordeev et al. 1989; Martinelli, pers. comm.).

They are:

1. The stochastic part of the tremor data is superimposed by intensity peaks of small bandwidth, indicating a quality factor of the involved resonators exceeding at least 25. The site of the resonators is at the source (Cosentino et al. 1989).
2. The center frequencies of dominating peaks are exceptionally stable, and not much influenced by the strength of the eruption (Schick 1981).
3. Significant integer harmonics or subharmonics corresponding to center frequencies of pronounced spectral peaks cannot be found (Fig. 1a–c).
4. The tremor envelopes can often be modeled well with solutions of non-linear differential equations, or simply by exponential functions (Schick et al. 1982;

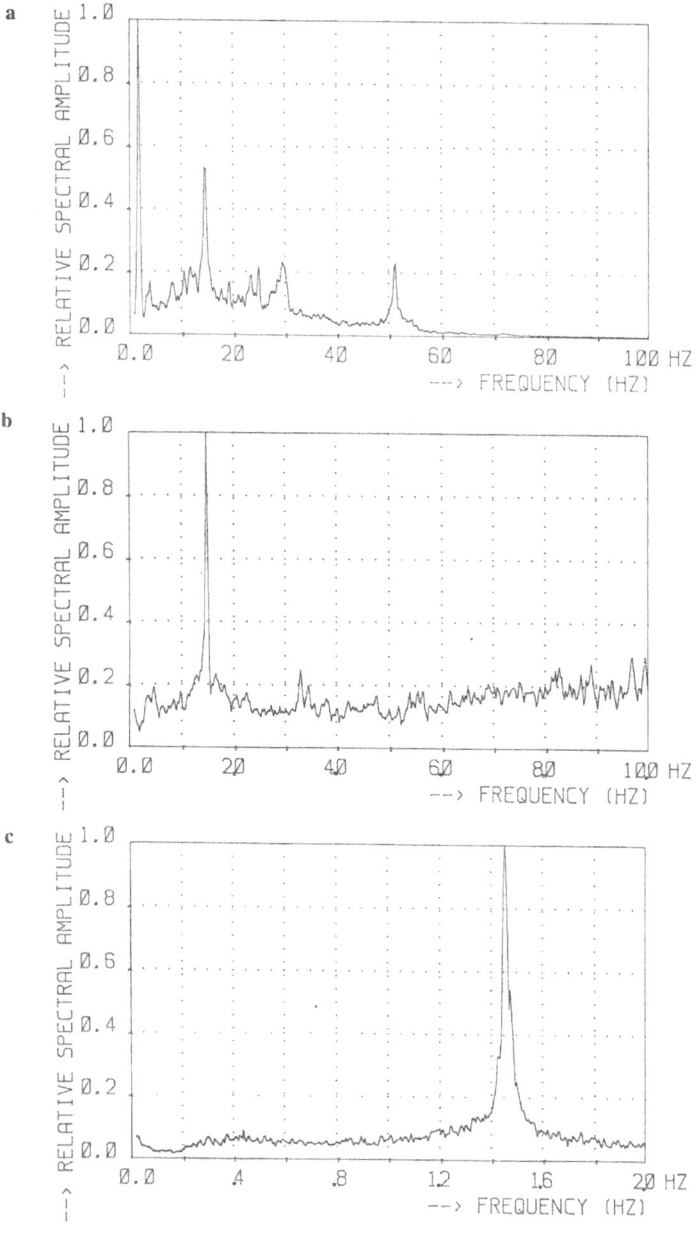

Fig. 1. a Spectrum of volcanic tremor recorded at station KLT, Merapi, Java, on June 25, 1988. Distance to crater 1.5 km. Time window 1 h. Seismometer 1 Hz, velocity output. Maximum value in frequency axis: 100 Hz. Frequency resolution 0.25 Hz. Exponential averaging over about 1000 spectral estimates. Note absence of integer harmonics. **b** Maximum value in frequency axis: 10 Hz. Frequency resolution 0.025 Hz. Linear averaging over 64 spectral estimates. Time window 2560 s. **c** Maximum value in frequency axis: 2 Hz. Frequency resolution 0.005 Hz. Linear averaging over 16 spectral estimates. Time window 3200 s

Kirbani 1983; Schick 1988). Volcanic tremor on Etna and Stromboli can show, over weeks and months, a quasi-stationary signal characteristic. Especially in times of elevated eruptive activity, pronounced beating with beating periods up to a few minutes is observed (Imbò 1954, Figs. 21–22).

5. A correlation between smoothed tremor envelope and visual volcanic activity does not lead to clear results (Sassa 1936; Schick 1988). Strong tremor need not necessarily be accompanied by strong visual eruptive activity. On the other hand, strong degassing of the volcano does coincide with strong tremor. Rhythmic eruptive activity, which is visually seen in the ball-like structures of the ash-clouds, is a significant indicator of stronger tremor. Volcanic tremor amplitudes follow well the strength of ash and gas or vapour eruptions, while lava fountaining or lava outflow is associated more with a lower tremor amplitude level (Schick 1988).
6. The tremor amplitude level shows, similar to the occurrence of shocks, often a significant correlation with changes in barometric pressure, rainfall loading on the volcano and tidal loading. Diurnal and semi-diurnal periods can often be detected in time series of volcanic tremor (Sassa 1936; Kirbani 1987; Schick 1988, Fadeli this Vol.).
7. The locations of the tremor sources are shallow, from close to the surface to not much deeper probably than 2–3 km. This is within the zone of Rittmann's pyromagma (Rittmann 1960).
8. There seems to exist a certain similarity in the generation of seismic signals associated with activity at geysers and at volcanoes (Kieffer 1984).

2.1 Considerations and Restrictions in Finding a Tremor Source Model

There seems now to exist a general agreement between investigators of volcanic tremor that, even considering the manifold possible source processes, the complexity of the recorded signals and the signal disturbance by the wave path from source to seismometer, the generation of the tremor can be seen in the magmatic process itself (e.g., Ferrazzini and Aki this Vol.). The behavior of pressure waves and vibrations within the source, generating seismic waves which can be recorded on the surface of the volcano, will tell much about the dynamics of the fluid system. The largely non-impulsive character of the tremor signals implies an analysis and interpretation with theoretically assumed source models in the frequency domain. "If one has succeeded in finding a prediction for the power spectral density of a flow variable, and the predicted power spectral density appears to be very much like the observed spectrum, one tends to feel a flush of triumph and momentarily deceive oneself into thinking that because the predicted spectrum matches the observation, the prediction must rest on correct physical insight"; these words by Landahl and Mollo-Christensen (1986) apply very much to the study of volcanic tremor. As an example, one may refer to the application of solutions of van-der-Pol's differential equation to the amplitude envelope of volcanic tremor on Etna and Merapi (Kirbani 1983; Fadeli et al. 1987). Although the similarities between model graphs and observations are striking, one must take great care in deducing a physical model from this correlation. Solutions of the van-der-Pol equation apply to many non-linear oscillatory and relaxation processes in nature and engineering, and Van der Pol (1926) has enumerated many instances of relaxation oscillations, from a pneumatic hammer over the humming noise of a water tap to the beating of the heart.

As there exists strong evidence that the source of tremor is connected with the non-steady flow of pyromagma, two-phase fluid flows (gas/vapour and liquid magma) have to be considered in any model. The complexity of two-phase flows, however, is extremely great. The amount of knowledge which is needed to perform a detailed analysis of the flow-regimes and flow-patterns of magma inside the volcano, far exceeds the information content of volcanic tremor. The problem is further complicated, because with the upward flow of the magma more and more magmatic melt is converted into vapour or gas. Thus, different parts of the vertical feeding conduits require different methods of analysis, let alone the problems of regime transitions of the different flow-patterns.

Further difficulties exist in the instrumental recording of the tremor. The great majority of the modern observations are performed with seismometers of a low cut-off frequency at 1 Hz. It certainly must not be expected that, the volcano as a fluid induced sound source, takes any notice of our instrumental recording techniques. As illustrated in Fig. 2, future studies may show that the frequency content below 1 Hz bears additionally essential information, in the study of volcanic tremor, for the understanding of volcano dynamics.

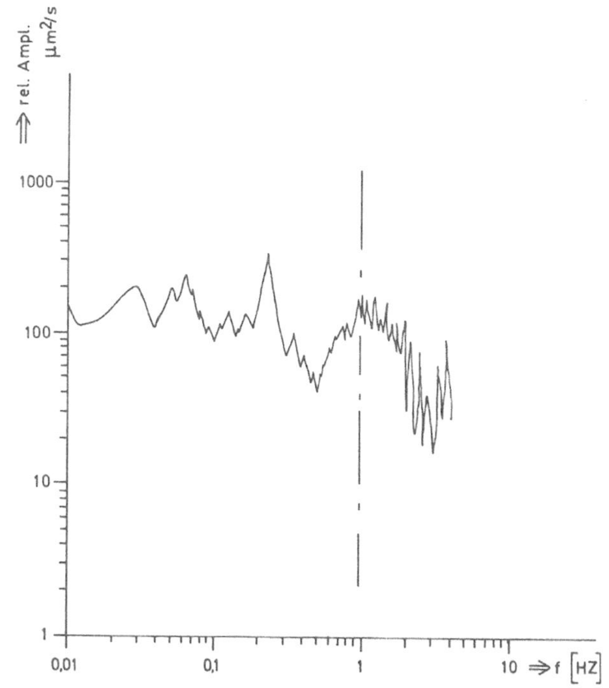

Fig. 2. Volcanic tremor spectrum observed with Wielandt-Streckeisen seismometer (LP-output) on Etna. Average over spectra from two stations in about 1.5 km crater distance (Torre del Filosofo and Pizzi Deneri). Time window 2 h. Linear averaging over 16 spectral estimates

2.2 Discussion about Possible Source Models

In summarizing these enormous difficulties, it requires great efforts to construct even a qualitative model which may reliably explain the generation of volcanic tremor. It is undisputable that the volcano, at least in comparison to the released seismic energy, contains an unlimited amount of potential thermal energy. This thermal energy leads, through buoyancy and convection, to a reservoir of static or slowly varying pressure energy. The question is, which mechanism converts the static potential energy into oscillatory kinetic energy? In model terms, the volcano bears some relationship to a thermodynamic machine, for instance to a steam-engine. Pichler and Schick (1985) describe a fluid-mechanical model, in which the introduction of a constriction in a linear pipe connected to a vapour emitting reservoir, leads from continuous to rhythmic steaming (Fig. 3).

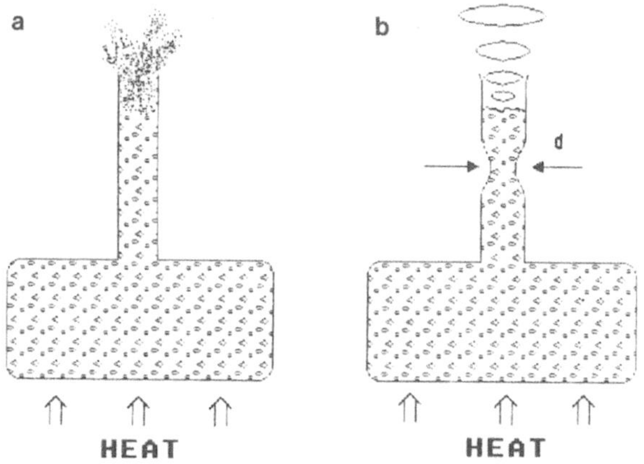

Fig. 3a,b. Experiment with boiling water in a tank-pipe system. **a** Continuous, quiet steaming; **b** the introduction of a constriction in the pipe leads, at a certain threshold value of d, to discontinuous, rhythmic steaming

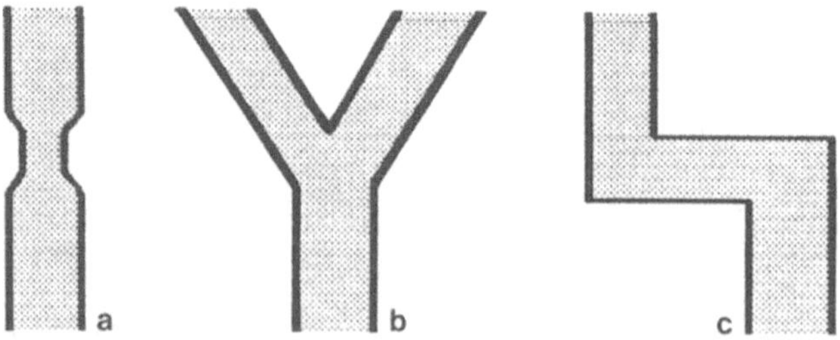

Fig. 4a–c. Frequently observed geometries of volcanic dikes at Etna. **a** Constriction; **b** joint; **c** horizontal offset

Exposed feeding conduits (dikes) in the Valle del Bove at Etna and the Somma crater rim at Vesuvius, just to mention two volcanoes, clearly show numerous constrictions, triple-junctions and horizontal offsets (Fig. 4). Abrupt contraction and abrupt expansion in a conduit lead to an exchange between potential and kinetic energy (Schaedel 1979, p. 31).

The formal description of the characteristics of volcanic tremor as the response of a relaxation oscillator has already been mentioned. Intensity – time functions of volcanic tremor, as for instance illustrated by Schick et al. (1982), show striking similarities to the behavior of self-excited and self-sustained feedback systems. Numerous instabilities in two-phase (gas/liquid) flows may, with a proper feed back, lead to periodic flow and pressure pulsations (Huhn and Wolf 1975). A major instability in fluid flow represents the transition from subsonic to supersonic flow and vice versa. In a modification of a model, presented originally by Steinberg and Steinberg (1975), Schick (1988) has observed the possibility that this instability could be an efficient source for the origin of tremor. The necessary feedback for the oscillating process could be seen in the gas/vapour escape (Fig. 5). The introduction of an active, undamped oscillator would also more easily explain the high Q-value of the observed spectral peaks than some other assumptions. In several tremor models, the high Q-value of the spectral peaks is seen as a response of a low damped, passive resonator, activated by impulsive signals of unknown origin. In order to keep radiation damping low, a high impedance contrast between source region and surrounding medium must be assumed. This introduction of the required impedance contrasts is, however, not easy to explain.

An observation very difficult to understand, and a serious criterion for the applicability of any source model, is the stability of the center frequencies of

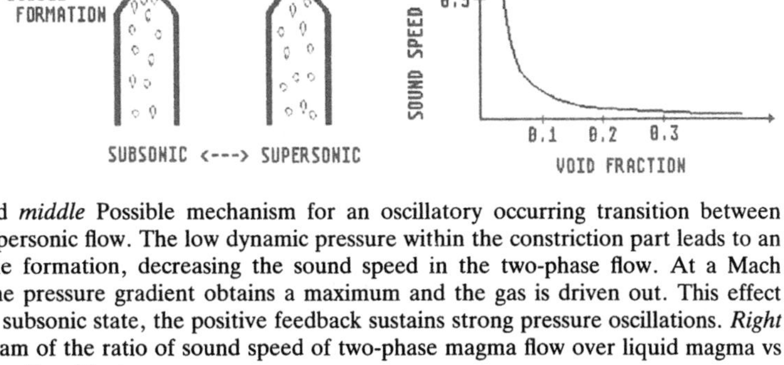

Fig. 5. *Left* and *middle* Possible mechanism for an oscillatory occurring transition between subsonic and supersonic flow. The low dynamic pressure within the constriction part leads to an increased bubble formation, decreasing the sound speed in the two-phase flow. At a Mach number of 1, the pressure gradient obtains a maximum and the gas is driven out. This effect again leads to a subsonic state, the positive feedback sustains strong pressure oscillations. *Right* Schematic diagram of the ratio of sound speed of two-phase magma flow over liquid magma vs void fraction. Small ambient pressure

dominant spectral peaks in regard to changing eruptive strengths. It is reasonable to assume an increase in the gas to liquid ratio with increasing eruption strength (Rittmann 1960). The speed of sound, however, decreases dramatically with increasing void ratio (e.g., Kieffer 1984) and one would expect a strong frequency-modulation of dominant peaks with varying eruptive strength. No matter whether we look at those dominant peaks as the impulsive response of magma conduits or as eigenfrequencies of an undamped oscillator, the speed of sound of the fluid should greatly affect the resonant frequency.

Perhaps a solution to this problem could be found in the clear separation of bubbly flow and liquid flow over the cross section of magma conduits. Many of the exposed dikes in the Valle del Bove on Etna show a clear separation of different flow regimes across their profile. The bubbly flow is always found in the center of the dike, the liquid flow borders the surrounding rock (Fig. 6). Thus, one could imagine that the central bubbly flow is the active vibrating part of the flow system, with a broad "noisy" spectrum. Because of the low speed of sound in this region, the dominant spectral energy will be found at low frequencies, and possibly at frequencies below the present recording range.

The liquid outer part, possibly passive in the generation of sound, is strongly coupled to the center part and responds with forced oscillations. As the sound velocity of the degassed, liquid part remains more constant over the time of the eruption, the eigenfrequencies of this part present possibly the stable peaks found in the tremor spectra. This assumption would also justify the use of these eigenfrequencies in constructing a structural model of the conduit and pipe sys-

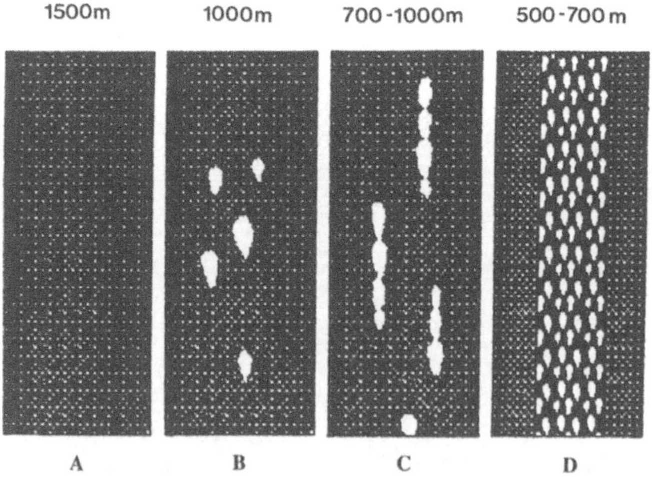

Fig. 6A–D. Flow structures of two-phase flow which are observed on exposed dikes in the Valle del Bove, Etna. The given values in meters refer to approximate heights of the previous magma level above the observed locations. **A** Magma is macroscopically free of bubbles (ca. 1500 m); **B** few bubbles, with relatively large volume (ca. 1000 m); **C** bubble columns, concentrated in the central part of the dike (ca. 700–1000 m); **D** pronounced phase seperation. Inner part of dike shows numerous bubbles, outer part is nearly free of bubbles. Transition between inner and outer part is very narrow (<500–700 m)

tem of a volcano, as was done for Etna by many investigators (e.g., Cosentino et al. 1989). The assumption of a passive resonator is, however, contrary to the above mentioned postulate of an active resonant circuit, as with the passive resonant circuit the problem with the high Q-value arises. A solution might be found in the integration of the liquid outer region and the bubbly inner region into one oscillating system. Such a system resembles in a certain way fluid oscillators built in engineering (Fig. 7), (Schaedel 1979). Here is a wide field for future theoretical, numerical and experimental investigations.

3 Conclusions

There exist, certainly, many routes to the origin of, what we call with the collective name, "volcanic tremor". It seems rather obvious that there is not one source model which can explain the generation of these signals. Furthermore, the information content of the tremor signals might be overemphasized. Dynamic gas speed measurements, as for instance performed by Tazieff (1970) or Andres (1988) could be of immense help in the understanding of volcanic tremor. As the hypomagma presents a plasma, mechanical vibrations of this electrically conducting system in the Earth's magnetic field must induce electromagnetic signals. As our recording sensors are situated outside the dynamo-field, these induced electromagnetic signals of very low frequency are certainly weak, and not easy to distinguish from the influence of non-volcanic sources. A cross-correlation between seismic and electromagnetic signals, and a comparison of spectra, might however prove to be a possibility for obtaining a significant electromagnetic response from the oscillating magma column.

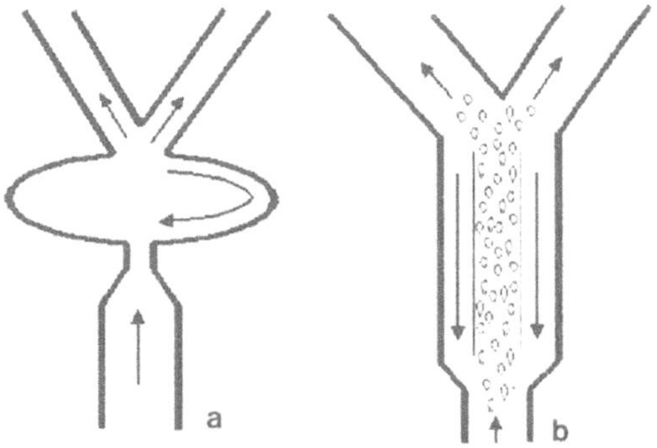

Fig. 7. a Model of a fluid oscillator used in engineering. Static input pressure converts to oscillating pressure in the output. The arrows indicate main flow directions. **b** Possible flow patterns in a volcanic conduit; attempt to find an analogy between the volcano and the fluid oscillator shown in **a**. The *arrows* indicate again main flow directions. The splitting of the dike may lead to flow reversals, hot and bubble rich magma flows upward in the central region of the dikes, whereas cooler, degassed magma flows downwards along the walls (forced convection). The model is highly speculative for the flow conditions in a volcanic dike. It is only used to illustrate the manifold possible flow patterns in dikes with a complex geometry

Nevertheless, even now, the recording of volcanic tremor is a powerful tool to describe volcano dynamics and even volcanic risk. Perhaps we are in a comparable situation as Otto von Guericke was in Magdeburg in the 17th century, when he used the then newly developed Torricelli instrument to predict weather. Guericke had only a faint understanding of the correlation between barometric pressure and weather fronts, but he was succesful in predicting, even on a purely heuristic basis, severe thunderstorms.

Acknowledgements. I thank Susanna Falsaperla, Abdulrahman Fadeli, Sri B. Kirbani, Bruno Martinelli and Gordon Woo for many helpful discussions, suggestions and comments. This research was supported by the Deutsche Forschungsgemeinschaft, Bonn, Kernforschungsanlage (KFA) Juelich and Badan Penerapan dan Pengkajian Teknologi (BPPT), Jakarta.

References

Andres RJ (1988) Sulfur dioxide and particle emissions from Mount Etna, Italy. Thesis, Inst Mining Tech, Socorro, New Mexico, USA

Cosentino M, Lombardo G, Privitera E (1989) A model for internal dynamical processes on Mt. Etna. Geophys J 97:367–379

Fadeli A, Kirbani SB, Mugiono R, Schick R (1987) Preliminary analysis of seismic data gathered with a new seismic station network on the Merapi volcano. 1987 Contrib Merapi Res Group, Gadjah Mada Univ, Sekip Unit III, Yogyakarta, Indonesia

Gordeev EI, Melnikov YY, Sinitsyn VI, Chebrov VN (1989) Volcanic tremor of Kliuchevskoi Volcano. IAVCEI Proc Volcanol (Volcanic Hazard), Springer Berlin Heidelberg New York, 1:486–503

Huhn J, Wolf J (1975) Zweiphasenstroemung. VEB Leipzig

Imbò G (1954) Sismicità del parosissmo vesuviano del marzo 1944. Ann Oss Vesuviano, Sesta Ser 1:1–211

Kieffer SW (1984) Seismicity at Old Faithful Geyser: an isolated source of geothermal noise and possible analogue of volcanic seismicity. J Volcanol Geotherm Res 22:59–95

Kirbani SB (1983) Analysis and interpretation of volcanic tremors at Etna (Sicily, Italy). Publ Inst Geophys Univ Stuttgart 210:106, Inst Geophys, Richard-Wagner-Str 44, D-7000 Stuttgart 1, FR Germany

Kirbani SB (1987) Maximum entropy method applied to volcanic tremors of Merapi volcano (central Java, Indonesia). 1987 Contrib Merapi Res Group, Gadjah Mada Univ, Sekip Unit III, Yogyakarta, Indonesia

Landahl MT, Mollo- Christensen E (1986) Turbulence and random processes in fluid mechanics. Cambridge Univ Press, Cambridge

Pichler H, Schick R (1985) Der Atna. In Vulkanismus. Spek Wiss, Heidelberg, pp 16–28

Rittmann A (1960) Vulkane und ihre Tätigkeit. Enke, Stuttgart

Sassa K (1936) Micro-seismometric study on eruptions of the volcano Aso. Mem Coll Sci Kyoto Imp Univ A19:11–56

Schaedel HM (1979) Fluidische Bauelemente und Netzwerke. Vieweg, Braunschweig

Schick R (1981) Source mechanism of volcanic tremor. Bull Volcanol 44-3:491–497

Schick R (1988) Volcanic tremor – source mechanism and correlation with eruptive activity. Nat Hazard 1:125–144

Schick R, Kirbani BP (1987) Two phase flow of magma and the origin of volcanic tremor. In: Int Workshop Forced Fluid Flow, Comm Eur Comm EUR 11164/1

Schick R, Riuscetti M (1973) An analysis of volcanic tremors at South Italian volcanoes. Z Geophys 39:247–262

Schick R, Lombardo G, Patane G (1982) Volcanic tremors and shocks associated with Etna (Sicily), September 1980. J Volcanol Geotherm Res 14:261–279

Seidl D, Schick R, Riuscetti M (1981) Volcanic tremors at Etna – a model for hydraulic origin. Bull Volcanol 44:43–56
Steinberg GS, Steinberg AS (1975) On possible causes of volcanic tremor. J Geophys Res 80:1600–1604
Tazieff H (1970) New investigations on eruptive gases. Bull Volcanol 34:421–438
Van der Pol B (1926) Relaxation oscillations. Phil M 2:928–945

Preliminary Results from a Field Experiment on Volcanic Events at Kilauea Using an Array of Digital Seismographs

Abstract

We carried out a field experiment at Kilauea Volcano for two months in 1988, during which the volcano was active and generated gas-piston events, long-period events and tremors. Six three-component digital instruments recorded 200 h of useful signal and 12 GEOS 6-channel recorders collected more than 200 events with a total amount of information of about 10 gigabits. Preliminary analyses of the data revealed that seismic spectra of all these volcanic events show many narrow spectral peaks, and we were able to establish that these peaks are not due to propagation path or recording site effect but due to source effect. We also found that, the movement of gas bubbles plays an important role in the excitation of gas-piston events, and possibly also long-period events because of the similarity in spectral feature between them. We propose preliminary source models for gas-piston events, long-period events and volcanic tremor based on fluid-gas filled crack models, and discuss outstanding problems with these models.

1 Introduction

With the cooperation of the U.S. Geological Survey, we carried out a field experiment for two months in early 1988 on Kilauea Volcano, Hawaii and collected an enormous amount of high quality seismic data which are directly relevant to the understanding of volcanic tremors.

We were extremely fortunate in the choice of the time and site of our experiment; the continuing eruption, since 1983, of the east rift zone of Kilauea volcano persisted and we observed, within the period of the experiment, a full episode of the magma dynamics giving rise to many kinds of seismic events. Koyanagi et al. (1987) and Wolfe et al. (1987) have described the history of this eruption. In early January 1983, the first outbreak occurred at Napau Crater (see Fig. 1). Within a few hours this outbreak produced a 6-km-long discontinuous line of erupting fissures along the east rift, between Napau crater and Pu'u Kahaualea, with Pu'u O'o as central vent. In the early stages of episodic eruptions at Pu'u O'o, a nearly vertical 20-m radius pipe-like conduit could be observed within a growing pyroclastic cone of 100-m diameter formed by deposits from lava fountains. The height of the fountains varied from a few meters to 400 m. At that time, according to Wolfe et al. (1987), intense harmonic tremors in the vent area did correspond to general activity of eruption, but the amplitude of tremor did not correspond directly with changes in the height of the fountain. Instead, the amplitude showed a positive correlation with

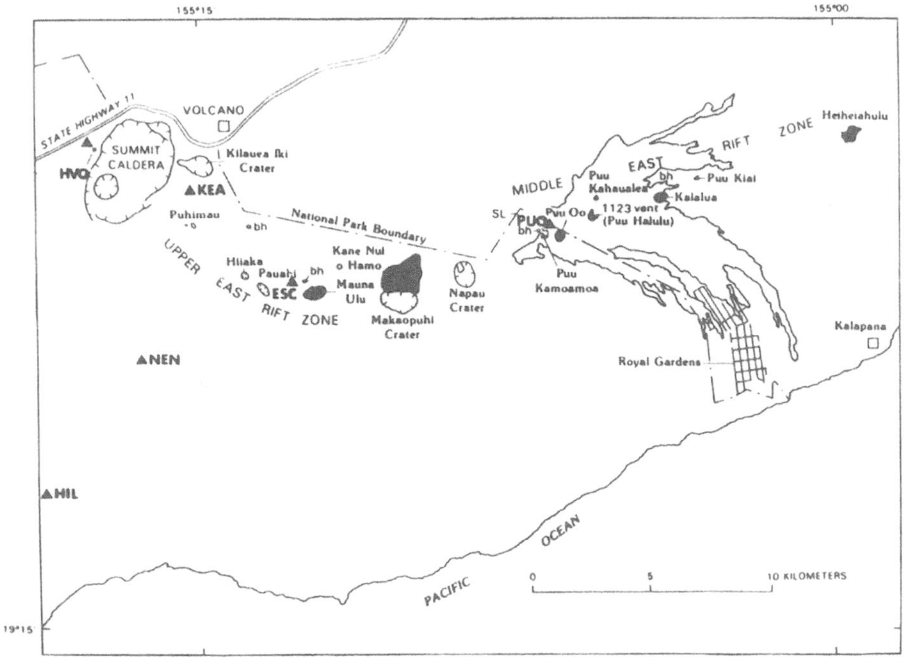

Fig. 1. Map of Kilauea volcano showing the U.S.C. instrument locations

the discharge of lava and suggests that part of the tremor is generated by a process deeper than those from which the fountaining originates. Also, the tremor attenuation pattern was symmetric with respect to the Pu'u O'o vent, which may indicate that the source of the tremor and the vent are vertically aligned, although the depth of this source is unknown.

During the first month of our experiment, we recorded a continuous high amplitude tremor originating from Pu'u O'o. At night, a glow from the crater could be observed and the lava activity could be heard, but smoke prevented us from seeing the crater floor during the day. Four kilometers downrift from Pu'u O'o, the lava pond of Kupaianaha showed an almost constant lava flow from the pond into a network of tubes leading to the ocean.

A decrease in the tremor amplitude was first noticed on January 29 and the first gas-piston event occurred on January 30. Gas-piston events occurred during episodes of ponding and withdrawal of a lava lake inside the crater due to the rise of gas slugs in the underlying volcanic conduits. This type of activity was observed during the 1969 Mauna Ulu eruption and named "gas-piston activity" by Swanson (1979). The gas-piston activity stopped on February 4, then started again on February 12 and stopped 6 days later.

Toward the end of this period a swarm of long-period events began on February 17 under the summit area of Kilauea and subsided by noon on February 19. This activity was followed by shallower and very small amplitude long-period events.

We operated two type of instruments; six U.S.C. three-component event-recorders produced 200 h of useful signals with a total amount of information on the order of 6 gigabits, and 12 U.S.G.S. GEOS 6-channel recorders collected more than 200 events with a total amount of information of about 3.5 gigabits.

In addition to the seismic observations, when all the instruments set up in Pu'u O'o were recording in a continuous mode, we were particularly fortunate to have the chance to visually observe the movement of lava and gas bubbles in the crater of Pu'u O'o which caused gas-piston events. This observation greatly facilitated our interpretation of seismograms, not only for gas-piston events, but also for long-period events and volcanic tremors.

In the present paper, we shall describe the field experiment, the collected data, and some preliminary results obtained from the analysis of about one-third of the U.S.C. data.

2 Field Experiment

The U.S.G.S. GEOS instruments (a total of 72 channels) were set up on a relatively flat and smooth basaltic lava flow on the west side at the foot of Pu'u O'o which has been the most seismically active vent of the east rift zone for most of this continuing eruption since 1983. The instruments were operated in two different array geometries. Initially, they were deployed over an area about 1 × 1 km. After 1 week of operation, we found that signals were incoherent between neighboring stations. The GEOS instruments were then redeployed in a smaller semi-circular array with a diameter of 120 m. Eleven three-component geophones were placed on the circumference and one at the center of the semi-circular array, and 36 vertical component geophones were distributed in seven radial directions. This array was located at a distance of approximately 860 m from the Pu'u O'o crater.

The semi-circular geometry was chosen to delineate the wave-type, polarization and phase velocity of a random wave field by the correlation method of Aki (1957) who demonstrated its effectiveness by applying it to traffic noises in Tokyo. The GEOS seismographs were operated for a week starting on Jan. 18 in the first configuration and from Jan. 29 to Feb. 21 in the semi-circular array. A description of the data recorded by the GEOS instruments may be found in Dietel et al. (1989).

The U.S.C. three-component broad-band event recorders were set up at six locations covering the area of Kilauea extending from the summit and south flank to the east rift zone down to Pu'u O'o as shown in Fig. 1. The array was operated from Jan. 17 to Mar. 1. One of the instruments was located 200 m north-west of the GEOS semi-circular array, at 1000 m from the Pu'u O'o crater. The other five stations were located near the Hawaiian Volcano Observatory (HVO), at Keanakakoi (KEA), Escape Road (ESC), Kipuka Nene (NEN) and Hilina Pali (HIL). The long-period response of the U.S.C. recorder has been improved by P. Roberts who designed, and field-tested in Valles Caldera, New Mexico, a stable equalizer circuit which gave a nearly flat velocity response from

0.1 to 40 Hz. A detailed description of this circuit is given in Roberts (1989). The sampling rate of the instrument is set at 240 samples/s and one field tape can record up to 3 h of three-component signals.

In the early period of our field experiment, a high-amplitude tremor was observed continuously near Pu'u O'o. The GEOS instrument records indicated that the amplitude of the tremor decreased as a function of the distance from the Pu'u O'o crater and did not depend on the distance from the rift. The tremor amplitude began to decrease very slowly and gradually on Jan. 29 and, after a seismically unusually quiet 15-min period the first gas-piston event occurred at 8:36 a.m. on Jan. 30. In the field, the first sign of the gas-piston activity observed was the sudden emission of a brown cloud accompanied by Pele's hair. An MEQ 800 seismometer that provided a visible record at the field site indicated that each cloud correlated with a seismic burst.

Gas-piston events occur repeatedly at nearly regular time intervals for several hours and have a very peculiar seismic signature with a spindle-shaped amplitude envelope and no clear shear-wave arrivals. A typical example of a gas-piston event recorded at Pu'u O'o on Feb. 4 is shown in Fig. 2. The seismograms represent the three components of ground velocity that have been high-pass filtered at 0.5 Hz to eliminate the oceanic tremor. We observed two periods of gas-piston activity; each lasted 5 days separated by 1 week of continuous tremor. The first period of gas-piston activity stopped on Feb. 4 at 15:50 and the second period started sporadically from about 1:00 a.m. on Feb. 12. and developed into vigorous activity on Feb. 13. On Feb. 3, 4 and 14, we made visual observation of the lava activity in the crater, and later correlated them with the seismic records.

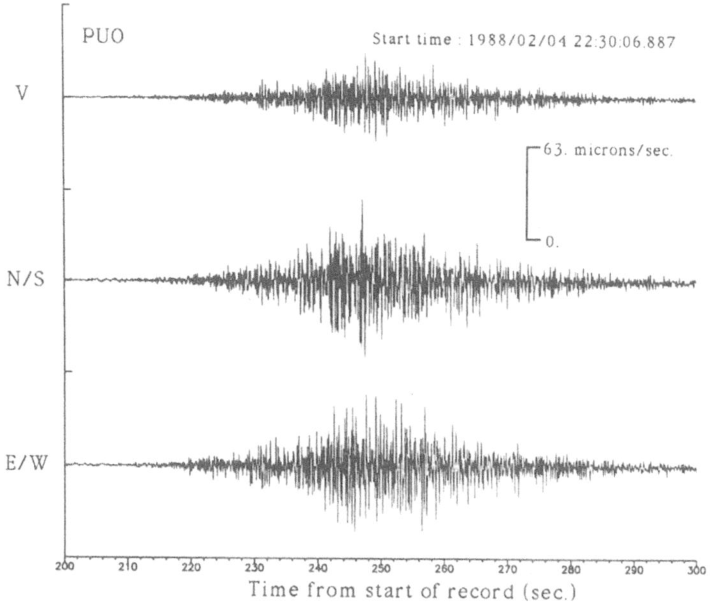

Fig. 2. The three components of ground velocity of a gas-piston event recorded at the site PUO

Pu'u O'o is presently a cinder-spatter cone 300 m high, its crater is approximately 150 m in diameter and 150 m deep with vertical walls. Inside the crater two open vents with an estimated diameter of 10 to 15 m were visible. They were located in the crater walls opposite to each other but at different levels, in the southwest and northeast directions, approximately in line with the trend of the east rift zone. A nearly circular pond about 30 m deep covering half of the crater area was supplied by lava from the northeast vent. The southwest vent was located at the highest level of the crater floor and a canyon allowed the lava to flow from the lower pond into the southwest vent or vice versa. Both vents were acting as outlet and inlet to the circulating lava in the crater.

Figure 3 represents the three components of ground velocity at the station PUO for a time window of 1 h taken from a 3-h long record on Feb. 4, while visual observations were made from the crater rim. A cycle of the gas-piston event starts with the rise of a lava lake at the bottom of the crater. The rise of the lava lake may take 20 min or so to reach a critical level, during that time which corresponds to location A in Fig. 3, seismic amplitudes are very low. When the critical level is reached, small upwelling domes of lava began to appear at the surface of the lake, breaking apart the black crust of the lake, and immediately followed by bursting of the gas bubbles under expansion. At location B, the size of these upwelling domes and gas bursts increases progressively, marking the beginning of the gas piston event as well as the beginning of the withdrawal of the lake. Finally the upwelling domes reach a diameter of tens of meters and a slug of gas bursts violently into a fully developed storm

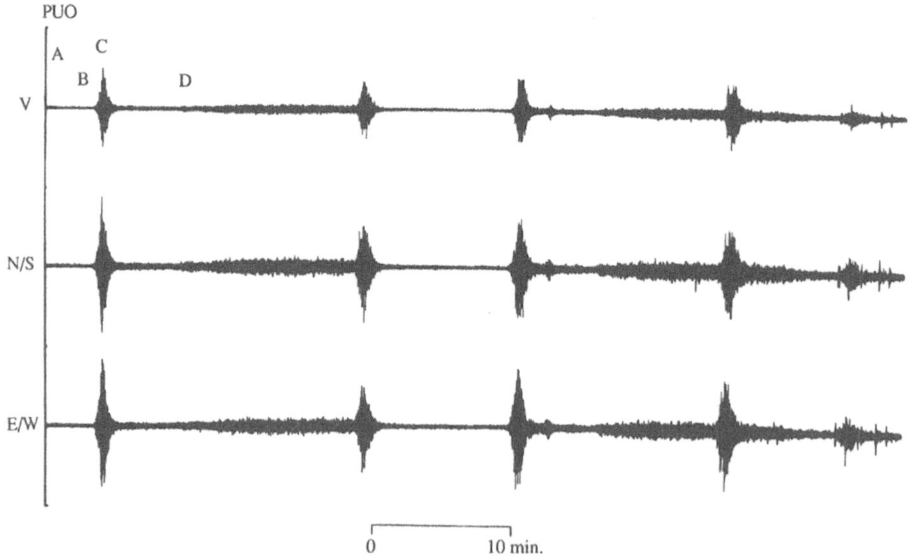

Fig. 3. Sample of the three components of ground velocity at the site PUO, from a 3-h continuous record done on Feb. 4

in which the lake is immediately and extremely rapidly, completely drained (in about 10 s), marking the time of the maximum seismic amplitude (location C) of the gas-piston event. Rather suddenly, everything becomes quiet again, and the lava lake rises again slowly. This is sometimes accompanied by cascading of lava from one vent to the other that correlates with marking the coda part of the gas-piston event as shown in D. To illustrate that rapidity of withdrawal of the lava lake, Fig. 4 shows four photographs taken a few seconds apart, from the north side of the crater rim during the gas-piston event shown in Fig. 2 which corresponds also to the first gas-piston event represented in Fig. 3. The northeast vent is in the bottom left of the photographs and the southwest vent is just outside the upper right corner of the photographs. In the first photograph a slug of gas has just burst on the northeast side of the lake and the lake is drained back into the northeast vent. The level of the lake before the drain-back corresponds to the darkest band on the crater walls. The second photograph shows the continuing withdrawal of the lake and the two different levels of the crater floor which begin to be apparent. In the third photograph the lake is nearly drained. The draining of the lake by one vent seems to trigger a violent outflow of the other vent, causing the torrent of lava running across the crater and suggesting that the two vents are connected at depth. In the last photograph, the pond is empty and the lower level of the crater floor can be fully observed. The flow rate of lava from the southwest vent is still very impressive here (note that the stream lines do not follow the wall of the lower pond). At the end of this event, the flux of lava from the southwest vent decreased gradually and completely stopped, while lava supplied by the northeast vent filled up the lower pond causing a cascade of lava into the southwest vent.

The magmatic activity inside the crater appeared to vary from day to day. During the second period of gas-piston events on Feb. 14, on the northeast side of the crater, the ponding and withdrawal of a lake which never reached the highest part of the crater floor, occurred repeatedly at intervals of 3 to 4 min. The gas-piston events were of lesser amplitude than in the first period presumably due to the smaller volume of gas slugs and of lava involved in the drain of the lake.

Following the second period of gas-piston activity at Pu'u O'o, a swarm of long-period events originated near the summit of Kilauea, beginning on Feb. 17 and ending by noon on Feb. 19. These long-period events were unusually strong and could therefore be recorded at several of our U.S.C. digital seismograph sites shown in Fig. 1. A preliminary location by the H.V.O. assigns their hypocenters ranging from 1 to 10 km deep about 2 km south of Kilauea Caldera. Some examples of long-period events recorded at the station KEA are shown in Fig. 5. They have very characteristic waveforms dominated by a long-period component which is sometimes superimposed by a high frequency signal. Although long-period events have been observed on numerous volcanoes in other regions such as Japan (Minakami 1974; Okada et al. 1988), New Zealand (Dibble 1974; Latter 1979), Kamchatka (Gorshkov 1959), Hawaii (Koyanagi 1968; Aki and Koyanagi 1981), Sicily (Guerra et al. 1976), Mexico (Haskov et al. 1983), Alaska (MuNutt 1986) and Washington (Endo et al. 1981; Fehler and

Fig. 4. Pictures taken from Pu'u O'o crater rim during a gas-piston events on Feb. 4. The four pictures have been taken a few seconds apart and show the withdraw of the lava lake

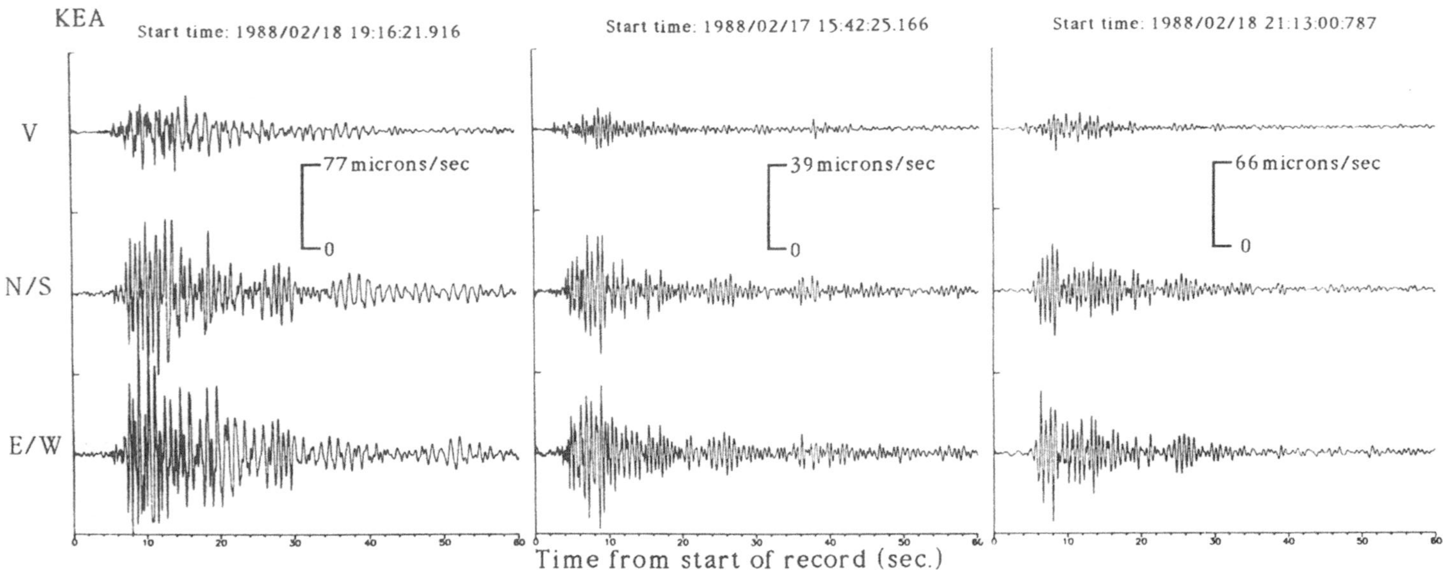

Fig. 5. The three components of ground velocity of three long-period events recorded at the location KEA

Chouet 1982) there have been only a few attempts of simulation of the waveforms associated with these events (Chouet 1985, 1986, 1988; Crosson and Bame 1985; Mahrer and Mauk 1987; Julian 1986; Ferrazzini et al. 1990).

As we shall describe later, we found a striking similarity of spectra between the long-period event and the gas-piston events, suggesting a similar source mechanism for both type of events.

In addition to long-period events, gas-piston events and tremors, we also recorded usual volcanic earthquakes. Together, they will help to separate the source, path and recording site effects on each of them.

3 Preliminary Results

Figure 6 shows typical examples of energy spectra of a gas-piston event recorded at the Pu'u O'o station, of tremor at Pu'u O'o and of a long-period event recorded at HVO. The most striking observation which applies generally to all these events is that the spectra are composed of many narrow peaks.

The first question coming to mind is: are those peaks caused by some random noise or have they physical reality? This question has been readily answered by observing the same peak at the same frequency repeatedly over many events at least for some of the peaks.

The next basic question is whether these peaks are produced at the source, along the propagation path, or near the recording site. It is easier to address this question for the long-period events than for gas-piston events, because the former was observed at widely spaced stations for which the path and site effects must be distinctly different.

Figure 7a,b shows individual energy spectra for the three components of a long-period event recorded at five stations, HVO, ESC, NEN, KEA and PUO. In Fig. 7a, all the spectra have been normalized to the same value to show the relative energy spectral density among different sites. In Fig. 7b, all the spectra have been normalized separately to emphasize the spectral shape for each channel. In Fig. 8 we compare two stacks of energy spectra for one long-period event. The first one is the stack of the energy spectra of all components from all the stations except Pu'u O'o. The stacked spectrum clearly show four distinct peaks. The second one represents the stack of the three-component spectra of the same long-period event observed at Pu'u O'o. The comparison of these two spectra clearly demonstrate that the same spectral peaks show up in both cases. The propagation path to Pu'u O'o station is considerably longer and different (nearly along the east rift zone) than those to the other stations and the recording site at Pu'u O'o is also uniquely different, located on newly formed (1983) lava sheets. The observed agreement in spectral peaks between Pu'u O'o and the other stations is impossible to explain as due to a path or recording site effect, but easily explained as due to a source effect. Thus, we conclude that at least some of the narrow peaks observed on the energy spectra of long-period events should be attributed to a process that takes place at the source.

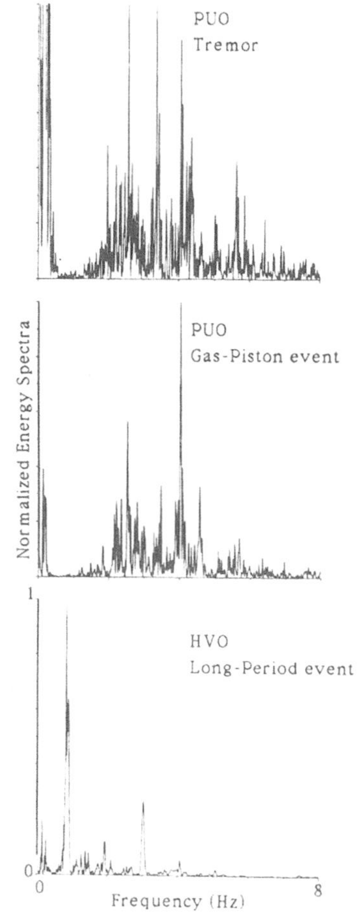

Fig. 6. Energy spectra of 60 s of tremor, of a gas-piston event and a long-period event

For gas-piston events, all our recorders are located within a relatively small area. Thus our argument may not be as compelling as in the case of long-period events, but we again find that different stations show the same peaks supporting the idea that the source effect is responsible for the observed narrow peaks. For example, Fig. 9 shows identical spectral peaks between the GEOS site and the U.S.C. recorder site near Pu'u O'o, which are separated by 200 m, enough distance to make the wave-field uncorrelated.

The general similarity of spectral signature among long-period events, gas-piston events, and tremor strongly suggests that the spectral peaks of tremor may be also attributed to the source effect.

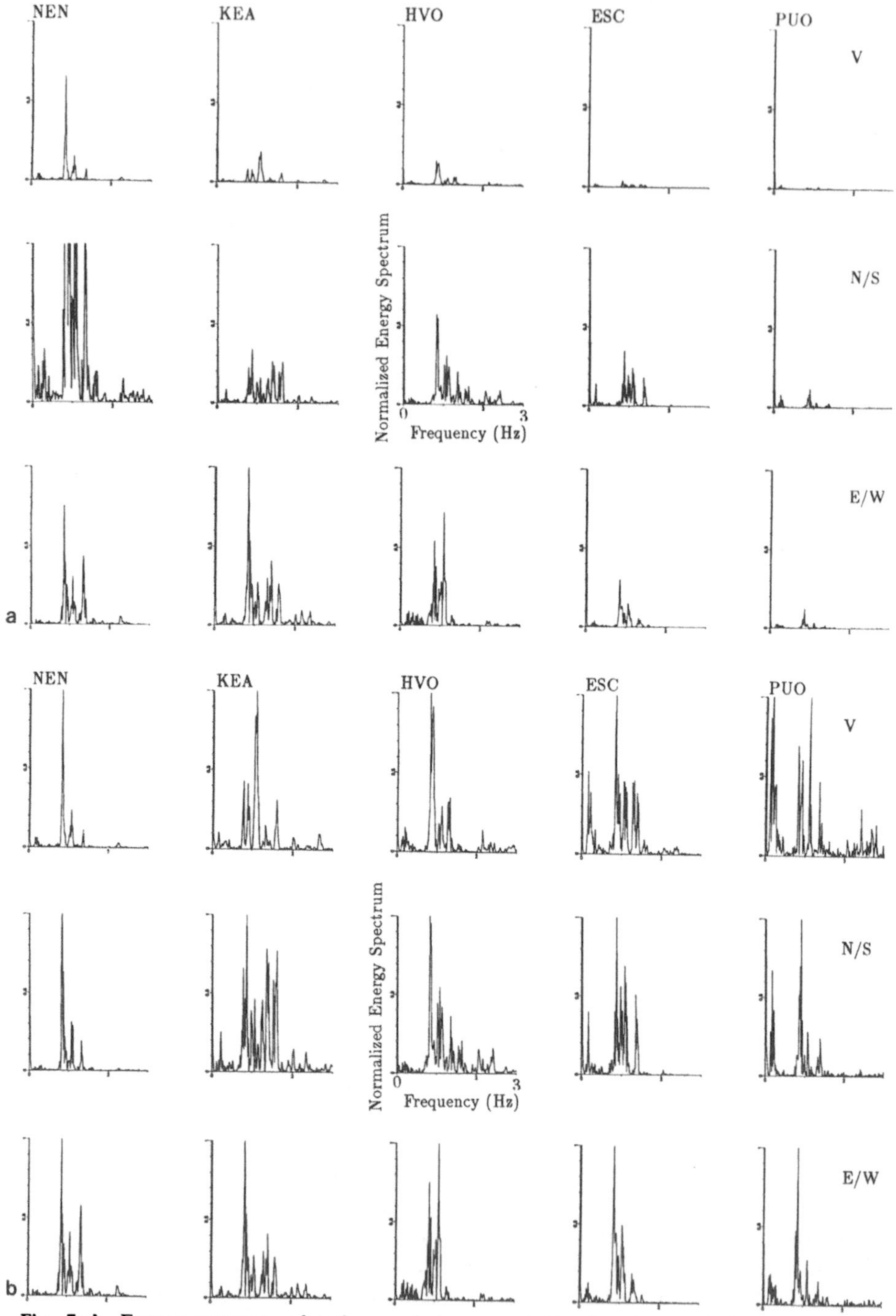

Fig. 7a,b. Energy spectrum of a long-period event at five locations and for the three components of ground velocity. **a** All the stack have been normalized to the same arbitrary value; **b** each stack has been normalized individually

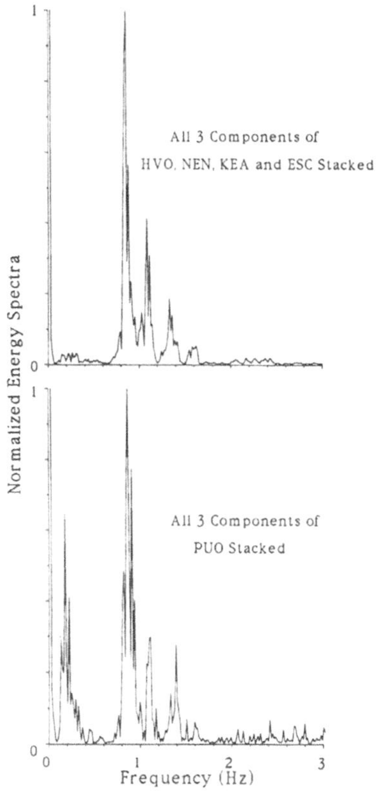

Fig. 8. Stack of the energy spectra of all components and of all locations except PUO, compare to the stack of energy spectra of the three components of PUO

3.1 Working Hypothesis on the Source Processes

Since the spectral peaks observed for the gas-piston and long-period events are due to the source effect, we shall proceed to a more specific source model which can quantitatively account for the observed spectral signature.

No real comparison of theoretical prediction with observed data from volcanoes has ever been made even though time domain solutions for displacement have been calculated for three kinds of three-dimensional models with different geometries; a spherical magma chamber (Crosson and Bame 1985), a fluid-filled pipe (Chouet 1985), and the fluid-filled crack (Chouet 1986, 1988). These three models have their own spectral characteristics that could potentially be used to identify them at the origin of the tremor. This task is still rather difficult today for two reasons. The first difficulty is due to the lack of versatility of the models, which stems from the need of simplified equations to obtain an analytical solution, or the need to reduce the computation time in the case of numerical solutions. Therefore the models are constrained to have a high degree of symmetry and the spectral characteristics of the source are strongly affected by such

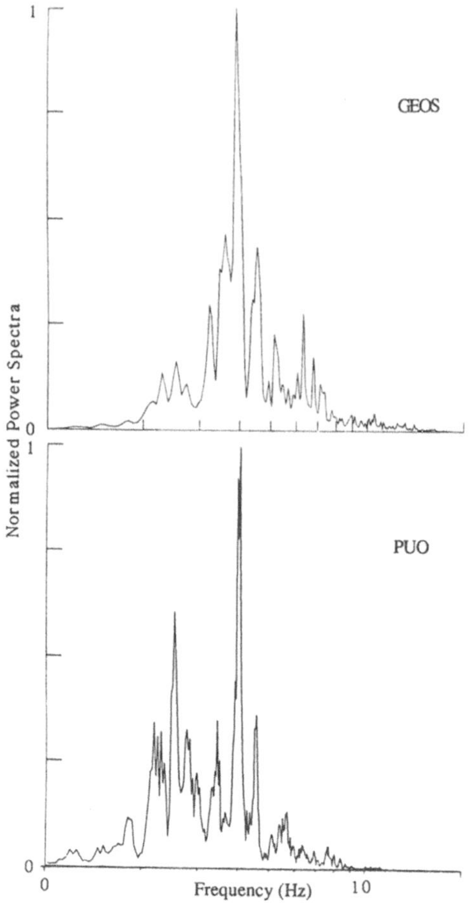

Fig. 9. Power spectra of the same gas-piston event recorded by the U.S.C. station PUO (*top*) and by a GEOS instrument (*bottom*)

simplifications, although other features of the spectrum should remain unaffected and could possibly be used as a tool. The second difficulty is that available observations suggest that tremor may result from the superposition of signals due to several sources rather than a single source, a situation that requires sophisticated models that will satisfy not only the observed wave field but also the mass balance of magmatic transport. At this point, it is still uncertain which source model may be better suited for the interpretation of long-period events, gas-piston events, and volcanic tremor. For example, the pipe model as well as the spherical chamber model have been invoked to explain the tremor observed at Mount St. Helens (Chouet 1985; Crosson and Bame 1985). We prefer, following the line of works by Aki et al. (1977), Chouet (1986, 1988), Aki and Koyanagi (1981) and Ferrazzini et al. (1990), a fluid-filled crack model in which vibration is set up by a sudden application of excess

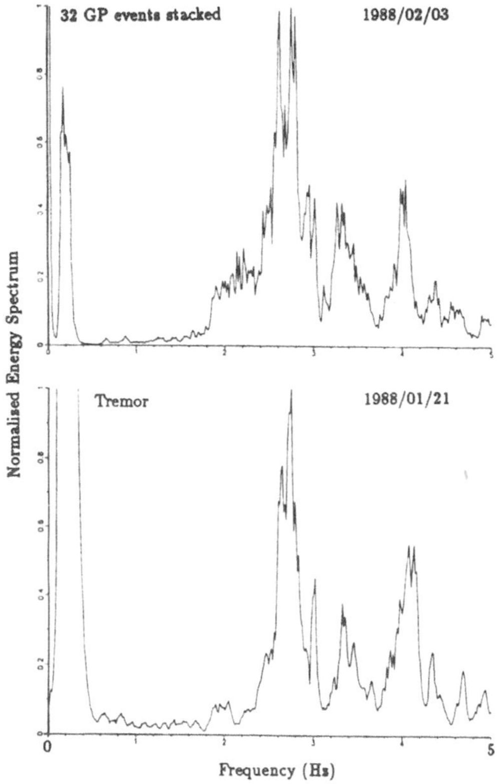

Fig. 10. Stack of energy spectra of gas-piston events recorded on Feb. 3 compared to the energy spectrum of the tremor recorded on Jan. 21

pressure in the fluid. The crack geometry seems to be the most appropriate for the east rift zone where fissure eruptions are the basic mode of eruption.

In Fig. 10, a stacked spectrum for gas-piston events is compared with a tremor spectrum observed 10 days before the gas-piston events started. The spectra are very similar. There is no evidence for a frequency shift in spite of the change in the level of magmatic activity. The same observation can be made in Fig. 11 where the stacked energy spectra for the three components of gas-piston events occurring on Feb. 3, during the first period of gas-piston activity, are compared with those occurring 10 days later during the second period. The stability of the frequency independent of amplitude suggest that the system is not only stable, but also linear.

For long-period events, the general shape of the energy spectra suggests a similar source as for gas-piston events and tremor. Figure 12 shows for ten long-period events, the stack of energy spectra from all components of the four stations located around the summit caldera. In Fig. 12a each stacked spectra has been normalized to its maximum value above 0.75 Hz, independently of the others. In Fig. 12b all the spectra have been normalized to a common arbitrary

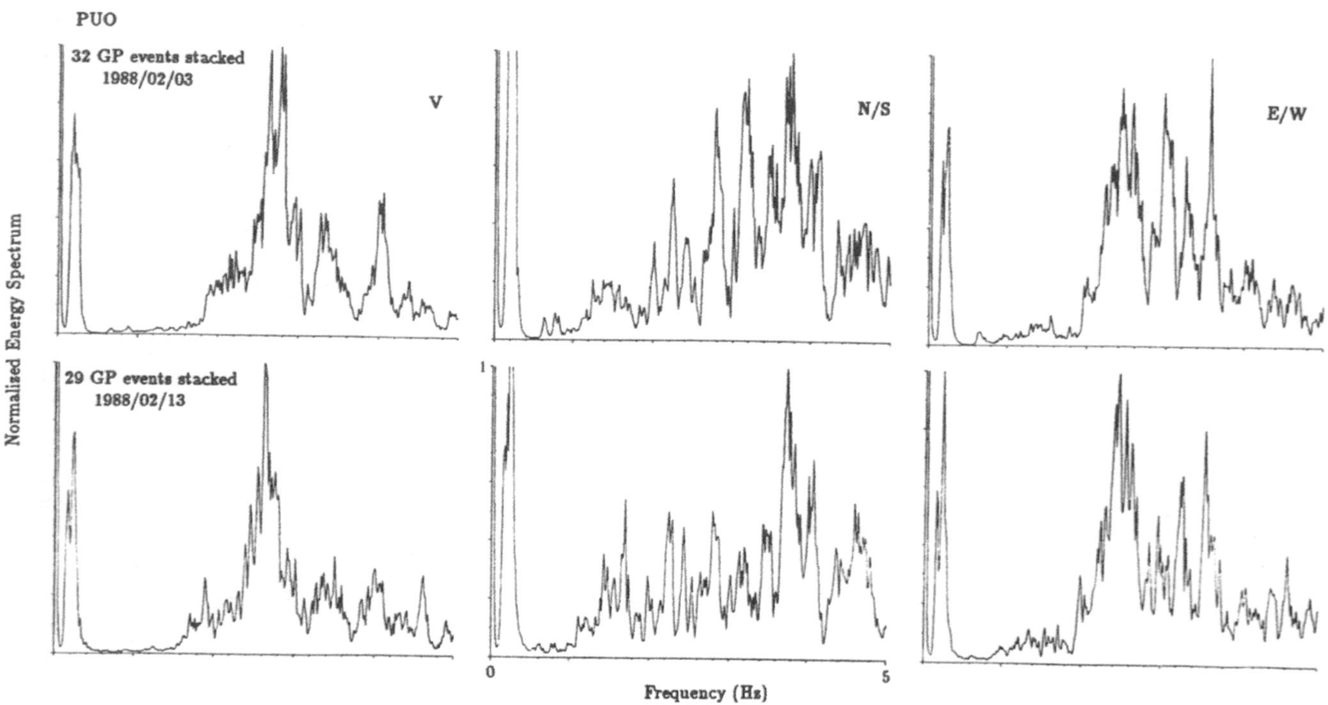

Fig. 11. Stack of energy spectra for the three components of 32 gas-piston events recorded on Feb. 3 compared to the stack of energy spectra for the three components of 29 gas-piston events recorded on Feb. 13

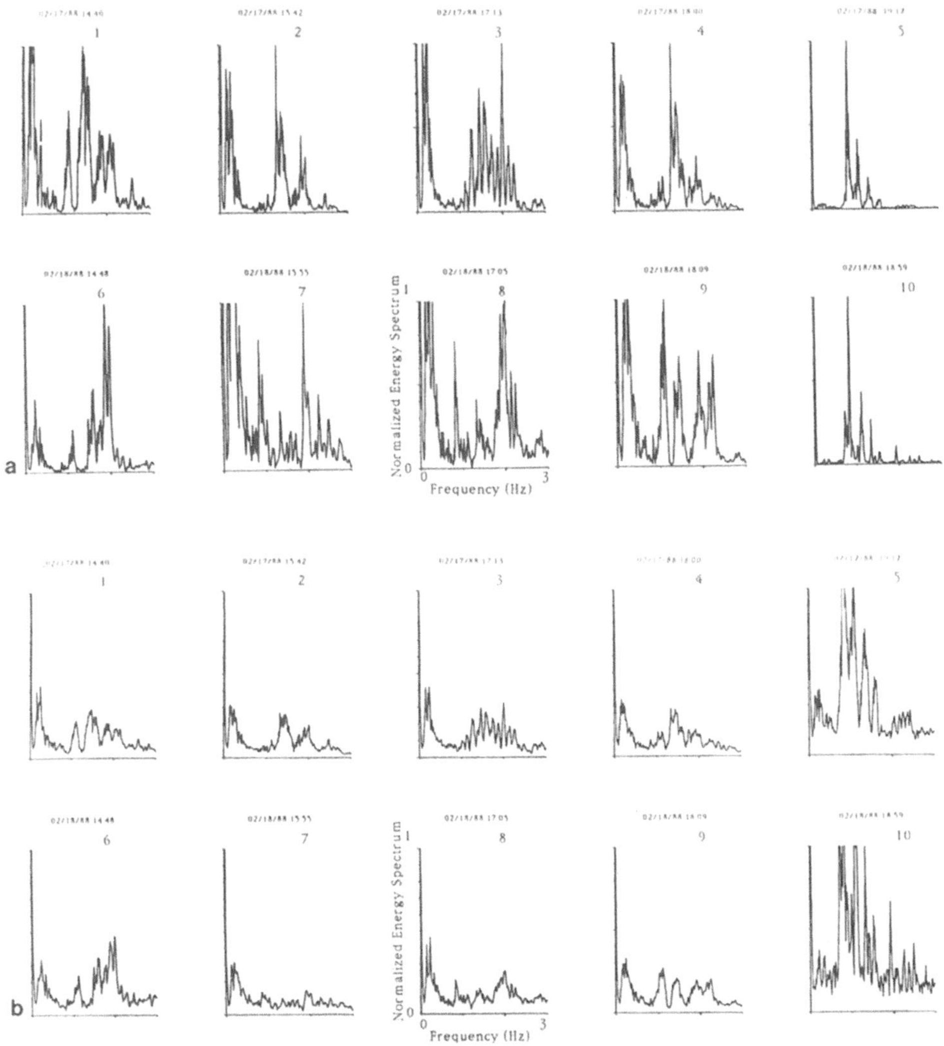

Fig. 12a,b. Same as Fig. 8 for ten long-period events. **a** Each spectrum has been normalized individually; **b** all the spectra have been normalized to the same arbitrary value

value. These figures show a variety of spectra, although some of them are identical and we can distinguish four kinds of spectra. In the first kind, the spectra have a very simple spectral shape, with four peaks and the dominant peak occurs at the lowest frequency, 0.85 Hz. This is the observed spectra for cases 5 and 10. In the second kind, which includes the observed spectra cases 2 and 4, the spectral shape is very similar to the first kind, but the spectra seems to be shifted toward higher frequencies and the spectral peaks as well as the interval between them becomes broader. The third kind corresponds to case 3 with a dominant peak at 2 Hz, surrounded by many other peaks regularly spaced.

The fourth kind shows the energy spectrum which may be a superposition of the above three kinds. For example, the spectrum for case 8 seems to be the superposition of the spectra for cases 5 and 3.

It is interesting to note that events with lower dominant frequency peaks are greater in absolute seismic amplitude. This is in agreement with the observations that the tremor frequency becomes lower with the increase in seismic amplitude for the deep tremor in Hawaii (Aki and Koyanagi 1981), and for a long-period tremor of Mount Aso (Kubotera 1974). In these cases, the lower frequency seems to reflect the greater volume of magma involved.

Another interesting observation is that the spectral shape of the first and second kind are very similar to the spectral shape of a stack of gas-piston events. The spectra are characterized by a dominant peak at low frequency (0.85 Hz for the long-period events of the first kind, around 2.5 Hz for gas-piston events and volcanic tremor), followed by two or three peaks with decreasing amplitude at frequencies which do not correspond to a simple multiple of the dominant frequency. Neither the spectrum of a resonating fluid-filled pipe (Chouet 1985) or a spherical magma chamber (Crosson and Bame 1985) will display such features which seems to be due to a three-dimensional effect at the source. The 3-D fluid-filled crack model developed by Chouet (1986, 1988, this Vol.) may be more appropriate, but the model does not seem to generate the spectral features of long-period events and gas-piston events mentioned above. The inability may be due to the approximation used in solving the equations of motion and continuity for the fluid in the plane of the crack i.e., in two dimensions, which may be responsible of the overall strong excitation of a very slow wave analogue to the tube wave, called "crack wave" (Ferrazzini and Aki 1987; Tang and Cheng 1988; Chouet, this Vol.). The excess pressure applied being symmetrical with respect to the crack plane, in the thickness of the crack, only symmetrical modes can exist thus favouring the crack-wave development. However, we noticed that the spectral peak distribution of the long period events seems to follow those of the normal modes of vibration of a rectangular room. Considering for simplicity a rectangular room with rigid walls of dimensions L_x, L_y, L_z, the corresponding eigenfrequencies are given by (Morse and Bolt 1944):

$$\nu = \frac{c}{2}\sqrt{\left(\frac{n_x}{L_x}\right)^2 + \left(\frac{n_y}{L_y}\right)^2 + \left(\frac{n_z}{L_z}\right)^2},$$

where the quantities n_x, n_y, n_z are integers greater or equal to zero and each trio of values (n_x, n_y, n_z) corresponds to a normal model of the room. In the case of the long-period event shown in Fig. 8, assuming that the dominant peak at 0.85 Hz corresponds to the normal mode (1, 0, 0) and the acoustic velocity equals 1 km/s, we find the largest dimension of the room on the order of L_x = 600 m. The second peak occurs at 1.08 Hz, frequency less than twice the frequency of the first mode thus, this peak should correspond to the normal mode (0, 1, 0) which leads to the second largest dimension of the room L_y = 470 m. The third or fourth peaks may be related to the third dimension of the room if this dimension is not too small compare to L_x and L_y, or may correspond to the normal modes (1, 1, 0) or (2, 0, 0). The eigenfrequencies found for those two modes are respectively 1.37 and 1.70 Hz which are very close to frequencies of 1.34 and 1.6 Hz of the observed third and fourth peaks, and suggest that the third dimension of the room L_z must be smaller than $L_x/2$. The same reasoning apply for the gas-piston events, the tremor and the long-period events of the second kind although the spectral peaks are broader and the resonant frequencies less accurate. Still assuming an acoustic velocity of 1 km/s, the two largest dimensions of the source are estimated to be on the order of 385 and 265 m for long-period events of the second kind and, 190 and 150 m for gas-piston events and tremor. In both cases, the third peak may be attributed to the normal mode (1, 1, 0).

For gas-piston events, as described earlier, we have some direct observations about the source process. Clearly, the burst of the slug of gas and the impact of lava movement due to this degassing is causing the vibration. From the observed movement of lava, we can make an estimation of the impact force and energy pumped into the system and directly relate them to the seismic observation. For example, a rough estimate of the gravitational work can be made by the volume of lava moved during the event that the energy rate at the peak of gas-piston may be about 10^{17} erg/s. As mentioned earlier, the spectra of gas-piston events and tremor are characterized by many narrow peaks. In order to explain the sharpness of the spectral peaks observed, let us consider the Q of resonator corresponding to a spectral peak and may be defined as

$$Q = \frac{f}{\Delta f},$$

where f is the peak frequency and Δf is the width of the peak at half its level in power spectrum. The observed spectral peaks for gas-piston events and tremor indicates a Q value as high as 11. It is well known that the damping of the fundamental mode of isotropic oscillation of a liquid sphere embedded in a solid is determined by the impedance contrast between solid and liquid (e.g., Aki et al. 1977). For high Q, this relation is simplified and Q becomes directly proportional to the impedance contrast. In the case of a fluid-filled crack model (Ferrazzini et al. 1990; Chouet, this Vol.), numerical experiments indicate that Q is also proportional to the impedance contrast for large Q. The only way to obtain the high impedance contrast required for Q as high as 11 is to have bubbles in the magma (Aki et al. 1978). Q is also affected by the viscosity of the gas-fluid mixture. To obtain high Q, viscosity must be low. Although bubbles tend to increase the viscosity (Sibree 1933; Wallis 1969), the increase may be negligible. Thus we find that gas may be playing an important role in gas-piston events and tremor, not only for exciting vibration but also for sustaining vibration.

The association of tremor with degassing activity has already been proposed by numerous authors. It was first suggested in Japan by Sassa (1935) on Aso Volcano. Sassa noticed that one kind of volcanic tremor with a dominant frequency of 2 Hz was observed only during the active periods of the volcano and that the changes in signal amplitude was related to the emission of volcanic smokes. More recently, on Mount Etna, Schick and Riuscetti (1973) suggested that the tremor source was related to the presence of gases in the upper part of the magmatic column rather than to the oscillation of a magma chamber at depth. Based on the idea of vibrating gas bubbles rising in the magma, Seidl et al. (1981) proposed a model used by Schick et al. (1982) and Gresta et al. (1987) to obtained a scheme for the feeding conduits.

This hypothesis agrees well with the recent work done by Jaupart and Vergniolle who consider gas as a major factor for triggering and governing the dynamics of eruptions. Jaupart and Vergniolle (1986, 1988) perform a simple laboratory experiment to explain Hawaiian fountains as well as Strombolian eruptions. They consider a reservoir of oil topped by a thin vertical conduit. Bubbles are generated at the bottom of the reservoir and rise into the oil. A foam begins to form at the roof of the reservoir and the oil level in the pipe increases as long as the foam can be accumulated beneath the roof. The thickness of the foam layer is limited by a critical value at which the foam becomes unstable and collapses, resulting in the formation of one or several large bubbles, depending on the fluid viscosity. These bubbles rise through the conduit and burst at the surface and the oil level inside the conduit drops to its minimum level. The size of the bubble rising in the conduit, and the difference between maximum and minimum oil level inside the conduit depend on the amount of gas and the fluid viscosity. As the oil viscosity decreases the size of the largest bubble created increases and the time interval between two bubbles increases. The gas-piston events that we observed and their cyclicity, are well explained by Jaupart and Vergniolle's model which should be helpful in constraining our seismic source model parameters.

In the case of the long-period events, we also find a very high Q resonator, and the similarity of spectra leads us to assume a similar source mechanism as for gas-piston events. The difference between gas-piston events and long-period events may be then, whether the degassing activity occur at depth, in a close system or in the atmosphere, in an open system.

The presence and composition of gas at depth of 1 to 5 km is a subject of controversy among scientists. Recently Tait et al. (1989) have presented a numerical model simulating the effects of fractional crystallization and oversaturation of volatile species in a shallow (1 to 5 km) magma chamber. They show that in the presence of magma saturated with volatiles, only a few percent of crystallization are required to build up overpressures equal to twice the effective tensile strength for a pure H_2O gas phase and that a pure CO_2 gas phase in the magma chamber is unlikely to exist. Our capability to determine the mass fraction of gas in magma by simulating the long-period events may then be crucial in resolving the controversy on the existence and composition of gas in magma chamber.

The sharply peaked spectra of volcanic events that we recorded require a source process that is inherently of high Q. The spectral peaks are attributed to the normal modes of vibration of a magma filled-body and because of their stability over the period of observation, we assume that the system is linear. The high Q value of the resonator may be due to a high contrast of impedance between the fluid and the surrounding medium that we justify by the obvious presence of gas and we suggest that gas may be a necessary ingredient in the generation of the tremor. Although, a quantitative model must be developed in order to determine if gas is also sufficient to explain the high Q resonator, or, if a non linear source mechanism such as a turbulent two-phase flow or, such as in the flow-induced vibration model proposed by Julian (1986), is required.

We have, however, a serious puzzle in our source model. Why is the peak frequency so stable if bubbles are involved in vibration? We rather expect that the bubble contents may easily fluctuate from time to time. This is an open question.

Acknowledgements. We would like to thank Peter Roberts, Chris Dietel, Robert Koyanagi, Bernard Chouet, Michelle Robertson and Aaron Martin for their help in the field experiment. We also acknowledge Tom Wright, Reggie Okamura, Christina Heliker, James Griggs and many others at the Hawaiian Volcano Observatory for their invaluable assistance in the experiment. Mr. and Mrs. Dallas Jackson provided excellent accommodations. John McRaney of the University of California provided all the logistics. Special thanks are due to David Okita of Hilo, Hawaii, who always provided us with superb helicopter assistance, advices concerning the conditions at Pu'u O'o and a constant helping hand during our experiment. The present work was supported in part by the National Science Foundation under grant EAR-8618107, and in part by funding from the University of Southern California.

References

Aki K (1957) Space and time spectra of stationary stochastic waves, with special reference to microtremors. Bull Earthq R Inst 25:415–457

Aki K, Koyanagi R (1981) Deep volcanic tremor and magma ascent mechanism under Kilauea, Hawaii. J Geophys Res 86:7095–7109

Aki K, Richards PG (1980) Quantitative seismology: theory and methods. WH Freeman, San Francisco

Aki K, Fehler M, Das S (1977) Source mechanism of volcanic tremor: fluid-driven crack models and their application to the 1963 Kilauea eruption. J Volcanol Geotherm Res 2:259–287

Aki K, Chouet B, Fehler M, Zandt G, Koyanagi R, Colp J, Hay R (1978) Seismic properties of a shallow magma reservoir in Kilauea Iki by active and passive experiments. J Geophys Res 83:2273–2282

Chouet B (1985) Excitation of a buried magmatic pipe: a seismic source model for volcanic tremor. J Geophys Res 90:1881–1893

Chouet B (1986) Dynamics of a fluid driven crack in three dimensions by the finite difference method. J Geophys Res 91, 13:967–992

Chouet B (1988) Resonance of a fluid-driven crack: radiation properties and implications for the source of long-period events and harmonic tremor. J Geophys Res 93:4375–4400

Chouet B, Julian BR (1985) Dynamics of an expanding fluid-filled crack. J Geophys Res 90, 11:184–198

Crosson RS, Bame DA (1985) A spherical source model for low-frequency volcanic earthquakes. J Geophys Res 90, 10:237–247

Dibble R (1974) Volcanic seismicity and accompanying activity of Ruapehu Volcano, New Zealand. In: Civetta L, Gasparini P, Luongo G, Rapolla A (eds) Physical volcanology. Elsevier, Amsterdam, pp 49–86

Dietel C, Chouet B, Aki K, Ferrazzini V, Roberts P (1989) Data summary for dense GEOS array observations of seismic activity associated with magma transport at Kilauea Volcano, Hawaii. US Geol Surv Open-File Rep 89–113

Endo E, Malone S, Noson L, Weaver C (1981) Locations, magnitudes and statistics of the March 20–May 18 earthquakes sequence. USGS Prof Pap 1250:93–108

Fehler M, Chouet B (1982) Operation of a digital seismic network on Mount St. Helens volcano and observations of long-period seismic events that originate under the volcano. Geophys Res Lett 9:1017–1020

Ferrazzini V, Chouet B, Fehler M, Aki K (1990) Quantitative analysis of long-period events recorded during hydrofracture experiments at Fenton Hill, New Mexico. J Geophys Res 95:21871–21884

Ferrazzini V, Aki K (1987) Slow waves trapped in a fluid-filled infinite crack: implication for volcanic tremor. J Geophys Res 92:9215–9223

Ferrick MG, Qamar A, St. Lawrence WF (1982) Source mechanism of volcanic tremor. J Geophys Res 87:8675–8683

Gorshkov G (1959) Gigantic eruption of the volcano Bezymianny. Bull Volcanol 20:77–109

Gresta S, Imposa S, Patane D, Patane G (1987) Volcanic tremor at Mt Etna: state of the art and perspectives. Pageoph 125:1079–1095

Guerra I, Bascio AL, Luongo G, Scarpa R (1975) Seismic activity accompanying the 1974 eruption of Mt. Etna. J Volcanol Geotherm Res 1:347–362

Havskov J, De la Cruz-Reyna S, Singh SK, Medina F, Gutierrez C (1983) Seismic activity related to the March–Apirl, 1982 eruptions of El Chichon volcano, Chiapas, Mexico. Geophys Res Lett 10:293–296

Jaupart C, Vergniolle S (1986) Separated two-phase flow and basaltic eruptions. J Geophys Res 92:13715–13719

Jaupart C, Vergniolle S (1988) Laboratory models of Hawaiian and Strombolian eruptions. Nature (Lond) 331:58–60

Julian B (1986) Volcanic tremor: flow-induced vibration? EOS 67, 44 (Abstr) VA32A-08

Koyanagi RY (1968) Earthquakes from common sources beneath Kilauea and Mauna Loa volcanoes in Hawaii from 1962 to 1965. US Geol Surv Prof Pap 600-C:120–125

Koyanagi RY, Chouet B, Aki K (1987) Origin of volcanic tremor in Hawaii. US Geol Surv Prof Pap 1350, 2:1221–1257

Kubotera A (1974) Volcanic tremor at Aso volcano. In: Civetta L, Gasparini P, Luongo G, Rapolla A (eds) Physical volcanology. Elsevier, Amsterdam, pp 29–47

Latter JH (1979) Volcanological observations at Tongariro National Park, 2. Types and classification of volcanic earthquakes, 1976–1978. N Z Dep Sci Ind Res Geophys Div Wellington New Zealand 150:60

Mahrer KD, Mauk FJ (1987) Seismic wave motion for a new model of hydraulic fracture with induced low velocity zone. J Geophys Res 92:9293–9309

Matsushima N, Nishimura Y, Suzuki A, Okada H (1987) Spectral analysis of volcanic earthquakes at Mt. Tokachi. Bull Volcanol Soc JPN 32:317–328

McNutt SR (1986) Observations and analysis of B-type earthquakes, explosions, and volcanic tremor at Pavlov volcano, Alaska. Bull Seismol Soc Am 76:153–175

Minakami T (1974) Seismology of volcanoes in Japan. In: Civetta L, Gasparini P, Luongo G, Rapolla A (eds) Physical volcanology. Elsevier, Amsterdam, pp 1–28

Morse PM, Bolt R (1944) Sound waves in rooms. Rev Modern Phys 16, 2:69–150

Nishi K, Nakamura S, Sodona T et al. (1988) Seismic activity of the volcano Suwanose-Jima, Tokara Islands, Japan. In: Kamo K (ed) Report of joint geophysical and geochemical observation of Suwanose-Jima. Kyoto Univ, July–August 1984, 1:1–51

Okada H, Nishimura Y, Maekawa T (1988) Earthquake family with low-frequency nature observed at Volcano Suwanose-Jima, Tokara Islands, south of Kyushu. In: Kamo K (ed) Report of joint geophysical and geochmical observation of Suwanose-Jima. Kyoto Univ, July–August 1984, 1:1–51

Roberts PM (1989) A versatile equalization circuit for increasing seismometer velocity response below the natural frequency. Bull Seismol Soc Am 79:1607–1617

Sassa K (1935) Volcanic micro-tremors and eruption-earthquakes. Mem Coll Sci Univ Kyoto Ser A, 18:255–293

Schick R, Riuscetti M (1973) An analysis of volcanic tremors at south Italian volcanoes. Z Geophys 39:247–262

Schick R, Cosentino M, Lombardo G, Patané G (1982) Volcanic tremor at Mount Etna. A brief description. Mem Soc Geol It 23:191–196

Seidl D, Schick R, Riuscetti M (1981) Volcanic tremors at Etna: a model for hydraulic origin. Bull Volcanol 44:43–56

Sibree JO (1933) The viscosity of froth. Trans Faraday Soc 30:325–331

St. Lawrence W, Qamar A (1979) Hydraulic transients: a seismic source in volcanoes and glaciers. Science 203:654–656

Swanson DA, Duffield W, Jackson D, Peterson D (1979) Chronological narrative of the 1969–71 Mauna Ulu eruption, Kilauea Volcano, Hawaii. US Geol Surv Prof Pap 1065:55

Tait S, Jaupart C, Vergniolle S (1989) Pressure, gas content and eruption periodicity of a shallow, crystallizing magma chamber. Earth Planet Sci Lett 92:107–123

Tang XM, Cheng CH (1988) Wave propagation in a fluid-filled fracture – an experimental study. Geophys Res Lett 15:1463–1466

Wallis GB (1969) One dimensional two phase flow. McGraw Hill, New York, 408 pp

Wolfe EW, Garcia MO, Jackson DB, Koyanagi RY, Neal CA, Okamura AT (1987) The Pu'u O'o eruption of Kilauea Volcano, episodes 1–20, Jaunary 3, 1983, to June 8, 1984. US Geol Suv Prof Pap 1350, 1:471–508

Numerical Studies of the Dynamics of Fluid-Filled Cracks Placed in Series: A Model for Inharmonic Peaked Tremor Spectra

T. Dahm

Abstract

A volcanic tremor is characterized by peaked spectra with frequencies ranging from 0.15 to 10 Hz, which may be explained by resonance set up in magmatic conduits. Temporal changes of volcanic tremor periods have been observed on various volcanoes. Another feature of peaked tremor spectra is that, especially on Mount Etna, higher modes do not appear to be harmonics of the fundamental mode. These observations cannot be explained by a simple model consisting of a single resonating dike, which can only generate harmonic spectra. As an extension to the single dike model, I have considered two or more dikes placed in series with a different characteristic impedance. The characteristic impedance of the dike is determined by the cross-sectional area, the acoustic velocity, and the density of the fluid.

Numerical results obtained using two dikes in series differ considerably from the single dike solution. In particular, spectral peaks are obtained at frequencies which are not harmonics of the fundamental mode. Moreover, the frequency of the fundamental mode may be much lower than that obtained using a single dike with the same length as the two dikes in series.

The symmetry in the inharmonic peaked spectrum, which is independent of the impedance contrast between the dikes, is a common characteristic of all models with complex geometry.

1 Introduction

Volcanic tremors are mostly linked to volcanic eruptions and, in general, are linked to volcanic activity. As seismic observations on active volcanoes can be made quite easily, there is increasing interest in using seismic data to characterize the state of volcanic activity and to forecast possible eruptions.

Although recently many tremor models have been established (see Chouet et al. 1987; Chouet 1988; Schick 1988 for latest reviews), the source mechanism of a volcanic tremor is still controversial. Based on the long duration of a volcanic tremor (from a few minutes to several days), and the relatively stable, sharply peaked tremor spectra, their generating mechanism is often associated with resonance effects linked to conduit, crack, or magma chamber geometry.

Several models have been proposed in which fluid vibration plays an active role (St. Lawrence and Quamar 1979; Seidl et al. 1981; Ferrick et al. 1982 and others). Ferrick et al. studied analytically the hydraulic transients of a magma-filled conduit with constant cross-sectional area. They showed that this system can generate a harmonically peaked spectrum which can be a first approximation to observed tremor spectra.

As an extension to this model, I have studied the hydraulic transients of magma-filled dikes with more complex geometry. I have used a finite element method (Kirschke 1974) and compared the numerical results with analytical results.

2 Geometry and Boundary Conditions

In order to facilitate the presentation I have used only two dikes placed in series. The geometry is illustrated in Fig. 1, which shows two dikes with length l_1 and l_2 and characteristic impedance Z_{c1} and Z_{c2} placed in series. The system has two ends with boundary conditions Z_{01} at $x = 0$ and Z_{l2} at $x = l_1 + l_2$.

In this example I have always used a reservoir boundary condition at $x = 0$ ($Z_{01} = 0$) and a reservoir or a closed-end boundary condition at the other end ($Z_{l2} = 0$ or $Z_{l2} = \infty$). With these boundary conditions no outflow of oscillatory energy occurs, i.e., the reflection coefficient at the ends is +1 or −1.

I have supposed an abrupt characteristic impedance contrast between the dikes where a partial reflection of a propagating pressure pulse occurs. The characteristic impedance Z_c of a dike is determined by the cross-sectional area A, the acoustic velocity α, and the density ρ of the fluid. For fluids without viscosity the characteristic impedance is $Z_c = \frac{\rho\alpha}{A}$. In this case α and ρ are held constant and only the cross-sectional area A is varied.

In order to simplify the system, the following assumptions are introduced:

1. One-phase flow without friction. The effect of viscous damping is discussed in Dahm (1989).
2. Adiabatic propagation of plane waves. This condition is true only if the sectional area of the conduit system is small compared with the wave length considered, and if the pressure amplitude is not too large.
3. Perfect rigid surrounding rock. The interaction between the fluid and the surrounding material is small if the impedance contrast between them is strong.

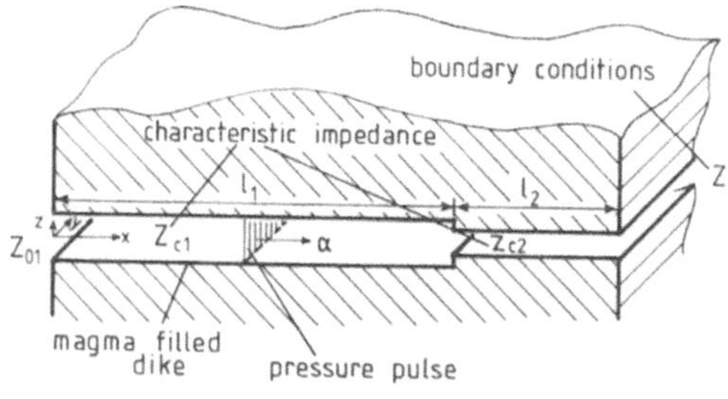

Fig. 1. Geometry of the dikes placed in series

3 Excitation

In this study I wanted to analyze the eigenfrequencies of the system and, therefore, an impulse exitation at $x = 0$ and $t = 0$ is considered. Then the synthetic impulse response can be observed. It is obvious that at an active volcano the excitation force for volcanic tremor is more complex.

4 Numerical Results

In Fig. 2 the well-known single dike solution for one closed end, with a reservoir at the other end, is shown. The dike is divided into finite elements and the program calculates, for every element, the pressure oscillation over time and the spectrum belonging to each time series.

In this case, each element has pressure oscillations with a constant amplitude and each spectrum shows the same eigenfrequencies, which are the odd harmonics of the fundamental eigenfrequency. It is easy to recognize the shape of the different modes by looking at the spectra.

The impulse response of the single dike is a harmonic peaked spectrum which may be inverted to yield the geometry of the generating system.

In Fig. 3a–d only one element of the system is considered (element 3) and the time series and the spectrum of this element are shown. Two dikes are placed in series, both having the same length l_0 but different cross-sectional areas. The boundary conditions are the same as in Fig. 2.

Due to partial reflections between the dikes, the time series do not have this simple shape as in Fig. 2. In the case of a strong characteristic impedance contrast (Fig. 3d), a long beating period occurs, which enters the spectrum as a peak-splitting effect. In this case the spectral peaks approximate the harmonic solutions of the two single dikes.

In Fig. 3c,d the fundamental frequency is much lower than that obtained by using a single dike (Fig. 3b). For example, the extremely long-period volcanic tremor (periods up to 7 s) reported by Sassa (1935) on Mount Aso may be generated by a dike of modest size. With $\alpha = 1200\frac{m}{s}$ the total length of the dike in the model in Fig. 3b would be $l \approx 2\,\text{km}$, in that of Fig. 3c $l \approx 800\,\text{m}$ and in that of Fig. 3d $l \approx 200\,\text{m}$.

Due to the different boundary conditions at $x = 0 (Z_{01} = 0)$ and $x = l_2(Z_{l2})$, the model of Fig. 3a does not have the same eigenfrequencies as that of Fig. 3c. The systems generally generate inharmonic peaked spectra and the fundamental frequency may be much lower than that obtained using a single dike of the same length as the two dikes in series.

In Fig. 4 the calculated eigenfrequencies of two dikes in series (both of length l_0) are plotted against the characteristic impedance contrast $\log\frac{Z_{c2}}{Z_{c1}}\left(= \log\frac{A_1}{A_2}\right)$.

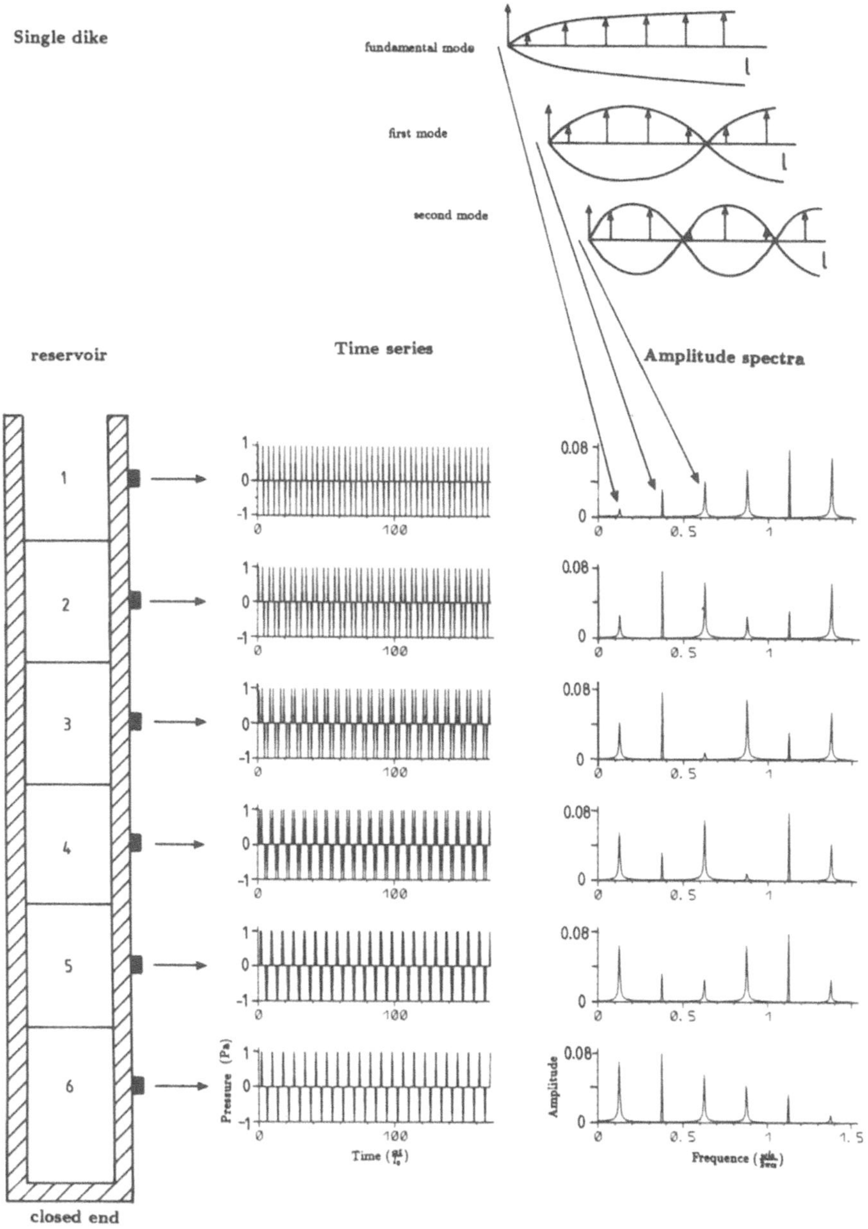

Fig. 2. Single dike solution

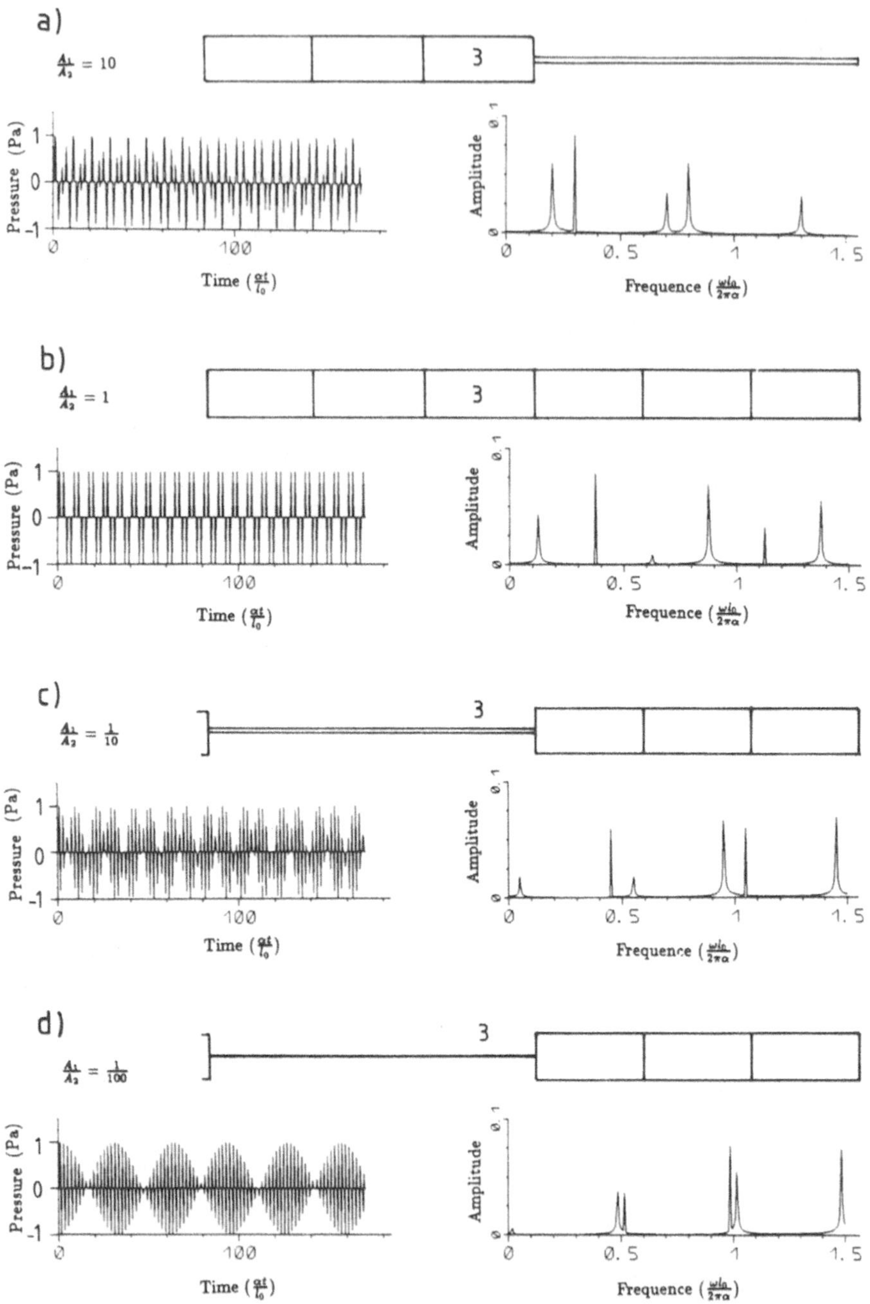

Fig. 3a–d. Time series and spectra of two dikes with length l_0 and the characteristic impedance contrast $\frac{Z_{c2}}{Z_{c1}} = \frac{A_1}{A_2}$ placed in series

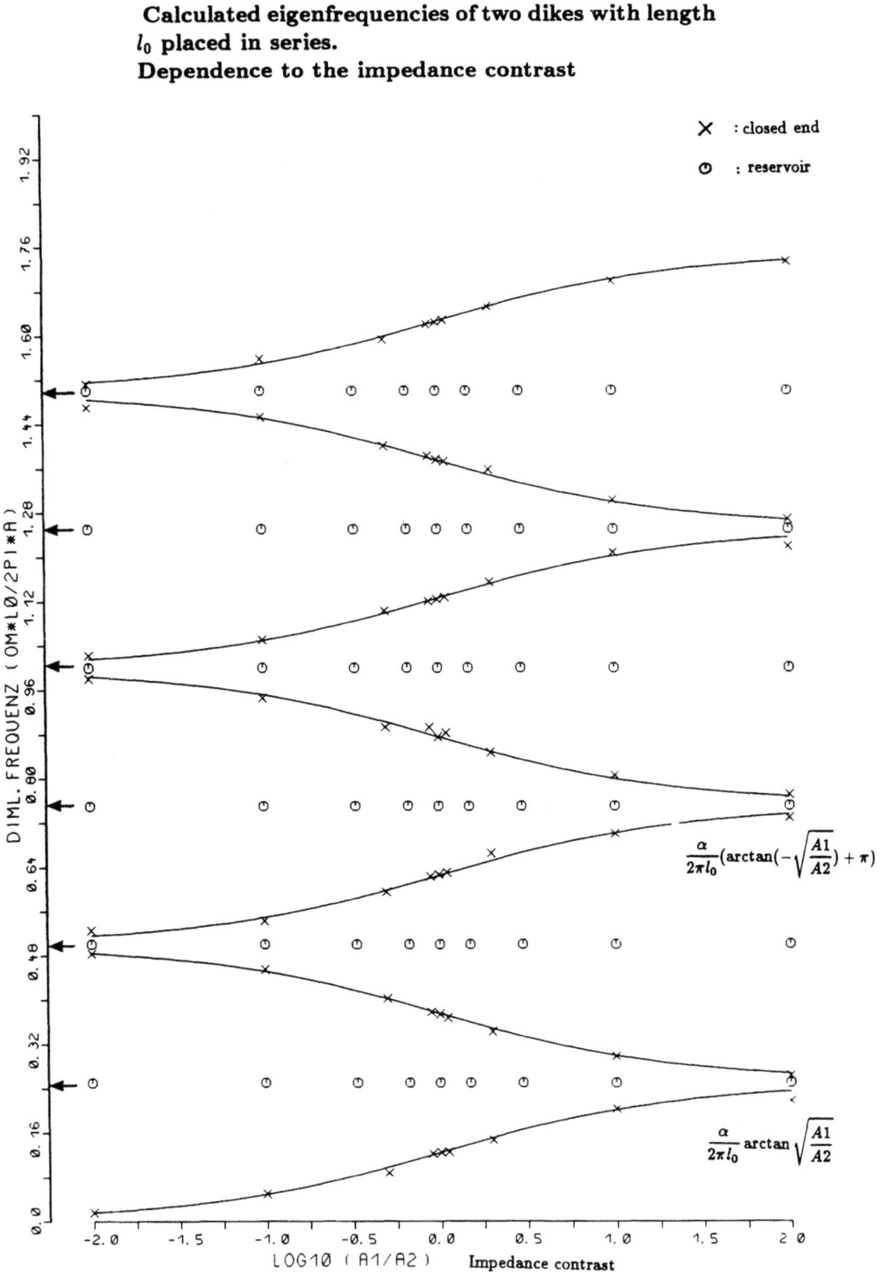

Fig. 4. Relation between the eigenfrequencies of two dikes with length l_0 and the characteristic impedance contrast $\frac{Z_{c2}}{Z_{c1}} = \frac{A_1}{A_2}$ between them

The crosses represent synthetic solutions for $Z_{l2} = \infty$ (closed end) and the circles solutions for $Z_{l2} = 0$ (reservoir). In the case of a frictionless system it is possible to find an analytical solution and this solution is plotted as a continuous line.

Looking at the crosses it is immediately clear that the eigenfrequencies are shifted symmetrically to periodic points on the frequency axis, which are marked with small arrows. These points of symmetry are a function of $\frac{l_0}{\alpha}$ (see mathematical description) and allow us to solve the inverse problem and to determine the length of both dikes placed in series.

As for the circles, even changing only one parameter is sufficient to significantly alter the solution. Here, the location of the eigenfrequencies is independent of the characteristic impedance contrast. The reason is that all eigenfrequencies occupy points of symmetry which are independent of the characteristic impedance contrast. A common characteristic of all models with two dikes placed in series is the symmetry in the shifted peaked spectra.

5 Mathematical Description

5.1 Single Dike Solution

Wylie and Streeter (1978) and Meyer and Neumann (1975) developed transient equations to describe the dynamics of a single pipe. These equations may also be used for the dike systems considered here.

The pressure P and the hydraulic discharge $Q = uA$ (u is the fluid velocity, A the cross-sectional area) can be divided into the mean values $\bar{P}$ and $\bar{Q}$ and oscillatory components p' and q'. $P(x,t) = \bar{P}(x) + p'(x,t)$, $Q(x,t) = \bar{Q}(x) + q(x,t)$ with:

$$p' = e^{st}P(x) \tag{1}$$

$$q' = e^{st}Q(x) \tag{2}$$

$$P(x) = P_{x=0}\cosh(\gamma x) - Z_c Q_{x=0}\sinh(\gamma x) \tag{3}$$

$$Q(x) = \frac{-1}{Z_c}(P_{x=0}\sinh(\gamma x) - Z_c Q_{x=0}\cosh(\gamma x)) \tag{4}$$

with $\gamma^2 = \frac{s}{\alpha^2}(S + \frac{R}{\rho})$, the linearized resistance R, the characteristic impedance $Z_c = \frac{\gamma\rho\alpha^2}{As}$, and the complex frequency $s = \sigma + i\omega$.

Remark: Wylie and Streeter (1978) introduced the hydraulic gradline $H = \frac{P}{\rho g}$. Here, I have used p' instead of h' and therefore the relation for the characteristic impedance Z_c differs from that given by Wylie and Streeter. With

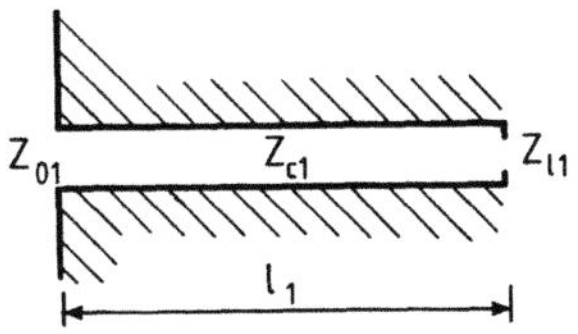

Fig. 5. Single dike with characteristic impedance Z_{c1}, length l_0 and boundary conditions Z_{01} and Z_{l1}

$Z_c = \dfrac{\gamma\rho\alpha^2}{As}$ the pressure reflection coefficient between two dikes with characteristic impedances Z_{c1} and Z_{c2} is $r = \dfrac{Z_{c2} - Z_{c1}}{Z_{c1} + Z_{c2}}$.

By introducing the acoustic impedance $Z(x) = \dfrac{P(x)}{Q(x)}$, the single dike can be described by one equation (cf. Fig. 5):

$$Z_{01} = Z_{c1}\frac{Z_{l1} + Z_{c1}\tanh\gamma l_1}{Z_{c1} + Z_{l1}\tanh\gamma l_1} \tag{5}$$

with $Z_{01} = \dfrac{P_{x=0}}{Q_{x=0}}$ and $Z_{l1} = \dfrac{P_{x=l_1}}{Q_{x=l_1}}$ as boundary conditions.

The solutions for $R = 0$ (fluids without viscosity) are given by Ferrick et al. (1982). The implicit form of Eq. (5) is now:

$$z\left(Z_{01}, Z_{l1}, Z_{c1}, \tanh\left(\frac{\sigma + i\omega}{\alpha}l_1\right)\right) = 0, \tag{6}$$

where Z_{01}, Z_{l1}, Z_{c1} are constant.

From Eq. (6) one sees that the solution must be periodic with $\dfrac{i\omega l_1}{\alpha} = i\pi$ since $\tanh z = \tanh(z + i\pi)$.

5.2 Solution for Two Dikes in Series

The impedance function for two dikes in series is (Wylie and Streeter 1978; cf. Fig. 6):

$$Z_{01} = Z_{c1}\frac{A_0 + A_1 f(\gamma l_1) + A_2 f(\gamma l_2) + A_{12} f(\gamma l_1) f(\gamma l_2)}{A_1 + A_0 f(\gamma l_1) + A_{12} f(\gamma l_2) + A_2 f(\gamma l_1) f(\gamma l_2)} \tag{7}$$

with

$f(\gamma l) = \tanh\gamma l$
$A_0 = Z_{l2}Z_{c2}$
$A_1 = Z_{c1}Z_{c2}$
$A_2 = Z_{c2}^2$
$A_{12} = Z_{l2}Z_{c1}$

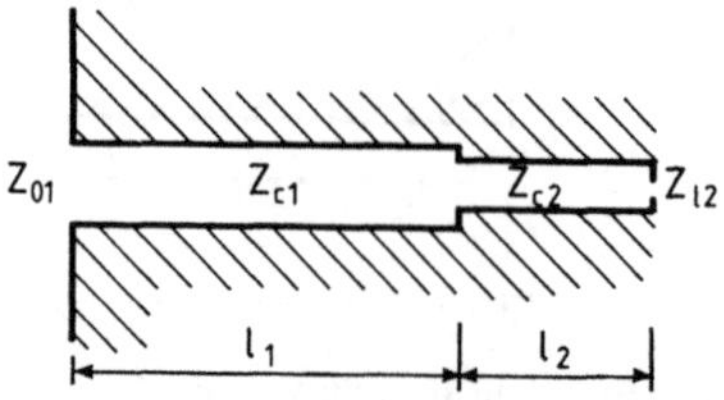

Fig. 6. Two dikes placed in series with characteristic impedances Z_{c1}, Z_{c2}, lengths l_1, l_2 and boundary conditions Z_{01}, Z_{l2}

The implicit form is:

$$Z(Z_{01}, Z_{l2}, Z_{c1}, Z_{c2}, f(\gamma l_1), f(\gamma l_2)) = 0. \tag{8}$$

Setting $l_1 = al_0$ and $l_2 = bl_0$ where a, b are integers and using the addition theorems, the number of independent variables can be reduced.

Equation (8) may now be written as:

$$Z(Z_{01}, Z_{l2}, Z_{c1}, Z_{c2}, f(\gamma l_0)) = 0. \tag{9}$$

For $R = 0$ a comparison with the single dike solution shows that the solution must be periodic with $\frac{i\omega l_0}{\alpha} = i\pi$. With:

$$\binom{n}{n-k} = \binom{n}{k} = \begin{cases} \frac{n!}{k!(n-k)!} & 0 \leqslant k \leqslant n \\ 0 & 0 \leqslant n \leqslant k' \end{cases}$$

where $n, k \in N$, Eq. (9) can be written as:

$$Z_{01} = Z_{c1} \sum_{p=0}^{i > \frac{a+b}{2}} \frac{f(\gamma l_0)^{2p}(A_0 S^0_{pq} + A_{12} S^{12}_{pq}) + f(\gamma l_0)^{2p+1}(A_1 S^1_{pq} + A_2 S^2_{pq})}{f(\gamma l_0)^{2p}(A_1 S^0_{pq} + A_2 S^{12}_{pq}) + f(\gamma l_0)^{2p+1}(A_0 S^1_{pq} + A_{12} S^2_{pq})}$$

with

$$S^0_{pq} = \sum_{q=0}^{p} \binom{a}{2q} \binom{b}{2p-2q}$$

$$S^{12}_{pq} = \sum_{q=0}^{p} \binom{a}{2q+1} \binom{b}{2p-2q-1}$$

$$S^2_{pq} = \sum_{q=0}^{p} \binom{a}{2q} \binom{b}{2p-2q+1}$$

$$S^1_{pq} = \sum_{q=0}^{p} \binom{a}{2q+1} \binom{b}{2p-2q}.$$

Example:

$l_1 = al_0$, $l_2 = bl_0$, a and b are odd and reservoir boundary conditions at both ends ($Z_{01} = 0$, $Z_{l2} = 0$).

This leads to:

$$Z_{c1}\frac{C_{a+b-1}f^{a+b-1} + C_{a+b-3}f^{a+b-3} + \ldots + C_3f^3 + Cf}{C_{a+b}f^{a+b} + C_{a+b-1}f^{a+b-2} + \ldots + C_2f^2 + C_0} = 0$$

with coefficients C_n and $f = \tanh\left(\frac{i\omega l_0}{\alpha}\right)$:

$$f = \pm\infty \Rightarrow \frac{\omega l_0}{\alpha} = n\frac{\pi}{2}, \text{n} = 1,3,5,\ldots \text{ is a solution;}$$

$$f = 0 \quad \Rightarrow \frac{\omega l_0}{\alpha} = m\frac{\pi}{2}, m = 2,4,6,\ldots \text{ is a solution.}$$

These solutions are independent of the coefficients C_n, i.e., independent of the characteristic impedance contrast. Due to $\tanh z = \tanh(z + i\pi)$, the solution for $\frac{i\omega l_0}{\alpha}$ is periodic with $i\pi$ and symmetric to $\frac{\omega_n l_0}{\alpha} = n\frac{\pi}{2}, n = 1,3,5,\ldots$ The nominator has only odd powers of f, i.e., the roots of the nominator show symmetry to $f = 0\left(\frac{\omega_m l_0}{\alpha} = m\frac{\pi}{2}, m = 2,4,\ldots\right)$. Between any two poles of f you find $a + b - 1$ roots of the norminator, which means $a + b - 1$ eigenfrequencies.

In this example one sees that all points of symmetry are occupied by eigenfrequencies and the number of eigenfrequencies between the odd harmonic points of symmetry determine the length $l = l_1 + l_2$ of the two dikes in series. This principle of reducing the number of independent variables in the impedance function of the system can also be used for more complex geometries such as more dikes placed in series, or branching systems (Dahm 1989, 1991). The solution, therefore, is always periodic with $\frac{\omega l_0}{\alpha} = \pi$.

6 Conclusions

The presented work shows that even for the simple model of two dikes placed in series, the solution for the eigenfrequencies is complex and in general different from the single dike solution. A single dike or a single pipe approximation in tremor modeling may lead, therefore, to large errors.

From the analytical description it appears that for any complex geometry (more than two dikes in series or branching systems) the solution is always periodic with $\frac{\omega l_0}{\alpha} = \pi$. If all eigenfrequencies of a system are known, it is possible to determine the characteristic length l_0 and the whole length of the two dikes placed in series.

In this analysis many simplifications are made (i.e., propagation of plane waves, one phase fluid, etc.) which are not very likely to be realistic for an active volcano.

Future developments call for the dropping of the simplifying assumptions and the development of solutions for more realistic models.

Acknowledgements. I would like to thank Prof. Dr. Rolf Schick and Dr. Walter Zürn for their helpful assistance and discussions throughout this work and the European Science Foundation and the Italian National Research Council's Gruppo Nationale di Vulkanologia (GNV) for supporting my participation at the workshop in Capri. I also gratefully appreciate the careful reviews of Walter Zürn, Calum Macdonald, and Heiner Igel. This work is a part of a diploma thesis at the Geophysical Institute at the University of Karlsruhe, FRG.

References

Chouet B, Koyanagi R, Aki K (1987) The origin of volcanic tremor in Hawaii, Part 2: theory and discussion. US Geol Surv Prof Pap 1350:1259–1280

Chouet B (1988) Resonance of a fluid-driven crack: radiation properties and implications for the source of long-period events and harmonic tremor. J Geophys Res 93:4375–4400

Dahm T (1989) Eigenschwingungen magma-gefüllter Klùfte mit Komplexer Geometric. Diplomarbeit, Geophys Inst Univ Kavlsruhe

Dahm T (1991) Eigenvibrations of magma-filled dike systems with complex geometry. In: Schick R (ed) Volcanic tremor and magma flow. KFK Jùlich (in press)

Ferrick MG, Quamar A, St. Lawrence St. (1982) Source mechanism of volcanic tremor. J Geophys Res 87:8675–8683

Kirschke K (1974) Druckstossvorgänge in wassergefüllten Felsklüften Veröff Inst Bodenmech Felsmech, Univ Karlsruhe, 61:164

Meyer E, Neumann EG (1975) Physikalische und technische Akustik. Vieweg, Braunschweig

Sassa K (1935) Volcanic micro-tremors and eruption-earthquakes. Mem Coll Sci Univ Kyoto A, 18:255–293

Schick R (1988) Volcanic tremor-source mechanism and correlation with eruptive activity. Nat Hazard 1:125–144

Seidl D, Schick R, Riuscetti M (1981) Volcanic tremors at Etna: a model of hydraulic origin. Bull Volcanol 44:43–56

St. Lawrence W, Quamar A (1979) Hydraulic transients: a seismic source in volcanous and glaciers. Science 203:654–656

Wylie E, Streeter VL (1978) Fluid transients. McGraw-Hill, New York

Seismic Image of the Volcanic Tremor Source at Izu-Oshima Volcano, Japan

MUNEYOSHI FURUMOTO, TAKAHIRO KUNITOMO, HIROSHI INOUE, and KOSHUN YAMAOKA

Abstract

The 1986 fissure eruption of Izu-Oshima volcano was followed by a strong activity of volcanic tremor. To investigate the characteristics of the activity and its relation to the magma system beneath the volcano, two seismological surveys were carried out; one was an array observation of volcanic tremor and the other was an investigation of the crustal structure using seismic signals generated by an air-gun. The first study revealed that two separated sources radiated high frequency tremor and that they were located beneath the northwestern and the southeastern parts of the caldera floor. The analysis of the air-gun data by the method of the semblance depicted the three-dimensional image of a magma-filled crack elongating in the NW-SE direction beneath the caldera. From these results, we deduced a kind of fluid-driven crack model of volcanic tremor in which the outflow of magma through the crack tips generates high frequency tremor and resonates the magma-filled crack.

1 Introduction

Generation of volcanic tremor, a non-impulsive seismic signal related to volcanism, is characteristic of many volcanos in the world. Although substantial progress has been made, our knowledge of the mechanisms and the processes of the source is still rudimentary. Since Sassa (1935) suggested the free oscillation of a magma chamber for explaining the observed features of volcanic tremor at Mount Aso, Japan, numerous models of the sustained resonance of a fluid (magma)-filled body have been developed to interprete volcanic tremor; e.g., a spherical magma chamber (Crosson and Bame 1985), a magma-filled pipe (Chouet 1985), and a magma-filled crack (Chouet 1986, 1988; see also Chouet et al. 1987, for a review). As shown by the line-up of almost all important elements of a magma system expected in a volcano (a magma chamber, a magma conduit, and a dyke), seismologists are still unable to identify the true resonator at present. Although these models have their own spectral characteristics that could be used to discriminate the differences among the source mechanisms, this task is still rather difficult (Ferrazzini et al. 1990).

An alternative approach to identify the source mechanism of volcanic tremor is to obtain information of fine structures of the source such as the source geometry and the shape of the relevant magma cavity by analyzing volcanic tremor and other seismic signals. For this purpose, extensive seismic surveys conducted on Izu-Oshima volcano, Japan, where a fissure eruption took place in

November 1986, provided an invaluable opportunity (Ida et al. 1987). Although various analyses of the data acquired in the surveys are still in progress, we have preliminarily obtained interesting results to understand the source mechanism of volcanic tremor.

The aim of this chapter is to report some of these results and to show a rather definite image of tremor source beneath Izu-Oshima volcano deduced from them. We will start with a brief description of two studies. The first one is the investigation of tremor sources by analyzing volcanic tremor observed with a portable seismic array (Furumoto et al. 1990). The second one is an active survey to depict a three-dimensional image of a magma chamber using seismic waves which were generated by a large air-gun and observed by a large array system (Furumoto et al. 1992). Finally, we will attempt to make a source model of volcanic tremor by combining the results of these studies.

2 Sources of Volcanic Tremor

Izu-Oshima volcano is a volcano island located on the volcanic front along the Izu-Bonin trench off the southern coast of Central Japan (Fig. 1). The island is approximately elliptical in shape, about 15 km long along the major axis in the northwest-southeast direction. A caldera about 3 km in diameter lies at the center of the island. The 1986 fissure eruption across the caldera was preceeded and followed by strong activities of volcanic tremor.

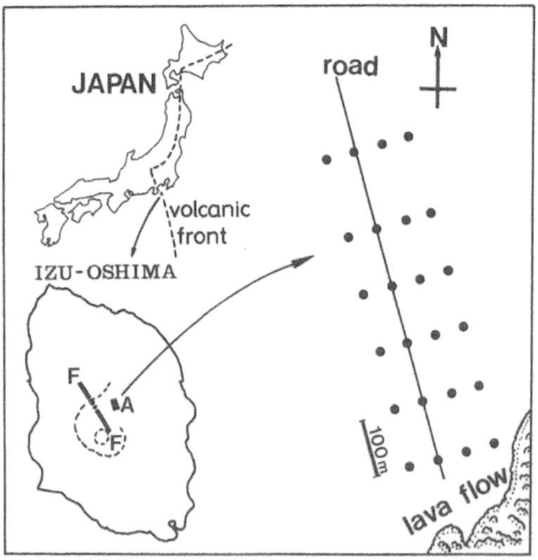

Fig. 1. Location map showing Izu-Oshima volcano and the seismic stations of the array. *Solid circles* indicate the locations of the 24 seismic stations. The *top inset* shows the location of Izu-Oshima and its relation to the volcanic front. The *bottom inset* shows the location of the array (*A*) and the trace (*F-F*) of the 1986 fissure eruption. The caldera rim and the central crater of the volcano are indicated by *dotted curves*

Temporal variation of the tremor activity was summarized by Watanabe (1987, 1988). In July 1986, the beginning of the intermittent activity of tremor was recorded at seismic stations installed on the caldera floor. The amplitudes of tremor gradually increased with time. Since October 24, 1986, the intermittent activity of tremor shifted to a continuous one with smaller amplitudes which lasted until the summit eruption on November 15, 1986. The source of volcanic tremor was estimated at depths 400–600 m just beneath the central crater using small seismic arrays. Although the eruption and the tremor activities stopped on November 23, new tremor activity has been continuing since January 1, 1987. For this new activity, Watanabe (1988) and Fukuyama (1988) observed a very interesting phenomenon that each episode of volcanic tremor after mid-January, 1987, was associated with short-term crustal deformation which indicated the deflation of the volcanic edifice.

To record volcanic tremor, a small portable 24-element array of 2-Hz vertical component seismographs (Furumoto et al. 1988) was deployed on the north-eastern floor of the caldera on March 21, 1987. Figure 1 shows the location of the array and the positions of the 24 stations. The stations were arranged on 4 × 6 grid points with spacings of 50 and 100 m in the nearly east-west and north-south directions respectively. After amplifying each channel, the signals were digitized at 125 Hz by an A/D unit under the control of a microcomputer.

We observed three episodes of volcanic tremor, each of which lasted for about ten minutes after a quiescent interval of about one hour. We here analyzed high frequency components of the recorded tremor. Since the aperture of the array is small, the spatial resolution of the tremor source becomes worse with the increase of the wavelength of the tremor signal. The recorded tremor signals were band-pass filtered in a range from 5 to 25 Hz. Although such high frequency components may be supposed as due to surface activities of the volcano (e.g., the fourth kind tremor on the Volcano Aso, Sassa 1935), we observed also a dominant peak of the spectra of the tremor with a period of about 1 s which must be attributed to the internal activity of Izu-Oshima volcano. Therefore, it is expected that the high frequency components constitute the higher side of the observed tremor signals whose source is embedded beneath the volcano. Furthermore, we can expect that a finer image of the source is revealed by the analysis of higher frequency signals.

To depict sources of volcanic tremor, the method of the semblance (e.g., Neidel and Tarner 1971) was applied to the array data. The semblance is a measure of the coherency existing among multichannel data and very useful to detect signal sources from data contaminated by large noises (Nikolaef and Troitskiy 1987). The semblance S is defined as follows:

$$S = \frac{\sum_{j=s(i)}^{e(i)} \left(\sum_{i=1}^{M} f_{ij(i)} \right)^2}{M \cdot \sum_{j=s(i)}^{e(i)} \sum_{i=1}^{M} f_{ij(i)}^2} \tag{1}$$

Here M is a total number of stations, $f_{ij(i)}$ is the $j_{(i)}$-th data at the i-th station, and $s_{(i)}$ and $e_{(i)}$ are starting and final points of the i-th station data, respectively.

These sample orders $j_{(i)}$, $s_{(i)}$, and $e_{(i)}$ represent the times when some signals from a source arrive at the i-th station. Since they are functions of the distances between the source and the individual stations, they can be calculated from the travel-time curves.

The value of the semblance is computed by summing the data along the times when waves from an assumed source arrive at the stations of the array. If there is a certain seismic source under the ground, the existence of spatial coherence among seismic waves radiated from the source provides a large value of the semblance. In other words, any value summed along inappropriate source-station relations will lead to a small semblance due to the equal probability of positive and negative data in the time series.

Changing the position of the assumed source, or scanning, in a target space of the Earth, we can depict a map of the spatial distribution of the semblance which characterizes the emmissive properties in the space. The scanning is carried out on the area of $4 \times 4\,\mathrm{km}^2$ which encompasses the outer caldera rim. We assumed that the wave velocity of tremor signal in the country rock is 2000 m/s, which is the P wave velocity of the top layer on the volcano (Ikami et al. 1989).

Figure 2 shows the contour map of the resultant semblance distribution on a horizontal section at a depth of 1500 m under the caldera floor (h = 500 m). The contour interval is 2.5 percent of the maximal value of the semblance. On the map, the two regions with the highest semblance values are shaded. Although the shaded regions spread on the map, we assume that small sources are located at the centers of the individual regions. The apparent spreading is due to the lack of resolution caused by a insufficient aperture of the seismic array. As shown in the figure, volcanic tremor was generated from the two separated sources rather than a single source as might be expected. Since the two sources

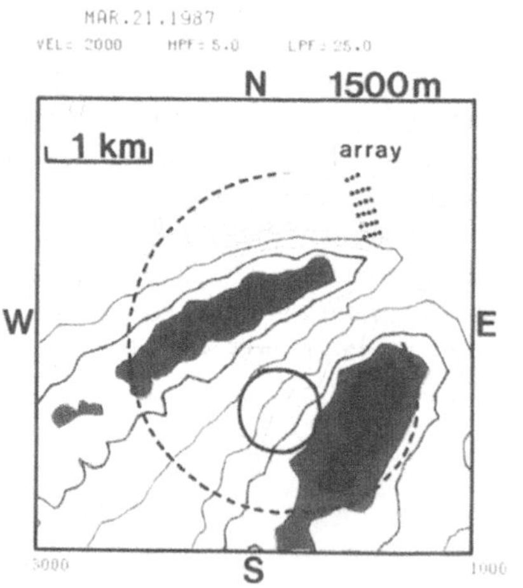

Fig. 2. Contour plot of the semblance distribution for volcanic tremor. The seismic stations of the array are indicated by *small dots*. The caldera rim and the central crater are shown by a *broken* and a *heavy line*, respectively

were simultaneously activated, they must be mechanically connected through some dynamic linkage such as the fluid dynamics of magma.

Note that the two sources are aligned along the NW-SE direction. As we will show later, the alignment of the sources has important implications for the relation between the sources and an assumed magma cavity.

2.1 Shape and Location of the Magma Chamber

To investigate the structure of Izu-Oshima volcano in detail, a seismic survey was carried out on March 21–23, 1987, employing a large air-gun and seismic arrays (Ida et al. 1987). The air-gun was shot at locations on the sea, around the volcanic island, with a spatial interval of about 100 m and the total number of the shots was about 1000. The signals from the air-gun were recorded by 15 permanent seismological stations installed by Izu-Oshima Volcano Observatory (OVO), Earthquake Research Institute, University of Tokyo, and a tentative linear seismic array comprising about 60 stations.

In the present study, among the total data, the signals from about 650 shot points along four shot lines encircling the island recorded at the 15 stations of OVO were used. The locations of the stations and the air-gun shot points used here are indicated in Fig. 3. Figure 4 shows an example of the recorded signals.

To detect magma bodies, we analyzed the coda parts of the recorded signals which are mainly composed of waves scattered from underground heterogeneities (Aki 1969). It would be expected that a magma body caused strong scattering, if any magma exists, since the acoustic impedance of the magma is very different from that of the country rock. However, it is difficult to delineate by eye, a signal trajectory of a scattered wave on a record section such as in Fig. 4; we can

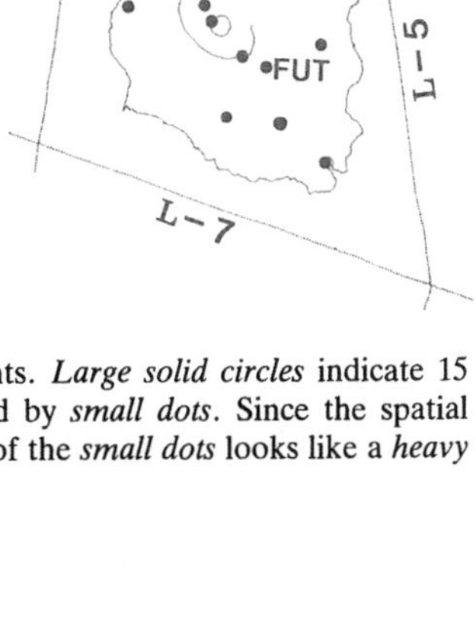

Fig. 3. Location map showing seismic station and shot points. *Large solid circles* indicate 15 stations of OVO. Shot points along four lines are indicated by *small dots*. Since the spatial interval of the shot points is small (100 m), each linear array of the *small dots* looks like a *heavy line*

Fig. 4. Example of the record section. The section shows the seismic waves from the shots along the shot line-1 recorded at FUT station (Fig. 4). The 170 traces are depicted from the *bottom to the top* of the figure according to the order of shot point from the southern end of the shot line. The amplitude of each trace is normalized by the maximum amplitude of the trace

hardly recognize any particular phases from a scattering (secondary) source on a record section. Therefore, we applied the same method that was used in the previous analysis of volcanic tremor. In the case of the semblance for the tremor sources, the functions of the data order $j_{(i)}$ etc. are calculated by using the travel time curves for the direct waves from a source to the stations. However, we must use here the travel time curves for scattered waves, which travel the paths from a source to a subsurface scattering point and subsequently to the stations. Calculating the semblance for an assumed scatterer in a target space using the coda parts, we can depict the spatial distribution of the scatterer of the seismic waves (Furumoto et al. 1992).

We calculated the semblance for each scattering point in a spatial grid with an interval of 500 m beneath the area encompassing the island. The travel times of the waves were computed for ray paths through a scattering point in a homogeneous half-space model. The effects on travel times of the sea depth under the shot points and the station heights were also computed and added to the travel times. Although the method used here makes it possible to depict three-dimensional images of target bodies, the computation for the full space is very time-consuming. In this study, therefore, we computed two-dimensional profiles of the semblance distribution.

In Figs. 5 and 6, the distributions of the semblance on a horizontal section at a depth of 3 km and a vertical section along the SW-NE direction are illustrated by shading. The dark parts of the profiles indicate the regions where strong

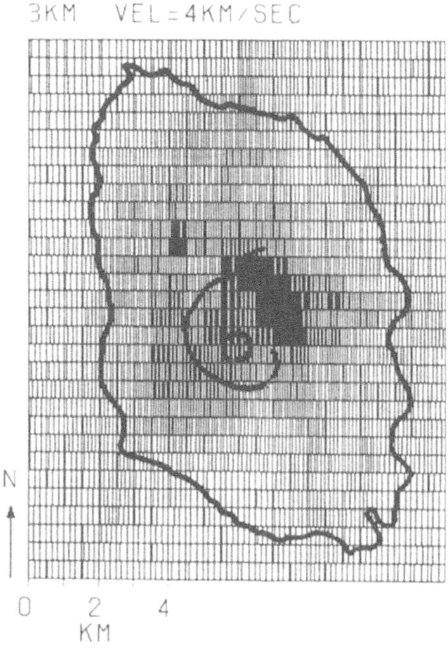

Fig. 5. Semblance distribution at a depth of 3 km. The half-space with the seismic wave velocity of 4 km/s is assumed for the target space

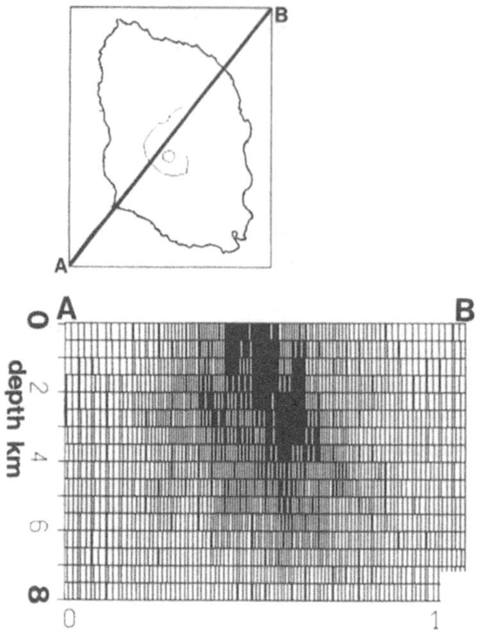

Fig. 6. Vertical corss-section of the semblance distribution. The line *A-B* in the index map indicates the trace of the section

scattering of seismic waves took place. We believe that the dark part appearing under the caldera represents a magma system beneath the volcano, mainly for the reasons (1) that a magma body should be one of the strongest scatterers, and (2) that the region of the strongest scatterer stands just beneath the caldera, where a magma system is expected.

One of the most interesting results of the semblance analysis is that the image of the magma cavity sectioned at a depth of 3 km is elongated in the northwest-southeast direction (Fig. 5). Since the coverage of the target region by the seismic waves is fairly uniform, it is expected that the shape of the image is similar to the original one of the scatterer, even if the computed image of the scatterer may be blurred through the lack of resolution. From the morphology, therefore, we assume that the magma is reserved in an open crack of the country rock elongated in a northwest-southeast direction.

2.2 Source Model of Volcanic Tremor

We have shown the features of the sources of volcanic tremor and the magma-filled crack. Among the features, the elongation direction of the magma-filled crack deserves more attention in order to understand the relation between the sources and the magma body. Geological observations show that fissures of fissure eruptions and dikes in the island preferentially develop along the NW-SE

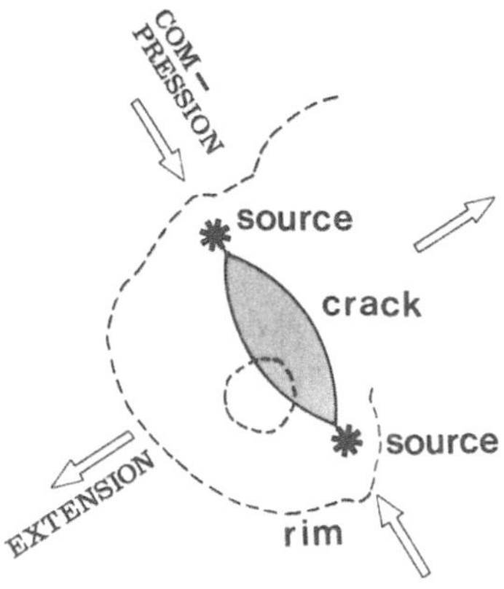

Fig. 7. Cartoon sketch showing the source model of volcanic tremor. The *shaded region* and the *two asterisks* indicate the magma-filled crack and the sources of the high frequency tremor. The caldera rim and the central crater are drawn by *broken lines. Large arrows* show the principal axes of the tectonic stress

direction (Nakamura 1961); the most spectacular evidence of this preference is the orientation of the 1986 fissure eruption of the volcano itself developed along the NW-SE direction from the eastern rim of the sumit crater to the outside of the caldera (Fig. 1). Such a preferred orientation of fissures can be well explained by the rock fracture mechanics: an extensional crack develops along the direction of the maximum compressional axis of the ambient tectonic stress (Nakamura 1969, 1977). Since the magma is likely to be stored in a fissure or a crack, the observed elongation of the scatterer gives another piece of evidence to support that the scatterer represents a magma-filled crack.

This preferred orientation of cracks also provides the key to explain the two sources of volcanic tremor. As mentioned earlier, the two sources align in the same direction as the preferred orientation of cracks, the NW-SE direction. This suggests that the sources are associated with a magma body elongated along the preferred orientation. We assume that this magma body is identical to the magma-filled crack discovered by the seismic imaging. Although there is some discrepancy in the horizontal locations between the imaged crack (Fig. 5) and the tremor sources (Fig. 2), this discrepancy can be attributed to the depth difference between the two maps; the tremor sources and the magma-filled crack are depicted at depths of 1000 and 3000 m below sea level, respectively. The magma-filled crack, as shown in Fig. 6, is inferred to dip up southwestwards. Therefore the position of the magma-filled crack at a depth of 1000 m below the sea level is concordant with the locations of the tremor sources.

Figure 7 schematically illustrates the composite image of the total source of volcanic tremor deduced from the results described above. The magma-filled crack trending along the preferred orientation is the parent body of the two sources of high frequency volcanic tremor and the dynamic linkage between the sources is maintained by the magma in the crack. In this model, we further assume that the tremor sources are loated at the tips of the magma-filled crack.

It may be perceived that the model is related to the fluid-driven crack models of volcanic tremor, which have been studied by many scientists (e.g., Aki et al. 1977; Chouet 1986, 1988). Aki et al. (1977) proposed a model to explain shallow harmonic tremor observed at Kilauea volcano, Hawaii, regarding the tremor as the result of the crack oscillation excited by the jerky opening of channels in responses to excess magma pressure. More recently, the model of the resonance of such a fluid-driven crack induced by an impulsive pressure transient was found to explain many of the observed characteristics associated with volcanic tremor (Chouet 1986, 1988).

Based on their results, the observation of the correlation between volcanic tremor and the crustal deflation (Fukuyama 1988; Watanabe 1988) and the source configuration illustrated in Fig. 7, the source process, in an episode of volcanic tremor, can be interpreted as follows. The increase of magma pressure in the crack causes outflow of magma through the crack tips, the most fragile parts of an extensional crack. The high frequency volcanic tremor analyzed here is generated by such a magma motion, though the detailed mechanism is not known yet. The magma motion, at the crack tips, also resonates the free oscillation of the magma-filled crack and generates longer period volcanic tremor whose frequency (1 Hz) is much lower than the components analyzed here (Chouet 1988). When the pressure of the magma falls, the outflow of magma through the tips and then, volcanic tremor ceases.

3 Summary

We have reported the results of the analyses about the source of volcanic tremor and the magma system of Izu-Oshima volcano, showing the ability of the array observations and the semblance method for seismological studies of a volcano. The obtained source model of volcanic tremor is related to the fluid-driven crack models that have been predicted and analyzed by many volcano-seismologists since the pioneering works by Sassa (1935) and Aki et al. (1977). The salient features of the tremor source of Izu-Oshima volcano in March, 1987 are summarized as follows: the two tremor sources beneath the caldera radiated high frequency components and they were probably located at the magma-filled crack tips; this magma-filled crack develops along the preferred orientation of fissures (the NW-SE direction) expected from the configuration of the ambient tectonic stress. The outflow of magma through crack tips, driven by the increase of magma pressure, resonated the crack oscillation with the frequency of the predominant component (1 Hz).

Acknowledgements. We are indebted to Kiyoshi Suyehiro, laboratory colleagues and the staff of the Regional Center of Earthquake and Volcanoes of Nagoya University for their enthusiastic assistance of the field survey.

References

Aki K (1969) Analysis of the seismic coda of local earthquakes as scattered waves. J Geophys Res 74:615–631

Aki K, Fehler M, Das S (1977) Source mechanism of volcanic tremor: fluid-driven crack models and their application to the 1963 Kilauea eruption. J Volcanol Geotherm Res 2:259–287

Chouet B (1985) Excitation of a buried magma pipe: a seismic source model for volcanic tremor. J Geophys Res 90:1881–1893

Chouet B (1986) Dynamics of a fluid-driven crack in three dimensions by the finite difference mentod. J Geophys Res 91:13 967–13 992

Chouet B (1988) Resonance of a fluid-driven crack: radiation properties and implications for the source of long-period events and harmonic tremor. J Geophys Res 93:4375–4400

Chouet B, Koyanagi RY, Aki K (1987) Origin of volcanic tremor in Hawaii. US Geol Surv Prof Pap 1350:1259–1280

Crosson RS, Bame DA (1985) A spherical source model for low frequency volcanic earthquakes. J Geophys Res 90:10 237–10 247

Ferrazzini V, Chouet B, Fehler M, Aki K (1990) Quantitative analysis of long-period events recorded during hydrofracture experiments at Fenton Hill, New Mexico. J Geophys Res 95:21 871–21 884

Fukuyama E (1988) Saw-teeth-shaped tilt change associated with volcanic tremor at the Izu-Oshima volcano. J Volcanol Soc Jpn 33:S128–S135 (in Japanese)

Furumoto M, Kunitomo T, Inoue H, Yamada I, Takagi Y (1988) Portable seismic array system: its versatile application. J Seismol Soc Jpn 41:419–429 (in Japanese)

Furumoto M, Kunitomo T, Inoue H, Yasmada I, Yamaoka K, Ikami A, Fukao Y (1990) Twin sources of high-frequency volcanic tremors of Izu-Oshima volcano, Japan. Geophys Res Lett 17:25–27

Furumoto M, Inoue H, Yamaoka K, Kunitomo T, Yamada I, Fukao Y, Suehiro K (1992) Holography of Izu-Oshiam volcano. (in prep)

Ida Y and the group of the structure survey of Izu-Oshima (1987) Survey of the velocity structure of Izu-Oshima. Progr Abst Seismol Soc Jpn 1987, 2:151 (in Japanese)

Ikami A, Kunitomo T, Furumoto M, Inoue H, Suyehiro K (1989) Shalow seismic profile of Izu-Oshima Island as derived from air-gun blasts observation. J Seismol Soc Jpn (in Japanese) 42:219–226

Neidel NS, Tarner TM (1971) Semblance and other coherency measure for multichannel data. Geophysics 36:482–497

Nikolaev AV, Troitskiy PA (1978) Lithospheric studies based on array analysis of P-coda and microseisms. Tectonophysics 140:103–113

Nakamura K (1961) Stratigraphic studies of the pyroclastics of Oshima Volcano Izu, deposited during the last fifteen centuries II. Sci Pap Coll Gen Educ Univ Tokyo 11:281–319

Nakamura K (1969) Arrangement of parasitic cones as a possible key to regional stress field. J Volcanol Soc Jpn 1:8–20

Nakamura K (1977) Volcanoes as possible indications of tectonic stress orientation – principle and proposal. J Volcanol Geotherm Res 2:1–16

Watanabe H (1987) Eruption mechanism of Izu-Oshima volcano inferred from the observation of volcanic tremor. The Earth Monthly 9:475–481 (in Japanese)

Watanabe H (1988) Seismic activity. In: Aramaki S (ed) The eruption of Izu-Oshima volcano. Earthq Res Inst Univ Tokyo 1986–1987

Sassa K (1935) Volcanic micro-tremors and eruption-earthquakes. Mem Fac Sci Kyoto Univ 18:255–293

Volcanic Tremor and Low-Frequency Earthquakes in Iceland

BRYNDÍS BRANDSDÓTTIR and PÁLL EINARSSON

Abstract

Most earthquakes in Iceland are of the usual high-frequency type, reflecting brittle failure of the crust. Earthquakes lacking energy in the higher frequencies (low-frequency earthquakes) also occur, particularly in volcanic regions. Low-frequency earthquakes in the volcanic systems of Iceland span a broad spectrum with respect to amplitude/duration ratios. They often accompany magmatic intrusions or extrusions, and in the cases of Krafla and Hekla, they have been found to correlate in time with the opening of surface fissures. In other cases, such as at SW-Mýrdalsjökull and Torfajökull, low-frequency earthquakes are associated with central volcanoes that have not erupted for centuries. Volcanic tremor has been recorded during all recent eruptions in Iceland. In Krafla, at least two types of tremor have been recorded, one associated with intrusions, the other with eruptions. The two types have distinctly different characteristics, and are produced by different physical processes.

1 Introduction

Volcanic eruptions in Iceland since 1966 have been studied in some details with portable seismographs and the local seismic network that began expanding rapidly after 1970. During this time, no fewer than 14 eruptions have occurred, providing the opportunity to study various seismic phenomena associated with volcanism. In addition, several volcanoes have been seismically active, even though no magma has erupted at the surface. In this paper we summarize our present experience and understanding of two of these phenomena, i.e., volcanic tremor and the so-called low-frequency earthquakes.

In Iceland, the Mid-Atlantic Ridge is superimposed on a hotspot which is responsible for the excess volcanic productivity. This combination of hotspot activity and rifting gives rise to diverse types of volcanism. Volcanic activity in Iceland is concentrated around central volcanoes or volcanic systems, 25 of which have been active in Postglacial time (Fig. 1). A volcanic system consists of a central volcano, where the productivity is highest, and a fissure swarm, where rifting structures such as volcanic fissures, normal faults and open fissures are concentrated. A central volcano may have a caldera, silicic volcanism, and a geothermal system associated with it. Fissure swarms are particularly well developed in volcanic systems along the divergent plate boundary.

The epicentral map of Iceland is characterized by earthquake clusters (Fig. 1). Some of the clusters are the result of persistent activity, and others are single

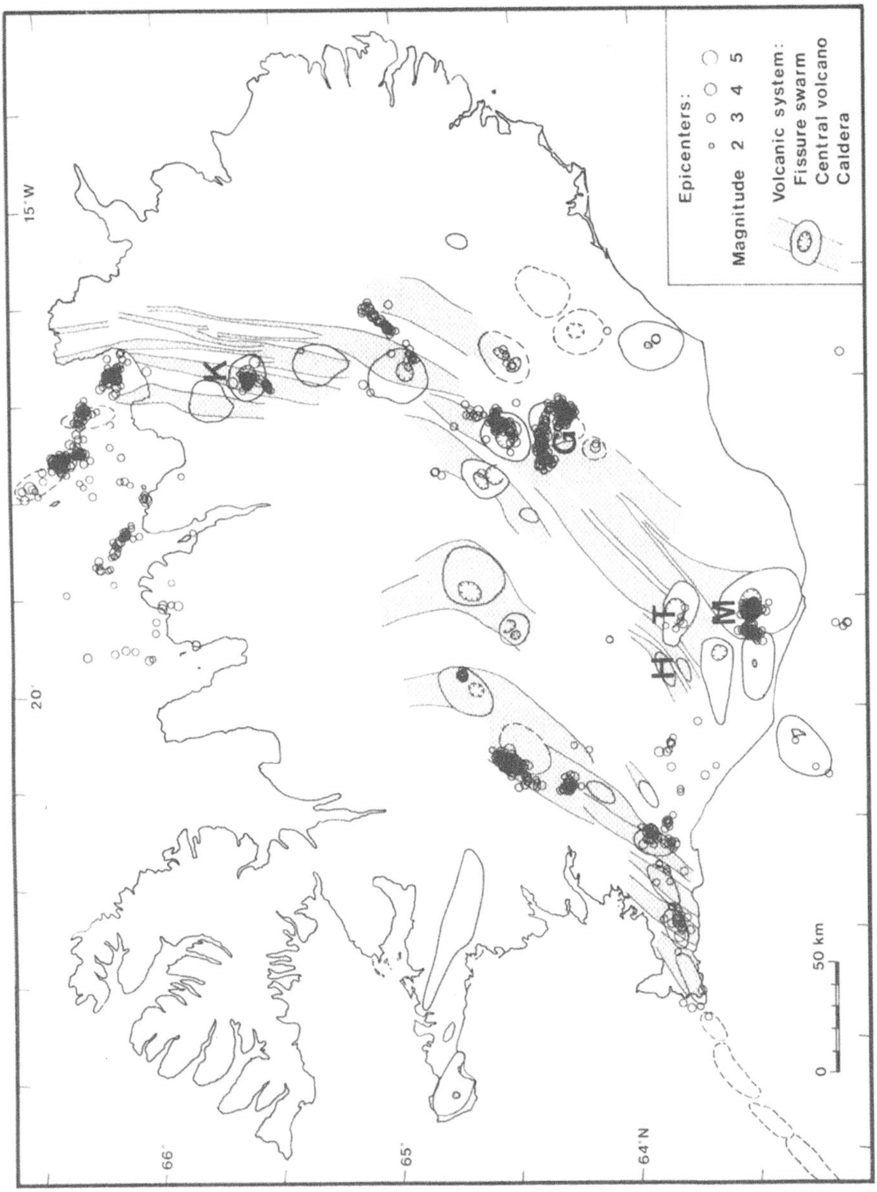

Fig. 1. Earthquake epicenters in Iceland, with M > 2, 1982–1985, and the main volcanic systems of the volcanic zones (after Einarsson and Sæmundsson 1987). Central volcanoes are indicated by a closed curve around the area of maximum volcanic production. Of the structural elements within the central volcanoes only calderas are shown and transecting fissure swarms, which are *hatched*. Krafla (*K*), Grímsvötn (*G*), Hekla (*H*), Mýrdalsjökull (*M*) and Torfajökull (*T*)

swarms. Many of these clusters are confined to central volcanoes, and each central volcano exhibits its own seismic characteristics. Fissure swarms of the rift zones are mostly aseismic, except during episodes of rifting and magmatism, like the events presently occurring in Krafla, NE-Iceland. Most earthquakes in Iceland are of the usual high-frequency type, reflecting brittle failure of the crust. Earthquakes lacking energy in the higher frequencies (low-frequency earthquakes) also occur, particularly in volcanic regions. The low-frequency earthquakes do not necessarily always reflect the same source process. In some cases, they seem to be associated with fissures opening at the surface, in other cases the low frequency characteristics may be the result of attenuation. These earthquakes have a quite variable appearance on the seismograms, some resemble Minakami's b-type earthquakes (Minakami 1974), others look like the type m and type l earthquakes at Mount St. Helens. In the case of Krafla intrusions the whole range of low- to high-frequency events seems to be present. Similarly, at least two types of volcanic tremor have been identified, reflecting different source processes.

2 Krafla

The structure of the volcanic rift zone in NE-Iceland is dominated by NNE-trending fissure swarms that are arranged en echelon within the N-S trending zone. The Krafla central volcano (K in Fig. 1) has a caldera about 10 km in diameter and is transected by a fissure swarm 100 km in length. The Krafla volcanic system became active in 1975 after a quiet interval of more than two centuries. The present activity has involved inflation and deflation of a shallow magma reservoir beneath the caldera, lateral intrusions into the fissure swarm, fissure eruptions, rifting and transform faulting, (Björnsson et al. 1977; Björnsson 1985; Brandsdóttir and Einarsson 1979; Einarsson 1978; Einarsson and Brandsdóttir 1980; Tryggvason 1980, 1984). Since 1975, a total of twenty inflation-deflation cycles have taken place in the Krafla area and nine deflations have resulted in volcanic eruptions.

Shear wave attenuation of local earthquakes has been used to delineate a shallow crustal magma reservoir at a depth of 3–7 km near the center of inflation in the Krafla caldera (Einarsson 1978). This attenuation zone, within the Krafla caldera, is localized, as shear waves generally propagate with little anomalous attenuation through the crust in the volcanic zone. During inflation periods, that may last from several weeks up to several years, magma steadily accumulates in the magma reservoir. Increasing magma pressure leads to strains and seismic activity in the chamber roof. The inflation is interrupted by short periods of rapid deflation when magma is erupted or laterally intruded into the transecting fissure swarm. Inflation seismicity stops abruptly when the magma reservoir deflates and stress in the chamber roof is relaxed. A dike propagates away from the chamber along a plane that is perpendicular to the least compressive stress. In the extensional regime of the rift zone, the least compressive stress is horizontal and oriented E-W. It was considerably lower than the overburden pressure at the beginning of the rifting episode. The magma pressure required to open up a N-S striking dike was, therefore, lower than the pressure necessary to bring magma to the surface. Therefore, during the first four years of the Krafla activity most of the magma was intruded laterally into the fissure swarm. Lateral intrusions to distances of 30–60 km have been documented in association with the Krafla rifting episode (Brandsdóttir and Einarsson 1979; Einarsson and Brandsdóttir 1980). Extensive surface rifting, including fissuring and normal faulting, is observed in the fissure swarm where the magma is emplaced. Each new dike, however, increases the least compressive stress, which eventually approaches the overburden pressure. Then the pressure required to open up a dike is also sufficient to sustain an eruption. This transition in the Krafla events, from primarily intrusive to eruptive activity, occurred in 1980. Since then, five substantial lava eruptions have taken place in the caldera and the fissure swarm to the north of it.

The intrusion and eruptive events are accompanied by characteristic seismic patterns. Intrusions are generally accompanied by intrusion tremor mixed with swarms of small earthquakes. Most of these earthquakes occur near the front of the migrating dike tip, for which propagation speeds of 0.5–1.2 m/s have been determined (Brandsdóttir and Einarsson 1979; Einarsson and Brandsdóttir 1980). The majority of earthquakes associated with the dike intrusion are of the high-frequency type. But low-frequency events also occur and seem to be correlated in time with the opening of surface fissures, eruptive or noneruptive. The low-frequency events range in character between the type m and type l Mount St. Helens events described by Malone (1983). Low- and high-frequency events occur simultaneously in the same area, so the low-frequency characteristics are not the result of attenuation. They must reflect the source process in one way or another.

At least two types of continuous tremor are also associated with the deflation events at Krafla. The intrusion process is accompanied by continuous, spasmodic tremor, the amplitude of which correlates with the rate of deflation. The intrusion tremor appears to have a relatively broad spectrum with predominant frequencies above 3 Hz. Intrusion tremor and earthquakes usually stop when the magma reaches the surface and the dike stops propagating. The resulting eruption is accompanied by eruption tremor which is distinctly different from the intrusion tremor. It has lower predominant frequencies (<3 Hz) and a narrow spectrum. Its amplitude is sustained and is related to the vigor of the eruption. The eruption tremor is most likely produced by processes at the eruption site, such as turbulent flow, lava fountaining or gas release.

A typical seismic pattern during a deflation event was recorded at the Reynihlíð seismograph station on January 30, 1981 (Fig. 2). Reynihlíð is located 10 km SSW of the center of inflation. Tiltmeter located at the Krafla power plant, 2 km SE of the center of inflation, indicated a rapid deflation starting around 7:00 on January 30, 1981 (Einarsson et al. 1981). At that time all disturbances seen on the seismogram were caused by an occasional car passing by (location A in Fig. 2). High-frequency intrusion tremor appears on the seismogram at about 7:40 (location B in Fig. 2). The gain is lowered in order to observe the time signals (location C in Fig. 2). During the next 2 hours the

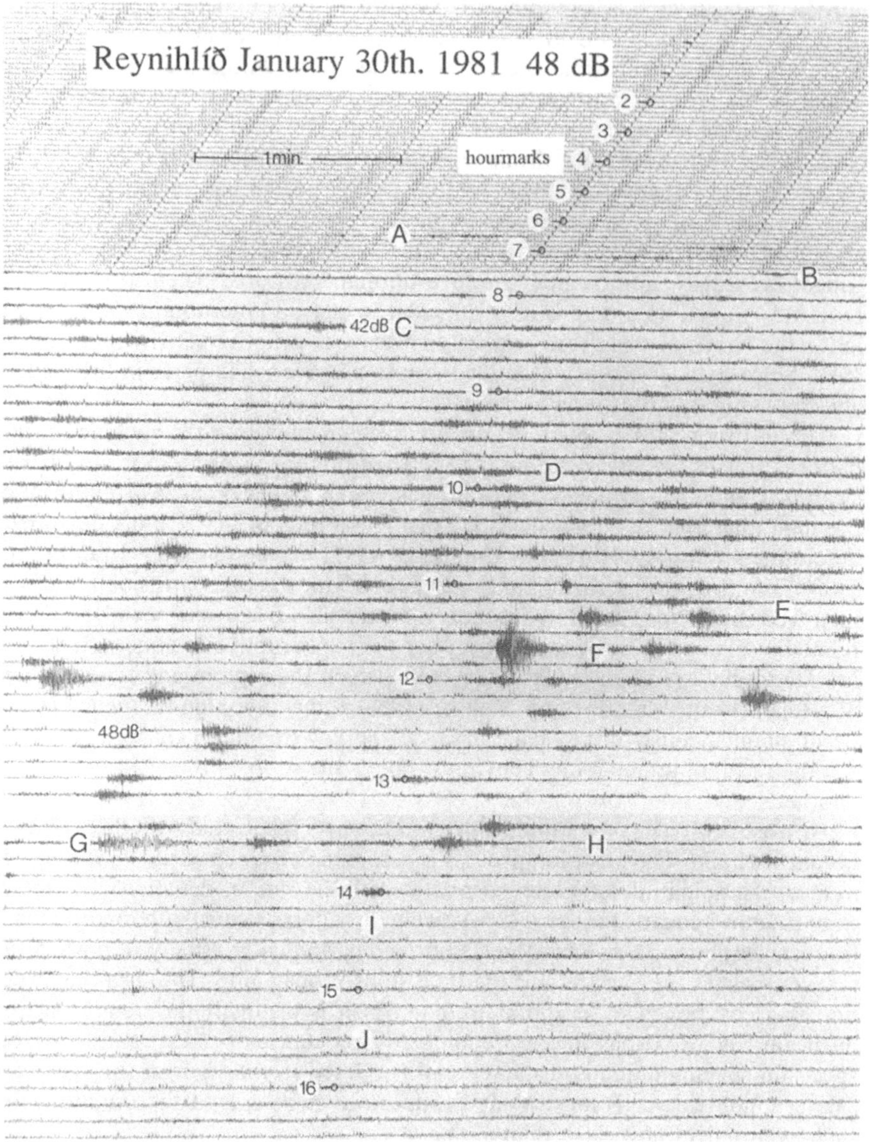

Fig. 2. Seismic record from the Reynihlíð station showing the beginning of the Krafla eruption on January 30, 1981

tremor amplitude steadily increases and culminates around 9:50 (location D in Fig. 2). Superimposed on the intrusion tremor are many small high-frequency earthquakes. The earthquake activity increases as the tremor amplitude decreases (locations E and F in Fig. 2). Epicenters are in the fissure swarm north of the center of deflation. The first low-frequency earthquake was recorded at 13:13, followed by another at 13:29 (location G in Fig. 2). The low-frequency earthquakes are characterized by less power in the higher frequencies and longer coda duration making their amplitude/duration ratio considerably lower than for the high-frequency events. A marked drop in the high-frequency seismicity is observed at about 13:40 (location H in Fig. 2) and very few earthquakes were recorded after 14:03. At about that time eruption tremor was first observed on a seismograph just west of the caldera. The eruption tremor is characterized by a lower frequency content and a more even amplitude than the intrusion tremor. The eruption tremor gained amplitude and became visible on the Reynihlíð station at about 14:20 (location I in Fig. 2). At 14:10 a fissure eruption started 6–8 km north of the center of deflation. The eruption lasted five days and a maximum deflation of 32 cm was observed in the caldera region (Einarsson et al. 1981; Grönvold and Sæmundsson 1981).

3 Grímsvötn

On May 28, 1983 an eruption started in Grímsvötn (G on Fig. 1). The eruption lasted until June 2 and it is the first confirmed eruption in Grímsvötn, after the large increase in seismograph coverage in Iceland that began in the early seventies. The Grímsvötn volcano, with its 6–8 km wide caldera, is situated beneath the Vatnajökull ice sheet, and is one of a group of central volcanoes that characterizes the structure of this part of the volcanic rift zone. The eruption was preceded by a significant increase in seismic activity for about three months and an intense earthquake swarm on the day of the presumed outbreak. The swarm on May 28 was most intense between 7:00 and 10:00, and then gradually faded out. The last earthquake occurred at 11:47 and after that no locatable earthquake occurred in Grímsvötn for more than 3 months, (Einarsson and Brandsdóttir 1983). The earthquakes on May 28 exhibited great variations with respect to maximum amplitude and coda length. All had low-frequency characteristics with respect to coda duration but clear P arrivals and some had clear S arrivals as well. Shortly after the earthquakes ceased on May 28, continuous tremor became visible on the nearest seismographs. The tremor was first seen at approximately 12 h on the Kvísker station at a distance of 65 km from Grímsvötn. The amplitude was relatively small and constant during the first three hours. It then became variable, as if bursts of tremor, were superimposed on the background tremor. The bursts were of different duration, from less than one up to 9 minutes. The interval between the bursts was also quite variable during the first 12 hours, but became more regular later. The tremor activity was highest in the period from 15:00 on May 28 to 17:00 on May 29. After that the activity subsided rapidly, but small bursts could be seen until 01:38 on June 2. Spectral characteristics of the tremor are difficult to discern in the analog records but it seems as if the background tremor has a more monochromatic character than the tremor bursts. The predominant frequency is slightly lower than 3 Hz, except during two short periods on May 28 and 29, when the frequency appeared to be slightly higher than 3 Hz. The predominant frequencies of the tremor bursts were between 2 and 5 Hz (Einarsson and Brandsdóttir 1983). The long-term increase in seismicity, prior to the eruption, bears close resemblance to increased activity observed in Krafla before some of the eruptions and rifting events, both with respect to time scale, seismicity rate, and maximum magnitude. By analogy to Krafla, we suggest that the earthquakes during the last few months, before the Grímsvötn eruption, were associated with an inflating magma chamber. This interpretation is strengthened by the fact that the earthquakes ceased when the eruption began.

4 Hekla

Whereas some volcanoes exhibit a prolonged seismic activity prior to eruptions, others erupt without pronounced precursory activity. Hekla, one of the most active volcanoes in Iceland, erupted without a warning in May 1970, August 1980, April 1981 and January 1991.

In 1970, three permanent seismic stations were operated in Iceland, at distances of 200, 110, and 80 km from Hekla. No precursory earthquake activity was observed in the last months prior to the May 1970 eruption. During the last half hour before the eruption continuous tremor was recorded on all three stations. Superimposed on the tremor were many earthquakes, with magnitudes between 3.6 and 4 (Einarsson and Björnsson 1977).

Few small earthquakes were recorded at several seismic stations in south Iceland prior to the August 17, 1980 eruption. The first earthquake, of magnitude about 1.5, was recorded at 13:04 followed by another earthquake of similar magnitude at 13:08. At about 13:10 continuous shaking resembling a series of small high-frequency quakes was recorded, but no definite volcanic tremor was recorded until 13:27, at about the same time as the beginning of the eruption (Grönvold et al. 1983). The volcanic tremor reached maximum amplitude during the next hour but started to decrease again at about 16 h (Fig. 3). The tremor amplitude followed the tephra production rate. Earthquake activity decreased after the initial outbreak of the eruption, but between 18 and 23 h it increased again. The largest earthquake, magnitude 3, occurred at 20:19 (Fig. 3). It has clear P and S phases but a low-frequency character with respect to coda duration. The earthquake coincided with the extension of the eruption fissure down the SW slope of the mountain. When the eruption fissure had reached its maximum length, the earthquake activity decreased and only minor seismic events were detected after the first day. Volcanic tremor continued, however, and its amplitude appeared to follow the vigor of the eruption. This eruption lasted four days.

Volcanic tremor appeared on a seismometer 22 km west of Hekla shortly after 22 h on April 8, 1981. The tremor had similar characteristics as the tremor associated with the August 1980 eruption. The tremor amplitude increased

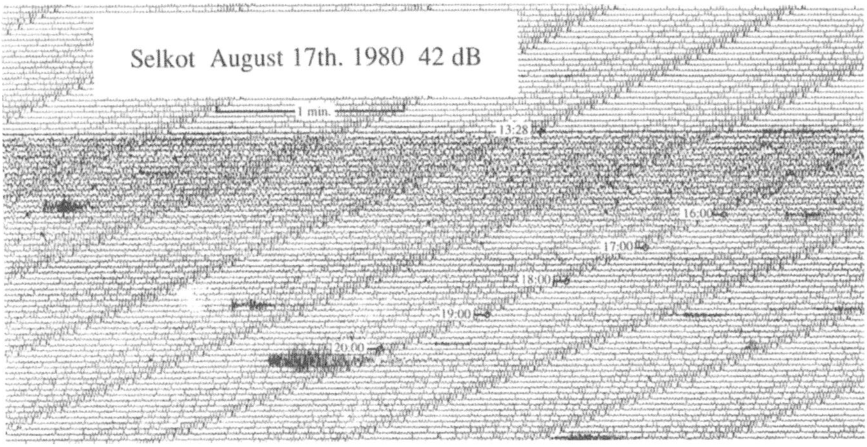

Fig. 3. Seismic record from the station Selkot, at a distance of about 50 km south of Hekla, showing the beginning of the Hekla eruption on August 17, 1980

rapidly between 02 and 05 h in the morning of April 9, and during this time the first tephra fall was noted. Most likely the lava eruption began simultanously. The tremor started to decrease in the morning of April 10 and finally disappeared below the level of detection on April 16 with the last signs of the eruption. Few earthquakes accompanied this eruption and all these earthquakes were smaller than magnitude 2.

An eruption started in Hekla at 17 h on January 17, 1991. The eruption was preceded by a small earthquake swarm that began at 16:32. The earthquake intensity increased at 14:42 but none had magnitude exceeding 3 on the Richter scale. A volcanic tremor, with dominant frequencies between 1.5 and 3 Hz, started around 17 h at the same time as the eruption. The eruption column was observed from nearby farms at 17:05 and 17:07. The tremor amplitude declined exponentially between 18 h, January 17, and 4 h on January 18, when it levelled off. The tremor slowly declined throughout the eruption, excluding brief intervals of increased intensity. Intermittent observations indicate that the tremor amplitude is directly related to the vigour of the eruption. A small array, consisting of five three-component instruments was installed 200–1600 m from the active crater. Spectral analyses of the volcanic tremor indicate that it is made up of a narrow-band random noise with a peak amplitude around 2.5 Hz. Individual tremor phases propagate with extremely low velocities, ranging from 500–1000 m/s. These phases seem to be made up of both Rayleigh and Love waves. The slow phase velocities are attributed to the tephra layers and highly cracked lava flows that make up the volcano. The tremor disappeared below the level of detection around 5 h on March 11 and was followed by a small earthquake swarm, only observed at a seismometer located on the mountain. The eruption was last sighted around noon March 11.

The low seismicity preceding and accompanying the 1970, 1980, 1981 and 1991 eruptions of Hekla may indicate that the eruptions were fed by storage chambers in the lower crust. Pressure changes at this depth cause crustal strains

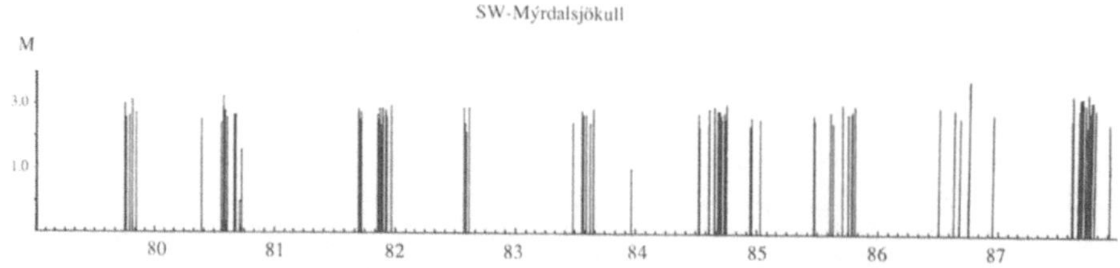

Fig. 4. Time sequence of 113 earthquakes with epicenters in SW-Mýrdalsjökull. Magnitude is plotted as a function of time, data from 1979–1987

over a wide area above the magma storage volume but the stresses may not reach values sufficient to cause earthquakes. This is consistent with the results of deformation measurements made in 1981, which give a source of pressure variations at a depth of 7–8 km beneath Hekla (Kjartansson and Grönvold 1983).

5 Mýrdalsjökull

Two subglacial volcanic centers in Mýrdalsjökull exhibit persistent seismic activity (M in Fig. 1) even though no eruptive activity has taken place in this area for decades. A pronounced annual cycle is observed in the frequency of earthquakes in the western system where nearly all earthquakes are low-frequency events. The earthquakes are of a similar size, seldom exceeding magnitude 3, and occur in the latter half of the year (Fig. 4). This volcanic system has probably not erupted for at least six centuries, and the seismicity has been interpreted as the result of failure, within the magma chamber roof, in response to long term pressure fluctuations in the magma. The seasonal cycle, however, is interpreted as the result of a superimposed triggering effect of the glacier. We suggest that meltwater from the glacier leads to elevated pore pressure in the crust beneath it in the latter half of the year. Thus, the frictional strength of faults is reduced, leading to increased earthquake activity.

6 Torfajökull

Swarms of small, low-frequency earthquakes originate near the Torfajökull central volcano (T in Fig. 1). This activity was first observed in late 1985 shortly after a seismograph was installed near the area, and was most intense in 1986 (Fig. 5) when four swarms occurred. The first swarm began in early March and lasted into April 1986 with a daily number of earthquakes around five. The next swarm began in early May with more than 20 earthquakes per day for a couple of days. Unfortunately, no data is available from the next 2 months, due to interference from a newly installed portable telephone system. The daily num-

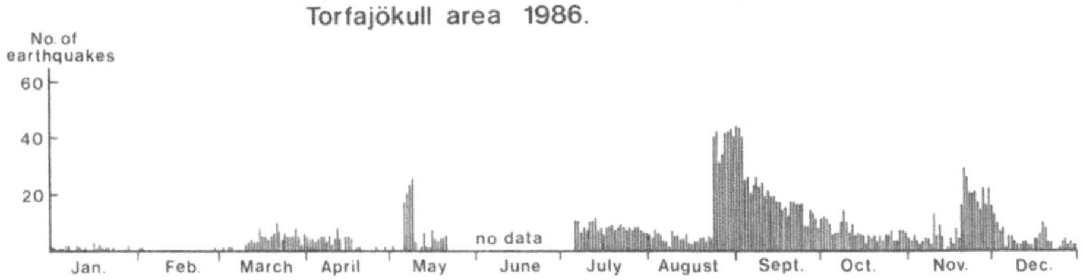

Fig. 5. Daily number of low-frequency earthquakes in the Torfajökull area during 1986

ber of earthquakes was fairly constant during July with five to twelve quakes per day, but jumped suddenly on August 24 to more than 40 quakes per day. No surface fault movements were observed in the Torfajökull region during this swarm. The fourth swarm took place in late November. The August–September swarm is the biggest low-frequency earthquake swarm in the area so far. All these earthquakes were too small to be located. This seismicity is thought to reflect minor magmatic movements in the roots of the volcano, since no eruptive activity has occurred in the Torfajökull area for a few centuries.

References

Björnsson A (1985) Dynamics of crustal rifting in NE Iceland. J Geophys Res 90:10151–10162

Björnsson A, Sæmundsson K, Einarsson P, Tryggvason G, Grönvold K (1977) Current rifting episode in North Iceland. Nature (Lond) 266:318–323

Brandsdóttir B, Einarsson P (1979) Seismic activity associated with the September 1977 deflation of the Krafla central volcano in NE Iceland. J Volcanol Geotherm Res 6:197–212

Einarsson P (1978) S-wave shadows in the Krafla caldera in NE-Iceland, evidence for a magma chamber in the crust. Bull Volcanol 41:1–9

Einarsson P, Björnsson S (1977) Seismic activity associated with the 1970 eruption of Volcano Hekla in Iceland. Jökull 26:8–19

Einarsson P, Brandsdóttir B (1980) Seismological evidence for a lateral magma intrusion during the July 1978 deflation of the Krafla volcano in NE-Iceland. J Geophys 47:160–165

Einarsson P, Brandsdóttir B (1983) Seismic activity preceding and during the 1983 volcanic eruption in Grímsvötn, Iceland. Jökull 34:13–23

Einarsson P, Sæmundsson K (1987) Earthquake epicenters 1982–1985 and volcanic systems in Iceland. In: Sigfússon P (ed) Map accompanying the festschrift "Í Hlutarins Eðli". Menningarsjóður, Reykjavík, p 433

Einarsson P, Grönvold K, Brandsdóttir B (1981) Eldgosið í Gjástykki í janúar og aðdragandi pess (The eruption in Gjástykki in January). Skjálftabréf, Earthq Lett 46:4–10 (in Icelandic)

Grönvold K, Sæmundsson K (1981) Hraungos í Kröflueldum (Lavaflows during the Krafla eruptions). Skjálftabréf (Earthq Lett) 47:6–14 (in Icelandic)

Grönvold K, Larsen G, Einarsson P, Thorarinsson S, Sæmundsson K (1983) The Hekla eruption 1980–1981. Bull Volcanol 46:349–363

Kjartansson E, Grönvold G (1983) Location of a magma reservoir beneath Hekla Volcano, Iceland. Nature (Lond) 301:139–141

Malone SD (1983) Volcanic earthquakes: Examples from Mount St. Helens. In: Kanamori H, Boschi E (eds) Earthquakes: observation, theory and interpretation. LXXXV Corso, Soc Ital Fisica, Bologna, Italy pp 436–455

Minakami T (1974) Seismology of volcanoes in Japan. In: Civetta L, Gasparini P, Luongo G, Rapolla A (eds) Physical volcanology. Dev Solid Earth Geophys 6:1–27, Elsevier, Amsterdam

Tryggvason E (1980) Subsidence events in the Krafla area, North Iceland, 1975–1979. J Geophys 47:141–153

Tryggvason E (1984) Widening of the Krafla fissure swarm during the 1975–1981 volcano-tectonic episode. Bull Volcanol 47:47–69

Non-Double Couple Earthquake Focal Mechanisms and the Accretionary Tectonic Cycle

Gillian R. Foulger and Roger E. Long

Abstract

This chapter discusses the results of recent investigations of small magnitude earthquake activity in geothermal areas in Iceland. Instrument deployment was sufficient to constrain source mechanisms and mixed suites of double couple and non-double couple events were reported. The focal mechanisms of these continuous events allow us to study the cyclic pattern of the stress field with time, close to the accretionary plate boundary. During periods of tectonic quiescence, plate movements build up an extensional stress field close to the boundary and extensional failure occurs. Immediately after rifting, the extensional component is removed. The resulting stress field is roughly triaxially symmetric and variable on a small scale in both magnitude and orientation.

1 Introduction

Ridge-centred volcanism, in excess of that required to facilitate sea floor spreading, has resulted in the building up of the Icelandic basalt dome in the north Atlantic. Thus, 100,000 km^2 are exposed above sea level, along with 600 km of spreading plate boundary. The boundary is characterized by en-echelon volcanic systems, 30 to 120 km long, the majority of which display a swarm of open fissures, a central volcano which may or may not have a caldera, and a high temperature geothermal area (Saemundsson 1978).

The seismicity of the spreading plate boundary in Iceland has been monitored for approximately 15 years by a regional seismograph network with a detection threshold of approximately $M_t = 2.0$ (M_t = duration magnitude, normalized to M_s). The majority of the activity may be divided into 3 groups: firstly, intense seismic activity accompanies volcanism and tectonism associated with crustal accretion processes. Such activity occurred recently in the Krafla volcanic system in response to inflation of the magma chamber and dyke emplacement in the fissure swarm (Brandsdottir and Einarsson 1979; Einarsson and Brandsdottir 1980), and similar tectonic episodes are historically documented (Thoroddsen 1899; Bjornsson et al. 1979). On average, this kind of activity probably occurs somewhere in Iceland once every few decades, when rifting occurs in one of the volcanic systems. Earthquake magnitudes may not exceed $M_t = 4.5$.

Secondly, volcanic earthquakes occur in the central volcanoes (Tryggvason 1973; Einarsson and Bjornsson 1979). In some cases they occur accompanying eruptive activity, but in others they occur fairly regularly, without eruptive activity, and may even be seasonal, e.g., Katla. Magnitudes may exceed $M_t = 5$, e.g., Grimsvotn. Such activity is related to subsurface magmatism. Thirdly, continuous, small magnitude earthquake activity is associated with the geothermal areas (Foulger 1982, 1988a). This activity represents the release of thermal stress, and results from geothermal heat loss. Examples where such activity is observed are the Hengill, Krafla, and Krisuvik geothermal areas (Ward and Bjornsson 1971; Einarsson and Bjornsson 1979). Such activity rarely exceeds $M_t = 4$.

This chapter summarizes and discusses three recent investigations where detailed observations were made of the third type of seismicity, i.e., the small magnitude geothermal activity. In all these cases, instrument deployments were sufficient to constrain the source mechanisms, and mixed suites of double couple and non-double couple events were reported (Klein et al. 1977; Foulger and Long 1984; Foulger 1988b; Foulger et al. 1989). The double couple events represent shear failure, whereas the non-double couple events are interpreted by some of the authors, and in this paper, as representing volumetric failure and resulting from the opening or closure of cracks or cavities. The mode of volumetric failure, and the preferred orientation of the shear fractures, are governed by the stress field at the accretionary plate boundary, which varies according to the stage of the tectonic cycle of the particular volcanic system under investigation.

The study of earthquake focal mechanisms, and in particular the non-double couple events, may give insight into the current stage in the tectonic cycle of other Icelandic volcanic systems, by analogy with those systems already studied. Perhaps these principles may also be applied to other extensional environments, e.g., the submarine accretionary plate boundary and certain volcanic areas.

2 Non-Double Couple Earthquakes at the Accretionary Plate Boundary in Iceland

Three seismological surveys on the accretionary plate boundary in Iceland have been reported recently, where sufficiently dense instrument networks have been deployed, to study the focal mechanisms of the continuous small magnitude earthquake activity. All report mixed suites of double couple and non-double couple events.

2.1 Reykjanes

The submarine accretionary plate boundary comes onshore at Reykjanes, and traverses the peninsula with an orientation of N 70° E (Fig. 1). The trace of the plate boundary crosses a series of 3 en-echelon volcanic systems (Einarsson and Saemundsson 1987) and high temperature geothermal areas occur at the intersections of the plate boundary with these systems (Klein et al. 1977). In 1972, the seismicity of a 10 × 15 km area on the tip of the peninsula was investigated

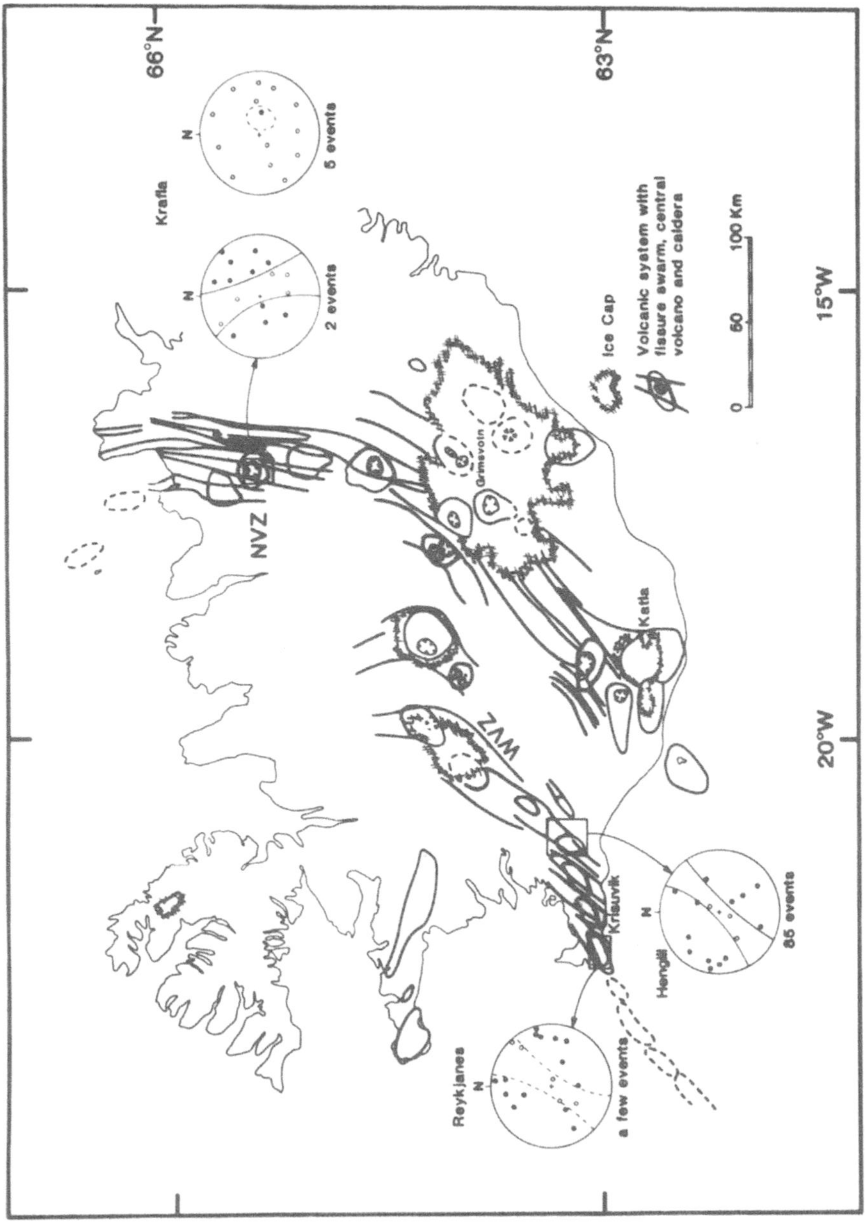

Fig. 1. Major volcanic systems of the neovolcanic zones of Iceland. WVZ, Western Volcanic Zone; NVZ, Northern Volcanic Zone. The sites of three recent local earthquake monitoring projects are shown as *boxes*. Alongside, typical upper hemisphere stereographic projections of p-wave polarities for the anomalous events are shown. *Dots* represent compressional arrivals and *open circles* represent dilatational arrivals. In the case of the Reykjanes event, Klein et al. (1977) suggested that the nodal lines were great circles crossing at an angle of 55°, representing a reduced dilatational "quadrant" of the kind appropriate to double couple events (where the nodal lines would cross at an angle of 90°). The *dotted small circular nodal lines* indicated for this event represent an alternative interpretation in terms of the tensile crack model of Foulger and Long (1984). Compiled from Klein et al. (1977), Foulger and Long (1984), Einarsson and Saemundsson (1987) and Foulger et al. (1989)

(Klein et al. 1973; Klein et al. 1977). The experimental target was the plate boundary where it crosses the most westerly, – the Reykjanes system (Fig. 1).

A 23-station network of short period geophones was deployed, data were recorded analogue on magnetic tape, and the network was operated for 5 weeks. During the deployment period a tectonic swarm of over 17,000 events within the network was recorded. Of these 2500 were located in the depth range 2–5 km, and focal mechanisms for 433 events were obtained by studying the distributions of p-wave polarities. For the majority of the events, double couple solutions were easily fitted to the data, and these showed that shear faulting ranged from right lateral strike slip on NS or EW orientated fault planes, to normal faulting on NE orientated planes (Fig. 2).

The orientations of the tension (t), null and pressure (p) axes of the double couple earthquakes do not exactly coincide with the orientations of σ_1, σ_2 and σ_3, the greatest, intermediate and least principal stresses, but they nonetheless

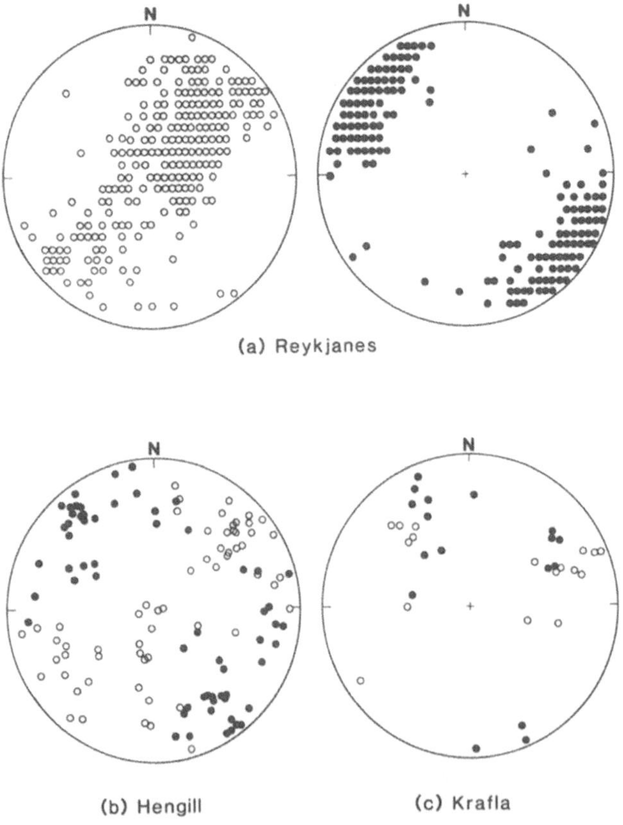

Fig. 2a–c. Stereographic projections of the p- and t-axes of double couple events. *Open circles*: t-axes, *dots*: p-axes. The t- and p-axes do not coincide exactly with the orientations of σ_1 and σ_3, the greatest and least principal stresses, but they nonetheless give an indication. **a** Reykjanes earthquakes; **b** Hengill earthquakes, **c** Krafla earthquakes

give an approximate indication (McKenzie 1969). The orientations of the t- and p-axes of the Reykjanes double couple events are shown in Fig. 2a. Thus, it may be deduced that σ_3 was orientated roughly NW, horizontal and perpendicular to the fissure swarm, and σ_1 and σ_2 varied from vertical to NE horizontal and parallel to the fissure swarm.

In addition, several events displayed radiation patterns with dominant compressional components that clearly did not permit double couple solutions (Fig. 1). It was suggested that these anomalous events were caused by either small scale crustal velocity heterogeneities or by explosive seismic sources with shear components, caused by rapid dyke injection and cracking (Klein et al. 1977).

2.2 Hengill

The Hengill volcanic system marks the eastern boundary of the Reykjanes peninsula zone, and the southernmost part of the Western Volcanic Zone that extends for 100 km in a N 30° E direction, into the interior of Iceland (Fig. 1). As a result of recent volcanic migration, the Hengill area contains a large, double volcanic system, containing two central volcanoes, one active and one extinct (Foulger 1988a). An extensive geothermal area extends over both systems, and displays a high level of small magnitude seismicity (approximately 1 $M_t = 0$ event per day) with a constant spatial distribution. During a 3-month period in 1981, a 20 × 20 km, 23-station network of 1 Hz seismometers was deployed that encompassed the geothermal area (Foulger 1988a,b). Recording was analogue on magnetic tape. Accurate locations were calculated for 2000 events inside the network, in the depth range 1–8 km, and well constrained focal mechanisms were obtained for 178 events using 1st p-wave polarities. Of these, 93 fitted a double couple interpretation. The range of mechanisms was similar to that observed for the events from the Reykjanes area, which indicates a similar orientation of the axes of principal stress for both volcanic systems (Fig. 2).

The remaining 85 were found to be non-double couple in type, and exhibited predominantly compressive arrivals (Fig. 1). They were mixed together with the double couple events, which demonstrated that their anomalous p-wave radiation patterns were not artifacts of some systematic error, e.g., in the crustal structure used by the location program. The anomalous events were interpreted as resulting from tensile crack formation induced in the heat sources of the geothermal area by cooling and contraction of the rock body (Foulger and Long 1984; Foulger 1988b). The tensile crack type events occurred only in proximity to the accretionary plate boundary, and showed a systematic crack orientation parallel to it. This indicated an orientation of σ_3 of horizontal and perpendicular to the plate boundary, in agreement with the results from the double couple events. Away from the boundary earthquake mechanisms were exclusively double couple.

It was concluded that the continuous small magnitude activity occurred in response to thermal cracking within the heat source of the geothermal area, as a result of continual surface heat loss, and thereby represented the release of thermal stress. The *mode* of this stress release, however, and therefore the focal mechanisms, were governed by the regional stress field. This was of constant orientation over the whole area, but permitted volumetric extensional fracture close to the accretionary plate boundary but shear fracture only away from it (Foulger 1988b).

2.3 Krafla

The Krafla volcanic system constitutes one of the five en-echelon systems of the Northern Volcanic Zone. The 150-km-long fissure swarm contains the Krafla central volcano which has a 10×7 km caldera, elongated in the spreading direction. Two discrete geothermal areas exist, one within the caldera, and one 9 km further south. This volcanic system recently experienced intense tectonic activity where a shallow magma chamber underlying the caldera repeatedly inflated and deflated, injecting magma laterally along the fissure swarm both to the north and to the south, forming subsurface dyke intrusions and surface lava flows. This activity resulted in a total of 8 m of crustal spreading over a 10 year period (Bjornsson et al. 1977, 1979; Brandsdottir and Einarsson 1979; Einarsson and Brandsdottir 1980; Johnsen et al. 1980; Bjornsson 1985).

In 1985, during a period of temporary quiescence, a 26 station network of 1 Hz seismometers was installed in a 15×15 km area and operated for 3 months (Foulger et al. 1989). The experimental target was the continuous small magnitude earthquake activity associated with the two geothermal areas. 500 locatable earthquakes were recorded and the seismic rate was 1 $M_t = -0.4$ event per day. Approximately 100 events were located in a preliminary analysis. The majority occurred at depths of 1–3 km directly beneath the two geothermal areas, and in the fissure swarm along the trace of recent dyke injections. The focal mechanisms of 25 events were well constrained by p-wave polarities. Of these 17 agreed with a double couple interpretation, but with variable orientations, that indicated that σ_1 and σ_3 were horizontal and their directions varied from perpendicular to parallel to the fissure swarm (Fig. 2). In addition, three events showed predominately compressional arrivals, and were interpreted as tensile cracks. Five events showed predominantly dilatational arrivals and were interpreted as the collapse of cavities or closure of cracks.

It was hypothesized that the earthquakes were induced by a similar process to that occurring in the Hengill area, i.e., thermal cracking in the heat sources of the geothermal areas and in the fresh dyke intrusions in the fissure swarm. Again, the mode of fracture was governed by the regional stress field. This was highly heterogeneous, permitting shear, extensional and compressional events in close juxtaposition. It was postulated that this type of stress field existed in the Krafla volcanic system because the recent dyke intrusions and crustal spreading had modified the previously uniform extensional stress field along this part of the plate boundary.

3 Genesis of Non-Double Couple Earthquakes

Non-double couple focal mechanisms have been reported for small earthquakes from 3 locations along the accretionary plate boundary in Iceland. In all cases events are reported that have a majority of compressional arrivals, but a few dilatational arrivals also occur which fall within a narrow band across the focal sphere. In the Krafla area, events with predominantly dilatational p-wave radiation patterns also occur, which are more numerous than those with a majority of compressional arrivals. In all cases, the non-double couple events are mixed spatially with double couple events, which rules out the possibility that the unusual radiation patterns observed are caused by distortion of the ray paths by lateral variations in crustal structure.

The events that display a predominantly compressional radiation pattern may be interpreted on the basis of a tensile crack model, the radiation pattern of which is omni-compressional (Aki and Richards 1980). The few dilatational arrivals therefore require further explanation, and these may be generated by a sudden pore pressure drop at the source (e.g., Foulger and Long 1984; Foulger 1988b), or a shear component in the mechanism (e.g., Shimizu et al. 1987). In an analogous way, the predominantly dilatational events observed in the Krafla area may be explained by crack or cavity closure with an accompanying pore pressure increase or a shear component.

We suggest here that in all these three cases, small magnitude earthquakes, both double couple and non-double couple, were recorded that were induced by the reduction of the confining stresses by cooling and contraction of the rock mass in the way hypothesized by Foulger and Long (1984) and Foulger (1988b). The depth distributions of the events indicates the depths of the cooling intrusions. In the case of the Reykjanes survey, the majority of the activity was a tectonic swarm (Klein et al. 1973, 1977), and the seismic rate of the geothermal activity in that study area is relatively low (Ward and Bjornsson 1971), such that probably only a few of the events recorded during the 5 weeks were thermal events.

The mode of failure, and therefore the focal mechanisms, are governed by the regional stress regime. The orientations of the axes of principal stress may be deduced from the orientations of the shear and tensile crack events. In addition, the nature of the non-double couple events indicates the mode of volumetric crustal response to the stress field. In the cases of the Reykjanes and Hengill areas, this was extensional, and in the case of the Krafla area it was variable on a local scale, and both extensional and compressional failure occurred.

4 The Accretionary Tectonic Cycle

Recent observations of the crustal accretion process in the Krafla volcanic system, N. Iceland, have been linked with older geodetic data, historical information and petrological evidence from other systems to arrive at a conceptual picture of the accretionary tectonic cycle in Iceland (Fig. 3; e.g., Niemczyk 1943; Einarsson 1978; Gronvold and Makipaa 1978; Sigurdsson and Sparks 1978a,b; Einarsson and Brandsdottir 1980; Johnsen et al. 1980; Sigurdsson 1980; Torge and Kangiesser 1980; Tryggvason 1980, 1983, 1986; Bjornsson 1985; Wendt et al. 1985).

The accretionary plate boundary comprises a series of discrete volcanic systems, each of which behaves as an independent spreading unit. These systems are quiescent for periods of centuries, during which time little or no crustal widening occurs, and extensional stress builds up (Fig. 3, stages 1 and 2). The Krafla volcanic system had been quiescent for 250 years prior to the most recent spreading episode, and quiescent periods of 100–150 years are reported for other systems (Bjornsson et al. 1979).

Periodically, a volcanic system will become activated and, for periods of the order of a few years, dyking accompanied by crustal widening on a scale of metres and release of extensional stress will occur in the fissure swarm (stage 3). The onset of such an episode is probably governed by fracturing in the lower crust. This allows the upward migration of melt from a partially molten subcrustal layer (Beblo and Bjornsson 1978, 1980; Beblo et al. 1983) into a shallow crustal magma chamber or chambers beneath a central volcano. Activation of the volcano ensues, that triggers rifting in the upper crust and at the surface. Given a sufficient supply of magma, dyking continues until all the extensional

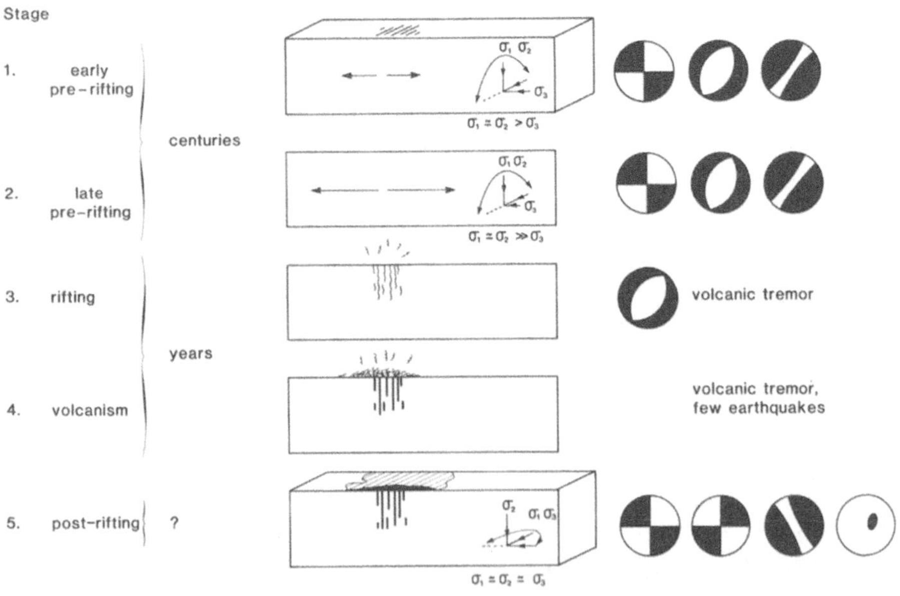

Fig. 3. Schematic diagram illustrating the stages of the accretionary tectonic cycle (inspired by Bjornsson 1985, Fig. 8). For each stage, the time scale, orientation of the principal axes of stress and expected suite of small earthquake focal mechanisms are shown. See text for detailed discussion

stress built up during the previous quiescent period has been released, such that further crustal widening is prohibited by the modified crustal stress field.

The total amount of crustal widening is governed by the amount of time elapsed since the last spreading episode, and the time-averaged full spreading rate. In Iceland, this is 2.2 cm/year (Talwani and Eldholm 1977). If magmatism continues, surface lava eruption then occurs (stage 4) and when this ceases the volcanic system enters the post rifting stage (stage 5). Subsequently, extensional stress once more begins to build up along the boundary, and the tectonic cycle begins again (stage 1). Close to the boundary, the stress field therefore varies cyclicly.

The study of earthquake focal mechanisms is one of very few ways in which the crustal stress field may be studied. In general, well-constrained focal mechanisms are only available for the occasional large event, which may represent an exceptional situation, e.g., magma chamber deflation, and only provides one data point. For Iceland, however, a fortuitous circumstance enables the method to be applied more widely. Where geothermal areas occur within the rift zones, the withdrawal of heat results in contraction of the rock and systematic reduction of the confining stresses. Thermal stress is then released as continuous small magnitude earthquakes, the focal mechanisms of which are controlled by the regional stress field. This activity may easily be monitored in a purpose designed experiment.

The focal mechanisms of the small earthquakes studied in the three cases discussed above give information about the stress state of the crust in three particular spreading segments at the times of the surveys. Historical evidence is available for the Reykjanes and Hengill areas, that gives insight into the tectonic stages of those systems at the times of the surveys. In the case of Reykjanes, the swarm of 17 000 events recorded in 1972 probably does not represent a spreading episode since no volcanic activity or surface deformation is reported as accompanying the swarm. The last eruption in the Reykjanes fissure swarm occurred in the year 1300 ca. (Jonsson 1983), which may indicate that a spreading episode occurred then, and therefore extensional stress may have been building up in this system for 700 years. Because of the en-echelon configuration of the Reykjanes peninsula series of volcanic systems, the rate of stress buildup across each system is less than where a single system occurs parallel to the plate boundary. A relatively long quiescent period between spreading episodes for the Reykjanes system is therefore to be expected.

The most recent tectonic episode in the Hengill system occurred in 1789 (Thorodssen 1899). Over 30 km of the fissure swarm were activated in an intense earthquake swarm and over 60 cm of subsidence observed. Subsurface dyke injections and crustal extension probably accompanied this activity, with attendant release of extensional stress. The Hengill system had therefore undergone 192 years of extensional stress build up at the time of the 1981 experiment.

Both the Reykjanes and Hengill systems were therefore in pre-rifting tectonic stages at the times of the surveys (Fig. 3, stages 1 or 2). In the cases of the Reykjanes and Hengill surveys, the mode of shear faulting observed indicates that at the times of the surveys, σ_3 was horizontal and normal to the fissure swarms. σ_1 and σ_2 were roughly equal and ranged from vertical to horizontal and parallel to the fissure swarms. The occurrence of tensile crack type events indicates that the stress fields were such as to permit volumetric extension. These observations suggest the characteristic suite of failure modes for the pre-rifting tectonic stages.

The tectonic stage of the Krafla system at the time of the experiment contrasted with those of the other two areas. The system was currently undergoing a rifting episode. Magmatic activity over the previous decade had caused dyking and crustal extension in the fissure swarm to the point where all the extensional stress was released, further dykes could not be accommodated and continued activity had resulted in surface eruptions. The system was therefore in a post rifting stage (Fig. 3, stage 5). The mode of shear faulting observed indicated that in general σ_2 was vertical, and that σ_1 and σ_3 had variable horizontal orientations. The occurrence of tensile crack and implosive events indicated that the stress field permitted both extensional and compressional volumetric sources, the latter being the most common. These observations testify to the absence of a strong uniform extensional stress regime, but indicate an inhomogenous stress field where the average deviatoric stress is small, locally variable in orientation, and such as to allow both compressional and extensional volumetric failure. These observations suggest the characteristic seismic signature of the post rifting tectonic stage.

The combined observations from three spreading segments give a detailed picture of the crustal stress fields that characterize different stages of the tectonic cycle. These stress fields are indicated in Fig. 3. During the pre-rifting stages, σ_3 is orientated horizontal and normal to the fissure swarm and its magnitude decreases with time. σ_1 and σ_2 range from vertical to horizontal and parallel to the fissure swarm. The stress field is such that extensional volumetric failure can occur. During stage 3, the horizontal compressive stress normal to the plate boundary progressively increases until it equals and may even locally exceed the former values of σ_1 and σ_2. The orientations of the axes of principal stress are thereby rearranged as shown in stage 5, with σ_2 vertical, and σ_1 and σ_3 with variable horizontal orientations. There is little difference between the magnitudes of σ_1, σ_2 and σ_3, – the triaxial stress ellipsoid is almost a sphere and it follows that the orientation of the axes of principal stress may vary on a small scale. This stress field is such that both extensional and compressional volumetric failure can occur, is inconsistent with the regional tectonic fabric, and probably short lived.

5 Summary

The cyclic pattern of spreading at the accretionary plate boundary in Iceland results in a crustal stress field, that also varies cyclicly with time, close to the boundary. This stress field may be studied by examining the focal mechanisms of the continuous, small magnitude earthquakes that are commonly associated with geothermal areas within the neovolcanic zone. The evidence that has so far accumulated from three such studies shows that during periods of tectonic quiescence, plate movements build up an extensional stress field close to the boundary where σ_3 is horizontal and perpendicular to the fissure swarm and tensile failure occurs. Immediately after rifting, however, the extensional component is removed. The resulting field is roughly triaxially symmetric, variable on a small scale in both magnitude and orientation, and may even be such, locally, that σ_1 is horizontal and normal to the fissure swarm. Implosive failure then occurs.

These results provide a seismic diagnostic for the various stages of the tectonic cycle and suggest that the stage of individual volcanic systems may be discovered using earthquake focal mechanisms. It is possible that these principles may also be applied to other environments where cyclic tectonism is expected. In the case of subduction zones and transform faults, it is likely that all failure is shear, and therefore volumetric sources whose type varies cyclicly with the stress field would be unlikely. However, along the submarine accretionary plate boundary and in volcanic regimes where the crust periodically accommodates volumes of intrusive material, similar volumetric, non-double couple types of earthquakes as are observed in Iceland may occur, and the value of monitoring this parameter to obtain useful information about the volcanic cycle should be explored.

Acknowledgements. Prof. Paolo Gasparini stimulated this paper.

References

Aki K, Richards PG (1980) Quantitative seismology. Freeman, San Francisco

Beblo M, Bjornsson A (1978) Magnetotelluric investigation of the lower crust and upper mantle beneath Iceland. J Geophys Res 45:1–16

Beblo M, Bjornsson A (1980) A model of electrical resistivity beneath NE-Iceland, correlation with temperature. J Geophys Res 47:184–190

Beblo M, Bjornsson A, Arnason K, Stein B, Wolfgram P (1983) Electrical conductivity beneath Iceland – constraints imposed by magnetotelluric results on temperature, partial melt, crust and mantle structure. J Geophys 53:16–23

Bjornsson A (1985) Dynamics of crustal rifting in NE Iceland. J Geophys Res 90:10151–10162

Bjornsson A, Saemundsson K, Einarsson P, Tryggvasson E, Gronvold K (1977) Current rifting episode in North Iceland. Nature (Lond) 266:318–323

Bjornsson A, Johnsen G, Sigurdsson S, Thorbergsson G, Tryggvason E (1979) Rifting of the plate boundary in North Iceland 1975–1978. J Geophys Res 84:3029–3038

Brandsdottir B, Einarsson P (1979) Seismic activity associated with the September 1977 deflation of the Krafla central volcano in north-eastern Iceland. J Volcanol Geotherm Res 6:197–212

Einarsson P (1978) S-wave shadows in the Krafla caldera in NE-Iceland, evidence for a magma chamber in the crust. Bull Volcanol 41:187–195

Einarsson P, Bjornsson S (1979) Earthquakes in Iceland. Jokull 29:37–43

Einarsson P, Brandsdottir B (1980) Seismological evidence for lateral magma intrusion during the July 1978 deflation of the Krafla volcano in NE Iceland. J Geophys 47:160–165

Einarsson P, Saemundsson K (1987) Earthquake epicenters 1982–1985 and volcanic systems in Iceland (map). In: Sigfusson Th. (ed) I Hlutarins Edli Festschrift for Thorbjorn Sigurgeirsson. Menningarsjodur, Reykjavik

Foulger GR (1982) Geothermal exploration and reservoir monitoring using earthquakes and the passive seismic method. Geothermics 11:259–268

Foulger GR (1988a) Hengill triple junction SW Iceland. 1 Tectonic structure and the spatial and temporal distribution of local earthquakes. J Geophys Res 93:13 493–13 506

Foulger GR (1988b) Hengill triple junction SW Iceland. 2 Anomalous earthquake focal mechanisms and implications for process within the geothermal reservoir and at accretionary plate boundaries. J Geophys Res 93:13 507–13 523

Foulger GR, Long RE (1984) Anomalous focal mechanisms: evidence for tensile crack formation on an accreting plate boundary. Nature (Lond) 310:43–45

Foulger GR, Long RE, Einarsson P, Bjornsson (1989) A Implosive earthquakes at the active accretionary plate boundary, Northern Iceland. Nature 337:640–642

Gronvold K, Makipaa H (1978) Chemical composition of Krafla lavas 1975–1977. Nordic Vulcanol Inst Rep 7816, 50 pp

Johnsen GV, Bjornsson A, Sigurdsson S (1980) Gravity and elevation changes caused by magma movement beneath the Krafla caldera, north east Iceland. J Geophys Res 47:1–3

Jonsson J (1983) Eldgos a sogulegum tima a Reykjanesskaga. (Historic eruptions on the Reykjanes Peninsula). Natturufraedingurinn 52:127–139

Klein FW, Einarsson P, Wyss M (1973) Microearthquakes on the Mid-Atlantic plate boundary on the Reykjanes Peninsula in Iceland. J Geophys Res 78:5084–5099

Klein FW, Einarsson P, Wyss M (1977) The Reykjanes Peninsula, Iceland, earthquake swarm of September 1972 and its tectonic significance. J Geophys Res 82:865–888

McKenzie DP (1969) Earthquakes and the directions of principal stresses. Bull Seismol Soc Am 59:591–601

Niemczyk O (1943) Spalten auf Island. Wittwer, Stuttgart

Saemundsson K (1978) Fissure swarms and central volcanoes of the neovolcanic zones of Iceland. Geol J Spec Issue 10:415–432

Schouten H, Klitgord KD, Whitehead JA (1985) Segmentation of mid-ocean ridges. Nature (Lond) 317:225–229

Shimizu H, Ueki S, Koyama J (1987) A tensile-shear crack model for the mechanism of volcanic earthquakes. Tectonophysics 144:287–300

Sigurdsson H, Sparks RSJ (1978a) Rifting episode in N Iceland in 1874–1875 and the eruptions of Askja and Sveinagja. Bull Volcanol 41:149–167

Sigurdsson H, Sparks RSJ (1978b) Lateral magma flow within rifted Icelandic crust. Nature (Lond) 274:126–130

Sigurdsson O (1980) Surface deformation of the Krafla fissure swarm in two rifting events. J Geophys 47:1–3

Talwani M, Eldholm O (1977) Evolution of the Norwegian-Greenland sea. Geol Soc Am Bull 88:969–999

Thoroddsen T (1899) Jardskjalftar a sudurlandi (Earthquakes of southern Iceland). Hid Islenzka Bokmenntafelag, Copenhagen

Torge W, Kanngieser E (1980) Gravity and height variations during the present rifting episode in northern Iceland. J Geophys Res 47:125–131

Tryggvason E (1973) Seismicity, earthquake swarms and plate boundaries in the Iceland Region. Bull Seismol Soc Am 63:1327–1348

Tryggvason E (1980) Subsidence events in the Krafla area, North Iceland 1975–1979. J Geophys 47:141–153

Tryggvason E (1983) The widening of the Krafla fissure swarm during the 1975–1981 volcano tectonic episode. Nordic Volcanol Inst Rep 8304:48

Tryggvason E (1986) Multiple magma reservoirs in a rift zone volcano: ground deformation and magma transport during the September 1984 eruption of Krafla, Iceland. J Volcanol Geotherm Res 28:1–44

Ward PL, Bjornsson S (1971) Microearthquakes, swarms and the geothermal areas of Iceland. J Geophys Res 76:3953–3982

Wendt K, Moller D, Ritter B (1985) Geodetic measurements of surface deformation during the present rifting episode in NE Iceland. J Geophys Res 90:10,163–10,172

Non-Double Couple Earthquake Mechanisms in Volcanic Environments

GEOFFREY R. ROBSON

Abstract

Earthquake mechanisms involving movement of pore fluids accompanying tensile and shear failure in volcanic environments are described. Three mechanisms: extension failure, extension-shear failure and shear failure are identified. The role of pore fluid pressure in determining the type of failure mechanism is described and the factors which determine the relative frequency of occurrence of the three types of failure are commented upon.

1 Introduction

In a volcanic environment fluids such as magma, hydrous solutions and steam occupy spaces within solid rock. It is well known that when fluid fills an interconnected network of pores within a solid, the mechanical properties of the solid-fluid system are modified as the system is stressed to the point of failure. It has been proposed that movement of fluid at the point of failure in such solid-fluid systems generates characteristic earthquake radiation patterns. Such mechanisms were proposed to explain the radiation pattern of the Ise Bay earthquake of 3 June 1929 by Ishimoto (1932), and Robson et al. (1968) have described an earthquake mechanism, termed "extension failure" which generates radiation patterns similar to those observed for the Ise Bay earthquake. Knopoff and Randall (1970) have termed earthquake source moments of the kind that would be generated by the extension failure mechanism, the compensated linear vector dipole (CLVD).

The theory of failure in fluid filled rocks is well established and widely used in engineering applications, but the possible application of the theory to volcanic environments where crustal and mantle rocks may contain rock melt as a pore fluid is less well established (though see Shaw 1980). The present chapter sets out to show that when rocks which contain a pore fluid are stressed to the point of failure they may exhibit displacements unlike those which would occur in dry rocks. Application of failure theory leads to the conclusion that three types of source displacement are possible and that earthquake waves may be generated in some circumstances.

A requirement of the theory is that the pore system be interconnected, but otherwise it imposes no special requirements on the pore system, either as to the ratio of pore space to host rock, the dimensions of the pores and their spacing, or on the nature or the viscosity of the fluid occupying the pores. If the displacements accompanying failure are to generate elastic waves, however, they must be completed within a time interval which is comparable to the period of the waves which are generated, a few seconds, at most, in duration. In this case parameters such as source volume, fluid viscosity and pore dimension assume a critical importance. Before the application of failure theory to earthquake mechanisms is considered in detail, therefore, the effect of changes in these critical parameters is discussed, in terms of a general model.

1.1 Source Volume

Suppose that a solid of volume v contains a fluid occupying an interconnected pore system which forms a small fraction f of the total volume of the rock, that a single pressure p extends throughout the fluid and that the system is acted upon by principal external stresses which are σ_1 the greatest, σ_2 the intermediate and σ_3 the least. Suppose further that a fracture intersects the pore system with orientation such that the stress acting across the fracture is smaller than the fluid pressure. In this case some of the fluid will flow from the pores into the fracture and the pore pressure will fall to some new and lower value. In practice the principal external stresses will change to new values but here the case is considered where these changes are small and may be neglected. The fluid volume will increase as a result of the fall in fluid pressure by an amount Δvf given by

$$\Delta vf = \beta_e fv\Delta p,$$

where β_e is the compressibility of the fluid and Δp is the reduction in fluid pressure. When f is small, the compressibilities of the solid and fluid are similar, and the process is adiabatic, Δvf is small and may be neglected. The effect on the solid of the reduction in fluid pressure will be equivalent to an increase in each of the principal external stresses by an amount equal to the reduction in fluid pressure. Such changes will cause a reduction in the volume of the solid Δv given by:

$$\Delta v = \beta_s(1 - f)v\Delta p, \tag{1}$$

where β_s is the compressibility of the solid. When β_s is 10^{-6} bar^{-1} and Δp is 10 bar, the ratio of Δv to v is 10^{-5}. Since the volume of fluid which migrates into the fracture is determined by the reduction in the volume of the porous solid, it follows that for each cubic kilometre of source rock 10^{-5} km^3 of fluid will migrate into the fracture.

The energy E which may be available for the generation of earthquake waves can be estimated using the relation from Robson et al. (1968):

$$E = 4\,\beta_s v(\Delta p)^2/3. \tag{2}$$

If β_s is 10^{-6} bar^{-1} and Δp 10 bar, as before, the energy available is 10^{10} J km^{-3}. The source volume required for the generation of an earthquake of magnitude 6 (with $E = 10^{14}$ J) is therefore 10^4 km^3.

1.2 Viscosity of Pore Fluids

Water and steam, under subcritical or supercritical conditions, are commonly present in the rocks of the upper mantle and crust in volcanic environments. Their viscosities are well established by experiment; that for water at 100 °C is 3 10^{-4} Nsm^{-2} and that for steam is about two orders of magnitude lower than this. The viscosities of rock melts, as they may occur in the crust and mantle, are less well established; partly because experimental determinations at high temperature and pressure are difficult to carry out, partly because the major element composition of rock melts in the crust and mantle is uncertain and partly because their content of volatile components, in particular water, is even more uncertain.

Ultrabasic melts, where they occur in the mantle, would have very low viscosities. Data for slag viscosities summarised by Rosenqvist (1974) suggest that melt with 45% silica and 5% alumina would have a viscosity of 0.01 Nsm^{-2} at 1700 °C and atmospheric pressure. Viscosity data for basaltic melts have been published by Ryan and Blevins (1987) who also quote from Euler and Winkler (1957), Murase and McBirney (1973) and Kushiro et al. (1976). The data are in broad agreement that at 1200 °C and at atmospheric pressure an anhydrous basaltic melt would have a viscosity about 70 Nsm^{-2} which would fall to about 1 Nsm^{-2} at 1600 °C. At 1200 °C and atmospheric pressure anhydrous andesitic melt would have a viscosity about 10^3 Nsm^{-2} which would fall to about 30 Nsm^{-2} at 1500 °C.

Both basaltic and andesitic melts show a small reduction in viscosity as pressure increases; from 3.5 Nsm^{-2} to 3.0 Nsm^{-2} in the case of a basaltic melt as the pressure increased from 15 kbar to 20 kbar at 1350 °C and from 140 Nsm^{-2} to 120 Nsm^{-2} in the case of an andesitic melt as the pressure increased from 7 kbar to 20 kbar at 1350 °C. We may note, however, that in the case of ultrabasic melts, where the ratio of nonbridging oxygens to tetrahedrally coordinated cations in the melt is greater than unity Scarfe et al. (1987) suggest that an increase in viscosity will occur as pressure increases. On the other hand, the presence of a small amount of water in an andesitic melt (4% by weight) caused a reduction in viscosity from 120 to about 5 Nsm^{-2} at 15 kbar and 1350 °C (Kushiro et al. 1976).

Navon et al. (1988) have studied inclusions of mantle-derived fluids in diamonds and conclude that at depths of 150–200 km upper mantle fluids or melts may contain up to 40% by weight of carbon dioxide and water. It is not known what the viscosity of such melts may be, but it is likely that they would be many orders of magnitude lower than those of basaltic lavas. Much recent evidence, then, suggests that rock melts in the crust and mantle may be considerably less viscous than lava flows and rock melts at the surface.

1.3 Pore Dimensions and Speed of Fluid Migration

The preceding discussion has shown that a source volume of the order of 10^4 km^3 would be required to generate an earthquake of magnitude 6 in fluid filled rock. It seems unlikely, however, that water or steam-containing rock-fluid systems would have dimensions extending to distances of the order of 20 km horizontally and vertically. If, on the other hand, the volume of such hydraulic systems in the upper crust is much smaller than this, of the order of 1 km^3, then the largest earthquake which could be generated by the adiabatic mechanisms discussed in this paper would be relatively small, of the order of magnitude 3 or less.

No such spatial limitations seem likely to apply to the volumes of crustal or mantle rock that contain interconnected systems of rock melt, which may be very large, tens of kilometres in linear extent judged by the areal extent of currently active volcanic systems. Geological evidence as to the characteristics of pore systems occupied by rock melt in the upper crust is found in regions of former volcanic activity which are now deeply eroded, as, for example, the Tertiary igneous province in western Scotland where there are areas extending over tens of kilometres in which multiple systems of intrusions penetrate older rocks, forming complex networks within them. The largest intrusions here have thicknesses measured in hundreds of metres but there is a range in scale down to very small thicknesses.

Evidence of the existence of pore networks of substantial dimensions in the mantle is not directly available, but if we suppose that mantle rocks melt along grain boundaries, then, in the first instance, the dimensions of the pores occupied by the melt will be determined by the ratio of melt to solid and by the size of the crystals of which the solid is formed. It is of interest, therefore, to note that Orowan (1967) considers that crystal grains in the mantle must be extremely large (perhaps metres or even hundreds of metres in size) as a result of annealing and extremely slow creep over millions of years. If, to estimate the pore dimensions, we take a grain diameter of 100 m and a melt: solid ratio of 1:100, the pores will be of the order of one metre in thickness and we may suppose that eventually they may be substantially larger than this as melt migration and segregation takes place. The migration of melt, once formed, would in any case lead to the formation of pore networks of substantial dimensions, but Orowan's estimate of grain size helps to confirm the concept that substantial pore networks may exist in the mantle where the conditions for melting are achieved.

To estimate the time taken by rock melt to transit complex structures of the kind described, because of its difficulty, is not attempted here, instead a simplified model is adopted which, it is believed, will give results of the right order of magnitude.

We suppose a series of plane parallel-sided pores are occupied by a rock melt and are intersected at right angles by a fracture. The time T for a volume $\beta_s v \Delta p$ to flow along these pores into the fracture under a pressure gradient $2\Delta p/L$ is, from Robson et al. (1968),

$$t = (6\eta\beta_s/f)(L/d)^2, \tag{3}$$

where η is the viscosity of the fluid, d the pore thickness, L the length of the pore system and f the ratio of rock melt to solid rock. Consider a source volume v with sides of length L. If v is $10^4\,km^3$ (for a magnitude 6 earthquake), L is 20 km and if β_s is $10^{-6}\,bar^{-1}$, f is 10^{-2} and t is 1 s, rearranging, we may write

$$d^2 = 0.24\eta, \tag{4}$$

as the relation between pore thickness and viscosity when flow is completed in one second. If η is 0.01 Nsm^{-2}, representing an ultrabasic melt in the mantle, the pore thickness d is found to be 0.15 m; when η is 10 Nsm^{-2}, representing basaltic or hydrous andesitic melt in the upper mantle or crust, d is found to be 1.5 m; when η is 10^4 Nsm^{-2}, representing anhydrous andesitic melt in the crust, d is found to be 49 m. These estimates are thought to be conservative, since, when the volume of the pore fluid is one percent of the source volume, only one thousandth of the the pore fluid migrates into the fracture after failure occurs (see Sect. 1.1 above). The fluid movement at the boundary of the source volume will be very small, therefore, and only where the fracture intersects the pore system will the flow equal that assumed to occur throughout the pore system in the model described above. The flow paths taken by the fluid within the fracture are relatively small (about half the distance between pores in the solid) and are neglected in estimating the time taken by the fluid flow to be completed.

The preceding discussion has shown that if a system with pores of the order of metres in thickness is occupied by a rock melt and extends through a volume of $10^4\,km^3$, pressure changes of the order of 10 bar can occur in a time interval of the order of one second as a result of fluid migration into a newly formed fracture. The conditions necessary for the generation of elastic waves on the scale of an earthquake of magnitude 6 are therefore met by the model described.

A similar conclusion may be reached for a pore system containing water. If v is $1\,km^3$, L 1 km and t 0.01 s and with values for β_s and f unchanged we have,

$$d^2 = 0.06\eta,$$

and when η is $3\ 10^{-4}\,Nsm^{-2}$, d is found to be 25 mm. The conditions necessary for the generation of high frequency elastic waves typical of earthquakes of magnitude 3 or less are also met, therefore, when the pore fluid consists of water. An analogous result, with smaller pore thicknesses, can be reached in the case of steam.

2 Failure in Fluid-Filled Rocks

The model adopted here is the same as that described in earlier paragraphs; a rock contains a fluid which occupies an interconnected pore network forming a small fraction f of the total volume of the rock. Pressure gradients in the fluid are neglected and it is assumed that (1) a single pressure p extends throughout the fluid; (2) the principal external stresses are σ_1 the greatest, σ_2 the intermediate, and σ_3 the least; (3) that when failure occurs fractures intersect the pore system and migration of fluid into the fractures occurs when the pore pressure exceeds the pressure acting across the fractures.

2.1 Extension Failure

The relationships between fluid pressure, the principal stresses and the conditions for failure are given by the Griffith theory. The conditions for failure are:

$$\sigma_3 - p + T = 0, \text{ when } p > \sigma_3 + (\sigma_1 - \sigma_3)/4, \tag{5}$$

and:

$$(\sigma_1 - \sigma_3)^2 - 8T(\sigma_1 + \sigma_3 - 2p) = 0, \text{ when } p < \sigma_3 + (\sigma_1 - \sigma_3)/4. \tag{6}$$

If the inequality of Eq. (5) is satisfied and the fluid pressure p equals the minimum principal stress plus the tensile strength of the solid T, tensile fractures form on a plane normal to σ_3. Since the fluid pressure exceeds the stress acting across the plane of failure, fluid flows from the pore system into the fractures. If after failure the change in the minimum principal stress is small then the fluid pressure will have been reduced by T. The effect on the solid of this reduction in pore pressure will be equivalent to an increase in each of the principal external stresses by an amount equal to the fall in pore pressure. Such changes will cause a reduction in the volume of the porous solid [Sect. 1.1 Eq.(1)].

If the linear dimensions of the volume before failure are L_1 L_2 and L_3 in the directions parallel to σ_1 σ_2 and σ_3 these dimensions will change after failure by an extension equal to:

$$\beta_s v \Delta p/(L_1 L_2) - \beta_s v \Delta p/(L_1 L_2 3),$$

which is equal to:

$$2\beta_s \Delta p L_3/3,$$

in the 3 direction and contractions equal to:

$$\beta_s \Delta p L_1/3 \quad \text{and} \quad \beta_s \Delta p L_2/3,$$

in the 1 and 2 directions respectively. When L_1 L_2 and L_3 are effectively equal, the displacements at the boundary of the source volume due to the movement of fluid will be in the proportions −1, −1, +2 in the 1, 2 and 3 directions.

The time required for fluid flow to occur has been estimated (Sect. 1.3) where it is shown that fluid flow can proceed to completion in a time short enough for earthquake waves to be generated. This is the mechanism described by Robson et al. (1968). It corresponds in terms of moments to the compensated linear vector dipole of Knopoff and Randall (1970).

Earthquakes exhibiting radiation patterns characteristics of this mechanism appear to be rare. Ishimoto (1932) has described the P wave radiation pattern of the Ise Bay earthquake of 3 June 1929 and attributes it to a source mechanism analagous to the one described here, though Foulger (pers. comm. 1989) has pointed out that the Ise Bay earthquake can also be interpreted as a double-couple thrust. The focus of the Ise Bay earthquake was at a depth of 300 km and it was well recorded throughout Japan. When the sense of the P wave first motions are plotted on a stereogram of the upper hemisphere of the focal sphere, it is found that a circular nodal line may be drawn which separates the compressional and dilatational fields, indicating, if the extension failure interpretation is adopted, that displacements at the source occurred in three mutually perpendicular directions and that displacements, in two of these directions, were equal in magnitude and sign. The ratios of the displacements corresponding to the circular nodal lines are −1, −1, $+1\frac{1}{2}$ rather than −1, −1, +2 as predicted by the model. Uncertainty in the determination of the focal depth of this earthquake may account for this discrepancy since the principal axis corresponding to the 3 direction is almost vertical. That earthquakes exhibiting the extension failure mechanism are rare, may follow from the fact that after failure, the fluid pressure throughout the source volume will fall to a value roughly equal to the least principal stress. Under these conditions the inequality of Eq. (5) cannot be satisfied and further earthquakes will not take place by this mechanism until the fluid pressure rises once more to satisfy the inequality and failure conditions of Eq. (5).

The field observations by Foulger, reported in Foulger and Long (1984) and Foulger et al. (1989), which are discussed later, represent an important step forward in the study of the focal mechanisms of earthquakes in volcanic areas. Foulger and Long (1984) and Foulger (1988) have described unusual micro earthquakes from the Hengill geothermal area in southern Iceland which have principal moments -1, -1, $+30$; the unbalanced principal moments indicating a volume change at the source. They propose that this radiation pattern is generated by the formation of a tensile crack in the presence of a pore fluid; a process which they would expect to occur in the tensile stress regime of an accreting plate boundary. The magnitudes of these micro earthquakes vary from -2 to $+1.2$ and their focal depths range from 2 to 7 km, the median depth being about 5 km. At a depth of 5 km the lithostatic pressure is about 1.5 10^3 bar so that it is unlikely that an absolute tension can occur. Tensile fracturing, therefore, may be the result of a process similar to the extension failure mechanism described above. Extension failure would not give rise to uncompensated moments, however, and so some additional factor must have been operating within the source volume. Foulger and Long (1984) proposed that this factor is restricted fluid flow at the moment of fracture, which would diminish the amplitude of the dilatational component of the source motion. This amounts to the assumption that fluid migration is restricted to a relatively narrow zone on either side of the fracture, from which it follows that the proposed source volume will be substantially inequidimensional, with dimensions in the ratios 30, 30 and 1 in the 1, 2 and 3 directions. If the energy released by an earthquake of magnitude zero is 10^5 J, then from Eq. (2) the source volume for such an earthquake according to the Foulger-Long proposal will be about 10^4 m^3 and its principal dimensions will be about 70, 70 and 2.5 m. Foulger (1988), commenting further on the Hengill non-double-couple events, has suggested that they may be caused by tensile cracks with a shear component, but, as will be shown in Section 2.2, "extension-shear" failure would not give rise to a volume change at the source and therefore would not generate the unbalanced moments which have been observed.

An alterative to explaining why the amplitude of the dilatational component may be restricted, is to explain why the compressional component may be augmented. Such explanation may be found if a non-adiabatic mechanism is considered which would add thermal energy to the fluid. Suppose, for example, that the tensile fracture took place in water-filled rock overlying high temperature rock or magma which was also penetrated by the fracture. In this case, that part of the fluid which penetrated the hot rock would undergo a rise in temperature and pressure which, if sufficiently rapid, would add to the energy radiated as elastic waves in a direction normal to the sides of the fracture. Relatively small volumes of rock would need to be cooled to generate the earthquakes observed. For a magnitude zero earthquake, if extension failure took place and some of the water in the fracture, say $10^{-2}\,m^3$, penetrated the hot rock, 10^7 J would be required to raise the temperature of this water from 250 °C to 450 °C at 500 bar pressure. If the wetted surface of the rock was $6\,m^2$ and was cooled to a depth of 10^{-3} m from 650 °C to 450 °C, approximately 2×10^7 J would be available to raise the temperature of the water and generate elastic waves. A few percent only of this energy would be required to generate an earthquake of magnitude zero.

In a second paper on non-double-couple earthquake mechanisms, Foulger et al. (1989) have described micro earthquakes from the vicinity of the Krafla volcano in northern Iceland. These earthquakes took place during a quiescent period following the intrusion of basaltic magma along a fissure system on which many of the micro earthquakes were located. Among many earthquakes recorded most showed double-couple mechanisms and some showed mechanisms similar to those discussed above, but there was a small number of others which showed principal moments varying from −1, −1, +0.3 to −1, −1, +0.7, magnitudes ranging from −0.2 to +0.4 and depths between 2.7 and 3.5 km. Foulger et al. (1989) conclude that the large dilatational field exhibited by these earthquakes implies cavity collapse along the edges of cracks. The closing of cavities between the edges of cracks appears, however, to imply other fracturing (normal faulting, parallel to the crack) to permit the crack edge to move inward and close the cavity. Such fracturing would also generate elastic waves which would be likely to mask the first motions postulated by Foulger et al. (1989).

An alternative to cavity collapse to explain the large dilatational field, is to consider another non-adiabatic mechanism such as extension failure accompanied by cooling and contraction of the fluid in the fracture. If the fluid in the rock undergoing extension failure is supercritical steam and the tensile crack when it formed penetrated rock containing cooler fluid, condensation of the steam would generate a dilatational pulse that would be superimposed on the displacements caused by the extension failure. Meyer (1963) has commented that condensing steam is a very efficient generator of acoustic waves. The volume of steam required by this mechanism is relatively small. If $0.1\,m^3$ of steam were cooled from 900 to 250 °C at 300 bar pressure, the volume would be reduced to one tenth its former value and 10^7 J would be released and would be available to heat the cooler fluid and to generate earthquake waves. If we assume, conservatively, that only a few percent of the available energy is used to generate earthquake waves, a magnitude zero earthquake is easily achieved.

2.2 Extension-Shear Failure

The inequality of Eq. (6) may in principal be satisfied for a wide range of values of p as it diminishes towards zero. In the case of a rock with pore cavities, the extent to which p can fall below the maximum principal stress is limited by the strength of the rock. In rock where the stresses approach the values at which failure occurs, p cannot be significantly smaller than σ_3. If σ_3 is substituted for p in Eq. (6) then, at failure

$$(\sigma_1 - \sigma_3) = 8T, \quad \text{when } p = \sigma_3.$$

It follows that 8T represents the maximum stress difference required to cause shear failure in a fluid-filled rock, whatever the actual values of the principal stresses.

On the other hand, it also follows from Eq. (6) that the fluid pressure in a fluid-filled rock which undergoes shear failure cannot be greater than

$$\sigma_3 + (\sigma_1 - \sigma_3)/4.$$

Substituting for p in Eq. (6) we have, at failure,

$$(\sigma_1 - \sigma_3) = 4T, \quad \text{when } p = \sigma_3 + (\sigma_1 - \sigma_3)/4.$$

The stress difference required to cause failure is halved, therefore, as the fluid pressure increases to a limiting value.

When the inequality and the conditions for failure of Eq. (6) are satisfied and shear failure occurs in the host rock, the shear fractures which form will intersect the pore system and when $p > \sigma_3$, fluid will flow from the pores into the fracture. If we assume, as before, that the change in the minimum principal stress which accompanies failure is small, then after failure, $p \approx \sigma_3$.

The effect of the fluid movements described above is to add to the displacements which accompany shear failure those which accompany extension failure. This type of failure is here termed "extension-shear" failure. The possibility that certain types of failure are possible which involve a shear component and an extensional component has been discussed by Shaw (1980). Shaw's extensional shear fracture regime corresponds, in terms of failure criteria, to the extension-shear failure mechanism discussed above. The relative magnitudes of the extension and shear components of extension-shear failure may be estimated if we assume that the shear strain is reduced to zero after failure; if the ratio between the bulk and rigidity moduli is known; and if the fall in fluid pressure can be specified. If the ratio of the moduli is 2:1 and the fall in fluid pressure after failure is $(\sigma_1 - \sigma_3)/4$, the displacement in the 1 direction due to the shear component is roughly twice that due to the extension component if all shear strain is released. When the fluid pressure is relatively high, however, the most likely plane of failure, according to the Griffith theory, is inclined at a relatively small angle to the 1 direction, in which case the shear strain may not be reduced to zero after failure. On the other hand, if extension shear failure occurs with smaller fluid pressures, the extension failure component will diminish relative to the shear component. It seems possible, therefore, that a considerable range in the ratio of the extension and shear components of extension-shear failure may occur in practice. Earthquakes exhibiting radiation patterns characteristic of extension-shear failure are uncommon, though well observed earthquakes at Mammoth Lakes on 25 and 27 May 1980 exhibit the radiation patterns predicted by this model. Comprehensive analysis of the available data for the Mammoth Lakes earthquakes has been made by Julian and Sipkin (1985) who, concluding

Table 1. Mammoth Lake earthquakes

Earthquake of 25 May 1980			
Observed source moments	−1.5	−0.6	+2.1
Extension failure component	−0.6	−0.6	+1.2
Residual moment	−0.9	0	+0.9
Earthquake of 27 May 1980			
Observed source moments	−1.0	−0.4	+1.4
Extension failure component	−0.4	−0.4	+0.8
Residual moment	−0.6	0	+0.6

that no substantial volumetric change took place at the source, use the scheme of Knopoff and Randall (1970) to decompose the source moments, which are found to have a large CLVD component. Typical values of the source moments quoted by Julian and Sipkin (1985) are set out in Table 1. It will be observed that when moments in the ratios −1, −1, +2 are extracted, the ratio of the extension and shear components is 2:3 for each earthquake; a value smaller than that estimated for the case when all shear strain is removed by the failure mechanism. Julian and Sipkin (1985) have considered the case of dyke intrusion along a tensile crack as a possible mechanism for the earthquake of 25 May. The estimated moment of the principal dipole of the CLVD normal to the tensile crack is used to calculate the volume of fluid intruded. This volume, together with the linear dimension of the fracture surface, estimated from the distribution of ground breakage and aftershocks, is used to calculate a minimum value for the crack width. The crack dimensions and the duration of the earthquake (assumed to be the filling time of the crack) are then used to calculate a viscosity for the fluid, which is found to be 0.3 Nsm^{-2}. The authors conclude that this viscosity is too low for magma, though when smaller aspect ratios for the dyke (which are also considered by the authors to be plausible ones) are used to estimate the viscosity of the intrusive magma, values for the viscosity range from 17 to 1.7 $10^4 Nsm^{-2}$. The Julian-Sipkin model, explicitly, does not take into account the displacements which would occur at the source as magma migration takes place. The Julian-Sipkin model also differs from the model discussed here, inasmuch as it is concerned with a tensile fracture filled from below, whereas the extension-shear model is concerned with a shear fracture which intersects a rock containing a pore fluid and is filled with fluid (magma) from the pores which are intersected.

2.3 Shear Failure

The theories of brittle fracture, such as the Griffith theory and the Coulomb-Navier theory, have the difficulty that they cannot be used to explain fracture in dry rocks at depths greater than, at most, a few kilometres because the magnitudes of the differences between the principal stresses required to cause failure become implausibly large. At 20 km depth, for example, a stress difference of several thousand bars is required if shear failure is to occur. Yet it is commonly accepted that at depths below a few kilometres the rocks of the lower crust and mantle respond to long term stress differences of the order of a few hundred bars as a result of the action of creep mechanisms. Yet brittle fracture, if we define it as the abrupt release of elastic strain energy accompanied by earthquake radiation patterns characteristic of shear failure, occurs at depths down to 700 km below the surface. The theoretical difficulty is removed if the rocks that undergo fracturing contain a pore fluid such as a rock melt. The discussion in Section 2.2 has shown, from the Griffith theory, that stress differences from 4T to 8T (at most a few hundred bars) cause shear failure (or extension-shear failure) in rock containing a pore fluid. The question arises, therefore, whether any of the many earthquakes which exhibit radiation patterns characteristic of shear failure take place in the presence of a pore fluid. This question cannot be answered easily from studies of the radiation patterns of the earthquakes concerned, since rocks containing pore fluids and dry rocks will exhibit the same pattern when shear failure occurs. It seems possible, moreover, that if extension-shear failure occurs and the extension failure component is small this component may not have been recognised as such and a double-couple solution accepted in spite of discordant data. A re-examination of focal mechanisms which exhibit discordant data with the extension-shear mechanism in mind might reveal new examples of this mechanism. If the earthquakes exhibiting the extension-shear mechanism were commonly found in association with earthquakes exhibiting the shear mechanism this would be circumstantial evidence that pore fluids were present in both cases.

3 Conclusions

If mantle and crustal rocks which contain an interconnected pore fluid are compressible, are rigid, at least for periods of time of the order of seconds, and are subject to mechanical failure according to conventional criteria, it is found that when failure occurs, three types of source displacement are possible, which are determined by the relationship between the fluid pressure and the least principal stress.

The condition that earthquake waves may be generated as a consequence of failure, depends critically upon the viscosity of the fluid and the dimensions of the spaces occupied by the fluid. Experimental and field data on viscosities and possible pore dimensions in the upper mantle and crust, indicate that earthquake wave generation may accompany failure.

Extension failure takes place when the fluid pressure exceeds the least principal stress by an amount equal to the tensile strength of the rock. In this case displacements at the source are in three mutually perpendicular directions, which are equal inward displacements in the 1 and 2 directions and an outward displacement in the 3 direction of twice the amplitude of the displacements in the 1 and 2 directions. Earthquakes exhibiting the extension failure mechanism seem likely to be rare, since, after failure occurs, the fluid pressure falls to a value close to the least principal stress, consequently further earthquakes will not take place by this mechanism unless the fluid pressure rises once more, or the least principal stress falls, by an amount equal to the tensile strength of the rock. It seems likely therefore that the occurrence of such earthquakes will be limited to regions of the mantle or crust where melting is actively in progress.

Extension-shear failure takes place when the fluid pressure exceeds the least principal stress by an amount limited by the tensile strength of the rock, and when the difference between the greatest and least principal stresses reaches a critical value. The displacements at the source following extension-shear failure are a combination of those of extension failure with those of shear failure. The contributions of the extension and shear components to the combined mechanism appear to be capable of wide variation; the extension component being small when the difference between the fluid pressure and the least principal stress is small and the shear component being small when the difference between the fluid pressure and the least principal stress is great. The source displacements in extension-shear failure, therefore, tend towards those of pure shear failure as the pore pressure, and with it the extension contribution, diminish. In regions of the upper mantle and crust where the tectonic regime tends to restore shear stresses rapidly, even though rock melts are generated and a pore fluid exists, extension-shear earthquakes are likely to have small extension components.

Shear failure takes place in the presence of a pore fluid when the fluid pressure is equal to the least principal stress and the difference between the greatest and least principal stresses reaches a critical value. Whether shear failure has occurred in rock containing a fluid or in dry rock is not easily determined, although the discussion in Section 2.3 presents an argument based on the theory of failure that below a few kilometres from the surface all shear failure takes place in the presence of a pore fluid. A re-examination of double-couple focal mechanism solutions with a view to the identification of any extension component present, and the investigation of a possible association between extension-shear mechanisms and pure shear mechanisms would help to resolve this question.

The three failure mechanisms discussed are adiabatic, yet in the upper crust in volcanic areas strong thermal gradients occur which facilitate the transfer of thermal energy by fluids within zones undergoing failure, making possible non-adiabatic source mechanisms. Examples of such mechanisms are provided by the observations made by Foulger in Iceland.

Acknowledgements. The author is much indebted to K.G. Barr, G.R. Foulger, B.R. Julian, R.E. Long and A.R. McBirney for helpful comments.

References

Euler R, Winkler HGF (1957) Über die Viscositäten von Gesteinsund Silicatschmelzen. Glastech Ber 30:325–332

Foulger GR (1988) Hengill Triple Junction SW Iceland II: anomalous earthquake focal mechanisms and implications for process within the geothermal reservoir and at accretionary plate boundaries. J Geophys Res 93:13 507–13 525

Foulger GR, Long RE (1984) Anomalous focal mechanisms: tensile crack formation at an accreting plate boundary. Nature (Lond) 310:43–45

Foulger GR, Long RE, Einarsson P, Bjornsson A (1989) Implosive earthquakes at the active accretionary plate boundary in northern Iceland. Nature (Lond) 337:640–642

Ishimoto M (1932) Existence d'une source quadruple au foyer sismique d'apres l'etude de la distribution de mouvements initiaux des secouses sismiques. Earthquake Res Inst Bull 10:449–471

Julian BR, Sipkin SA (1985) Earthquake processes in the Long Valley caldera area. J Geophys Res 190:11 155–11 169

Knopoff L, Randall MJ (1970) The compensated linear-vector dipole, a possible mechanism for deep earthquakes. J Geophys Res 75:4957–4963

Kushiro I, Yoder HS, Mysen BO (1976) Viscosity of basaltic and andesitic liquids at high pressures. Carnegie Inst Wash Year Book 75:615–618

Meyer E (1963) High-intensity sound in liquids. In: Albers VM (ed) Underwater acoustics. Plenum, New York

Murase T, McBirney AR (1973) Properties of some common igneous rocks and their melts at high temperatures. Geol Soc Am Bull 84:3563–3592

Navon O, Hutcheon ID, Rossman GR, Wasserburg GJ (1988) Mantle-derived fluids in diamond micro-inclusions. Nature (Lond) 335:384–389

Orowan E (1967) Mechanical properties of crust and mantle. In: Runcorn K (ed) International Dictionary of Geophysics. Pergamon, Oxford

Robson GR, Barr KG, Canales Luna L (1968) Extension failure, an earthquake mechanism. Nature (Lond) 218:28–32

Rosenqvist T (1974) Principles of extractive metallurgy. McGraw-Hill, New York

Ryan MP, Blevins JYK (1987) The viscosity of synthetic and natural melts and glasses at high temperatures and 1 bar (10 pascals) pressure and higher pressures. US Geol Surv Bull 1764

Scarfe MC, Mysen BO, Virgo D (1987) Pressure dependence of the viscosity of silicate melts. Geochem Soc Spec Publ 1:59–67

Shaw HR (1980) The fracture mechanisms of magma transport from the mantle to the surface. In: Hargraves RB (ed) Physics of magmatic processes. Princeton, Princeton, New Jersey

Nonlinear Inversion of S-Wave Polarization for Constraining the Source Mechanism of Small Earthquakes

ALDO ZOLLO and PASCAL BERNARD

Abstract

An accurate estimate of micro-earthquake source mechanisms in volcanic areas may be obtained by using three-component S waveforms from near source records. S-wave polarizations recorded at the earth's surface are here used to infer the parameters of a point-source double couple fault model (strike, dip and slip angles of the fault plane) for an earthquake. We define an error function associated to polarization misfits, and a related probability density function which is computed in the whole model space. Because of the small number of parameters, the probability density function can be represented on plane view projections. This gives a global image of the resolution of model parameters by actual data. This method has been applied to estimate the source mechanism of an $M_L = 2$ earthquake which occurred at the Campi Flegrei caldera during the recent seismic crisis of (1983–1984).

1 Introduction

The deployment of three-component seismic stations in volcanic areas provides a more complete data set for constraining the earthquake source parameters. In particular, the three-dimensional S-wave motions bring additional information on fault mechanisms and greatly improve the accuracy of its determination. Great accuracy on fault mechanism estimates in volcanic areas is particularly needed because of the possible relation between microseismicity and local stress field induced by magmatic processes. Most of the inversion procedures applied to the study of earthquake mechanisms work on P and/or S wave polarities or, more generally on the three components of ground motion. Ground motion amplitudes depend simultaneously on source (fault mechanism and location, source strength, rupture directivity) and medium effects (i.e., reflections and refractions, geometrical divergence, anelastic attenuation, anisotropy, site amplification effect).

This explains the usual complexity of near-source amplitude waveforms, and makes these data difficult to interpret.

The three-dimensional ground motion (acceleration, velocity or diplacement) can also be parametrized by its modulus and direction (polarization unit vector). For direct S-waves, the modulus is a function of mechanism and location of the source and slip distribution on the fault plane. The value of this parametrization is that the S-wave polarization from point double-couple sources does not depend on the complexity of the source history. Its observed stability in near-source records may therefore be the consequence of a stable mechanism at the source. Bernard and Zollo (1988, 1989) used S-wave polarization data from near-source, high frequency records of the $M_S = 6.9$ Irpinia 1980 earthquake (southern Italy) to constrain the geometrical features of the sequentially activated fault segments. They proposed a method to statistically analyze the S polarizations observed at a set of stations and to invert source parameters.

In this paper, we generalize the nonlinear inverse method by proposing a probabilistic approach to estimate the free-parameters of a point-source double-couple fault model. A point source is described by its location and its fault mechanism. The free-parameters of the model are the strike (ϕ_s), dip (δ), and slip (λ) of the fault assuming that the position of the radiating point source is known. The inverse method is applied to a set of three-component digital waveforms from a $M_L = 2$ ($M_o = 1.1 \times 10^{12}$ Nm) 1984 (depth = 3 km) earthquake which occurred in the Campi Flegrei volcanic area (southern Italy) on 1 April 1984 during the last seismic crisis of 1982–1984.

2 Method of Inversion

In a homogeneous medium, the far-field approximation to the radiation from a point double-couple source can be expressed as:

$$u(t) = \Delta v(t)\, A\, s \quad \text{(e.g., Aki and Richards 1980)}, \tag{1}$$

where u = displacement vector at a given site, A = matrix depending on the radiation pattern, Δv = slip velocity function at the source, and s = the unit slip vector. Equation (1) can be generalized for spherically symmetric and layered media by computing the appropriate spreading matrix and plane wave reflection and transmission coefficients. Similarly, the recorded velocity or acceleration ground motion a is given by

$$a(t) = S(t)\, A\, s, \tag{2}$$

where S is the source time function. Let us write $a = |a|\, \hat{a}$, where $|a|$ is the modulus and $\hat{a}$ is the unit polarization vector of the ground motion time function. Given a set of N observations $\hat{a}_i$ ($i = 1, \ldots, N$), Eq. (2) can be written in the form of a system:

$$d = G(p), \tag{3}$$

where $d = \{\hat{a}_1, \ldots, \hat{a}_N\}$ is the vector of data and $p = \{\phi_s, \delta, \lambda\}$ is the vector of parameters. G is generally a nonlinear functional form, relating parameters to data. It depends on the medium's elastic parameters (S wave velocity and density), and on the relative locations of the point source and stations.

According to Tarantola and Valette (1982) we define the joint probability density function (pdf) P(d, p) of data d and parameters p assuming that the distribution of expected errors is Gaussian:

$$P(d, p) = P(d|p)\frac{\rho(p)}{\mu(d, p)} \propto \rho(p)\prod_{i=1}^{N} \exp[-(\theta_i(\hat{a}_i, G(p)))^2/2\theta_{oi}^2]$$

$$= \text{const}\ \rho(p) \exp\left[-\sum_{i=1}^{N}(\theta_i(\hat{a}_i, G(p)))^2/2\theta_{oi}^2\right] \qquad (4)$$

where $P(d|p)$ is the *conditional* pdf of data d given the parameters p;

$\mu(d, p)$ is the *null information* pdf and it is constant in our case because we chose cartesian reference systems for both data and parameters;

$\rho(p)$ is the a priori pdf of parameters. In the following, we set $\rho(p) = 1$ in the considered range of variation of the parameters, and 0 elsewhere;

θ_i is the angular difference between the observed and theoretical S polarization vectors at station i, and θ_{oi} is the expected variance of θ which accounts for errors in modeling and data quality (signal to noise ratio). The definition of θ_i in Eq. (4) takes into account that for a given direction of a synthetic S polarization, one cannot distinguish between two opposite polarizations. This means that if one mechanism is found to be consistent with the data, the mechanism obtained by exchanging P and T axis provides exactly the same fit. In particular, for a given fault plane, a thrust event cannot be distinguished from a normal fault one, since they correspond to opposite vectors; nevertheless, the slip vectors would be different.

The probability function (4) contains the complete information on the inverse problem. Its maxima correspond to the "most likely" values of parameters in the sense of Eq. (4) and its topography in the M-dimensional parameter space (M = number of parameters) directly gives a global image of the resolution on the model parameters. In order to compute P(d, p), the model parameter space is sampled using a constant step for each parameter. Angular distances θ_i are evaluated for each model. Synthetic waveforms at the free surface are generated by a ray-tracing program.

The *marginal* pdfs associated to each model parameter is computed from the P(d, p):

$$\sigma(\phi_s) = \iint P(d, \delta, \lambda)\, d\delta\, d\lambda)$$
$$\sigma(\delta) = \iint P(d, \phi_s, \lambda)\, d\phi_s\, d\lambda \qquad (5)$$
$$\sigma(\lambda) = \iint P(d, \phi_s, \delta)\, d\phi_s\, d\delta$$

These functions give the probabilities for each parameter disregarding any possible correlation with the others. Their maxima correspond to the maximumum likelihood parameter values. A flat marginal pdf means that the related parameter is completely unresolved.

In order to have a more complete image of the pdf over the parameter space we projected it over the three orthogonal planes defined by the fault angles. On each plane a model point is represented by a symbol whose size is proportional to its probability density (see Fig. 4). On these plots a group of large size symbols indicates a peak of probability around the best fit values of the parameters. The areal extension and the shape of these *patches* gives infomation on the expected errors and correlation between the parameters. This kind of representation enables us to get at the same time the best estimates of the parameters and a global image of resolution obtained with the actual data.

3 Inversion with Synthetic Data

Synthetic S-wave polarizations have been computed at a set of seven stations for a given fault mechanism model. Station configuration, hypocenter location, and velocity model are the same as in the real case considered later. In Fig. 1, the theoretical S-vectors (closed circles) are represented by projections on the horizontal plane defined by their azimuth when the motion is upward (upper hemisphere, east is on right, north towards the top) and by their distance to the center of the circle which is proportional to the angle from the vertical. When the motion is downwards, the azimuth is changed by π. Theoretical S vectors at the free-surface of a vertically hetereogeneous medium have been computed by a dynamic ray tracing program written by V. Farra (Farra et al. 1986). The positions of the stations on the focal sphere (lower hemisphere Wulff projection) are also shown in the figure. Values of θ_{oi} are assumed to range from 15° to 30° depending on the source-receiver distance.

The inversion of seismic data for constraining the double-couple fault parameters is a highly nonlinear problem. Most of the adopted inversion techniques are iterative methods which assume a linearized model and are, thus, strongly dependent on the assumed starting solution.

In Fig. 2 are shown the cross-sections of the function $\sum_{i=1}^{N} (\theta_i \hat{a}_i, G(p))^2/\theta_{oi}^2$ (norm function) around the three smallest minima found. Minima of this function correspond to maxima of the probability (Eq. 4). We point out that more than 50 minima were found with varying amplitude. For each minimum, one of the three parameters is allowed to vary, the others remain fixed at the value they have at the minimum.

Deep holes are clearly observed in the vicinity of the *true* parameter values. Note that the topography of this function is locally very irregular, presenting numerous secondary minima. Such an observation questions the use of standard inversion techniques with gradient evaluation because they may converge to some irrelevant minimum and, in any case, will not give a global image of the resolution of parameters.

The marginal pdf for the three parameters has been computed by integrating the *joint* pdf over the model parameter space ($0° < \phi_s < 360°$, $0° < \delta < 90°$, $-180° < \lambda < 0°$) (Fig. 3). Peaks are centered at the expected values of the parameters. The two maximum likelihood points found for strike and slip are associated with the orthogonal planes of the double-couple fault model. The dip *marginal* pdf presents a single peak due to the almost similar values of dip for

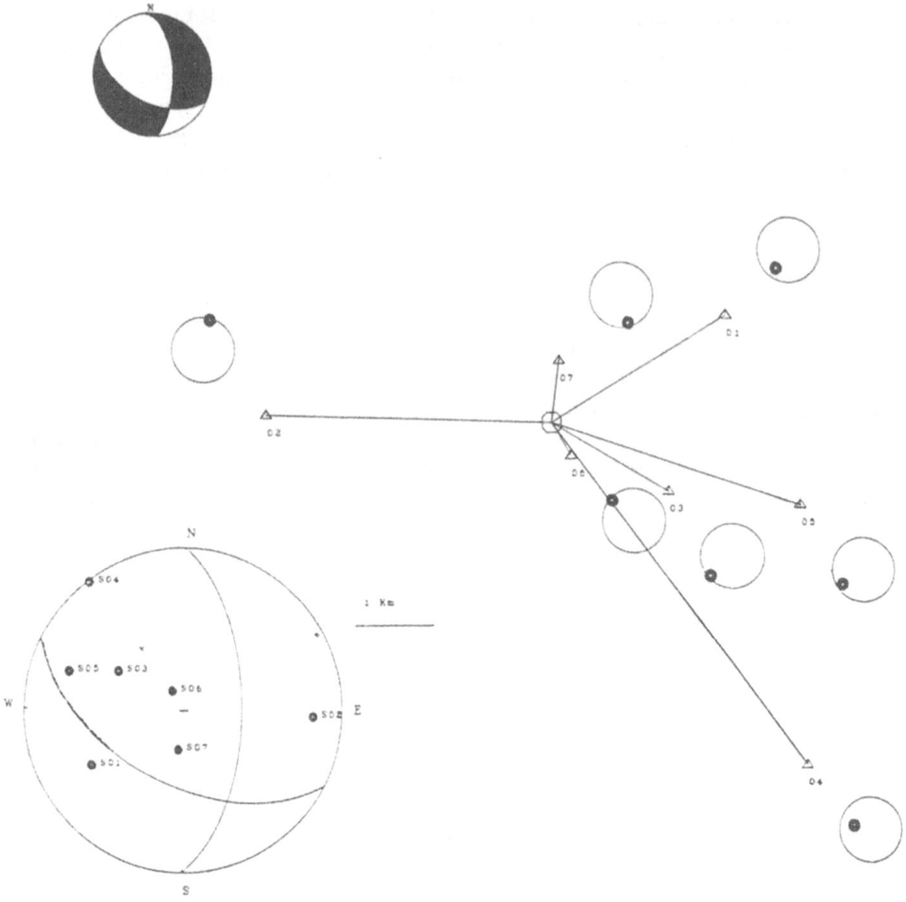

Fig. 1. Polarization data used for the synthetic test. For each station S vectors are projected on the (north east) plane, north is at the top of the figure (see the text). Theoretical S polarizations have been computed by using the mechanism shown on the *bottom left* of the figure ($\phi_s = 0°$, $\delta = 60°$, $\lambda = -45°$). Station positions on the focal sphere (lower hemisphere equal area projection) are also indicated

the two fault planes. The estimated uncertainty is of the order of 15°–20°, roughly corresponding to the mean error θ_{oi}. Larger values of θ_{oi} produce a flattening of the curves and a consequent decrease of resolution of the peaks. In order to look in detail at the shape of the pdf. Over the whole parameter space, we represent on diagrams the projection of the probability function on the planes (ϕ_s, δ), (ϕ_s, λ) and (δ, λ) (Fig. 4).

Each model is represented by a symbol whose size is proportional to its probability as indicated at the top of the figure. The 99% probability level

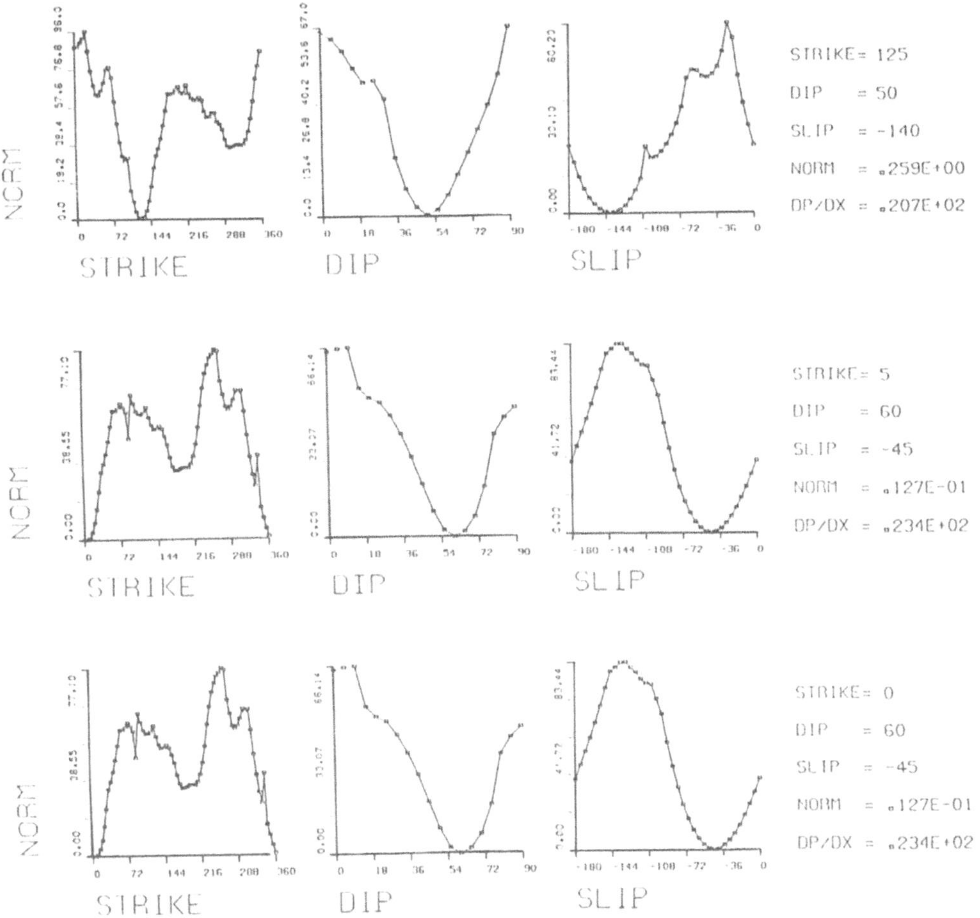

Fig. 2. Cross-section of the norm function around the three lowest minima for the synthetic case

models are represented (i.e., the probability that the true model is not represented is smaller than 1%).

This graphic representation gives a global image of the shape of the probability and allows one to identify the correlation among the parameters. Very localized *patches* indicate good model resolution. An estimate of the errors on the parameters may be visually inferred by the areal extension of the patches. A decrease of resolution on the model parameters corresponds to a scattered distribution of points on the planes. A diagonal distribution of points on the plane indicates a correlation between the parameters. For a poorly resolved parameter, points are distributed parallel to the corresponding axis.

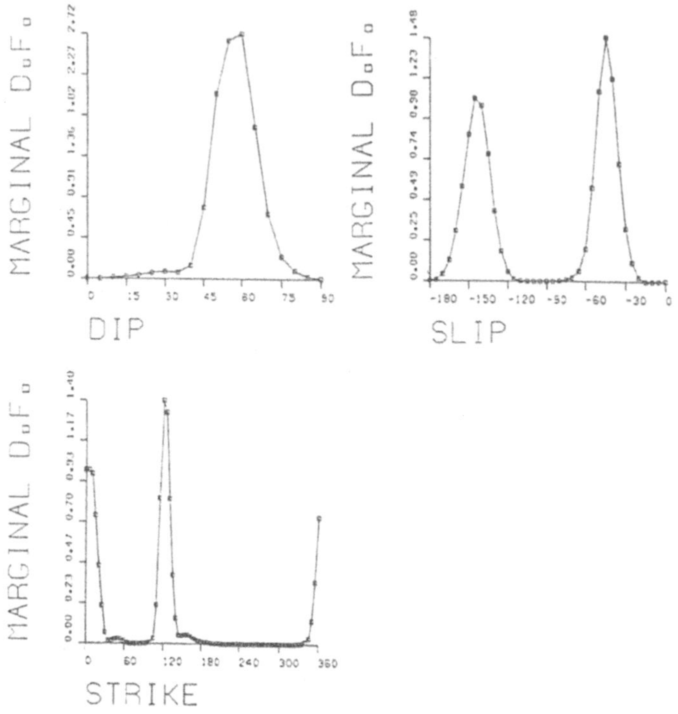

Fig. 3. Plots of the marginal pdfs of the three model parameters for the synthetic case

4 Inversion of Real Data: Source Mechanism of a Micro Earthquake in Campi Flegrei

The method has been used to estimate the fault parameters of a micro-earthquake which occurred within the caldera of Campi Flegrei during the last seismic crisis (1983–1984). Short period (1–20 Hz) three-component digital data were recorded by a set of stations operated in the area by the University of Wisconsin, Madison and Osservatorio Vesuviano (Aster et al. this Vol.) An intense seismic activity (more than 10 000 events in 2 years) accompained a large ground deformation episode with a total uplift of about 2 m which affected the central part of the caldera (De Natale et al. 1987). Figure 5 shows the epicenter location of this earthquake and the stations used for the analysis. The event occurred on 1 April 1984 within a large swarm of events having magnitudes smaller than 3 (about 500 events in less than 7 h). The 40 best-located events of this swarm are also shown in the same figure. P and S wave readings on digital waveforms have been used for locating the events by using a *master event* location technique (del Re and La Volpe 1988). Events appear confined within a small volume of $2 \times 1 \times 1\,km^3$ at an average depth of 3 km.

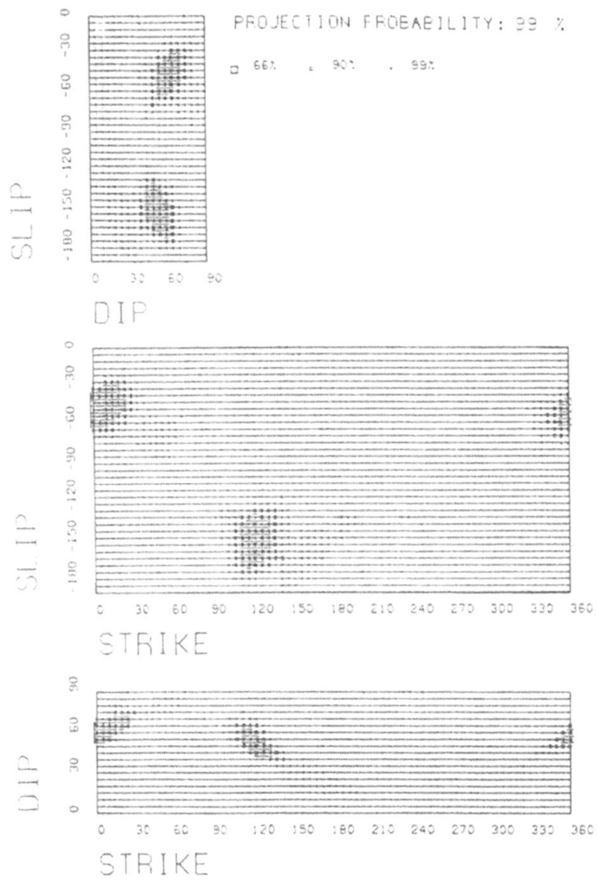

Fig. 4. Plane projections of the probability density function (4) (see text). Angles are expressed in degrees

Three-component digital waveforms were filtered at several frequency bands and the mean P and S polarization vectors were measured in the selected windows (Fig. 6). The mean polarization p_m is defined as the vector which minimizes the weighted sum:

$$S = \sum_i w_i \theta_i^2 \Big/ \sum_i W_i,$$

where w_i is the squared velocity ground motion amplitude (v), and θ_i the angular difference between v and p_m. Figures show some examples of filtered records and the projections of polarization vectors on the plane (N, E). The size of circles is proportional to the squared amplitude. The filled circle indicates the measured mean polarization. (TETA) and (FI) are, respectively, the angle with the vertical direction and the azimuth of the measured vector. Stability of the S vector in time is generally observed for a time window lasting 0.2–0.3 s (on average) and in the frequency band 1–9 Hz. At higher frequencies, S vectors

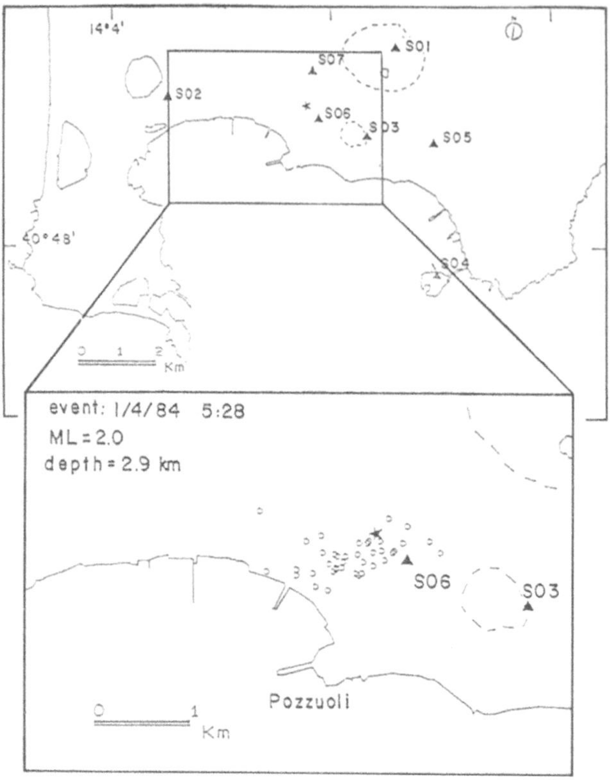

Fig. 5. 1 April 1984, 5^h 28′ epicenter location (*black star*) and stations used for computing the fault mechanism. Epicenters of the best located events of the 1 April swarm are shown in the enlarged view of the area

show random distribution over the sphere. The analysis of S polarization stability on the records as a function of the frequency confirms that coherently polarized signals have a dominant frequency range as already observed for the Irpinia (1980) earthquake records (Bernard and Zollo 1988). At higher frequencies (f > 10 Hz for the Campi Flegrei earthquake, f > 5 Hz for the Irpinia 1980 earthquake) waves are much more perturbed by the medium that hides the source effects. Errors in polarization measurements (ALFA in the figure) range from 18° to 30° depending on the distance source-to-receiver.

Fig. 6a–c. Filtered S velocity waveforms at station S02 and estimate of the mean P and S polarization vectors. A zero-phase shift Butterworth filter (36 db oct^{-1}) has been used to analyze the signals. *Top*: Vertical, north-south, east-west filtered records.
Bottom: Projection on the (north, east) plane of the P and S vectors computed in the chosen time window indicated by *lines* over the time series. The *size* of *circles* is proportional to the square modulus. The *filled circle* gives the position of the computed mean polarization vector (see text). **a** Polarization analysis of the 1–3Hz filtered time series; **b** polarization analysis of the 3–9 Hz filtered time series; **c** polarization analysis of the 6–18 Hz filtered time series

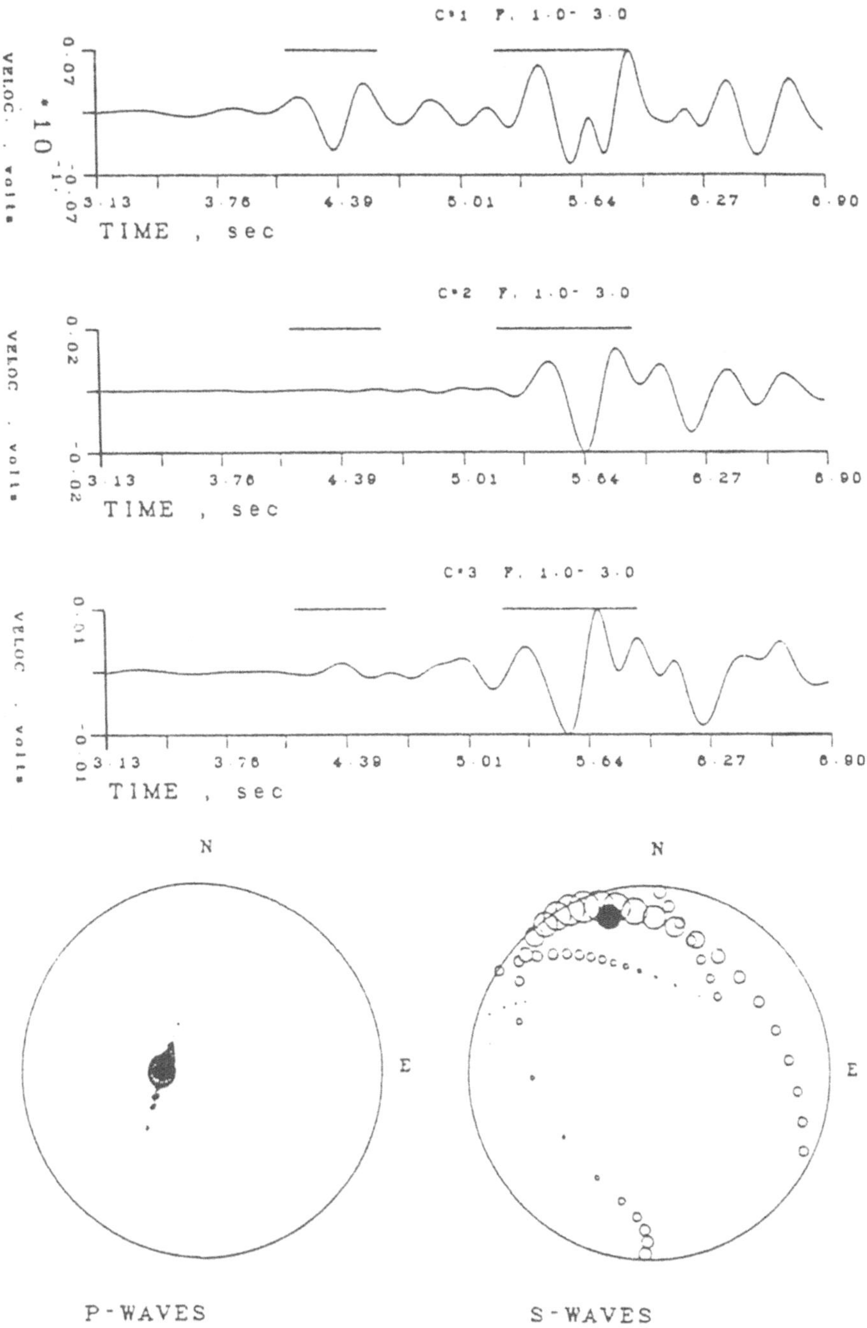

C*1 F. 1.0- 3.0
VELOC. , volts
*10
0.07
-0.07
3.13
3.76
4.39
5.01
5.64
6.27
6.90
TIME , sec
C*2 F. 1.0- 3.0
VELOC. , volts
0.02
-0.02
3.13
3.76
4.39
5.01
5.64
6.27
6.90
TIME , sec
C*3 F. 1.0- 3.0
VELOC. , volts
0.01
-0.01
3.13
3.76
4.39
5.01
5.64
6.27
6.90
TIME , sec
N
E
P-WAVES
N
E
S-WAVES

a

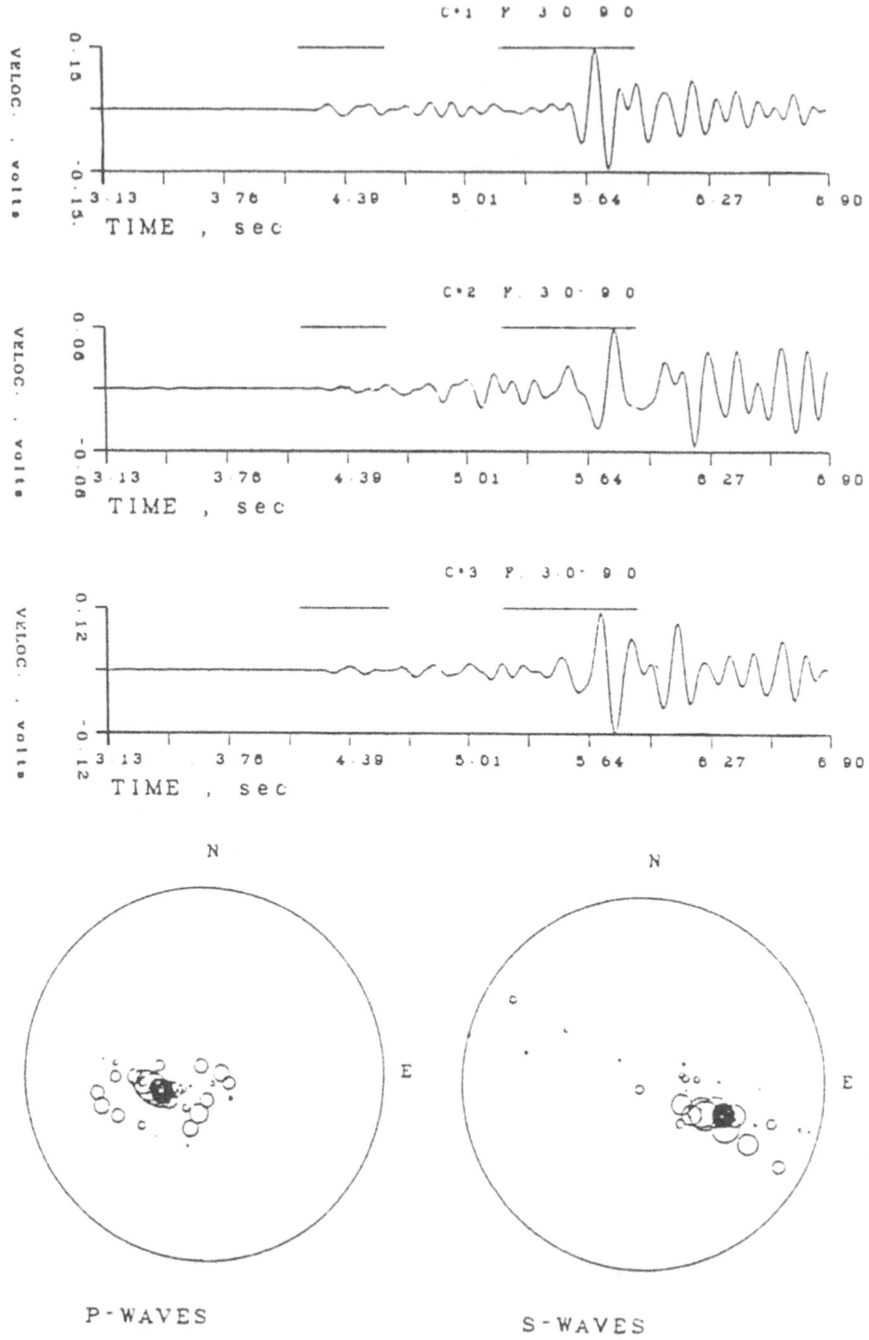

b

Fig. 6b. *Continued*

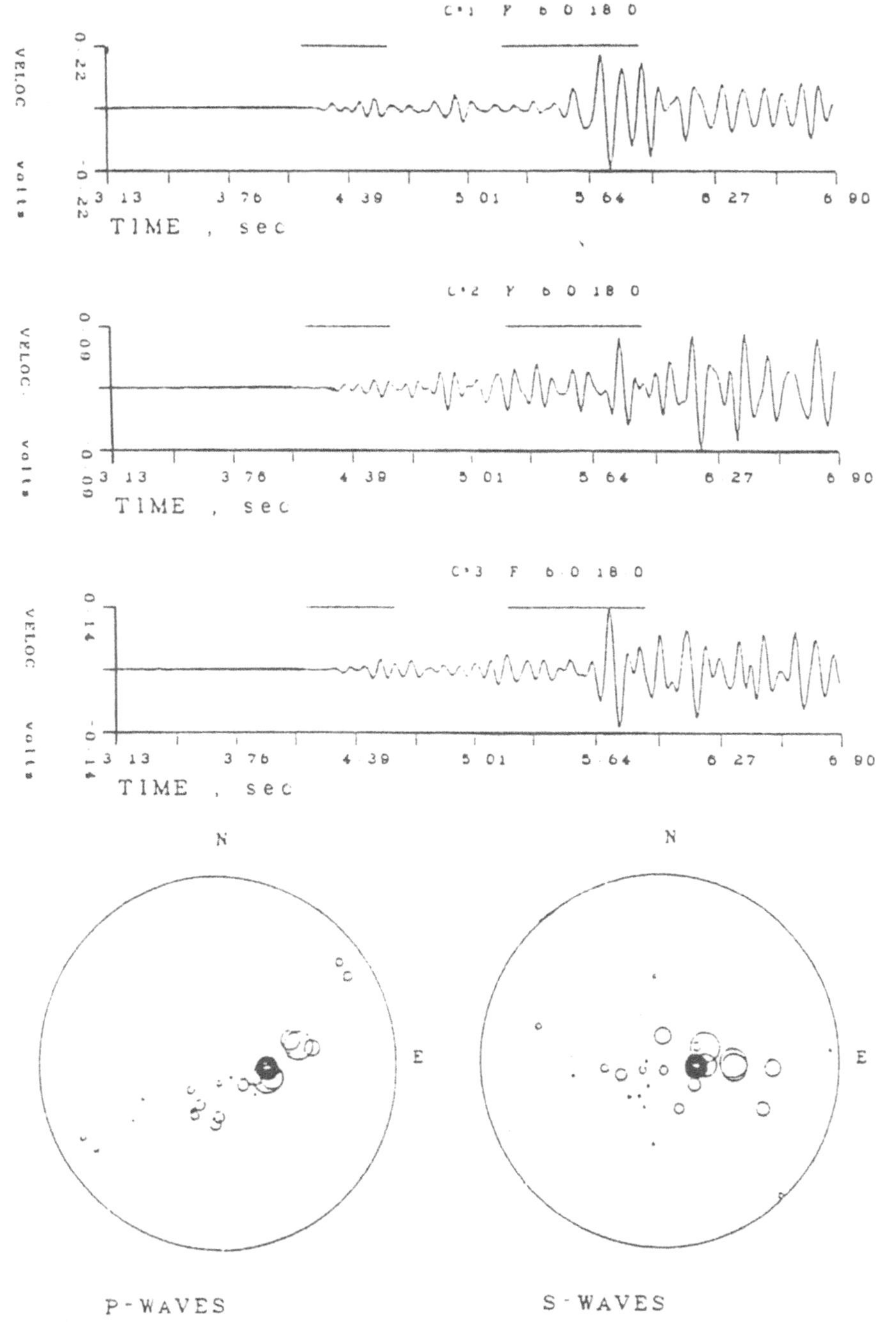

c

Fig. 6c. *Continued*

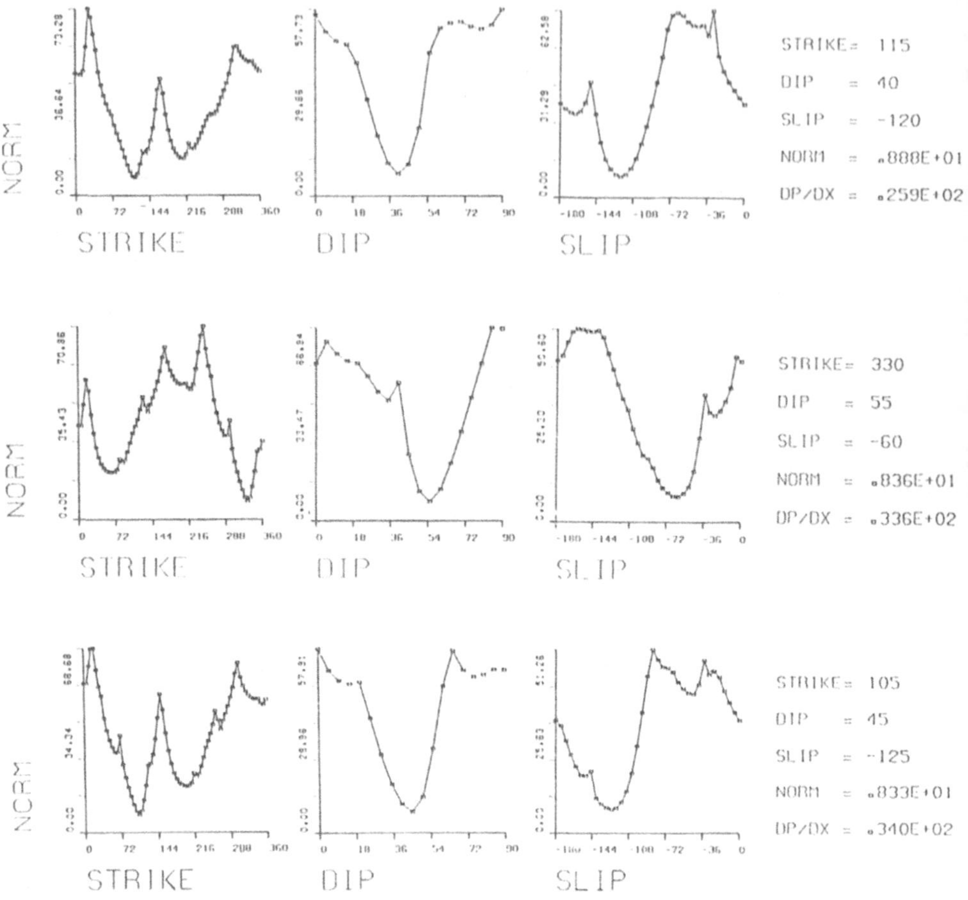

Fig. 7. Cross-section of the norm function around the three lowest minima for the real case

4.1 Results of the Inversion

A gradient velocity model was used for generating ray-theoretical synthetic seismograms ($v_s = 1.27 + 0.31\,z$; $v_p = 1.7\,v_s$). The adopted model takes into account near-surface P velocity measurements in the area (*CNR* [1987]) and results from a 3-D inversion of body wave arrival times (Aster and Mayer 1988) which show no evidence for a strong depth dependence of velocity in the upper 3 km of the central caldera. Furthermore, the chosen gradient is able to reproduce, at the surface, near-vertical P arrivals (incidence angle <40°) as observed from P-wave polarization analysis.

As for the synthetic case, the cross-section of the smaller minima, the marginal pdfs and the plane projections of probability are shown in Figs. 7, 8, and 9. Although the local topography of the norm around the best found minima is somewhat irregular, the *marginal* pdfs show that best parameter

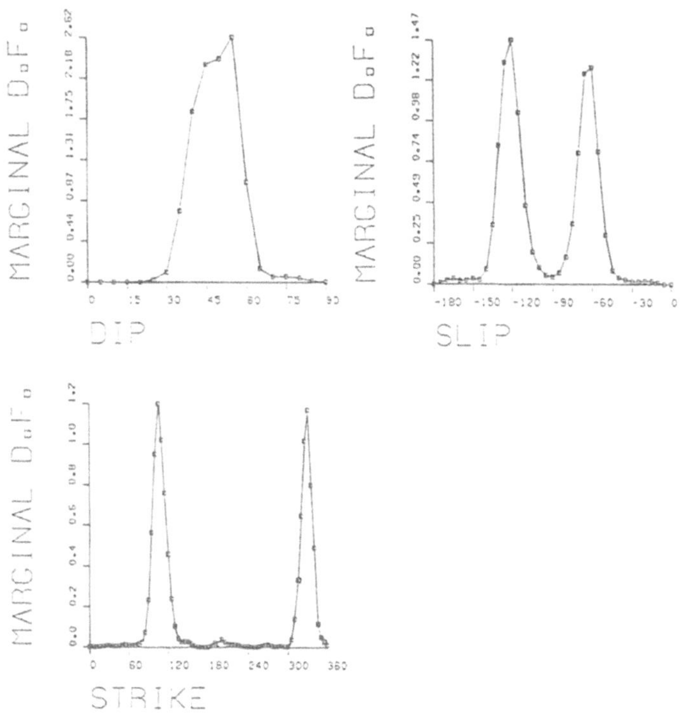

Fig. 8. Plots of the marginal pdfs of the three model parameters for the real case

values are well resolved. Errors on parameters are around 15°–20°. The single peak of the marginal pdf *for* dip is due to the fact that the two found plane solutions share an equal (within the errors) dip value. The projections of 99% probability models evidence a slight trade-off between dip and strike parameters although the highest probability models are all confined in 25° × 25° error limit areas.

The *most likely* fault plane solutions are plotted on an equal area lower hemisphere projection together with stations in Fig. 10. In the lower part of the Fig. 10, synthetic polarizations (open circles) generated by the found solution are compared to the observed ones by representing the projections of vectors on the (N, E) plane (see above). The values under the circles give the measure of the angular difference between the two vectors. Distant stations show larger divergences than expected from the increase of the noise level with distance in the S-wave window.

Finally, synthetic polarization diagrams at each station have been computed in an attempt to resolve the ambiguity on the slip sign. This plots show the polarization of the ground motion versus time in the (south, east) plane. No significant S-motion is observed in the vertical plane. In Fig. 11 the band-pass filtered horizontal records for each station, the (east, south) polarization diagram and the theoretical S-vector orientation in the same plane are plotted.

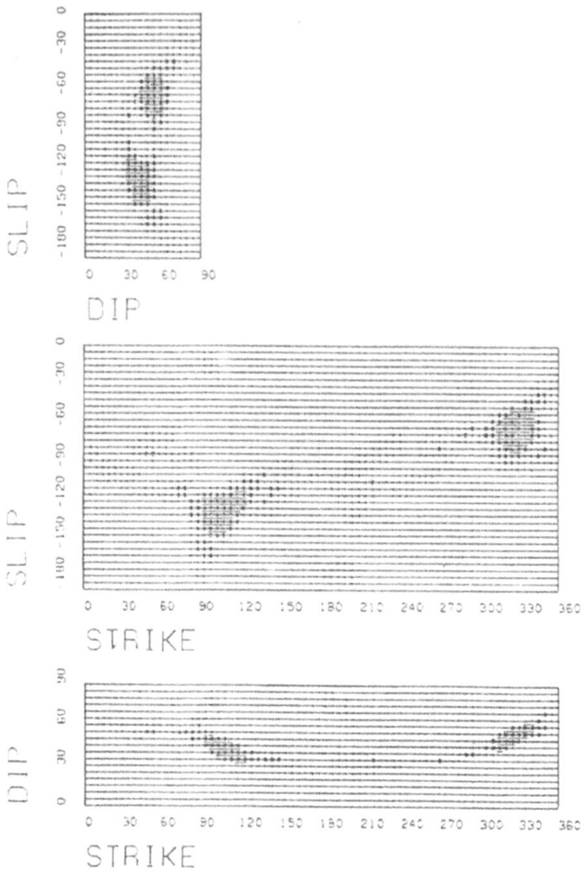

Fig. 9. Plane projections of the probability density function for the real case. Same representation as in Fig. 4

Arrows indicate the first S arrivals. Synthetics are computed by using ray-theory, using the fault plane solution obtained from inversion (center of Fig. 11). If a positive slip value is considered (thrust fault mechanism), the observed first S-pulse polarizations in the selected frequency bands, show the same orientation of synthetics for most of the stations.

5 Discussion and Conclusions

Microseismicity in volcanic areas is the only precise indicator of local stress field. Its spatial distribution and change with time may be related to magma chamber activity, or fluid induced cracks in dikes. The detailed study of earthquake source mechanisms may provide additional evidence for a causal relation be-

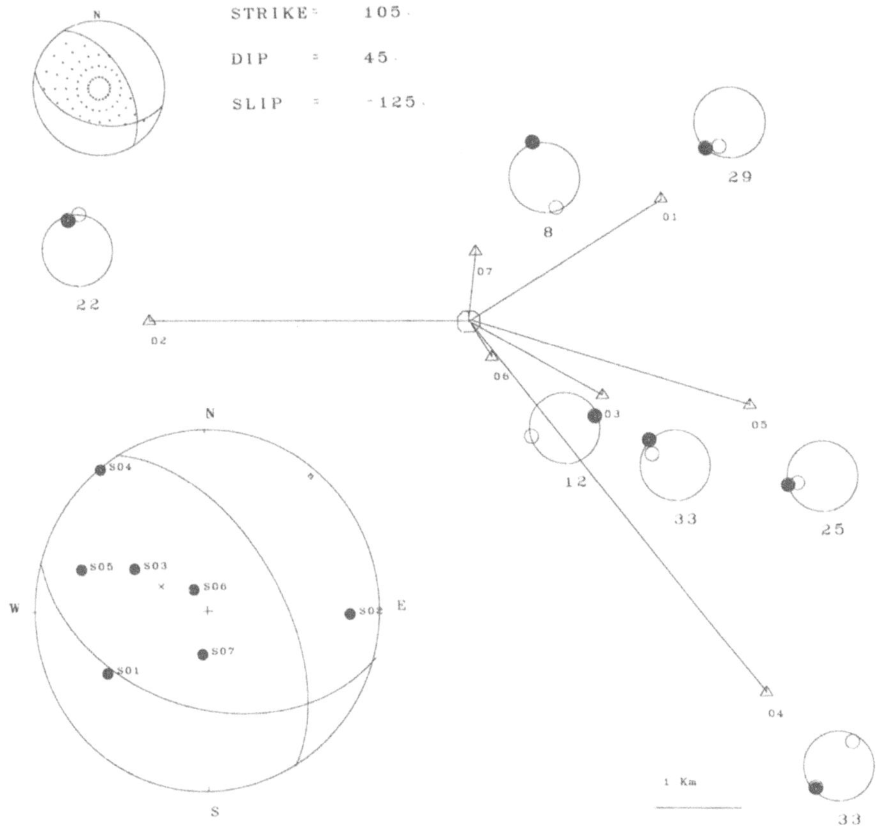

Fig. 10. Comparison between observed and computed S vector directions. Theoretical S polarizations are obtained from the *most likely* fault mechanism model. Same representation as in Fig. 1

tween magma activity and seismicity in volcanic areas. For instance, the principal stress axes inferred from micro-earthquakes may be related to the local stress field induced by the deformation processes in the magma chamber, and their spatial distribution may allow one to estimate the size and shape of the deforming volume. Alternatively, the stress axes may be oriented along the directions of tectonic regional stress or they may randomly follow very local failure conditions. A correct estimation of the fault mechanisms of micro-earthquakes is therefore needed for a proper physical understanding, which has obvious implications on the assessment of volcanic hazard. In both cases, reliable source mechanism estimates are clearly needed to discriminate between different models. Moreover, precise focal mechanism estimates are needed for a detailed study of rupture phenomena at the micro earthquake scale: this can provide useful insights on the dynamical properties of rupture processes in volcanic environments (estimate of local stress drop, seismic moment, geometry and time history of the rupture).

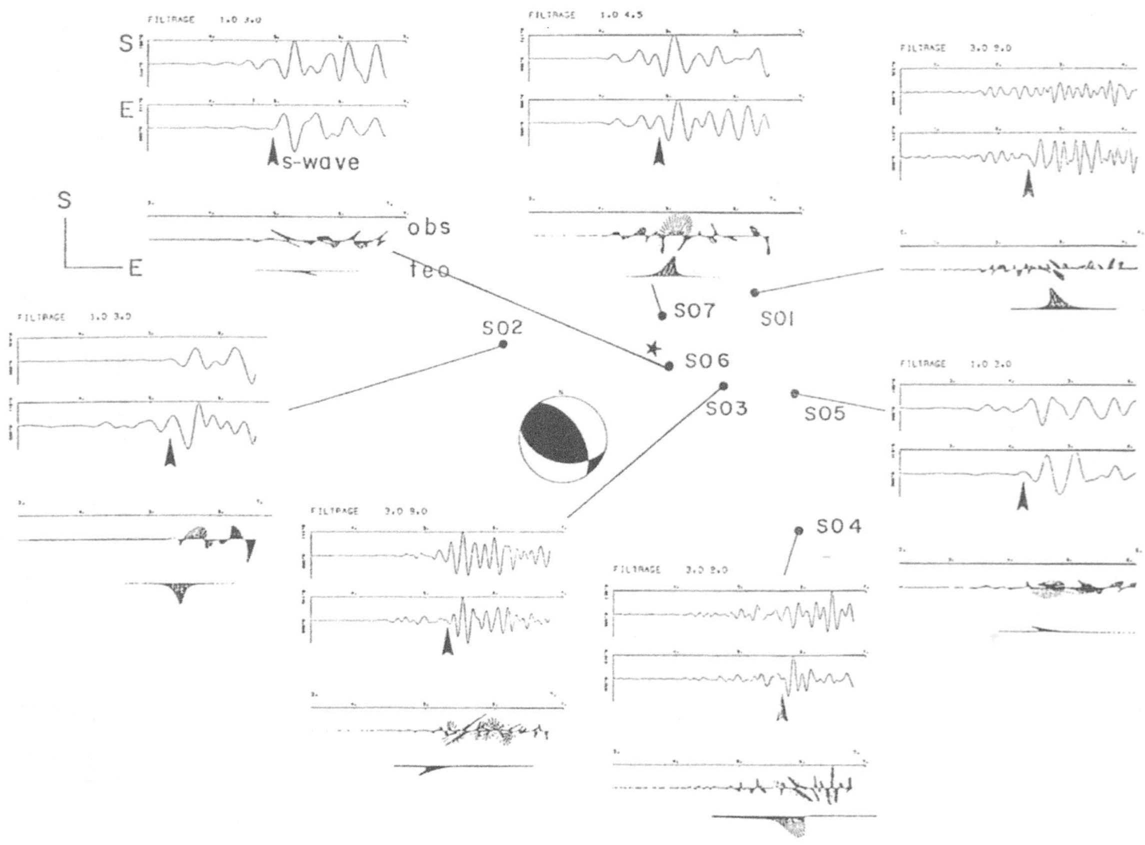

Fig. 11. Plots of the filtered horizontal records and associated north-east polarization diagrams (see text). Synthetic polarizations (*bottom trace* of each station plot) are obtained by imposing a thrust fault mechanism (*center* of Fig)

The increasing number of three-component digital instruments deployed in volcanic areas has now made available more complete data sets which bring new information on faulting processes. In particular, the use of three-component S waveforms certainly provide additional constraints on the focal mechanism and improve the accuracy of its estimate. This is not the case of usually applied techniques which consider only part of the available information as provided by P polarities and amplitudes.

In the present study, we proposed a method of fault mechanism determination which is based on the inversion of S wave polarizations recorded at a set of three-component instruments. The observed stability with time of S polarization in a proper frequency and source to receiver distance ranges for the Irpinia (Bernard and Zollo 1988, 1989), Campi Flegrei (this work), Imperial Valley (Zollo and Bernard 1989) cases (despite the high variability of amplitudes in the same time and frequency window) makes the S polarization a suitable parameter for studying the source geometrical parameters of earthquakes, as long as the details of the propagation medium are known at the wavelength of interest.

Micro-earthquakes which are recorded by short period seismometers (1 Hz of natural frequency) at a few kilometers from the sources may be reasonably modeled by point double-couple fault mechanisms. In this case, the locations are generally well constrained by P- and S-wave arrival times read on records from the local network. Therefore using S vector data from micro-earthquakes the only free geometrical source parameters are the three fault angles (strike, slip and dip) assuming that the S-velocity model is known. In a recent paper (Bernard and Zollo 1989) we proposed a method to compute the error function associated to the S polarization misfit between near source data snd synthetics generated by a double couple point source. In this paper, we generalized this method by defining and computing, in the whole model space, the joint probability of parameters and data associated to the polarization misfit function. This approach allows us to relax the general hypothesis of linearity, to estimate the best fit parameters and have a global image of the resolution on the model parameters. This is not the case of standard inversion techniques based on gradient evaluation or least squares criteria which can rarely provide reliable estimates of uncertainties and resolution on best fit parameters.

The method has been applied to estimate the fault mechanism of a $M_L = 2$ micro-earthquake recorded at the Campi Flegrei caldera. Results show that the best fitting model parameters are associated to a well defined maximum of the probability density function. The uncertainty on the inverted angular parameters is about 15°. Finally, the comparison of the plane view projections of the probability density functions shows that the resolution on the model parameters for the real case is practically the same as using synthetic data.

Acknowledgements. We would like to thank R. Madariaga for helpful comments and suggestions. K. Aki and an anonymous referee are also acknowledged for their critical review of the manuscript. This work was supported by the Commission of European Comunities and I.N.S.U. (France). This is an I.P.G.P. publication.

References

Aki K, Richards P (1980) Quantitative seismology. Freeman, San Francisco, 913 pp

Aster R, Meyer RP (1988) 3-Dimensional velocity structure and hypocenter distribution in the Phlegraean Field, Italy. Tectonophysics, 199:195–218

Bernard P, Zollo A (1989) Inversion of near source S polarization for parameters of double couple point sources. Bull Seism Soc Am, 79:1779–1809

Bernard P, Zollo A (1988) Inversion of S polarization from near source accelerograms: application to the records of 1980 Irpinia earthquake (Italy). Bonnin J et al. (eds) In: Seisuric hazald in reditenulan regions. Kluwer, Dotdrecht, The Netherlands

CNR (1987) Phlegrean Fields. Rosi M, Sbrana A (ed) Pubbl CNR m. 114, 9

Del Re G, La Volpe G (1988) Lo sciame sismico del 1 Aprile 1984 ai Campi Flegrei: intrusione magmatica o rottura di asperitá? Tesi di Laurea in Fisica Terrestre, Univ Napoli, Italia

De Natale G, Iannaccone G, Martini M, Zollo A (1987) Seismic sources and attenuation properties at the Campi Flegrei volcanic area. Pageoph 125:884–917

Farra V, Bernard P, Madariaga R (1986) Fast near source evaluation of strong ground motion for complex source models. Ewing Symp 6, AGU, Wash

Tarantola A, Valette B (1982) Inverse problem = quest for information. J Geophys 50:159–170

Zollo A, Bernard P (1989) Near source S wave polarization inversion of the 15 October 1979, 23:19 Imperial Valley aftershock records: evidence for anisotropy and for a single double-couple mechanism. Geophys Res Lett. 16:1047–1050

Seismic Activity at Stromboli (Southern Italy) for the Period 1983–1986

S. Falsaperla, B. Martinelli, and R. Schick

Abstract

A set of seismic data recorded at Stromboli during 1983–1986 is analyzed. The events, which are mainly connected to Strombolian activity, were collected at two seismic stations. Consideration of the analyzed data is made in respect to the following correlation functions: autocorrelation functions of released seismic energy and number of eruptions per hour; cross-correlation functions between the hourly number of eruptions and barometric pressure. Analysis of these correlation functions and comparison of eruption sequences with Poisson distributions suggest the existence of critical activity phases, in which a small disturbance force may trigger the transition from one state to another. The seismic energy flux seems to present a useful parameter to quantify the dynamical eruptive processes of Stromboli.

1 Introduction

The persistent eruptive activity at Stromboli represents an important and readily available source of information to help understand dynamic volcanic processes. Strombolian eruptions have been observed for over 2000 years and the activity has always been concentrated in a remarkably well-confined three-crater region (Washington 1917; Schick and Mueller 1988). In spite of the stationary nature of the eruption sites in space and time, the frequency of the "Strombolian" eruptions depends strongly on time. The volcanic eruptions are associated with seismic activity and several authors refer to the shocks as explosion-quakes. The origin of the seismic sources is assumed to be few hundred meters below the crater terrace and inside the magma conduits. The seismic magnitude is low, seldom exceeding a value of 2. The reasons for the observed fluctuations in activity might be found inside the volcanic system and are possibly due to a change in physical parameters such as temperature, pressure, liquid-to-gas ratio, viscosity, etc. External factors such as barometric pressure and earth tides, for instance, may generate a detectable influence in the behaviour of the activity. This response can be used to study some dynamic properties of the processes leading to the eruptions. A critical point in the analysis of the signals is given by the choice of the length and stationary nature of the considered data set. Our data set includes periods of highly explosive activity in 1983–1984, as well as the latest effusive episodes from December 1985 to April 1986, which were among the longest this century. The present work deals with an analysis of seismically

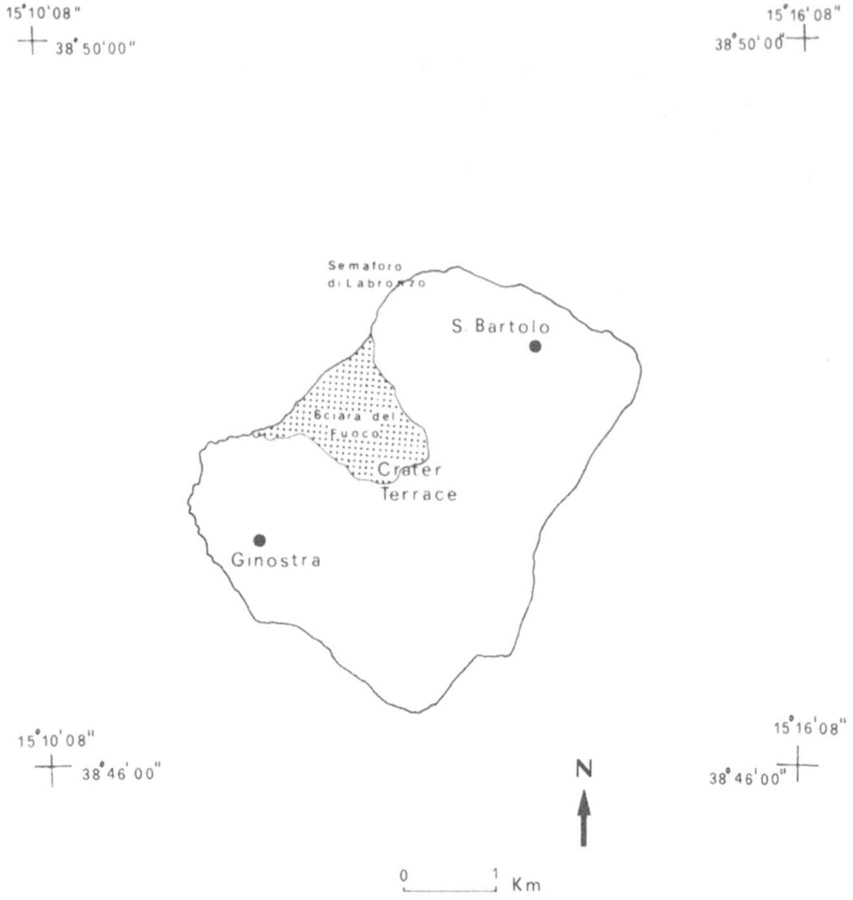

Fig. 1. Schematic map of the island of Stromboli. The *circles* show the location of the Ginostra and S. Bartolo seismic stations

recorded eruption sequences, with an emphasis on the seismic energy flux and on the correlation between eruption frequency and barometric pressure.

2 Analysis of Data

The analyzed seismic signals were recorded at two stations, San Bartolo and Ginostra. The station to crater distances of 1.8 km are approximately equal. The stations are located to the northeast and southwest of the crater area (Fig. 1). The Ginostra station is part of the Aeolian Island Seismic Network (Falsaperla and Neri 1986). Figure 2 shows the relative maximum amplitudes of the two stations for different shocks. The shocks have similar waveforms, with an emergent

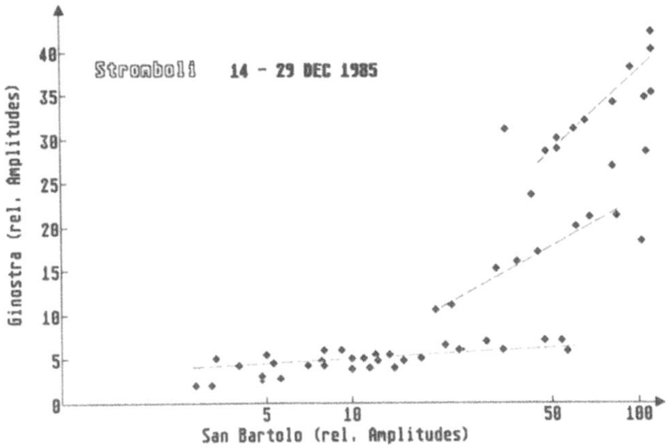

Fig. 2. Comparison of amplitudes of the shocks recorded on seismograms by the S. Bartolo and Ginostra stations. The *dashed lines* show three alignments of points referring to three possible sources

onset and a relatively long duration. Furthermore, the frequency content does not change with the amplitude (Falsaperla and Pellegrino 1988), which leads one to suppose that the wave radiation field was generated from similar source volumes. Moreover, this absence of change is not in agreement with the usual radiation characteristics of tectonic, i.e., shear fracture, shocks. Figure 3a illustrates an amplitude-distance profile from the crater region to Semaforo Labronzo (Fig. 1). The measurements were performed in October 1974 with four portable stations. A comparison with theoretically calculated amplitude distance relations (Fig. 3b) shows that for distances of less than 1000 m from the crater area, a strong wave path influence exists (Del Pezzo et al. 1974), whereas for distances greater than that there would seem to be an absence of such an influence. Referring again to Fig. 2, three branches in the amplitude distributions can be clearly distinguished. Assuming a homogeneous wave propagation model, the amplitudes at the stations depend only on source distances and source strengths. The three branches fit well to linear functions, the slopes of which depend on the source location only. An increase of the slope means a source location closer to Ginostra. Unfortunately, on the basis of the available recordings, one may not attribute the various recorded events, with any certainty, to one of the three craters.

The values for the released seismic energy were obtained by squaring the integrated absolute values of the ground velocity of the seismic signals, and thus, do not represent the exact value of the kinetic energy of the ground motion. In the following report we have called these values pseudo energy (PE). An example is shown in Fig. 4a,b, where a comparison between the number of events per hour is made with the hourly released energy. For a limited time window (Fig. 4a), an inverse relationship can be observed. A comparison between the autocorrelation functions for representative energy and for number-of shocks time series (Figs. 5 and 6, respectively) shows that the energy is more

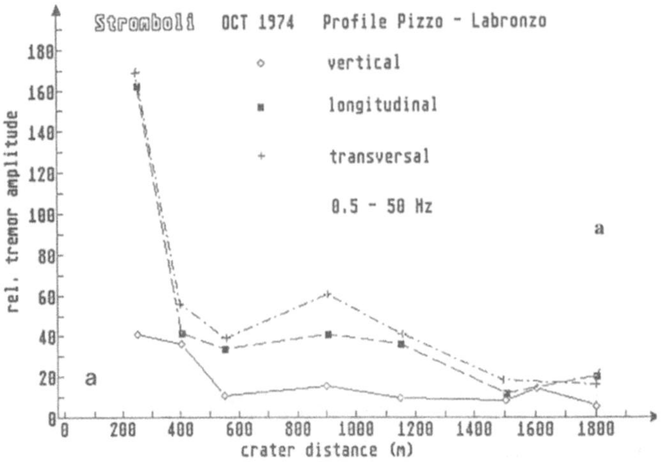

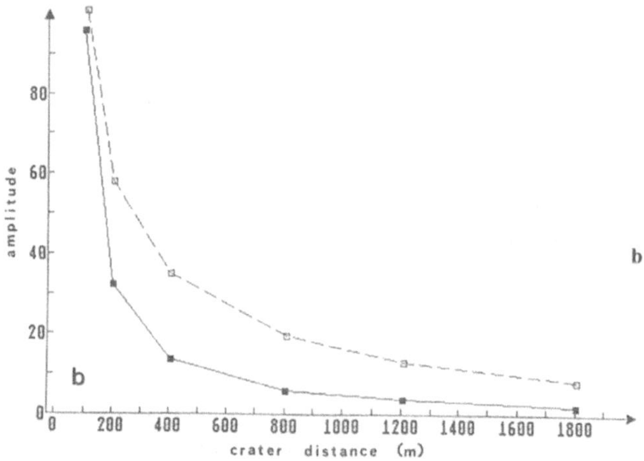

Fig. 3a,b. Tremor amplitude (in arbitrary units) from the crater terrace to Semaforo Labronzo. **a** Real profile in a 0.5–50 Hz frequency range; the *three symbols* refer to the three components; **b** theoretical profile (the *empty* and *full squares* refer to body and surface waves, respectively)

strongly correlated than the number of shocks. The cumulative seismic energy release can be surprisingly well approximated by piecewise linear segments (Fig. 7). The changes in the slopes of these relations apparently appear more frequently in times of fluctuating Strombolian activity. In February 1983, for example, an activity period characterized by many high amplitude events was followed by one in which a greater number of low amplitude events dominated. This transition is associated with a marked and sudden change in the slope of the linear trend for PE. Similar observations in the cumulative number of events were also reported by Schick and Müller (1988) for Stromboli and by Fadeli (this Vol.) for Merapi, Indonesia.

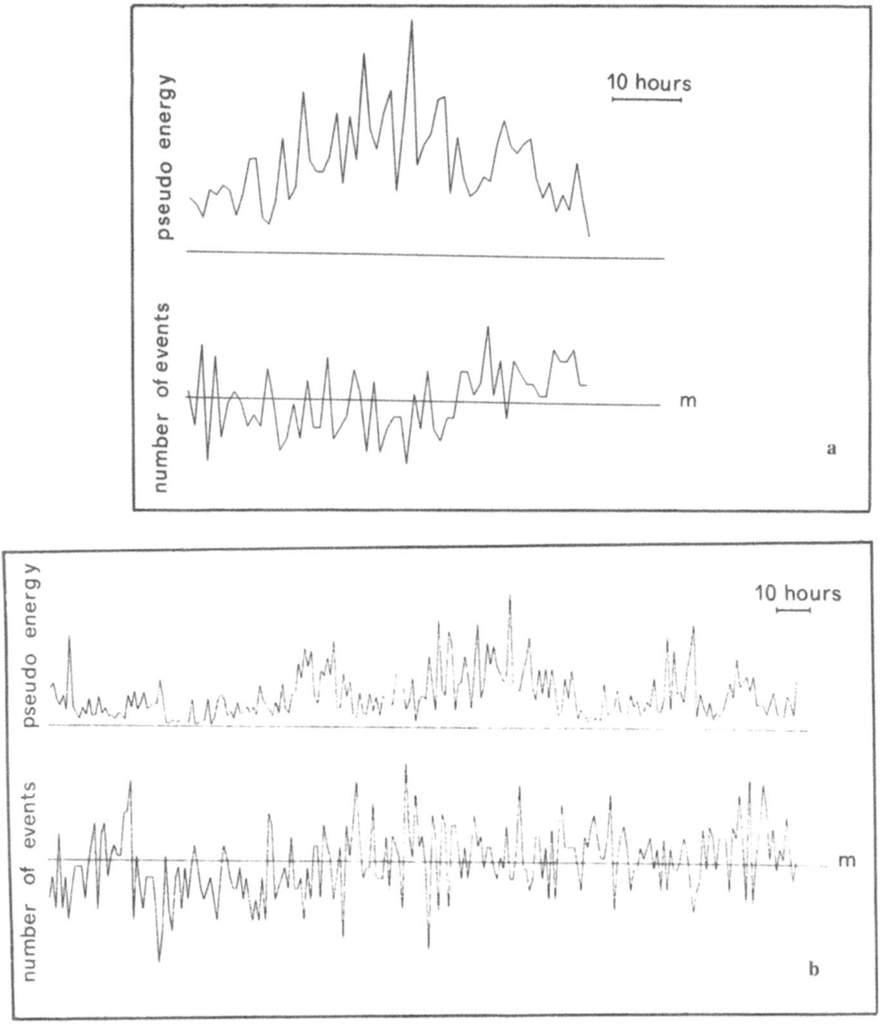

Fig. 4a,b. Comparison between numbers of events per hour (the graph shows how the shocks are distributed around the mean value *m*) and pseudo energy: **a** 24 September–4 October 1984; **b** 14–21 December 1985

An analysis concerning autocorrelation functions and Poisson distributions suggests the existence of "critical" activity phases, in which a small internal or external disturbance may trigger the transition from one state of equilibrium to another. The autocorrelation functions obtained from time series consisting of number of events/hour lead to different results for different observation times (Fig. 8a,b). Significance at the 5, 1 and 0.1% level is shown. It is interesting to see that the autocorrelation functions change with the externally observed volcanic activity. The series referring to Fig. 8b was observed during a time in

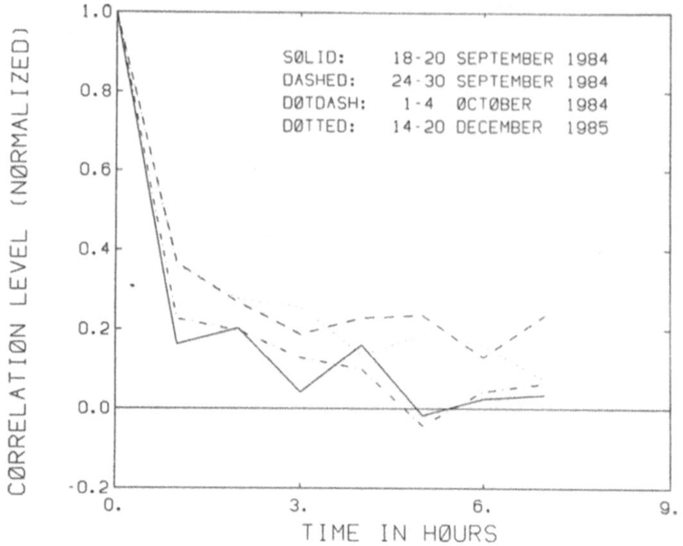

Fig. 5. Normalized autocorrelation functions of pseudo-energy

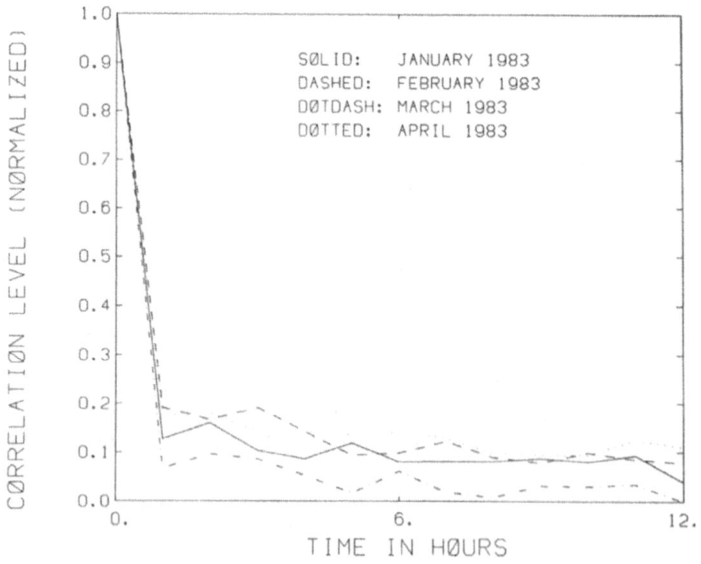

Fig. 6. Normalized autocorrelation functions of the number of events

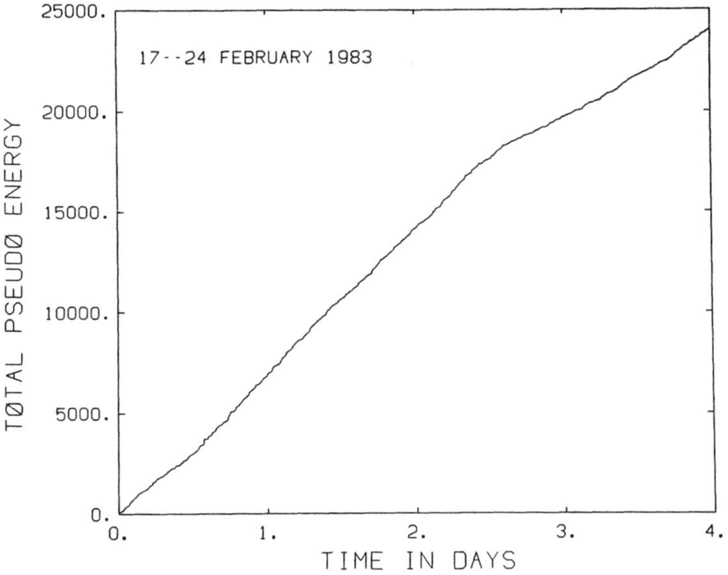

Fig. 7. Temporal pattern of cumulative pseudo-energy PE (PE is expressed in arbitrary units) during 17–24 February 1983

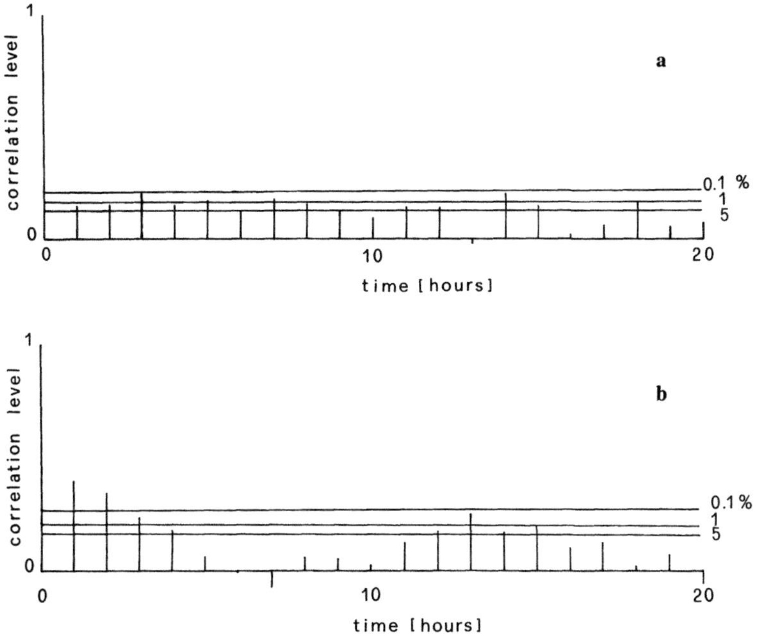

Fig. 8a,b. Autocorrelation functions of the number of events: **a** 24 September–4 October 1984; **b** 14–21 December 1985. The functions have been normalized. The *horizontal lines* show the significance levels deduced by the chi-square test at 5, 1 and 0.1%

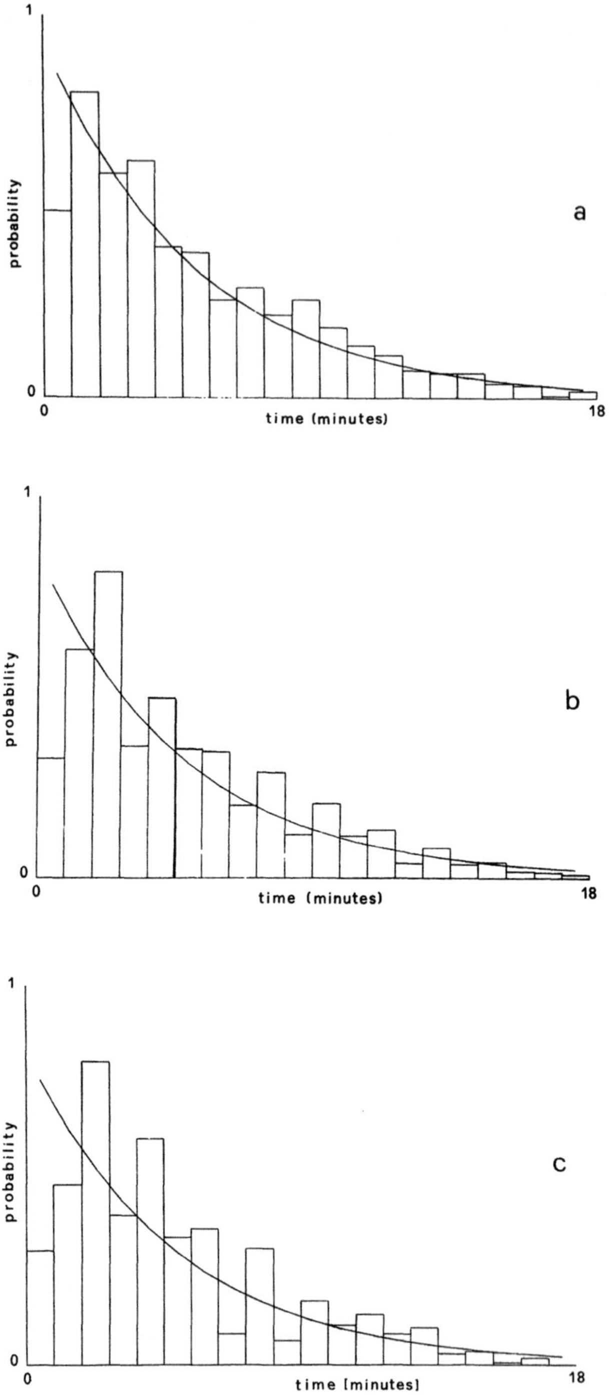

Fig. 9a–e. Theoretical Poisson distributions matched with the observed ones (each time interval for the experimental distribution is equal to 0.9 min): **a** 1–4 September 1984; **b** 12–17 September 1984; **c** 18–20 September 1984; **d** 24 September–4 October 1984; **e** 14–21 December 1985

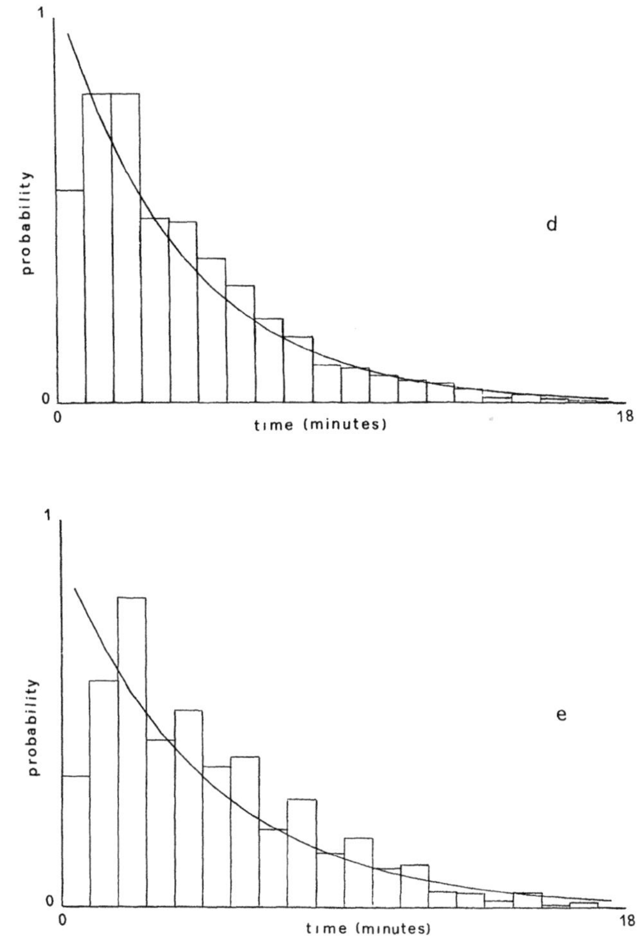

Fig. 9d,e. *Continued*

which a lava flow occurred. Figure 8b shows a significant approximately semidiurnal periodicity.

The distribution of the time intervals between two shocks, which occur during periods of Strombolian acitivity, deviates segnificantly from a Poisson distribution at a repose times of less than 2 min (too few shocks) and at those of 2–3 min (too many shocks) (Fig. 9a–d). Similar deviations are observed in Fig. 9e which refers to a period of effusive activity in December 1985.

A comparison between barometric pressure and the hourly occurrence of shocks has been attempted. The aim of this study was to determine if a stationary time delay exists between the two data series (Fig. 10a,b), and if so, to characterize this time delay. Our analysis refers to the period from January to April 1983. In determining the relative time delay, cross-correlation methods

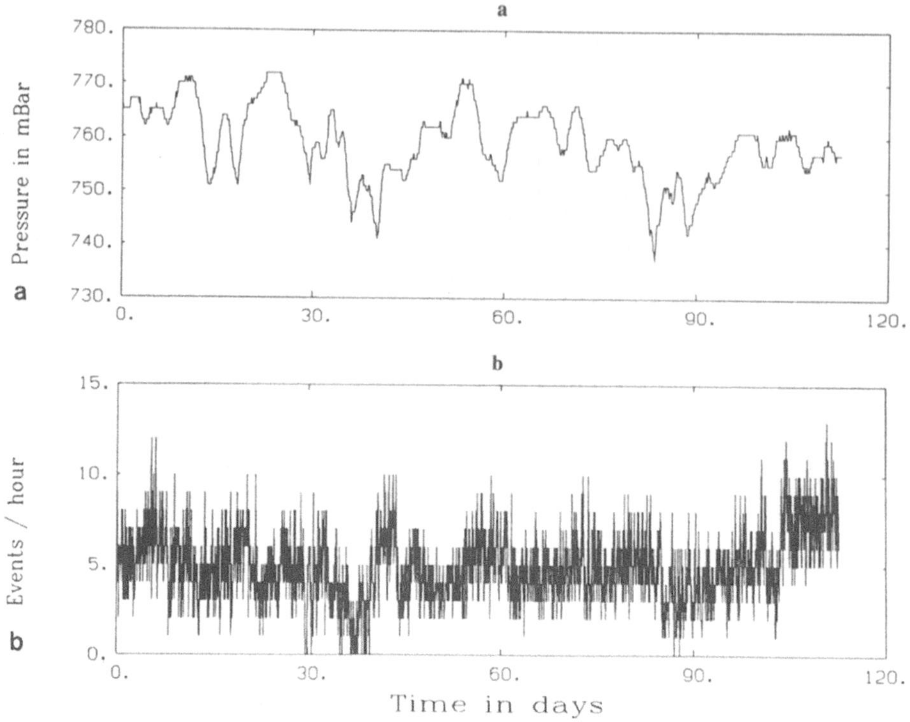

Fig. 10. **a** Barometric pressure and **b** temporal pattern of number of shocks/h. The time span considered is from 1 January to 28 April 1983

were used. A data window of duration D = 512 h (about 21 days) is applied successively to the available data in the overlapping interval (O, D), (Δ, Δ + D), . . . , ((N − 1)Δ, (N − 1)Δ + D). Δ = 336 h is the shift that each adjacent data window undergoes and N = 14 is the total number of segments employed. Figure 11 shows that barometric pressure and hourly occurrence of shocks are not joint stationary processes. To model the possible influence of barometric pressure on the seismic activity as a stochastic dynamic process an intermediary variable, the state, must be involved in an explicit way. Deeper insight in the fluid dynamic behaviour of magma will provide the necessary relations. The influence of other external forces (e.g. tides) should be considered at the same time.

3 Discussion and Conclusions

The analyzed data show that the seismic energy flux represents an additional useful function to quantify eruptive activity. A similar observation was made for Strombolian eruptions on Krakatau by Decker and Hadikusumo (1961). The

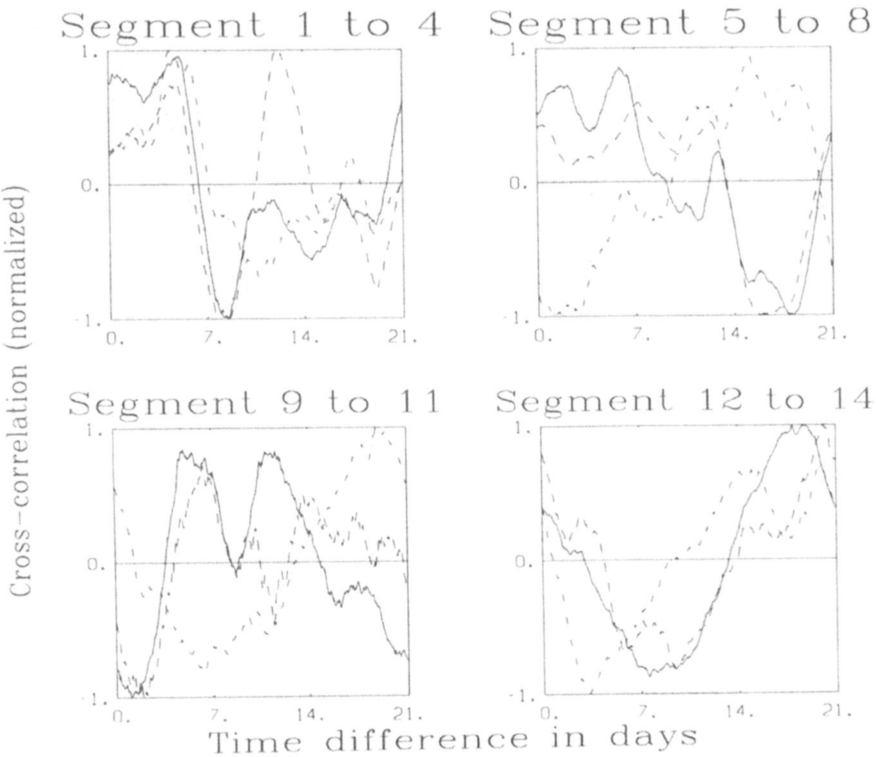

Fig. 11. Cross-correlation for the number of shocks/h and barometric time series over periods of 21 days (1 January–28 April 1983). The *lines* in each diagram represent different time segments shifted by 1 week in relation to one other. *Solid lines* correspond to the first week, *dashed lines* to the second week, *dot-dashed* lines to the third week and *dotted* lines to the fourth week

energy flux has a significantly higher tendency of conservation than, for instance, the shock rate. It means that the energy flux can be modelled using stochastic Markov-chain processes which allow a better estimation of future conditions of the states of the volcano. Furthermore, it will be very interesting to follow the seismic energy flux for a longer period of time. Future observations may provide evidence, if histograms of the energy flux in the time domain show discrete energy bands. The investigation of these energy bands would allow a much deeper insight into the dynamics of a volcano than is presently possible. A similar proposal was made by Carta et al. (1981) who studied the eruption sequences of Mt. Vesuvius volcano.

The triggering action between different states of volcanic activity by the influence of small external forces deserves careful analysis in future. The influence of barometric pressure on volcanic activity should not be considered independently because other phenomena of the same size (e.g. Earth tides) could have a similar effect. It will be interesting to study the mutual influences of internally and externally acting forces, which effect the dynamic conditions of the magma flow, by using a more comprehensive data set. Among externally acting forces are: tidal forces, barometric pressure, wind, temperature-induced stress, rainfall loading, sea-wave surges and oceanic loading. The observation of the seismic energy level transitions (state) and their correlation to known external factors (input) might lead, with the methods of the mathematical system theory, to a modelling of volcanic activity. The field of flow acoustics should help in providing the necessary basic equations.

Acknowledgements. The authors are grateful to Dieter Seidl and Walter Zuern for helpful discussions and contributions. This work was supported by Grant No. 203.05.12 of the Consiglio Nazionale delle Ricerche.

References

Carta S, Figari R, Sartoris G, Sassi E, Scandone R (1981) A statistical model for Vesuvius and its volcanological implications. Bull Volcanol 44-2:129–151

Decker RW, Hadikusumo D (1961) Results of the 1960 expedition to Krakatau. JGR 66, 10:3497–3511

Del Pezzo E, Guerra I, Lo Bascio A, Luongo G, Nappi G, Scarpa R (1974) Microtremors and volcanic explosions at Stromboli, Part 2. Bull Volcanol 38, 4:1023–1036

Fadeli A (1984) A study of the eruption mechanism of volcano Stromboli (Italy). Inst Geophys, Univ Stuttgart, FRG, Publ 213

Falsaperla S, Neri G (1986) Seismic monitoring of volcanoes: Stromboli (southern Italy). Per Mineral 55:153–163

Falsaperla S, Pellegrino A (1988) Analisi dei transienti sismici dello Stromboli (Isole Eolie) del settembre 1987. Open-file Rep 27/88, IIV, CNR, Catania

Schick R, Mueller W (1988) Volcanic activity and eruption sequences at Stromboli during 1983–1984. Modeling of volcanic processes. Vieweg, Germany

Washington HS (1917) Persistence of vents at Stromboli and its bearing on volcanic mechanism. Bull Geol Soc Am 26:249–278, 6–9

Wave Polarization and Location of the Source of the Explosion Quakes at Stromboli Volcano

E. Del Pezzo, C. Godano, A. Gorini, and M. Martini

Abstract

Portable 1 Hz, three component, digital seismic stations were used to investigate the spectral content and the wave composition of the volcanic earthquakes at Stromboli, Aeolian Islands, - Southern Italy. Two successive surveys were carried out in spring 1986 and summer 1988 and more than 1000 seismograms were recorded. Two distinct phases are visible on the seismograms. The first phase is that of the first arrival and is peaked at 1 Hz. The second phase begins two seconds later and shows two peaks at 2 and 5 Hz. Two different techniques were used to investigate the properties of the wave polarization. The first is based on the pattern of the wave vector along the seismogram, while the second one analyzes the properties of the covariance matrix among the three components of the signal. Both methods lead to the same results. Linear polarization with a mainly horizontal ground particle motion is found for the first phase of the seismograms. This phase is presumed to be composed mainly of P waves coming from the source. The pattern of polarization shows a possible source area located some hundreds of meters north of the crater with linear dimensions of about several hundreds of meters. The existence of two phases on the seismograms cannot be explained using simple step pressure models as, for example, that recently developed by Crosson and Bame. The present experimental evidence has to be used as a new constraint for future development of theoretical models.

1 Introduction

Stromboli Volcano, Aeolian Islands, southern Italy, is characterized by a persistent moderate explosive activity, accompanied by low energy seismic events (not greater than 10^{12} ergs, Capaldi et al. 1978). The seismic activity at Stromboli has an approximately constant rate of occurrence (about 10–12 events per hour) providing the opportunity for volcanologists to obtain much valuable data in short time intervals to develop knowledge of explosive volcanic processes (e.g., Settle and McGetchin 1980).

Volcanic earthquakes are a distinct class of earthquakes which are directly generated in some way by the volcanic process itself. Explosion quakes are a subset of volcanic earthquakes, and are directly associated with the explosive process. Beneath active volcanoes earthquakes also occur, with dominant frequencies of 1–5 Hz, and a lack of clear phase arrivals. These events are called low frequency events and are similar to the B type events classified by Minakami (1974). Their spectra present most power in a frequency range which is narrower than that of the tectonic events. Fehler and Chouet (1982) demonstrate that the narrowness of the seismic spectrum in low frequency events is due to the source rather than the path. Ferrick et al. (1982) propose that unsteady fluid flow in volcanic conduits is the source mechanism of low frequency volcanic earthquakes. Schick (1981) reports that basic concepts of fluid transient theory can be applied to analysis of tremor at Mt. Etna. Crosson and Bame (1985) developed a theoretical model for the source of the low frequency volcanic earthquakes (hereafter called LF events through this paper). LF events are modelled by a spherical cavity filled with magma which is excited by the occurrence of a pressure step inside. The motivation of this model was an observation carried out by L. Bascio et al. (1973) at Stromboli, where they found LF events originating before the volcanic explosion (a few seconds). The expansion of gas in a volcanic conduit is believed, by the same authors, to be responsible for the source mechanism. Del Pezzo et al. (1974) carried out a successive seismic survey at Stromboli to investigate the spectral content of the volcanic quakes. They found that spectral peaks persisted around 5 Hz and confirmed the existence of double shocks, about 10 s apart from each other. The second shock of this pair is linked to the explosion, as shown by the same authors using a microphone connected to the magnetic tape of the seismic traces. Anyway not only pairs, but also a great number of single shocks are visually connected to the explosions.

Are shocks recorded at Stromboli real explosion-quakes or LF events? A doubt arises because the spectral content and the waveform suggest similarity with the LF waveforms recorded with similar instruments at Mt. S. Helens (Malone 1983) but on the other hand, strict correlation occurs between the explosion itself and a great number of the events recorded at Stromboli.

To investigate further the experimental aspect of the quakes recorded at Stromboli, two seismic surveys were carried out in spring 1986 and summer 1988 using high-dynamic range, three-component digital equipment.

The aim of this paper is to study the wave composition of the explosion-quakes, the spectral content of the seismic phases that compose the seismograms and the properties of the polarization of the signals.

These elements may put constraints on the definition of the theoretical models of the source.

2 Data

Data were collected using seismic stations equipped with three 1 Hz geophones (Mark L4c). This equipment (Lennartz, model 5800) consists of high-dynamic range (120 dB) digital instruments with local PCM recording onto magnetic tape. The sampling rate was held constant at 125 Hz.

In the first survey, carried out in May 1986, the stations were positioned along a profile oriented from the base of the volcano toward the craters (Fig. 1). One of the stations was in operation almost continuously over three days (location S99 of Fig. 1). The other two stations recorded contemporaneously at a couple of locations along the profile. After two hours of continuous recording the two of stations were moved to occupy two other point locations. One of the new point locations was coincident with the previous one. In this way, nine

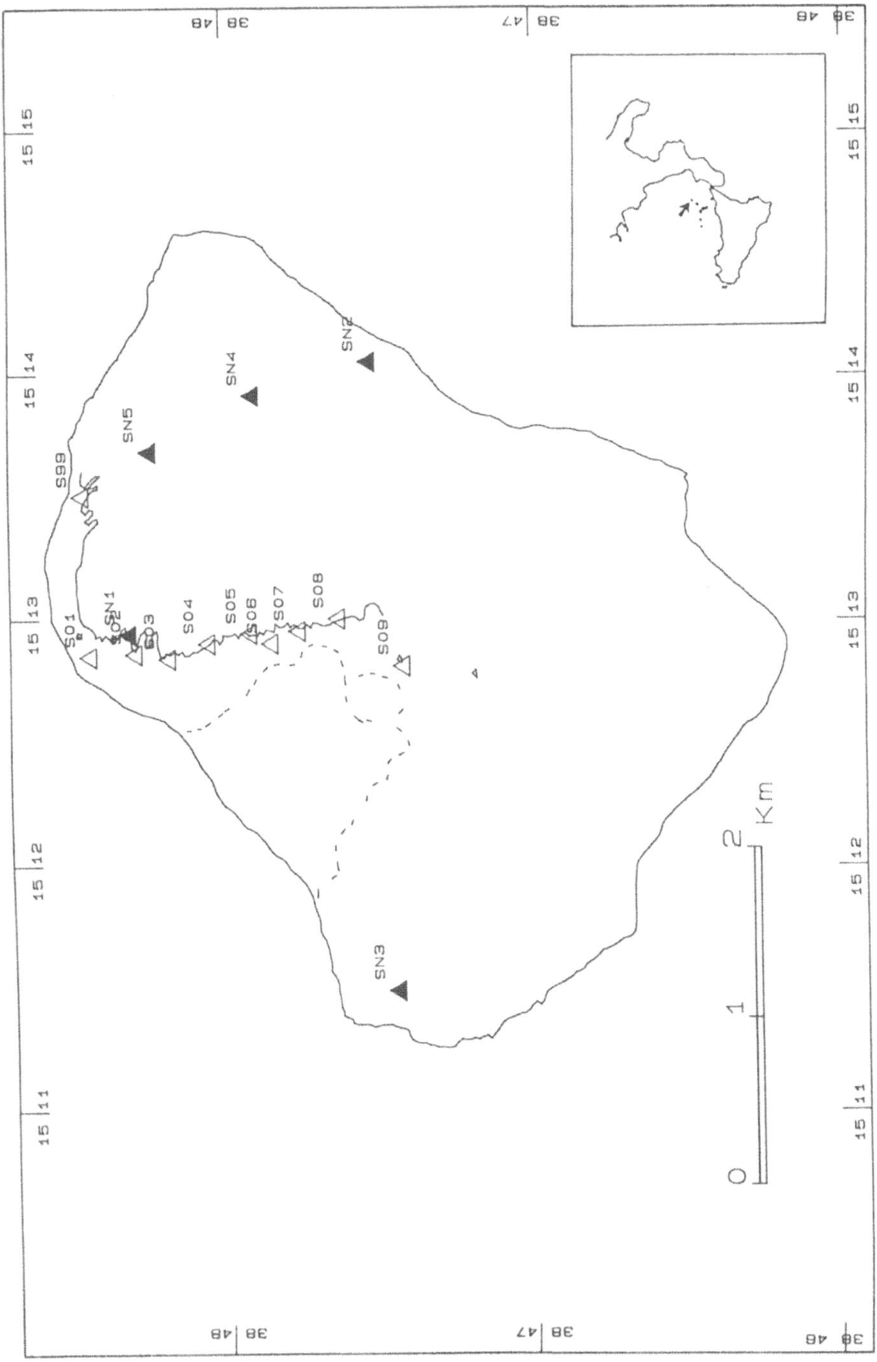

Fig. 1. Map of the island of Stromboli. *Triangles* show the location of the seismic stations. *Black symbols* show the stations location of the second survey. *Dashed lines* show the contours of "Sciara del Fuoco"

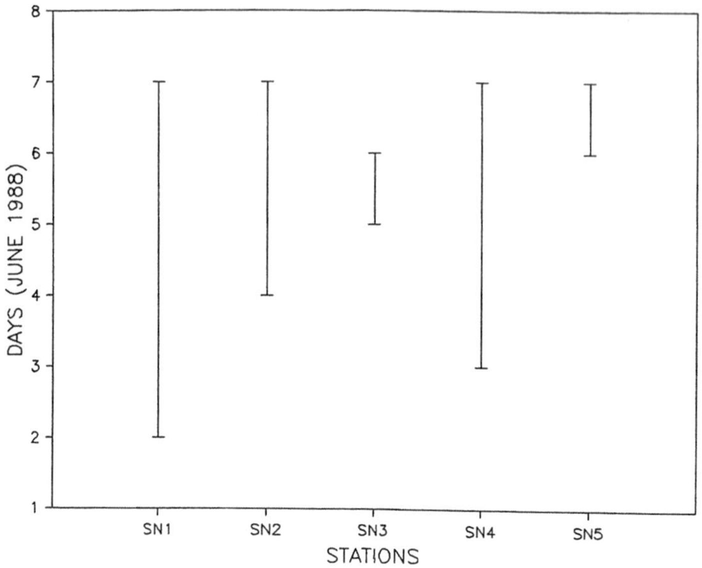

Fig. 2. Time functions of the stations for the survey of summer 1988

points were covered in eight steps and in each step, a couple of stations recorded for a period of 2 h.

The second survey was carried out in June 1988 and the stations were positioned radially with respect to the crater area. The southern zone of the island was not covered due to some logistic difficulties.

The time operation table of the seismic stations is indicated in Fig. 2. Calibration of this portable equipment was fully carried out in laboratory prior to deployment in the field. During the surveys, a step-like current pulse was sent to the input of the calibration coil at each change of tape (1 h) for a continuous check of the calibration. This operation was considered necessary in order to eliminate variations in the response of the horizontal components due to possible defects in setting geophones horizontal by means of a bubble level. Some tilting of the components occur frequently when geophones are buried in loose volcanic materials like those composing the terrain at Stromboli.

More than 1000 events were recorded in both surveys. One station point was mantained in the same location during both surveys (see SN1 and SO2 of Fig. 1).

Visual observations of the shape of events recorded at SN1 and SO2 show that the seismic activity is apparently not changed over time. No variations in the shape of the waveforms, recorded at Stromboli, was found by Falsaperla et al. (1986) during long time periods of seismic monitoring, characterized by "normal" volcanic activity.

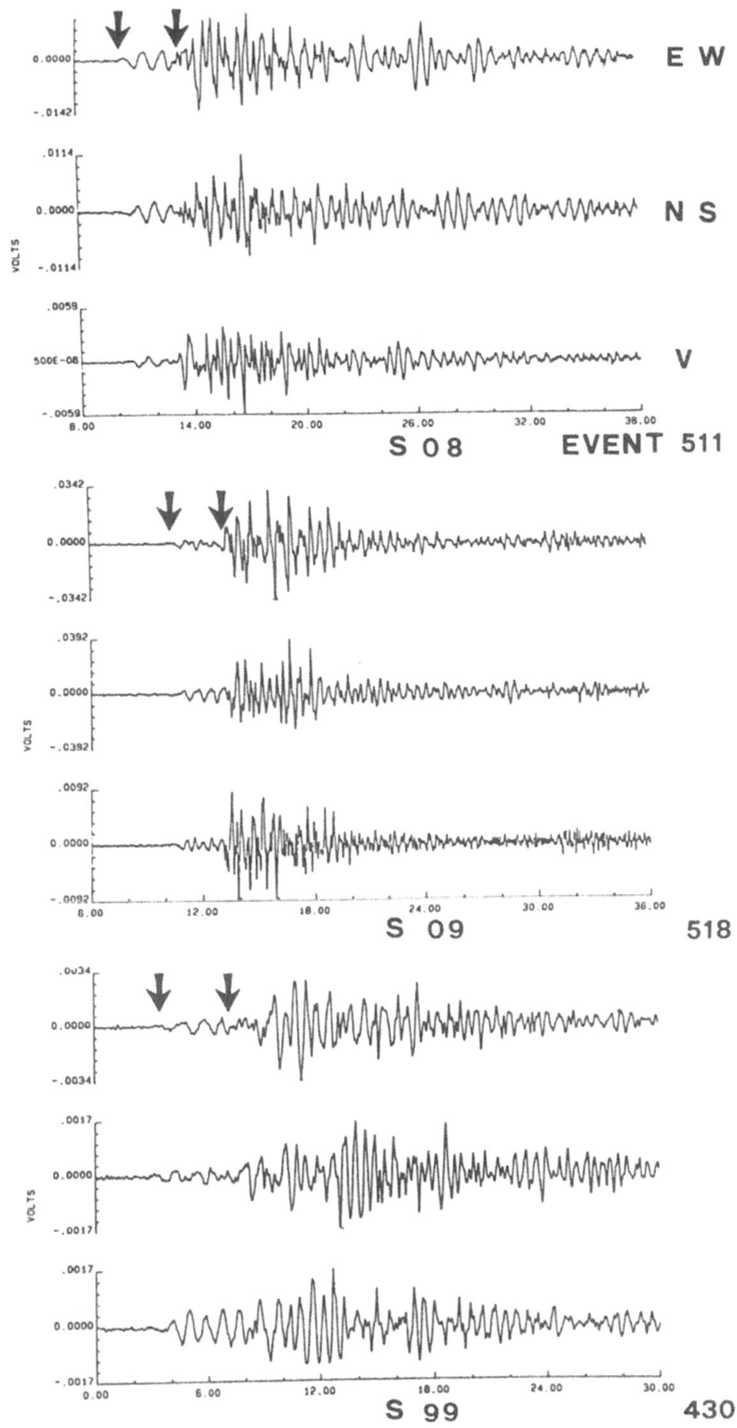

Fig. 3. Some examples of the quakes recorded at Stromboli on the three components of the ground motion. Two clear seismic phases are visible in the first 5 s of each seismogram (*arrows*)

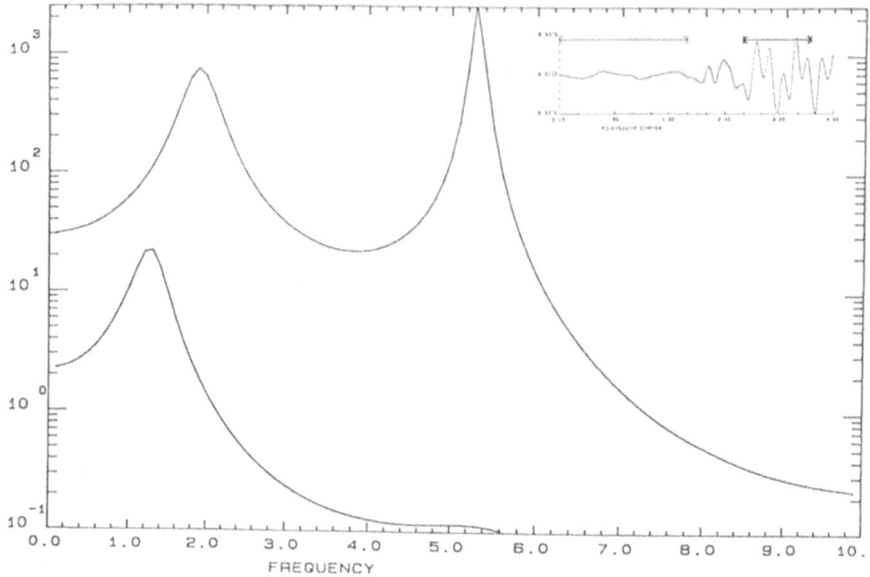

Fig. 4. Typical power spectrum of first and second phase, obtained using MEM technique. Distinct peaks are evident on 2 and 5 Hz for the second phase. 1 Hz is predominant for the first phase. Vertical axis is the spectral power in arbitrary units

2.1 Spectral Composition of the Quakes

On the greater part of the seismograms along the profile, two seismic phases are clearly visible. The first one is the phase of the first arrival and is characterized by a duration of about 2 s and a predominant frequency of about 1 Hz. The second phase starts 2 s after the first arrival with wider amplitudes and higher frequency content (Fig. 3).

At stations SN2, SN3, SN4 and SN5 these two phases are not always clear or distinct, probably due to the highest noise level.

The spectral content of the first 5 s of the seismograms was investigated using both Maximum entropy (MEM) techniques (Ulrich and Bishop 1975) and classical FFT methods. The MEM method was used to obtain the best resolution on the spectral peaks.

Three distinct peaks are evident at 1, 2 and 5 Hz. The 5-Hz peak is less pronounced. 1 Hz is predominant in the first phase, 2 and 5 Hz in the second phase. Figure 4 shows a typical MEM spectrum.

Riuscetti et al. (1977) report spectra of tremor waveforms at Etna and Stromboli which exhibit a two-peaked nature, with values at 2.5 and 4 Hz. Tremor power spectra obtained during the present survey show approximately the same spectral content of the quakes (Fig. 5) according to the results of

Fig. 5. Example of Fourier spectrum of tremor and quake. The signal has maximum power in the 1–5 Hz band in both cases

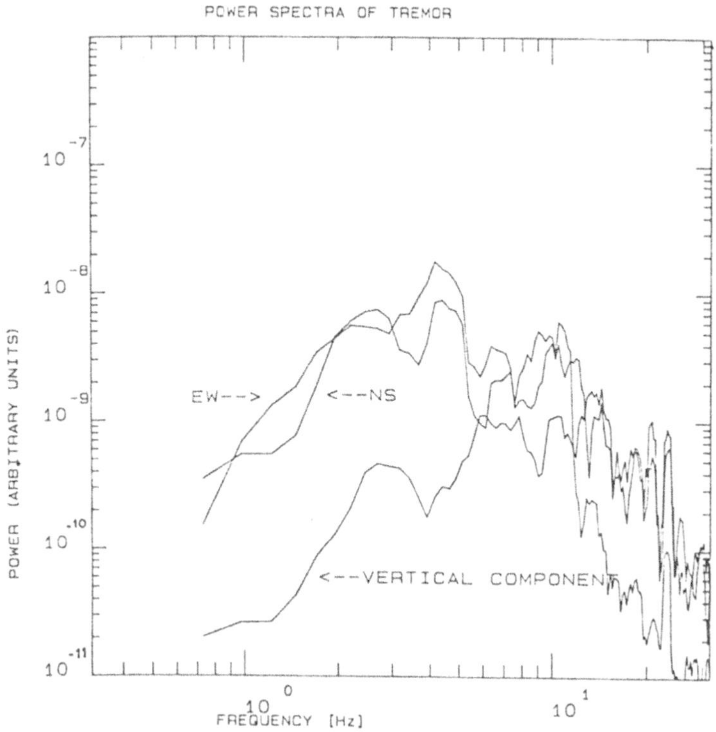
POWER SPECTRA OF TREMOR
POWER (ARBITRARY UNITS)
10^{-7}
10^{-8}
10^{-9}
10^{-10}
10^{-11}
EW-->
<--NS
<--VERTICAL COMPONENT
10^{0}
10^{1}
FREQUENCY [Hz]

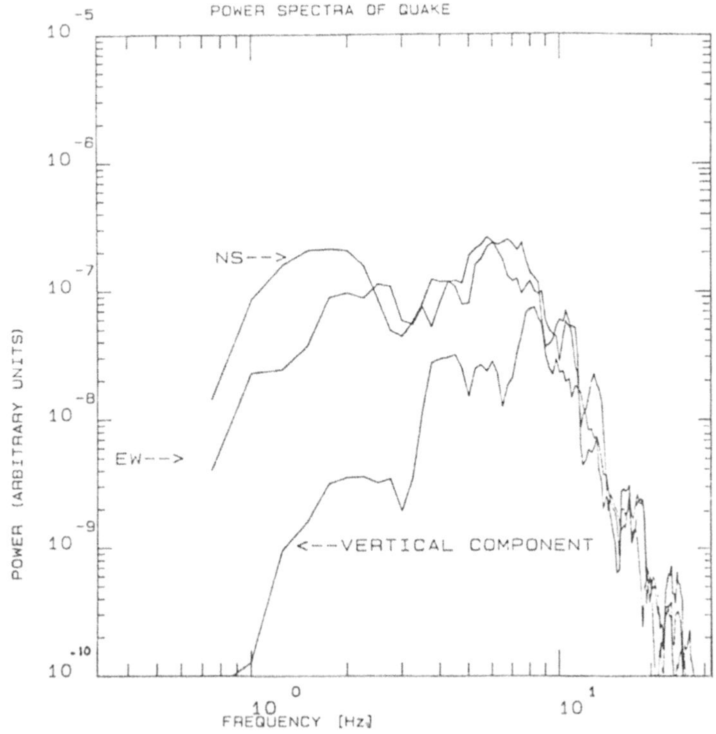
POWER SPECTRA OF QUAKE
POWER (ARBITRARY UNITS)
10^{-5}
10^{-6}
10^{-7}
10^{-8}
10^{-9}
10^{-10}
NS-->
EW-->
<--VERTICAL COMPONENT
10^{0}
10^{1}
FREQUENCY [Hz]

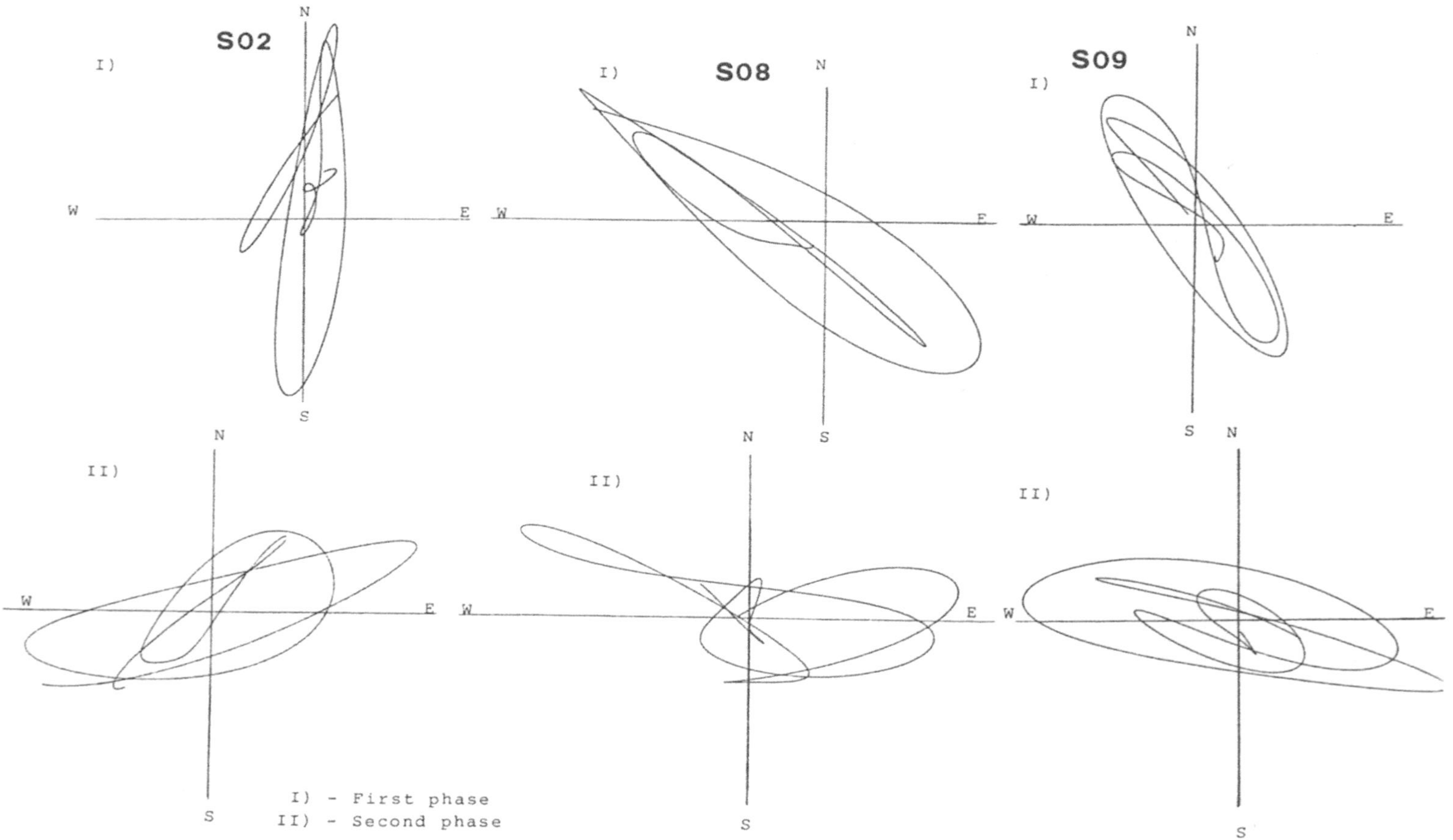

Fig. 6. Examples of particle orbit diagrams in the horizontal phane. A predominant direction can be easily noted

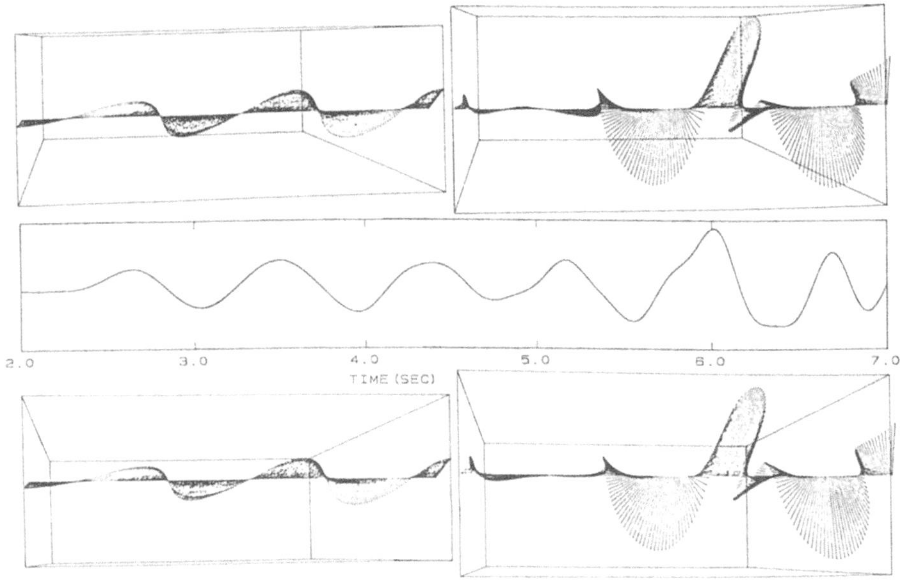

Fig. 7. Stereo plot of the ground particle velocity vector. First phase shows focused energy in an approximately radial direction toward craters. A wider distribution of vector directions is evident on the second phase

Riuscetti et al. (1977). This supports the hypothesis that harmonic tremor and volcanic earthquakes may have a common source. This hypothesis is supported by many authors (Fehler 1983; Crosson and Bame 1985). Chouet (1985) proposes that LF earthquakes are the "Green functions" of the harmonic tremor.

3 Polarization of the Waves and Location of the Source

An estimate of the source location that generates the explosion-quakes at Stromboli is possible by analyzing the polarization pattern of the seismic waves.

A great number of the earthquakes recorded at Stromboli are directly linked to the explosive process. The visible explosions occur at the crater area (Fig. 1) with throwing of pyroclastic materials, so it is reasonable that the source of seismic signals could be close to the crater area. Capaldi et al. (1978) using a tripartite array of seismic stations showed that most seismic activity is shallow and comes approximately from the direction of craters, but we have no information as to the size of the seismic source. A check of the results of Capaldi et al. (1978) is impossible using classical methods of earthquake location.

Three component recording allows the reconstruction of the particle motion of the ground during the quakes. The seismograms obtained were band-pass filtered in the 0.5–2.5 Hz window in order to eliminate high-frequency seismic noise and to isolate the frequency band of the first phase. A zero-phase shift filter was used.

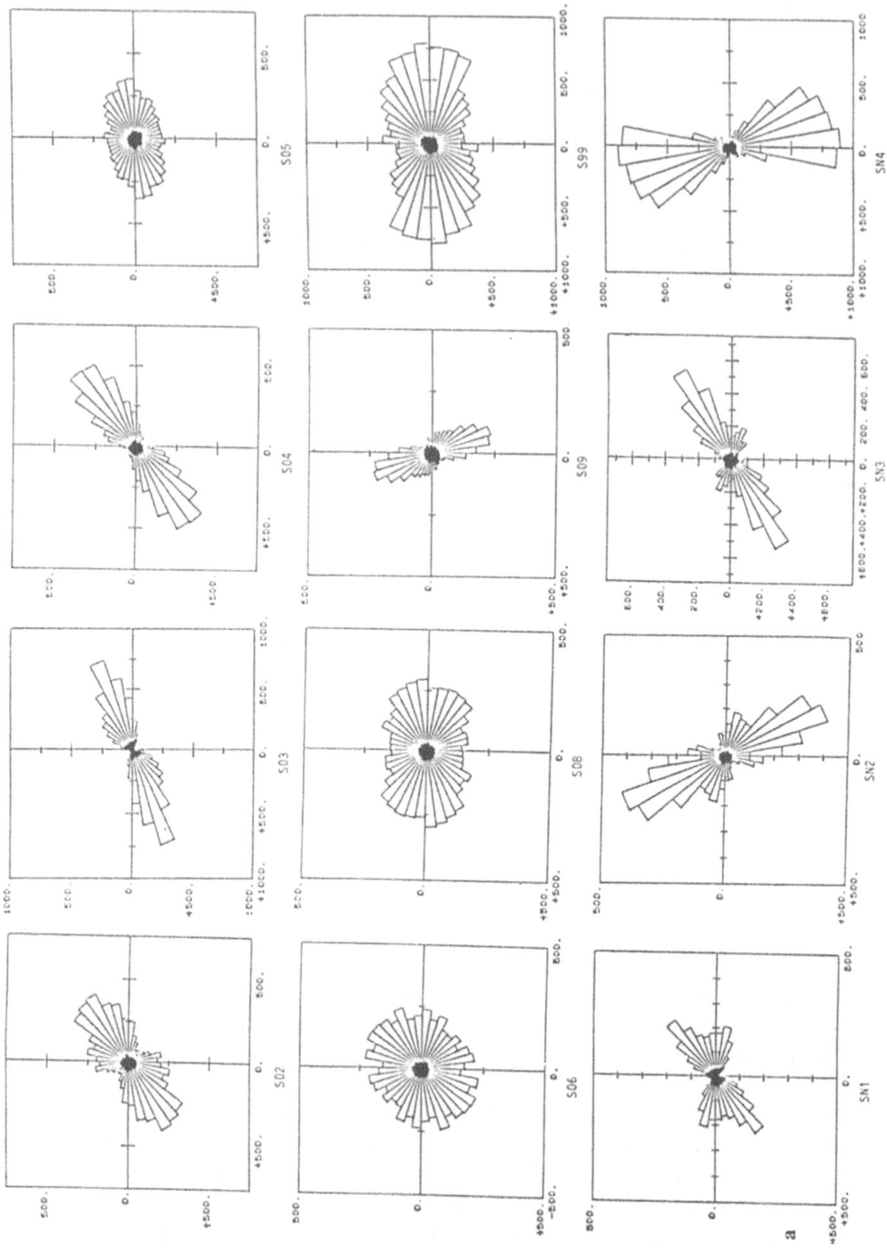

Fig. 8a,b. Predominant azimuths for each station of the seismic profile, for **a** first and **b** second phase

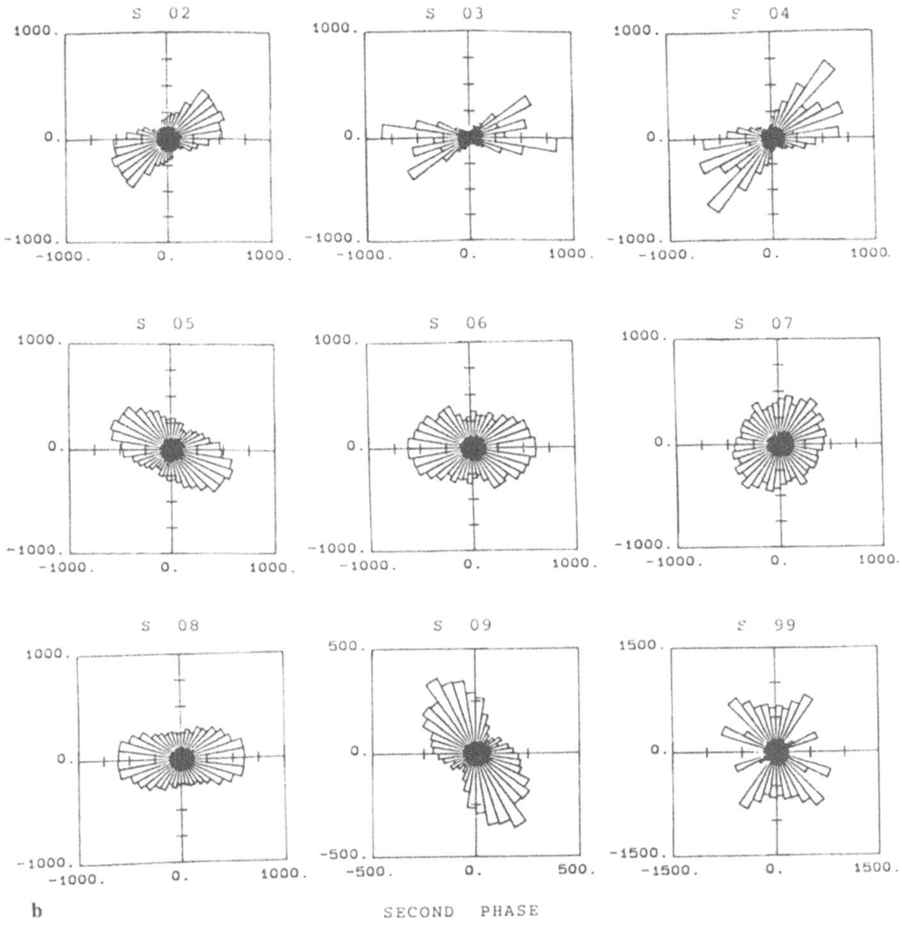

Fig. 8.b. *Continued*

A strong linear polarization of the waves is evident by the particle orbit diagrams in Fig. 6. The ground motion is represented by plotting the position of the particle in the plane containing the ground motion vector. In Fig. 6 a marked, horizontal particle motion is clear for the first phase.

The hypothesis that the first of the seismograms is composed by compressional P waves, coming from the source, is also confirmed by the direction of the vector which describes the ground motion in space. This vector, for Stromboli explosion quakes, is shown in Fig. 7, where a stereo-plot of the particle velocity vector is shown as a function of time for the first 2.5 s (first phase) and for a successive window of 2.5 s (second phase).

Once more, the first phase shows focused energy with a characteristic P wave motion, in an approximately radial direction toward the crater area, while the second phase, shows a wider distribution of the vector directions.

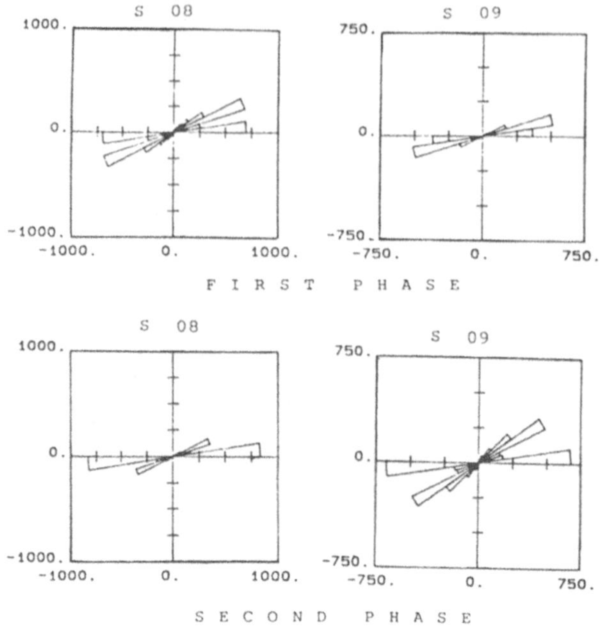

Fig. 9. Incidence angles for stations 8 and 9, the closest to the crater area

To evaluate possible differences between the polarization of the two phases, data were filtered in two narrow bands 0.8–1.2 and 1.8–2.2 Hz, respectively for first and second phase. Values falling below 30% of the maximum amplitude were arbitrarily considered noise-contaminated and rejected.

Predominant directions of the ground motion for each station of both surveys were obtained from the distributions in Fig. 8. The distributions of the azimuth angles for the first phase are reported in these plots. The peaks were considered as an estimate of the predominant directions.

Incidence angles are shown in Fig. 9 for stations 8 and 9.

For the stations of the second survey, without a clear phase separation probably due to a strong microseismic noise caused by many causes (wind, sea, etc.), the azimuth distributions are calculated from the first 1.8 of the pass-band (0.8–1.2 Hz) filtered signal.

In the plots of Fig. 10 predominant azimuths are shown inferred by the plots of Fig. 8, pointing to a zone which is located north of the crater area.

The second phase of the seismograms show much scattering of the data, but in some cases, an azimuth different from that of the first phase can be noticed.

It is unclear whether the waves that compose the second phase have the same nature of those that compose the first. Probably, they are not S-waves because their motion is not normal to the direction of the wave vector.

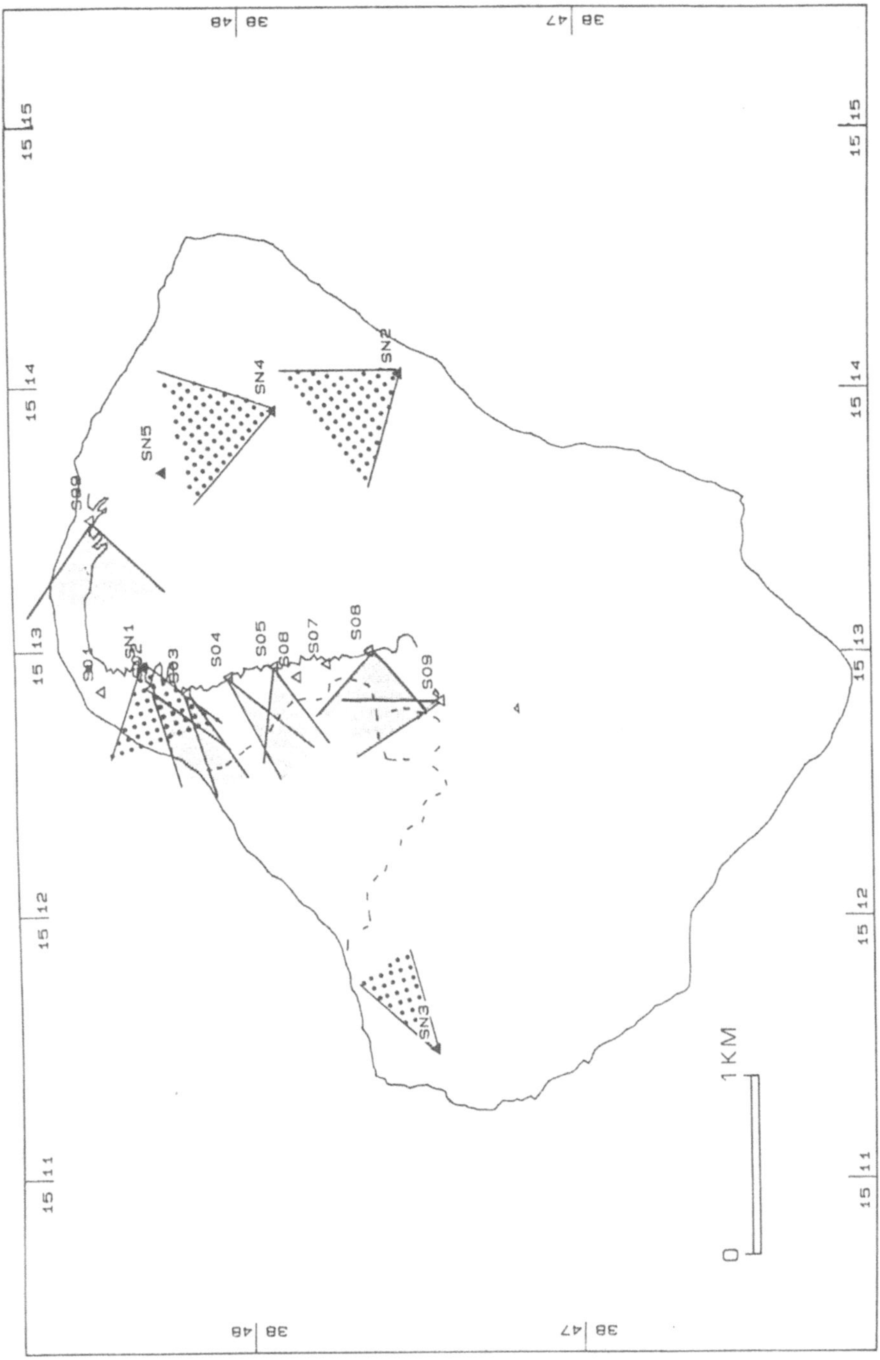

Fig. 10. Direction of polarization for the first phase inferred from the plots of Fig 8. *Dashed angles* show the direction of the wavefronts to the stations for the first survey, while the *dotted angles show* the directions for the second survey. The source seems to be located north of the crater area

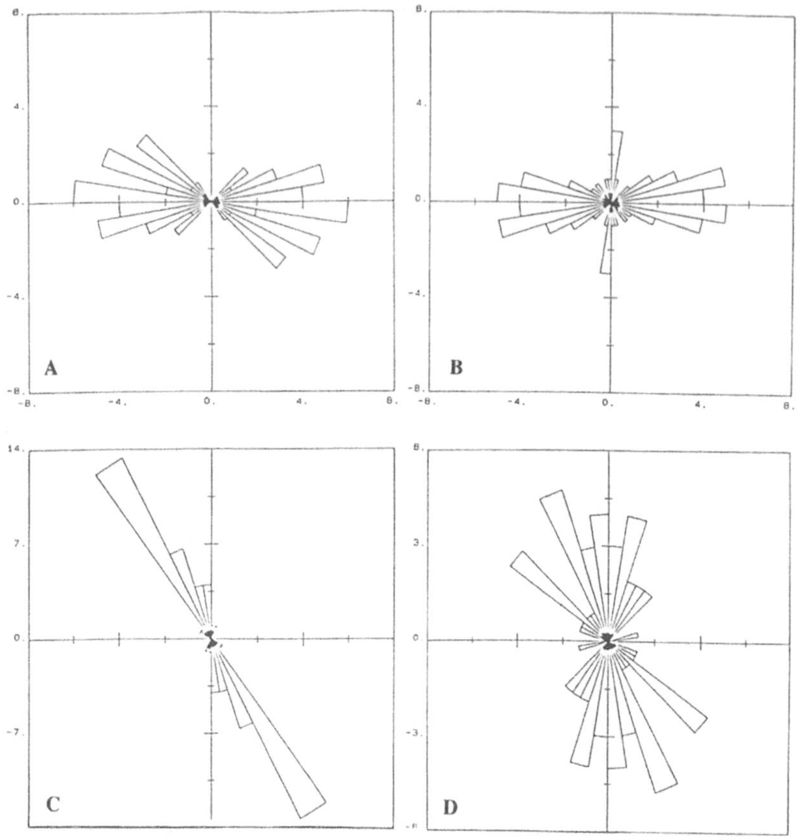

Fig. 11A–D. Incidence angles at station 8 for first **A** and second **B** phase. Azimuths at the same station for first **C** and second phases **D** (covariance matrix method)

3.1 Rectilinearity Filter

In order to obtain measures of rectilinearity and direction of the particle motion, a method described in Montalbetti and Kanasevich (1970) was applied.

The method is based on the evaluation of the covariance matrix for a set of N points of the signal; in our case N is equal to 300, corresponding to 2.4 s;

$$C = \begin{pmatrix} \mathrm{var}(x) & \mathrm{cox}(x, y) & \mathrm{cov}(x, z) \\ \mathrm{cov}(y, x) & \mathrm{var}(y) & \mathrm{cov}(y, z) \\ \mathrm{cov}(z, x) & \mathrm{cov}(z, y) & \mathrm{var}(z) \end{pmatrix}.$$

If the matrix is diagonalized, an estimate of the rectilinearity of particle motion trajectory over a specified time-window can be obtained by the ratio of the principal axes of the matrix.

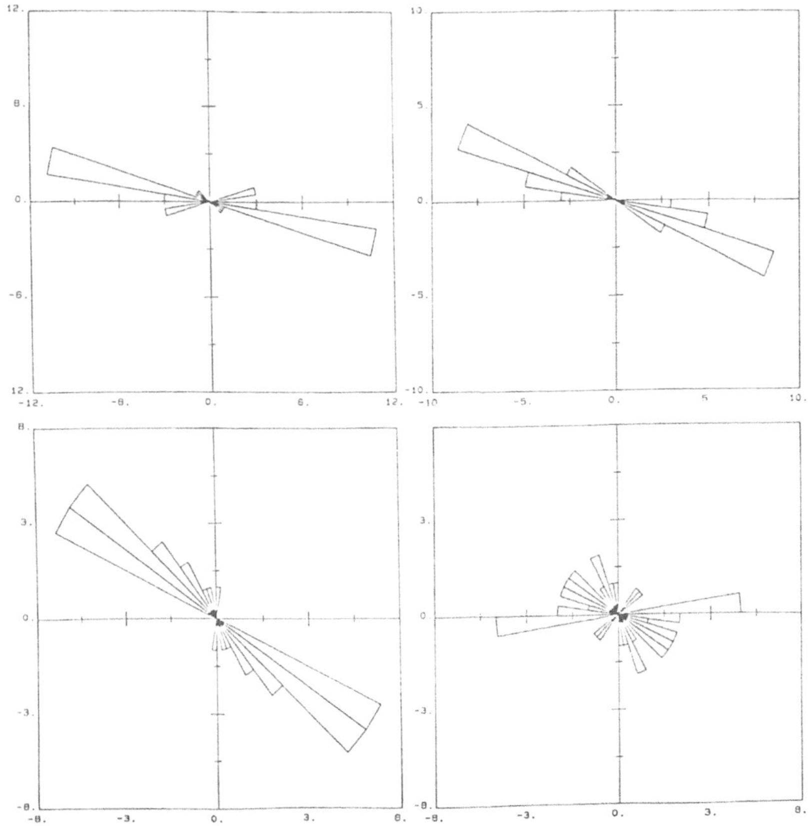

Fig. 12. The same as Fig. 11 at station 9

The direction of polarization can be measured by considering the eigenvector of the largest principal axes. If λ_1 is the largest eigenvalue and λ_2 the next largest one of the covariance matrix, then a coefficient of the form:

$$RL = 1 - \frac{\lambda_1}{\lambda_2}$$

would be close to unity when rectilinearity is high ($\lambda_1 \gg \lambda_2$) and close to zero when the two principal axes approach one another in magnitude (low rectilinearity). The direction of polarization can be determined by considering the components of the eigenvector associated with the largest eigenvalue with respect to the coordinate directions x, y, z.

Azimuth and incidence angles for stations 8 and 9 are shown in Figs. 11 and 12 for the first and second phase, respectively.

As can be seen, the angles are approximately in agreement with the results of the other method. (Fig. 13).

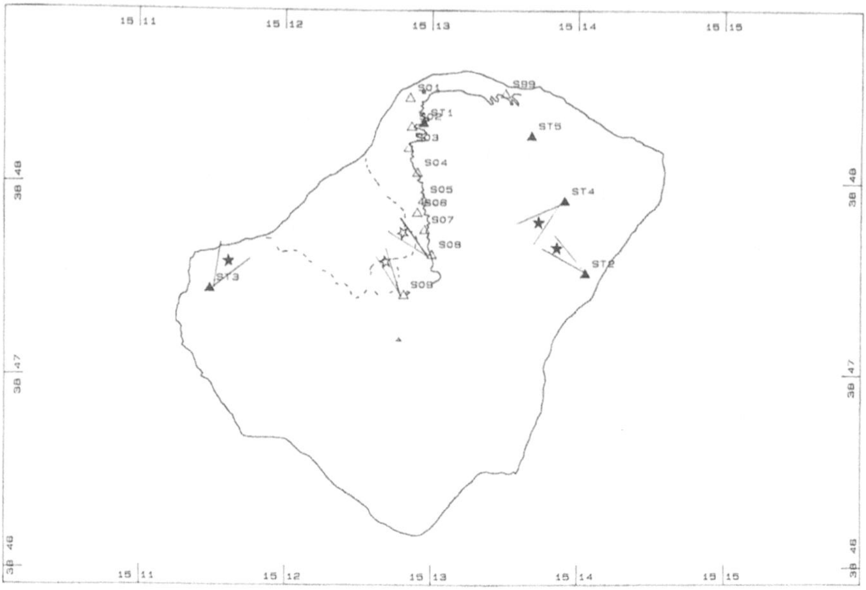

Fig. 13. Direction of polarization obtained using the covariance matrix. *White stars* show the directions for the first survey, while *black stars* are relative to the second survey

The distribution of the RL coefficient for the first and second phases are shown in Fig. 14. The first phase is strongly rectilinearly polarized, confirming the hypothesis that the seismograms are composed of compressional waves.

Although the value of the RL coefficient for the second phase is smaller than that calculated for the first phase, we can say that this phase also presents a rectilinear polarization.

4 Conclusions

The quakes recorded at Stromboli are visibly correlated with the volcanic explosions which occur at the craters. Spectra of tremor and explosion-quakes share the same shape and confirm the conclusion that tremor and quakes have the same source (Riuscetti et al. 1977; Chouet 1985; Crosson and Bame 1985).

The polarization of the seismic waves shows that the quakes seem to originate in a zone north of the crater area. The source is shallow but no precise estimates of its size are yet possible.

An indication of such dimensions can be inferred by the peak frequencies which are about 1.2 and 5 Hz, giving a rough estimate of the dimension of the source. If we assume a velocity of 1.0 km/s, the source is of the order of several hundred meters.

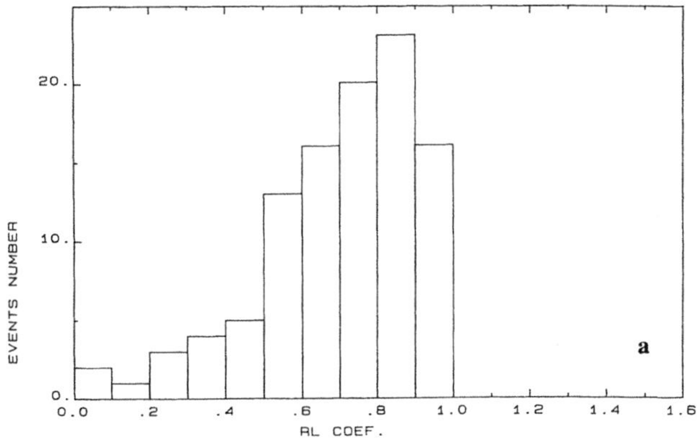

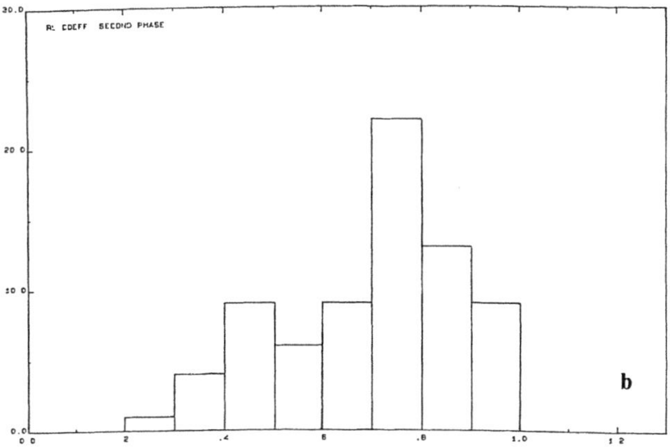

Fig. 14. **a** Distribution of rectilinearity coefficient for the first phase. **b** The same for the second phase

The source probably radiates linearly polarized waves which are evident in the first phase of the seismograms and partly in the second. From the direction of the polarization, a P wave nature for these waves can be inferred.

The possibility of applying simple pressure step models, for example, Crosson and Bame's model, still seems in doubt. This is mainly due to the fact that the two well-defined seismic phases, apparently not generated by differences in wave velocity or path, as stated, are in disagreement with this model, suggesting a multiple source.

The basic idea of magma oscillation in resonant cavities, triggered by expansion of gases in volcanic conduits is however the most appropriate, simply because quakes and explosions are visibly correlated in time. The existence of two phases can be explained with some other phenomena which accompany the gas expansion, as for example the turbolency generated by a change in velocity of the gas flow (Steinberg and Steinberg 1975).

References

Capaldi G, Guerra I, Lo Bascio A et al. (1978) Stromboli and its 1975 eruption. Bull Volcanol 41:259–285

Chouet B (1985) Excitation of a buried magmatic pipe: a seismic source model for volcanic tremor. J Geophys Res 90:1881–1893

Crosson RS, Bame D (1985) A spherical source model for low frequency volcanic earthquakes. J Geophys Res 90:10237–10247

Del Pezzo E, Guerra I, Lo Bascio A, Luongo G, Nappi G, Scarpa R (1974) Microtremors and volcanic explosions at Stromboli, part 2. Bull Volcanol 38, 4:1023–1036

Falsaperla S, Frazzetta G, La Volpe L, Neri G (1986) Stromboli: attivita' e sismicita' associata. Boll Gruppo Nazionale Vulcanol 2:233–242

Fehler M (1983) Observations of volcanic tremor at Mount St. Helens volcano. J Geophys Res 88:3476–3484

Fehler M, Chouet B (1982) Operation of a digital seismic network on Mount St. Helens volcano and observations of long period seismic events that originate under the volcano. Geophys Res Lett 9:1017–1020

Ferrick MG, Qamar AF, St. Lawrence WF (1982) Source mechanism of volcanic tremor. J Geophys Res 87, 10:8675–8683

Lo Bascio A, Luongo G, Nappi G (1973) Microtremors and volcanic explosions at Stromboli (Aeolian Islands). Bull Volcanol 37:596–606

Malone SD (1983) Volcanic earthquakes: Examples from Mount St. Helens. In: Kanamori H. Boschi E (eds) Earthquakes: observations, theory and interpretation. Elsevier Amsterdam, North Holland, pp 436–455

Minakami T (1974) Seismology and volcanoes in Japan. In: Civetta L, Gasparini P, Luongo G, Rapolla A (eds) Developments in solid Earth geophysics, physical volcanology. Elsevier, New York, pp 1–27

Montalbetti JF, Kanasevich ER (1970) Enhancement of teleseismic body phase with a polarization filter. Geophys J R Astr Soc 21:119–129

Riuscetti M, Schick R, Seidl D (1977) Spectral parameters of volcanic tremors at Etna. J Volcanol Geotherm Res 2:289–298

Schick R (1981) Source mechanism of volcanic earthquakes. Bull Volcanol 44, 3:491–497

Settle M, Mc Getchin TR (1980) Statistical analysis of persistent explosive activity at Stromboli, 1971. Implications for eruptive prediction. J Volcanol Geotherm Res 8:45–58

Steinberg GS, Steinberg AS (1975) On possible causes of volcanic tremor. J Geophys Res 80:1600–1604

Ulrich TJ, Bishop TN (1975) Maximum entropy spectral analysis and auto-regressive decomposition. Rev Geophys Space Phys 13:183–200

4 Tomography and Detection of Magma Chambers

Seismological Detection and Delineation of Magma Chambers: Present Status with Emphasis on the Western USA

H.M. Iyer

Abstract

During the past two decades, rapid strides have been made in the science of detection and delineation of magma chambers. Various seismological techniques including velocity and attenuation tomography, seismicity mapping, and refraction and reflection, and potential field techniques such as gravity, geodesy and electromagnetic surveys have provided useful information on the shape, size, and physical state of magma chambers beneath several volcanoes. Results from these techniques supplement the basic models of magmatic systems derived from geological, geochemical, hydrological, and thermal data. One or more of these techniques have been used to study volcanoes in many parts of the world, notably in the USA, USSR, Italy, Japan, and a few segments of the mid-ocean ridges. However, there are only a few volcanic centers where several of these techniques have been brought to bear simultaneously to image in detail the deep source of magma as well as the intermediate-depth and shallow storage chambers. The available case histories from such studies emphasize the fact that even though each of these techniques is a powerful tool to probe magmatic systems, the techniques differ in resolving power and they do not necessarily see the same parts of the magmatic systems. In this chapter, I discuss the various techniques, and give an overview of results from available case histories in western USA. I show how these techniques provide information on the regional setting of volcanoes, deep source of magma, intermediate storage chambers, small upper-crustal chambers, and fine structure of magma conduits. Significant structural differences between magmatic systems of different types of volcanoes will also be discussed. Appendices give partial listing of case histories of geophysical studies of volcanoes around the world.

1 Introduction

Determining how magma chambers form, change, and move in the crust in volcanic areas is of considerable importance for understanding volcano-building processes, the eruptive behavior and hazards of volcanoes, and for the exploration of magmatic and hydrothermal heat available in the form of geothermal energy. Also, many of the fundamental geologic and tectonic processes of the earth are related to the generation, transport, and storage of magma. While there is considerable speculation and some progress in understanding the generation and migration of magma, significant progress has been made, in recent years, in understanding the storage of magma. In the past, the bulk of our knowledge on the origin, storage, and transport of magma has been derived from geology, petrology, and geochemistry. However, recently, geophysical techniques, especially seismological techniques, have been particularly useful for probing magma chambers in the crust and the uppermost upper mantle of the earth. Many seismological and other geophysical techniques, such as electromagnetic methods, magnetics, gravity, and geodesy have provided useful information on the location, size, and physical properties of magma chambers in diverse volcano-tectonic environments.

In this review, I briefly describe the geophysical methods for magma chamber detection with particular emphasis on the newer seismological techniques. This is followed by a discussion of our understanding of magma chambers in a few tectonically and volcanically different areas (mostly in the western USA) where multiple techniques have been used. In this discussion, I summarize our present knowledge of magmatic systems in diverse tectonic environments based on geophysical studies of magma chambers. Finally, I include appendices listing known geophysical techniques and case histories in which these techniques have been used to study magma chambers around the world. Note that, the restricted literature search dictated by considerations of time, accessibility to published work, and language barriers cause the list to be incomplete.

In view of the existence of several previous reviews on detection of magma chambers (Piermattei and Adams 1973; Sanford and Einarsson 1982; Iyer 1984a, 1988) and the vastness of the subject, the treatment here will be brief. My main aim is to emphasize the magnitude of our recent progress in the art and science of imaging magma chambers and to prove that we now have available, a suite of experimental and analytical techniques, to image magma chambers of diverse dimensions and shapes.

Historically, Gorshkov (1958) was probably the first to use seismic methods to locate magma chambers. He showed that S-waves passing under the Kliuchevskaia group of volcanoes in Kamchatka were sharply attenuated. About a decade later, Kubota and Berg (1967) and Matumoto (1971) applied a similar technique to map magma chambers beneath the Katmai volcano in Alaska. Another decade later, seismic tomography using waves from local and distant seismic sources became available (for reviews see Thurber and Aki 1987; Iyer 1989). The introduction of seismic tomography together with rapid advances in other seismological and geophysical techniques and an enhanced interest in the anatomy of volcanoes, have contributed to a spectacular progress in the geophysical study of magma chambers during the past ten years. The measure of our progress can be judged by the fact that the U.S. Department of Energy (USDOE) has initiated a program to drill a deep hole into the magmatic environment in Long Valley, California, and in a few years we may be sampling the proximity of a live magma chamber (Rundle et al. 1986). The choice of the USDOE drill site was heavily influenced by results from the seismic studies carried out in the area.

2 Magma Chambers and Their Physical Properties

What is a magma chamber? According to the definition in the Glossary of Geology (Bates and Jackson 1980), it is a "reservoir of magma in the shallow part of the lithosphere (to a depth of a few km or tens of km), from which volcanic materials are derived". Magma itself is defined as "mobile rock material" with or without "suspended solids (such as crystals and rock fragments) and/or gas phases". In actuality, geological and physical models of magma chambers cited in literature visualize them in general to be small volumes (2–3 km dimensions) of total melt. On the other hand, geophysically inferred magma chambers seem to have volumes of tens to hundreds of km^3, and the measured average seismic velocities and densities within them correspond to rock with only a few percent of melt. The difference between the geological and geophysical models may be due to the fact that the anomalous volumes delineated by the geophysical techniques are large thermally disturbed zones comprised of hot rock, partial melt, crystals, and inclusions of small volumes of total melt. The modeling processes invariably yields only average seismic properties and densities of the total volumes, values which are nowhere near those expected for total melt.

2.1 Physical Properties Measured

2.1.1 Seismic Properties

Seismic techniques model, in two or three dimensions, velocities and attenuation of seismic waves within a volume of rock. The presence of a magma chamber within the volume is indicated by zones in which the values of one or more of these parameters are anomalous. To interpret the extent of melt in the magma chamber from the measurements one must know the change in P- and S-wave velocities (V_p and V_s) and the corresponding attenuations (Q_p^{-1}, and Q_s^{-1}) for physical states between the solidus and liquidus for common rock types. Also, effects of variations in mineralogy and the presence of cracks or voids with or without fluids which affect the above parameters have to be understood. Thus, it is obvious that it is hard to uniquely interpret observed seismic anomalies (Iyer and Stewart 1977; Iyer et al. 1981a). The situation is further complicated by the fact that practically no laboratory measurements of seismic properties of partial melts are available. One of the few measurements on seismic properties of rock melts was made by Murase and McBirney (1973). They showed that for a suite of rocks ranging in composition from mafic to silicic, as melting commenced the compressional wave velocities decreased rapidly by about 50%, the shear velocities became nearly zero, Q_p decreased by about 3–4 orders of magnitude, and Q_s was indeterminate. These results are supported by measurements in non-geologic systems (Iyer and Stewart 1977).

The theoretical behavior of rock with liquid inclusions, however, is fairly well understood. The general inference from a wide variety of calculations is that the wave velocity and attenuation in partial melt depends upon wave velocity in the solid, compressibility and viscosity of the fluid, the relative volume of the fluid, and the shapes of the fluid inclusions. An estimate of the order of magnitude of the variations of the four seismic parameters (V_p, V_s, Q_p, Q_s) and their dependence on the model assumptions are given by Mavko (1980) for olivine and pyroxene at upper mantle conditions of temperature and pressure. Mavko shows that the four seismic parameters can vary over a wide range (see Figs. 6 and 7 of

Mavko). For example, for a 10% melt, V_p and V_s can vary by 10–40% and 20–100%, respectively, and Q_p and Q_s by several orders of magnitude, compared with values for solid rock, depending on the assumptions on the melt geometry. The above discussion shows that with the available laboratory and theoretical data, no exact calculations on partial melt can be made even if good field measurements are available. In practice, most of the tomographic experiments show that P-velocity (for which the largest number of measurements are available) decreases by about 5–20% in magma chambers, implying a few percent of melt.

2.1.2 Other Geophysical Properties

A density decrease would also be expected in partial melt as can be seen from a simple application of Birch's law (Birch 1961) relating to change in compressional wave velocity and density, or due to change in mineralogy and chemistry. Change in density is reflected as a change in Bouguer gravity. For example, a 30 mgal change in Bouguer gravity can be approximately modeled as a sphere of radius 10 km with a density contrast of 0.5 gm/cc buried at a depth of 10 km. From the modeled density contrast, a rough estimate of the melt fraction can be made assuming density values for solid and molten rock.

Electrical resistivity decreases dramatically on commencement of melting. In contrast with seismic data, good laboratory data are available for the interpretation of electrical measurements (Shankland et al. 1981). However, the field application of electrical resistivity measurements for magma chamber detection are few.

Rocks lose their magnetic field at the Curie point temperature (about 560 °C for crustal rocks). Therefore, geomagnetic data can provide information on the existence of high temperatures at shallow depths if a shallowing of the Curie isotherm can be established. Techniques to compute the depth the Curie point isotherm are discussed by Bhattacharyya and Leu (1975), Smith et al. (1974, 1977), and Blakeley (1988).

Anomalous uplifts and tilts in volcanic areas may be caused by injection of magma at depth or inflation of pre-existing magma chambers. Extensive literature exists on modeling observed deformation in volcanoes (Mogi 1958; Davis 1986; Wu and Wang 1988). Modeling deformation and tilt provides information on the volume and pressure of the magma, the depth of injection, and the shape of the intrusion.

3 Techniques for Detection and Delineation of Magma Chambers

In the following sections I describe the various seismological techniques which have yielded models of magma chambers (with emphasis on the newer techniques such as seismic tomography). A brief summary of other geophysical techniques are also included for the sake of completion. Summary of techniques are given in Appendix 1, case histories using specific techniques are listed in Appendix 2, and results using multiple techniques in different volcanic environments are synthesized in Appendix 3.

3.1 Seismic Techniques

3.1.1 Seismic Tomography

Seismic tomography is a technique of imaging complex zones of velocity and/or attenuation structure within the Earth using seismic waves from earthquakes and explosions. The technique, borrowed from medical science, has been developing at a rapid rate during the past decade, and many versions of it are now in use to image seismic anomalies of global to purely local dimensions within the Earth. Seismic tomography is ideally suited for imaging magma chambers. Its applications include, modeling anomalous structures in V_p, V_s, Q_p, Q_s, and the V_p/V_s ratio, in volcanic areas using seismic waves from near earthquakes, explosions and distant earthquakes (teleseisms). All these techniques rely on illuminating the target volume with seismic waves traversing the volume in different azimuths at different angles of emergence at the surface. Such criss-crossing seismic waves are recorded by a dense array of seismic instruments deployed on the Earth's surface above the target volume. Anomalous structures that may be present in the volume produces travel-time changes for velocity anomalies and amplitude or spectral changes for attenuation anomalies in the recorded seismic waves. Modeling the anomalous structures is achieved by assuming that unknown velocity or attenuation perturbations are distributed in a parameterized volume (usually block parameterization), writing linear equations equating sums of the effects due to the perturbations with the observed travel-time or amplitude changes, and inverting for velocity- or attenuation-perturbation models. Even though there are significant differences in theoretical approaches to the various modeling schemes used in seismic tomography, the following features are common to all of them.

1. The final models give only perturbations (from an assumed initial model) of the measured quantities (V_p, V_s, Q_p, Q_s, V_p/V_s).
2. The smallest resolvable anomaly is dependent on the average instrument spacing in the array used in the experiment, the wavelength of the seismic wave, and resolution (see later).
3. Resolution depends upon how well designed the experiment is and the extent of cross-fire of the rays.

3.1.1.1 Teleseismic Compressional-Wave-Velocity Tomography Aki et al. (1977) were the first to introduce teleseismic tomography to seismology. In their technique, now popularly known as the ACH technique, teleseismic P-wave residuals are inverted using a least-squares-inversion technique to yield three-dimensional velocity models of heterogeneous volumes within the earth. The ACH technique has proved to be extremely useful for modeling complex structures within the Earth, especially magma chambers. Numerous publications on the theory and application of the technique to seismic modeling in volcanic environments are available and are listed in Appendices 2 and 3. The largest number of case histories, on the application of the ACH tomography to image magma chambers, are based on work carried out by the U.S. Geological Survey in diverse volcanic environments in the western USA and Hawaii using carefully designed field experiments. These studies include modeling continental and oceanic hot spots, intraplate silicic volcanic centers, and subduction-related volcanoes (see Sect. 4 and Appendices 2 and 3 of this presentation, and Iyer 1984a, 1988). Additionally, other researchers have applied the ACH technique to the study of other volcanic areas including the Rhinegraben in Germany (Raikes and Bonjer 1983), Larderello in Italy (Foley et al., this Vol.), Pavlov Volcano in Alaska, USA (McNutt and Jacob 1986), and the Rio Grandé Rift, USA (Parker et al. 1984). Together, these studies indicate significant differences in the sizes and shapes of magma chambers in different volcanic environments and provide clues on the magma sources and the mechanisms of magma transport in these environments.

Other tomographic techniques using teleseismic waves and applicable for magma chamber studies include the back-projection technique developed by Humphreys and Clayton (1988), which is analogous to the medical tomographic technique, and a hybrid technique using a combination of seismic waves from teleseisms and regional earthquakes within the modeled volume developed by Hirahara (1981). Also, Sharp et al. (1980) have used a forward-modeling technique, involving random search to model teleseismic residuals, to delineate a magma chamber beneath Mt. Etna in Italy.

3.1.1.2 Local Earthquake Wave Tomography Aki and Lee (1976) were the first to show that earthquake waves from within a modeled volume (unlike in teleseismic tomography where the sources are outside the volume) can be used for tomography. Since the source locations are not precisely known, perturbations in origin times and hypocentral parameters of the earthquakes must also be calculated, in addition to velocity perturbations associated with the anomalous structure within the volume. The simultaneous solution for source and path parameters becomes computationally impractical for large data sets and a method called parameter separation is used to tackle this problem. Thurber (1983) used a combination of parameter separation and an approach similar to back-projection, to develop an elegant tomographic algorithm, which is most commonly in use today for magma chamber studies (Thurber 1984; Amato and Berge 1988). Local earthquake wave tomography can in principle resolve smaller (km-size) anomalies than teleseismic-wave tomography (resolution about 6 km for short-period P-waves). However, application of local earthquake tomography is limited to the availability of a good distribution of earthquakes within the modeled volume. Local earthquake tomography using the Aki and Lee (1976) or Thurber (1983) techniques or their variations has been applied to the study of several volcanic areas. (See Sect. 5 and Appendices 2 and 3).

3.1.1.3 High-Resolution Tomography – the NeHT Technique The smallest size of an anomaly that can be resolved using teleseismic tomography is limited by the seismic wavelength and is about 6 km. Local earthquake tomography, on the other hand, can resolve much smaller bodies, but its application is limited to areas where the seismic activity is quite high. In order to overcome these limitations, Nercessian et al. (1984) developed a high-resolution tomographic imaging technique using controlled explosions as sources. Known by the acronym NeHT after the names of the authors, this experimental technique is based on a modification of the teleseismic tomographic technique. The experimental procedure in the NeHT technique consists of setting off a ring of large explosions around the target area, over which a dense array of seismographs are deployed. The distances between the array and the shots are so chosen as to illuminate the volume to be imaged by the upward-travelling seismic waves. As the rays do not have turning points within the volume, the problem is more linear than in local earthquake tomography, and at the same time the high-frequency waves provide much higher resolution than teleseismic tomography. In practice, cost considerations limit the size of the shots to less than 2000 kg, and the size of the seismic array to 100–150 instruments with an average spacing of about 1 km. These constraints limit the volume imaged to a few hundred km^3 and resolution to about 1 km.

The NeHT technique was first used to make a high-resolution image of the Mont Dore volcano in France by Nercessian et al. (1984) using an inversion technique without blocks developed by Tarantola and Nercessian (1984). It has since been successfully used at the Newberry volcano in Oregon (Achauer et al. 1988) and the Medicine Lake volcano in California (Evans and Zucca 1988) using an inversion algorithm based on the ACH technique.

3.1.1.4 V_p/V_s Tomography The spatial variation of the P-wave velocity to S-wave velocity ratio (V_p/V_s) in a volcanic area is a useful diagnostic to discriminate between abnormally low P-velocity anomalies caused by partially molten rock and lowering of P-velocity associated with fractured, vapor-dominated hydrothermal systems. Though both systems produce low P-velocities, V_p/V_s is higher than normal for partial melt, but lower than normal for vapor-dominated hydrothermal systems. Walck (1988) developed a technique of directly solving for V_p/V_s perturbations for local earthquake data using an iterative tomographic back-projection technique. V_p/V_s for each event-station pair is simply the ratio of the travel times of S-waves to P-waves (Chatterjee et al. 1985). Using this technique, Walck (1988) identified different zones with high and low V_p/V_s values in the Coso region, California.

Aster and Meyer (1988) using a different approach based on Thurber's (1983) tomographic technique of using local earthquakes, have computed tomographic models of P-velocity, S-velocity and V_p/V_s perturbations in the Campi Flegri caldera, Italy.

3.1.1.5 Teleseismic P-Wave Attenuation Tomography Young and Ward (1980) developed a Q_p tomographic technique involving use of spectral ratios of the first few seconds of teleseismic waves. The spectral ratios at an array of seismic stations in the test area with respect to a reference station, contain the differential attenuation factor $\delta t*$. The differential attenuation factor is the integral over the seismic ray path of the quantity $\delta Q^{-1}dT$, where δQ^{-1} is the perturbation in attenuation and dT is the elemental travel time. From the $\delta t*$s computed using spectral ratios over an array of seismographs for earthquakes at various azimuths and distances, a three-dimensional model of the perturbation of Q^{-1} over the modeled volume is computed. Young and Ward used a linearized inversion technique similar to the ACH technique for inverting the $\delta t*$ data.

3.1.1.6 High-Resolution P-wave Attenuation Tomography Evans and Zucca (1988) used a modified version of the Young and Ward (1980) technique to model the P-wave spectra from the high-resolution imaging data at the Medicine Lake volcano, California, and computed a three-dimensional Q_p^{-1} perturbation model for the region.

3.1.2 Teleseismic S-Velocity Modeling Using Residuals and Rayleigh-Wave Phase Velocity Inversion

Estimation of S-velocity perturbations simultaneously with P-velocity perturbations provides important constraints on partial-melt interpretations. However, very little S-velocity measurements and no teleseismic S-wave tomography have been done in magmatic environments. Daniel and Boore (1982) carried out a few S-wave residual and Rayleigh-wave phase velocity measurements in the Yellowstone caldera using a small portable seismic array. They modeled the S-residuals using a simple ray-plotting technique and the Rayleigh-wave phase velocities using the inversion technique of Backus and Gilbert (1967) to derive the comparative S-velocity structure of the lithosphere outside and beneath the Yellowstone caldera.

3.1.3 S-Wave Attenuation – Mostly Local Earthquakes

Strong attenuation occurs as S-waves travel through magma. Therefore, observations of S-wave absorption are often cited as evidence for the presence of magma chambers in discussions on detection of magma chambers. The first reported application of this technique was by Gorshkov (1958) who used it qualitatively to detect magma chambers in Kamchatka volcanoes. Basically the technique is very simple. Raypaths from several seismic stations which exhibit abnormal S-wave attenuation are traced back towards the source, and the magma chamber is interpreted to be the zone around the ray-intersection points. The technique has been applied to locate magma chambers in volcanoes in the USA, Iceland, New Zealand, and the USSR. Sanders (1984) has made the technique more rigorous by applying a scheme of assigning ratings to observed S-attenuation together with ray-plotting and has used it to delineate zones of high S-wave attenuation in Long Valley, California. Sanders et al. (1988) have further enhanced the usefulness of the technique by using a simplified version of back-projection tomography to model relative attenuation in three dimensions.

3.1.4 Reflected Phases from Earthquakes

Reflected phases from subcrustal magma boundaries, if they can be unambiguously identified, are useful to map the spatial extent and to infer the properties of magma chambers. The most successful application of the technique is in the Socorro area of the Rio Grande Rift, USA, where Sanford and Long (1965) and Sanford et al. (1973) found that seismograms from local earthquakes frequently contained two reflected phases. They identified these to be P to S and S to S conversions and showed that they could be from a high to low rigidity interface at depth in the crust. Several papers have been written on this discontinuity which is now believed to be the upper boundary of a magma chamber (see later section). Ake and Sanford (1988) using synthetic seismogram modeling of the reflected phases showed the magma body to be thin and sill-like and to have a layered structure.

3.1.5 Seismic Reflection Using Artificial Sources

The deep seismic reflection technique using explosions and vibrators has proved to be highly effective for delineating the fine structure of the crust in many areas of the world. However, perhaps due to the presence of complex reflectors and the high attenuation of seismic waves in the upper few kms of the upper crust in volcanic areas, the technique has met with only limited success for actual detection of magma chambers. Nevertheless, the capability of seismic reflection technique for the study of magma chambers has been demonstrated with impressive results in the Socorro region of the Rio Grande Rift, New Mexico (Brown et al. 1980; de Voogd et al. 1988). Reflection imaging by the COCORP consortium in this region has delineated a reflector in the upper crust coinciding with the reflector identified by Sanford et al. (1973) using converted phases identified in local earthquake seismograms (see Sect. 3.1.4). Brocher (1981) modeled the reflectivities of the reflected waves and argued for a layered structure for the magma body. Reflection seismology has also yielded useful results on magma chambers in Death valley, California (de Voogd et al. 1988) and several sections of the East Pacific Rise (Detrick et al. 1987; Detrick and Langmuir 1988).

3.1.6 Seismic Refraction Including Fan Shooting

Though the seismic refraction technique has in the past provided detailed crustal models of many volcanic environments (e.g., Yellowstone and Snake River plain: Smith et al. 1982, and Braile et al. 1982; Hawaii: Zucca et al. 1982; Long Valley: Hill 1976; Hill et al. 1985a; Lassen Peak, California: Berge and Stauber

1987), its application to actual delineation of magma chambers has been limited. The few successful applications include observations of reflections believed to be from a magma chamber interface in Long Valley, California (Hill 1976; Hill et al. 1985a); inferences on the location and size of magma chambers using fan shooting in Kamchatka volcanoes (Utnasin et al. 1976) and in parts of the Yellowstone caldera (Schilly et al. 1982); and time-term modeling of refracted arrivals to delineate a magma chamber in the Yellowstone caldera (Lehman et al. 1982). Standard refraction processing techniques are used in all these studies.

3.1.7 Seismicity

Mapping seismic activity in the vicinity of a volcano, particularly the intense seismicity associated with a volcanic crisis, can provide useful information on the location and size of magma chambers beneath the volcano. The assumption is that the magma chamber represents a zone of low rigidity and hence a seismically quiet zone within a seismically active region in the body of the volcano. Such quiet zones, interpreted to be magma chambers, have been identified in the upper 5 km beneath the Mauna Loa and the Kilauea volcanoes in Hawaii (Decker et al. 1983; Klein et al. 1987). Ryan et al. (1981) and Ryan (1988) on the other hand used a different concept to model seismicity data, particularly the deeper earthquakes, at the Kilauea volcano, Hawaii. They assumed the magma conduit to be the envelope of interconnected fracture networks which form a honeycomb-like structure, the earthquakes occurring in the walls of the honeycomb enclosing a zone of low rigidity.

3.1.8 Finite Element Modeling of Rayleigh Wave Data

Schlue et al. (1987) used finite-element modeling of Rayleigh-wave phase-velocity and spectral data and identified a vertical dike-like, low-velocity body in the Rio Grande Rift. In this technique, they recorded nuclear explosions from the Nevada Test Site at two seismic stations on either side of the Albuquerque-Belen basin of the Rio Grande Rift. Pure path phase velocity-dispersion curves and amplitude ratios as a function of frequency between the two stations were then computed and compared with several laterally heterogeneous models of the basin including one with a long, thin, vertical magma dike emplaced in the middle of the basin. The model with the dike fitted the phase velocity and amplitude data best.

3.2 Other Geophysical Techniques

Among non-seismological techniques, gravity has provided useful models of anomalous low-density bodies, some of which are interpreted to be magma chambers, in a few volcanic areas including the Clear Lake volcanic field (Isherwood 1976) and the Long Valley caldera in California (Carle 1988) and in the Yellowstone caldera, Wyoming (Eaton et al. 1975). In all of these areas pronounced gravity lows were present, and forward modeling techniques taking into consideration near-surface low-density structure yielded useful information on the size of and density contrast within anomalous low-density bodies in the crust. More sophisticated techniques, involving direct inversion of gravity data (Carle 1988), joint inversion of gravity and teleseismic residuals (Evoy 1978), or combined interpretation of gravity and seismic refraction data (Lehman et al. 1982) have also been attempted to model magma chambers.

Electromagnetic techniques, especially magnetotellurics, have provided useful models of regional structures associated with the presence of partial melt in Yellowstone-Snake River plain (Stanley et al. 1977) and Iceland (Thayer et al. 1981). However, application of the technique to delineate individual magma chambers has not met with great success.

Geomagnetic data has been used to map the depth to the Curie isotherm (about 560 °C) which, when it occurs at abnormally shallow depths is an indicator of hot rock and partial melt (Bhattacharyya and Leu 1975; Smith et al. 1974, 1977).

Uplift and tilt measurements form a vital part of the ongoing hazard monitoring program of active volcanoes and therefore the largest number of available case histories on detection of magma chambers are based on modeling these measurements. In most cases inferences from such measurements are related to the short-term behavior of volcanoes. However, occasionally abnormal uplifts together with high seismic activity continue in volcanic areas for a long time without eruptions and are used to model magma chambers. Recent examples are, Long Valley in California (Rundle and Hill 1988), the Campi Flegri caldera in Italy (Berrino et al. 1984), and the Rabaul caldera in New Britain (Archibald et al. 1988). Also, ongoing uplift has been observed in the Yellowstone caldera (Dzurisin and Yamashita 1987), and in the Socorro region of the Rio Grande Rift (Larsen et al. 1986). Most of the models which have appeared in published literature to explain volcanic deformation are based on the well known Mogi model (Mogi 1958) in which a cavity with a specified internal pressure is embedded in elastic half space. Variations of this model, too numerous to discuss here, exist in the literature (for e.g., Davis 1986; Wu and Wang 1988). The important point that I would like to make in this discussion is that, the available data and modeling techniques are inadequate to delineate the full extent of magma chambers, and even spatially extensive uplifts (as in Long Valley for example) are invariably modeled as localized inflation centers involving only very small volumes (of the order of $0.1\,km^3$) of magma.

4 What Have We Learned: Synthesis of Results

In this section, I discuss the application of multiple geophysical techniques to the delineation of magma chambers in diverse volcanic environments such as hot spots, subduction zones, rifts and spreading centers, and in intraplate volcanoes which are not associated with these environments in any obvious fashion. The findings are also summarized in Appendix 3.

4.1 Hot Spots

4.1.1 Yellowstone-Snake River Plain

The detailed study of the Yellowstone-Snake River Plain volcanic system using a variety of geophysical techniques reveal the anatomy of a continental hot spot. A large volume of information on geophysical mapping of magma chambers in the Yellowstone region exists in published literature. Reviews can be found in Iyer (1984b, 1988), Smith and Christiansen (1980), and Smith and Braile (1984).

Volcanism migrated along the eastern Snake River plain during the past 14 m.y. culminating in 2.2 m.y. of volcanism in Yellowstone. The most catastrophic eruption in Yellowstone occurred 0.63 m.y. ago producing a caldera with horizontal dimensions of 80 km by 40 km. Key seismological information on the magma chambers beneath Yellowstone caldera and the eastern Snake River plain is provided by teleseismic P-residual tomography (Iyer 1979; Iyer et al. 1981a; Evans 1982a). The results show the existence of a large volume (approximately $100 \times 100 \times 150\,km^3$) of low-velocity rock in the crust and upper mantle beneath the Yellowstone caldera, and a similar volume but only in the upper mantle, in the adjacent part of the eastern Snake River plain where the major silicic volcanism occurred 3–6 m.y. ago. Assuming that the low-velocity volume is hot partially-molten rock, these findings support the concept that a series of explosive volcanic centers erupted sequentially along the eastern Snake River plain-Yellowstone volcanic system.

Additional evidence for a large volume of partial melt beneath the Yellowstone caldera comes from the existence of abnormally low S-wave velocities (Daniel and Boore 1982) and very low resistivities (Stanley et al. 1977) in the crust and upper mantle. Existence of a very high temperature at depths as shallow as 10 km beneath the caldera is shown by the shallowing of the Curie isotherm (Smith et al. 1974, 1977; Bhattacharyya and Leu 1975), low resistivity (Stanley et al. 1977), shallowing of seismicity (Fournier and Pitt 1985), and the massive thermal flux implied by the surface geothermal manifestations. Detailed structure of the upper 20 km of the crust beneath the Yellowstone caldera is provided by seismic refraction and fan-shooting data (Smith et al. 1982; Schilly et al. 1982), a joint interpretation of seismic refraction and gravity data (Lehman et al. 1982) and local earthquake tomography (Benz and Smith 1984). These studies reveal that in addition to relatively low P-wave velocities, the upper 10 km of the crust in the Yellowstone caldera harbors two major zones of abnormally low velocities. At least one of these anomalies, in the northeastern part of the caldera, could be a magma chamber with a high percentage of partial melt. These results taken together with the ongoing caldera-wide uplift of nearly 20 mm/yr (Pelton and Smith 1979, 1982; Dzurisin and Yamashita 1987) show that the Yellowstone magma chamber is still active. Figure 1 shows some of the deep and shallow geophysical anomalies related to the Yellowstone magma chambers.

The above data taken together, yield a conceptual model of the magmatic system beneath the Yellowstone caldera, namely, a large deep zone of partial melt (P-velocity decrease of 5–10%; S-velocity decrease of 10–40%; density decrease of 0.1–0.2 g/cc; very low resistivity) with its upper surface at a depth of 4–8 km (based on shallowing of Curie isotherm and earthquake hypocenters). The uplift data imply an inflation center of 10–15 km depth beneath the caldera (Dzurisin and Yamashita 1987). We infer the propagating nature of the volcanism from the presence of a magma chamber in the upper mantle superposed by a healed crust in the eastern Snake River plain where the major volcanic activity occurred 3–6 Ma. Superposed on this large-scale model are the smaller more intensive molten volumes, inferred from seismic refraction modeling and local earthquake tomography.

4.1.2 Hawaii

Teleseismic tomography of Hawaii, an oceanic hot spot, shows a broad band of low-velocities deeper than 75 km beneath the island, approximately coinciding with the direction of the Hawaiian island chain (Ellsworth 1977). Teleseismic data provide no evidence for any substantial (several hundred km^3 volume) magma chamber in the crust and upper mantle beneath the Hawaiian volcanoes (Ellsworth and Koyanagi 1977). However, clear evidence for an extensive magma conduit system with probable small storage chambers within it comes from modeling of seismicity data. Assuming that the seismicity patterns delineate the magma pathways, Ryan et al. (1981) and Ryan (1988) have constructed a detailed three-dimensional model of the magmatic system of the Kilauea volcano to a depth of 34 km (Fig. 2a). The main feature of this model is a primary magma conduit, a few kilometers in diameter, reaching into the upper mantle. The upper 6 km of the conduit supplies the complex rift zones and dikes of the volcano. Klein et al. (1987) interpret the seismicity beneath the Kilauea volcano to a depth of about 60 km in different ways. An aseismic zone in the depth range of 3 to 7 km beneath the Kilauea caldera is interpreted to be a magma reservoir characterized by hot and ductile rock incapable of sustaining seismic strain. Beneath this is a narrow vertical zone of earthquakes extending to a depth of 20 km. Klein et al. (1987) interpret this feature as a magma conduit containing pressured and stressed regions in brittle fractures. A quiescent zone (interpreted to be a stress-free brittle zone rather than magmatic zone) occurs around 20 km depth beneath which is a large cluster of earthquakes along a south dipping zone merging at greater depths with the diffused seismicity associated with the present location of the Hawaii hot spot (Fig. 2b). The association of the deep diffused seismicity with magma is not clear. Thurber (1984) using local earthquake tomography finds a small volume of low-velocity rock (about 30 km^3) with 20% velocity decrease beneath the Kilauea summit

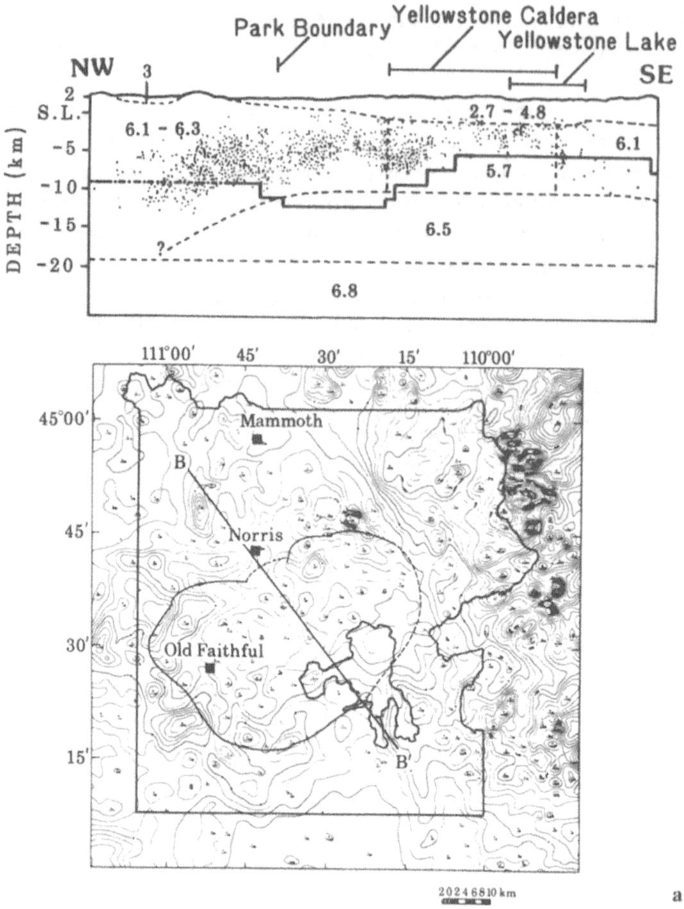

a

Fig. 1a–c. Geophysical anomalies associated with the Yellowstone hot-spot magma chambers. **a** Shallow structure based on refraction seismology. *Dotted lines* indicate different velocity blocks with P-wave velocities shown in km/s within each block. The velocity structure and figure outline is adapted from Smith and Braile (1984). *Dots* show locations of micro-earthquakes provided by A.M. Pitt, U.S. Geological Survey. The boundary indicated by the *dark line* is the Curie isotherm, (depths from Bhattacharyya and Leu 1975). Bottom part of the figure shows the Yellowstone Park boundary, Yellowstone caldera, Yellowstone Lake and other landmarks. The *contours* are residual magnetic anomalies (plotted at 50 gamma intervals) used in the computation of the Curie isotherm. *Line BB'* indicates the direction along which the cross-section is shown. **b** Velocity structure as above together with the two zones of ultra-low velocities at the northeast and southwest extremities of the caldera identified by Lehman et al. (1982) using time-term analysis of seismic refraction data (*dark shade*), and by Benz and Smith (1984) by local earthquake tomography (*light shade*). **c** Deep structure beneath Yellowstone caldera. *Left* P-velocity variations (Iyer et al. 1981a); *middle* density variations in units of 0.01 g/cc (Evoy 1978); courtesy of the author; *right* S-velocity variations (Daniel and Boore 1982)

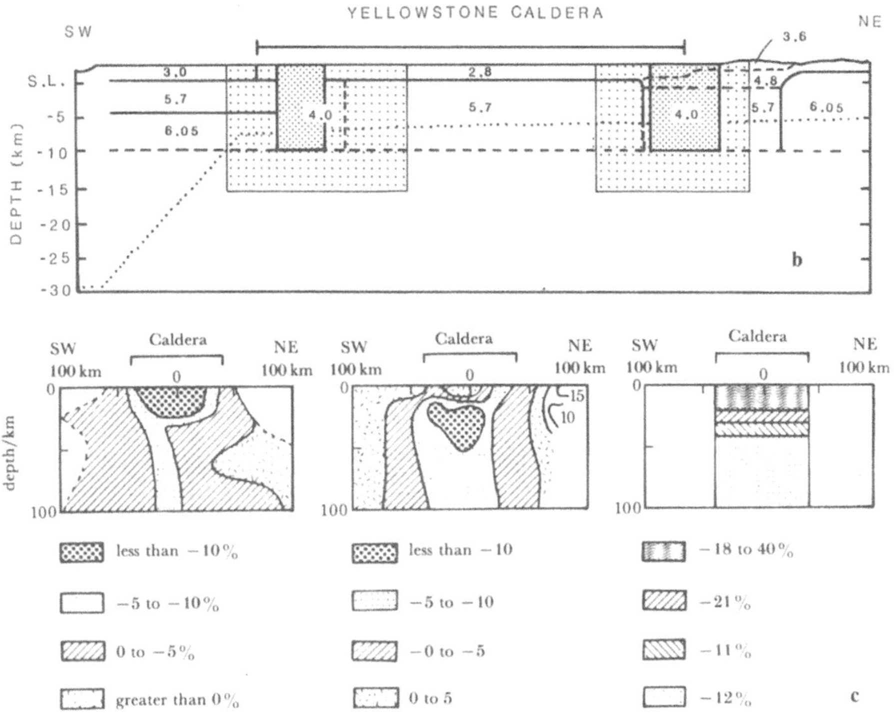

Fig. 1b–c. *Continued*

complex and interprets it as the roof of the main magma body. The pre-eruption migration of magma into small chambers near the surface and along rift zones is indicated by seismicity and inflation (Dzurisin et al. 1980; Duffield et al. 1982).

Thus the overall magma model for the Kilauea volcano, Hawaii, based on the above studies is remarkably similar to an early model proposed by Eaton and Murata (1960) based on deep and shallow seismicity data (Fig. 2c). The model has a deep upper mantle magma reservoir with narrow conduits and storage chambers connecting the reservoir to the volcano summit.

4.2 Continental Rifts

4.2.1 Rio Grande Rift

Parker et al. (1984) and Davis et al. (1984) used teleseismic arrivals from a 1000-km long profile with 50-km instrument spacing across the Rio Grande rift, New Mexico, and computed a two-dimensional P-velocity perturbation model of the

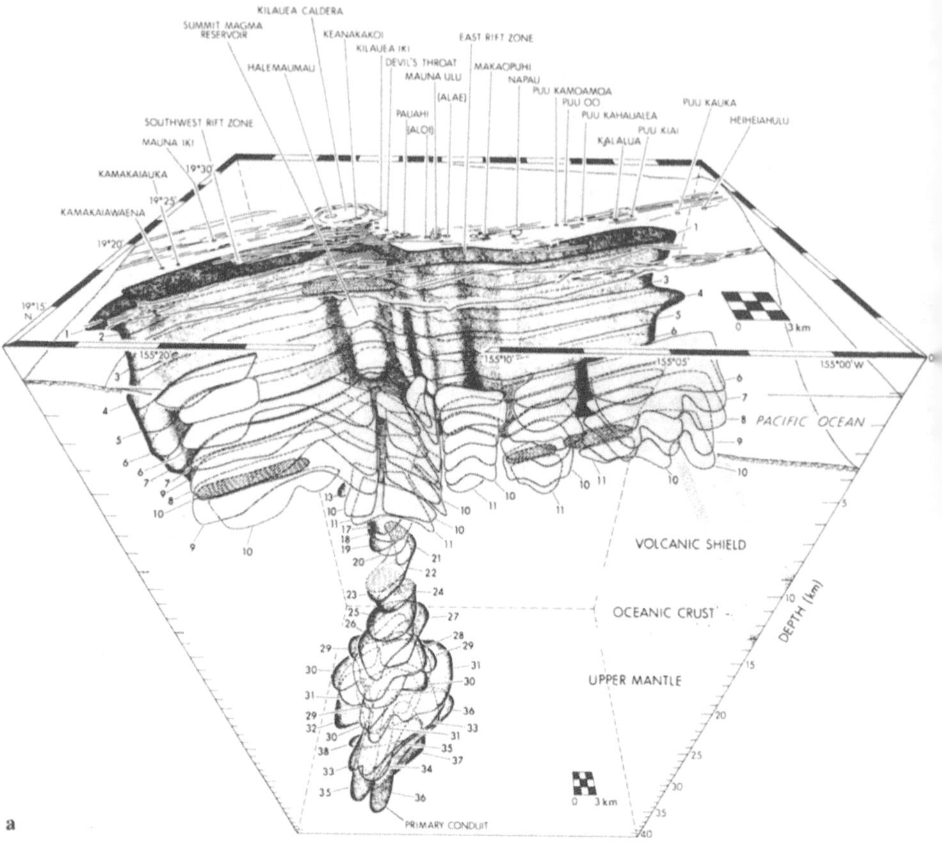

a

Fig. 2a–c. Internal structure of Kilauea volcano, Hawaii, derived primarily from seismicity data. **a** Ryan's (1988) three-dimensional model. The *numbers* denote depth in kilometers of stacks used in modeling (Ryan 1988). **b** Cross-section of the seismicity beneath Kilauea volcano (Klein et al. 1987). **c** Conceptual model of the Kilauea magma system (Eaton and Murata 1960)

upper mantle beneath the rift. The main feature of the model is a large sub-crustal zone of lower-than-normal velocity, several hundred kilometers across, and within the resolution of the experiment extending to the base of the crust. Parker et al. (1984) argue that the low-velocity zone is upwarped asthenosphere characterized by high temperature, high viscosity, and possibly partial melt. The velocity decrease in the upwarped mantle is about 5% (Fig. 3a). Their model is supported by a long-wavelength gravity low extending about 500 km across the rift (Olsen et al. 1987). Additional evidence for higher-than-normal sub-crustal temperatures beneath the rift comes from abnormally low P_n velocities (Olsen et al. 1987).

Other geophysical data show that crustal magma chambers may be present within the rift itself. The best evidence for such a chamber comes from the

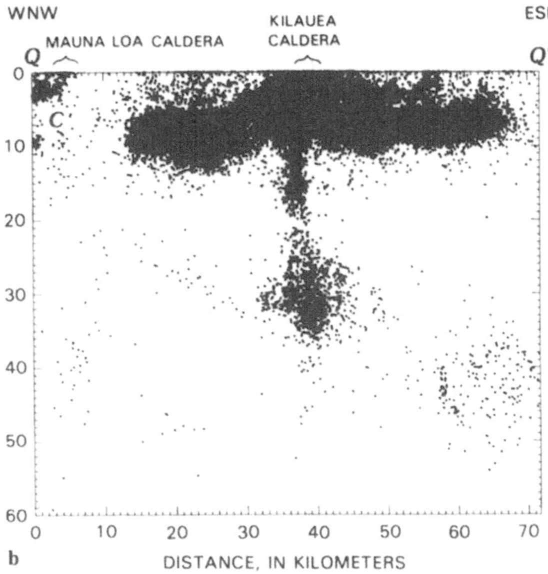

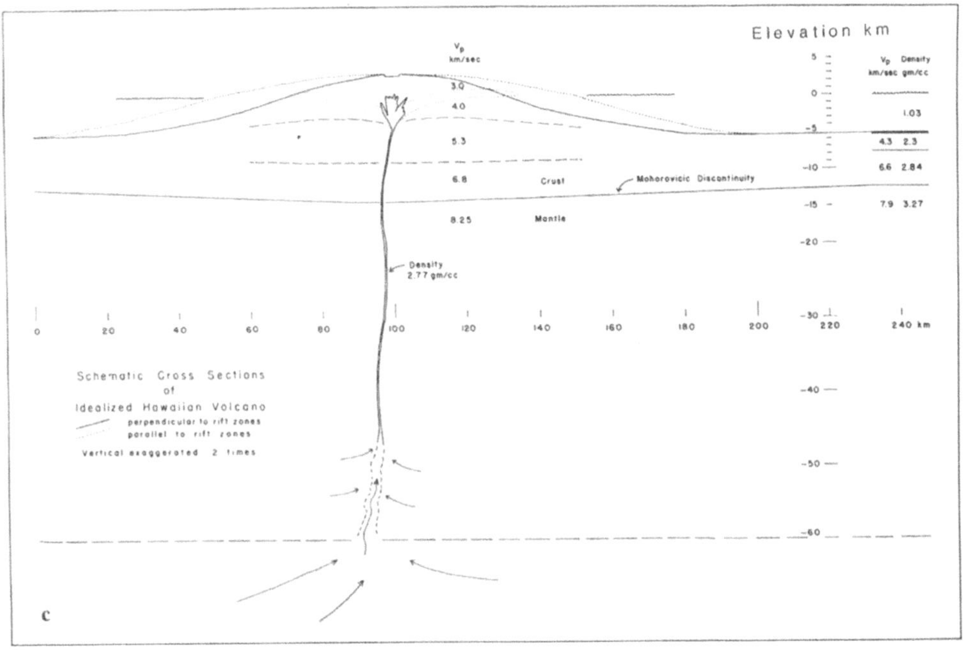

Fig. 2b–c. *Continued*

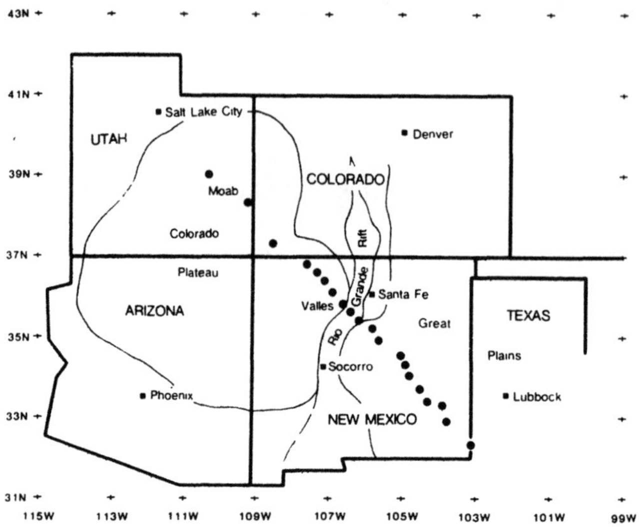

a

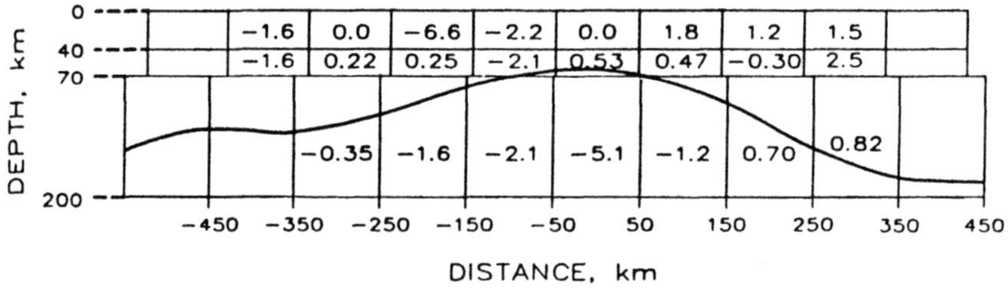

Fig. 3a–c. Rio Grande rift anomalous structure. **a** geographic location of profile of seismographs (*dots*) used for P-residual tomography of the rift, and block model of velocity perturbations across the rift (Parker et al. 1984). **b** Inferred outline of the magma body (*dashed line*) at a depth of about 19 km in the Socorro area of the Rio Grande rift based on reflected earthquake waves and seismic reflection profiling. *Triangles* are seismic wave reflection points. COCORP seismic reflection profiles are also shown (Rinehart et al. 1979). **c** Line drawing of prominent seismic reflectors from line 1 and 1A of the COCORP profiles (Brown et al. 1980)

Socorro region of the Rio Grande rift (about 200 km south of the center of the long profile), from what is now well known from extensive published literature as the "Socorro magma body".

Sanford and Long (1965) and Sanford et al. (1973), using S-wave reflected phases from local earthquakes, identified a discontinuity within the crust near Socorro. They showed the discontinuity to be approximately at a depth of 18 km and to have a very strong S-velocity contrast. Subsequently, by the analysis of an extensive data set, Rinehart et al. (1979) showed that the discontinuity could be the upper surface of a magma body. They estimated the body to be about 45 km long and 35 km wide, centered on the Rio Grande Rift (Fig. 3b). In a later study using a micro-earthquake swarm in the Socorro area, Ake and Sanford (1988) identified P to P reflections from the magma body, and by synthetic seismogram

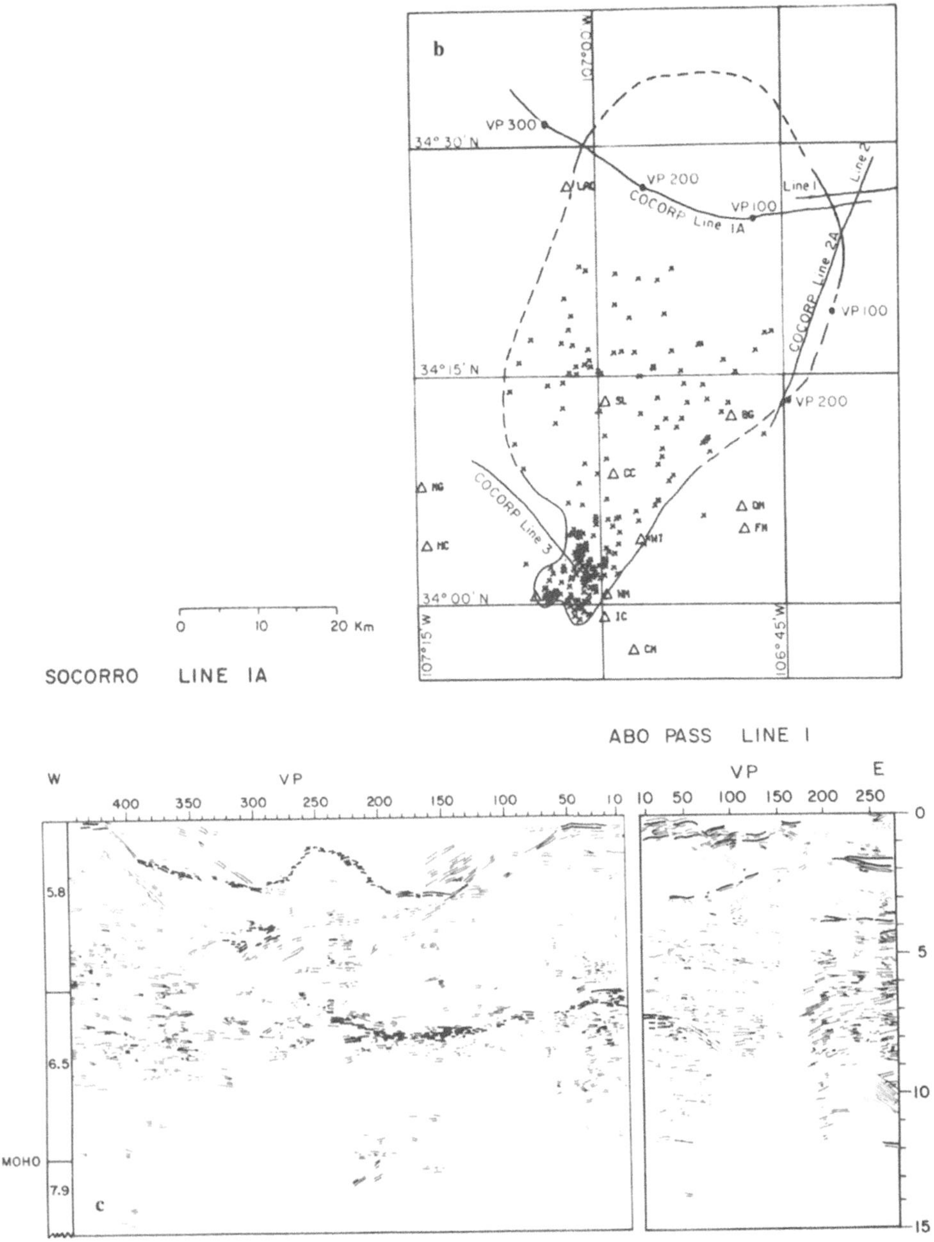

Fig. 3b–c. *Continued*

modeling of these and the S to S reflections, showed the magma body to be thin and sill-like, with a 70-m layer of magma (low velocity) underlain by about 60 m layer of slightly higher velocity interpreted to be crystalline mush. Additional evidence for the magma body comes from a strong reflecting horizon in the COCORP deep seismic reflection data in the region (Brown et al. 1980; de Voogd et al. 1988) (Fig. 3c). Brocher (1981) modeled the reflectivities from the COCORP data and showed that the magma body inferred from the reflector is probably composed of one to several layers of magma, approximately 40 m thick, separated by solid rock. In summary, modeling of reflected seismic phases from local earthquakes and from active reflection profiling provide a very good three-dimensional view of the Socorro magma body. It is located at a depth of 19–20 km, is approximately 1500 km^2 in areal extent, sill-like and a few tens of meters thick, and may be composed of alternating layers of molten and solid rock or crystalline mush. Evidence that the Socorro Magma body may be active comes from leveling data which indicate that uplift is occurring at a rate of about 18 mm/year over Socorro, consistent with magmatic movements at a depth of 19 km (Larsen et al. 1986).

Seismic techniques have identified a few other magma-chamber-like bodies within the Rio Grande rift. Ward et al. (1981), using local earthquake tomography, found a 10 by 10 km low-velocity body with a P-velocity decrease of about 15%, in the depth range of 4–10 km, in the southern extremity of the "Socorro magma body". It is not clear whether this is connected to the main body. Immediately to the north of the Socorro magma body and approximately at the center of the Rio Grande rift, Schlue et al. (1987) have identified a 28-km long, 20-km deep, narrow magma-filled dike by modeling spectral amplitude ratios of reflected Rayleigh waves. Further to the north, in the Valles Caldera of the Jemez Mountain volcanic field, Ankeny et al. (1986) have delineated using local earthquake tomography, a low-velocity body of about 15 km diameter at a depth of 10 km, with velocity decrease of 7%. They interpret this body to be a partially molten magma chamber.

All the above data, though not adequate, provide a conceptual model of a continental rift, namely, a large magma source in the upper mantle (asthenospheric upwarp) together with smaller accumulations of magma in the crust associated with specific volcanic centers (see conceptual model, Fig. 2 of Olsen et al. 1987).

Several magnetotelluric models have been computed for different areas within the Rio Grande rift. Though all the models invariably show the presence of a conducting layer within the crust, there are considerable variations in their depths and thicknesses making it difficult to make comparisons with seismic models (Jiracek et al. 1979).

4.2.2 Imperial Valley Spreading Center, California, and Icelandic Rift Zone

P-residual tomographic imaging of southern California (Humphreys et al. 1984) reveals a large (200 km wide, 100 km deep) low-velocity zone (2–4% velocity decrease) beneath the Imperial Valley spreading center. No definitive evidence for smaller magma chambers in the crust itself have so far been published for this region. Employing local earthquake tomography, Ho-Liu et al. (1988) have identified a zone with low P-velocity, and high S-wave attenuation in the depth range of 8–12 km beneath the Brawley seismic zone in Imperial Valley. But it is not clear whether the anomalies are related to partial melt or to intense fracturing associated with the high level of seismicity. However, in the Icelandic rift zone where active spreading is currently occurring, small magma chambers have been found in two areas within the rift zone. Einarsson (1978) mapped two small magma chambers (1–2 km diameter) at 3 km depth beneath the Krafla inflation center by mapping S-wave shadows from local earthquakes. In the Hengill-Grensdalur volcanic complex, Toomey and Foulger (1989) and Foulger and Toomey (1989) imaged three high-velocity bodies in the older volcanic complexes and a low-velocity body beneath the currently active Hengill volcano. They interpret all of these bodies to be magma chambers in different stages of crystallization.

4.2.3 Mid-Oceanic Ridge

A detailed discussion of the extensive search for mid-ocean ridge magma chambers is beyond the scope of this review. The results are best summarized by Detrick and Langmuir (1988): "observation of travel time delays and attenuation of P- and S-arrivals in refraction data at a number of locations in the East Pacific Rise is the best evidence for a ridge crest magma chamber (Detrick et al. 1987) . . . Additional evidence for an axial magma chamber comes from strong shallow reflector in reflection surveys often present in the 9° to 13° well-surveyed segment of the East Pacific Rise . . . also in a few places along the Juan de Fuca Ridge . . . no evidence for magma chambers in the slower-spreading Mid-Atlantic Ridge."

4.3 Subduction Zones

Seismic tomography has successfully delineated subducting lithospheric plates as relatively high-velocity features, in Japan (Hirahara 1981) and the Cascades, USA (Michaelson and Weaver 1986; Rasmussen and Humphreys 1988; Harris et al. 1989). In addition, in a tomographic model of the Tohoku district in the northeastern section of the Japan arc, Hasemi et al. (1984) found a 50–100 km wide zone of low P-wave velocity (3–5%) extending from the surface to 65 km depth, beneath the volcanic front. Similarly, using teleseismic tomography, Harris et al. (1989) found low-velocity zones (presumably partial melt) above the subducting lithosphere in the crust and upper mantle beneath the Crater Lake volcano in the Cascades, USA.

4.3.1 Subduction-Related Volcanoes

Modeling magma chambers at individual volcanoes in subduction zones have yielded a variety of magma chamber models. In Mt. Katmai, Alaska, Kubota and Berg (1967) and Matumoto (1971) using S-wave attenuation identified several small magma chambers in a 100 × 100 km area in the crust (see Fig. 10 of Iyer 1984a). In the Kamchatka volcanoes seismic refraction and fan-shooting data reveal the presence of small crustal magma chambers of assorted sizes in the crust, and conduits and larger chambers in the upper mantle, beneath several volcanoes (Utnasin et al. 1976). By carefully observing attenuation characteristics of S-waves from local earthquakes along numerous criss-crossing ray paths in the Ruapehu and Ngauruhoe volcanoes in New Zealand, Latter (1981) identified several small, anomalously attenuating zones, interpreted to be magma chambers, in the shallow crust beneath these volcanoes.

Detailed study using teleseismic tomography and the NeHT high-resolution imaging in the Cascade volcanoes reveal three basic types of structures:

1. No identifiable (larger than 5 × 5 × 5 km^3) magma chamber: e.g., Mt. Hood (Weaver et al. 1982); Mt. Shasta (Stauber and Berge 1985).
2. Possibly a small (200–1000 km^3) magma chamber: e.g., Mt. Lassen (Amato and Berge 1988)
3. High-velocity intrusion in the upper crust with a small (few km^3) magma chamber in the upper 5 km of the crust (e.g., Newberry volcano: Achauer et al. 1988; Stauber et al. 1988. Medicine Lake volcano: Evans 1982b; Evans and Zucca 1988)

Tomographic imaging studies at the Newberry and Medicine Lake volcanoes are of special interest because they provide some insight into the mechanisms of magma migration and storage at these volcanoes (Fig. 4). Teleseismic tomography at both these volcanoes (Stauber et al. 1988; Evans 1982b), though revealing a high-velocity anomalous body in the upper crust, showed no evidence for a low-velocity magma chamber. High-resolution NeHT tomography, on the other hand, showed a small low-velocity body 2–3 km beneath the summit calderas of both the volcanoes in addition to other anomalies related to older volcanic flows, sediments, faults, and hydrothermal effects. The velocity decrease in both the low-velocity bodies was of the order of 10%. The high-velocity core appears to be the cooled remnant of repeated magmatic intrusions of small volumes of magma. The low-velocity body is, probably, one such intrusion, which is currently in a partially molten stage. Remember that the Kilauea volcano in Hawaii, though chemically different from the Newberry and Medicine Lake volcanoes, has a somewhat similar structure.

4.4 Intraplate Silicic Volcanic Centers

Teleseismic tomographic technique has been most successful in imaging low-velocity anomalies, interpreted to be magma chambers, in several intraplate silicic volcanic centers in western USA (Iyer 1988 and Appendices 2, 3). Of these, the Long Valley-Mono craters region of California has been probed in detail employing multiple geophysical techniques, and the Coso Volcanic field, also in California, has been studied using several seismological techniques.

Several reviews (Hermance 1983; Hill et al. 1985b; Rundle and Hill 1988), and two companion papers in this volume, summarize our knowledge on the

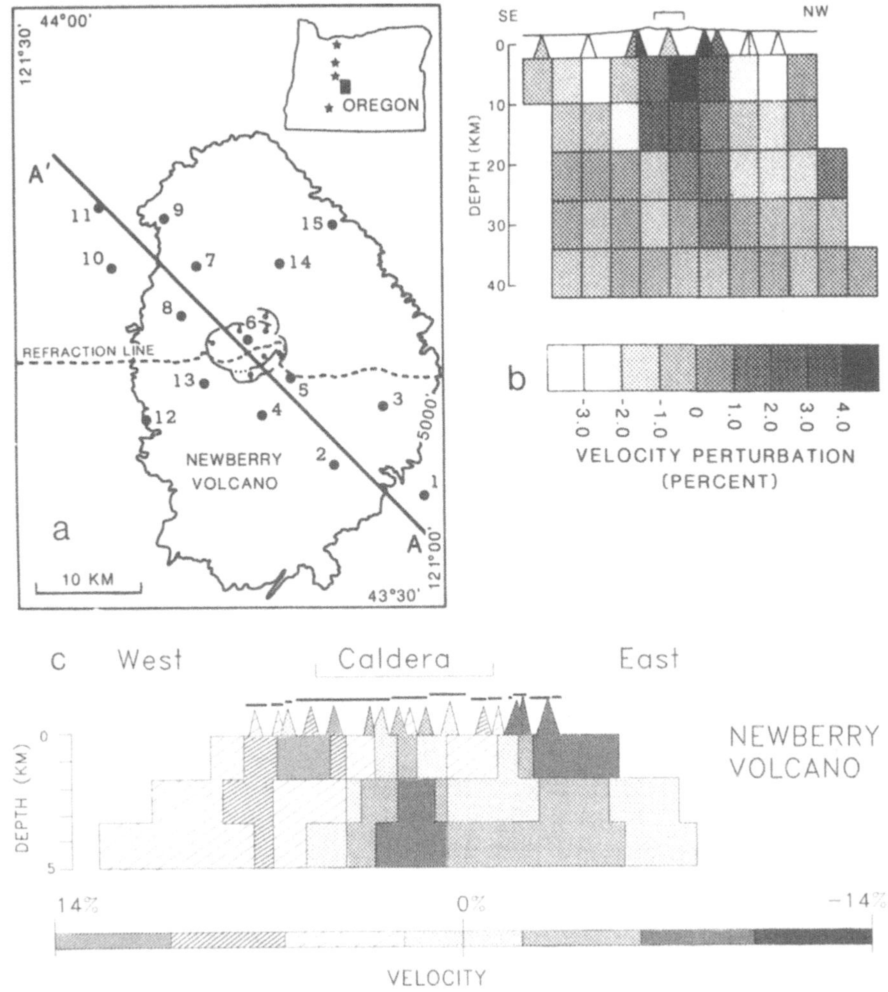

Fig. 4a–c. Tomography of Newberry volcano, Oregon, USA. **a** Geographic location and locations of seismic stations used in teleseismic-residual tomography. **b** Cross-section of deep P-velocity model along profile AA' (Stauber et al. 1988). **c** West-east cross-section of shallow P-velocity model based on high-resolution NeHT tomography (Achauer et al. 1988)

Long Valley magma chamber. Long Valley caldera is one of several young volcanic centers at varying stages of evolution, located along the eastern border of central California. A catastrophic eruption occurred in Long Valley 0.7 million years ago creating a 30-km-long, 15-km-wide caldera. This eruption has been followed by several episodic extrusions of lava to as recent as 50000 years ago. About 10 km to the north of Long Valley lies Mono Craters where the most recent volcanism occurred only 600 years ago. In addition to the young volcanism, geologic, heat-flow, and hydrothermal modeling have indicated that a crustal magma chamber may exist beneath the Long Valley caldera and

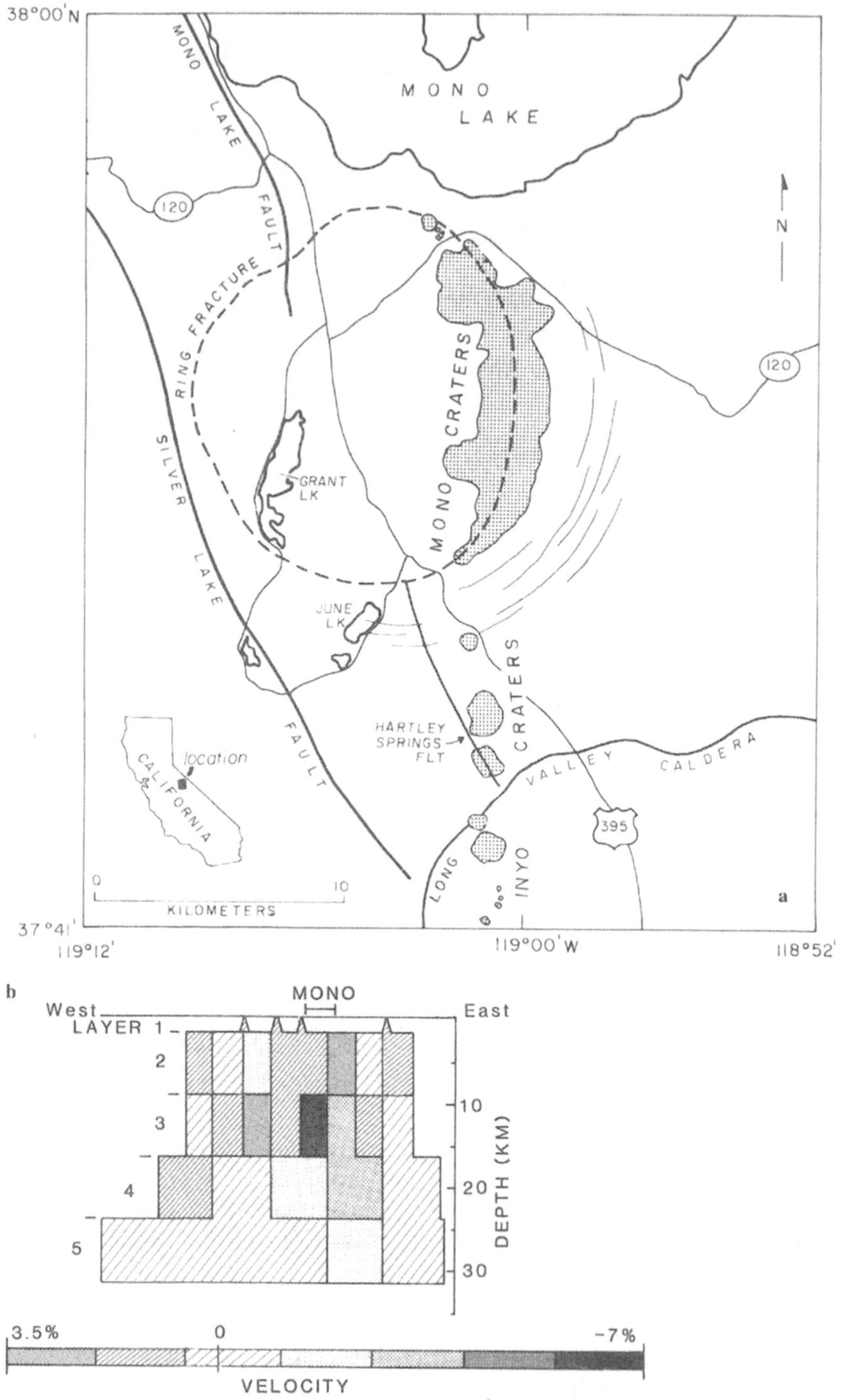
38°00'N
MONO LAKE
MONO LAKE FAULT
120
N
RING FRACTURE
MONO CRATERS
120
SILVER LAKE FAULT
GRANT LK
JUNE LK
HARTLEY SPRINGS FLT
CRATERS
LONG VALLEY CALDERA
INYO
395
location
CALIFORNIA
0
10
KILOMETERS
a
37°41'
119°12'
119°00'W
118°52'
b
MONO
West
East
LAYER 1
2
3
4
5
10
20
30
DEPTH (KM)
3.5%
0
-7%
VELOCITY

probably a smaller chamber in the relatively younger and hence not probably fully evolved Mono Craters. Steeples and Iyer (1976) found teleseismic delays with maximum values around 0.5 s in Long Valley and using simple rayplotting techniques modeled it to be a magma chamber of a few hundred km^3 volume located beneath the caldera.

Since 1980 several intense swarms of seismic activity have occurred in the southern part of Long Valley caldera following a few moderate sized earthquakes to the south of the caldera. The seismic activity has created fears that a volcanic eruption may be in the offing in this region and has led to focused multi-disciplinary investigations in Long Valley. One of these studies is a detailed teleseismic residual tomography study using an array of 90 seismic stations with an average spacing of about 5 km encompassing the whole of the Long Valley-Mono Craters region. Detailed results and interpretation of the teleseismic tomography are given in Dawson et al. (1990) and is summarized by the authors (this Vol.). Essentially, they imaged two low-velocity bodies: one, about 150–600 km^3 volume in the 7–20 km depth range beneath the resurgent dome of Long Valley caldera, and a similar body (imaged earlier by Achauer et al. 1986) in the 10–20 km depth range beneath the Mono Craters (Fig. 5). Other studies provide additional clues on the size and physical state of the Long Valley magma chamber. Sanders (1984) delineated zones of high S-wave attenuation, interpreted as melts, at depths of 5–10 km. Hill (1976) and Hill et al. (1985a) identified reflecting horizons in seismic refraction experiments and interpreted them to be from the roof of a magma chamber. Luetgert and Mooney (1985) recorded reflected earthquake phase on a long seismic profile, identified a reflection boundary at a depth of about 20 km depth beneath the Long Valley caldera, and interpreted it to be a dipping lower boundary of a magma chamber. The ongoing uplift, occurring at the rate of about 50 mm/year, probably since 1980, has been generally interpreted to be due to stresses generated by magma emplacement and variously interpreted as inflation centers or dikes at depth (Rundle and Hill 1988). Finally, Carle (1988) modeled the gravity low in the Long Valley caldera as a low-density anomaly in the 5–12 km depth range beneath the resurgent dome. Even though Carle interprets the low-velocity body as a solidified pluton, it coincides remarkably with the upper part of the magma chamber identified using teleseismic tomography (Dawson et al., this Vol.). All of these results taken together imply the presence of a large volume of partially molten rock beneath the resurgent dome in Long Valley with a few pockets of total melt in the upper part of the volume. [Local earthquake tomography of the Long Valley region by Kissling (1988) does not reveal the anomalous low-velocity body identified using teleseismic tomography. The reason for this discrepancy is not understood at present.] The only available geophysical model of the magma chamber beneath the relatively youthful Mono Craters is based on teleseismic tomography. Apart from showing that the Mono Craters magma

←

Fig. 5a,b. Teleseismic P-residual tomography of Mono Craters, California, USA (Achauer et al. 1986). **a** Tectonic and geographic features of Mono Craters. *Shaded areas* denote Holocene silicic flows and domes. **b** East-west cross section of tomographic P-velocity perturbation model chamber has approximately the same volume as the Long Valley magma chamber the data are inadequate to make further comparisons.

The Coso volcanic field in southern California comprises another group of young volcanic centers which lie along the western boundary of the Basin and Range Province. Continued rhyolitic and basaltic eruptions during the past 0.5 m.y. at a more or less constant rate, high heat flow, a high level of seismic activity, and geothermal manifestations, suggest that a small magma chamber may be present in this area. Because of this, the Coso region has been a target of several seismological imaging techniques (teleseismic P-velocity tomography: Reasenberg et al. 1980; teleseismic P-attenuation tomography: Young and Ward 1980; local earthquake P-velocity tomography: Walck and Clayton 1987; V_p/V_s tomography: Walck 1988; S-wave attenuation tomography: Sanders et al. 1988 and Ho-Liu et al. 1988). Teleseismic P-velocity tomography has imaged a small magma chamber of about 5 km diameter, with about 5–8% velocity contrast, extending from 5 to about 20 km depth in the Coso area. Teleseismic P-attenuation tomography marginally corroborates the deeper part of this model by indicating the presence of low-Q zones in the 12–20 km depth range. Local earthquake tomography has delineated a complex distribution of velocity anomalies in the upper crust of the Coso region and the Indian Wells Valley, 30 km to the south of Coso (Walck and Clayton 1987). The main anomaly in this group is a narrow elongated zone of 2–4% velocity decrease in the depth range 3–5 km in the Indian Wells Valley, several kilometers to the south of the P-anomaly found by Reasenberg et al. (1980) at Coso. Walck and Clayton (1987) found no significant low-velocity zones in the crust corresponding to the Coso magma chamber of Reasenberg et al. (1980). Sanders et al. (1988) found a zone of severe S-wave attenuation in the depth range of 3–7 km beneath the Indian Wells Valley, more or less in the vicinity of the P-wave anomaly of Walck and Clayton (1987). V_p/V_s tomography encompassing both the Coso and Indian Wells Valley regions showed a complex pattern of highs and lows in the Coso region and a high V_p/V_s in the depth range of 2–5 km in the Indian Wells Valley which could be related to partial melt (Walck 1988). Ho-Liu et al. (1988) used a back-projection tomographic technique to image an attenuation anomaly in the Indian Wells Valley coinciding more or less with the P-velocity anomaly of Walck and Clayton (1987) and the S-attenuation anomaly of Sanders et al. (1988).

The results from Coso and Indian Wells illustrate the difficulties in modeling magma chambers using multiple imaging techniques, namely, that the anomalies do not converge. A possible way of reconciling all the different results in the Coso region is to assume that the anomalies are related to two partially molten bodies, one at Coso and the other in the Indian Wells region, and the disparity in their distribution is due the different techniques sensing different physical properties of the associated magma-hydrothermal systems.

5 Conclusions

During the past decade, a whole suite of geophysical, and especially seismological techniques, have become available to image magma chambers of dimensions ranging from tens of kilometers to tens of meters. These techniques have begun to yield plausible models of the magma storage and conduit systems associated with diverse volcano-tectonic environments. For the first time, they provide the capability to carry out well-designed, multidisciplinary experiments to image magma chambers. Future directions include: (1) improving modeling techniques using state-of-the-art analytical and computational tools; (2) generating models with sufficient resolution to image variations in physical properties within the modeled volume; and (3) collecting more laboratory data on partial melts to interpret the observations in terms of the physical and thermal state of melt systems.

Acknowledgements. I thank Professors Roberto Scarpa and Paolo Gasparini for inviting me to attend the International Workshop on Volcanic Seismology and to contribute this paper. I am extremely grateful to Tim Hitchcock for extensive help during the preparation of this work including drafting figures, and to Walter Mooney and Chris Sanders for critical reviews.

Appendix 1. Manifestations of Magma Chambers and Geophysical Techniques to Detect Them

1. Broad deep source
a) Body-wave and surface-wave tomography
b) Seismic-wave attenuation (Q)
c) Gravity
d) Magnetics (Curie isotherm)
e) Electromagnetic studies

2. Conduits
a) Seismicity
b) High-resolution (NeHT) tomography

3. Substantial crustal or sub-crustal magma chambers
a) Teleseismic P-wave tomography
b) Seismic-wave attenuation
c) Seismic reflection and refraction
d) Gravity
e) Electromagnetic studies

4. Small crustal chambers including dikes and sills
a) Local earthquake tomography
b) High-resolution (NeHT) tomography
c) Seismicity
d) Seismic refraction (fan shooting)

5. Magma interfaces
a) Reflected earthquake phases
b) Seismic reflection

Appendix 2. Techniques with Best Examples

Seismic Techniques

Teleseismic P-Wave Tomography

1. Methods
a) ACH method (Aki et al. 1977)
b) Combined use of teleseismic and regional earthquakes (Hirahara 1981)
c) Humphreys-Clayton (HC) method (Humphreys and Clayton 1988)
d) Forward modeling using random search (Sharp et al. 1980)

2. Examples
a) Yellowstone-Snake River plain, USA (Iyer 1979; Iyer et al. 1981a; Evans 1982a)
b) Hawaii, USA (Ellsworth 1977; Ellsworth and Koyanagi 1977)
c) Rio Grande rift, USA (Parker et al. 1984)
d) Imperial Valley spreading center, USA (Humphreys et al. 1984)
e) Rhinegraben, F R G (Raikes and Bonjer 1983)
f) Japan subduction zone (Hirahara 1981)
g) Long Valley-Mono Craters, California, USA (Achauer et al. 1986; Dawson et al. 1990 and this Vol.)
h) Coso Volcanic field, California, USA (Reasenberg et al. 1980)
i) Clear Lake volcanic field, California, USA (Oppenheimer and Herkenhoff 1981)
j) Mineral Mountains, Utah, USA (Robinson and Iyer 1981)
k) San Francisco Mountain, Arizona, USA (Stauber 1982)
l) Pavlov Volcano, Alaska, USA (McNutt and Jacob 1986)
m) Mt. Lassen, California, USA (Amato and Berge 1988)
n) Mt. Hood, Oregon, USA (Weaver et al. 1982)
o) Newberry volcano, Oregon, USA (Stauber et al. 1988)
p) Medicine Lake volcano, California, USA (Evans 1982b)
q) Larderello, Italy (Foley et al. this Vol.)
r) Mt. Etna, Sicily (Sharp et al. 1980)

Local Earthquake-Wave Tomography

1. Method
a) Aki-Lee (AL) method and modifications (Aki and Lee 1976; Thurber 1983; Hirahara 1981)
b) Back-projection tomography (Walck and Clayton 1987)

2. Examples
a) Kilauea volcano, Hawaii, USA (Thurber 1984)
b) Valles caldera, New Mexico, USA (Ankeny et al. 1986)
c) Yellowstone caldera, Wyoming, USA (Benz and Smith 1984)
d) Clear Lake volcanic field, California, USA (Eberhart-Phillips 1986)
e) Coso volcanic field, California, USA (Walck and Clayton 1987)

f) Long Valley, California, USA (Kissling 1988)
g) Socorro in the Rio Grande rift, New Mexico, USA (Ward et al. 1981)
h) Campi Flegri caldera, Italy (Aster and Meyer 1988)
i) Hengill-Grensdalur volcanic complex, Iceland (Foulger and Toomey 1989; Toomey and Foulger 1989)
j) Mid-ocean ridge (Toomey et al. 1988)

High-Resolution Tomography

1. Method
a) Modified ACH method (NeHT technique) using experimental method developed by Nercessian et al. 1984 (Achauer et al. 1988; Evans and Zucca 1988)
b) NeHT experiment using inversion without blocks (Tarantola and Nercessian 1984)

2. Examples
a) Mont Dore, France (Nercessian et al. 1984)
b) Newberry volcano, Oregon, USA (Achauer et al. 1988)
c) Medicine Lake volcano, California, USA (Evans and Zucca 1988)

V_p/V_s Tomography Using Local Earthquakes

1. Method
a) Method of Walck (1988)

2. Examples
a) Coso volcanic field, California, USA (Walck 1988)

Teleseismic P-Wave Attenuation Tomography

1. Method
a) Method of Young and Ward (1980)

2. Examples
a) Coso volcanic field, California, USA (Young and Ward 1980)
b) Clear Lake volcanic field, California, USA (Ward and Young 1980)

High-Resolution P-Wave Attenuation Tomography

1. Method
a) Method of Evans and Zucca (1988)

2. Examples
a) Medicine Lake volcano, California, USA (Evans and Zucca 1988)

Teleseismic S-Wave Velocity Modeling

1. Method
a) S-wave residuals (Daniel and Boore 1982)
b) Rayleigh-wave dispersion (Daniel and Boore 1982)

2. Examples
a) Yellowstone caldera, Wyoming, USA (Daniel and Boore 1982)

S-Wave Attenuation – Using Mostly Local Earthquakes

1. Methods
a) Qualitative mapping (Kubota and Berg 1967; Matumoto 1971; Latter 1981; Sanders 1984)
b) Tomography (Sanders et al. 1988)

2. Examples
a) Mt. Katmai, Alaska, USA (Matumoto 1971; Kubota and Berg 1967)
b) Krafla, Iceland (Einarsson 1978)
c) Long Valley, California (Sanders 1984)
d) Indian Springs, California, USA (Sanders et al. 1988)
e) Ruapehu and Ngauruhoe volcanoes, New Zealand (Latter 1981)
g) Kamchatka volcanoes, USSR (Utnasin et al. 1976)

Reflected Phases

1. Method
a) Method of Sanford et al. (1973)

2. Examples
a) Socorro, New Mexico, USA (Sanford and Long 1965; Sanford et al. 1973; Rinehart et al. 1979; Ake and Sanford 1988)

Seismic Reflection

1. Method
a) Using explosions or vibrators

2. Examples
a) Socorro region of the Rio Grande rift, New Mexico, USA (Brown et al. 1980; de Voogd et al. 1988)
b) Death Valley, California, USA (de Voogd et al. 1988)
c) Mid-ocean ridges (Detrick and Langmuir 1988; Detrick et al. 1987)

Seismic Refraction Including Fan Shooting and Time Terms

1. Examples
a) Yellowstone, Wyoming, USA (Smith et al. 1982; Schilly et al. 1982; Lehman et al. 1982)
b) Long Valley, California, USA (Hill 1976; Hill et al. 1985a)
c) Hawaii, USA (Zucca et al. 1982)
d) Mt. Lassen, California, USA (Berge and Stauber 1987)
e) Kamchatka volcanoes, USSR (Utnasin et al. 1976)

Seismicity

1. Examples
a) Kilauea volcano, Hawaii (Eaton and Murata 1960; Ryan et al. 1981; Ryan 1988; Klein et al. 1987)
b) Rabaul, New Britain (Mori and McKee 1987)

Finite Element Modeling of Rayleigh-Wave Data

1. Method
a) Method of Schlue et al. (1987)

2. Examples
a) Rio Grande rift, USA (Schlue et al. 1987)

Nonseismic Techniques

Gravity

1. Examples
a) Yellowstone, Wyoming, USA (Eaton et al. 1975; Evoy 1978; Lehman et al. 1982)
b) Clear Lake volcanic field, California (Isherwood 1976)
c) Long Valley, California (Carle 1988)

Magnetics – Curie Isotherm

1. Examples
a) Yellowstone, Wyoming, USA (Smith et al. 1974, 1977; Bhattacharyya and Leu 1975)

Electromagnetic (EM) Techniques

1. Examples
a) Iceland (Thayer et al. 1981)
b) Yellowstone and Eastern Snake River plain, USA (Stanley et al. 1977)
c) Socorro, Valles caldera, New Mexico, USA (Jiracek et al. 1979)

Geodetic Measurements

1. Examples
a) Hawaii (Dzurisin et al. 1980; Duffield et al. 1982)
b) Long Valley, California (Hermance 1983; Hill et al. 1985b; Rundle and Hill 1988)
c) Rabaul, New Britain (Archibald et al. 1988)
d) Yellowstone, Wyoming, USA (Dzurisin and Yamashita 1987)
e) Socorro, New Mexico, USA (Larsen et al. 1986)
f) Campi Flegri caldera, Italy (Berrino et al. 1984)

Appendix 3. What Have We Learned: Synthesis of Results

Hot Spots

1. Yellowstone-Snake River plain

a) Teleseismic P-residual tomography (Iyer 1979; Iyer et al. 1981a; Iyer 1984b; Evans 1982a)
b) Teleseismic S-wave residuals (Daniel and Boore 1982)
c) Seismic refraction including fan shooting (Smith et al. 1982; Schilly et al. 1982; Lehman et al. 1982)
d) Local earthquake wave tomography (Benz and Smith 1984)
e) Seismicity (Fournier and Pitt 1985)
f) Gravity (Eaton et al. 1975; Evoy 1978; Lehman et al. 1982)
g) Magnetics – Curie isotherm (Smith et al. 1974, 1977; Bhattacharyya and Leu 1975)
h) Electromagnetic techniques (Stanley et al. 1977)
i) Uplift (Pelton and Smith 1979, 1982; Dzurisin and Yamashita 1987)

2. Hawaii

a) Teleseismic P-residual tomography (Ellsworth 1977; Ellsworth and Koyanagi 1977)
b) Local earthquake wave tomography (Thurber 1984)
c) Seismicity (Eaton and Murata 1960; Ryan 1988; Ryan et al. 1981; Klein et al. 1987)
d) Uplift and tilt (Dzurisin et al. 1980; Duffield et al. 1982)

Continental Rifts

1. Rio Grande rift

a) Teleseismic P-wave tomography (Davis et al. 1984; Parker et al. 1984)
b) Local earthquake wave tomography (Ward et al. 1981; Ankeny et al. 1986)
c) Reflected phases (Sanford and Long 1965; Sanford et al. 1973; Rinehart et al. 1979; Ake and Sanford 1988)
d) Seismic reflection (Brown et al. 1980; de Voogd et al. 1988; Brocher 1981)
e) Surface wave reflection modeling (Schlue et al. 1987)
f) Electromagnetics (Jiracek et al. 1979)
g) Uplift (Larsen et al. 1986)

2. Imperial Valley, California

a) P-wave tomography (Humphreys et al. 1984)

Oceanic Rifts

a) Seismic refraction, reflection and attenuation of seismic waves (Detrick et al. 1987; Detrick and Langmuir 1988)
b) Local earthquake tomography (Toomey et al. 1988)

Subduction Zones – Regional

1. Cascades

a) Teleseismic P-wave tomography (Michaelson and Weaver 1986; Rasmussen and Humphreys 1988; Harris et al. 1989)

2. Japan

a) Teleseismic and regional earthquake P-wave tomography (Hirahara 1981; Hasemi et al. 1984)

Subduction-Related Volcanoes

1. Newberry volcano, Oregon, USA

a) Teleseismic P-wave tomography (Stauber et al. 1988)
b) High-resolution P-wave tomography (Achauer et al. 1988)

2. Medicine Lake volcano, California, USA

a) Teleseismic P-wave tomography (Evans 1982b)
b) High-resolution P-wave tomography: velocity (Evans and Zucca 1988)
c) High-resolution P-wave tomography: attenuation (Evans and Zucca 1988)

3. Mt. Hood, Oregon, USA

a) Teleseismic P-wave tomography (Weaver et al. 1982)

4. Mt. Lassen, California, USA

a) Teleseismic P-wave tomography (Amato and Berge 1988)
b) Local earthquake P-wave tomography (Amato and Berge 1988)

5. Mt. Katmai, Alaska, USA

a) Attenuation (Kubota and Berg 1967; Matumoto 1971)

6. Pavlov volcano, Alaska, USA

a) Teleseismic P-wave tomography (McNutt and Jacob 1986)

7. Mt. Etna, Sicily

a) Teleseismic P-wave tomography (Sharp et al. 1980)

8. Kamchatka volcanoes, USSR

a) Refraction, fan shooting, and attenuation (Utnasin et al. 1976)

Intraplate Silicic Volcanic Centers

1. Clear Lake volcanic field, California, USA

a) Teleseismic P-residual tomography (Iyer et al. 1981b; Oppenheimer and Herkenhoff 1981)
b) Local earthquake wave tomography (Eberhart-Phillips 1986)
c) Teleseismic P-wave attenuation tomography (Ward and Young 1980)
d) Gravity (Isherwood 1976)

2. Coso volcanic field, California, USA

a) Teleseismic P-residual tomography (Reasenberg et al. 1980)
b) Local earthquake wave tomography (Walck and Clayton 1987)
c) V_p/V_s tomography (Walck 1988)
d) Teleseismic P-wave attenuation tomography (Young and Ward 1980)
e) S-wave attenuation using local earthquakes (Sanders et al. 1988; Ho-Liu et al. 1988)

3. Mineral Mountains, Utah, USA

a) Teleseismic P-residual tomography (Robinson and Iyer 1981)

4. San Francisco Mountain, Arizona, USA

a) Teleseismic P-residual tomography (Stauber 1982)

5. Long Valley-Mono Craters, California, USA

a) Teleseismic P-residual tomography (Steeples and Iyer 1976; Dawson et al. 1990 and this Vol.)
b) Local earthquake tomography (Kissling 1988)
c) S-wave attenuation using local earthquakes (Sanders 1984)
d) Gravity (Carle 1988)
e) Uplift and tilt (Hermance 1983; Hill et al. 1985b; Rundle and Hill 1988)

References

Achauer U, Greene L, Evans JR, Iyer HM (1986) Nature of the magma chamber underlying the Mono Craters area, eastern California, as determined from teleseismic travel time residuals. J Geophys Res 91:13 873–13 891

Achauer U, Evans JR, Stauber DA (1988) High-resolution seismic tomography of compressional wave velocity structure at Newberry volcano, Oregon Cascade Range. J Geophys Res 93:10 135–10 147

Ake JP, Sanford AR (1988) New evidence for the existence and internal structure of a thin layer of magma at mid-crustal depths near Socorro, New Mexico. Seismol Soc Am Bull 78:1335–1359

Aki K, Lee WHK (1976) Determination of three dimensional velocity anomalies under a seismic array using first P arrival times from local earthquakes. Part I. A homogeneous initial earth model. J Geophys Res 81:4381–4399

Aki K, Christoffersson A, Husebye ES (1977) Determination of the three dimensional seismic structure of the lithosphere. J Geophys Res 82:277–296

Amato A, Berge PA (1988) Teleseismic and local earthquake tomography of the Lassen volcanic area, northern California (Abs). EOS Trans Am Geophys Union 69:1472

Ankeny LA, Braile LW, Olsen KH (1986) Upper crustal structure beneath the Jemez Mountains Volcanic Field, New Mexico, determined by three-dimensional simultaneous inversion of seismic refraction and earthquake data. J Geophys Res 91:6188–6198

Archibald MJ, McKee CO, Talai B, Mori J, de Saint Ours P (1988) Electronic distance measuring network monitoring during the Rabaul seismicity/deformational crisis of 1983–1985. J Geophys Res 93:12 123–12 136

Aster RC, Meyer RP (1988) Three-dimensional velocity structure and hypocenter distribution in the Campi Flegri caldera, Italy. Tectonophysics 149:195–218

Bates RL, Jackson JA (eds) (1980) Glossary of geology. Am Geol Inst, Falls Church, Virginia

Backus G, Gilbert F (1967) Numerical application of a formalism for geophysical inverse problems. Geophys J R Astron Soc 13:247–276

Benz HM, Smith RB (1984) Simultaneous inversion for lateral velocity variations and hypocenters in the Yellowstone region using earthquake and refraction data. J Geophys Res 89:1208–1220

Berge PA, Stauber DA (1987) Seismic refraction study of upper crustal structure in the Lassen Peak area, northern California. J Geophys Res 92:10 571–10 579

Berrino G, Corrado G, Luongo G, Toro B (1984) Ground deformation and gravity changes accompanying the 1982 Pozzuoli uplift. Bull Volcanol 47:187–200

Bhattacharyya BK, Leu L (1975) Analysis of magmatic anomalies over Yellowstone National Park: Mapping Curie Point isothermal surface for geothermal reconnaissance. J Geophys Res 80:4461–4465

Birch F (1961) The velocity of compressional waves in rocks to 10 kilobars Part 2. J Geophys Res 66:2199–2224

Blakeley RJ (1988) Curie Temperature isotherm analysis and tectonic implications ot aeromagnetic data from Nevada. J Geophys Res 93:11 817–11 832

Braile LW, Smith RB, Ansorge J et al. (1982) The Yellowstone-Snake River plain seismic profiling experiment: crustal structure of the eastern Snake River plain. J Geophys Res 87:2597–2609

Brocher TM (1981) Geometry and physical properties of the Socorro, New Mexico, magma bodies. J Geophys Res 86:9420–9432

Brown LD, Chapin CE, Sanford AR, Kaufman S, Oliver J (1980) Deep structure of the Rio Grande rift from seismic reflection profiling. J Geophys Res 85:4773–4800

Carle SF (1988) Three-dimensional gravity modeling of the geological structure of Long Valley caldera. J Geophys Res 93:13 237–13 250

Chatterjee SN, Pitt AM, Iyer HM (1985) V_p/V_s ratios in the Yellowstone National Park region, Wyoming. J Volcanol Geotherm Res 26:213–230

Daniel RG, Boore DM (1982) Anomalous shear wave delays and surface wave velocities at Yellowstone caldera, Wyoming. J Geophys Res 87:2731–2744

Davis PM (1986) Surface deformation due to inflation of an arbitrarily oriented triaxial ellipsoidal cavity in an elastic half-space, with reference to Kilauea volcano, Hawaii. J Geophys Res 91:7429–7438

Davis PM, Parker EC, Evans JR, Iyer HM, Olsen KH (1984) Teleseismic deep sounding of the velocity structure beneath the Rio Grande rift. In: Baldridge WS, Dickerson PW, Riecker RE, Zidek J (eds) New Mexico Geol Soc Guidebook, 35th Field Conf, Rio Grande rift, northern New Mexico, pp 29–38

Dawson PB, Evans JR, Iyer HM (1990) Teleseismic tomography of the compressional wave velocity structure beneath the Long Valley region, California. J Geophys Res 95:11 021–11 050

Decker RW, Koyanagi RY, Dvorak JJ et al. (1983) Seismicity and surface deformation of Mauna Loa volcano, Hawaii. EOS Trans Am Geophys Union 64:545–547

Detrick RS, Langmuir CH (1988) Geometry and dynamics of magma chambers. In: The Mid-Oceanic Ridge-a dynamic global system. Available from Ocean Stud Board, Nat Res Counc, Washington, DC, 20418:123–150

Detrick RS, Buhl P, Vera E et al. (1987) Multi-channel imaging of a crustal magma chamber along the East Pacific Rise. Nature (Lond) 326:35–41

De Voogd B, Serpa L, Brown L (1988) Crustal extension and magmatic processes: COCORP profiles from Death Valley and Rio Grande rift. Geol Soc Am Bull 100:1550–1567

Duffield WA, Christiansen RL, Koyanagi RY, Peterson DW (1982) Storage, migration, and eruption of magma at Kilauea volcano, Hawaii, 1971–1972. J Volcanol Geotherm Res 13:273–307

Dzurisin D, Yamashita KM (1987) Vertical surface displacements at Yellowstone caldera, Wyoming, 1976–1986. J Geophys Res 92:13 753–13 766

Dzurisin D, Anderson LA, Eaton GP et al. (1980) Geophysical observations of Kilauea volcano, Hawaii, 2. Constraints on the magma supply during November 1975–September 1977. J Volcanol Geotherm Res 7:241–269

Eaton JP, Murata KJ (1960) How volcanoes grow. Science 132:925–938

Eaton GP, Christiansen RL, Iyer HM et al. (1975) Magma beneath Yellowstone National Park. Science 188:787–796

Eberhart-Phillips D (1986) Three-dimensional velocity structure in northern California Coast Ranges from inversion of local earthquake arrival times. Seismol Soc Am Bull 76:1025–1052

Einarsson P (1978) S-wave shadows in the Krafla caldera in NE-Iceland, evidence for a magma chamber in the crust. Bull Volcanol 41–3:187–195

Ellsworth (1977) Three-dimensional structure of the crust and mantle beneath the island of Hawaii. PhD Thesis, Massachusetts Inst Tech

Ellsworth WL, Koyanagi RY (1977) Three-dimensional crust and mantle structure of Kilauea volcano, Hawaii. J Geophys Res 82:5379–5394

Evans JR (1982a) Compressional-wave velocity structure of the upper 350 km under the eastern Snake River Plain near Rexburg, Idaho. J Geophys Res 87:2654–2670

Evans JR (1982b) Compressional-wave velocity structure of the Medicine Lake volcano from teleseismic relative residuals (abs). In: Tech Progr Abstr Biographies, 52nd Int Meet Exposition Soc Exploration Geophys, USA

Evans JR, Zucca JJ (1988) Active high-resolution seismic tomography of compressional wave velocity and attenuation structure at Medicine Lake volcano, northern California Cascade Range. J Geophys Res 93:15 016–15 036

Evoy JA (1978) Precision gravity re-observations and simultaneous inversion of gravity and seismic data for sub-surface structure of Yellowstone. MS Thesis, Univ Utah

Foulger GR, Toomey DR (1989) Structure and evolution of the Hengill-Grensdalur central volcano complex, Iceland: geology, geophysics, and seismic tomography. J Geophys Res 94:17 511–17 522

Fournier RO, Pitt AM (1985) The Yellowstone magmatic-hydrothermal system, USA In: Stone C (ed) Int Symp Geotherm Energy, Int Vol, Geotherm Resour Counc, pp 319–327

Gorshkov GS (1958) On some theoretical problems in volcanology. Bull Volcanol 19:103–113

Harris RA, Iyer HM, Dawson PB, Berge PA, Michaelson CA (1989) Imaging the Juan de Fuca slab beneath southern Oregon using teleseismic residuals (abs). Seismol Res Lett 60:1
Hasemi AH, Ishi H, Takagi A (1984) Fine structure beneath the Tohoku district, northeastern Japan arc, as derived by an inversion of P-wave arrival times from local earthquakes. Tectonophysics 101:245–265
Hermance JF (1983) The Long Valley-Mono Basin volcanic complex in eastern California: status of present knowledge and future research needs. Rev Geophys Space Phys 21:1545–1565
Hill DP (1976) Structure of Long Valley caldera, California, from a seismic refraction experiment. J Geophys Res 81:745–753
Hill DP, Kissling E, Luetgert JH, Kradolfer U (1985a) Constraints on the upper crustal structure of the Long Valley-Mono Craters volcanic complex, eastern California from seismic refraction measurements. J Geophys Res 90:11 135–11 150
Hill DP, Bailey RA, Ryall AS (1985b) Active tectonic and magmatic processes beneath Long Valley caldera, eastern California: an overview. J Geophys Res 90:11 111–11 120
Hirahara K (1981) Three-dimensional seismic structure beneath southwest Japan: the subducting Philippine sea plate. Tectonophysics 79:1–44
Ho-Liu P, Kanamori H, Clayton R (1988) Applications of attenuation tomography to Imperial Valley and Coso-Indian Wells region, southern California. J Geophys Res 93:10 501–10 520
Humphreys E, Clayton RW (1988) Adaptation of back projection tomography to seismic travel time problems. J Geophys Res 93:1073–1085
Humphreys ER, Clayton RW, Hager BH (1984) A tomographic image of mantle structure beneath southern California. Geophys Res Lett 11:625–627
Isherwood WF (1976) Gravity and magnetic studies of the Geysers-Clear Lake Region, California. Proc UN Symp Dev Use Geotherm Resour, San Francisco, California, USA, 20–29 May 1985, pp 1065–1073. Available from US Government Printing Office, Washington, DC
Iyer HM (1979) Deep structure under Yellowstone National Park, USA, a continental "hot spot". Tectonophysics 56:165–167
Iyer HM (1984a) Geophysical evidence for the locations, shapes and sizes, and internal structures of magma chambers beneath regions of Quaternary volcanism. In: Moorbath S, Thompson RN, Oxburgh ER (eds) Relative contributions of mantle, oceanic crust, and continental crust to magma genesis. Phil Trans R Soc Lond A310:473–510
Iyer HM (1984b) A review of crust and upper mantle structure studies of the Snake River plain-Yellowstone volcanic system: a major lithospheric anomaly in western USA. Tectonophysics 105:291–308
Iyer HM (1988) Seismological detection and delineation of magma chambers beneath intraplate volcanic centers in western USA. In: King C, Scarpa R (eds) Modeling volcanic processes. Vieweg, Braunschweig Wiesbaden, pp 1–56
Iyer HM (1989) Seismic tomography. In: James DE (ed) Encyclopedia of geophysics. Van Nostrand Reinhold, pp 1133–1151
Iyer HM, Stewart RM (1977) Teleseismic technique to locate magma in the crust and upper mantle. In: Dick HJB (ed) Magma genesis. Proc Am Geophys Union Chapman Conf Partial melting in the upper mantle, State of Oregon, Dep Geol Miner Ind, Portland, Oregon, pp 281–299
Iyer HM, Evans JR, Zandt G, Stewart RM, Coakley JM, Roloff JN (1981a) A deep low-velocity body under the Yellowstone caldera, Wyoming: delineation using teleseismic P-wave residuals and tectonic interpretation. Geol Soc Am Bull 92: Part I (Summary) 792–798; Part II (full text in Microfische edition) 1471–1646
Iyer HM, Oppenheimer DH, Hitchcock T, Roloff JN, Coakley JM (1981b) Large teleseismic P-wave delays in The Geysers-Clear Lake geothermal area, California. US Geol Surv Prof Pap 1141:97–116
Jiracek GR, Ander ME, Holcombe HT (1979) Magnetotelluric soundings of crustal conductive zones in major continental rifts. In: Riecker ER (ed) Rio Grande Rift: tectonics and magmatism. Am Geophys Union, Washington, DC, pp 209–222
Kissling E (1988) Geotomography with local earthquake data. Rev Geophys 26:659–698

Klein FW, Koyanagi RY, Nakata JS, Tanigawa WR (1987) The seismicity of Kilauea's magma system. In: Decker RW, Wright TL, Stauffer PH (eds) Volcanism in Hawaii. US Geol Surv Prof Pap 1350, US Gov Printing Office, Washington, DC, pp 1019–1185

Kubota S, Berg E (1967) Evidence for magma in the Katmai Range. Bull Volcanol XXXI: 175–214

Larsen S, Reilinger R, Brown L (1986) Evidence of ongoing crustal deformation related to magmatic activity near Sócorro New Mexico. J Geophys Res 91:6283–6292

Latter JH (1981) Locations of zones of anomalously high S-wave attenuation in the upper crust near Ruapehu and Ngauruhoe volcanoes, New Zealand. J Volcanol Geotherm Res 10:125–156

Lehman JA, Smith RB, Schilly MM, Braile LW (1982) Upper crustal structure of the Yellowstone caldera from seismic delay time analyses and gravity correlations. J Geophys Res 87:2713–2730

Luetgert JH, Mooney WD (1985) Crustal refraction profile of the Long Valley caldera, California, from the January 1983 Mammoth Lakes earthquake swarm. Seismol Soc Am Bull 75:211–221

Matumoto T (1971) Seismic body waves observed in the vicinity of Mount Katmai, Alaska, and the evidence for the existence of molten chambers. Geolog Soc Am Bull 82:2905–2920

Mavko GM (1980) Velocity and attenuation in partially molten rocks. J Geophys Res 85:5173–5189

McNutt SR, Jacob KH (1986) Determination of large-scale velocity structure of the crust and upper mantle in the vicinity of Pavlof volcano, Alaska. J Gephys Res 91:5013–5022

Michaelson CA, Weaver CS (1986) Upper mantle structure from teleseismic P wave arrivals in Washington and northern Oregon. J Geophys Res 91:2077–2094

Mogi K (1958) Relations between eruptions of various volcanoes and the deformations of the ground surfaces around them. Bull Earthq Res Inst Univ Tokyo 36:99–134

Mori J, McKee C (1987) Outward-dipping ring-fault structure at Rabaul caldera as shown by earthquake locations. Science 235:193–195

Murase T, McBirney AR (1973) Properties of some common igneous rocks and their melts at high temperatures. Geol Soc Am Bull 84:3563–3592

Nercessian A, Hirn A, Tarantola A (1984) Three-dimensional seismic transmission prospecting of the Mont Dore volcano, France. Geophys J R Astr Soc 76:307–315

Olsen KH, Baldridge WS, Callender JF (1987) Rio Grande Rift: an overview. Tectonophysics 143:119–139

Oppenheimer DH, Herkenhoff KE (1981) Velocity-density properties of the lithosphere from three-dimensional modeling at The Geysers-Clear Lake region, California. J Geophys Res 86:6057–6065

Parker EC, Davis PM, Evans JR, Iyer HM, Olsen KH (1984) Upwarp of anomalous asthenosphere beneath the Rio Grande rift. Nature (Lond) 312:354–356

Pelton JR, Smith RB (1979) Recent crustal uplift in Yellowstone National Park. Science 206:1179–1182

Pelton JR, Smith RB (1982) Contemporary vertical surface displacements in Yellowstone National Park. J Geophys Res 87:2745–2761

Piermattei R, Adams WM (1973) Historical review of seismic studies on structure of volcanoes. Ann Geofis XXVI:525–559

Raikes S, Bonjer KP (1983) Large-scale mantle heterogeneity beneath the Rheinish Massif and its vicinity from teleseismic P-residuals measurements. In: Fuchs K et al. (eds) Plateau uplift. Springer, Berlin Heidelberg New York, pp 315–331

Rasmussen J, Humphreys E (1988) Tomographic image of the Juan de Fuca plate beneath Washington and western Oregon using teleseismic P-wave travel times. Geophys Res Lett 15:1417–1420

Reasenberg P, Ellsworth W, Walter A (1980) Teleseismic evidence for a low-velocity body under the Coso geothermal area. J Geophys Res 85:2471–2483

Rinehart EJ, Sanford AR, Ward RM (1979) Geographic extent and shape of an extensive magma body at mid-crustal depths in the Rio Grande rift near Socorro, New Mexico. In: Riecker RE (ed) Rio Grande Rift: tectonics and magmatism. Am Geophys Union, Washington, DC, pp 237–251

Robinson R, Iyer HM (1981) Delineation of a low-velocity body under the Roosevelt Hot Springs geothermal area, Utah, using teleseismic P-wave data. Geophysics 46:1456–1466
Rundle JB, Hill DP (1988) The geophysics of a restless caldera – Long Valley, California. Annu Rev Earth Planet Sci 16:251–271, Annu Rev Inc Palo Alto California
Rundle JB, Carrigan CR, Hardee HC, Luth WC (1986) Deep drilling to the magmatic environment in Long Valley, California. EOS Trans Am Geophys Union 67:490–491
Ryan MP (1988) The mechanics and three-dimensional internal structure of active magmatic systems: Kilauea volcano, Hawaii. J Geophys Res 93:4213–4248
Ryan MP, Koyanagi RY, Fiske RS (1981) Modeling the three-dimensional structure of macroscopic magma transport systems: application to Kilauea volcano, Hawaii. J Geophys Res 86:7111–7129
Sanders CO (1984) Location and configuration of magma bodies beneath Long Valley, California, determined from anomalous earthquake signals. J Geophys Res 89:8287–8302
Sanders C, Ho-Liu P, Rinn D, Kanamori H (1988) Anomalous shear wave attenuation in the shallow crust beneath the Coso volcanic region, California. J Geophys Res 93:3321–3338
Sanford AR, Einarsson P (1982) Magma chambers in rifts. In: Pálmason G (ed) Continental and oceanic rifts. Geodyn Ser 8, Am Geophys Union, Washington, DC, pp 147–168
Sanford AR, Long LT (1965) Microearthquake crustal reflections, Socorro, New Mexico. Seismol Soc Am Bull 55:579–586
Sanford AR, Alptekin O, Toppozada, TR (1973) Use of reflection phases on microearthquake seismograms to map an unusual discontinuity beneath the Rio Grande rift. Seismol Soc Am Bull 63:2021–2034
Schilly MM, Smith RB, Braile LW, Ansorge J (1982) The 1978 Yellowstone-eastern Snake River plain seismic profiling experiment: data and upper crustal structure of the Yellowstone region. J Geophys Res 87:2692–2704
Schlue JW, Hostettler KK, Edwards CL (1987) Evidence from Rayleigh wave data for magma in an upper crustal dike in the Albuquerque-Belen Basin of the Rio Grande rift, New Mexico. J Geophys Res 92:9281–9292
Shankland TJ, O'Connell RJ, Waff HS (1981) Geophysical constraints on partial melt in the upper mantle. Rev Geophys Space Phys 19:394–406
Sharp ADL, Davis PM, Gray F (1980) A low velocity zone below Mt. Etna and magma storage. Nature (Lond) 287:587–591
Smith RB, Braile LW (1984) Crustal structure and evolution of an explosive silicic volcanic system at Yellowstone National Park. In: Explosive volcanism: inception, evolution, and hazards. Geophysics Study Committee, Natl Res Counc, Natl Acad Press, Washington, DC, pp 96–109
Smith RB, Christiansen RL (1980) Yellowstone Park as a window on the Earth's interior. Sci Am 242:104–117
Smith RB, Shuey RT, Freidline RO, Otis RM, Alley LB (1974) Yellowstone hot spot: new magnetic and seismic evidence. Geology 2:451–455
Smith RB, Shuey RT, Pelton JR, Bailey JP (1977) Yellowstone hot spot: contemporary tectonics and crustal properties from earthquakes and aeromagnetic data. J Geophys Res 82:3665–3676
Smith RB, Schilly MM, Braile LW et al. (1982) The 1978 Yellowstone eastern Snake River plain seismic profiling experiment: crustal structure of the Yellowstone region and experimental design. J Geophys Res 87:2583–2596
Stanley WD, Boehl JE, Bostick FX, Smith HW (1977) Geothermal significance of magnetotelluric sounding in the eastern Snake River plain-Yellowstone region. J Geophys Res 82:2501–2514
Stauber DA (1982) Two-dimensional compressional wave velocity structure under San Francisco volcanic field, Arizona, from teleseismic P residual measurements. J Geophys Res 87:5451–5459
Stauber DA, Berge PA (1985) Comparison of the P-velocity structures of Mt. Shasta, California, and Newberry Volcano, Oregon (Abstr). EOS Trans Am Geophys Union 66:25
Stauber DA, Green SM, Iyer HM (1988) Three-dimensional P velocity structure of the crust below Newberry volcano, Oregon. J Geophys Res 93:10095–10107

Steeples DW, Iyer HM (1976) Low-velocity zone under Long Valley as determined by from teleseismic events. J Geophys Res 81:849–860

Tarantola A, Nercessian A (1984) Three-dimensional inversion without blocks. Geophys J R Astr Soc 76:299–306

Thayer RE, Bjornsson A, Alvarez L, Hermance JF (1981) Magma genesis and crustal spreading in the northern neovolcanic zone of Iceland: telluric-magnetotelluric constraints. Geophys J R Astr Soc 65:423–442

Thurber CH (1983) Earthquake locations and three-dimensional crustal structure in the Coyote Lake area, central California. J Geophys Res 88:8226–8236

Thurber CH (1984) Seismic detection of the summit magma complex of Kilauea volcano, Hawaii. Science 223:165–167

Thurber CH, Aki K (1987) Three-dimensional seismic imaging. Annu Rev Earth Planet Sci 15: Annu Rev Inc Palo Alto California, pp 115–139

Toomey DR, Foulger GR (1989) Tomographic in version of local earthquake data from the Hengill-Grensdalur central volcano complex, Iceland. J Geophys Res 94:17497–17510

Toomey DR, Solomon SC, Purdy GM (1988) Microearthquakes beneath the median valley of the Mid-Atlantic ridge near 23° N: tomography and tectonics. J Gephys Res 93:9093–9112

Utnasin VK, Abdurakhmanov AI, Anosov GI, Budyansky Yu A, Fedorchenko VI, Markhinin Ye K (1976) Types of magma foci of island arc volcanoes and their study by the method of deep seismic sounding in Kamchatka. In: Aoki H, Iizuka S (eds) Volcanoes and tectonosphere. Tokai Univ Press, Tokyo, JPN, pp 123–137

Walck MC (1988) Three-dimensional V_p/V_s variations for the Coso region, California. J Geophys Res 93:2047–2052

Walck MC, Clayton RW (1987) P wave velocity variations in the Coso region, California, derived from local earthquake travel times. J Geophys Res 92:393–405

Ward RW, Young C (1980) Mapping seismic attenuation within geothermal systems using teleseisms with application to The Geysers-Clear Lake region. J Geophys Res 85:5227–5236

Ward RM, Schlue JW, Sanford AR (1981) Three-dimensional velocity anomalies in the upper crust near Socorro, New Mexico. Geophys Res Lett 8:553–556

Weaver CS, Green SM, Iyer HM (1982) Seismicity of Mount Hood and structure as determined from teleseismic P wave delay studies. J Geophys Res 87:2782–2792

Wu M, Wang HF (1988) Deformation and inferred stress field for ellipsoidal sources at Long Valley, California, 1975–1982. J Geophys Res 93:13285–13296

Young C, Ward RW (1980) Three-dimensional Q^{-1} model of the Coso Hot Springs known geothermal resource area. J Geophys Res 85:2459–2470

Zucca JJ, Hill DP, Kovach RL (1982) Crustal structure of Mauna Loa volcano, Hawaii, from seismic refraction and gravity data. Seismol Soc Am Bull 72:1535–1550

Three-Dimensional Imaging of the Crust and Upper Mantle in the Long Valley-Mono Craters Region, California, Using Teleseismic P-Wave Residuals

P.B. Dawson, H.M. Iyer, and J.R. Evans

Abstract

Teleseismic travel time residuals measured at 90 seismic stations centered on the Long Valley caldera in eastern California were inverted to obtain a three-dimensional image of the velocity structure beneath the array. Direct inversion of these data indicate that the 2- to 4-km thick low-velocity caldera fill contaminates the signal from mid-crustal velocity anomalies beneath the caldera. Thus, two methods were used to strip the effects of the upper crust from the travel time residuals: (1) ray tracing through upper crustal velocity models provided by seismic refraction experiments and gravity surveys, and (2) an iterative stripping scheme using the inversion process itself. These methods adequately remove the effects of the shallowest crustal structures, including the caldera fill. The resulting "stripped" models show two well resolved mid-crustal low-velocity bodies in the Long Valley region. The first body is centered between 7 and 20 km depth beneath the resurgent dome of the Long Valley caldera and has a volume of 150 to 600 km^3. The second, with a similar volume, is centered between 10 and 20 km depth beneath the Mono Craters, about 10 km north of Long Valley. Velocity contrasts in both of these bodies are about 6 to 10%, and the features are interpreted as silicic magma chambers. Upper mantle velocities are lower than average beneath the Mono Craters, and higher than average beneath Long Valley. The high eruptive rate of the Mono Craters and these upper mantle structures suggest that the focus of volcanism is shifting north from Long Valley to the Mono Craters.

1 Introduction

Long Valley caldera, located in eastern California, is one of several calderas worldwide which currently exhibit recurring earthquake swarms and intra-caldera uplift. The processes responsible for the physical unrest of these calderas are not well understood and, as is the case at Long Valley, major regional tectonism further complicates their observation and interpretation. Significant geophysical evidence indicates that magma is present beneath the Long Valley caldera, perhaps as shallow as 4 or 5 km beneath the southern part of the resurgent dome. Delineating the magmatic systems in the Long Valley region, and observing their interaction with the regional seismicity and tectonics is tantamount to understanding the unrest of the caldera, and by extrapolation, to calderas worldwide such as Campi Flegrei, Italy, Rabaul, Papua New Guinea, and Yellowstone, Wyoming.

The recognized seismic and volcanic hazards related to Long Valley have prompted numerous geological and geophysical studies focused primarily on the caldera. Both active and passive seismic experiments have defined the regional seismic structure and, in particular, the structure beneath the caldera to depths of about 10 km. Steeples and Iyer (1976) described a low-velocity anomaly beneath the resurgent dome of the caldera centered at about 10- to 15-km depth and subsequent geophysical experiments have confirmed the upper portion of their interpreted magma chamber (Rundle and Hill 1988). Various dislocation models based on precise geodetic data indicate up to 0.25 km^3 of magma injection into the upper crust beneath the caldera occurred between 1975 and 1985. While nearly all of the geophysical studies conducted to date image the upper portion of the magma chamber beneath the resurgent dome of the caldera, they provide little constraint on the depth of the chamber. It appears that the upper portion of the chamber is comprised of several small batches of melt rather than a large contiguous magma chamber (Goldstein and Stein 1988). Achauer et al. (1986) describe a magma chamber in the mid-crust beneath the southern Mono Craters using teleseismic tomography. Reviews of the geological and geophysical processes observed at Long Valley include Goldstein and Stein (1988), Rundle and Hill (1988), Goldstein (1987) Rundle et al. (1986), Hill et al. (1985a,b), Hermance (1983), Bailey (1982) and Muffler and Williams (1976).

The purpose of this chapter is to describe the three-dimensional P-wave velocity structure of the lithosphere in the Long Valley region using teleseismic tomography. This work was designed to compliment the detailed knowledge of the upper crust in the region providing a description of the crustal magmatic systems, and a characterization of the region of mafic intrusion and mafic-silicic melt interaction beneath the Long Valley area. We will also comment on the mafic source region which we believe drives the crustal magmatic systems.

1.1 Geologic Setting

Long Valley is located about 350 km east of San Francisco, California. It is one of several Quaternary volcanic centers located along the eastern front of the Sierra Nevada. Long Valley is a 17 × 32 km long (long axis striking east) resurgent caldera which lies in an east-west offset of the northwest trending Sierra Nevada frontal fault system. The Inyo and Mono Craters and the Mono Lake volcanic complexes are the most recent volcanic centers in the region and form a linear north striking zone from the west moat of the caldera to the north side of Mono Lake (Fig. 1). The seismic and volcanic activity at Long Valley is directly related to the boundary between the Basin and Range and Sierra Nevade provinces. Lachenbruch and Sass (1978) suggest that silicic volcanic systems along this boundary are localized in zones where the lithosphere pulls apart, causing upwelling of mafic magmas into the lower crust and accompanying faulting in the upper crust.

The regional volcanic system suggested by Bailey (1982) includes a mafic magma body common to the Long Valley, Mono Craters, and Mono Lake volcanic centers which supplies heat to the mid-crustal silicic magma chambers. The Long Valley caldera was formed 700 000 years ago with the eruption of 600 km^3 of rhyolitic magma (Bailey et al. 1976). The residual magmatic system has been recharged with mafic magmas at intervals of every few hundred

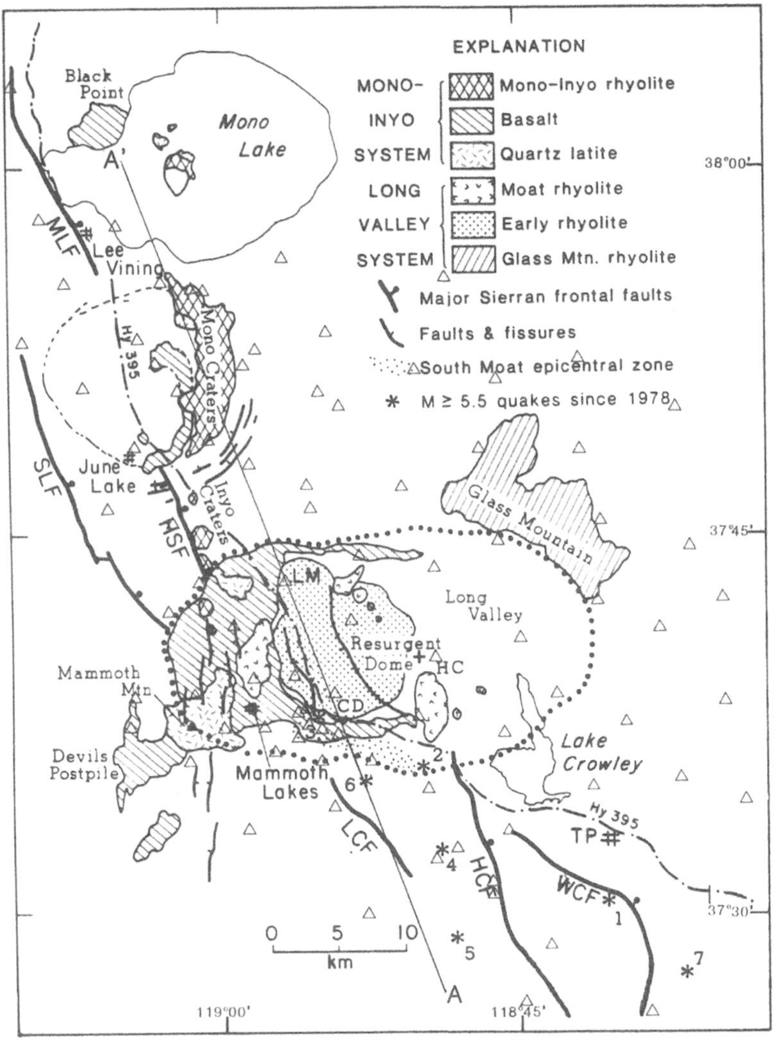

Fig. 1. Geologic and tectonic setting of the Long Valley region (after Rundle and Hill, 1988) and location key. *Dots* indicate the topographic boundary of the caldera. *Triangles* indicate the seismic station locations. *A-A'* refers to the cross-section shown in Fig. 6. Sierran frontal faults shown as *bold lines* are: *HCF* Hilton Creek fault; *HSF* Hartley Springs fault; *LCF* Laurel Canyon fault; *MLF* Mono Lake fault; *SLF* Silver Lake fault; and *WCF* Wheeler Crest fault. *CD* Casa Diablo; *CV* Crestview; *HC* Hot Creek; *LM* Lookout Mountain; *TP* Tom's Place. *Numbers* refer to the following earthquakes: *1* 4 October 1978, M = 5.7; *2* 25 May 1980, M = 6.1; *3* 25 May 1980, M = 6; *4* 25 May 1980, M = 6.1; *5* 27 May 1980 M = 6.2; *6* 30 September 1981, M = 5.7; and *7* 23 November 1984, M = 5.8

thousand years, feeding numerous rhyolitic eruptions. The Mono Craters volcanic system may be in the stages of pre-caldera formation, with a silicic magma chamber being formed in the mid-crust. The more mafic volcanic system at Mono Lake may represent the first intrusion of mafic magmas into the crust in that area. A general decrease in age of the volcanic centers from Long Valley north to Mono Lake supports this model. This trend may indicate that the magmatic source is shifting to the north, however, some of the latest eruptions in the Long Valley region have been within the caldera.

1.2 Design of Experiment

This teleseismic tomography experiment was designed to model the three-dimensional compressional-wave velocity structure of the upper mantle and the crust in the Long Valley region. The impetus for the experiment was to constrain the extent of the low-velocity anomalies seen by Steeples and Iyer (1976) beneath the Long Valley caldera and the Mono Craters as described by Achauer et al. (1986) and to compare the results with the inferred geologic model of Bailey (1982). We use the term "tomography" to describe the imaging of seismic velocity distribution within a target volume estimated from the integral of the seismic velocity along each of many different paths through the volume. In this case, the line integral values are travel times along the ray paths.

The damped least-squares inversion upon which this work is based is described by Aki et al. (1977). The advantages of using this method are the stability of the solution by using rays with no turning points in the target volume and the generally even sampling of the target volume by using large earthquakes from around the world as energy sources. Disadvantages of this method are the loss of absolute velocity information derived from the need to make assumptions about structure outside the target volume, difficulty in resolving the depth of horizontal structures, and the relatively poor resolution inherent in the 1-Hz compressional phases used. We recorded teleseisms at 90 stations forming a 75- by 45-km oval-shaped network centered on the Long Valley caldera (Fig. 1). This allows the resolution of the velocity structure in a target volume which extends to about 60 km depth beneath the network.

2 Compressional Wave Velocity Structure

Over 3200 usable readings from 190 events were timed at a distinct peak or trough in the first cycle using digital band-pass filters and visual phase correlation. Steeples and Iyer (1976) review the accuracy of the visual correlation method. Timing accuracy for this experiment is approximately 0.05 s. Theoretical travel times for each event were calculated using the Herrin (1968) travel time tables. Absolute travel time residuals at each station were calculated (the difference between the theoretical travel time and the observed travel time) and are due to source errors, heterogeneity along the ray path, and structure beneath the array. Relative residuals were calculated by subtracting the weighted mean residual for each event from the absolute residuals of every station for that event to isolate the part of the residual due to structure beneath the array. This procedure minimizes source and path effects at the expense of losing absolute velocity information. The relative residuals, hereafter called residuals, form the data set which is modeled in this study. For a formal presentation of the computation of relative residuals, see Achauer et al. (1986).

2.1 Interpretation of Residuals

The interpretation of the residuals is based on changes in spatial residual patterns with approach incidence angle and azimuth. These changes are due to velocity anomalies beneath the array. The magnitude of the residuals indicates the intensity and size of the anomalies and the distance the patterns move with changes in azimuth indicates their depth. Negative residuals imply the presence of high velocity structure and positive residuals indicate low-velocity structure exists beneath the array.

Figure 2a is a contour map of the mean residual at each station for events from all azimuths averaged together. The mean residual is due to velocity structure to a depth of approximately 1 to 1.5 station spacings (5 to 8 km), is azimuthally independent, and is called the invariant part of the residual. The invariant residual reflects the upper crustal velocity structure of the Long Valley region. The significant feature of this map is that stations within the caldera are about 0.1 to 0.4 s late, relative to stations outside of the caldera. Delays are greatest in an east-west band, centered on the north-central part of the caldera. The caldera fill structure determined from seismic refraction experiments (Hill et al. 1985b) corresponds with this pattern. Stations along the west margin of the network have generally earlier arrivals than stations to the east. This pattern reflects the higher velocity Jurassic and Cretaceous plutons and the Paleozoic and Mesozoic roof pendants of the Sierra Nevada batholith relative to the Cenozoic volcanics to the east. Another feature is a region of delays (0.1–0.2 s late) along the border of Mono Lake. Pakiser (1976) observed a similar delay pattern in time terms which he related to an extremely thin layer of low-velocity sediments in the Mono Basin.

To see the effect of structures below the upper crustal structure the teleseismic events were divided into five groups according to ray orientation beneath the network (Fig. 2b–f). In each group, the average residual for each station was calculated and the invariant residual for that station removed. The resultant changes in residual patterns between azimuth groups can be thought of as "shadows" cast by anomalous velocity structure at depth below the network. The dominant patterns are regions of 0.1–0.2 s late arrivals which migrate about the resurgent dome and the Mono Craters along the back azimuth of event propagation (Fig. 2b–e). We interpret these patterns as being caused by anomalous low-velocity zones centered beneath the resurgent dome of the Long Valley caldera and the Mono Craters. The fifth group of rays consists of PKIKP events (Fig. 2f). The wave fronts for these events are nearly vertically incident

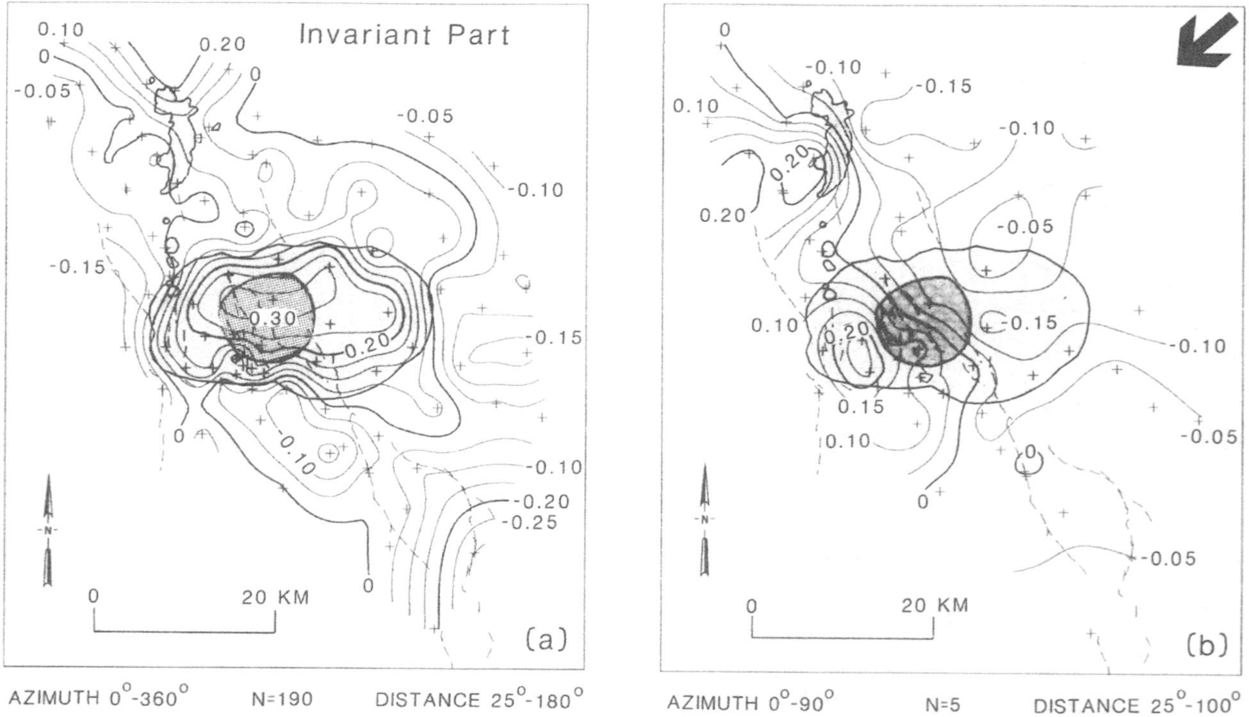

Fig. 2a–f. Relative residuals for six distance-azimuth groups. Long Valley caldera, Inyo-Mono Craters, and the resurgent dome are shown as *bold lines* and stippled. Contour interval is 0.05 s. *Positive values* indicate travel-time delays. *AZ* Azimuth group; *DI* distance of events in degrees; *N* number of events contributing to the figure. *Small plusses* indicate station locations. **2a** is the mean relative residual, or "invariant" part of the residual. This value has been subtracted from the mean value of the residuals for the subsequent plots. *Large arrows* indicate the direction of wave propagation for the primary quadrants

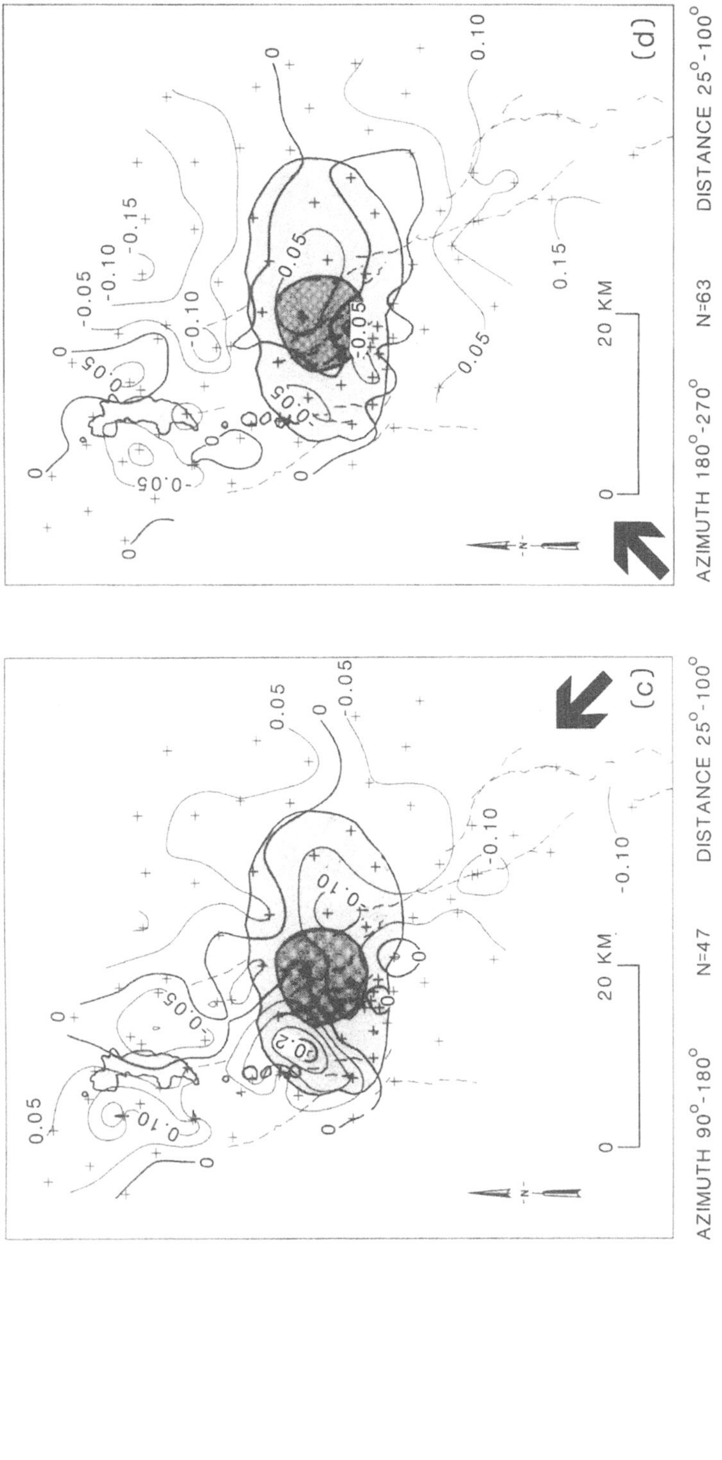

Fig. 2c,d. *Continued*

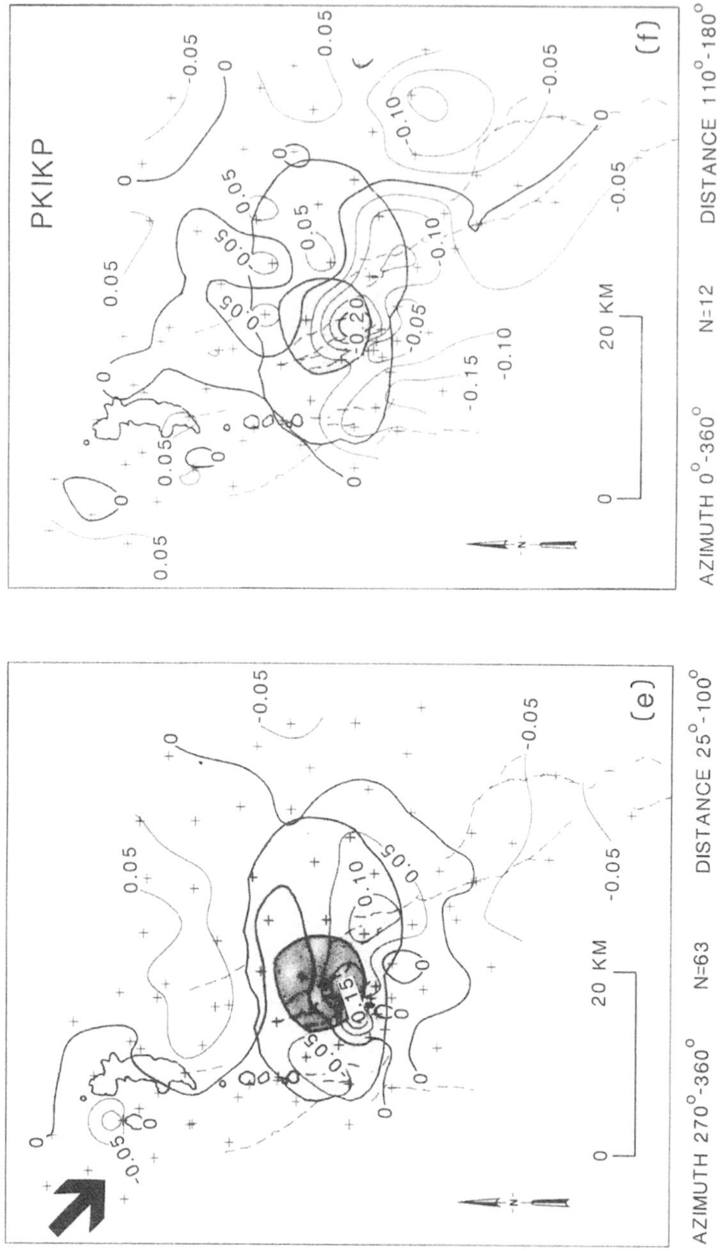

Fig. 2e,f. *Continued*

and show a region of early arrivals centered south of the resurgent dome. This feature may represent a zone of high-velocity residuum in the upper mantle or lower crust related to the long period of volcanism at Long Valley.

2.2 Inversion for Velocity Structure

To quantify the residual patterns discussed above the residuals were inverted using the "ACH" damped least squares method (Aki et al. 1977). A model, consisting of several plane layers divided into discreet blocks, is perturbed resulting in travel time changes in the blocks equal to the observed travel time at each station for each block. The solution contains velocity perturbations in each block relative to the mean layer velocity. Absolute velocities are indeterminate because relative travel time residuals are used in the solution. The first layer of the model is an exception to the block structure of the other layers. Rays arriving at each station generally do not intersect in the first layer and are most sensitive to the velocity structure immediately beneath a seismograph. Thus, the separate blocks for each station are essentially static terms.

As shown, significant low-velocity sediments and caldera fill are present in the Long Valley region. The poor vertical resolution inherent in the inversion causes the low-velocities to be smeared into layers below the first layer. Two attempts were made to remove the near surface low-velocity signature from the residuals by applying corrections to the first layer static terms.

In the first attempt, teleseismic rays were traced from a common depth of 8 km through all of the available seismic refraction profiles in the Long Valley region using the two-dimensional ray tracing package RAY84 (Luetgert 1988) following the method of Cerveny et al. (1977). Traveltimes through these models were normalized to the mean travel time, contoured, and are shown in Fig. 3. A correction factor in seconds was taken from this map and applied to the first layer static terms prior to inverting the data. These time terms are referred to as geologic corrections as they are derived from the well known upper crustal velocity structure in the Long Valley region. This method indicates that up to 0.3 s of the teleseismic travel time residual can be attributed to the upper few kilometers of the crust in the caldera while about 0.2 s delay is caused by the sediments of the Mono basin.

The second method is an iterative scheme using the inversion. A single layer model (Layer 1 of Fig. 4) was inverted five times, each time adding the calculated velocity perturbation back into the first layer perturbation input to that iteration as a corrective term. By the fifth iteration, these first layer corrections changed by no more than 0.9%. The final velocity perturbations from this method were then used as a correction to the static terms of the layered model.

For the most part, the geologic corrections and iterative corrections produce similar results and are compatible with the reading errors of this experiment. The most significant difference between the methods is the increase in magnitude of the low-velocity anomaly beneath the resurgent dome using the geologic corrections. The geologic corrections are due to structure within 5 km of the surface and are affected primarily by near surface sediments and caldera fill

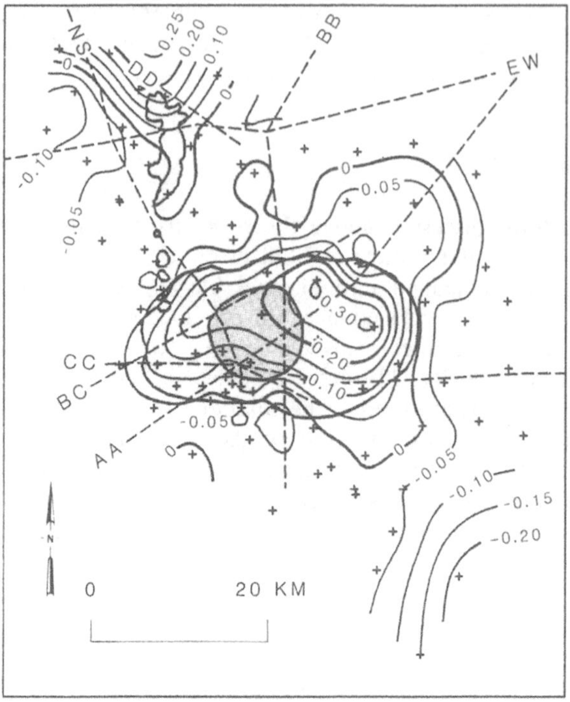

Fig. 3. Map of geologic corrections. Stations and physiographic outlines are the same as in Fig. 2. *Dashed lines* indicate seismic refraction profiles. Contour interval is 0.05 s. *Positive values* indicate increased travel time

structure. Other geophysical evidence (Rundle and Hill 1988) indicates that significant anomalous volumes as shallow as 4 or 5 km beneath the resurgent dome may be the cause of the observed decrease in velocity. We conclude that most of the low-velocity sediments are adequately removed when using these stripping methods, but that smearing of these low velocities (on the order of a few percent) may be occurring into the second layer using the geologic corrections.

Figure 4 shows six layers of a model derived from the residual data in Fig. 2, with corrections applied to the first layer using the geologic terms described above. A damping factor of $\theta^2 = 0.0020\,s^2/\%^2$ was found to provide the best trade-off between model smoothing and the fit to the observations. A ratio of 5:1 is maintained between observations and unknowns with this model. The model has been smoothed by spatially averaging four horizontally offset versions of the model to minimize the artificial effects of block boundaries and to maximize the spatial resolution of the inversion. The variance of the data is reduced 82% by each of the offset model inversions. Standard deviations of the velocity perturbation estimates are about 0.8% and diagonal elements of the resolution matrix range from about 0.5 to 0.8 in the areas interpreted below. Columns of the resolution matrix are compact in general and features about

5 km across or larger with velocity perturbations significantly larger than the standard deviations are well resolved and interpretable.

Significant features in Fig. 4 are the compact low-velocity anomalies beneath the Mono Craters and the resurgent dome of Long Valley caldera, and high-velocity anomalies coinciding with the Sierra Nevada batholith, the Benton Range east of Long Valley, and at depth beneath the caldera. Of primary interest to this work is the low-velocity anomaly beneath the resurgent dome. This anomaly is seen in layers 2 and 3 of Fig. 4 with the lowest velocity (6% lower than the mean layer velocity) occurring in layer 2. By applying a damping factor within the inversion, velocity estimates in the models are reduced, so the actual velocity may be lower. The feature dips slightly to the east with depth, starting beneath the western edge, and ending beneath the central portion of the resurgent dome. The volume of this anomaly can be estimated using the model from Fig. 4 and both stripping techniques. If a decrease from the mean layer velocity of 3% or greater indicates the presence of melt, the volume of the anomaly is approximately 1000 to 1500 km^3 using the iterative and geologic stripping techniques, respectively. Similarly, if a 5% decrease is required, the volume is approximately 150 to 600 km^3.

The low-velocity anomaly in the mid-crust beneath the Mono Craters has been discussed in detail by Achauer et al. (1986). They suggested that the eastward dipping low-velocity anomaly beneath Mono Craters may extend through the crust, but did not have a large enough array diameter to see below about 30 km. We can confirm that the low-velocity extends to about 27 km and is contained in a volume of 200 to 600 km^3. Below about 27 km the velocity

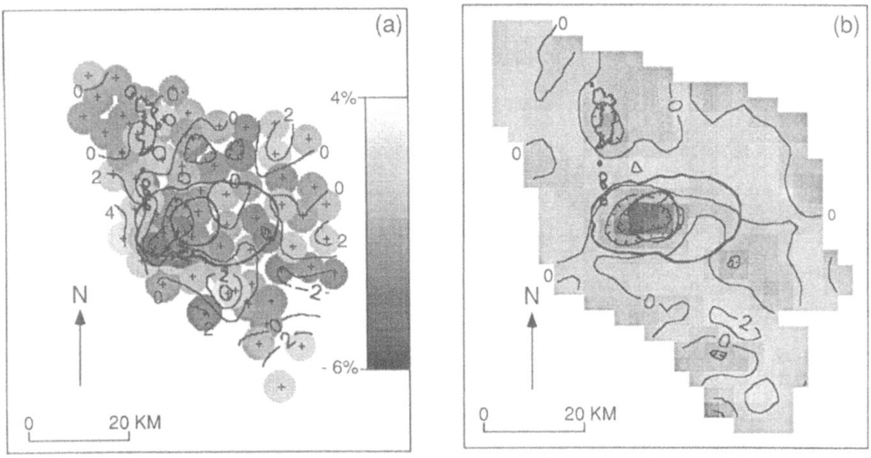

LAYER 1 0.0 - 8.0 KM VELOCITY 5.01 KM/S

LAYER 2 8.0 - 17.7 KM VELOCITY 6.07 KM/S

Fig. 4a–f. Tomographic model showing the velocity structure for six layers in the Long Valley region. Long Valley caldera, the resurgent dome, and Mono Craters are indicated by a bold line. Contour interval = 2% perturbation from the starting layer velocity. The first layer **a** has been corrected using the geologic corrections from Fig. 3. *Shading* indicates velocity perturbations, low velocities are dark

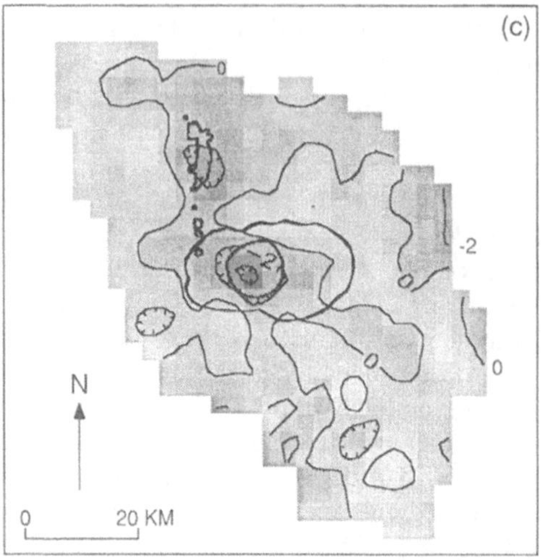

LAYER 3 17.7 - 27.6 KM VELOCITY 6.17 KM/S

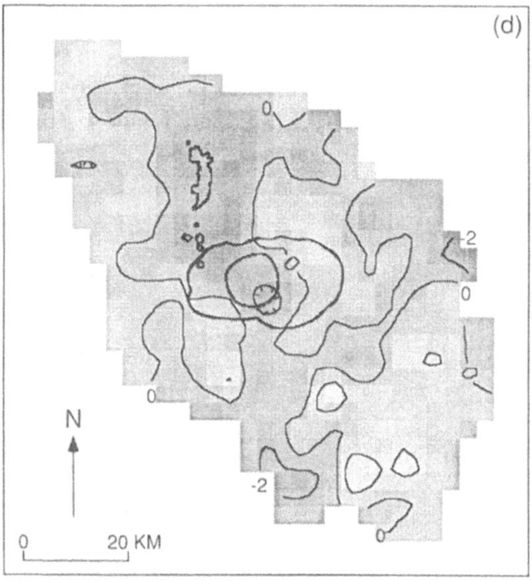

LAYER 4 27.6 - 38.1 KM VELOCITY 6.55 KM/S

Fig. 4c,d. *Continued*

structure beneath the Mono Craters is generally a few percent lower than average.

Another feature of the model is the distinct separation of the Long Valley and Mono Craters low-velocity anomalies. A zone of relatively high velocity material, on the order of a few percent increase in mean layer velocity, clearly

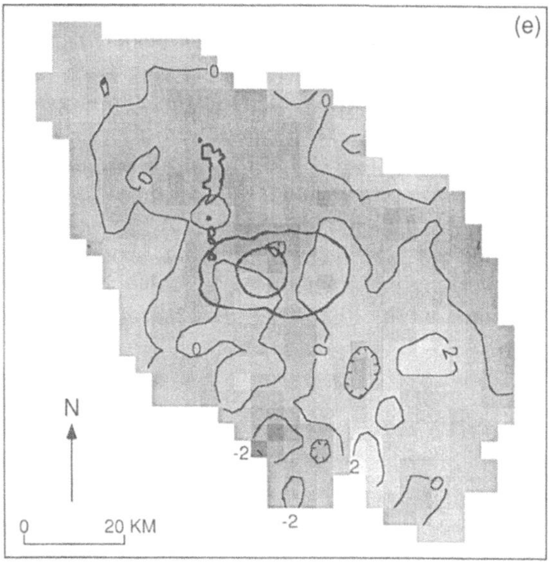

LAYER 5 38.1 - 48.8 KM VELOCITY 6.67 KM/S

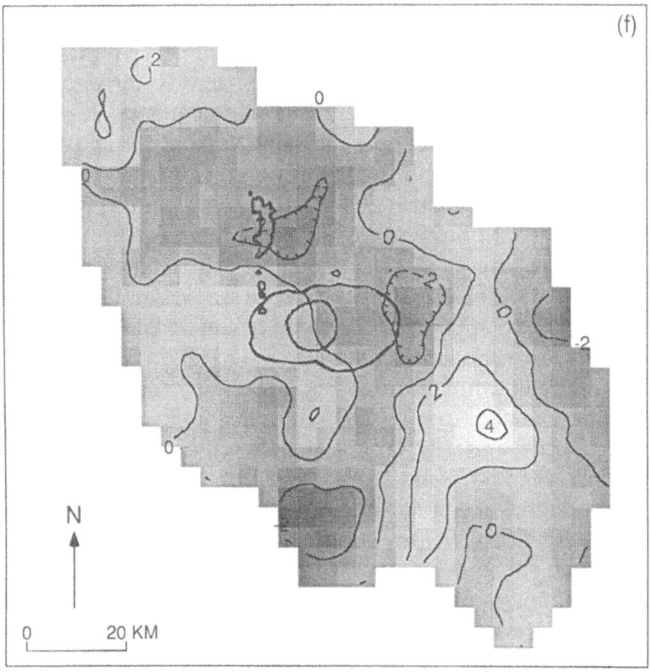

LAYER 6 48.8 - 61.5 KM VELOCITY 7.89 KM/S

Fig. 4e,f. *Continued*

extends from the surface to the bottom of the anomalies. Below layer 3 the magnitude of the low-velocity anomalies decreases sharply, with broader zones of marginally low-velocity underlying both anomalies in layer 4. The northwestern trend of low-velocity at this depth coincides with the structural trend of the Sierra Nevada. The separation of both low-velocity zones by high velocity material is also reduced at this depth. Regions of high velocity on the order of a few percent occur through the first three layers of the model west and northwest of the Mono Craters and south of the Long Valley caldera. These anomalies coincide with the Sierran batholith. Another weak and laterally broad zone of high velocity underlies the Long Valley caldera in layers 5 and 6. This zone was predicted by the PKIKP residual pattern (Fig. 2f).

3 Discussion

Figure 5 shows the geophysical phenomenon which support the presence of magma beneath the Long Valley caldera. Seismic reflections seen by Hill (1976) and Hill et al. (1985b) coincide with a zone of shear wave attenuation (Sanders, unpublished data, 1991) beneath the northwestern part of the resurgent dome at about 6 to 8 km depth. Luetgert and Mooney (1985) interpret a reflection seen at about 18 km depth beneath the southern part of the resurgent dome as coming from the bottom of a magma chamber. Throughout a volume of approximately 125 km^3 centered between 5 and 10 km depth beneath the resurgent dome are numerous point source and dike inflation centers (see Rundle and Hill 1988). Over approximately the same volume Carle (1988) has imaged a volume of low-density material by inverting gravity data. His interpretation of the low-density volume is that it is a mostly cooled pluton, with the possible inclusion of melt. Steck and Prothero (1988) show that teleseismic P-waves are converting to shear-wave energy across strong velocity contrasts in the same region. The correlation of these features, in so far as the resolution of each technique allows, is direct and implies that the same magma chamber or parts of the chamber are seen by the various methods.

The significant feature in Fig. 5 is the extension, to about 28 km depth, of the magma chamber as observed by the teleseismic data. Smearing of an intense low-velocity anomaly occurring between 5 and 10 km depth may contribute to the observed low-velocity anomaly which is observed between about 18 and 28 km depth, however, the resolution kernel for layer 3 indicates that the low-velocity is a real feature, and is not an artifact of vertical resolution. This suggests that the upper portion of the magma chamber consists of more anomalous material, perhaps significant batches of melt, while the deeper part of the chamber is composed of partial melt of a form which does not have as a large delaying signature.

Seismicity in the Long Valley region has direct implications as to the nature of the magma chamber beneath the caldera. The sequence of seismic events which began in 1978 and is continuing to the present (see Rundle and Hill 1988) has been characterized by quasiperiodic episodes of seismic activity occurring

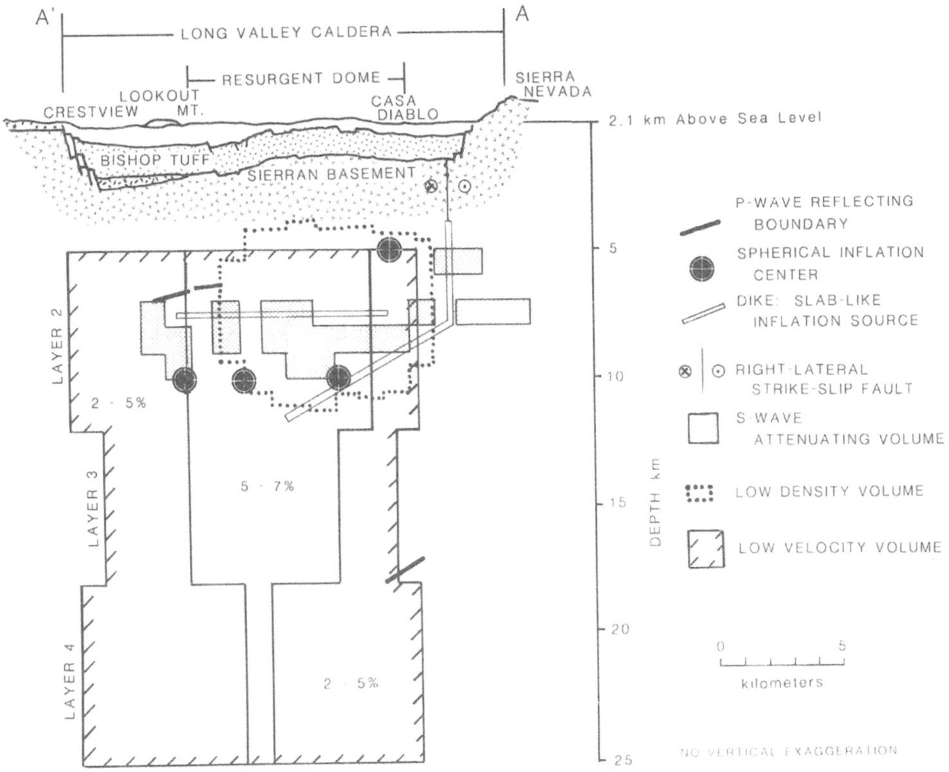

Fig. 5. Vertical profile through the Long Valley caldera (after Rundle and Hill 1988). Section is approximately coincident with A-A′ of Fig. 1. Shallow structure is from seismic refraction, deeper structure is from the listed geophysical measurements. Low velocity (in %) from Fig. 4 indicated by *large boxes*. No vertical exageration

over a 2400 km^2 region. The majority of events in this time period were located to the south of the caldera within the Sierran block in an area of higher than average velocity and to the east of the caldera along the White Mountains fault zone (Fig. 6). An exception to this pattern are events associated with moderate to intense swarms beneath the south moat of the caldera. In this Volume, Hill discusses in detail the seismicity and thickness of the seismogenic crust beneath Long Valley. He shows that the seismogenic crust thins from depths of 18- to 20-km south of Long Valley to 8 to 9 km beneath the south moat of the caldera and less than 5 km beneath the resurgent dome. Temperature estimates using Hill's method for the depth of the seismogenic crust beneath the south moat and resurgent dome of the caldera are consistent with the presence of magma at depths of 5 to 10 km.

Figure 7 is a depth cross-section (A-A′, Fig. 6) showing the seismicity which has occurred within 4 km of the line for the period 1980–1987. The revealing aspect of this figure is the deepening of events to the SE of the caldera. The M_L

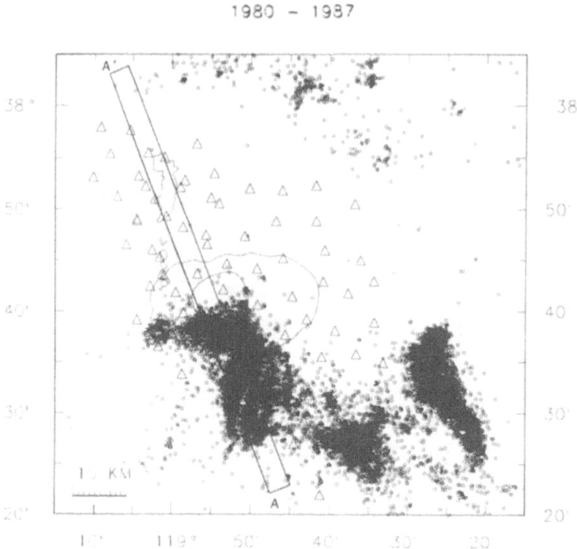

Fig. 6. Map of the regional seismicity from 1980 through 1987 (plot courtesy of F.W. Lester) *Triangles* indicate station locations. Box *A-A'* refers to the cross-section shown in Fig. 6

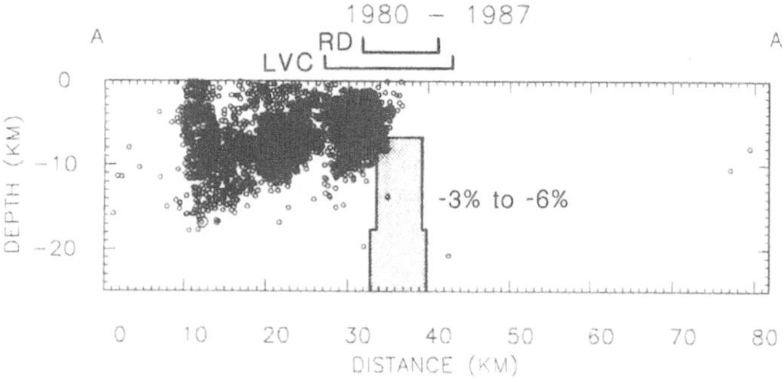

Fig. 7. Vertical section along profile A-A' in Fig. 6. *Stippled area* indicates the region of low velocity from Fig. 4. *RD* Resurgent dome; *LVC* Long Valley caldera. No vertical exageration

> 6 events are located near the lower extent of the earthquake zones and represent the nucleation centers for the upward release of stress along the Sierra Nevada frontal fault system. The shallowing of seismicity towards the caldera suggests that the brittle to plastic transition beneath the Sierra Nevada south of the caldera has been perturbed by the Long Valley magma chamber, which implies a long period of residence of magma in the mid-crust beneath the caldera. Another significant feature is the rapid shallowing of hypocenters beneath the south moat of the caldera. The presence of brittle fracture next to our observed low-velocity anomaly, but not into it, further supports the

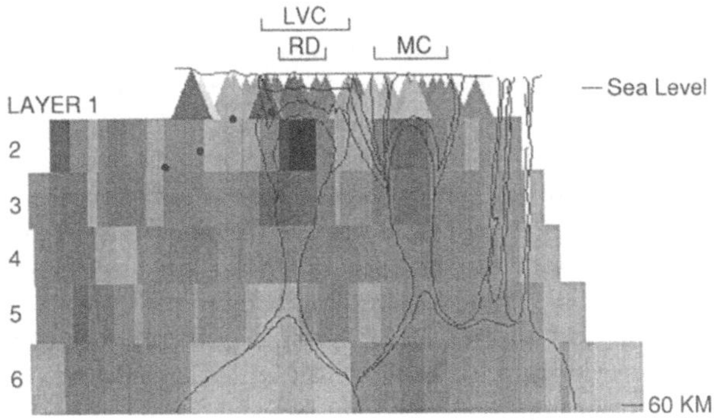

Fig. 8. Vertical section through the tomographic model (Fig. 4) along A-A′ of Fig. 1. *Shading* is the same as in Fig. 4. *Dark circles* are M > 6 earthquakes from 1980. *LVC* Long Valley caldera; *RD* resurgent dome; *MC* Mono Craters. *Black lines* represent an interpretation of Bailey's (1982) conceptual geologic model of the magmatic system. No vertical exageration

presence of magma beneath the resurgent dome. The near vertical cessation of earthquakes 15 km south of the caldera is a real feature. The Sierran frontal fault system steps eastward south of the caldera, as do the seismic events and this cross-section passes to the west of the frontal faults.

Figure 8 is a depth cross-section (A-A′, Fig. 1 and coincident with A-A′ of Fig. 6) through the model shown in Fig. 4. A modified version of the inferred geological model of the magmatic system of the Long Valley region forwarded by Bailey (1982) has also been projected onto this cross-section. This allows the direct comparison of a conceptual geologic model of the magmatic systems to the three-dimensional model of the compressional-wave velocity structure imaged over the same region. It must be remembered that since our model is composed of blocks it exhibits a rough, blocky picture of the true structure. The northwest trending zone of marginally low-velocity anomalies observed in layer 4 of Fig. 4 (which appears in cross-section as a zone of a few percent low velocity) may correspond to the region of mafic intrusion into the lower crust. The depth to the Moho in the Long Valley region is not constrained and layer 4 may approximate the maximum extent of the crust. Beneath the Long Valley caldera at depths greater than 38 km is a funnel-shaped feature of higher than average velocity, while the region beneath Mono Craters, at similar depths, exhibits lower than average velocity. This suggests a fundamental difference in the upper mantle beneath the two systems. The high velocity at depth beneath Long Valley is probably reflecting the age of the system, in that heat has been episodically supplied from this area for over 1.2 million years. Perhaps the withdrawal of heat from this area has produced a residuum with a high-velocity signature in the upper mantle. The Mono Craters, on the other hand, are only about 40 000 years old. The deeper low-velocity structure may indicate that this system has the potential for continued injection of basaltic magma into the lower crust.

The relatively high velocities between the Mono Craters and Long Valley magma chambers in the mid to upper crust severely restrict the size of any local magma chambers directly beneath the Inyo Craters to less than a few km^3. As suggested by Sampson and Cameron (1987), the approximately 400- to 600-year-old Inyo Craters were supplied by at least two separate magmatic systems, one similar in composition to older extruded volcanics within the caldera, the other similar to the recent Mono Craters eruptives. From the viewpoint of teleseismic tomography the magma chambers which supplied the Inyo Craters eruptions are most likely the low-velocity bodies we have imaged. The mixing of these magmas prior to eruption leads to the unlikely situation that both systems fortuitously erupted along intersecting paths at the same time. It is conceivable that major seismic events on this portion of the Sierra Nevada front triggered the eruptions and the frontal faults provided travel paths for the magma. This interpretation has significant bearing on the volcanic hazards of the Long Valley region, suggesting that a large seismic event near the Inyo-Mono Craters could trigger an eruption near this dike zone.

The problem of estimating the bulk melt in the magma chamber is quite difficult. As discussed by Achauer et al. (1986), a great variety of attenuating mechanisms affect the possible velocity perturbations which we image. For example, a crystal wetting melt on the order of 1% of the bulk material could produce the observed velocity perturbations, while field descriptions of recent extrusions indicate that some of the magma chamber beneath Mono Craters may be nearly 100% melt.

Achauer et al. (1986) give a simple estimate of percent melt within the Mono Craters magma chamber caused by a 7% velocity anomaly as on the order of 20%. Whether the 20% is spread through the volume of 200 to 600 km^3 imaged beneath Mono Craters, or occurs as discrete pockets of magma is indeterminate. A similar interpretation of the low-velocity anomaly beneath Long Valley suggests that within the anomalous volume the amount of melt could be on the order of 20 to 25%. An important clue as to the nature of the magma chamber is given by the porphyritic poor composition of the recently extruded magmas in the region which suggests that there must be zones of nearly complete melt. Because of the resolution inherent with our method, we can only average velocity changes on the order of 5 km. This does not restrict the magma chambers to being one continuous body; it is quite likely that the volumes we image as magma chambers are a complicated system of partial silicic melts, ranging from discreet batches of melt to crystal wetting melt. The current volumes of both the Mono Craters and the Long Valley caldera magma chambers are approximately the same as the erupted volume of the Bishop Tuff (Bailey 1982), but the amount of melt is on the order of 20%. This suggests that a major eruption such as the one which produced the Long Valley caldera is not likely, but that continuing Inyo-Mono Craters type eruptions are probable.

4 Conclusions

Patterns in teleseismic body wave travel time residuals suggest the presence of magma chambers beneath the Mono Craters volcanic complex and the resurgent dome of Long Valley caldera. A moderately resolved three-dimensional velocity image of the Long Valley region has been derived by inverting the residuals using the ACH method. The derived model matches the residuals within the reading errors and reveals the presence of two mid- to upper-crustal magma chambers in the region. One occurs beneath the southern portion of the Mono Craters, the other beneath the resurgent dome of the Long Valley caldera. Both magma chambers exhibit eastward dipping trends with depth which suggests structural control by the Sierra Nevada frontal fault system. A northwest striking zone of a few percent low velocity between 28 and 38 km depth is interpreted as an upwelling zone of mafic intrusion into the lower crust, controlled by regional tectonic structures. Below about 38 km depth the mantle source for the magmas is markedly different beneath Long Valley than the Mono Craters. Seismic velocities in the mantle beneath Long Valley are higher than below the Mono Craters, suggesting the focus of volcanism may be shifting north from Long Valley to the Mono Craters and, reflecting the greater age of the Long Valley magmatic system.

The volume of magma available in both chambers limits near future eruptions to the order of those which formed the Inyo-Mono Craters. It is possible that large seismic events occurring along the Sierran front near the Inyo-Mono Craters may trigger eruptions such as those which formed the Inyo Craters.

The physical unrest of the Long Valley region continues and so do our efforts to understand the reasons for the unrest. The information gained from this experiment shows that teleseismic tomography is an important part of a comprehensive, multi-disciplinary study needed to understand the nature of the volcanism in the Long Valley region.

References

Achauer U, Greene L, Evans JR, Iyer HM (1986) Nature of the magma chamber underlying the Mono Craters area, eastern California, as determined from teleseismic travel time residuals. J Geophys Res 91, B14:13873–13891

Aki K, Christoffersson A, Husebye ES (1977) Determination of the three-dimensional seismic structure of the lithosphere. J Geophys Res 82:277–296

Bailey RA (1982) Other potential eruption centers in Claifornia: Long Valley, Mono Lake, Coso, and Clear Lake volcanic fields. Calif Div Mines Geol Spec Publ 63:17–28

Bailey RA, Dalrymple GB, Lanphere MA (1976) Volcanism, structure and geochronology of Long Valley caldera, Mono County, California. J Geophys Res 81:725–744

Carle SF (1988) Three-dimensional gravity modeling of the geologic structure of Long Valley caldera. J Geophys Res 93, B11:13237–13250

Cerveny V, Molotkov IA, Psencik I (1977) Ray method in seismology. Chambers Univ Press, Praque, 214 pp

Goldstein NE (1987) Proceedings of the symposium on the Long Valley caldera: A predrilling data review. Rep *LBL-23940*, Lawrence Berkeley Lab, Berkeley, Calif, 195 pp

Goldstein NE, Stein RS (1988) What's new at Long Valley. J Geophys Res 93:13187–13190

Hermance JF (1983) The Long Valley/Mono basin volcanic complex in eastern California and status of present knowledge and future research needs. Rev Geophys Space Phys 21:1545–1565

Herrin E (1968) 1968 Seismological tables for P-phases. Bull Seismol Soc Am 58:1193–1241

Hill DP (1976) Structure of Long Valley caldera from seismic refraction experiments. J Geophys Res 81:745–753

Hill DP, Bailey RA Ryall AS (1985a) Active tectonic and magmatic process beneath Long Valley caldera, eastern California: an overview. J Geophys Res 90, B13:11 111–11 120
Hill DP, Kissling E, Luetgert JH, Kradolfer U (1985b) Constraints on the upper crustal structure of the Long Valley-Mono Craters volcanic complex, eastern California, from seismic refraction measurements. J Geophys Res 90:11 135–11 150
Iyer HM, Evans JR, Zandt G, Stewart RM, Coakley JM, Roloff JN (1981) A deep low-velocity body under the Yellowstone caldera, Wyoming: delineation using teleseismic P-wave residuals and tectonic interpretation. Geol Soc Am Bull 92: Part I 792–798, Part II 1471–1646
Lachenbruch AH, Sass JH (1978) Models of an extending lithosphere and heatflow in the Basin and Range province. Geol Soc Am Mem 152:209–250
Luetgert JH (1988) User's manual for RAY84/R83PLT interactive two-dimensional ray tracing/synthetic seismogram package. US Geol Surv Open File Rep 88–238:80
Luetgert JH, Mooney WD (1985) Crustal refraction profile of the Long Valley caldera California, from the January 1983 Mammoth Lakes earthquake swarm. Bull Seismol Soc Am 75:211–221
Muffler LJP, Williams DL (1976) Geothermal investigations of the U.S. Geological Survey in Long Valley, California, 1972–1973. J Geophys Res 81:721–744
Pakiser LC (1976) Seismic exploration of Mono Basin, California. J Geophys Res 81:3601–3618
Rundle JB, Carrigan CR, Hardee HC, Luth WC (1986) Deep drilling to the magmatic environment in Long Valley caldera. EOS Trans Am Geophys Un 67:490–491
Rundle JB, Hill DP (1988) The geophysics of a restless caldera-Long Valley, California. Annu Rev Earth Planet Sci 16:251–271
Sampson DE, Cameron KL (1987) The geochemistry of the Inyo volcanic chain: multiple magma systems in the Long Valley region, eastern California. J Geophys Res 92, B10:10 403–10 421
Sanders CO (1984) Location and configuration of magma bodies beneath Long Valley, California, determined from anomalous earthquake signals. J Geophys Res 89, B10:8287–8302
Steck LK, Prothero WA (1988) P to S converted wave observations at Long Valley caldera. Geophys Res Lett 15, 12:1433–1436
Steeples DW, Iyer HM (1976) Low-velocity zone under Long Valley as determined from teleseismic events. J Geophys Res 81:849–860

Three-Dimensional Inversion of Teleseismic Travel Time Residuals for Crust and Upper Mantle Velocity Structure in the Larderello Geothermal Field, Italy

J.E. Foley, M.N. Toksoz, and F. Batini

Abstract

The analysis of teleseismic travel time residuals observed on the Larderello seismic network has revealed a sharp low velocity zone in the center of the geothermal production area. Residuals from 101 teleseismic events, automatically determined with the implementation of a simulated annealing optimization procedure, are inverted to produce an image of an anomalous zone interpreted to be a series of intrusive bodies. The top of this zone is constrained to be below 6 km depth from local earthquake seismicity patterns and extends to a depth of greater than 20 km. The zone extends into the upper mantle and broadens with increasing depth. The observed travel time residual anomaly corresponds with observed gravity lows, high heat flow, the shallowing of a dominant upper-crustal reflection, and diminished teleseismic P-wave amplitudes. This body has elevated the temperatures in the region to over 350 °C at 2 km and is the origin of the geothermal energy.

1 Introduction

The deep structure of the Larderello geothermal field of central Italy is investigated by inversion of teleseismic travel time residuals. The information this study can provide is necessary to further the basic understanding of the origin of the geothermal phenomenon observered in Larderello.

We determine the size, extent and magnitude of the low velocity zone (LVZ) within the geothermal region which has given rise to the observed teleseismic travel time anomalies, temperature anomalies, gravity anomalies, P-wave teleseismic amplitude anomalies and variation of a deep crustal reflection horizon observed in the region. This information helps us understand the regional processes which control this important economic resource.

A wide range of studies of Earth velocity structure in volcanic regions utilizing anomalous travel times and amplitudes of seismic waves have been made, examples include Hawaii (Ellsworth 1977), Yellowstone (Iyer et al. 1981), the Geysers (Oppenheimer and Herkenhoff 1981) and Long Valley, California (Sanders 1984; see Iyer 1988 for a review).

In this paper we first review the geologic background of the Larderello geothermal area (Sect. 1.2), then describe the seismic network and data used in the analysis (Sect. 2.1). An automated technique for determination of travel time residuals is presented in Section 2.2, followed by a brief description of travel time residual calculations (Sect. 2.3) and two data reduction schemes implemented to improve the inversion (Sect. 2.4). The general azimuthal pattern in the data are discussed in Section 2.5. The inversion procedure is outlined and results presented in Sections 3.1 and 3.2. Discussion of various geophysical observations made in the area and interpretation of the data are discussed in Sections 4.1 and 4.2.

1.2 Geologic Background

Most Italian geothermal areas are located, from the geological point of view, in the complex structures of the Apennines, which are tectonically characterized as overlapping folds (Fig. 1) Boccaletti et al. 1983).

The inner Apennines are characterized by high heat flow. The Bouguer anomalies and the data from deep seismic soundings show a thin crust with a Moho discontinuity at a depth of less than 20 km and a soft mantle (V_p = 7.8 km/s) (Giese et al. 1981). Moreover, tensile stresses are present in the area and cause the rise of hot masses from the mantle into the crust.

This geological/structural evolution gave rise to a local geothermal anomaly with heat flow values exceeding 300 mW/m^2 in correspondence to the largest geothermal fields. The Larderello field is a system of the hydrothermal type, and is the most important in Italy, with an installed capacity of over 500 MW. The cap rock is represented by sedimentary terrains (clays and conglomerates) deposited during the post-orogenic transgressive phase and a complex of tectonic wedges composed of mainly clayey and arenaceous flysches.

The complexes overthrust during the Alpine orogeny onto other tectonic wedges are made up intensely fractured Triassic evaporites (chiefly dolomitic limestones and anhydrites) and represent the main reservoir exploited so far. A tectonic surface separates the complexes and tectonic wedges from a crystalline basement including formations characterized by an increasing degree of metamorphism towards the bottom, with the phyllites and quartzites giving way to mica schists and gneisses. The recognition of the structural boundaries of the main units was made possible by the acquisition of seismic reflection data, which were compared with results from field geology and wells drilled in the areas of question.

Well defined on nearly all reflection lines are deep markers, sufficiently continuous and with clear amplitude and frequency characters. Horizon K has been followed in correspondence to these markers. It follows the major amplitude events and rests on a group of reflectors which are mostly parallel to it.

The structural trend of K (Fig. 2) generally reflects the structural behavior of the top of the metamorphic basement, with a main positive element placed in a roughly NNE-SSW direction. The minimum depth (about 3000 m) is observed in proximity to the San Pompeo 2 well, between seismic stations LAGO and FRAS.

The structural panorama is complicated by vertical discontinities, well evidenced on the seismic sections, which in general determine depths of the K at

Fig. 1. Larderello geothermal field, located in central Italy, is a system of the hydrothermal type, and is the most important in Italy with an installed capacity of over 500 MW

the boundaries of the areas now being explored. As far as the nature and the causes of the K reflections are concerned (Batini and Nicolich 1984), it should be stressed once again that they are determined by strong variations of the reflection coefficient and thus by the presence of a medium of a very different acoustic impedance from that of the embedding medium.

Inside the crystalline basement, strongly corrugated during the Hercynian orogenesis and subsequently subjected to tensile stresses caused by the rise of intrusive bodies during the late Alpine orogenic phase, densely fractured permeable zones are present which allow circulation of the geothermal fluids (Batini et al. 1978, 1985).

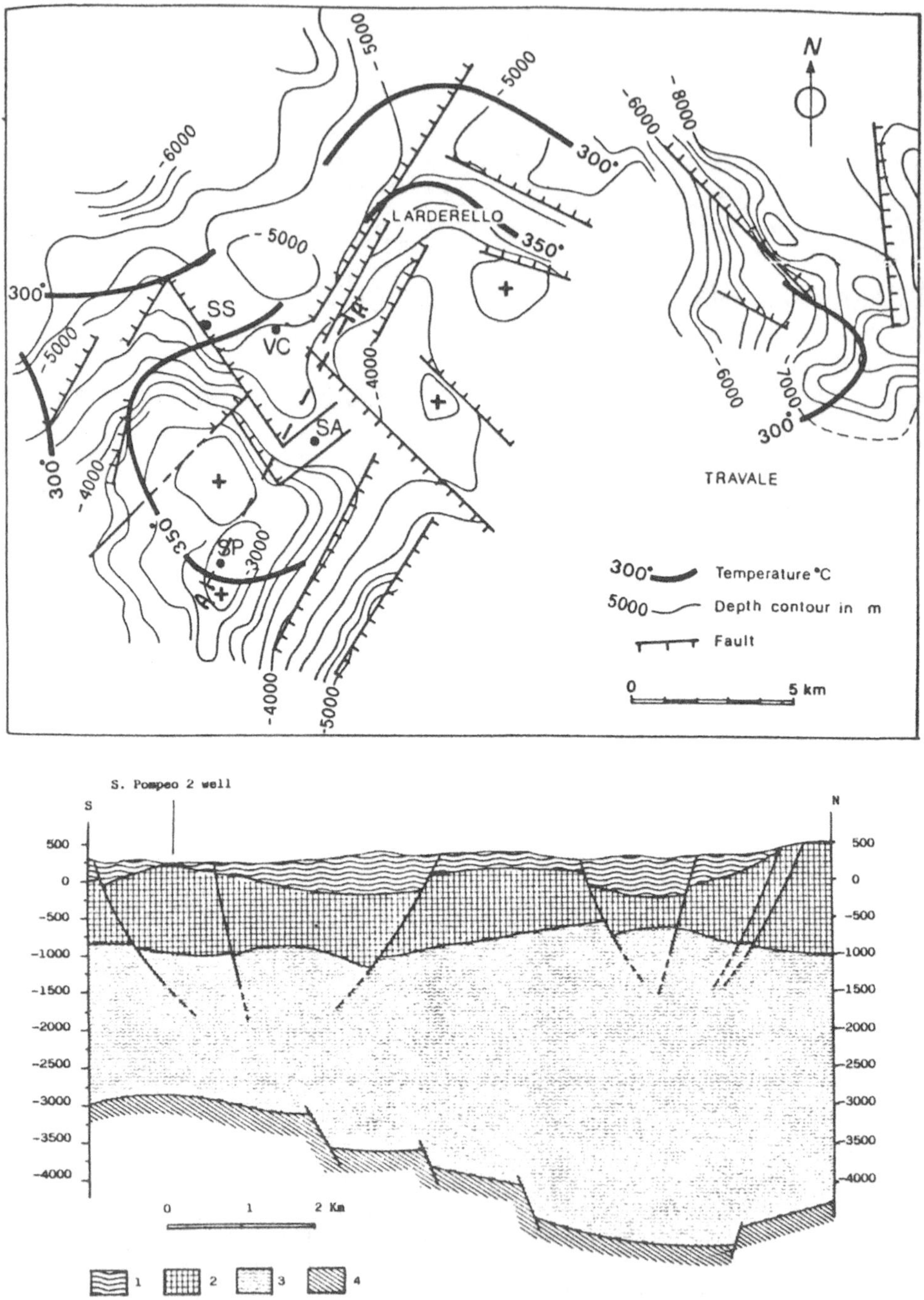

Fig. 2. Structural trend of the K reflection horizon. This strong marker is visible on nearly all reflection lines with depths of approximately 3 km in the center of the network to depth of 7 to 8 km on the periphery

2 Waveform Correlation for Travel Time Residuals

2.1 Data

The data for the teleseismic travel time inversion and the P-wave amplitude anomaly calculations come from the Italian National Energy Board (ENEL) seismic monotoring network located in Larderello, Italy (Fig. 3, Table 1). This seismic network began recording digital seismic data in 1985 and has recorded approximately 200 teleseismic events through June 1988. The network covers an area approximately 30 by 40 km and has 26 uniform and calibrated stations with 1 Hz vertical geophones (Batini et al. 1985). The average station spacing is 6.5 km. A typical teleseismic data set recorded on the network is shown in Fig. 4. This event is Number 44 from Table 2.

Of 200 teleseismic events, 101 selected on the basis of high quality were chosen for analysis (Table 2). Events with waveforms that have emergent onsets which makes accurate timing of the arrival at each station very difficult, and events recorded on less than 15 stations were not included.

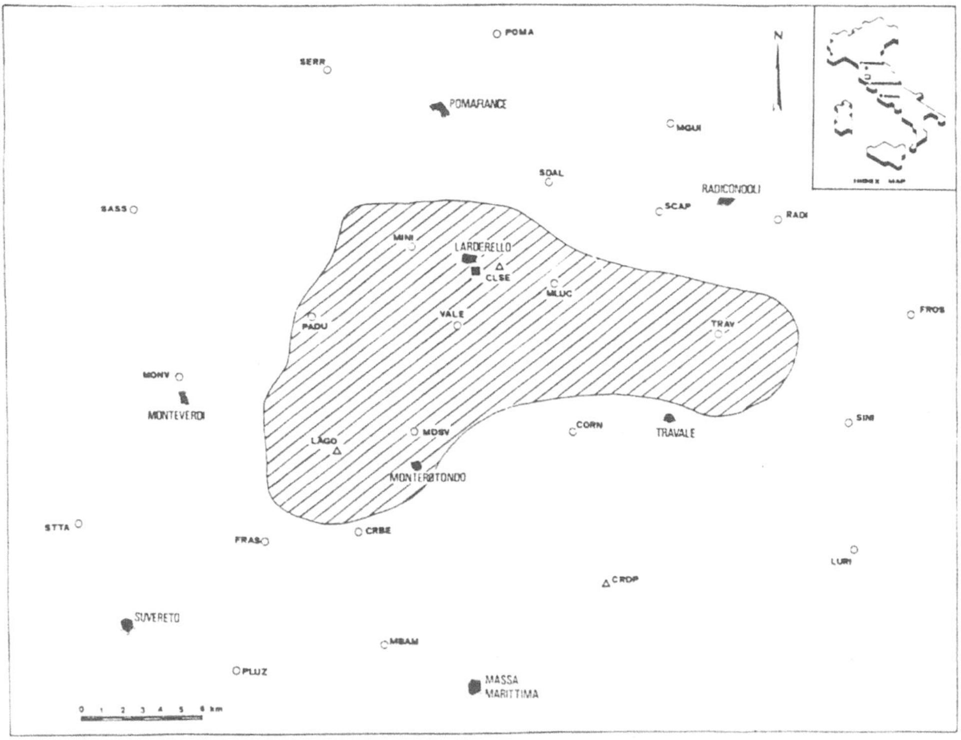

Fig. 3. ENEL seismic network, established in 1985, has 26 uniform and calibrated stations. Twenty-five of the stations were used in the travel time and amplitude calculations. One station was removed from the analysis due to unexplainable wave shapes and timing delays. Three stations have three-component data but were not utilized in this analysis

Table 1. Parameter values from the Larderello geothermal field seismic network

Station name	Lat	Long	Elev m	Resid s	Temp °C	K m	Grav mGal	Amp %[a]
MINI	43.23	10.85	444	0.37	260	4600	19	0.33
SERR	43.31	10.79	223	−0.23	150	–	26	2.60
POMA	43.33	10.90	231	−0.19	150	–	33	1.94
SDAL	43.26	10.93	380	−0.13	225	5000	23	0.22
SCAP	43.25	11.00	360	−0.25	175	8000	24	0.21
MGUI	43.29	11.01	341	−0.26	150	–	31	0.44
RADI	43.25	11.07	489	−0.30	150	8500	29	0.22
MLUC	43.22	10.94	572	0.83	240	4200	21	0.10
FROS	43.20	11.15	432	−0.38	150	–	26	0.92
SINI	43.16	11.11	350	−0.03	150	–	16	1.08
LURI	43.12	11.12	458	−0.08	150	–	18	1.09
TRAV	43.19	11.03	472	−0.05	300	7000	17	1.12
CORN	43.16	10.94	886	−0.03	250	4400	24	0.64
VALE	43.20	10.87	782	0.63	320	3800	20	2.01
CRDP	43.09	10.97	816	−0.21	150	–	18	0.62
MDSV	43.15	10.85	754	0.51	300	3600	19	0.56
CRBE	43.11	10.82	442	0.08	300	3200	25	0.99
MBAM	43.06	10.82	342	−0.03	250	3400	32	0.80
LAGO	43.15	10.81	319	0.40	300	3000	15	2.21
PADU	43.20	10.79	439	0.16	250	4600	22	0.26
SASS	43.25	10.69	438	−0.33	150	–	40	1.11
MONV	43.18	10.71	445	−0.20	255	5000	29	0.68
FRAS	43.11	10.76	220	0.00	200	4000	30	0.64
STTA	43.12	10.65	361	−0.28	150	–	37	0.55
PLUZ	43.05	10.74	156	0.00	150	4000	33	1.04

[a] Amp: waveform amplitude, see text for definition.

2.2 Waveform Correlation

A waveform correlation technique based on simulated annealing was applied to the data in order to automate the travel time residual calculations and to make more quantitative classifications of the arrival time pick quality.

The idea of simulated annealing, introduced by Kirkpatrick et al. (1983), is that one can make an analogy between optimization of an objective function and the physical process of crystal growth (annealing). In the teleseismic P-wave travel time problem we wish to align all traces to an equivalent initial phase. We recast this problem as a search for a set of teleseismic trace shifts that optimize the power of the stacked waveforms (the objective function) which has been windowed to contain the P-wave first arrival. A window of the expected arrival time is determined by calculating the total travel time from the event hypocenter to the station from a Herrin (1968) reference Earth model.

This optimization can be thought of as the determination of a set of trace shifts which minimizes a function containing local minima (Fig. 5). This function has an energy state at each possible combination of trace lags. We use simulated annealing to find the lowest energy level.

Attempting to locate this global minimum with conventional iterative improvement techniques can lead to solutions which rest in local minima. The

Table 2. Teleseismic event information

Event number	Lat.	Long.	Event date	Time	Depth (km)	Mag Mb	Back Azm	Incidence angle
001	−21.62	−176.49	Oct 12, 1985	02:12	157.0	5.8	18.3	1.8
002	40.32	69.84	Oct 13, 1985	15:59	33.0	5.8	72.2	38.1
003	46.30	146.29	Oct 18, 1985	04:19	291.0	6.0	29.2	23.2
004	36.72	54.80	Oct 29, 1985	13:13	33.0	6.0	85.8	42.1
005	51.77	175.55	Oct 30, 1985	19:05	33.0	5.6	9.4	22.5
006	53.26	−166.92	Oct 31, 1985	19:33	33.0	5.8	358.6	22.7
007	51.55	−173.11	Jan 18, 1986	01:59	33.0	5.8	2.5	22.1
008	−18.81	169.61	Mar 5, 1986	15:47	287.0	5.6	43.0	0.7
009	−7.13	−71.64	Mar 26, 1986	22:06	609.0	5.8	259.7	20.8
010	54.65	161.58	Apr 1, 1986	13:40	35.0	5.7	16.8	24.3
011	43.89	147.57	Apr 16, 1986	12:52	23.0	6.3	29.8	22.3
012	51.52	−174.78	May 7, 1986	22:47	33.0	6.4	3.5	22.1
013	36.37	70.71	May 7, 1986	23:25	223.0	5.6	77.4	37.1
014	51.50	−174.80	May 7, 1986	23:41	33.0	5.1	3.5	22.1
015	51.16	−176.57	May 8, 1986	01:18	33.0	5.4	4.7	22.0
016	51.16	−176.89	May 8, 1986	02:04	33.0	5.5	4.9	22.1
017	51.14	−176.44	May 8, 1986	04:03	33.0	5.8	4.6	22.0
018	51.34	−175.36	May 8, 1986	05:37	18.0	6.0	3.9	22.1
019	51.46	−174.24	May 9, 1986	19:04	33.0	5.8	3.2	22.1
020	51.49	−178.47	May 14, 1986	04:02	33.0	5.4	5.8	22.2
021	52.33	−174.50	May 17, 1986	16:20	26.0	5.8	3.3	22.4
022	43.68	148.42	May 21, 1986	05:47	39.0	6.1	29.4	22.1
023	−21.82	−179.08	May 26, 1986	18:40	583.0	6.1	24.4	1.6
024	−20.19	178.86	May 26, 1986	19:06	538.0	6.4	27.5	0.9
025	38.00	37.92	Jun 6, 1986	10:39	10.0	5.6	95.0	48.0
026	43.27	146.49	Jun 8, 1986	11:02	56.0	6.0	30.8	22.3
027	−22.04	−178.93	Jun 16, 1986	10:48	547.0	6.3	24.3	1.7
028	−22.04	−178.93	Jun 16, 1986	10:48	547.0	6.3	24.3	1.7
029	53.88	160.39	Jun 17, 1986	00:42	33.0	5.9	17.7	24.2
030	51.25	−179.75	Jul 5, 1986	03:01	33.0	5.6	6.6	22.2
031	10.39	56.83	Jul 7, 1986	16:26	08.0	6.4	115.4	34.8
032	29.96	51.58	Jul 12, 1986	07:54	10.0	5.7	98.6	41.6
033	16.06	−93.90	Jul 13, 1986	09:12	80.0	5.9	291.6	20.7
034	47.26	151.13	Jul 19, 1986	5:59	141.0	5.9	26.0	22.9
035	−25.85	−177.49	Jul 27, 1986	10:43	147.0	5.8	24.3	1.0
036	51.60	−174.10	Jul 28, 1986	04:06	33.0	5.4	3.1	22.2
037	36.47	71.08	Aug 21, 1986	01:34	235.0	5.4	77.1	37.0
038	−23.28	−176.75	Sep 6, 1986	08:52	122.0	5.6	20.1	2.3
039	36.71	71.09	Sep 15, 1986	21:42	89.0	5.8	76.8	37.1
040	−15.57	167.56	Oct 23, 1986	02:18	160.0	5.7	43.0	1.9
041	−14.34	167.68	Nov 12, 1986	00:26	33.0	5.3	41.7	2.2
042	51.43	−173.82	Nov 14, 1986	21:42	33.0	5.5	2.9	22.1
043	−16.27	167.55	Nov 20, 1986	13:14	60.0	5.7	43.7	1.7
044	29.98	51.64	Nov 20, 1986	20:08	32.0	5.2	98.5	41.6
045	−15.27	167.35	Nov 22, 1986	17:08	158.0	5.8	43.0	2.0
046	−3.34	−77.41	Nov 23, 1986	01:39	106.0	6.4	266.4	20.2
047	41.96	81.32	Jan 5, 1987	22:52	17.0	5.9	65.7	35.2
048	−19.49	−177.46	Feb 10, 1987	0:59	395.0	6.2	19.1	1.0
049	43.17	132.29	Feb 11, 1987	17:42	499.0	5.5	39.4	24.4
050	54.72	161.74	Feb 14, 1987	16:42	33.0	5.7	16.7	24.3
051	−19.68	168.76	Feb 17, 1987	4:19	33.0	5.6	45.3	0.6
052	51.30	−179.28	Feb 18, 1987	0:00	33.0	6.2	6.3	22.2

Table 2. *Continued*

Event number	Lat.	Long.	Event date	Time	Depth (km)	Mag Mb	Back Azm	Incidence angle
053	−15.85	167.89	Feb 23, 1987	15:49	234.0	5.9	42.8	1.8
054	38.10	91.18	Feb 25, 1987	19:56	26.0	5.7	65.2	31.9
055	53.03	−162.55	Feb 28, 1987	23:20	999.9	4.8	356.0	22.7
056	46.33	152.07	Mar 3, 1987	1:32	90.0	5.9	25.8	22.5
057	51.16	179.58	Mar 6, 1987	13:48	56.0	5.4	7.1	22.1
058	31.97	131.70	Mar 18, 1987	3:36	38.0	6.5	46.9	21.4
059	−14.81	167.24	Mar 19, 1987	17:14	132.0	5.3	42.8	2.2
060	−20.31	−176.27	Mar 19, 1987	22:51	231.0	5.8	17.0	1.4
061	52.01	−177.44	Mar 21, 1987	10:41	97.0	6.0	5.1	22.4
062	51.54	−173.49	Mar 22, 1987	2:49	22.0	5.9	2.7	22.1
063	−13.72	167.18	Mar 26, 1987	16:17	190.0	5.5	41.9	2.5
064	−22.93	−66.23	Apr 1, 1987	1:48	224.0	6.2	244.7	18.3
065	49.90	78.81	Apr 3, 1987	1:17	0.0	6.2	56.8	37.1
066	49.85	78.69	Apr 17, 1987	1:03	0.0	6.0	56.9	37.1
067	61.44	−150.83	Apr 18, 1987	2:01	70.0	5.6	351.0	26.0
068	49.78	78.09	May 6, 1987	4:02	0.0	5.5	57.1	37.2
069	−6.35	152.40	May 7, 1987	1:51	10.0	5.3	53.9	7.1
070	50.02	156.27	May 12, 1987	4:03	46.0	5.4	21.7	23.3
071	49.90	78.73	Jun 20, 1987	0:53	0.0	6.1	56.8	37.1
072	54.17	−162.54	Jun 21, 1987	5:46	33.0	6.2	356.1	23.1
073	53.43	158.32	Jul 6, 1987	23:22	149.0	5.3	29.1	24.2
074	49.87	78.79	Nov 15, 1987	03:31	0.0	6.0	56.8	37.1
075	41.36	89.64	Dec 22, 1987	00:16	22.0	5.9	62.6	32.9
076	43.29	142.42	Jan 3, 1988	12:42	177.0	6.0	33.3	22.8
077	54.78	161.66	Jan 11, 1988	21:07	43.0	5.8	16.7	24.4
078	50.81	173.47	Feb 7, 1988	18:15	33.0	6.2	10.9	22.3
079	49.95	78.91	Feb 13, 1988	03:05	0.0	6.1	56.7	37.1
080	46.66	152.59	Feb 20, 1988	22:11	54.0	5.9	25.4	22.6
081	57.27	−142.79	Mar 6, 1988	22:35	10.0	6.8	345.7	25.0
082	10.37	−60.58	Mar 10, 1988	06:17	56.0	6.2	264.9	27.9
083	54.70	161.61	Mar 12, 1988	08:38	23.0	5.7	16.7	24.3
084	49.89	78.96	Apr 3, 1988	01:33	0.0	6.0	56.7	37.0
085	10.67	−62.84	Apr 12, 1988	19:41	98.0	5.5	266.7	27.4
086	0.95	−30.24	Apr 20, 1988	04:25	10.0	5.8	232.6	33.1
087	−23.92	−176.98	Apr 25, 1988	01:19	127.0	5.5	21.2	2.4
088	−22.81	170.25	May 3, 1988	23:22	33.0	5.9	46.6	0.4
089	11.46	−85.93	May 6, 1988	14:46	100.0	5.7	282.9	21.6
090	−13.38	−76.21	May 6, 1988	16:34	51.0	5.9	258.2	18.2
091	42.61	143.72	May 7, 1988	01:59	79.0	6.1	32.9	22.4
092	73.35	54.47	May 7, 1988	22:49	0.0	5.6	19.6	41.0
093	−13.91	166.24	May 16, 1988	23:07	51.0	5.9	43.4	2.6
094	52.09	174.24	May 18, 1988	06:13	33.0	5.2	10.2	22.7
095	0.76	−30.30	May 21, 1988	15:15	10.0	5.5	232.6	33.0
096	50.56	−174.56	May 25, 1988	14:05	40.0	5.7	3.4	21.8
097	−7.47	128.36	May 30, 1988	21:11	66.0	6.4	76.1	12.1
098	−15.38	167.56	Jun 5, 1988	18:22	116.0	5.9	42.8	2.0
099	51.02	−177.60	Jun 18, 1988	16:15	42.0	5.1	5.3	22.0
100	−15.23	168.19	Jun 22, 1988	21:53	53.0	5.4	41.8	1.9
101	−17.68	−178.88	Jul 6, 1988	01:10	545.0	5.6	21.0	0.3

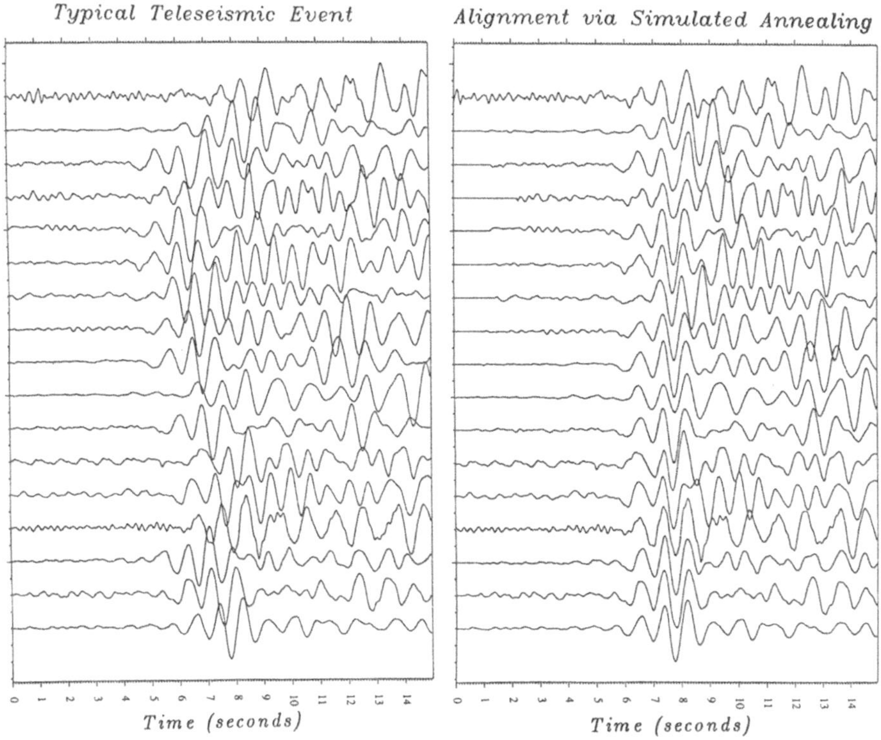

Fig. 4. Typical data set recorded of the Larderello Seismic Network (*left*). This event is number 44, see Table 2 for event information. The teleseismic data recorded on this network is dominated by 1 Hz energy. *Right* The events after it has been automatically aligned with the simulated annealing procedure

Fig. 5. A schematic view of an objective function which contains many local minima and one global minimum. Many optimization procedures, such as steepest descent, will get trapped in local minima. Simulated annealing allows for uphill motion of the objective function to escape local minima

Simulated Annealing Procedure Schematic

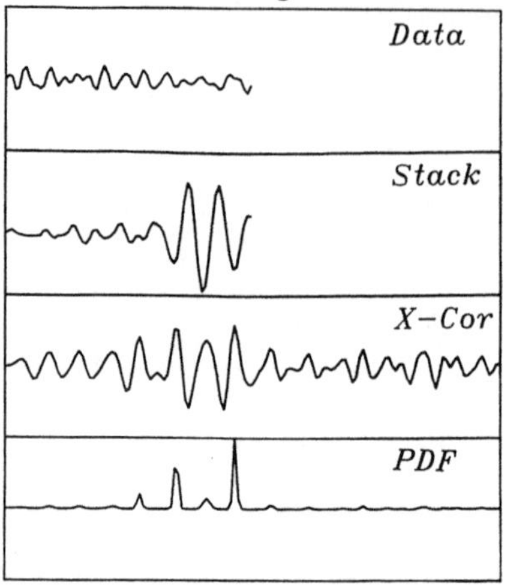

Fig. 6. Simulated annealing procedure schematic: A data trace is windowed to contain the phase to be correlated; this window is removed from the window stack which contains all the data windows; a cross-correlation function is made between the data window and the partial stack; the correlation function is converted into a probability distribution and a shift is determined. This procedure is followed for each trace to complete one iteration

false conclusion that optimal trace arrangement has been reached would come from a stable stack power level with successive iterations. We use simulated annealing to insure that the global minima is reached in each case. This is done by probabilistically allowing uphill motions on the objective function.

Standard waveform correlation procedures shift traces to the maximum cross correlation value between the waveform and a pilot trace. With simulated annealing, we use the stack as the pilot and probabilistically determine the trace shifts. We follow the procedure outlined by Rothman (1986) for event alignment; an outline of the process is described below and shown schematically in Fig. 6.

The procedure begins with a stack of the filtered trace windows without shifts. All seismograms are low-pass filtered to eliminate energy outside of the teleseismic P-wave frequency band (0.2–3 Hz). After the initial stack is made, a seismogram is removed from it and a cross correlation function is calculated between the seismogram and the partial stack. The cross correlation function is then converted into a probability distribution using the relation:

$$\mathrm{PDF(i)} = exp(\mathrm{XCOR(i)/T}) \qquad (1)$$

Here, PDF(i) is the probability distribution, XCOR(i) is the cross correlation function and T is the conversion *temperature*. The temperature parameter controls the conversion from cross correlation to probability distribution. A random selection is made from the PDF to determine the shift which is applied to the trace before it is added back into the stack. The procedure is executed for each seismogram in the stack to complete a single iteration. High temperatures are used at early iterations to allow for easy movement on the objective function. The temperature is systematically lowered with each iteration to near zero when the configuration state is near of the global minimum.

Many iterations are often required with simulated annealing (Rothman 1986); here stable trace configurations were found in 20 to 80 iterations. With this procedure we probabilistically allow for shifts not associated with the maximum cross correlation when defining the optimal trace alignment.

The pilot function (the stack) develops into an accurate representation of the teleseismic P-wave receiver function. On the last iteration, we retain the maximum pilot-trace correlation value as the quality index of the trace timing selection.

When classifying a successful phase pick as one which cannot be improved manually, this technique has an 80% success rate for the Larderello teleseismic data set. In general, the method fails when the teleseismic source is without distinctive waveshape character (harmonic with very narrow frequency band).

2.3 Residuals

After a set of relative travel times are determined with the simulated annealing procedure, we calculate the relative travel time residuals with respect to a Herrin Earth model. Elevation corrections are applied to the travel time data and then absolute residuals are calculated using the equation:

$$R_{ij}^{H} = T_{ij}^{obj} - T_{ij}^{H}, \tag{2}$$

where R^H is the absolute residual from the Herrin table for the ith station and j^{th} event; T_{ij}^{obj} is the observed time from the PDE event catalog; and T_{ij}^{H} is the theoretical time through the reference Earth.

Relative residuals RR_{ij} are calculated using:

$$RR_{ij} = R_{ij}^{H} - \frac{1}{N}\sum_{i=1}^{N} R_{ij}^{H}, \tag{3}$$

where N is the number of stations in event j; see Zandt (1978) for complete details.

Travel time residuals were calculated for 101 teleseismic earthquakes and nuclear explosions. The azimuthal and incidence angle distribution of these data (Fig. 7) can be categorized into five groups; events from the NNE, NE, East, West and events arriving with nearly vertical incidence (PKIKP). Table 2 gives the hypocentral and geometrical information for these events used in this study. Gaps in coverage are quite large from the SW (74°); however, there is sufficient azimuthal variation in the data to obtain a varied cross-fire of rays through the model. Average station residuals were calculated for the entire data set and for the five subdivisions (Table 3).

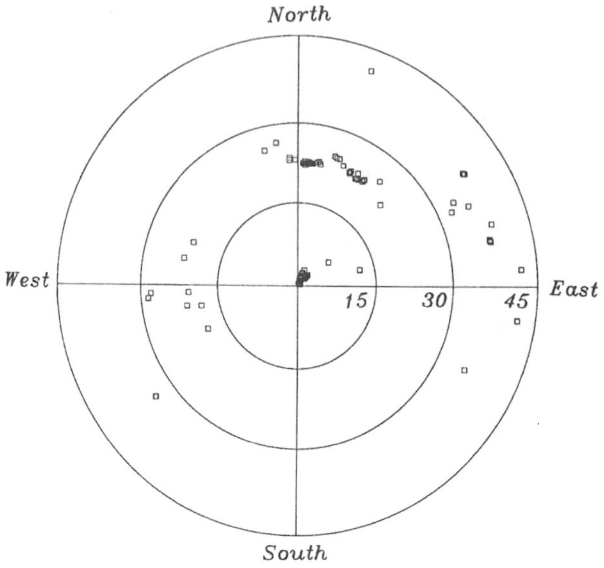

Fig. 7. Azimuthal coverage of the events used in the teleseismic travel time inversion. Concentric circles represent 15°, 30° and 45° of incidence angle. Data is subdivided into five groups: NNE, NE, East, West, and PKIKP events

2.4 Residual Reductions

Two corrections were applied to the relative residual data, the first was to reduce the residuals to a reference level of 2 km below sea level. This was accomplished by utilizing the well defined near surface velocity structure determined from the extensive reflection work done in the region (Batini and Nicolich 1984). One-dimensional, 2-km-deep, near-surface P-wave velocity profiles for each station were defined and the travel time of each ray was then reduced in accordance to that model and the angle of incidence of the ray to a depth of 2 km. This procedure helps to reduce the effect of back propagating near surface low velocity anomalies into the mid- and lower-crust.

After inversion of the data set reduced to 2 km was performed, we found that a second reduction of the data was necessary. The initial inversion with the 2 km data lowers the velocity of the upper most layer (between 2 and 7 km depth) by 25 to 30%. However, results from reflection analysis (Batini and Nicolich 1984) indicate that the strong LVZ in Larderello must be below the K reflection horizon which is found at depths of 3 to 5 km below the surface in the center of the network. We also feel that the LVZ must be below the level of intense earthquake activity observed in the region.

The second reduction applied to the data lowers the reference datum deeper into the crust and helps reduce upward smearing of residuals into the top layers

Table 3. Average teleseismic travel time residuals

Station	Lat.	Long.	NNE	n	NE	n	East	n	West	n	PKIKP	n	All	n
MINI	43.23	10.85	0.07	16	0.21	14	0.34	15	0.34	6	0.37	21	0.26	72
SERR	43.31	10.79	−0.20	11	−0.24	8	−0.32	10	−0.09	4	−0.23	13	−0.23	46
POMA	43.33	10.90	−0.23	25	−0.24	19	−0.26	20	−0.08	7	−0.19	23	−0.22	94
SDAL	43.26	10.93	−0.13	20	−0.19	15	−0.18	15	0.18	6	−0.13	18	−0.13	74
SCAP	43.25	11.00	−0.24	19	−0.29	13	−0.20	14	0.04	8	−0.25	17	−0.21	71
MGUI	43.29	11.01	−0.25	4	−0.28	3	−0.28	5	0.14	2	−0.26	8	−0.23	22
RADI	43.25	11.07	−0.39	17	−0.43	11	−0.16	12	−0.11	6	−0.30	10	−0.30	56
MLUC	43.22	10.94	0.82	5	0.77	7	0.62	1	0.57	7	0.83	2	0.71	22
FROS	43.20	11.15	−0.27	3	−0.53	2	−0.32	5	−0.47	5	−0.38	3	−0.39	18
SINI	43.16	11.11	0.05	15	−0.02	12	−0.10	8	0.07	6	−0.03	18	−0.00	59
LURI	43.12	11.12	−0.04	11	−0.10	11	−0.08	10	−0.02	6	−0.08	15	−0.07	53
TRAV	43.19	11.03	0.51	21	0.59	12	0.58	12	0.72	5	0.63	18	0.58	68
CORN	43.16	10.94	0.10	11	0.00	11	−0.01	9	−0.04	5	−0.03	14	0.00	50
VALE	43.20	10.87	−0.08	14	−0.11	10	−0.04	13	0.12	7	−0.05	11	−0.05	55
CRDP	43.09	10.97	−0.13	25	−0.20	18	−0.21	18	−0.16	9	−0.21	22	−0.18	92
MDSV	43.15	10.85	0.45	23	0.48	16	0.14	13	0.18	8	0.51	19	0.39	79
CRBE	43.11	10.82	0.24	11	0.28	6	0.26	5	−0.16	1	0.08	8	0.20	31
MBAM	43.06	10.82	0.18	9	0.07	13	0.09	10	−0.03	7	0.05	17	0.07	56
LAGO	43.15	10.81	0.35	13	0.41	16	0.49	19	0.32	4	0.40	23	0.41	75
PADU	43.20	10.79	0.12	5	0.07	9	0.27	7	−0.02	4	0.16	10	0.13	35
SASS	43.25	10.69	−0.35	10	−0.30	12	−0.26	12	−0.36	8	−0.33	17	−0.32	59
MONV	43.18	10.71	−0.20	20	−0.15	19	−0.03	17	−0.29	7	−0.20	19	−0.16	82
FRAS	43.11	10.76	0.05	19	0.07	16	0.12	16	−0.12	8	0.00	15	0.04	74
STTA	43.12	10.65	−0.20	18	−0.23	14	−0.18	16	−0.28	8	−0.28	20	−0.23	76
FLUZ	43.05	10.74	0.08	15	0.09	18	0.10	18	−0.08	10	−0.00	24	0.04	85

Azimuthal Ranges (degrees from north):
NNE: 346 to 16; NE: 17 to 47; East: 48 to 116; West: 232 to 292; PKIKP: All events with incidence angle less than 10°

of the model from the strong, deeper anomalies below. A contour map of the deepest earthquakes in the region (Fig. 8) was made by scanning the local seismicity catalog for the deepest events occurring on a 2 by 2 km grid, and a datum level of 6 km was selected by examination of the local earthquake depth distribution pattern. Earthquakes occur across the entire network to a depth of at least 6 km, justifying the datum level reduction. To correct the relative travel time residuals to 6 km, the same procedure is followed for the 2 km reduction using deep velocity estimates of Batini and Nicolich (1984).

2.5 Residual Distribution

Figure 9 shows the average station travel time residuals for the data divided into the four azimuth groups, the PKIKP events and all the data together. The average values and number of events are listed in Table 3. From these maps, we can see that the strongest low velocity area is concentrated in the center of the network. Strong positive residuals (slow) of up to 0.6 s are observed which

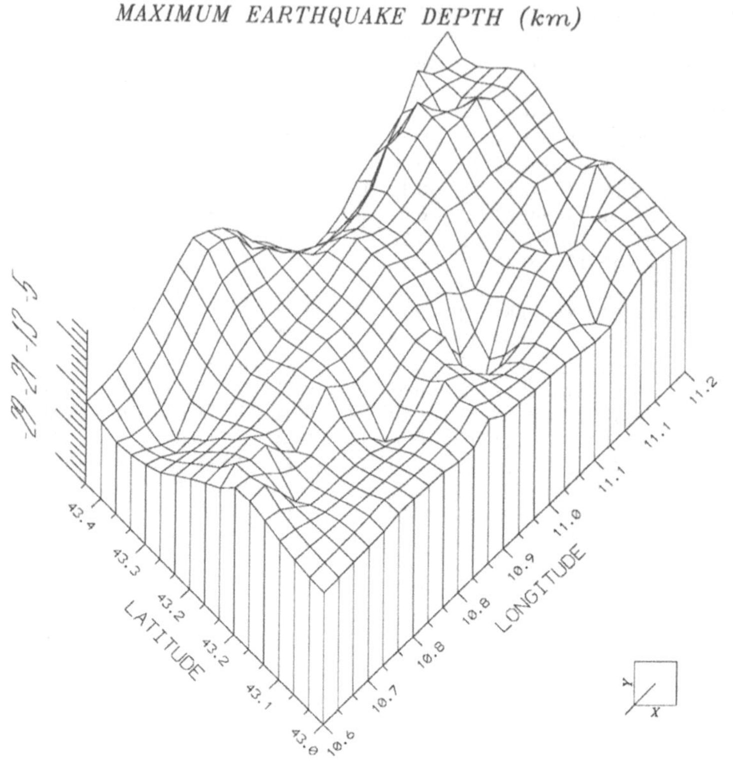

Fig. 8. Local earthquake depth distribution beneath the Larderello seismic network. Earthquakes occur throughout the entire region to depths of 6 km and greater

contrast to the negative (fast) residuals up to −0.4 s on the periphery. The contour of zero residual is aligned to the NE in each case with dimensions of approximately 30 by 40 km. The total travel time differences are about 1 s which represents a very strong travel time anomaly for a small region (approximately 25 km between average extremes). This range in average travel time residual is similar to those found in Geysers – Clear Lake Geothermal field in California, (Oppenheim and Herkenhoff 1981).

To produce a 1 s relative delay for a PKIKP event (near-vertical ray) from crustal effects requires a 35% velocity decrease along the entire crustal raypath. Velocities of rhyolite melts (Murase and McBirney 1973) and dacite melts (Hayakawa 1957) of about 4 km/s would represent a 38% velocity reduction of a 6.5 km/s country rock. This indicates that the low velocity anomaly either exists as a large body of melted magma in the crust or, more likely, that it penetrates the crust into the mantle to depths below 20 km.

Patterns exhibited in Fig. 9 indicate that average station travel time residuals calculated for events with rays passing through the center of the network have generally slower raypaths. Those paths which exclude the network center are generally faster. Stations located near the zero-second contour show this effect

Travel Time Residuals (All Events)

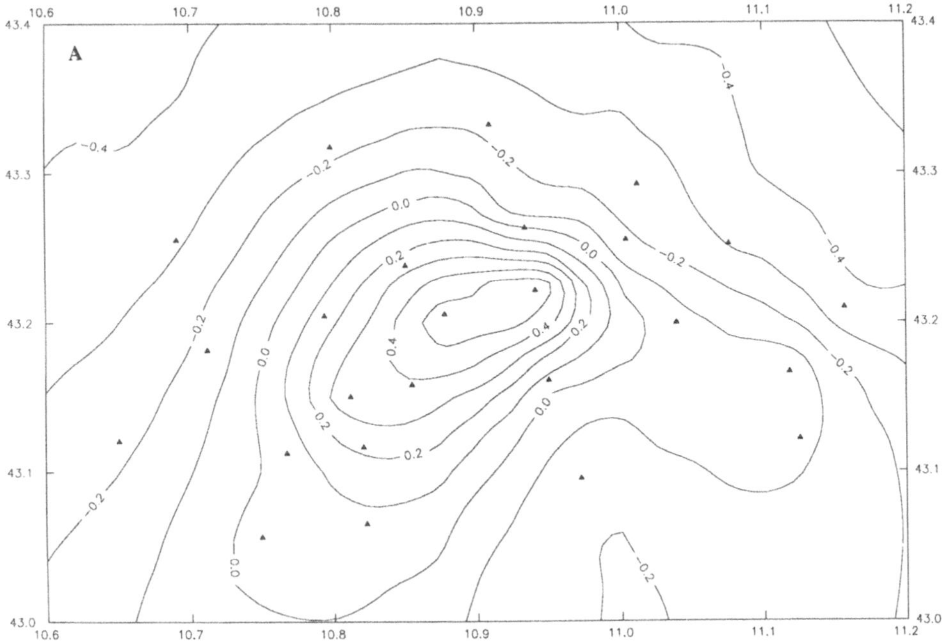

Travel Time Residuals (NNE)

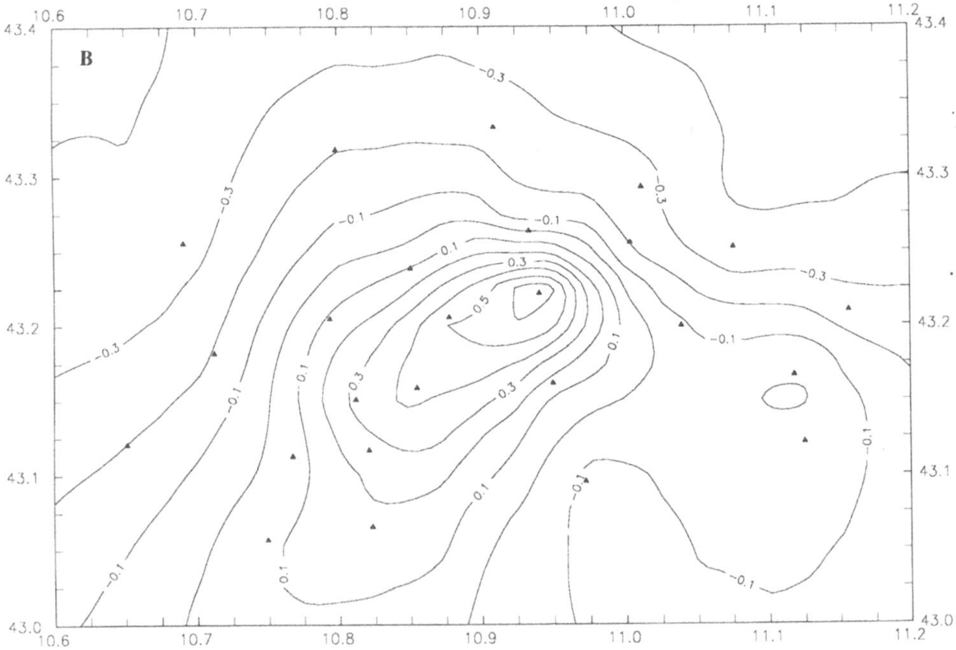

Fig. 9A–F. Average travel time residuals recorded on the Larderello seismic network. All residuals represent data reduced to 6 km. Contours are lines of equal residual plotted at 0.1 s intervals. See text for description of the residual patterns. **A** Data from all 101 events. **B** NNE, 28 events in the azimuth range of 346 to 16°. **C** NE, 18 events in the azimuth range of 17° to 47°. **D** East, 20 events in the azimuth range of 48° to 116°. **E** West, 10 events in the azimuth range of 232° to 292°. **F** PKIKP, 25 events arriving with incidence angles less than 5°

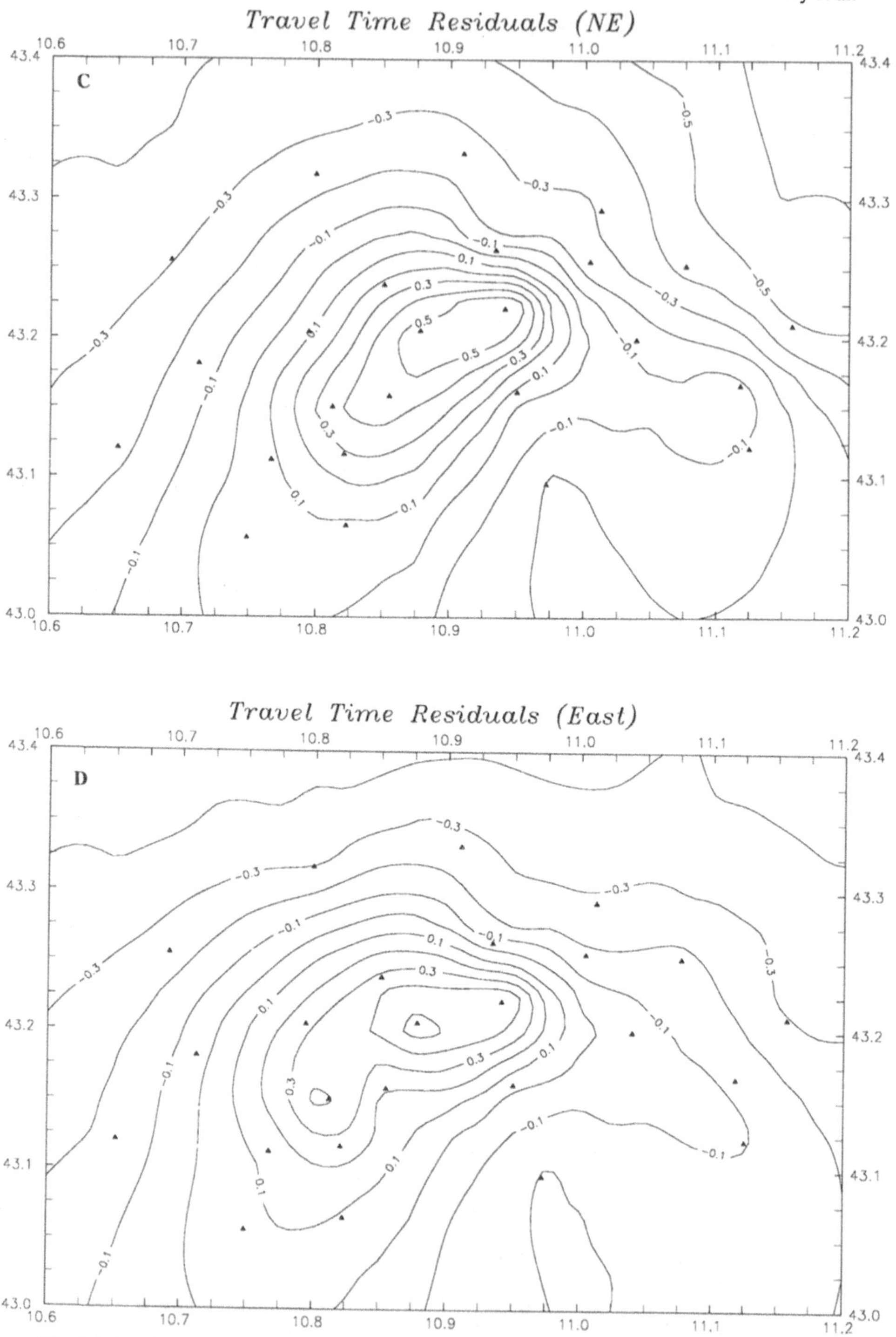

Fig. 9C,D. *Continued*

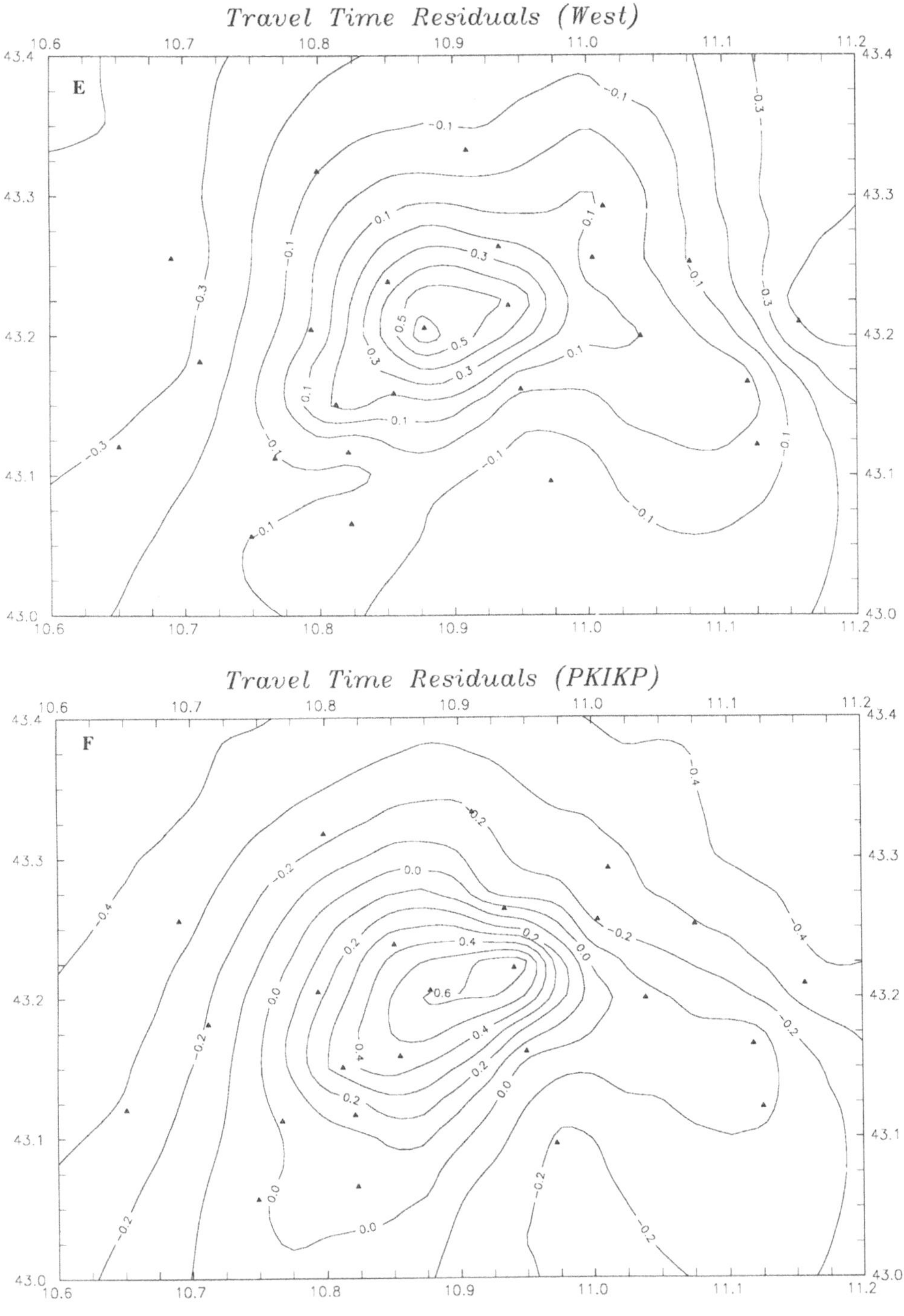

Fig. 9E,F. *Continued*

clearly. For events arriving from the east, stations MGUI and PADU (see Fig. 3 for station names) have residuals of −0.29 s (fast) and 0.28 s (slow), respectively. From the west the pattern is reversed; the average residual at MGUI is 0.14 and −0.02 s at PADU. Rays arriving at PADU from the west and MGUI from the east are advanced. Rays arriving at PADU from the east and MGUI from the west are delayed.

This pattern indicates that the first layer stripping procedure has effectively removed the near surface anomalies and that the data represents only mid-crust and mantle effects. It also implies that the velocity contrasts are quite strong and have rapid horizontal variations.

3 Inversion

3.1 Method

Travel time residuals are inverted for velocity perturbations using the method of Aki et al. (1977). This method of inversion has been used quite extensively in the past, particularly for the purposes of magma chamber delineation. See Iyer (1988) for a review of Aki inversions in volcanic regions. The details of the inversion are only sketched below, for a complete review of the inverse formulation see Zandt (1978), or Taylor and Toksoz (1979).

An Earth model consisting of a set of layers subdivided into right rectangular blocks is established as a starting model. For each event and each station, rays are traced through the model. Travel time residuals are distributed in the blocks along the raypath proportional to the length of the ray in each block. Since events are located are various distances and azimuths from each station, a dense mesh of distributed residuals is formed beneath the network. The Aki technique solves the system of equations which characterizes this travel time distribution using a damped least squares technique.

The starting model for the travel time inversion is the five layer model shown in Table 4. The crust is characterized by three layers each of which is 5 km thick with velocities of 6.5, 7, and 7.5 km/s, respectively. The upper mantle is parameterized by two 10 km thick layers at 8.0 km/s. This model is consistent

Table 4. Starting model

Layer No.	P velocity (km/s)	Layer thickness (km)	Horizontal dimensions (km)
1	6.5	5.0	5 × 5
2	7.0	5.0	5 × 5
3	7.5	5.0	5 × 5
4	8.0	10.0	5 × 5
5	8.0	10.0	5 × 5

Net center: 43°14′ 10°54′

with deep reflection work done in Larderello (Batini et al. 1985) and the refraction model of Giese et al. (1981) for central Italy.

3.2 Results

Figure 10 shows the results of the travel time inversion. These plots show the velocity perturbation for each of the five layers in percent of the background velocity as contours of equal velocity perturbation. For the crustal layers (1, 2, and 3; from 6 to 21 km), the region of low velocity is confined to the center of the network in an area 25 by 41 km in extent, with an elongate pattern to the NE. This pattern coincides with the region of low amplitude teleseismic P-waves, high heat flow, negative Bouguer gravity anomaly, and the elevated K reflection horizon and will be discussed below.

Layers 4 and 5 (depth from 21 to 41 km) also have a distinct low velocity pattern in the center of the network. The relative velocity perturbations are still quite large at these depths (10% in the center). The zone of maximum low velocity migrates to the northeast at greater depths and the total area of LVZ increases.

Slices through the final perturbed model (Fig. 11) show the pattern of the anomalous body in vertical cross section. Velocity contrasts are depicted with different shade patterns with cross hatched blocks representing velocity decreases and stippled blocks representing regions of increased velocity. Blocks without any pattern represent regions not sampled densely enough (at least 10 rays) for a parameter estimate to be made.

The SW to NE slice (Fig. 11a) shows that strong velocity reductions (to 18%) are necessary to satisfy the data from depth of about 6 to 16 km. The LVZ in the crustal part of the model is about 20 km wide. Deeper in the crust and in the upper mantle, the perturbations are smaller and cover a larger region. However, the results in the lower part of the model may be corrupted by smearing effects in the inversion. The residual redistribution technique tends to back propagate travel time residuals equally along each ray path. When criss-crossing ray coverage is not sufficient in an area, artificial velocity anomalies can be created. This effect may be accentuating the deep and widespread low velocity area of layers 4 and 5 where velocity reductions of up to 10% are required by the data, and the LVZ covers an area about 35 km wide.

The NW to SE slice shows a more uniform pattern of low velocity, with a reduction of about 10% existing from the top layer (6 km) to a depth of 41 km. The width of the LVZ is about 15 km in the crustal layers, and increases at depth. However, resolution of the LVZ is poor to the SE due to poor data coverage.

To quantify the confidence in the final model, resolution and standard error functions are generated for each model parameter (Zandt 1978). The resolution values presented here are a measure of how well each block velocity estimate is independently measured. A resolution value of 1 indicates a perfectly resolved parameter with no interdependence on other parameters. Lower values of resolution indicate that the final velocity should be thought of as an estimate of

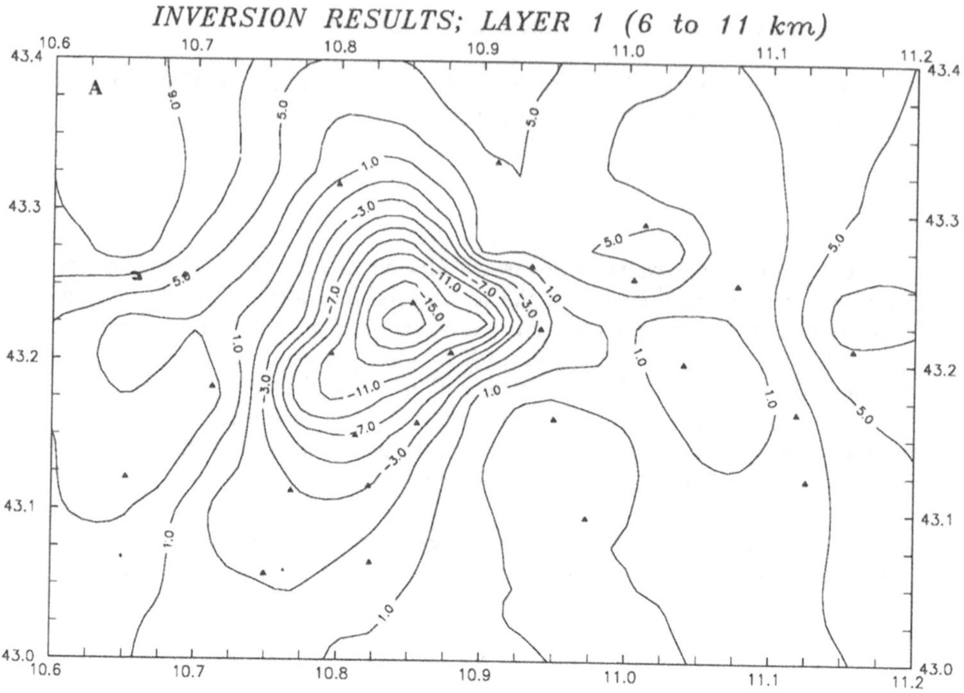

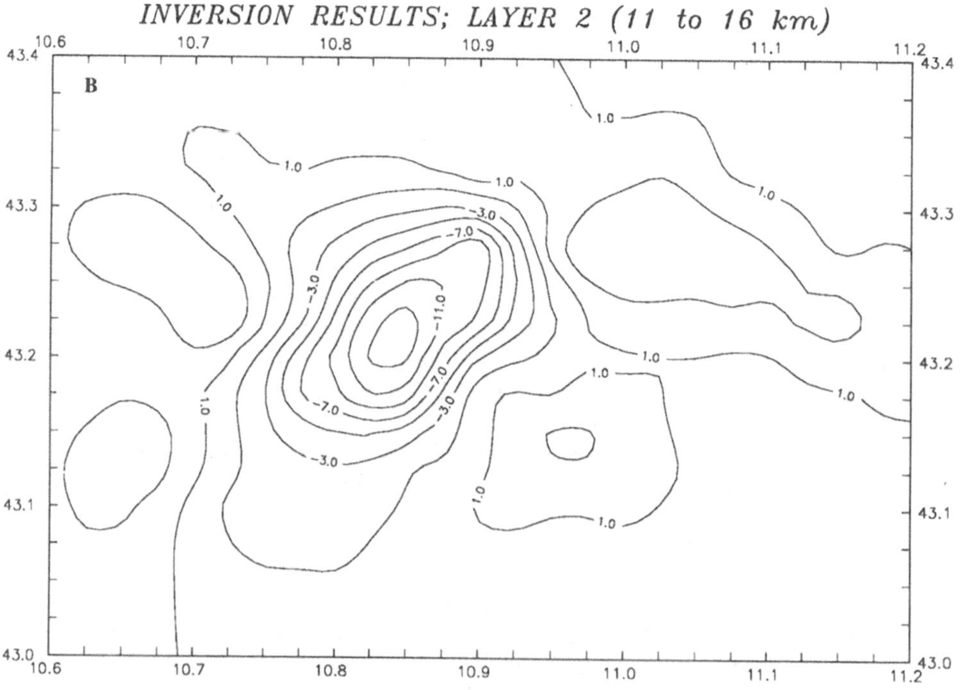

Fig. 10A–E. Inversion results for the five-layer input model described in Table 4. Plotted are contours of equal velocity perturbation. Contour interval is 0.2%. Layer 1 has station symbols plotted for reference

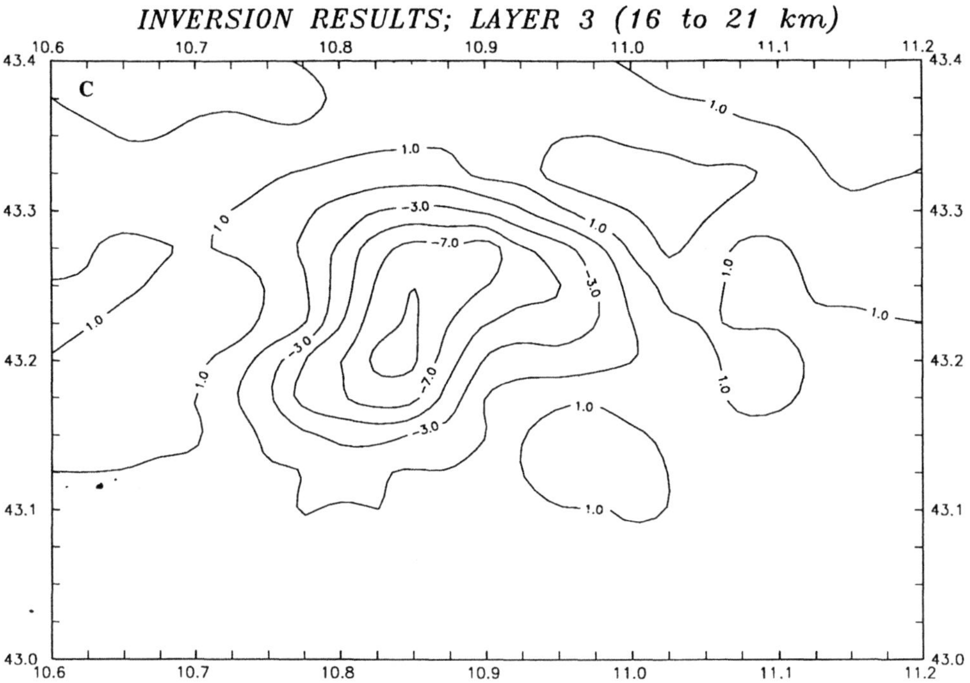

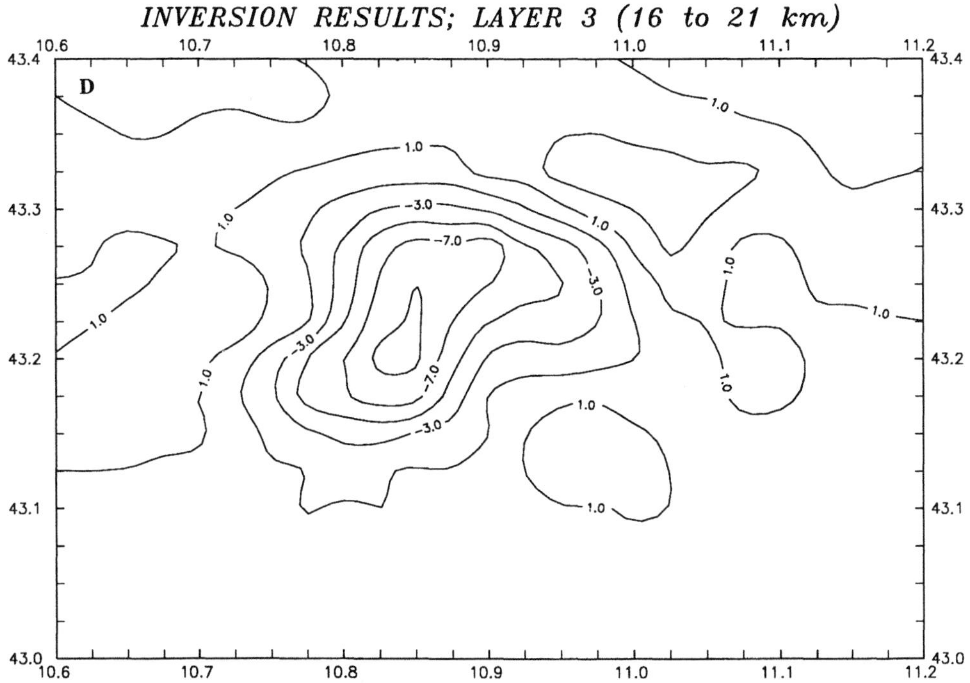

Fig. 10C,D. *Continued*

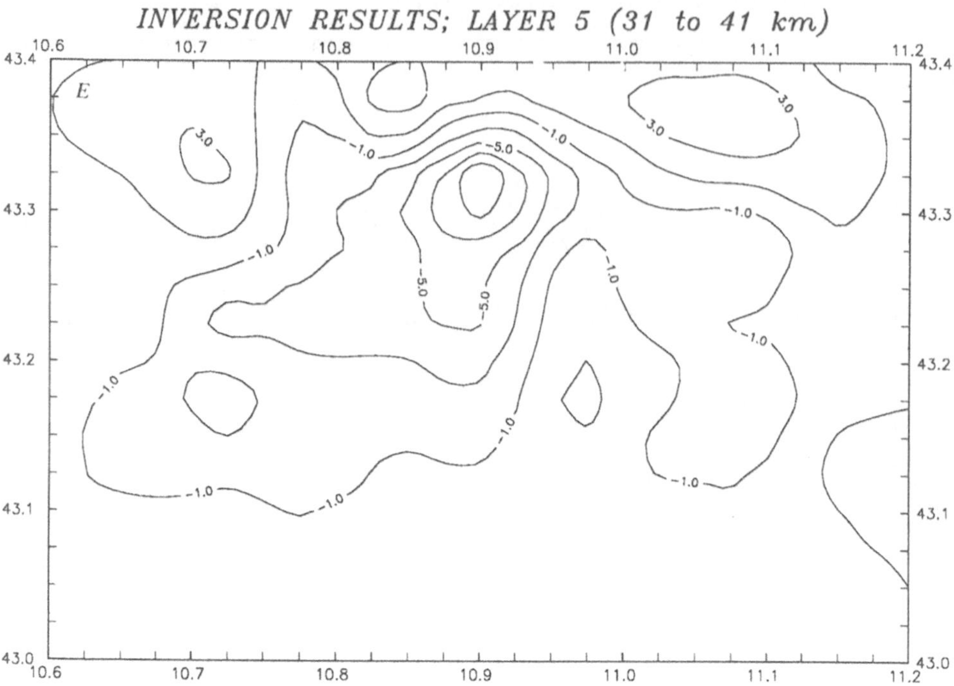

Fig. 10E. *Continued*

an average of the surrounding blocks. The standard error values are a measure of the relative accuracy of the parameter estimate.

Figure 12 shows the resolution and error calculations for the SW to NE slice. Resolution is best in the center of the model and to the NE at depth due to good azimuthal and incident angle coverage in that quadrant, and worst from the SW, where coverage is poor. Because the addition of the damping parameter used to solve the least squares problem discussed above tends to reduce the number of degrees of freedom of the problem, the trade-off between standard error and model resolution must be examined. Through trial and error, we find the best trade-off when damping is set to $0.003 \frac{s^2}{\%^2}$.

4 Discussion

The anomalous body which gives rise to the velocity structure discussed above also has strong effects on other geophysical observations made in the region. The temperature field measured at 2 km (Fig. 13a), Bouguer gravity field (Fig. 13b), average teleseismic body wave amplitudes (Fig. 13c), and the depth to the

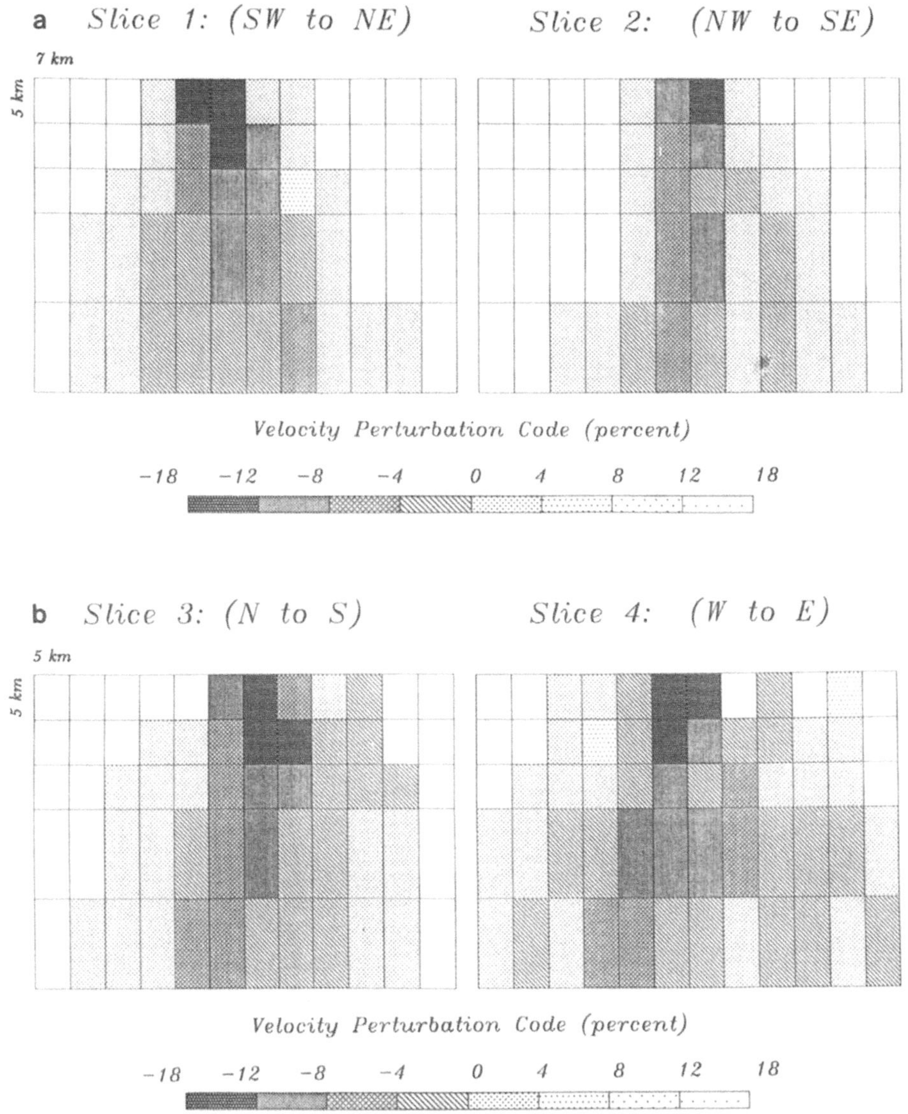

Fig. 11a,b. Inversion results for vertical slices through the region. **a** Diagonal slices from SW to NE and NW to NE. All block widths are 7 km. The three crustal layers are 5 km thick and the two mantle layers are 10 km thick. The *top* of the model represents a depth of 6 km below the surface. The velocity perturbation code is relative to the starting model described in Table 4. **b** NS and EW slices through the center of the final model. Blocks are 5 km wide

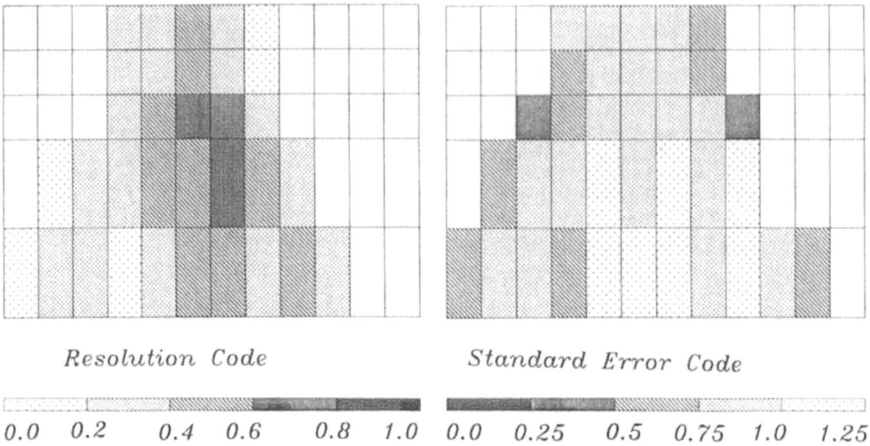

Fig. 12. Resolution and error plots of the vertical slice from SW to NE Blocks are all 7 km wide, upper three layer thicknesses are 5 km and the lower 2 layer thicknesses are 10 km. The *top* of the model represents a depth of 6 km

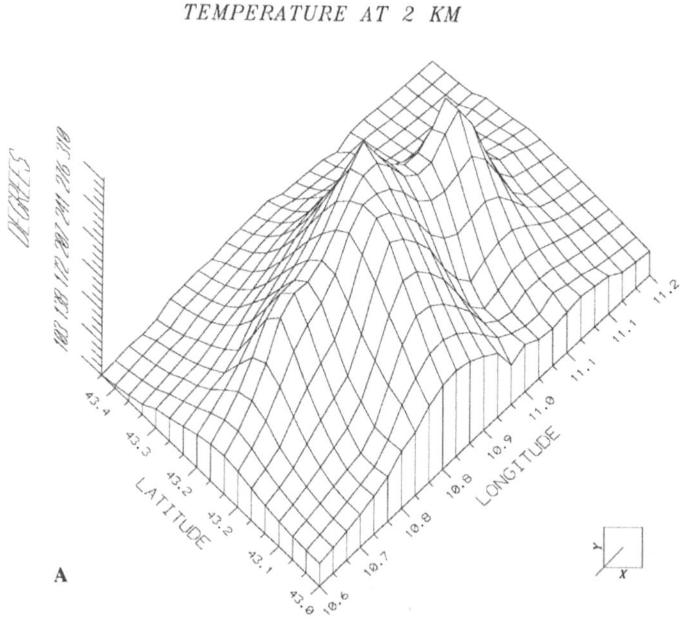

A

Fig. 13A–E. Geophysical parameter maps of data recorded within the Larderello geothermal field. All maps are made by making a parameter estimate at each of the seismic stations in the network and then contoured for display. **A** Temperature measured at 2 km in °C. **B** Bouguer Gravity anomalay in mGals. **C** Depth of the K reflection Horizon in m. **D** PKIKP amplitude residuals from teleseismic P-wave measurements. **E** PKIKP travel time residual averages

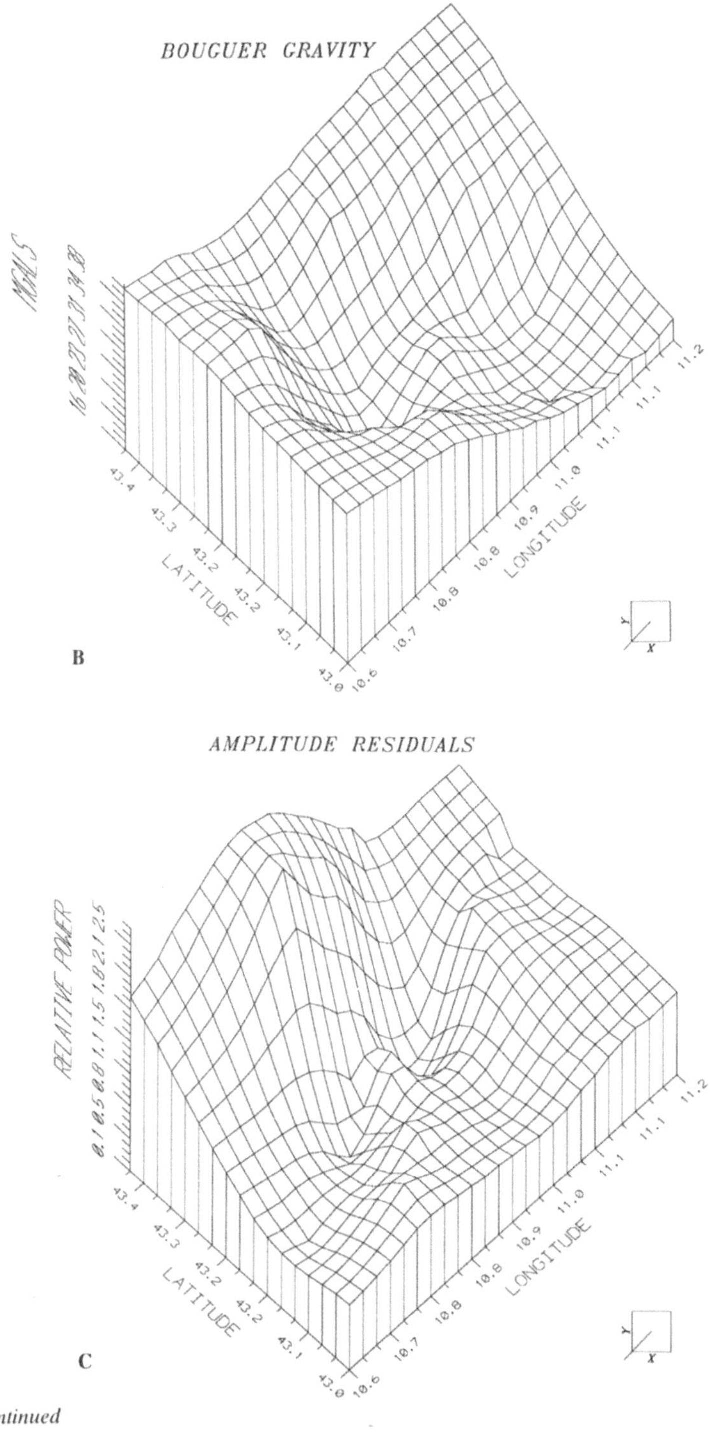

Fig. 13B,C. *Continued*

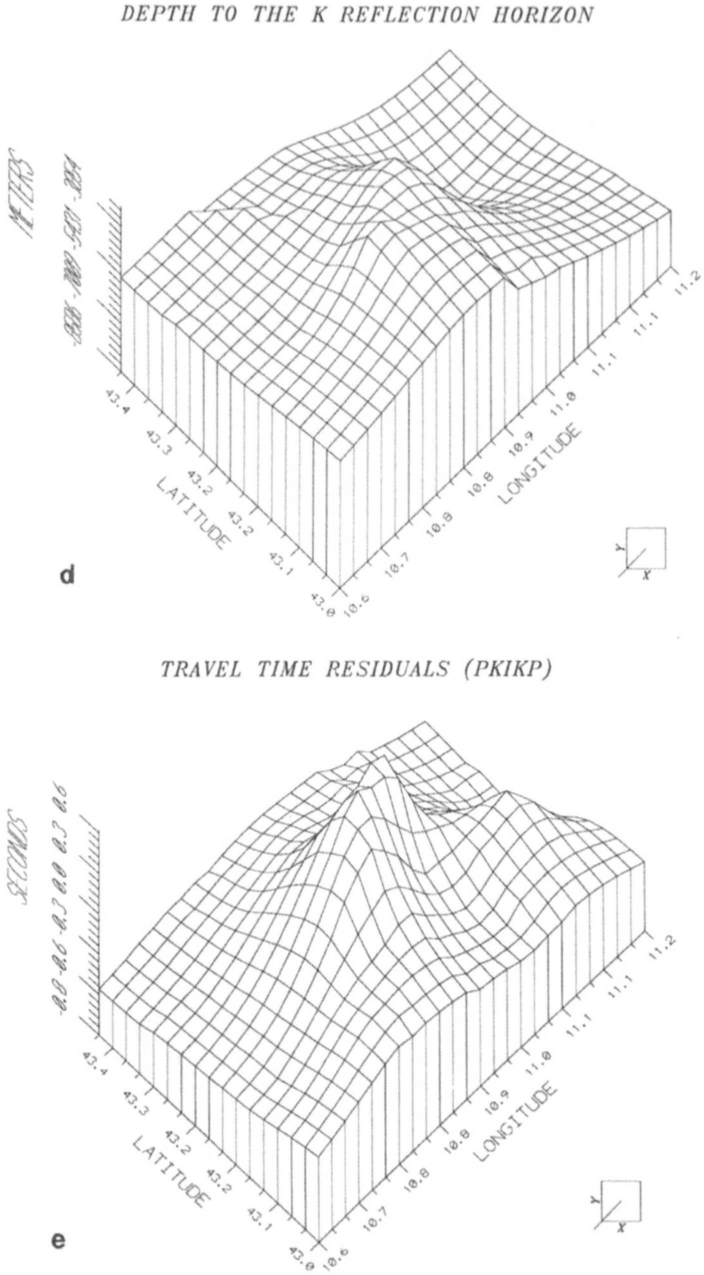

Fig. 13D,E. *Continued*

K reflection horizon (Fig. 13d) are strongly variable across the network. See Table 1 for parameter estimates at each station. The 2-km temperature values vary from about 150 to 350 °C with the highest values in the center of the field corresponding with an area of very active geothermal production. The Bouguer gravity anomaly (Cataldi et al. 1978) is also greatest in the center of the field, with a gavity low of 25 mGals. A low density, silica-rich magmatic intrusion has been proposed to be the source of this gravity feature (Puxeddu 1984). This gravity feature compares well with gravity lows of 30 mGals at the Geysers (Iyer 1988) and 50 mGals at Yellowstone (Ellsworth 1977).

The depth of the K reflection horizon (Batini et al. 1985) which varies in depth from about 3 to 8 km, is deepest on the periphery of the network and shallowest in the center of the network coincident with the area of high geothermal activity.

Teleseismic P-wave amplitude residuals patterns can be used to help understand the structure beneath seismic arrays (Thompson and Gubbins 1982). Relative amplitude residuals were calculated for all events used in the travel time residual calculations using the formula:

$$A(i) = \frac{1}{N}\left(\sum_{i=1}^{N} \frac{SWP}{AWP}\right). \tag{4}$$

Where SWP is the station window power for an individual event, AWP is the average window power for the event, and N is the number of events recorded at the station. Window power is defined as the sum of the squared amplitudes in the first 1 s of the event. The average amplitude residual variations for the PKIKP events (Fig. 13e) show a complicated pattern with generally smaller window power values in the center of the network and larger values to the north.

To view the correspondence of various geophysical parameters we calculated non-dimensionalized product maps (Davis 1973). These maps are made by first converting the single parameter maps into *standard* maps via the formula:

$$Z_i = \frac{P_i - P_i^{ave}}{S}. \tag{5}$$

Here Z_i is the standardized map parameter, P_i is the input parameter and S is the parameter standard deviation. Standard temperature and standard PKIKP travel time residual maps are shown in Fig. 14. Each point in the standard map now represents the parameter variation from the average weighted by the data standard deviation. The final maps are made from point by point multiplication of the non-dimensionalized parameters of two *standard* maps.

Two non-dimensionalized product maps are shown in Fig. 15. The first shows the correspondence between travel time residuals and temperature, with only PKIKP events used in the calculations to reduce the effect of gaps in the azimuthal data coverage. Two phenomena are visible on this map, the first of which clearly shows that the highest temperatures directly correspond with the most delayed PKIKP travel times. The peak is centered near station VALE.

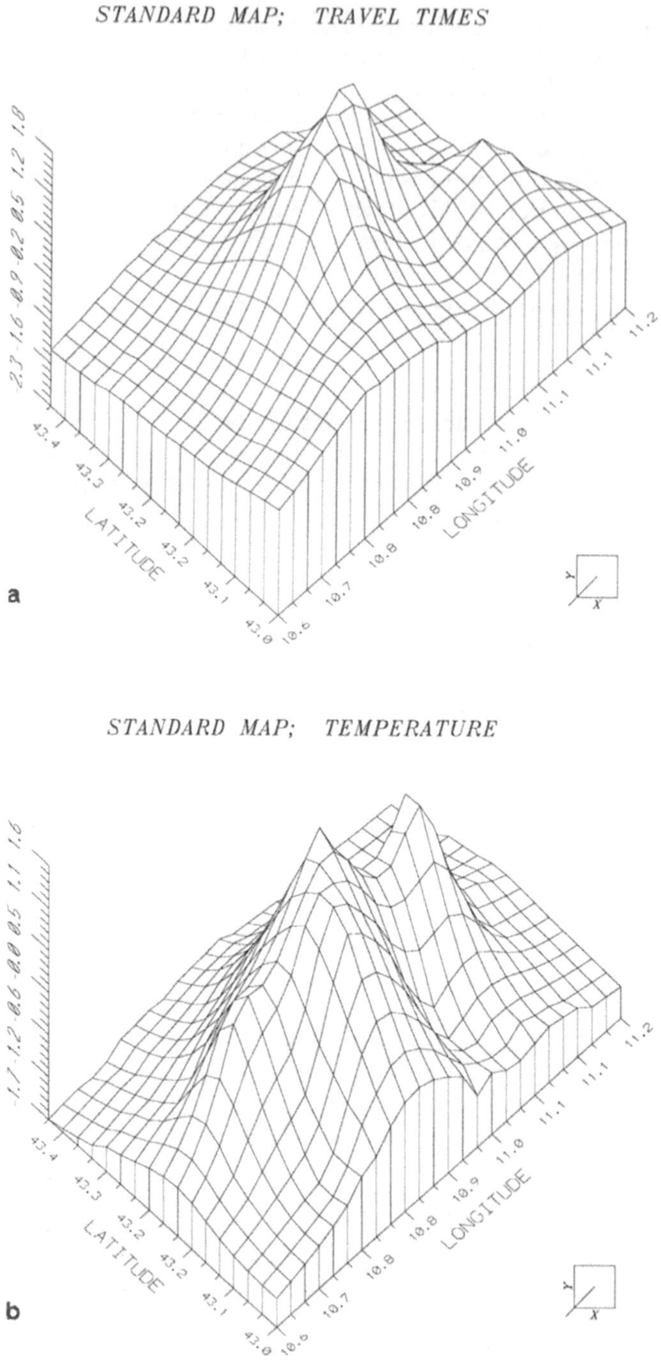

Fig. 14A,B. Non-dimensionalized *standard* maps for the parameters temperature and PKIKP average travel time residuals

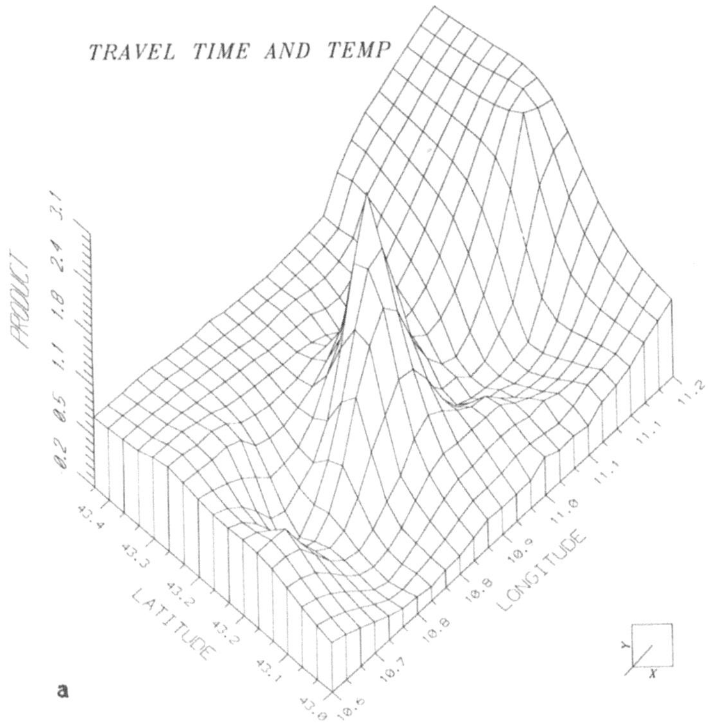

a

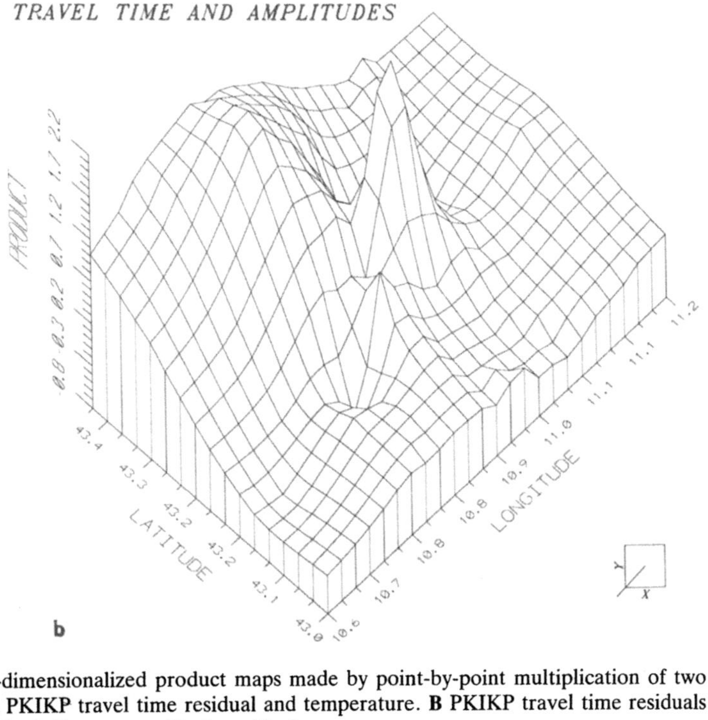

b

Fig. 15A,B. Non-dimensionalized product maps made by point-by-point multiplication of two *standard* maps. **A** PKIKP travel time residual and temperature. **B** PKIKP travel time residuals and PKIKP teleseismic P-wave amplitude residuals

The second observation is that directly to the west and east of the peak correspondence, we see distinct lows in the non-dimensionalized product which come from high temperatures which correspond geographically with normal to fast PKIKP travel time averages. This tells us that the anomalous body, which gives rise to the travel time delays and high temperatures in the center of the network, only effects the temperature observations on the periphery and not the travel times.

The product map between PKIKP travel times and averaged relative PKIKP amplitude anomalies (Fig. 15b) shows a complicated pattern of correspondence. Here, the sense of the amplitude residuals is reversed to produce a positive value when large amplitudes and delayed amplitudes match. From this map we can see that in the center of the network smaller than average amplitudes and low velocity regions coincide. To the northwest, larger than average amplitudes and relative high velocities correlate. This leads us to believe that the existence of the LVZ in this area is either scattering or attenuating energy to diminish the average window power calculations made in the region of low velocity. Very low Q values may exist in fully melted regions and significantly reduce amplitudes provided that ray travels in the material for a few wavelengths. A high degree of scattering may also exist in the region due to the large velocity and density contrasts between the competent country rock and partially of fully melted material. Forward modeling of the waveshapes and amplitudes is neccessary to separate these factors.

4.1 Interpretation and Conclusions

Our interpretation of the geophysical data described above is that a strong and extensive low-velocity zone exists beneath the Larderello geothermal field. The inversion of teleseismic P-wave travel time residuals has defined the area of low velocity to be roughly cylindrical in shape with a diameter of about 15 to 20 km. Given the high temperatures at 2 km (up to 350 °C), we believe that the LVZ may exist as a volume of partially or fully melted material with a bulk velocity reduction of 15 to 20%.

The depth of the top of the LVZ is assumed to be below the level of active seismicity at a depth of 6 km. Tomographic inversion of the relative travel time residuals indicates clearly that the LVZ penetrates to a depth of at least 21 km and suggests that the region may extend deeper into the crust or upper mantle to a depth of about 40 km. At depth the zone trends to the northeast and broadens to a width of about 40 km.

The K horizon, which is characterized by strong reflections at depths between 3 km in the center of the network to 8 km on the periphery, may represent a petrophysical boundary (Batini et al. 1985), marking a zone of highly fractured, fluid-filled rocks which may have resulted from the high stresses induced from implacement of a hot, low density, silica-rich material (Giese et al. 1981). High-angle normal faults on the K horizon, seen on many of the reflection lines and interpreted from local earthquake fault mechanism analysis (Batini et al. 1985), are consistent with this interpretation.

References

Achauer U, Greene L, Evans JR, Iyer HM (1986) Nature of the magma chamber underlying the Mono Craters area, eastern California, as determined from teleseismic travel time residuals. J Geophys Res 91:13873–13891

Aki K, Christoffersson A, Husebye ES (1977) Determination of the three-dimensional seismic structure of the lithosphere. J Geophys Res 82:237–296

Batini F, Nicolich R (1984) The application of seismic reflection methods to geothermal exploration. Seminar on Utilization of geothermal energy for electric power production and space heating. Florence, 11–17 May

Batini F, Burgassi PD, Cameli GM, Nicolich R, Squarci P (1978) Contribution to the study of the deep lithospheric profiles: deep reflecting horizons in Larderello-Travale geothermal field. Mem Soc Geol Ital 9:477–484

Batini F, Console R, Luongo G (1985) Seismological study of Larderello-Travale geothermal area. Geothermics 14:255–272

Boccaletti M, Coli M, Eva C et al. (1983) Considerations on the seismotectonics of the northern Apennines. Tectonophysics 47:7–38

Cataldi R, Lazzarotto A, Muffler LPJ, Squarci P, Stefani GC (1978) Assessment of geothermal potential of central and southern Tuscany. Geothermics 7:91–131

Davis JC (1973) Statistics and data analysis in geology. Wiley, San Francisco, 550 pp

Ellsworth WL (1977) Three dimensional structure of the crust and upper mantle beneath the island of Hawaii. PhD thesis, Massachusetts Inst Tech, 377 pp

Giese P, Morelli C, Nicolich R, Wigger P (1981) Seismische Studien zur Bestimmung der Krustenstruktur im Bereich der geothermischen Anomalie der Toskana. Comm Eur Commun EUR 75–78 de MF

Hayakawa M (1957) Prospecting of the underground structure at "Showa-Shinzan" by various geophysical methods, particularly seismic survey. JPN J Geophys 1:13–20

Herrin E (1968) Seismological tables of P-phases. Bull Seismol Soc Am 58:1193–1241

Iyer HM (1984) Geophysical evidence for the locations, shapes and sizes and internal structures of magma chambers beneath region of Quaternary volcanism. Phil Trans R Soc Lond Ser A, 310:473–510

Iyer HM (1988) Seismological detection and delineation of magma chambers beneath interplate volcanic centers in western USA. In: King C, Scarpa R (eds) Modeling of volcanic processes. Vieweg, New York

Iyer HM, Oppenheimer DH, Hitchcock T, Roloff LN, Caokley JM (1981) Large teleseismic P-wave delays in the Geysers Clear Lake geothermal region. US Geol Surv Prof Pap 1141:97–116

Kirkpatrick S, Gelatt C, Vecchi M (1983) Optimization by simulated annealing. Sience 220:671–680

Murase T, McBirney AR (1973) Properties of some common igneous rocks and their melts at high temperatures. Geol Soc Am Bull 84:3563–3592

Oppenheimer DH, Herkenhoff KE (1981) Velocity-density properties of the lithosphere from three-dimensional modeling at the Geysers-Clear Lake region, California. J Geophys Res 86:6057–6065

Puxeddu M (1984) Structure and late Cenozoic evolution of the upper lithosphere in southwest Tuscany (Italy). Tectonophysics 101:357–382

Rothman D (1986) Automatic estimation of large residual statics corrections. Geophysics 51:332–346

Sanders CO (1984) Location and configuration of magma bodies beneath Long Valley California, determined from anomalous Earthquake signals. J Geophys Res 89:8287–8302

Steeples DW, Iyer HM (1976) Low velocity zone under Long Valley as determined from teleseismic events. J Geophys Res 81:849–860

Taylor SR, Toksoz MN (1979) Three dimension crust and upper mantle structure of the northeastern United States. J Geophys Res 84:7627–7644

Thompson CJ, Gubbins D (1982) Three-dimensional lithospheric modeling at NORSAR: linerarity of the method and amplitude variations from the anomalies. Geophys J R Astr Soc 71:1–36

Zandt G (1978) Study of the three dimensional heterogeneity beneath seismic arrays in central California and Yellowstone, Wyoming. PhD Thesis, Massachusetts Inst Tech, USA

Physical Meaning of the Coda Envelopes

LONG-SHENG GAO

Abstract

This article shows a relatively complete solution for the coda envelopes in the time domain. The main theoretical basis is the energy transportation computation. After establishing the integral equation, the subsequential method of computation is used to find the solution. At this moment, the analytical solution for the multiple scattering in two-dimensional space has been obtained, whereas the solution for the three-dimensional space is only a fifth-order approximation which is from the digital computation. The solution shows that in the two-dimensional case the coda decay rate may reflect only the intrinsic absorption of the medium. In three-dimensional space the decay rate also reflects mainly the intrinsic absorption but with a certain percentage of scattering attenuation. This conclusion is to simplify matters: the receiver is set very close to the source. When the receiver is distant from the source, as the distance increases the scattering attenuation will effect the decay rate of codas more. Considering that intrinsic absorption is very sensitive to changes in temperature, microfractures and the content of fluid within the pores, coda Q should be a good detector of these changes. Therefore, the theoretical results show that precise analyses of coda envelopes and the discrimination of the different Q-factors are essential in detecting underground thermal-fluid activities that may be the precursor of imminent disasters.

1 Introduction

The Q factor has been recognized to be such an important parameter in seismology and geophysics, even though the precision of measurement has not been high and its physical substance has not been clear enough. Body waves, surface waves of different types and frequencies, even the Chandler wobble with a period of 434 days, all give their own Q factors while vibrated and attenuated in different attenuation mechanisms. These Q-factors may have quite different physical substances.

During this period, coda is a new source giving Q-factors for the earth medium. The coda Q is named to express the decay feature of the tails of seismograms.

Coda is assumed to be produced mainly from the scattering of body waves by heterogeneities distributed randomly in the upper part of the Earth (Aki 1969; Aki and Chouet 1975). Probably coda also composes certain components of surface waves (Malin 1980), reverberations in layered structure in the neighbourhood of source and receiver (see, e.g., Bouchon 1982; Richards and Menke 1983). In addition, coda can be generated by the conversion of body waves into surface waves by topography at the free surface or at buried interfaces at depth (Levander and Hill 1985).

Scattering, which is also called refraction, covers a variety of physical processes. It can be counted as two opposite mechanisms, attenuating and collecting energies. However, these processes should not dissipate the elastic energy but only spread or delay it arrival. The other category of attenuation is absorption or intrinsic dissipation which includes internal friction and other thermodynamic processes. Wave energies transform into heat due to those processes finally. The coda research provides a good chance to seek to discriminate between these two categories of attenuation, which both attenuate wave energies and are mixed together in most wave research.

Obviously, such effort, is not only of theoretical interest. This discrimination should be helpful in finding earthquake precursors, seismic prospecting, and also, it should have wide application in the field of general material research.

Herraiz and Espinosa (1987) made a comprehensive review on the theory of coda waves, emphasizing the scattering and attenuation research of high frequency seismic waves by means of codas. It showed a number of unsolved problems in scattering research. Kopnichev (1977); Gao et al. (1983a,b) worked on the multiple scattering model as an inevitable consequent development of the single back-scattering mode (Aki 1969) and single isotropic-scattering mode (Sato 1977). Matsunami (1987) carried out laboratory tests on the single- and multiple-scattering model for the generation of seismic coda waves. His results showed that the multiple-scattering model of Gao et al. for coda generation explains the time decay of coda that consists of waves scattered by small-scale crack like heterogeneities present in the Earth's crust very well. Peng et al. (1987) used the Monte-Carlo simulation for coda formation by discrete scatterers, their numerical results show an excellent agreement with the theoretical prediction by Gao et al. (1983) for uniform randomness, for single-, double- and triple-scattering, and found that multiple-scattering may be neglected when the lapsed time shorter than the mean free time. The previous theoretical research of multiple-scattering was limited to a special case, i.e., when the receiver is close to the source (Gao et al. 1983a,b). This paper avoids this previous limitation and attempts to obtain a better physical understanding of coda Q through the calculation of scattering energy.

This chapter introduces first, the solution of multiple scattering in two-dimensional space (Shang and Gao 1988) finally, then the development in three-dimensional space and discusses the physical meaning of the decay rate of codas.

2 Equations of Scattering

There are two simulations for two extremes, the Born approximation or single-scattering which simulates the scatterers being tenuously distributed (Sato 1977) and the diffusion approximation which deals with the very dense distribution of scatterers (Dainty and Toksoz 1977). To cover all the different cases, including the medium dense distribution, let us consider the energy transportation process, assuming that the scattered waves are isotropic and incoherent, thus only the additive energy is under concern. We calculate the energy due to multiple scattering based on the model of discrete scatterers with random but statistically homogeneous distribution. In general, the energy flux density along $\vec{k}$ direction $\vec{J}(\vec{x}, t, \vec{k})$ can be expressed as:

$$\vec{J}(\vec{x}, t, \vec{k}) = VE(\vec{x}, t, \vec{k})\vec{k}, \tag{2.1}$$

where $\vec{k}$ is a unit vector, $[\vec{k}] = 1$, $E(\vec{x}, t, \vec{k})$ denotes the energy density related with the energy flux in $\vec{k}$ direction, V is the velocity of transmitted waves.

Let us consider the collected energies at $(\vec{x}, t)$ from a small volume element d^3x' around $(\vec{x}', t')$.

The scattered energy per second of the wave travelling along $\vec{k}'$ direction at a given time t' from one scatterer is:

$$\sigma|\vec{J}'(\vec{x}', t', \vec{k}')| = \sigma VE(\vec{x}', t', \vec{k}'), \tag{2.2}$$

where σ is the mean effective cross section of scatterers. The total scattered energy flow from volume d^3x' per second $F_s(x', t')$, should be the sum of the energy from all scatterers in the volume element n_0d^3x', where n_0 is the mean density of scatterers, and from all $\vec{k}$ directions, that is:

$$\begin{aligned} F_s(\vec{x}', t') &= \sum_{k'} n_0\sigma|\vec{J}(\vec{x}', t', \vec{k}')|d^3x' \\ &= n_0\sigma Vd^3x' \sum_{k'} E(\vec{x}', t', \vec{k}') \\ &= n_0\sigma VE(\vec{x}', t')d^3x', \end{aligned} \tag{2.3}$$

where $E(\vec{x}', t')$ is the energy density at $(\vec{x}', t')$. The wave energies collected at the place $\vec{x}$ at time t is from all the medium elements at $(\vec{x}', t')$, which travel from point $\vec{x}'$ to $\vec{x}$ in a time of duration:

$$t - t' = \frac{|\vec{x} - \vec{x}'|}{V}. \tag{2.4}$$

The energies should be distributed homogeneously on a spherical area with the center at $\vec{x}'$ and radius of $|\vec{x} - \vec{x}'|$ due to the assumption of isotropy. Thus, at the time of $t = t' + |\vec{x} - \vec{x}'|/V$ at the point $\vec{x}$, the absolute value of the energy flux density from a volume d^3x' is:

$$\Delta|\vec{J}_s(\vec{x}, t)| = \frac{F_s(\vec{x}', t')}{4\pi|\vec{x} - \vec{x}'|^2} \exp\left[-\left(n_0\sigma + \frac{\omega}{Q_iV}\right)|\vec{x} - \vec{x}'|\right]. \tag{2.5}$$

Here the exponential factor takes the attenuation into account which includes not only scattering loss $n_0\sigma$ but also intrinsic absorption denoted by $\frac{\omega}{Q_iV}$, where Q_i is the quality factor correlated with intrinsic absorption. Substituting (2.3) into (2.5) and noting the relation $\Delta|\vec{J}(\vec{x}, t, \vec{k})| = VdE(\vec{x}, t, \vec{k})$, Eq (2.5) can be transformed for the increment in the energy density at $(\vec{x}, t)$ as:

$$dE_s(\vec{x}, t) = \frac{n_0\sigma E(\vec{x}', t')}{4\pi|\vec{x} - \vec{x}'|^2} \exp\left[-\left(n_0\sigma + \frac{\omega}{Q_iV}\right)|\vec{x} - \vec{x}'|\right] d^3x', \tag{2.6}$$

where the subscript $_s$ means scattering, dE_s is the energy increment due to scattering from the volume element. Integrate all contributions from all elements d^3x', then we have the whole scattered energy density at $(\vec{x}, t)$:

$$E_s(\vec{x}, t) = n_0\sigma \iiint_\infty \frac{E(\vec{x}', t')}{4\pi|\vec{x} - \vec{x}'|^2} \exp\left[-\left(n_0\sigma + \frac{\omega}{Q_i V}\right)|\vec{x} - \vec{x}'|\right] d^3x'. \tag{2.7}$$

If we know the whole wave field in the scattering medium, we could find the scattering energy density E_s easily. However, it is impossible to find $E(\vec{x}', t')$ before having the solution to multiple scattering. Obviously Eq (2.7) is only part of an integral equation. The complete integral equation for the energy density of wave fields, which includes both primary and scattered wave fields in three-dimensional space, is:

$$E(\vec{x}, t) = E_p(\vec{x}, t) + n_0\sigma \iiint_\infty \frac{E\left(\vec{x}', t - \frac{|\vec{x} - \vec{x}'|}{V}\right)}{4\pi|\vec{x} - \vec{x}'|^2} \times \exp\left[-\left(n_0\sigma + \frac{\omega}{Q_i V}\right)|\vec{x} - \vec{x}'|\right] d^3x', \tag{2.8}$$

where E_p is the energy density of the primary wave. For the two-dimensional case the difference is only in the geometry, thus (2.8) can be adapted as:

$$E(\vec{x}, t) = E_p(\vec{x}, t) + n_0\sigma \iint_\infty \frac{E\left(\vec{x}', t - \frac{|\vec{x} - \vec{x}'|}{V}\right)}{2\pi|\vec{x} - \vec{x}'|} \times \exp\left[-\left(n_0\sigma + \frac{\omega}{Q_i V}\right)|\vec{x} - \vec{x}'|\right] d^3x'. \tag{2.9}$$

2.1 Solution in Two- and Three-Dimentional Spaces

Define $\varepsilon(x, t)$ and $\varepsilon_p(x, t)$ as:

$$E(\vec{x}, t) = \varepsilon(\vec{x}, t) \exp\left[-\left(n_0\sigma + \frac{\omega}{Q_i V}\right) Vt\right],$$

and

$$E_p(\vec{x}, t) = \varepsilon_p(\vec{x}, t) \exp\left[-\left(n_0\sigma + \frac{\omega}{Q_i V}\right) Vt\right]. \tag{3.1}$$

Equation (2.9) can be rewritten without the exponential factor,

$$\varepsilon(\vec{x}, t) = \varepsilon_p(\vec{x}, t) + n_0\sigma \iint_\infty \frac{\varepsilon\left(\vec{x}', t - \frac{|\vec{x} - \vec{x}'|}{V}\right)}{2\pi|\vec{x} - \vec{x}'|} d^2x'. \tag{3.2}$$

Here we use the following equation while omitting the exponential factor in (2.9):

$$\iint\limits_{\infty} \frac{E\left(\vec{x}', t - \dfrac{|\vec{x} - \vec{x}'|}{V}\right)}{2\pi|\vec{x} - \vec{x}'|} e^{\left[-\left(n_0\sigma + \frac{\omega}{Q_i V}\right)|\vec{x} - \vec{x}'|\right]} d^2x'$$

$$= \iint\limits_{\infty} \frac{\varepsilon\left(\vec{x}', t - \dfrac{|\vec{x} - \vec{x}'|}{V}\right)}{2\pi|\vec{x} - \vec{x}'|} e^{\left[-\left(n_0\sigma + \frac{\omega}{Q_i V}\right)Vt\right]} d^2x'$$

$$= e^{\left[-\left(n_0\sigma + \frac{\omega}{Q_i V}\right)Vt\right]} \iint\limits_{\infty} \frac{\varepsilon\left(\vec{x}', t - \dfrac{|\vec{x} - \vec{x}'|}{V}\right)}{2\pi|\vec{x} - \vec{x}'|} d^2x'.$$

According to the assumptions of isotropy, the wave field is spherically symmetrical to its source. Equation (3.2) is adapted then in the form of:

$$\varepsilon(R, t) = \varepsilon_p(R, t) + n_0\sigma \iint\limits_{\infty} \frac{\varepsilon\left(R', t - \dfrac{|\vec{x} - \vec{x}'|}{V}\right)}{2\pi|\vec{x} - \vec{x}'|} d^2x', \tag{3.3}$$

where $R = |\vec{x}|$ and $R' = |\vec{x}'|$. Making use of the technique of δ-function, Eq. (3.3) can be written as,

$$\varepsilon(R, t) = \varepsilon_p(R, t) + n_0\sigma \iint\limits_{\infty} d^2x' \int_0^{\infty} \frac{\varepsilon(R', t')}{2\pi|\vec{x} - \vec{x}'|} \delta\left(t' - t + \frac{|\vec{x} - \vec{x}'|}{V}\right) dt. \tag{3.4}$$

It can be seen in Fig. 1 that:

$$|\vec{x} - \vec{x}'| = (R^2 + R'^2 - 2RR' \cos\theta)^{1/2},$$

where R is the distance between the source and the receiver, R′ is the distance between the scatterers and the receiver.

The computation for the radially symmetrical wave field, after taking the scattered wave field into account, shows that Eq. (3.4) can be rewritten as:

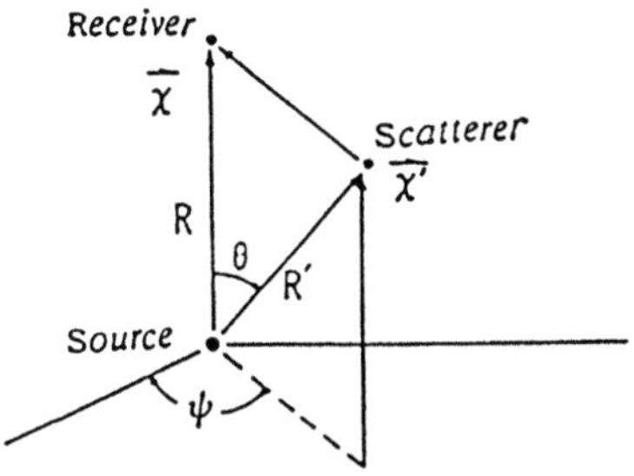

Fig. 1. The spherical polar coordinates of $\vec{X}$, $\vec{x}'$

$$\varepsilon(R,t) = \varepsilon_p(R,t) + \frac{2n_0\sigma V}{\pi}\iint_\Gamma \varepsilon(R',t')\frac{R'dR'dt'}{\sqrt{4R^2R'^2 - [R^2 + R'^2 - V^2(t-t')^2]^2}} \tag{3.5}$$

where Γ is the integration domain shown in Fig. 2.

This integration equation can be solved with the subsequential method. This mathematical subsequential solution approach just corresponds to the physical subsequential process. When we first put the primary wave field into the space, we obtain the first order approximation of the scattered wave field, the single scattering wave field. Then, we put the first order of the scattered wave field into the space and we obtain the second order approximation of the scattered wave field, the double scattering wave field. In this process the direct wave field is, of course, attenuated according to the exponential law in the space domain.

Mathematically, we express this process as follows:

substitute $\varepsilon(R', t')$ under the integration of Eq. (3.5) with the primary wave to obtain the first-order approximation of the solution. Thus, it can be written as:

$$\varepsilon_1 = \varepsilon_p + \iint \varepsilon_p f\, dR'\, dt' = \varepsilon_p + \varepsilon_{s,1}, \tag{3.6}$$

where:

$$f = \frac{2n_0\sigma V}{\pi}\,\frac{R'}{\sqrt{4R^2R'^2 - [R^2 + R'^2 - V^2(t-t')^2]^2}},$$

and $\varepsilon_{s,1}$ is the increment of wave energy due to the first-order scattering. The second-order approximation of the solution of energy density is:

$$\begin{aligned}\varepsilon_2 &= \varepsilon_p + \iint \varepsilon_1 f\, dR'\, dt' \\ &= \varepsilon_p + \iint (\varepsilon_p + \varepsilon_{s,1}) f\, dR'\, dt'.\end{aligned} \tag{3.7}$$

Thus we have:

$$\varepsilon_2 = \varepsilon_p + \varepsilon_{s,1} + \varepsilon_{s,2},$$

where

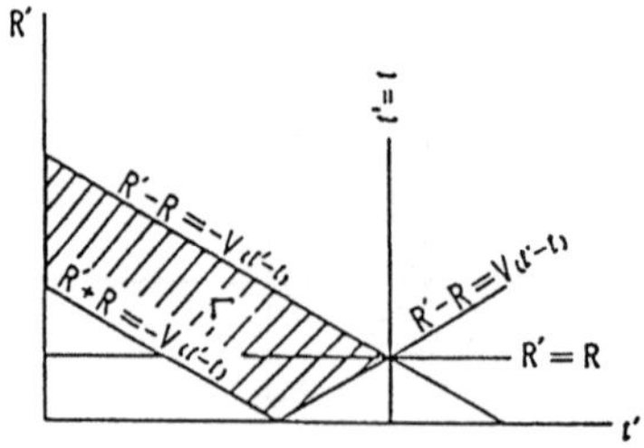

Fig. 2. Γ domain in the coordinates of t', R'

$$\varepsilon_{s,2} = \iint \varepsilon_{s,1} f \, dR' \, dt'.$$

The energy density of higher order scatterings can be obtained in a similar way. Therefore, if we find all the increments of energy density due to the different orders of scattering we will find the approximate solution for up to n-th order scąttering energy as:

$$\varepsilon_s^{(n)} = \sum_{i=1}^{n} \varepsilon_{s,i}, \tag{3.8}$$

then, the complete energy density should include the primary and scattered waves,

$$\varepsilon^{(n)} = \varepsilon_p + \sum_{i=1}^{n} \varepsilon_{s,i}. \tag{3.9}$$

Let us consider the primary wave as a pulse-like wave. The power $P_p(R, t)$ of the primary wave received at R can be written as:

$$P_p(R, t) = W_0 \frac{S_{\Delta t}\left(t - \frac{R}{V}\right)}{\Delta t} - \frac{1}{2\pi R}, \tag{3.10}$$

where:

$$S_{\Delta t}(x) = \begin{cases} 1 & \text{for } 0 < x < \Delta t \\ 0 & \text{otherwise,} \end{cases} \tag{3.11}$$

where W_0 is the total radiated energy, $S_{\Delta t}(x)/\Delta t$ is a simulation of Dirac's delta function, when Δt is very small. The solution of such type of primary wave is feasible and can be a basis for dealing with more complex sources. Thus, the energy density at R is:

$$\varepsilon_p(R, t) = \frac{W_0 \delta\left(t - \frac{R}{V}\right)}{2\pi RV}, \tag{3.12}$$

and

$$\varepsilon(R,t) = W_0 \frac{\delta\left(t - \frac{R}{V}\right)}{2\pi RV} + \frac{n_0 \sigma W_0}{\pi^2} \iint_{\Gamma} \frac{\delta\left(t' - \frac{R'}{V}\right) dR' \, dt'}{\sqrt{4R^2R'^2 - [R^2 + R'^2 - V^2(t-t')^2]^2}}. \tag{3.13}$$

After integrating, we have the analytical solution for two-dimensional space as:

$$E_s(R, t) = \frac{W_0}{2\pi} \frac{n_0 \sigma H\left(t - \frac{R}{V}\right)}{\sqrt{V^2t^2 - R^2}} \exp\left[n_0 \sigma \sqrt{V^2t^2 - R^2} - n_0 \sigma Vt - \frac{\omega}{Q_i} t\right). \tag{3.14}$$

Obviously, when R = 0, namely when the observer stands at the source, the collected energy density would be:

$$E_s(t) = \frac{W_0 n_0 \sigma}{2\pi V t} H(t)\, e^{-\frac{\omega}{Q_i} t}, \tag{3.15}$$

or the power of coda is:

$$P_s(t) = VE_s(t) = \frac{W_0 n_0 \sigma}{2\pi t} H(t)\, e^{-\frac{\omega}{Q_i} t}. \tag{3.16}$$

For practical use, we write the power of coda of surface waves in the form of:

$$P_c(t, 0) = \frac{W_0}{2\pi} \frac{\omega}{VQ_s t} e^{-\frac{\omega}{Q_i} t}. \tag{3.17}$$

This expression can be used approximately when the receiver is not far from the source, or for local earthquakes. This equation has just the same form of single-scattering but Qc is replaced by Qi. When the receiver is not close to the source, we have:

$$P_c(t, t_s) = \frac{W_0}{2\pi} \frac{\omega}{VQ_s t\left[1 - \left(\frac{t_s}{t}\right)^2\right]^{1/2}} \exp\left\{-\left[\frac{1}{Q_s}\left(1 - \sqrt{1 - \left(\frac{t_s}{t}\right)^2}\right) + \frac{1}{Q_i}\right]\omega t\right\}. \tag{3.18}$$

This expression shows that the early coda has more involvement of Q_s but the very late coda reflects only Q_i. However, due to the codas of different elapsed time scattered from different parts of Earth material, and the fact that the different parts have different quality factors, it will be very difficult to distinguish Q_s from Q_i.

The solution in three-dimentional space can be approached in a similar way. The only difference between these two different spaces is in the geometrical spreading factor. Namely we should use $(4\pi|\vec{X} - \vec{X}'|^2)^{-1}$ to substitute $(2\pi|\vec{X} - \vec{X}'|)^{-1}$ in the relevant equations. For the impulsive source, Eq. (3.4) is adapted as:

$$\varepsilon(R, t) = \varepsilon_p(R, t) + n_0\sigma \iiint_{\infty} d^3x' \int_0^{\infty} \frac{\varepsilon(R', t')\delta\left(t' - t - \frac{|\vec{x} - \vec{x}'|}{V}\right)}{4\pi|\vec{x} - \vec{x}'|^2} dt', \tag{3.19}$$

where $d^3x' = R'^2 \sin\theta\, d\psi'\, d\theta'\, dR'$ as in ordinary spherical polar coordinates, and $|\vec{x} - \vec{x}'| = |R'^2 + R^2 - 2RR' \cos\theta'|^{1/2}$.

The first order approximation of solution for Eq. (3.19) is:

$$E_{s,1}(R, t) = \frac{n_0\sigma W_0 H\left(t - \frac{R}{V}\right)}{4\pi R^2} \ln\frac{Vt + R}{Vt - R} \exp\left[-\left(n_0\sigma Vt + \frac{\omega t}{Q_i}\right)\right]. \tag{3.20}$$

This is just the solution of Sato (1977) for single isotopic scattering. It is very hard to find the analytical solution for higher-order scattering in three-dimensional space. However, we have the digital solution for the first several orders' approximation. The solutions of up to five order approximations give the following expression:

$$E_s = \frac{W_0}{4\pi} \frac{\omega}{Q_s V^3 t_s t} \ln \frac{t + t_s}{t - t_s} e^{\left(0.16 \frac{\omega t}{Q_s} - 0.86 \frac{\omega}{Q_s} \frac{t_s^2}{t} - \frac{\omega t}{Q_i}\right)}, \tag{3.21}$$

or for power spectrum:

$$Ps = \frac{W_0}{4\pi} \frac{\omega}{Q_s V^2 t_s t} \ln \frac{t + t_s}{t - t_s} e^{\left\{\left[0.16 - 0.86\left(\frac{t_s}{t}\right)^2\right] \frac{\omega t}{Q_s} - \frac{\omega t}{Q_i}\right\}}, \tag{3.22}$$

The solution (3.21) has some similarity with the results from the diffusion theory (Dainty and Toksoz 1977; Gao and Li 1988).

3 Discussion

Equations (3.18) and (3.22) obtained in this presentation showed that the decay rate of coda envelopes has a complex physical meaning; even in the simple model in which we suggested that the wavelets propagate in an inhomogeneous medium where numerous isotropic scatterers distributed in the medium stochastically and uniformly. In general, the decay rate reflects both intrinsic Q, Q_i, and scattering Q, Q_s. For certain special cases, for example when wavelets propagate in two-dimensional space and the receiver is at the same place as the source, the decay rate reflects only the intrinsic Q. A similar conclusion was obtained by Frankel and Wennerberg (1987) by means of finite difference simulation. As broadly recognized, coda Q, Q_c, is just based on the decay rate of the coda envelopes. If the recorded codas on the seismograms are from surface waves, we believe that the acquired coda Q_s in previous works are just close to intrinsic Q. However, as we have seen the recorded codas of local earthquakes may be a mixture of body waves and surface waves, therefore we need to have the solution both in two- and in three dimensional spaces. This article shows that the scattered waves in three-dimensional space have slightly more complex expressions than in two-dimensional space. In this case, even when the receiver is sited at the same place as the source, the decay rate is correlated with both Q_i and Q_s. However, we can find from Eq. (3.22) that the decay rate should reflect more Q_i effect than Q_s effect as the elapsed time increases, if these two quality factors are of similar magnitude. Of course, if these two quantities are of quite different magnitudes, one of the two could be ignored. Recently, the preliminary results on discriminating Q_s from Q_i show that the scattering Q-values are generally larger than the intrinsic Q-values by one order. Therefore, the main origin of attenuation in the actual Earth seems to be intrinsic absorption but not scattering. This reality enhances furthermore the suggestion that the coda Q-values are basically close to the intrinsic Q-values.

As many articles have discussed, coda Q is strongly dependent on frequency. If this dependency is due to the fact that coda Q mainly reflects scattering effects, then this frequency dependency may be explained in terms of the characteristics of the scattering process. Some theoretical works (Wu 1982; Sato 1982) show a frequency dependence of Q_s which has a peak around ka = 1, where k is the wavenumber and a is the correlation distance of the heterogeneous medium. They gave a good explanation on the frequency dependence: the calculation for the von Karman autocorrelation function (self-similar), which is one of the examples for random fractional velocity fluctuation with relatively more short wavelength heterogeneity (Frankel 1989) shows that in the low frequency band ($ka \ll 1$), for example, lower than 0.5–1.0 Hz, Q_s^{-1} is proportional to the third power of frequency, the Rayleigh scattering, whereas in the higher frequency band ($ka \gg 1$), for example higher than 1.0 Hz, Q_s^{-1} decreases with frequency of negative power $-2n$, where n is a featuring constant of the autocorrelation function. If we recognize that the coda Q is mainly scattering Q, the above theory can be deployed to explain the observed feature. However, if coda Q reflects mainly intrinsic Q, we need to find a more reasonable explanation. Aki cited different possible mechanisms of attenuation which may associate with absorption (Aki 1980a): the viscous damping due to liquid in the liquid-filled cracks (saturated and partially saturated), the dissipation due to flow between adjacent interacting cracks, etc. Those researchers did not pay enough attention to the frequency dependence of intrinsic Q, of the high frequency seismic waves. Therefore, explanation of the true origin that causes such a characteristic frequency dependency of coda Q is a physical or chemical environment in the area through which the coda waves passed.

To a certain extent, the frequency dependence feature of the Q-factor has its own special meaning when judging the material properties other than Q value itself. As the experience of measuring Q_c from the local earthquake seismograms shows (Gao et al. 1986) the Q value may change drastically among different stations in a small area, but the derivative of Ln Q_c, with respect to frequency, does not change very much in such an area.

At present, this article and the works on multiple-scattering have not given rise to obviously new, different treatment of the coda records, but only more concern about the true meaning of codas theoretically. However, this analysis of the physical meaning will be helpful in finding an approach toward discriminating Q_i from Q_s. This distinction may increase the precision of Q-measurement substantially.

If coda Q mainly reflects the intrinsic Q, it should be very sensitive to temperature and the content of liquid within the detected body. Therefore, if volcanic activities are caused by temperature change and/or the movement of melting materials, coda Q should be a good measure for monitoring the volcanic eruption. However, if the activities are caused by small stress variations, consideration of coda Q may be less sensitive to the change in stress in comparison to temperature; it may thus be less sensitive to the stress related precursors. Therefore, the coda Q monitoring approach may prove useful for watching the formation of the magma chamber, and thus, the first eruption of a volcano. However, as the work on monitoring the subsequent eruptions of Etna by means of coda approach (Del Pezzo et al. this Vol.) shows that there is no clear evidence between the correlation of the coda Q change and eruption, it seems that the simple coda Q monitoring is not effective for the prediction of subsequent eruptions. This may mean that the subsequent eruptions are caused by local stress readjustment and/or the small scale of geochemical reaction and so on. Of course, this situation does not mean that coda is not useful to eruption monitoring. More thorough research is needed to improve the precision of monitoring and to find a more comprehensive approach.

Though we may believe, in general, that Q_i is not very great in the actual Earth, by investigating the equations obtained, let us see if, in the medium, there is no any absorption at all to show what it would look like. From Eq. (3.17) the energy density close to the source in two-dimensional space may be expressed as:

$$P_c(t) = \frac{W_0}{2\pi}\frac{\omega}{Q_s}\frac{1}{Vt} \tag{5.1}$$

This means that codas will decay at a very slow rate. Consider that at time t the coda records all the waves that travel a distance r, obviously $r = Vt$. Therefore, Eq. (5.1) means that the wave energy attenuates as if affected only by a geometrical diffusivity factor. As time goes by, the impulsive energy spreads by a factor of r^{-1}. This fits the suggestion that Q_i causes real energy loss but Q_s causes only apparent attenuation, and also shows that the solution shown in this presentation is quite reasonable.

As is well known, from the calculation on scattering properties, every single-scattering process has very complex directivity features. Therefore, the single scattering model has problems when dealing with anisotropic considerations. However, in the case of the multiple-scattering model the anisotropic problem may be mitigated due to the stochastic direction distribution in scattering processes with every different order of scattering.

The collected energy by the codas, at every moment, includes all the different orders of scattered waves. Gao et al. (1983a,b) showed criteria of $\tau = 0.8$ for two-dimensional space, $\tau = 0.65$ for three-dimensional space, where τ is the time counted in terms of "mean free travel time" namely the ratio between time t and the mean free travel time. We should notice that this distinction is only for the predominant contribution of different orders of scattering. Physically, we can not distinguish which part of the coda is from single-scattering and which is from multiple-scattering. Therefore, the information acquired by the wavelets of codas at their propagating in the medium should be considered from both the simple- and multiple-scattering processes, no matter how long the elapsed time is.

Richards and Menke (1983) pointed out a logic inconsistency in the single-scattering model. This model first considers only single-scattering but then attributes some of the attenuation between the source and scatterers to scattering attenuation in addition to absorption. The multiple-scattering model (Gao et al. 1983a) considers that the scattering process occurs not only in the neighbouring scatterers but also in remote ones between which there are numerous scatterers. Therefore, when the wavelets move in the medium, their motion is attenuated by both scattering and absorption. Therefore, when we calculate the scattering energies due to every single-scattering process, we should take those two attenuation factors into account simultaneously.

In the present calculation, the source is an impulse which has a spectrum consisting of continuous and uniform frequency components over the whole frequency band. However, in practice we need to deal with the limited waveform which is neither an impulse nor a continuous sinusoidal signal. This gap between theory and practice needs to be fulfilled.

To make the measured Q-value more comparable with the results from other approaches and to make use of the measured coda Q in the prediction of earthquakes and volcanoes, we need to work further to understand the meaning of its physical substance. Also whether we can increase the precision of measurement of the Q-factor will depend on whether we have a deep understanding of the recorded waveforms of codas. Qc from the decay rate of the envelopes of coda was thought as being the Q-factor of shear waves, if the weak scattering hypothesis holds (Aki 1980b; Roecker et al. 1982). We face the serious task of recognizing the complexity of the wave transformation between body wave to surface wave (Frankel 1989) and also the modification of the solution from a delta function to the practical waveform calculation.

Acknowledgements. The author thanks Mr. Xu Yong-Shuan and Miss Yu Guan for their assistance in computer work. The author expresses his thanks to Professor C.Y. Fu for his firm support in the research on coda waves.

The author would like to express his thanks to Professor K. Aki for his encouragement in the author's theoretical work on scattering and his observation work on codas. The author also expresses his debt to Professor J.P. Poirier and all colleagues at the Department of Geomaterials of IPG in Paris for their enthusiastic help in preparing this presentation during his visit.

References

Aki K (1969) Analysis of the seismic coda of local earthquakes as scattered waves. J Geophys Res 74:615–663

Aki K (1980a) Attenuation of shear-waves in the lithosphere for frequencies from 0.05 to 25 Hz. Phys Earth Planet Interiors 21:50–60

Aki K (1980b) Scattering and attenuation of shear waves in the lithosphere. J Geophys Res 85:6496–6501

Aki K, Chouet B (1975) Origin of coda waves: source, attenuation and scattering effects. J Geophys Res 80:3322–3342

Bouchon M (1982) The complete synthesis of seismic crustal phases at regional distance. J Geophys Res 87:1735–1741

Dainty AM, Toksoz MN (1977) Elastic wave scattering in highly scattering mediums, a diffusion approach. J Geophys 43:375–388

Frankel A (1989) A review of numerical experiments on seismic wave scattering. 131:639–685

Frankel A, Wennerberg (1987) Energy-flux model of seismic coda: separation of scattering and intrinsic attenuation. Bull Seismol Soc Am 77:1223–1251

Gao LS, Li SL (1988) Time domain solution of multiple scattering and the coda envelopes. Pageoph (in press)

Gao LS, Lee LC, Biswas NN, Aki K (1983a) Comparison of the effects between single and multiple-scattering on coda waves for local earthquakes. Bull Seismol Soc Am 73:377–389

Gao LS, Biswas NN, Lee LC, Aki K (1983b) Effects of multiple scattering on coda waves in a three-dimensional medium. Pure Appl Geophys 121:3–15

Gao LS, Shi RB, Hua ZX, Li RX (1986) Q-factor as the function of frequency around different stations within Beijing Seismic Network. Acta Seismol Sinica 9:354–366
Herraiz M, Espinosa AF (1987) Coda waves: a review. Pageoph 125:499–577
Kopnichev YF (1977) The role of multiple scattering in the formation of a seismogram's tail. Izv Acad Sci USSR, Phys Solid Earth 13:394–398
Levander AR, Hill NR (1985) P-SV resonances in irregular low-velocity surface layers. BSSA 75:847–864
Malin PE (1980) A first-order scattering solution for modelling elastic wave codas, I. The acoustic ease. Geophys J, 63, 361–380
Matsunami K (1987) Laboratory tests of the single- and multiple-scattering model for the generation of seismic coda waves. Bull Dis Prev Res Inst 37:137–168
Peng JY, Jin A, Aki K (1987) Mente- Carlo simulation of coda formation by discrete scatterers. EOS 68:1362
Richards PG, Menke W (1983) The apparent attenuation of a scattering medium. BSSA 73:1005–1022
Roecker SW, Tucker B, King J, Hatzfeld D (1982) Estimates of Q in central Asia as a function of frequency and depth using the coda of locally recorded earthquakes. BSSA 72:129–149
Sato H (1977) Energy propagation including scattering effects. Single isotropic scattering approximation. J Phys Earth 25:27–41
Sato H (1982) Attenuation of S waves in the lithosphere due to scattering by its random velocity structure. J Geophys Res 87:7779–7785
Shang TL, Gao LS (1988) The transportation theory of multiple scattering and its application to seismic codas of impulsive source. Sci Sin B 1:85–94
Wu RS (1982) Attenuation of short-period seismic waves due to scattering. Geophys Res Lett 9:9–12
Wu RS (1984) Seismic wave scattering and the small scale inhomogeneities in the lithosphere. PhD Thesis, Massachusetts Inst Tech, Cambridge, Mass
Wu RS (1985) Multiple scattering and energy transfer of seismic waves, separation of scattering effect from intrinsic attenuation, I., Theoretical modelling. Geophys JR Astron Soc 82: 57–80

5 Seismic Activity in Resurgent Calderas and Geothermal Fields

Fracture Processes Induced by Forced Fluid Percolation

F.H. Cornet

Abstract

The theory of the rupture processes induced by excess fluid pressure is reviewed. After a brief overview of some fracture mechanics principles, the theory of hydraulic fracturing is summarized. Then, shear failure mechanisms are discussed. Some examples of microseismicity and ground deformation associated with forced fluid flow, including magma intrusions, are presented. It is argued that most seismic signals are associated with shear ruptures and not with tensile cracking. However, short wave length instabilities associated with mode I fractures may be partly at the origin of the long period events observed in some instances during high flow rates borehole injections.

1 Introduction

When a fluid is forced in a rock mass through a conduit, first the rock surrounding the conduit deforms elastically under the increasing pressure. As the pressure reaches a critical level, various rupture processes may initiate, depending on the fluid viscosity, the flow rate, the shape and morphology of the conduit(s), the regional stress field and the rock strength. The purpose of this chapter is to provide a schematic overview of the various rupture processes induced by excessive fluid pressures in a rock mass, i.e., by forced fluid percolation (all along the paper, compressive stresses are reckoned positive).

First, some elementary principles of fracture mechanics are recalled and applied to the analysis of hydraulic fractures, i.e., fractures normal to a principal direction of the local stress field. Thereafter, the possibility of developing shear fractures either through the rock matrix or along preexisting surfaces of weakness, because of excessive interstitial fluid pressures, is discussed.

These principles are then examined through practical examples in which the forced fluid percolation process has been investigated experimentally in situ through the monitoring of both induced micro-seismicity and surface deformation.

This paper concentrates on fracture processes induced by the fluid pressure alone. It does not address two equally important problems, namely thermal cracking and stress corrosion. Indeed, forced fluid percolation may induce large temperature gradients and therefore thermal stresses which may lead possibly to thermal cracking. Also, because of physicochemical effects, fluid percolation may affect the quantity of energy required to develop new surfaces and therefore may lead to various failure processes linked to corrosion.

2 Introductory Elements of Fracture Mechanics

Consider a solid submitted to mixed boundary conditions (body forces, time dependent surface tractions and/or surface displacements); a fracture corresponds to a discontinuity in the displacement field in the solid. The questions addressed in fracture mechanics concern the determination of the boundary conditions for which the fracture starts to extend, the prediction of fracture extension geometry and the analysis of the fracturing process stability. These questions can be tackled after a criterion of fracture has been defined. The basis of most fracture criteria are derived more or less directly from Griffith's energy criterion of fracture (Griffith 1921).

2.1 Griffith's Fracture Criterion

Griffith postulated that the development of fracture is an energy dissipation process and that the appearance of a new surface in the solid absorbs a certain amount of energy ΔD, called surface energy:

$$\Delta D = f(\mathbf{ds}), \tag{1}$$

where $\mathbf{ds}$ is the crack extension of area da, with normal $\mathbf{n}$.

During fracture extension, the principle of energy conservation applies:

$$\Delta W(\mathbf{ds}) + \Delta Q_0(\mathbf{ds}) = \Delta U(\mathbf{ds}) + \Delta Q_1(\mathbf{ds}) + \Delta T(\mathbf{ds}) + \Delta D(\mathbf{ds}), \tag{2}$$

where, for crack extension $\mathbf{ds}$:

$\Delta W(\mathbf{ds})$ is the external work done to the system,
$\Delta Q_0(\mathbf{ds})$ is the heat provided to the system,
$\Delta U(\mathbf{ds})$ is the change of internal (elastic) energy,
$\Delta Q_1(\mathbf{ds})$ is the variation of internal heat,
$\Delta T(\mathbf{ds})$ is the variation of kinetic energy,
$\Delta D(\mathbf{ds})$ is the variation of surface energy.

Griffith proposed that fracture propagation occurs if the creation of a new surface induces a decrease in potential energy for the system.

For an adiabatic process, this implies:

$$[\Delta W(\mathbf{ds}) - \Delta U(\mathbf{ds})] \geqslant \Delta D(\mathbf{ds}). \tag{3}$$

When there is strict equality between both terms of (3), the fracturing process is quasi-static, while if the left term is larger than the right one, some kinetic energy is generated. For quasi-static fracture processes, of all possible crack configuration **ds** of area da, that which actually occurs, $\mathbf{ds_0}$, is that which induces the largest decrease in potential energy of the system (Palaniswamy and Knauss 1974):

$$[\Delta W(\mathbf{ds}) - \Delta U(\mathbf{ds}) - \Delta D(\mathbf{ds})] - [\Delta W(\mathbf{ds_0}) - \Delta U(\mathbf{ds_0}) - \Delta D(\mathbf{ds_0})] \leqslant 0. \quad (4)$$

If the rock is isotropic with respect to its strength:

$$\Delta D(\mathbf{ds}) = \Delta D(\mathbf{ds_0}) = 2\gamma \, da, \quad (5)$$

where γ is Griffith's free surface energy per unit area (the factor 2 appears because a fracture has two faces). In this case, Eq. (4) reduces to:

$$[\Delta W(\mathbf{ds}) - \Delta U(\mathbf{ds})] - [\Delta W(\mathbf{ds_0}) - \Delta U(\mathbf{ds_0})] \leqslant 0. \quad (6)$$

Equation (6) implies that for isotropic materials, the fracture extension geometry can be determined without knowing the surface energy of the rock. For anisotropic materials the free surface energy depends on the crack extension orientation:

$$\Delta D(\mathbf{ds}) = 2\gamma\,(\mathbf{n}) \, da. \quad (7)$$

In such instances, fracture extension geometry can be determined only through Eq. (4) and the angular variation of surface energy must be known. The existance of weakness planes in the rock mass implies that, for points located on such surfaces, if γ_m is the free surface energy of the matrix (supposed here to be isotropic) and γ_w is the free surface energy along the weakness plane, with $\gamma_w \ll \gamma_m$:

$\mathbf{ds} = 2\gamma_m$ da for all **n** orientations not normal to the weakness plane,
$\mathbf{ds} = 2\gamma_w$ da for **n** normal to the weakness plane.

For a real rock mass, the free surface energy is not only a function of orientation but also a function of the crack tip location.

For isotropic materials, we get from Eqs. (3) and (5):

$$[\Delta W(\mathbf{ds_0}) - \Delta U(\mathbf{ds_0})]/da = 2\gamma. \quad (8)$$

Let us define the strain energy release rate G(**n**) by:

$$G(\mathbf{n}) = \lim_{da \to 0}[\Delta W(\mathbf{n}) - \Delta U(\mathbf{n})]/da. \quad (9)$$

Accordingly, for an isotropic material, the critical loading conditions for which fracture extends and the direction of fracture extension (normal to $\mathbf{n_0}$) satisfies, for all **n**, the following conditions:

$$G(\mathbf{n}) \leqslant G(\mathbf{n_0})$$
$$G(\mathbf{n_0}) = 2\gamma. \quad (10)$$

If $\delta G(\mathbf{n_0})/\delta a > 0$, some kinetic energy is generated and the fracturing process becomes unstable with generation of vibrations in the solid.

If $\delta G(\mathbf{n_0})/\delta a = 0$, the fracturing process remains quasistatic and no vibration is generated.

If $\delta G(\mathbf{n_0})/\delta a < 0$, the fracture stops. It may resume propagation only if additional work is supplied to the system. The fracture process is stable.

The strain energy release rate concept as expressed by (10) has revealed very convenient for analyzing, numerically, fracture processes in rock mechanics. It is readily adapted to anisotropic or heterogeneous conditions provided the function $\gamma(\mathbf{n})$ is known, which, unfortunately, is generally not the case.

2.2 Irwin's Theory of Fracture

As mentioned (Sect. 2), fractures may be regarded as surfaces of discontinuity for the displacement field in the solid. Irwin (1957) suggested that the relative displacements $\hat{\mathbf{u}}$ between points on the two faces of the fracture could be described completely in terms of three independent kinematic movements.

Consider the frame of reference shown in Fig. 1. OX_1, OX_2 lie in the plane tangent to the fracture surface Σ, at the fracture tip; OX_1 is perpendicular to the crack tip curve Σ, OX_2 is tangent to Ω. OX_3 is orthogonal to both OX_1 and OX_2. The fracture surface Σ is presumed to be smooth, with no singularity point.

The three fracture modes defined by Irwin are:

Mode I, or opening mode, for which $\hat{u}_3 \neq 0$; $\hat{u}_1 = \hat{u}_2 = 0$;
Mode II, or sliding (in plane shearing) mode, for which $\hat{u}_1 \neq 0$; $\hat{u}_2 = \hat{u}_3 = 0$;
Mode III, or tearing (anti-plane shearing) mode, for which $\hat{u}_2 \neq 0$; $\hat{u}_1 = \hat{u}_3 = 0$;

where $\hat{\mathbf{u}}$ is the displacement discontinuity vector near the fracture tip.

These three basic fracture modes are necessary and sufficient to describe all the possible crack propagation schemes in the most general state of stress. Each of the three crack movements is associated with a particular stress field, in the immediate vicinity of the crack edge. For linearly elastic materials, the three stress components S_{33}, S_{12}, S_{13} at points on the X_1 axis close to the origin take the simple forms (assuming that the radius of curvature at the fracture tip is infinitely small):

$$\begin{aligned} S_{33} &= (2X_1)^{-1/2} K_I + 0(1), \\ S_{13} &= (2X_1)^{-1/2} K_{II} + 0(1), \\ S_{12} &= (2X_1)^{-1/2} K_{III} + 0(1), \end{aligned} \tag{11}$$

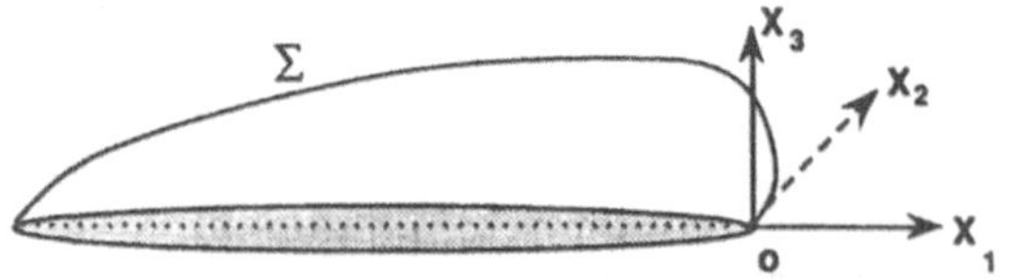

Fig. 1. Coordinate system used to define Irwin's fracture modes (Irwin 1957). In this diagram, the fracture is shown thicker than in reality; for very thin fractures, the radius of curvature at the fracture tip o tends to zero so that the O X_1 X_2 plane is tangent to the fracture surface

where the higher order terms in the variable X_1 have been neglected. Thus, these stress components tend to infinity as the point **X** gets closer and closer to the fracture tip. For points **X** not located on the X_1 axis, similar expressions are found except for additional trigonometric functions of the angular coordinate of **X**. Irwin defined the three quantities K_I, K_{II}, and K_{III}, as the stress intensity factors for mode I, II and III. Determination of these stress intensity factors is a problem of the mathematical theory of elasticity. Solution for many problems have been published (e.g., Sih and Liebowitz 1968; Sih 1973).

The important implication of these stress intensity factors stems from the fact that the region for which Griffith's energy balance must be satisfied for a crack to grow can be any portion of the body enclosed within a simple closed surface L, surrounding the crack tip, where energy is being dissipated. Following Sanders (1960), the Griffith's criterion of fracture may be stated as: "The rate at which work is being done by forces acting across L, equals the rate of increase of strain energy stored in the material inside L, plus the rate at which energy is dissipated by the growing crack". Accordingly, only the stress field in the vicinity of the crack tip needs to be examined.

For the opening mode, in plane strain conditions, Irwin showed that the strain energy release rate and the stress intensity factor are related by:

$$G_I = (1 - v^2)/E * K_I^2, \tag{12}$$

where E is Young's modulus and v is Poisson's ratio of the rock.

Accordingly, since a fracture propagates when G_I is equal to 2γ it can be also stated that it propagates if K_I reaches a critical value, called the critical stress intensity factor for mode I, also referred to as the fracture toughness of the rock. Thus, even though the stress components S_{33}, S_{13}, S_{12} may be very large in the immediate neighbourhood of the fracture tip, fracture extends only when the energy balance is satisfied. This shows that the concept of strength, as expressed in terms of maximum stress, is not an intrinsic property of the material and therefore is often of dubious value.

Equation (12) has been obtained on the assumption that the fracture extends in its own plane. The validity of this hypothesis has been verified, both experimentally and numerically, for mode I.

By extension, it has been postulated that fracture in mode II or III would occur when the stress intensity factor in mode II and/or III reached a critical value. However, for these two modes, the fracture generally does not extend in its own plane, i.e., propagation does not occur in pure mode II or pure mode III (e.g., Sih 1974; Swedlow 1974; Ingraffea 1977; Cornet 1979) so that there is no simple relationship between "critical stress intensity factors" and critical strain energy release rate, especially if the crack closes and sustains frictional forces. For these two modes of fracture, and therefore for general fracture propagation problems, the concept of fracture toughness is purely heuristic.

In more complicated configurations, such as fracture propagation in a layered medium, it can be shown (e.g., Comninou and Dundurs 1979) that, at the interface between two layers, the stress intensity factor concept does not apply. Indeed, the stress field near the fracture tip, when it reaches the interface, is not singular as the inverse square root of the distance between the point under consideration and the fracture tip so that the stress intensity factor is not defined, nor is the fracture toughness.

However, for all the above-mentionned configurations, the strain energy release rate concept remains always valid and this makes it a more attractive fracture criterion.

2.3 Fracture Propagation in Rock Materials

In many structural engineering situations, fracture propagates following the opening mode (mode I) and extensive experimental work has been conducted with this regard. In particular, tests for measuring fracture toughness (i.e., the critical stress intensity factor for mode I) has been standardized (ASTM 1978).

A similar approach has been followed initially for rocks but it soon appeared that even for pure mode I fracture extension, the energy dissipation process associated with macroscopic crack growth is not related only to the area of macroscopic fracture extension. For example, Hoagland et al. (1973) observed that microcracking develops in the minerals surrounding the tip of an artificial preset notch. At first the microfracture zone increases in size as the macroscopic crack grows, but this increase levels off after a certain macroscopic crack length has been reached so that the so called fracture toughness measurement becomes independent of crack length.

Important factors affecting the fracture toughness of rocks are environment and local stress field, and most noticeably the magnitude of the principal stresses which are not singular at the crack tip (see e.g. Henry et al. 1977; Schmidt and Huddle 1977; Ouchterlony 1980; Atkinson 1984; Kanninen and Popelar 1985; Atkinson and Meredith 1987).

These effects are best analyzed if the concept of free surface energy associated with macroscopic fracture extension is replaced by that of process zone. The process zone is that domain in the neighborhood of the fracture tip where energy is dissipated through microcracking. As a consequence of this material degradation, the elastic properties are modified and the stress field cannot be analyzed any longer by means of linear elasticity theory (see e.g. Madariaga and Villotte, this volume). From field observations at the tip of dykes, Pollard (1987) estimates that the radius of the process zone can range from 0.03 m to 370 m, depending on the fracture or dyke size.

Another important peculiarity of rock fracture is the tendency to generate tensile ruptures, even in pure mode II or mode III configurations (Brace and Bombolakis 1963; Hoek and Bieniawski 1965) so that *en echelon* tensile fractures are observed before the development of large scale shear ruptures (see e.g. Tapponier and Brace 1975). This is in contradiction with theoretical prediction based on the maximum energy release rate criterion, if it is assumed that the surface energy is the same for all fracture modes: numerical results show that the fracture should extend in pure shear, in an unstable manner (Cornet 1979). However if it is assumed that the surface energy associated with pure shearing modes is about four times larger than that implied by the mode I then *en echelon* fracturing is predicted (Fairhurst and Cornet 1981). The difficulty in this case is to define the procedure for identifying the optimum configuration of *en echelon* fractures. The exact mechanism controlling the development of shear zones is still an active research domain and will not be discussed further here (this is discussed by Segall and Pollard 1983).

3. The Mechanics of Hydraulic Fracturing

3.1 Hydraulic Fracture Orientation

In this paper, the expression "hydraulic fracturing" refers to the opening of fractures because of excessive fluid pressure. "Opening" of a fracture refers to the complete physical separation of the two faces of a fracture, because of a large enough normal component of the displacement occurring between the fracture faces, i.e., because of a large enough mode I component in the displacement discontinuity along the fracture surface. These fractures cannot transfer shear stresses from one face to the other (we will neglect the shear component induced by the fluid viscosity during flow) and therefore are normal to one of the principal directions of the local stress tensor. The local stress tensor is the stress near the faces of the fracture, it must not be confused with the regional stress tensor which is that prevailing in the domain before the opening of the fracture. If the hydraulic fracture is inclined, with respect to the principal directions of the regional stress field, the displacement discontinuity through the fracture exhibits a shear component (mode II and/or mode III), and the fracture induces a local rotation of the principal stress directions.

It will be shown later that, in some instances, high fluid pressures can also initiate shear fractures. Macroscopic shear fractures often exhibit a small normal component in the displacement discontinuity, because of the roughness of the surfaces (Barton et al. 1985). However, they retain numerous solid contacts so that they do transmit shear stresses after rupture. These hydraulically-induced shear fractures are not normal to a principal direction of the local stress field; they will not be referred to as hydraulic fractures.

The opening of a hydraulic fracture requires that the fluid pressure inside the fracture is larger than the normal stress supported by the fracture. For solids which are homogeneous and isotropic with respect to their strength, it is readily found from Eq. (6) that, for a given pressure distribution, the fracture orientation which corresponds to the largest decrease in potential energy is normal to the minimum principal stress direction, so that, hydraulic fractures should be normal to the minimum principal stress. This is well supported by laboratory experiments (e.g., Haimson 1968). Further, if for some reason the minimum principal stress direction changes in the rock mass, before the fracture propagates, so does the hydraulic fracture when it reaches the corresponding zone (von Schoenfeldt 1970; Daneshy 1971; Warren and Smith 1985).

However, according to Eq. (4), hydraulic fractures may initiate at an angle with respect to the minimum principal stress direction because, for example, of some preexisting weakness plane. Hayashi and Abe (1982) have discussed the conditions for which such inclined fractures can remain opened in a stable manner because of an internal pressure. A numerical investigation was carried out by Cornet (1982) to investigate, for an isotropic homogeneous solid, the growth of a crack inclined with respect to a uniaxial stress direction with a uniform pressure distribution inside the crack. The numerical model (Cornet 1979), based on Crouch's displacement discontinuity method (Crouch 1976), computes the strain energy release rate for various virtual crack increment configurations and picks up as actual crack extension configuration that for which the strain energy release is maximum.

Results are presented in Fig. 2. They correspond to a 10-cm-long crack, inclined by the angle θ with respect to the uniaxial stress ($T_1 = 10\,\text{MPa}$) direction. The change in crack orientation is β (see Table 1); the uniform internal pressure is P (in MPa) and the maximum strain energy release rate is G (in MPa·cm). The Young's modulus and Poisson's ratio were respectively chosen equal to 70 000 MPa and 0.25, values supposed to be that of a granite

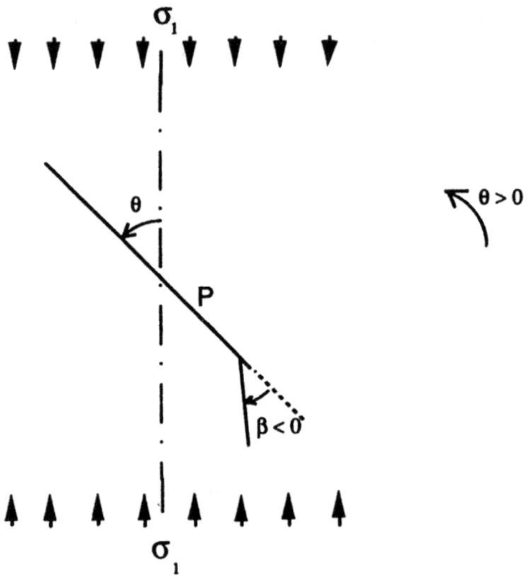

Fig. 2. Propagation of a pressurized fracture inclined at an angle θ with respect to a uniaxial stress field σ_1 ($\sigma_1 = 10\,\text{MPa}$); pressure P (MPa) is uniform up to the crack tip; fracture extension occurs at angle β; G is the strain energy release rate (MPa·cm)

θ degrees	P MPa	β degrees	G MPa·cm
30	8.9	−45	0.02
	18.8	−22.5	0.11
	100.	−7	2.01
	300.	0	18.7
45	16.	−30	0.02
	20.	−22.5	0.03
	40.	−15	0.14
	80.	−7.5	0.62

sample with surface energy of the order of 50 J/m^2 (critical strain energy release rate equal to 0.01 MPa·cm). These results show that for quasi-static crack growth ($G = G_c$) the fracture extension gets progressively parallel to the applied uniaxial stress direction. For large internal pressures, the effect of the uniaxial stress field on the crack tip stress field becomes less and less significant, so that fracture reorientation is less marked. However, this result is obtained only because it is assumed that the pressure remains uniform up to the crack tip. In fact, for large internal pressures the fracturing process is very unstable ($\delta G/\delta a \gg 0$) so that the fracture should propagate at a much faster rate than the fluid flow inside the crack, unless the fracturing fluid is a highly compressible gas. This should result in a pressure decrease near the crack tip which, in turn, should result in a different stress distribution near the crack tip. In the numerical model described, the fluid pressure is dealt with as an independant variable when, in fact, it should be computed from the coupled problem of fluid flow and fracture hydraulic aperture. This will be discussed later.

Let us only conclude now that experimental work, both in the laboratory and in the field (e.g., Aki et al. 1982; Cornet 1982; Rummel and Kappelmeyer 1983; Fehler et al. 1987; Jung 1987), has indeed demonstrated that hydraulic fractures can locally exhibit orientations different from that of the regional maximum principal stress and that these "abnormal" hydraulic fracture orientations depend on the anisotropy and heterogeneity of the rock strength as well as on the viscosity and the flow rate of the fracturing fluid. The lower the flow rate and the fluid viscosity, the more significant the influence of the heterogeneity and the anisotropy of the rock mass.

3.2 Simplified Two-Dimensional Models

Simplified two-dimensional models provide useful insight into the various factors controlling the driving pressure as well as the shape and volume of fractures.

As mentioned, the simplest model of a hydraulic fracture is that of a pure mode I fracture, i.e., normal to the least principal stress direction. If the pressure is assumed to be uniform up to the crack tip and if the fracture is penny-shaped (flat, thin disc) the stress intensity factor is (see e.g., Sih and Liebowitz 1968):

$$K = 2/\pi \cdot (P - \sigma_3)(\pi\, a)^{1/2}, \tag{13}$$

while if it is blade-shaped (infinite extension in the X_2 direction with elliptic cross-section of length 2a and width w ($w \ll 2a$) in the OX_1, OX_3 plane):

$$K = (P - \sigma_3)(\pi\, a)^{1/2}, \tag{14}$$

where P is the driving pressure and σ_3 is the minimum principal stress.

In either case, it is readily found that as soon as the fracture extends, i.e., when K_1 reaches the critical value K_{Ic}, K increases so that the fracturing process is unstable if the driving pressure P remains constant. But in fact, if fracture extension becomes unstable, the fluid cannot flow as fast as the fracture extends and a pressure gradient is generated near the fracture tip.

The stress intensity factor at the tip of a blade shaped fracture, normal to the minimum principal stress, loaded by a uniform pressure applied only along a small segment of length 2r ($r \ll a$) located at the center of the crack, is (Zoback and Pollard 1978):

$$K = 2(P - \sigma_3)r \cdot (\pi\, a/2)^{-1/2}, \tag{15}$$

so that fracture propagation is stable.

These examples outline the two following points:

1. The stability of the propagation process depends on the pressure distribution near the crack tip. In some laboratory experiments on Plexiglas, the fracture has been found to extend in successive jumps (Rummel 1987), however it is often believed that fracture extension is quasi-static because of a stable pressure gradient near the crack tip (e.g., Abe et al. 1976; Aki et al. 1977).
   Delanay and Pollard (1982) report average dike propagation rates of the order of 1 m/s. Since this is an average propagation velocity, it seems reasonable to consider that locally, and temporarily, the propagation velocity may be higher and therefore that the kinetic energy may not always be negligible.
2. Once the fracture has extended a certain distance, the pressure required to extend it further is very close to the minimum principal stress magnitude. The difference between the driving pressure and the minimum principal stress is essentially controlled by the pressure gradient near the fracture tip, i.e., by the fluid viscosity, the flow rate, the stiffness and the "strength" of the surrounding rock.

Various models have been developed in the oil industry for determining the geometry of hydraulic fractures as a function of injection flow rate and injection pressure, fluid viscosity, injected volume, flow losses through the walls of the fracture because of local permeability, effective driving pressure (difference between the normal stress to the fracture surface and the fluid pressure), elastic and strength characteristics of the rock mass. Because it is often observed that the vertical extension of hydraulic fractures in sedimentary rocks is limited (see Sect. 3.3), two dimensional models have revealed to be fairly satisfactory (Perkins and Kern 1961; Geertsma and Deklerk 1969; Nordgren 1972; Simonson et al. 1976; Cleary 1983; Nolte 1988).

The problems encountered in volcanology are quite different for none of the "pumping" parameters are known. It is only through interpretation of data, coming either from direct observation of old dykes or other volcanic intrusions (Pollard and Muller 1976; Rubin and Pollard 1987; Jackson and Pollard 1988) or from surface deformation observed during magma injections (Tarantola et al. 1979; Rubin and Pollard 1988) that these flow parameters can be evaluated.

In the former case, as the magma stops flowing, the pressure tends toward a hydrostatic state and the dyke walls adjust to this new loading condition. Thus, if it is assumed that the surrounding rock behaves elastically, analysis of the shape of the dyke can be used to infer the difference between the magma pressure and the normal stress supported by the dyke wall. Following this principle, Pollard (1987) concludes, after considering both penny-shaped and blade-shaped geometries, that a rule of thumb is that the thickness to length ratio for dykes is approximately equal to the ratio of the average driving pressure ($P - \sigma_3$) to the host rock stiffness:

$$w/2a \approx (P - \sigma_3)(1 - v^2)/E. \tag{16}$$

Values of (w/2a) are found to be of the order of 10^{-3}. This suggests that for Young's modulus in the order of 10^3 to 10^4 MPa, which are the values encountered in most rock masses in the upper 1 or 2 km of the crust, the driving pressure varies between 1 and 10 MPa.

This result is confirmed by Rubin and Pollard's interpretation of surface deformation observed during dyke propagation for the Kilauea volcano in Hawaii and for the Krafla volcano in Iceland. For these two volcanoes, these authors observe that the ratio of driving pressure to stiffness is of the order of 5×10^{-4} and they conclude that, for these two cases, the driving pressure is smaller than 4 MPa.

Accordingly, it seems reasonable to conclude that the driving pressure is somewhere between 1 and 10 MPa.

3.3 Factors Influencing Fracture Geometry

Hydraulic fractures have been observed to be approximately penny-shaped when developed in the laboratory in an isotropic homogeneous rock sample submitted to a uniform stress field (e.g., Haimson 1968). However, as soon as the size of the fracture becomes larger than a few meters, it is generally subjected to a non-uniform stress field (e.g., gravity effects) and this stress gradient is one of the prime factors affecting fracture geometry.

When the minimum principal stress is horizontal, as is the case in most oil fields, hydraulic fractures are vertical. Variations in minimum principal stress magnitude from one bed to another have been found in some instances to reach values of the order of 2 to 5 MPa, both in volcanic environments (Haimson and Rummel 1982) and in sedimentary rocks (e.g., Barton 1983; Teufel and Warpinski 1983; Cornet 1988). These variations are considered to be the main cause of vertical containment of hydraulic fractures: the fracture remains contained in the domain where the minimum principal stress is the smallest (e.g., Simonson et al. 1976; Thiercelin and Lemanczyk 1983). In fact, this parameter is one of the most important factors influencing the design of large hydraulic stimulations in oil reservoirs (Abou-Sayed 1982; Daneshy 1983; Cleary 1983).

In their analysis of factors influencing dyke propagation, Rubin and Pollard (1987) consider in addition to this effect of remote stresses, the magma pressure distribution within the dyke and geometrical factors such as the proximity of the Earth surface. These effects are analyzed through their influence on the stress intensity factor as computed along all the periphery of the dyke. The influence of some heterogeneities can also be analyzed along similar lines by considering various values for the fracture toughness of the rock mass. This rock strength heterogeneity may become very significant in layered formations because of blunting of the dyke tip caused by shearing along the interface between two layers (Weertman 1980). This has been observed in coal seams and also possibly occurs when a dyke reaches an interface between previous lava flows.

The Earth's surface is an important element controlling hydraulic fracture geometry for it constitutes, generally, a surface with no normal stress (except at the bottom of deep oceans) so that the minimum principal stress direction is normal to the Earth's surface. This implies that close to the Earth's surface, hydraulic fractures become parallel to it. This is well known in volcanology to be the cause for sill formation (see e.g., Jackson and Pollard 1988). However, if one of the other principal stress components remains tensile near the Earth's surface the fracture is normal to both the Earth's surface and the minimum principal stress direction (recall that compressive stresses are reckoned positive).

For a horizontal penny-shaped crack embedded in an infinite medium, the opening of the fracture caused by a uniform pressure P is given by Sneddon (1946):

$$2w = 4(P - \rho gz)(1 - \nu^2)(a^2 - r^2)^{1/2}/\pi E, \tag{17}$$

where:

w is the half thickness of the crack at distance r from the crack center,
a is the crack radius,
z is the depth at which the crack is embedded;
ρg is the vertical stress gradient.

As the sill radius a becomes large as compared to depth z, the influence of the Earth's surface cannot be neglected any longer and Eq. (17) is not valid any more. Pollard and Johnson (1973) analyze this geometric condition by considering the vertical deflection of a circular plate clamped around its edge and loaded by a uniform pressure. According to the elastic theory of plates, the vertical deflection at the center of the plate is given by:

$$w = 3(P - \rho gz)(1 - \nu^2)(a^2 - r^2)^2/16\,ET_e^{\,2}, \tag{18}$$

where T_e is the effective mechanical thickness of the plate. According to Pollard and Johnson, the effective mechanical thickness is the thickness of a single plate that has the same resistance to bending as does the stack of sedimentary layers over the magma intrusion. Accordingly, for the same values of the elastic parameters, the vertical deformation is found to be much larger when there is some influence by the Earth's surface, i.e., when the still length becomes larger than the distance to the Earth's surface.

4 Other Types of Failure Induced by Forced Fluid Percolation

In the previous Section we discussed the fracturing process in which a high fluid pressure induces the creation and the propagation of an opened fracture, i.e., a fracture with a predominant mode I component. This fracturing process requires specific conditions of flow rate and fluid viscosity so that the pressure can reach the required critical value at the fracture tip. These hydrodynamic conditions depend essentially on the characteristics of the hydraulic conductivity of the rock mass.

If the rock mass is permeable or fractured, fluids may percolate through these interstices. When the rate of fluid loss through the walls of a hydraulic fracture equals the injection flow rate, the fracture stops propagating. But this does not imply that all fracturing processes stop. The increase in interstitial pressure induced by the fluid percolation may still lead to some failure processes. These are observed when the interstitial pressure reaches a critical value which depends on the mechanical properties of the rock mass as well as on the regional stress field (see e.g., Raleigh et al. 1972; Lockner and Byerlee 1977; Pearson 1981).

Before discussing these effects, let us first recall some results concerning failure of rock masses submitted to compressive stresses. These failure processes can involve either the rock matrix or slip along preexisting fractures. Because of the complexity of the fracturing process in a homogeneous rock under compression (development of numerous microcracks with a predominant mode I component leading to dilatancy effects, i.e., increase in bulk volume prior to shear failure; see e.g., Brace et al. 1966; Crouch 1970; Cornet and Fairhurst 1974) the concept of strain energy release rate is not used generally. Fracture criteria applied in stability analysis are empirical and expressed in terms of ultimate stresses.

For example, Bieniawski (1974) found that the peak triaxial strengths of a range of rock types were well represented by:

$$\sigma_1/C_0 = 1 + A(\sigma_3/C_0)^k, \tag{19}$$

where σ_1 and σ_3 are respectively the maximum and the minimum principal stress, C_0 is the uniaxial compressive strenth of the rock and A and k are two scalars which depend on the rock type. A has been found to range from 5 for a norite to 3 for a mudstone and k from 0.8 for the norite to 0.7 for the mudstone.

Hoek and Brown (1980) found that the peak triaxial compressive strengths of a wide range of isotropic rocks could be described by:

$$\sigma_1/C_0 = \sigma_3/C_0 + (m\sigma_3/C_0 + 1.0)^{1/2}, \tag{20}$$

where σ_1 and σ_3 are the maximum and minimum principal stress magnitude and C_0 the uniaxial compressive strength. The parameter m varies with the rock type from 7 for carbonate rocks to 17 for fine grained polymanerallic igneous crystalline rocks (andesite, dolerite, diabase, rhyolite) and 25 for coarse grained polymineralic igneous and metamorphic rocks (amphibolite, gabbro, gneiss, granite, norite, quartz-diorite).

For preexisting fractures, the friction law has been found by Byerlee (1978) to be well represented by the Coulomb criterion, provided the normal stress supported by the slip plane is large enough. Sliding occurs when:

$$\tau = 0.85\,\sigma_n \qquad \text{if } \sigma_n \leqslant 200\,\text{MPa}; \tag{21.a}$$

$$\tau = 0.5 + 0.6\,\sigma_n \quad \text{if } \sigma_n \geqslant 200\,\text{MPa}; \tag{21.b}$$

where σ_n is the normal stress supported by the fracture plane and τ is the shear stress for which slip occurs.

However, as the normal stress reaches the order of 5 to 10 MPa, results are much more scattered and do not obey the simple linear law expressed by Eq. (21.a). As shown by Barton (1983), this apparent scatter is caused by the roughness of the surfaces which implies that, during sliding, some normal displacement occurs because of the asperities of the surfaces riding on one another. This normal displacement induces an increase in volume in the rock mass, which will be referred to as the joint dilatancy. This joint dilatancy must not be confused with the dilatancy induced by the opening of microcracks in the rock matrix because of large deviatoric stresses. Barton proposes the following equation, based on experimental results:

$$\tau = \sigma_n \tan[\mathrm{JRC} \log_{10}(\mathrm{JSC}/\sigma_n) + \phi], \tag{22}$$

where:

JRC is a scalar representing the roughness of the surface, defined empirically between 1 and 20;

JSC is the joint wall uniaxial compressive strength;

ϕ is the residual friction angle, i.e., the friction angle of the surface after it has slid for a distance long enough to stablize this friction angle.

All these failure processes are considered to depend usually on effective stresses (Handin et al. 1963; Brace and Martin 1968; Cornet and Fairhurst 1974), i.e., on the stress tensor defined by:

$$\sigma' = \sigma - P\mathbf{I}, \tag{23}$$

where σ' is the effective stress tensor, σ is the real stress, P is the interstitial pressure at the point where σ is defined and **I** is the unit tensor. Accordingly, the influence of an increase in interstitial pressure on the stability of a rock mass is generally analyzed simply by determining the consequences of the corresponding variation of effective stress magnitudes. Preexisting fractures which were in equilibrium before the interstitial pressure increase may become unstable because of the fluid percolation. Similarly, if the deviatoric stress is large enough, an increase in interstitial pressure may induce some failure in the rock matrix.

However, it will be mentioned in the next Section that a change in interstitial pressure in the natural fracture network of a rock mass is not satisfactorily represented by Eq. (23) when the hydraulic conductivity is not isotropic because, in fact, the interstitial pressure variation seems to induce a rotation of principal stress directions. The exact effective stress law for failure of saturated fractured rock masses remains to be properly defined.

5 Geophysical Monitoring of Forced Fluid Percolation

5.1 Microseismicity Induced by Man-Made Injections

When the failure process induced by forced fluid percolation becomes unstable, some kinetic energy is generated and this kinetic energy is then dissipated partly through elastic waves. Accordingly, these unstable failure processes may be studied by monitoring the mechanical vibrations propagating in the rock mass.

This has been performed in a number of large scale in situ experiments conducted mainly for investigating the possibility of developing "hot dry rock" geothermal energy (see e.g., Albright and Hanold 1976; Pearson 1981; Pine and Batchelor 1984; Fehler et al. 1987; Talebi and Cornet 1987; Baria et al. 1989) but also for controlling grout injections (Majer 1989). Some work has been conducted also in the laboratory by recording acoustic emmissions induced by fluid injections in rock samples submitted to various loading conditions (e.g., Lockner and Byerlee 1977; Zoback et al. 1977).

Different types of information are retrieved from such monitoring. The number of events per unit time has been used in the laboratory to investigate the onset of hydraulic fracturing (e.g., Zoback et al. 1977) or for defining the critical injection flow rate for which a hot dry rock criculation loop becomes unstable (e.g., Pine and Batchelor 1984). Location of impulsive events helps to identify regions where the local fluid pressure reaches the critical level for which unstable failure processes occur. Focal mechanisms based on first motion direction for both P- and S-waves help to characterize the type of rupture (pure shear, pure opening or combined mode fracture). Spectral analysis of time series may yield some information on the size of the source and, for shear events, on some magnitude of the stress drop. Hodograms are used either for identifying P wave polarization for the sake of event location (e.g., Pearson 1981) or for identifying different shear wave arrivals because of the anisotropy of the rock mass (Roberts and Crampin 1986).

These are standard seismological investigations and only some of their conclusions are discussed now.

5.1.1 Shear Events

The vast majority of events observed in field experiments of fluid injection are consistant with shear motion along preexisting fracture planes. This is evidenced by their shear wave magnitude which is generally too large to be caused by a tensile fracture (Pearson 1981). It is confirmed by focal mechanisms which are consistent with double-couple sources for most events (Cash et al. 1983; Cornet and Julien 1989). The fracture planes (as defined by one of the nodal planes) can be either sub-parallel to the main flow direction as determined from event location and borehole observations (Fehler et al. 1987; Julien 1987), or at an angle to it (Talebi and Cornet 1987). The former events are interpreted as corresponding to the shearing of asperities distributed along the fault which transmitted shear stresses through the fault, prior to their failure. Such shearing events cannot be observed with true hydraulic fractures. The latter events are interpreted as slip along fractures intersecting the main flow zone; they can be observed with true hydraulic fractures as well as during flow in fault zones. The former events show that, indeed, major flow can occur along fractures, or faults, inclined with respect to the principal stress directions.

For many large earthquake sources, the power spectrum of the signals recorded at some distance from the source is characterized by a low frequency constant spectral amplitude and a high frequency trend where the power decreases as $1/\omega^n$ with increasing frequency. The frequency for which the two trends intersect is called the corner frequency. It has been shown (e.g., Brune 1970; Brune et al. 1979; Madariaga 1979) that the corner frequency is somewhat related to the source dimensions whilst the seismic moment is directly proportional to the amplitude of the low frequency trend.

For example, the results of Brune et al. suggest that for circular or elliptic sources a rough source size estimate is given by:

$$K = f_c\, r/V_S, \tag{24}$$

where:

f_c is the far field S-wave corner frequency,
r is the radius of the "equivalent" circular source,
V_S is the shear wave velocity,
K is a scalar the value of which is quite model dependant; for a first order estimate of the source size, Brune et al. (1979) suggest a value of 0.3.

Pearson (1981) analyzed microseismic signals generated by a massive hydraulic stimulation as recorded by a three-component geophone package located some 300 m from the injection point. He observed signals with corner frequencies varying between 300 and 1400 Hz (corresponding to source diameters of the order of 3 m, if K is taken as equal to 0.37 in Eq. 24) and seismic moments ranging from 6 to 15 10^{13} dyn/cm (yielding a mean stress drop ranging from 0.5 to 10 bar according to Brune's model (1970) in which the mean stress drop τ is equal to $7M_0/16r^3$, where M_0 is the seismic moment). However, he observed that the corner frequencies seemed to increase as the locus of events migrated away from the injection point, toward the recording station and he attributed this observation to local variations in the natural stress regime.

Talebi and Cornet (1987) conducted a similar investigation but with a network of 15 three-component geophone packages set at various distances and in various azimuths from the source. They observed very clear corner frequencies in the 200–800 Hz range but noticed that, for the same event, the corner frequencies observed at various stations decreased as the distance from source to station increased. They concluded that these corner frequencies were only reflecting attenuation effects linked to propagation in the rock mass and could not be associated in a direct manner to source characteristics. Accordingly, given the observed corner frequencies, the mean dimension of these shear sources are very likely smaller than 1 m.

As mentioned, when enough recording stations are available, focal mechanisms of sources can be determined. When the source is a pure shear event, the focal mechanism yields two nodal planes, one of which is the actual fault plane. When many focal mechanisms are available, they may be used to advantage to determine the principal directions of the regional stress field (see e.g., Vasseur et al. 1983; Gephart and Forsyth 1984; Michael 1984; Julien and Cornet 1987). This requires that:

- the stress field is uniform throughout the volume under consideration;
- the focal mechanisms are independant from each other;

- the slip vector within the fault plane exhibits the same direction as the resolved shear stress supported by this plane.

During a 2500 m^3 water injection, Cornet and Julien (1989) observed about 70 micro-seismic events, 40 of which yielded well defined focal mechanisms. Considering only events close enough to one another so that they could be assumed to have occurred in the same stress field, these authors estimated the principal directions of the regional stress field. They compared these results with those obtained with the hydraulic tests on preexisting factures method (HTPF method described by Cornet and Valette 1984). In this case, one principal stress was assumed to be vertical and was found to be the maximum stress, for the depth range under consideration. Whilst the focal mechanisms inversion confirmed that the maximum principal stress was not far from being vertical (108° with the horizontal direction), the minimum sub-horizontal principal stress direction determined from the inversion of focal mechanisms made a 50° angle with that derived from the hydraulic tests data. Since the latter is consistant with other local stress determinations as well as with the main flow direction as determined from events location, it is concluded that the water injection has perturbed, at least temporarily, the regional stress field. Accordingly, this investigation suggests that, focal mechanisms of induced seismicity may not be representative of the regional tectonic stress field, because of the possibility of non isotropic stress field variations induced by the fluid percolation process.

5.1.2 Other Source Mechanisms

Events not associated with shear mechanisms have also been described (Bame and Fehler 1986; Baria et al. 1989). They correspond generally to so called long period events which are caused by a mechanical interaction between the fluid and the surrounding rock leading to trapped waves (Chouet and Julian 1985; Chouet 1986; Ferrazini and Aki 1987). For these long period events to occur, it is necessary that, firstly, some "open" fracture exists and, secondly, that a vibrating source excites the fracture.

Bame and Fehler (1986) observed such long period events during massive hydraulic stimulations of hot dry rock geothermal wells at the Fenton Hill site, in New Mexico. These events exhibit two dominant frequencies, the first in the 110–130 Hz range, the second between 700 and 800 Hz. They conclude that these frequencies are consistant with open fracture sizes of the order of respectively, 20 and 3 m. Since these events are observed from the beginning of pumping, these authors conclude that they are caused by the opening of tensile fractures in the rock.

It could be argued, however, that since there is no significant variation of dominant frequency as pumping proceeds, flow is not generating a true hydraulic fracture, since its size would grow with injected volume, but rather opens chanels in a major preexisting fracture, or fault zone, as suggested by Fehler et al. (1987).

It must be mentioned that other types of signals associated with deep fluid injections have been reported in the literature (Baria et al. 1988). In particular, high frequency signals (with frequencies in excess of 3000 Hz) have been described, but no specific discussion of source mechanism is given.

Thus, it appears that no event associated with a pure mode I fracture process has been reported unambiguously in the literature. The high frequency signals described by Baria et al. may well be caused by such sources but not enough data are available to ascertain this hypothesis. This lack of mode I induced signal could be taken as a strong support of the hypothesis that propagation of true hydraulic fractures is a stable quasi-static process. But it could be argued, as well, that for most of these documented examples, flow occurred along preexisting faults or fractures so that no real information is available concerning the propagation of true hydraulic fractures. Some answer to this question is given by the results observed during a massive hydraulic injection conducted in granite with high flow rate (over 100 l/s) and a very viscous fluid [reticulated gel with apparent viscosity (non-Newtonian fluid) larger than 300 cp] (Baria et al. 1989). This injection was shown later on (by direct observation of the borehole wall) to have developed a true hydraulic fracture in the direction of the maximum principal stress, yet very few microseismic signals were observed as compared to the microseismic activity recorded during other injections with lower flow rates and lower fluid viscosity. It will be concluded therefore that hydraulic fracture propagation is globally quasi-static, but that local instabilities may occur generating very high frequency signals which cannot be observed at too long distances from the fracture front because of attenuation effects.

5.2 Ground Surface Deformation

The opening of fractures in a rock mass induces displacements in the surrounding rock. Attempts have been made to map shallow man-made hydraulic fractures by interpreting measurements of ground deformation obtained during the hydraulic fracture development (Evans et al. 1980; Julien et al. 1984).

Various closed-form elastic solutions for the displacement field associated with the opening of fractures are available in the literature. For a uniform pressure distribution in the fracture, Sneddon (1946) discusses the case of a penny-shaped crack in an infinite body; Sun (1969) gives the solution for a penny-shaped fracture parallel to the free surface of an infinite half space; Pollard and Holzhausen (1979) give that for a slit, inclined with respect to this free surface but infinite in a direction parallel to it (plane-strain conditions) whilst Davis (1983) derives the solution for a rectangular fracture inclined with respect to the free surface. For the sake of mapping hydraulic fractures, these closed form solutions can be used to identify the fracture shape which best fits observations by applying inverse problem theory.

For example, Evans et al. (1980), interpreted tilt data recorded at five different stations during a $2\frac{1}{2}$ hour injection of 440 m^3 of nitrogene at a depth of 335 m in a Devonian sub-horizontal shale sequence. Results are consistant with an initial vertical fracture propagating upward for 183 m, nearly symmetrically with respect to the injection well in the horizontal direction. Although the further extension of the fracture is less well constrained by the data, it is consistant with the opening of a sub-horizontal fracture at about 153 m from the ground surface, i.e., at the top of the initial vertical extension.

For an injection of 800 m^3 of water (average injection rate of 1.4 m^3/h), at a depth of 224 m in a sub-horizontal oil shale formation, Julien et al. (1984) analyzed tilt measurements obtained at 16 stations distributed in various azimuths, at distances ranging from 100 m to 200 m from the injection well. Each station consisted of two inclinometers measuring radial and tangential inclinations (as defined with respect to the well head) with a sensitivity of the order of 10^{-2} µrad. These 32 tilt measurements were interpreted using Sun's solution of a penny-shaped horizontal crack. The vertical displacement w observed at ground surface (see Fig. 3), at distance r from the surface projection of the fracture centre, for a horizontal fracture of volume V, at depth h, with radius a, is given by (Sun 1969):

$$w = \left(\frac{3V}{2\pi a^2}\right)\left[\sqrt{k}\cdot \sin(\theta/2) - \frac{h}{a\sqrt{k}}\cdot \cos(\theta/2)\right], \qquad (25)$$

in which:

$k = [X^2 + (2h/a)^2]^{1/2}$;
$X = r^2/a^2 + h^2/a^2 - 1$;
$tg\theta = 2h/Xa$.

Equation (25) is accurate to within 2% if $h \geqslant (2a/5)$; it becomes nearly exact when $h \geqslant 5a$.

Interestingly, Eq. (25) is independant of the pressure in the fracture and of the elastic moduli of the rock mass. Assuming that there was no fluid loss, as suggested by the negligible pressure decay observed at the end of the test, Julien et al. used this formulation to interpret the tilt results, which reached 28 µrad at some stations at the end of injection. They found that results were consistant with a horizontal 150 m circular fracture centred on injection well and a thickness at the crack centre equal to 17 mm. This yields, for this fracture, a thickness to length ratio equal to about $0.5 \cdot 10^{-4}$. This value is quite significantly lower than that proposed by Pollard (1987) for dyke intrusions (recall the 10^{-3} value mentioned in Sect. 3.2), a result which is consistant with lower driving pressures (of the order of 1 MPa).

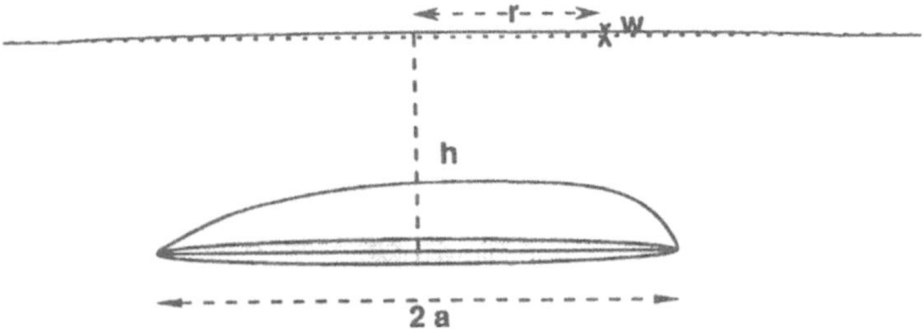

Fig. 3. Ground surface deformation caused by a horizontal penny-shaped hydraulic fracture of radius a, at depth h; w is the vertical displacement at ground surface at distance r from the projection of the fracture center

Using similar equipment to that of the experiment of Julien et al., tilts at ground level were monitored during various fluid injections, conducted through a straddle packer, in a granitic rock mass (Desroches 1988). For the first injection, 100 m^3 of water were injected at a flow rate of 1.3 m^3/min, at a depth of 440 m. Six subsequent injections were conducted between 645 m to 730 m with a viscous gel (viscosity ranging from 70 cp to 300 cp depending on the injection), at flow rates varying between 1 m^3/min for the low viscosity gel to 4 m^3/min for the high viscosity gel. The total injected volume varied between 125 m^3 to 320 m^3. For all these injections, no surface tilt was observed, i.e., if tilts occurred they never reached values larger than 0.5 μrad. However according to the classical hydraulic fracturing theory tilts of the order of a few microradians should have been observed. This implies that these injections did not open as large a fracture as expected and that most of the flow occurred in closed preexisting fractures. This conclusion is somewhat similar to that reached by Fehler et al. (1987) from their analysis of the micro-seismic activity observed during the Fenton Hill experiments, as reported in the previous section. Accordingly for crystalline rock masses, it seems reasonable to consider that large flow can occur through channels in preexisting faults, or fractures, without actually completely opening the fracture.

6 Conclusion

Forced fluid percolation in a rock mass can induce either hydraulic fractures, i.e., fractures for which both faces are completely separated by the fluid and which therefore transmit no shear stresses, or shear fractures either through the rock matrix or, more generally, by reactivation of preexisting fractures or faults. These shear fractures exhibit open channels and solid asperities; they transmit shear stresses but may also accomodate large fluid flows.

Inception and propagation of hydraulic fractures can be analyzed according to fracture mechanics principles. In a homogeneous, isotropic rock mass, a hydraulic fracture develops normal to the minimum principal stress direction; it can be at an angle to it when the surface energy of the material is anisotropic or/and heterogeneous. Its geometry is controlled essentially by regional stress variations and the rock mass stiffness, but also by variations in surface energy. From the analysis of induced seismicity, the fracturing process appears to be close to being quasi-static because of the development of a strong pressure gradient near the fracture tip. However, this pressure gradient is not well documented; it is conceivable, from both laboratory observations and high frequency signals observed in some specific instances during field experiments, that local instable fracture extensions do exist. Because of their short lengths, these instabilities generate only high frequency signals which attenuate very rapidly and can be observed only in the immediate vicinity of the fracture tip. They might be the source of the long period events which correspond to trapped waves in open fluid-filled fractures.

Development of shear movements is analyzed generally according to maximum effective stress principles rather than fracture mechanics theory. They are unstable and the source of most of the induced seismic activity observed during fluid injections. They can be generated either by slip along preexisting fractures intersecting the main flow path or by shearing of asperities distributed along the fault in which flow is occuring. In the latter case, surface tilt observations and analysis of micro-seismicity suggest that flow occurs along tortuous channels within the fracture or the fault, without significant normal displacement of the fracture walls, so that no detectable displacement is induced in the rock mass away from the fault zone. However, some small normal displacement may be induced by the shearing of the fault because of the roughness of the surfaces. Contrary to what is generally believed, injection of large amount of fluid in the fracture network of a rock mass does not seem to modify the regional stress field according to the classical effective stress concept: inversion of focal mechanisms of induced shear events suggests that both the magnitude and the direction of the principal stresses are affected by the fluid percolation process.

Factors controlling the type of rupture process, i.e., hydraulic fracturing or shearing, depend on the fluid pressure, the flow rate and the fluid viscosity as well as on the regional stress field and the rock mass characteristics. Both rupture processes may be observed simultaneously. The lower the flow rate and the fluid viscosity, the more pronounced the effects of heterogeneities and the more chances for shearing. The larger the flow rate and the fluid viscosity, the more chances for hydraulic fracturing. Similarly, the more isotropic the stress field, the more pronounced the effects of the rock mass characteristics whilst the more anisotropic the stress field, the more likely the development of true hydraulic fractures.

Acknowledgements. I wish to thank D. Pollard, L. Vernik, G. Shamir and J. Baumgartner for fruitful discussions on this paper.

References

Abe H, Mura T, Keer LM (1976) Growth rate of a penny-shaped crack in hydraulic fracturing of rocks. J Geophys Res 81, 29:5335–5340

Abou-Sayed AS (1982) The influence of subsurface conditions on hydraulically induced fractures in deep rocks. In: Nemat-Nasser S, Abe H, Hirakawa S (eds) Hydraulic fracturing and geothermal energy. Nijhoff, The Hague, pp 125–144

Aki K, Fehler M, Das S (1977) Source mechanism of volcanic tremor; fluid driven crack models and their application to the 1963 Kilauea eruption. J Volcanol Geotherm Res 2:259–287

Aki K, Fehler M, Aamodt RL et al. (1982) Interpretation of seismic data from hydraulic fracturing experiments at the Fenton Hill, New Mexico, hot dry rock geothermal site. J Geophys Res 87, nb B2:936–944

Albright JN, Hanold RJ (1976) Seismic mapping in basement rocks. 2nd Annu Symp enhanced oil and gaz recovery, Enery Res Dev Ass, Tulsa, Oklahoma

ASTM (1978) 1978 Annual book of standards. Rep ANSI/ASTM 399-18a, Am Soc Testing Materials, Philadelphia, USA

Atkinson BK (1987) Fracture mechanics of rock. Academic Press, Lond, New York, 528 pp

Atkinson BK, Meredith PG (1987) Experimental fracture mechanics data for rocks and minerals. In: Atkinson BK (ed) Fracture mechanics of rock. Academic Preee, Lond, New York pp 477–525

Bame DA, Fehler MC (1986) Observations of long period earthquakes accompanying hydraulic fracturing. Geophys Res Lett 13:149–152

Baria R, Green ASP, Jones RH (1989) Anomalous seismic events observed at the CSM HDR project. Int J Rock Mech Min Sci Geomech Abs 26, nb 3/4:257–270

Barton N (1983) Hydraulic fracturing to estimate stress and rock mass stability at a pumped hydro-project. In: Haimson BC, Zoback MD (eds) Hydraulic fracturing stress measurements. National Academy Press, Washington, DC, pp 61–67

Barton N, Bandis S, Bakhtar K (1985) Strength, deformation and conductivity coupling of rock joints. Int J Rock Mech Min Sci Geomech Abs 22, nb 3:121–140

Bieniawski ZT (1974) Estimating the strength of rock materials. J S Afr Inst Min Metall 74:312–320

Brace WF, Bombolakis EG (1963) A note on brittle crack growth in compression. J Geophys Res 68, nb 12:3709–3713

Brace WF, Martin RJ (1968) A test of effective stress law for crystalline rocks of low porosity. Int J Rock Mech Min Sci 5:415

Brace WF, Paulding BW, Scholtz C (1966) Dilatancy in the fracture of crystalline rocks. J Geophys Res 71, nb 16:3939–3953

Brune J (1970) Tectonic stress and the spectra of seismic shear waves from earthquakes. J Geophys Res 75:4997–5009

Brune JN, Archuletta RJ, Hartzell S (1979) Far field S-wave spectra, corner frequencies and pulse shapes. J Geophys Res 84, nb B5:2262–2272

Byerlee J (1978) Friction of rocks. Pure App Geophys 116:615–626

Cash D, Homuth EF, Keppler H, Pearson C, Sasaki S (1983) Fault plane solutions for microearthquakes induced at the Fenton Hill hot dry rock geothermal site: Implications for the state of stress near a Quaternary volcanic center. Geophys Res Let 10, nb 12:1141–1144

Chouet B (1986) Dynamics of a fluid driven crack in three dimensions by the finite difference method. J Geophys Res 91:13 967–13 992

Chouet B, Julian BR (1985) Dynamics of an expanding fluid filled crack. J Geophys Res 90:11 184–11 198

Cleary MP (1983) Theory and application of hydraulic fracturing technology. In: Nemat-Nasser S, Abe H, Hirakawa S (eds) Hydraulic fracturing and geothermal energy. Nijhoff, The Hague, pp 267–290

Comninou M, Dundurs J (1979) A closed crack tip terminating at an interface. J App Mech 46, nb 1:97

Cornet FH (1979) Comparative analysis by the displacement discontinuity method of two energy criteria of fracture. J App Mech 46, nb 2:349–355

Cornet FH (1982) Interpretation of hydraulic injection tests for in-situ stress determination. In: Haimson BC, Zoback MD (eds) Proc Worksh Hydraulic fracturing stress measurements. US Geol Surv Open File Rep 82-1075:414–443

Cornet FH (1988) Two examples of stress measurements by the HTPF method. In: Cundall PA, Sterling RL, Starfield AM (eds) Proc 29th US Symp Rock Mech. Balkema, Rotterdam, pp 615–624

Cornet FH, Fairhurst C (1974) Influence of pore pressure on the deformation behaviour of saturated rocks. 3rd Int Cong Rock Mech. Denver, Int Soc Rock Mech, II

Cornet FH, Julien P (1989) Stress determination from hydraulic test data and from mechanisms of induced seismicity. Int J Rock Mech Min Sci Geomech Abs 26, nb 3/4:235–248

Cornet FH, Valette B (1984) In situ stress determination from hydraulic injection test data. J Geophys Res 89, nb B13:11 527–11 537

Crouch SL (1970) Experimental determination of volumetric strains in failed rock. Int J Rock Mech Min Sci 7:589–603

Crouch SL (1976) Solution of plane elasticity problems by the displacement discontinuity method. Int J Num Meth Eng 10:301–343

Daneshy AA (1973) A study of inclined hydraulic fractures. Soc Petrol Eng J 13:61–68

Daneshy AA (1983) Propagation of hydraulic fracture and its conductivity in layered media. In: Nemat-Nasser S, Abe H, Hirakawa S (eds) Hydraulic fracturing and geothermal energy. Nijhoff, The Hague, pp 159–174

Delaney PT, Pollard DD (1982) Solidification of basaltic magma during flow in a dike. Am J Sci 282:856–885

Davis PM (1983) Surface deformation associated with a dipping hydrofracture. J Geophys Res 88, nb B7:5826–5834

Desroches J (1988) Rapport d'avancement sur l'inclinomettrie. In: Cornet FH (ed) Projet Mayet de Montagne – Etude in-situ de la percolation forcee d'eau en milieu granitique. Fin Rep Eur Econ Com, DG XII, contract nb EN 3G-0051-F
Evans K, Holzhausen GR, Wood MD (1980) The mapping of nitrogene gas induced hydraulic fractures in Devonian shale by observation of the associated surface deformation. Soc Petrol Eng Pap nb SPE/DOE 8933
Fairhurst C, Cornet FH (1981) Rock fracture and fragmentation. 22nd US Symp Rock Mech. Mas Inst Tech, pp 21–46
Fehler M, House L, Kaieda H (1987) Determining planes along which earthquakes occur: Method and application to earthquakes accompanying hydraulic fracturing. J Geophys Res 92:9407–9414
Ferrazini V, Aki K (1987) Slow waves trapped in a fluid-filled infinite crack: implication for volcanic tremor. J Geophys Res 92, nb B9:9215–9223
Geertsma J, de Klerk F (1969) A rapid method of predicting width and extent of hydraulically induced fractures. J Petrol Tech December: 107
Gephart JW, Forsyth DW (1984) An improved method for determining the regional stress tensor using earthquake focal mechanism data: application to the San Fernando earthquake sequence. J Geophys Res 89, nb B11:9305
Goodman RE, Ohnishi Y (1973) Undrained shear testing of jointed rock. Rock Mech 5:129–149
Griffith AA (1921) The phenomenon of rupture and flaw in solids. Phil Trans R Soc Lond Ser A, A221:163–198
Haimson BC (1968) Hydraulic fracturing in porous and nonporous rock and its potential for determining in situ stresses at great depths. Tech Rep 4-68 US Corps Eng, Omaha, Neb, 235 pp
Haimson BC, Rummel F (1982) Hydrofracturing stress measurements in the Iceland research drilling project drill hole at Reydarfjordur, Iceland. J Geophys Res 87, nb B8:6631–6649
Handin J, Hager RV Jr, Friedman M, Feather JN (1963) Experimental deformation of sedimentary rocks under confining pressure: pore pressure tests. Bull Am Assoc Petrol Geol 47:717–755
Hayashi K, Abe H (1982) Opening of a fault and resulting slip due to injection of fluid for the extraction of geothermal heat. J Geophys Res 87, nb B2:1049–1054
Henry JP, Paquet J, Tancrez JP (1977) Experimental study of crack propagation in calcite rock. Int J Rock Mech Min Sci Geomech Abs 14, nb 2:85–91
Hoagland RG, Hahn GT, Rosenfield K (1973) Influence of microstructure on fracture propagation in rock. Rock Mech 5:77–106
Hoek E, Bieniawski ZT (1965) Brittle fracture propagation in rock under compression. Int J Rock Mech Min Sci 1:137
Hoek E, Brown ET (1980) Underground excavations in rock. Instn Min Metall, London
Ingraffea AR (1977) Discrete fracture propagation in rock, laboratory tests and finite element analysis. PhD Thesis, Univ Colorado, Boulder
Irwin GR (1957) Analysis of stresses and strains near the end of a crack traversing a plate. J App Mech 24, nb 3:361–364
Jackson MD, Pollard DD (1988) The laccolith-stock controversy: new results from the southern Henry Mountains, Utah. Geol Soc Am Bull 100:117–139
Julien Ph (1987) De la contribution des mecanismes au foyer dans le calcul du champ de contrainte – application au cas de la sismicite induite. These de doctorat Univ Pierre et Marie Curie, Paris 6
Julien Ph, Saleh B, Blum PA, Despax D (1984) Inversion de donnees inclinometriques obtenues au cours de deux experiences de fracturation hydraulique. Rev Fr Geotech nb 31:53–63
Julien P, Cornet FH (1987) Stress determination from after-shocks of the Campania-Lucania earthquake of November 23, 1980. Ann Geophys 5, nb 3:289–300
Jung R (1987) Propagation and hydraulic testing of a large unpropped hydraulic fracture in granite. In: Kappelmeyer O, Rummel F (eds) Terrestrial Heat from Impervious Rocks-Investigations in the Falkenberg Granite Massif. Geol Jahrb E 39, Bundesanst Geowiss Rohstoffe, Hannover, FRG

Kanninen MF, Popelar CH (1985) Advanced fracture mechanics. Oxford Univ Press, Oxford 563 pp
Lockner D, Byerlee JD (1977) Hydrofracture in Weber sandstone at high confining pressure and differental stress. J Geophys Res 82, nb 14:2018–2026
Madariaga R (1979) On the relation between seismic moment and stress drop in the presence of stress and strength heterogeneity. J Geophys Res 84, nb B5:2243–2250
Majer EL (1989) The application of high frequency seismic monitoring methods for the mapping of grout injections. Int J Rock Mech Min Sci Geomech Abs 26, nb 3/4:249–256
Michael AJ (1984) Determination of stress from slip data: faults and folds. J Geophys Res 89, nb B13:11517
Nolte KG (1988) Principles for fracture design based on pressure analysis. Soc Petrol Eng Prod Eng feb: 22–33
Nordgren RP (1972) Propagation of a vertical hydraulic fracture. Soc Petrol Eng J August: 306–314, Trans AIME 253
Ouchterlony F (1980) Review of fracture toughness testing of rock. Rep DS 1980-15, Swedish Detonic Res Found, Sweden
Pearson C (1981) The relationship between microseismicity and high pore pressures during hydraulic stimulation experiments in low permeability granitic rocks. J Geophys Res 86, nb B9:7855–7864
Perkins TK, Kern LR (1961) Widths of hydraulic fractures. J Petrol Tech september: 937–949, Trans AIME 222
Pine RJ, Batchelor AS (1984) Downward migration of shearing in jointed rock during hydraulic injections. Int J Rock Mech Min Sci Geomech Abs 21, nb 5:249–263
Pollard DD (1987) Elementary fracture mechanics applied to the structural interpretation of dykes. In: Halls HC, Fahrig WF (eds) Mafic dyke swarms. Geol Assoc Canada Spec Pap 34:5–24
Pollard DD, Holzhausen G (1979) On the mechanical interaction between a fluid-filled fracture and the earth's surface. Tectonophysics 53:27–57
Pollard DD, Johnson AM (1973) Mechanics of growth of some lacolithic intrusion in the Henry Mountains, Utah. Tectonophysics 18:311–354
Pollard DD, Muller OH (1976) The effect of gradients in regional stress and magma pressure on the form of sheet intrusions in cross section. J Geophys Res 81, nb 5:975–984
Raleigh CB, Healy JH, Bredehoeft JD (1972) Faulting and crustal stress at Rangely, Colorado. In: Heard HC et al. (eds) Flow and fracture of rocks. Geophys Monogr Ser 16, Am Geophys Un, Washington, DC
Roberts G, Crampin S (1986) Shear wave polarizations in a hot dry rock geothermal reservoir: anisotropic effects of fractures. Int J Rock Mech Min Sci Geomech Abs 43, nb 4:291–302
Rubin AM, Pollard DD (1987) Origins of blade-like dikes in volcanic rift zones. In: Volcanism in Hawaii. US Geol Surv Prof Pap nb 1350, pp 1449–1470
Rubin AM, Pollard DD (1988) Dike-induced faulting in rift zones of Iceland and Afar. Geology 16:413–417
Rummel F (1987) Fracture mechanics approach to hydraulic fracturing stress measurements. In: Atkinson BK (ed) Fracture Mechanics of Rocks. Academic Press, Lond, New York
Rummel F, Kappelmeyer O (1983) The Falkenberg geothermal frac-project: concept and experimental results. In: Nemat-Nasser S, Abe H, Hirakawa S (eds) Hydraulic fracturing and geothermal energy. Nijhoff, The Hague, pp 59–74
Sanders JL (1960) On the Griffith – Irwin fracture theory. J Appl Mech 27, Ser E, nb 2:352–353
Schmidt RA, Huddle CW (1977) Effect of confining pressure on fracture toughness of Indiana Limestone. Int J Rock Mech Min Sci Geomech Abs 14:289–293
Sih GC (1973) Methods of analysis. In: Solutions of crack problems. Noordhof, Leyden, The Netherlands
Sih GC (1974) Strain energy density factor applied to mixed mode crack problems. Int J Fracture Mech 10:305–321
Sih GC, Liebowitz H (1968) Mathematical theories of brittle fracture. In: Liebowitz H (ed) Fracture – an advanced treatise, vol II. Academic Press, Lond New York, pp 67–190

Simonson ER, Abou-Sayed AS, Clifton RJ (1976) Containment of massive hydraulic fractures. SPE Pap nb 6089, 51st Annu Fall Meet Soc Petrol Eng AIME, New Orleans
Sneddon IN (1946) The distribution of stress in the neighbourhood of a crack in an elastic solid. Proc R Soc Lond Ser A, A187, pp 229–260
Sun RJ (1969) Theoretical size of hydraulically induced horizontal fractures and corresponding surface uplift in an idealized medium. J Geophys Res 74, nb 25:5995–6011
Swedlow JL (1974) Criteria for growth of the angled crack; cracks and fracture. ASTM STP 601:506–521
Talebi S, Cornet FH (1987) Analysis of the microseismicity induced by a fluid injection in a granitic rock mass. Geophys Res Let 14, nb 3:227–230
Tapponier P, Brace WF (1976) Development of stress induced microcracks in westerly granite. Int J Rock Mech Min Sci 23:103–112
Tarantola A, Ruegg JC, Lepine JC (1979) Geodetic evidence for rifting in Afar-vertical displacements. Earth Planet Sci Lett 48:363–370
Teufel LW, Warpinski NR (1983) An assessment of factors affecting hyraulic fracture containment in layered rock: observation from a mineback experiment. In: Nemat-Nasser S, Abe H, Hirakawa S (eds) Hydraulic fracturing and geothermal energy. Nijhoff, The Hague, pp 251–266
Thiercelin MJ, Lemanczyk ZR (1983) The effect of stress gradient on the height of vertical hydraulic fractures. SPE ?DOE Pap nb 11626 Proc Symp Low permeability, Denver, Marsh 14–16 Soc Petrol Eng Am Inst Min Eng
Vasseur G, Etchecopar A, Philip H (1983) Stress state inferred from multiple focal mechanisms. Ann Geophys 1, nb 4–5:291–298
Von Schoenfeldt H (1970) An experiment of open-hole hydraulic fracturing as a stress measurement method with particular emphasis on field trials. MRD-3-70, Crops of Engineers, Missouri River Division
Warren WE, Smith CW (1985) In situ stress estimates from hydraulic fracturing and direct observation of crack orientation. J Geophys Res 90, nb B8:6829–6839
Weertman J (1980) The stopping of a rising, liquid filled crack in the earth's crust by a freely slipping horizontal joint. J Geophys Res 85, nb B2:967
Zoback MD, Pollard DD (1978) Hydraulic fracture propagation and the interpretation of pressure time records for in-situ stress determination. 19th US Rock Mech Symp, Mackay School of Mines, Reno, Nevada
Zoback MD, Rummel F, Jung R, Raleigh CB (1977) Laboratory hydraulic experiment in intact and prefractured rock. Int J Rock Mech Min Sci 14:49–58

Temperatures at the Base of the Seismogenic Crust Beneath Long Valley Caldera, California, and the Phlegrean Fields Caldera, Italy

David P. Hill

Abstract

In seismically active volumes of the crust, the base of the seismogenic zone commonly corresponds with the temperature at which the first mineral species in the crustal rock enters the quasi-plastic domain. This sub-solidus temperature marks the onset of the transition from brittle (seismogenic) failure to plastic (aseismic) flow for a given rock type, and it varies with both strain rate and water content. Earthquake activity and deformation accompanying recent unrest in Long Valley caldera, California, and the Phlegrean Fields caldera, Italy, provide an opportunity to compare predictions of laboratory-based constitutive relations for the brittle-plastic transition in regions of differing rock type (quartzo-feldspathic in Long Valley and feldspathic in the Phlegrean Fields), for which strain rates (3×10^{-14} and $5 \times 10^{-12}\,s^{-1}$ for Long Valley and the Phlegrean Fields, respectively) and earthquake focal depths are well determined. Magma underlies sections of both calderas at depths somewhere between 4 to 10 km. Results indicate temperatures in the range 250 to 350 °C for the base of the seismogenic zone in the granitic crust beneath Long Valley caldera, the depths to which vary from less than 5 km beneath the resurgent dome and Mammoth Mountain to 8 or 9 km beneath the south moat. Temperatures could approach 500 °C at depths of 5 km beneath the resurgent dome, however, if strain rates increase significantly with depth toward the underlying inflation center and rocks in the plastic domain are dry. The higher temperatures (600 to 700 °C) predicted for the base of the seismogenic zone at depths of 4 to 5 km beneath the Phlegrean Fields caldera reflect a higher strain rate and a feldspathic (trachyte) composition of the crustal rocks. These temperatures are consistent with a linear extrapolation of geothermal gradients measured in adjacent, 3-km-deep wells. If, however, strain rates increase significantly with depth toward the inflation center at a depth between 3 and 5 km and rocks in the plastic domain are dry, the temperature at the base of the seismogenic zone could approach 800 °C beneath the central section of the Phlegrean Fields caldera.

1 Introduction

Knowledge of the volume and distribution of magma within the crust beneath a volcanic system is critical not only for understanding magmatic processes specific to the system, but also for making realistic assessment of the hazard posed by possible future eruptions from the magmatic system. Because earthquakes represent shear failure in brittle rocks, their distribution contains useful information on temperatures and the distribution of magma in the crust. At one extreme, for example, large, long-lived magma chambers form aseismic volumes because melts have vanishingly small shear strength, and at the other extreme, dikes intruded into cool, brittle crust generate intense earthquake swarm activity local to the intrusion (Hill 1977). A growing body of information on rock mechanics from laboratory experiments provides an ever clearer picture of the temperature- and rate-dependent transition in the mode of rock deformation from brittle (seismogenic) failure to plastic (aseismic) flow at subsolidus temperatures (Kirby and Kronenberg 1987). Using these results from laboratory measurements, we can begin to make more specific statements about temperatures and magma distribution in the crust based on the distribution of well-located earthquake hypocenters.

Here I examine constraints on subsurface temperatures and depths to magma implied by the three-dimensional distribution of earthquake hypocenters beneath two large, well-studied volcanic systems that have shown pronounced seismic unrest in recent years: Long Valley caldera in eastern California and the Phlegrean Field caldera near Naples in the Regione Campania of southern Italy. My approach follows closely the methods described by Sibson (1982) and Meisner and Strelau (1982) in their analysis of the relation between maximum focal depths and temperatures beneath active fault systems. After first reviewing the theoretical and experimental basis for the brittle-plastic transition, I describe the two calderas and their seismicity and conclude with a discussion of the implications for temperatures and magma distribution beneath the two calderas.

2 The Brittle-Plastic Transition

The theoretical and experimental basis for the strength of crustal rocks and the transition in failure mode from brittle fracture to plastic flow are clearly described in a number of papers (see, for example, Brace and Kohlstedt 1980; Meissner and Strelau 1982; Sibson 1982; Scholz 1988). Figure 1 summarizes the essence of these results in terms of the three-layer strength profile for the crust proposed by Scholz (1988). Here, strength carries the usual meaning of the maximum shear stress, τ, a rock volume can sustain without suffering permanent deformation; $\tau = 0.5(S_1 - S_3)$, where S_1 and S_3 are the greatest and least principal effective stresses, respectively, and compression is taken to be positive. Effective stress, S, is the rock matrix stress, σ, reduced by the pressure, p, in a fluid filling interconnected pore spaces within the rock matrix, or $S = \sigma - p$.

2.1 Brittle (Seismogenic) Layer

In the upper, brittle layer, shear strength is limited by frictional resistance to slip along preexisting fractures (joints or faults) in the crust. The static frictional strength for crustal rocks at depths of interest here (less than about 15 km and confining pressures less than about 500 MPa) is, to a good approximation, described by:

$$\tau = 0.75\,\mathrm{Sn}, \tag{1}$$

where Sn is the effective compressional stress normal to the local fracture surface, and 0.75 is the average coefficient of friction, μ, for confining pressures

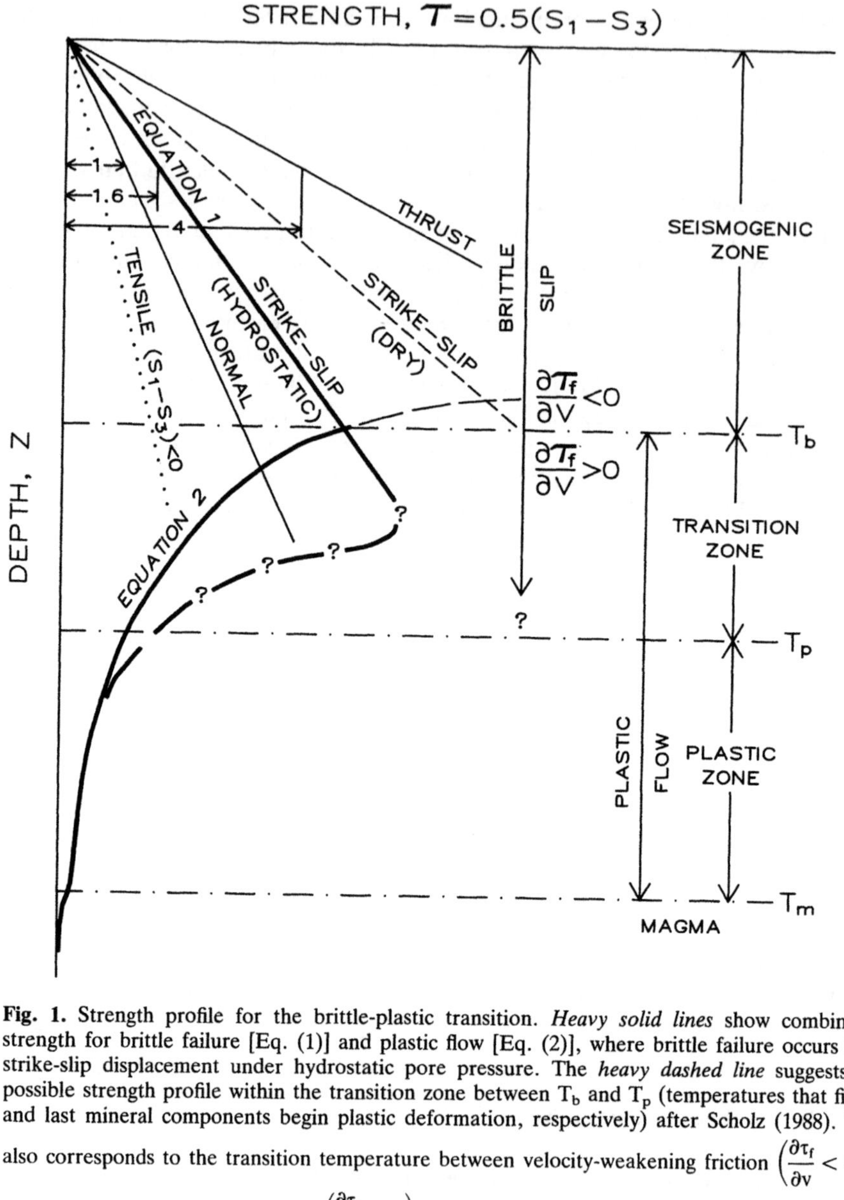

Fig. 1. Strength profile for the brittle-plastic transition. *Heavy solid lines* show combined strength for brittle failure [Eq. (1)] and plastic flow [Eq. (2)], where brittle failure occurs by strike-slip displacement under hydrostatic pore pressure. The *heavy dashed line* suggests a possible strength profile within the transition zone between T_b and T_p (temperatures that first and last mineral components begin plastic deformation, respectively) after Scholz (1988). T_b also corresponds to the transition temperature between velocity-weakening friction $\left(\frac{\partial \tau_f}{\partial v} < 0\right)$ and velocity-strengthening friction $\left(\frac{\partial \tau_f}{\partial v} > 0\right)$. T_m is the magma temperature. *Light solid lines* indicate brittle strength for normal and thrust stress regimes with hydrostatic pore pressures. The *light dashed line* indicates brittle strength for strike-slip faulting under dry conditions, and the *dotted line* indicates tensile strength $[(S_1 - S_3) < 0]$

less than 500 MPa (Byrlee 1978; Sibson 1983). Thus strength in the brittle layer is pressure-sensitive and increases as the effective lithostatic pressure increases with depth. Most commonly, water is the pore fluid and pore pressure follows the hydrostatic gradient. In this case, both Sn and the frictional strength, τ, will increase linearly with depth at a rate that depends on the ambient stress regime and faulting mode. The curves in Fig. 1 illustrate the case for classical Andersonian faulting regimes and hydrostatic pore pressure. Under this idealized case, the frictional strength for a vertical strike-slip fault increases with depth as $\tau \approx 10\,z$ (z is depth in km and τ is in MPa), and the respective strengths for normal, strike-slip, and thrust faulting at a given depth have the ratios 1:1.6:4 (see Sibson 1982). In reality, strike-slip faulting will occur in a pie-shaped zone on either side of the ideal Andersonian case and will grade into transition zone of mixed mode faulting toward the idealized "normal" and "thrust" cases in Fig. 1. Under some cirumstances, tensile failure may be seismogenic (Foulger this Vol.), and as indicated in Fig. 1, rocks are much weaker under tensional stress $[(S_1 - S_3) < 0]$ than under shear stress $[(S_1 - S_3) > 0]$.

2.2 Plastic (Aseismic) Layer

In the lower, plastic layer, strength is temperature-sensitive and follows a flow law of the form

$$\tau = \left[(\dot{e}/A)\exp\left(\frac{H^*}{RT}\right)\right]^{-n}, \tag{2}$$

where:

$\dot{e}$ = strain rate;
A, n = constants;
H^* = activation energy;
R = universal gas constant;
T = absolute temperature.

Table 1 lists the experimentally determined parameters in this flow law for several crustal rock types [see Kirby and Kronenberg (1987) for a more extensive compilation]. For a given rock type (fixed H^*, A, and n), strength in the plastic layer is dominated by the exponential dependence on temperature, T. The power-law dependence on strain rate ($\dot{e}^{-n}$) is weak by comparison and becomes significant only for order of magnitude variations in strain rate. Applying this flow law to strain rates typical of magmatic and tectonic processes, which range from 10^{-11} to $10^{-16}\,s^{-1}$, requires extrapolation over many orders of magnitude from the 10^{-4} to $10^{-6}\,s^{-1}$ rates typical of laboratory experiments. Mindful of the uncertainties associated with such an extrapolation, one can use this flow law to estimate the variations in plastic strength with depth in the crust given independent information on temperature as a function of depth, dominant crustal rock types, and ambient strain rates.

Table 1. Steady-state flow-law parameters

Rock type	A ($MPa^{-1/n} s^{-1}$)	H* (KJ/mol)	n	Reference
Westerly granite (wet)	1.9×10^{-03}	137	1.5	Hansen and Carter (1982)
Westerly granite 1	1.4×10^{-09}	106	2.9	Carter et al. (1981)
Westerly granite 2	2.6×10^{-09}	139	3.4	Hansen and Carter (1982)
Diabase	2.2×10^{-04}	260	3.4	Shelton and Tullis (1981)
Diabase	6.2×10^{-02}	276	3.0	Caristan (1982)
Anorthosite	3.3×10^{-04}	238	3.2	Shelton and Tullis (1981)
Yule marble (average)	1.9×10^{-04}	252	8.0	Heard (1976)

2.3 Brittle-Plastic Transition Zone

A transition zone of finite thickness separates the upper, brittle layer from the lower, plastic layer (Fig. 1). The upper and lower bounds to this transition zone represent the temperatures at which the first and last mineral components in a rock enter the plastic domain (Scholz 1988). For granitic (quartzo-feldspathic) rocks subject to typical tectonic strain rates of 10^{-14} to $10^{-16} s^{-1}$ for example, quartz first enters the plastic domain at temperatures around 300 °C and the feldspathic minerals enter the plastic domain at temperatures around 450 °C. Deformation within this transition zone involves both quasi-brittle and quasi-plastic processes, although the effective rheology of these combined processes remains poorly quantified (Hobbs et al. 1986; Scholz 1988).

As a basis for proceeding, I use the following two assumptions:

1. The base of the seismogenic zone (where it can be identified as a sharp drop in the density of hypocenters with depth) corresponds closely with the top of the brittle-plastic transition zone. In particular, spontaneous shear failure (earthquake generation) will be much more common in the brittle domain, which is characterized by unstable, velocity-weakening friction, than within the transition zone where velocity-strengthening friction and quasi-plastic processes dominate (Scholz 1988). Under this assumption, isolated earthquakes beneath the seismogenic zone represent instabilities within the brittle-plastic transition zone, the failure of which involves temperature- and rate-dependent processes as suggested by Hobbs et al. (1986). Brittle failure in tension (abrupt hydrofracture) may also occur deeper than the onset of plastic deformation for shear failure because the low tensile strength for rocks (Fig. 1). Foulger's (1988) analysis of tensile earthquakes near Hengill, Iceland, indicates that the foci of tensile events may indeed extend to systematically greater depths than shear events in the same crustal volume.
2. The top of the brittle-plastic transition zone corresponds closely with the intersection of the brittle shear failure and plastic flow laws [Eqs. (1) and (2)] as plotted in Fig. 1. Specifically, I assume that this intersection marks the onset of quasi-plastic deformation for the weakest mineral species in a specific rock (i.e., quartz in a quartzo-feldspathic rock) and the transition from velocity-weakening to velocity-strengthening friction. This assumption is the same as that used by Sibson (1982) and Meissner and Strehlau (1982).

2.4 Complications and Ambiguities

Given these assumptions, one must bear in mind the host of factors that can complicate the relatively simple strength profile illustrated in Fig. 1. Here, I list the most important of these factors as a cautionary reminder to those tempted to draw hard conclusions from an uncritical interpretation of this approach. More detailed discussions can be found in Chen and Molnar (1983), Kirby (1983), Smith and Bruhn (1984), and Sibson (1984).

1. *Experimental Uncertainties*. The lines representing the strength profile in Fig. 1 should in fact be bands, the width of which reflect the scatter in experimental results. Although the coefficient of friction, μ, is essentially independent of rock type, the experimentally determined values of μ at a given confining pressure show a scatter of roughly 10% about $\mu \approx 0.75$ for confining pressures less than 500 MPa (see Byrlee 1978). Chin and Molnar (1983) point out that the experimental uncertainties associated with the parameters for the flow law of crustal rocks (Table 1) result in an uncertainty of several tens of percent in the curves for plastic strength of crustal rocks.

2. *Water*. Water reduces strength in all rheological domains. The amount of water (or its pressure) in the respective domains, particularly in the lower part of the brittle layer and the upper part of the brittle-plastic transition zone, can significantly influence the depth to the base of the seismogenic zone (brittle zone) for a given geothermal profile.

In the brittle crust, fluid pressure gradients greater than the hydrostatic gradient reduce frictional strength rotating the frictional strength curves in Fig. 1 toward the vertical axis, and frictional strength becomes vanishingly small as pore pressures approach the lithostatic gradient. Fluid pressure gradients less than hydrostatic increase frictional strength by rotating the strength curves away from the vertical axis toward the limiting case of dry friction (for strike-slip faulting and $\mu = 0.75$, this leads to $\tau \approx 16z$ as indicated by the dashed line in Fig. 1). Note that, other factors remaining the same, decreasing strength in the brittle layer increases the depth to the base of the brittle layer and vice versa.

The flow parameters demonstrating the weakening effect of water in the plastic domain (see the curves for wet and dry grainite in Fig. 7) have been determined for relatively few rocks (Kirby and Kronenberg 1987). Fournier (1989) points out that the onset of plastic flow in a wet environment combined with pressure solution processes will weaken rock by reducing pore volume, thereby increasing pore pressure; he describes evidence that pore pressures commonly approach lithostatic values at depths in the crust where temperatures exceed 350 to 400 °C. Note that, other factors remaining the same, decreasing strength in the plastic or quasi-plastic layer decreases the depth to the base of the brittle layer and vice versa.

3. *Composition*. With the exception of certain clays, frictional strength in the brittle domain is virtually independent of rock type (Byrlee 1978). The coefficient of friction in clays with structural water (kaolinite, montmorillonite,

etc.) can be as low as 0.1 to 0.2, and a fault zone filled with gouge consisting of such clays would be relatively weak. Strength in the plastic domain depends markedly on rock type with strength increasing at a given temperature and strain rate in more mafic and more refractory rocks (see Fig. 7).

3 Earthquakes and the Base of the Seismogenic Zone

Long Valley caldera and the Phlegrean Fields caldera offer particularly attractive subjects for an investigation into the relation between the distribution of earthquake hypocenters, the brittle-plastic transition, and magma bodies. Both calderas have shown intense episodes of earthquake swarm activity and ground uplift in recent years, related to increases in subsurface volume commonly attributed to magma intrusion (Barberi et al. 1984; Hill et al. 1985a). Dense instrumental networks installed to monitor the unrest in the two calderas provide high-quality data for precise hypocentral locations of local earthquakes as well as for detailed definition of the deformation field and strain rates (Hill 1984; Lirer et al. 1987). Both calderas have produced eruptions in the last 500 to 600 years. The Phlegrean Fields caldera offers a further advantage: bore holes drilled within the caldera for geothermal exploration define temperatures extending to depths as great as 3 km. (The deepest holes drilled in the Long Valley caldera to date do not exceed 2.1 km.)

3.1 Long Valley Caldera, Eastern California

Long Valley caldera is a 17 by 32 km elliptical depression at the base of the Sierra Nevada eastern escarpment in east-central California (Fig. 2). The caldera was formed 730 000 years ago with the catastrophic eruption of 750 km^3 of rhyolitic magma, the ash flow deposits from which form the Bishop tuff (Bailey et al. 1976; Hildreth 1979; Carle 1988). The residual Long Valley magma chamber fed numerous smaller rhyolitic eruptions over the next 600 000 years, the last of which occurred in the west moat 100 000 years ago. Magma for the Bishop tuff as well as these post-Bishop rhyolites appears to have resided at depths of about 6 km (Bailey et al. 1976; Johnson and Rutherford 1989). The most recent eruptions in the caldera occurred 550 to 650 years ago with the intrusion of an 8- to 10-km long rhyolitic dike beneath the southern end of the Mono-Inyo volcanic chain. This mid-14th century intrusion vented in three places to form the three southernmost Inyo Domes (Fig. 2; Miller 1985; Eichelberger et al. 1988). Volcanism along the Mono-Inyo chain, which began some 200 000 years ago, has produced multiple eruptions of basaltic to rhyolitic lavas fed by sources within the crust that appear to have no simple connection with the residual Long Valley magma chamber (Bailey 1989).

The current episode of unrest in Long Valley caldera began on May 25–27, 1980, with four magnitude-6 earthquakes near the southern margin of the caldera followed by recurring earthquake swarms in the south moat of the

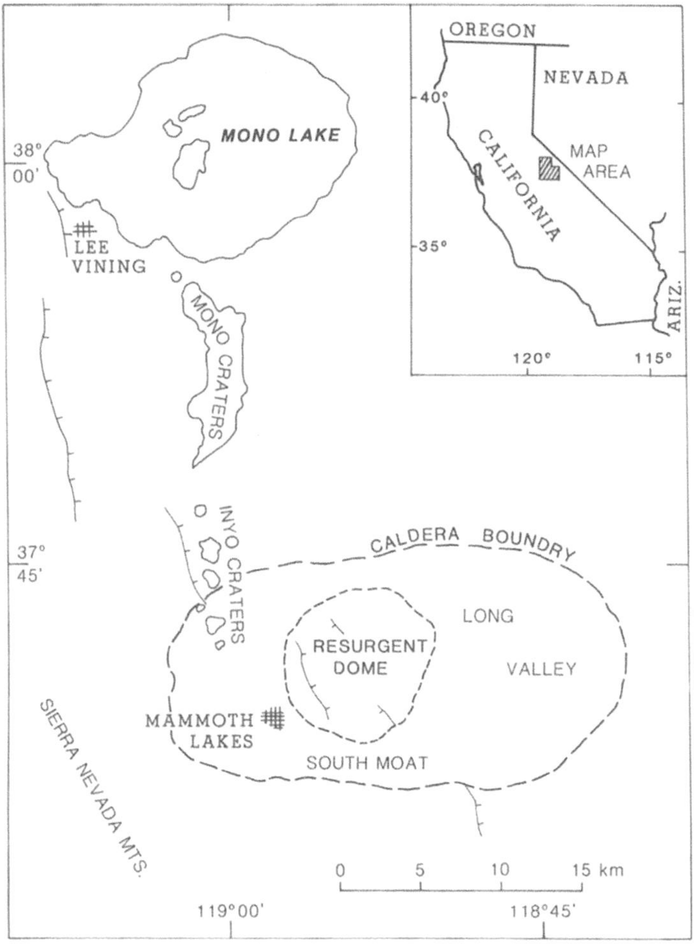

Fig. 2. Location map for Long Valley caldera and the Mono-Inyo volcanic chain. *Hachured lines* indicate principal normal faults

caldera (Hill et al. 1985a). Deformation accompanying this earthquake activity is characterized by a broad (diameter approximately 20 km), dome-shaped uplift centered over the southern section of the resurgent dome that began sometime after mid-1979 and had accumulated a maximum of 0.25 m by mid-1980 (Savage and Clark 1980), and 0.5 m by mid-1987 (Savage 1988). This deformation is consistent with a volume increase of roughly 0.2 km^3 since 1980 and is presumably associated with the injection of magma into one or more bodies at depths between 4 to 10 km beneath the resurgent dome (Savage et al. 1987). The most intense post-1980 earthquake swarm began on January 7, 1983, and included two $M = 5.3$ earthquakes. Deformation data suggest that this swarm may have been associated with the intrusion of a dike to depths approaching 4 km of the

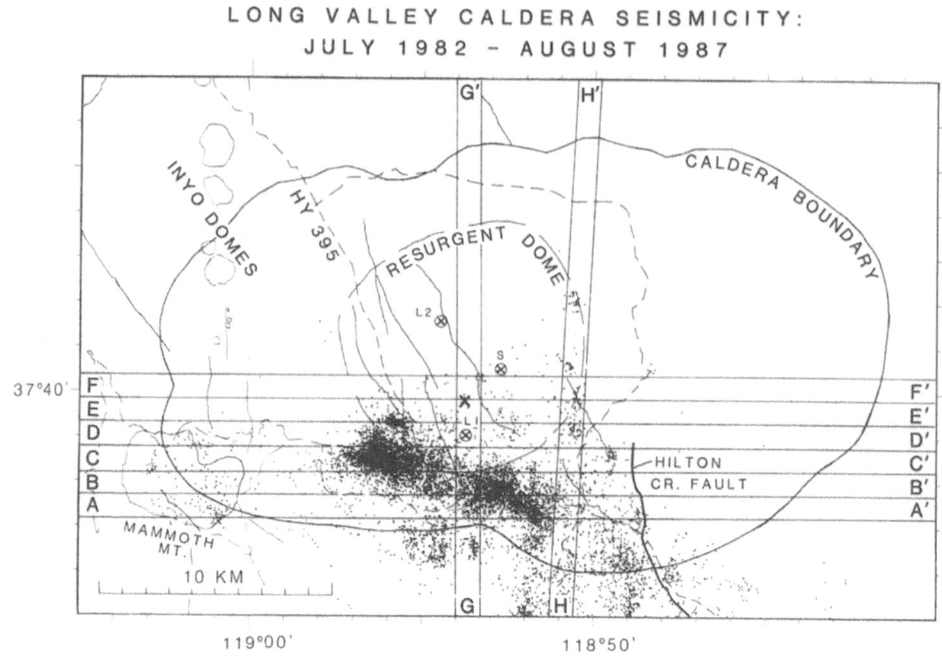

Fig. 3. Seismicity map for Long Valley caldera for the period July 1982 through August 1987. Faults indicated by *solid lines* and roads by *dashed lines*. *x* indicates location of inflation sources driving the uplift; *L1* and *L2* are Langbein's (1989) shallow (5.5 km) and deep (10 km) inflation centers, respectively, and *S* is Savage's (1988) single inflation center at 10 km. *X* indicates proposed site of 6-km-deep drill hole. *A-A'* ... *H-H'* show locations of 1-km-wide cross-sections plotted in Fig. 4

surface along the ringfracture beneath the south moat of the caldera (Savage and Cockerham 1984).

Figure 3 shows the distribution of earthquake epicenters in Long Valley caldera for the period July 1982 through August 1987. By July 1982, ten or more vertical component seismic stations with telemetered signals were operational within the caldera, permitting uniformly high quality hypocentral locations for $M \geqslant 1.5$ earthquakes occurring within the caldera. Locations for the 9527 well-recorded earthquakes plotted in Fig. 3 were obtained by R.S. Cockerham using a one-dimensional P-wave velocity model that incorporates the basic results from regional and local seismic- refraction profiles (Eaton 1966; Hill et al. 1985b) with a VELEST inversion of P-wave travel times from 2800 local earthquakes and 11 explosions. Cockerham (written comm. 1988) estimates the hypocentral locations for the set of earthquakes plotted in Fig. 3 are reliable to within ±0.5 km both horizontally and vertically.

Noteworthy aspects of the distribution of epicenters in the caldera (Fig. 3) include:

1. The dense, elongate cluster of earthquakes in the south moat. Most of the $M \geqslant 4$ earthquakes in the caldera have occurred within the western half of this

cluster, including the second M = 6 earthquake of May 25, 1980, and the two M = 5.3 earthquakes of the January 1983 swarm. Most of the earthquakes defining the eastern half of the cluster occurred during the January 1983 swarm. Focal mechanisms for earthquakes throughout this cluster are predominantly strike-slip with the right-lateral planes subparallel with the axis of the south moat (Savage and Cockerham 1984).

2. The diffuse distribution of epicenters centered beneath Mammoth Mountain at the southwestern edge of the caldera. This cluster has developed gradually since 1983 with flurries of small ($M \leqslant 3$) earthquakes showing predominantly normal focal mechanisms.
3. A scattering of epicenters over much of the southern sector of the caldera. Many of these events occurred as solitary earthquakes while others occurred in clusters. Spatial clustering within this diffuse distribution is most dense along the southern caldera boundary and in a broad zone that enters the caldera from the Sierra Nevada to the south and narrows northward along the east margin of the resurgent dome. This diffuse, north-trending seismicity zone generally coincides with the northward splaying of the Hilton Creek fault into the caldera. The Hilton Creek fault is a major east-dipping normal fault forming the eastern escarpment of the Sierra Nevada south of the caldera with more than 5 m of Holocene dip-slip offset.
4. An extensive area devoid of earthquakes includes most of the east, north, and west sections of the caldera moat as well as the northwestern half of the resurgent dome. This quiescent area extends northward to Mono Lake, and notably, it includes the 14th-century eruption vents along Mono-Inyo volcanic chain.

A series of east-west and north-south sections through the caldera illustrate the depth distribution of earthquakes in the caldera for the period from July 1982 through August 1987 (Fig. 4a,b). Each of the sections includes the hypocenters in a 1-km-thick, vertical slice through the caldera, the locations of which are indicated on the seismicity map in Fig. 3. The dashed line in each of the depth sections represents the approximate locus of the base of the seismogenic zone consistent with assumption 1. discussed in the section on the brittle-plastic transition zone; it represents a miminum depth for the base of the brittle zone ($\mathfrak{T}_b$) in Fig. 1 (we cannot conclude, for example, that the brittle zone has zero thickness beneath the seismically quiescent sections of the crust). The contours in Fig. 5 show the depths to the base of the seismogenic zone in map view based on the dashed lines in Fig. 4a,b.

Broadly considered, the base of the seismogenic zone for the 1982–1987 period shallows to the northwest from depths of 10 to 12 km beneath the southern-most extent of the caldera boundary to less than 4 km out beneath the western part of resurgent dome and the west moat where the seismicity dies out. Locally, the seismogenic zone maintains a thickness of 8 to 9 km beneath the most intense swarm activity and the largest (M = 5–6) earthquakes in the south moat. Further to the west, however, it shallows rather abruptly to less than 4 km beneath Mammoth Mountain. The base of the seismogenic zone also shallows abruptly to the north toward the center of the resurgent dome with steep gradients closely coinciding with the margin of the resurgent dome along its

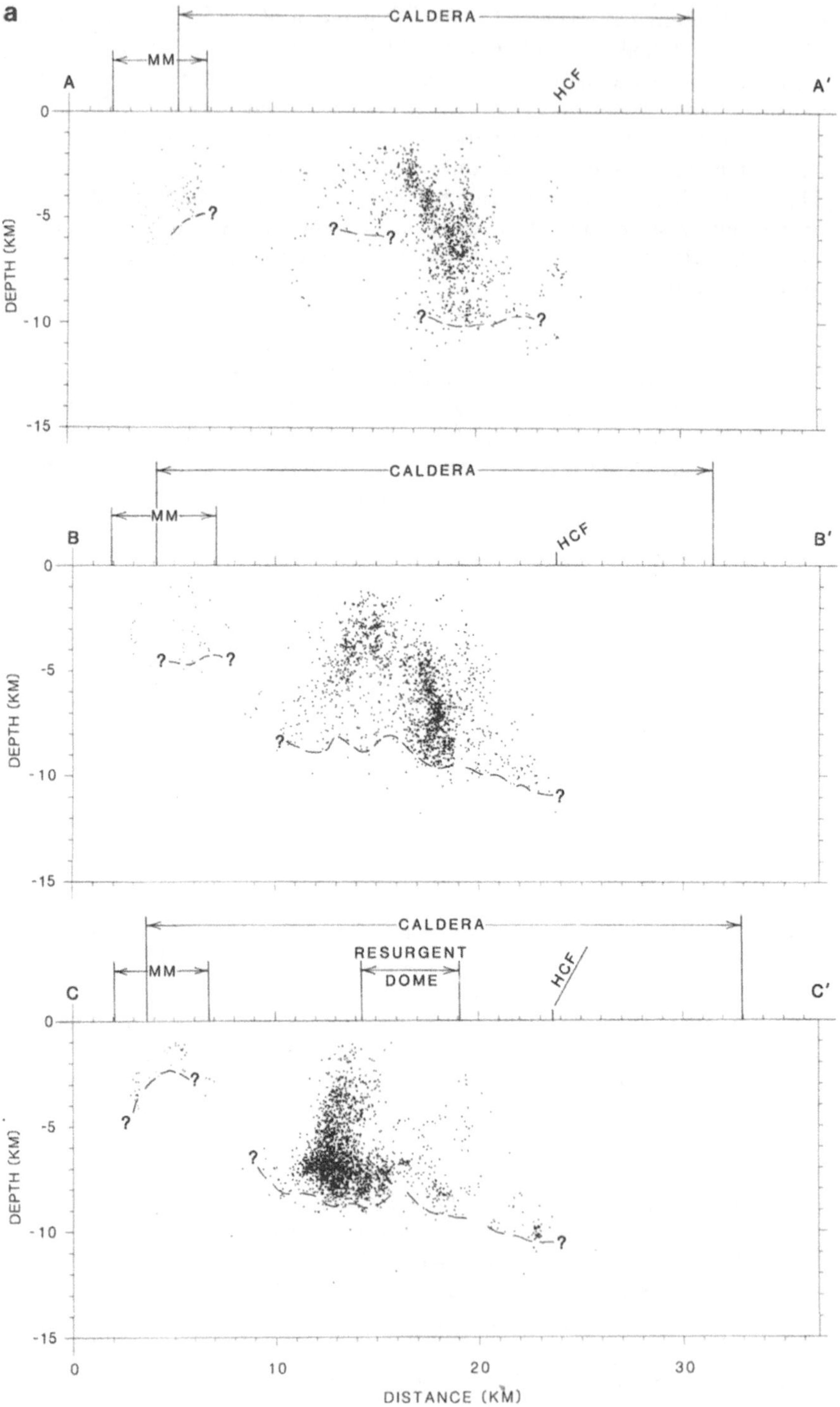
a
CALDERA
MM
A
HCF
A′
B
B′
C
RESURGENT
DOME
C′
DEPTH (KM)
0
-5
-10
-15
10
20
30
DISTANCE (KM)

Fig. 4a

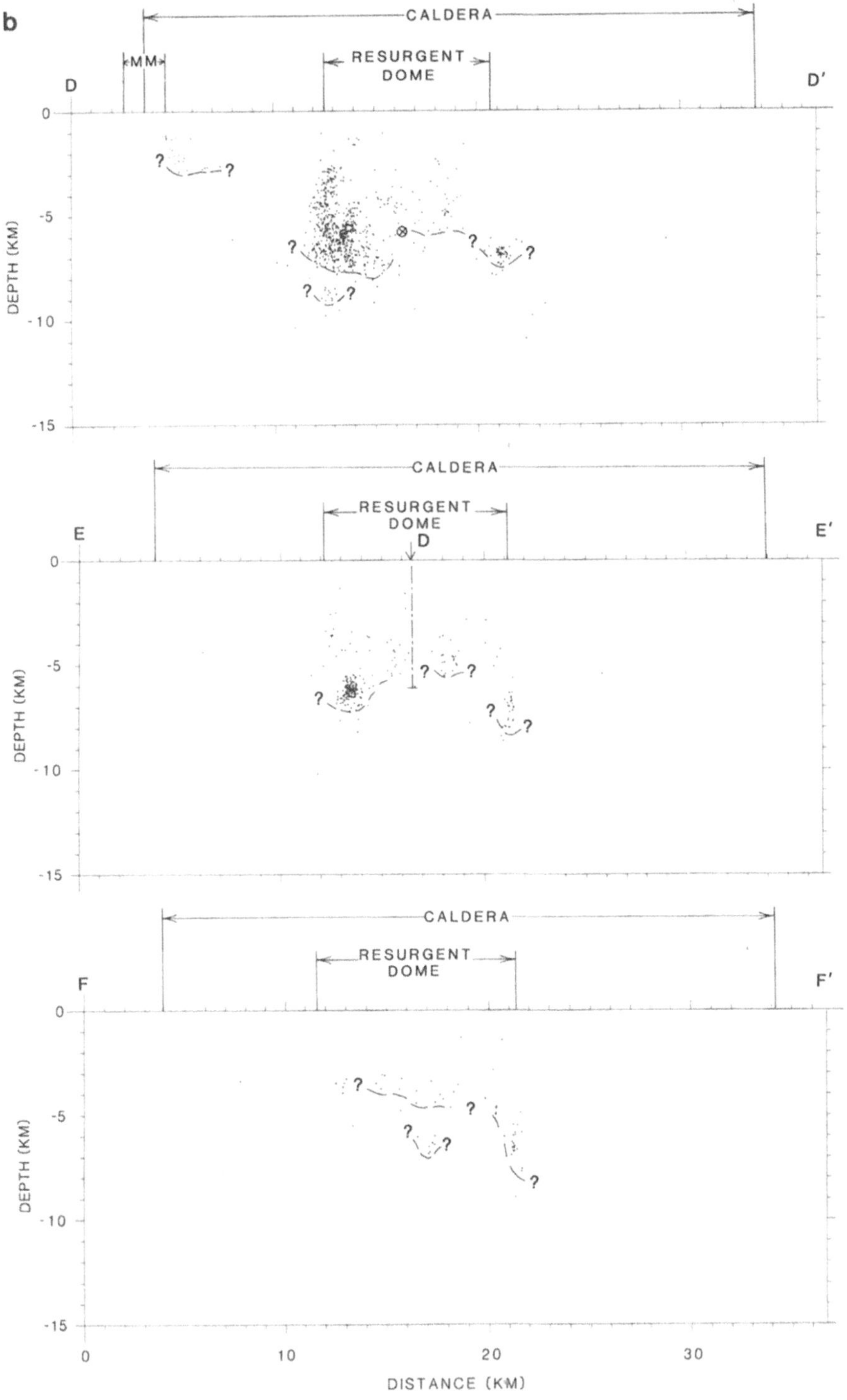

Fig. 4b. *Continued*

c

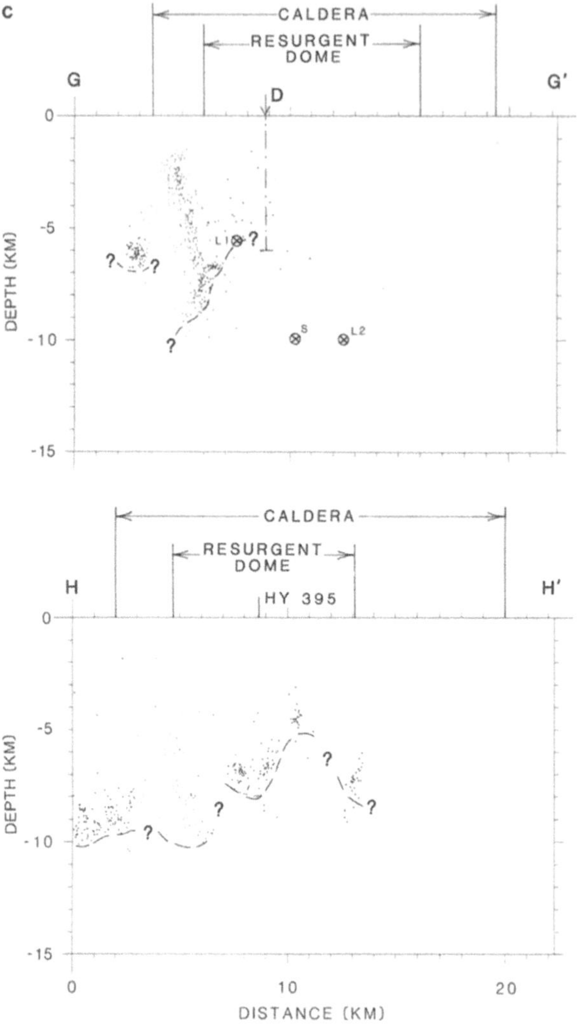

Fig. 4a–h. Seismicity cross-sections for Long Valley caldera with locations marked in Fig. 3. *Dashed line* indicates approximate base of seismogenic zone. *x* in *D-D′* and *G-G′* indicates positions of the 5- to 10-km-deep inflation source beneath the southern section of the resurgent dome proposed by Savage (1988) and Langbein (1989); other solutions for inflation sources have similar locations (see Rundle and Hill 1988; Goldstein and Stein 1988). The *dash-dot line* beneath *D* in *E-E′* and *G-G′* indicates planned drill hole

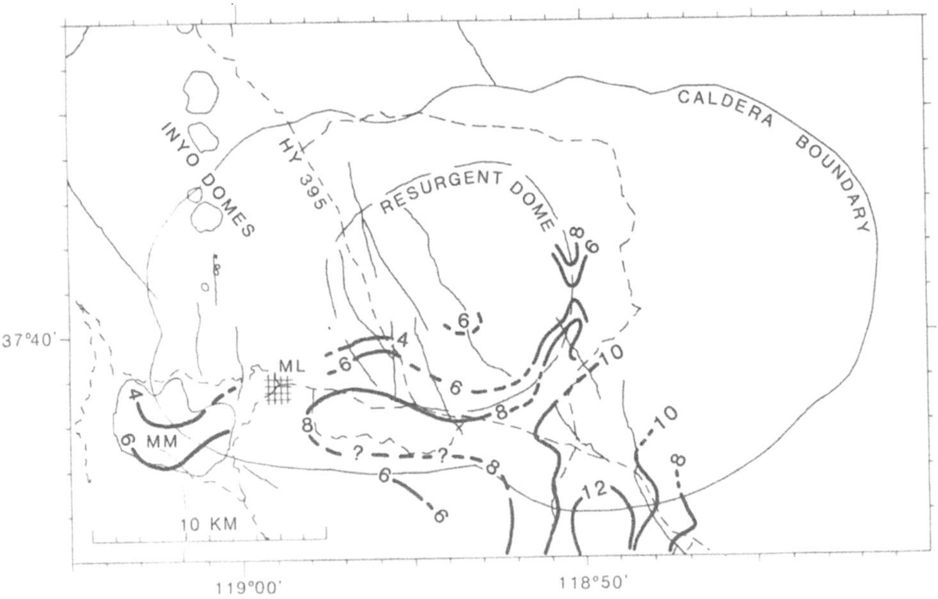

Fig. 5. Contour map for depth to base of seismogenic zone beneath Long Valley caldera. Contour interval 2 km

southeastern quadrant. Two details of particular interest related to magma bodies and heat sources in the crust beneath the caldera are: (1) the pronounced, dome-like upwarp in the base of the seismogenic zone to depths of less than 6 km beneath the resurgent dome and the center of uplift for recent ground deformation, and (2) shallow depths (less than 4 km) to the base of the seismogenic zone beneath the northern section of Mammoth Mountain and adjoining sections of the west moat, which appear to be the deep heat source for the current hydrothermal system in the caldera (Sorey 1985).

3.2 Phlegrean Fields Caldera, Italy

The Phlegrean Fields caldera lies at the western edge of the Naples metropolitan area in the Regione Campania of southern Italy (Fig. 6). The caldera-forming eruption 34 000 years ago ejected approximately 80 km^3 of trachytic lavas from a ring fracture system 10 km in diameter (Barbari et al. 1984; Lirer et al. 1987). Subsequent eruptions from monogenetic vents within the caldera have generally become smaller with time as the vents migrated toward the center of the caldera near Pozzuoli. Barberi et al. (1984) suggest that this eruptive pattern, together with chemical variations within the suite of trachitic lavas, indicate a gradually cooling magma chamber. The most recent eruption, which occurred in 1538 from a vent 3 km west of Pozzuoli, produced Monte Nuovo (Fig. 6).

Recent unrest in the Phlegrean Fields has occurred in two episodes, both dominated by pronounced uplift of the central section of the caldera (Lirer et al.

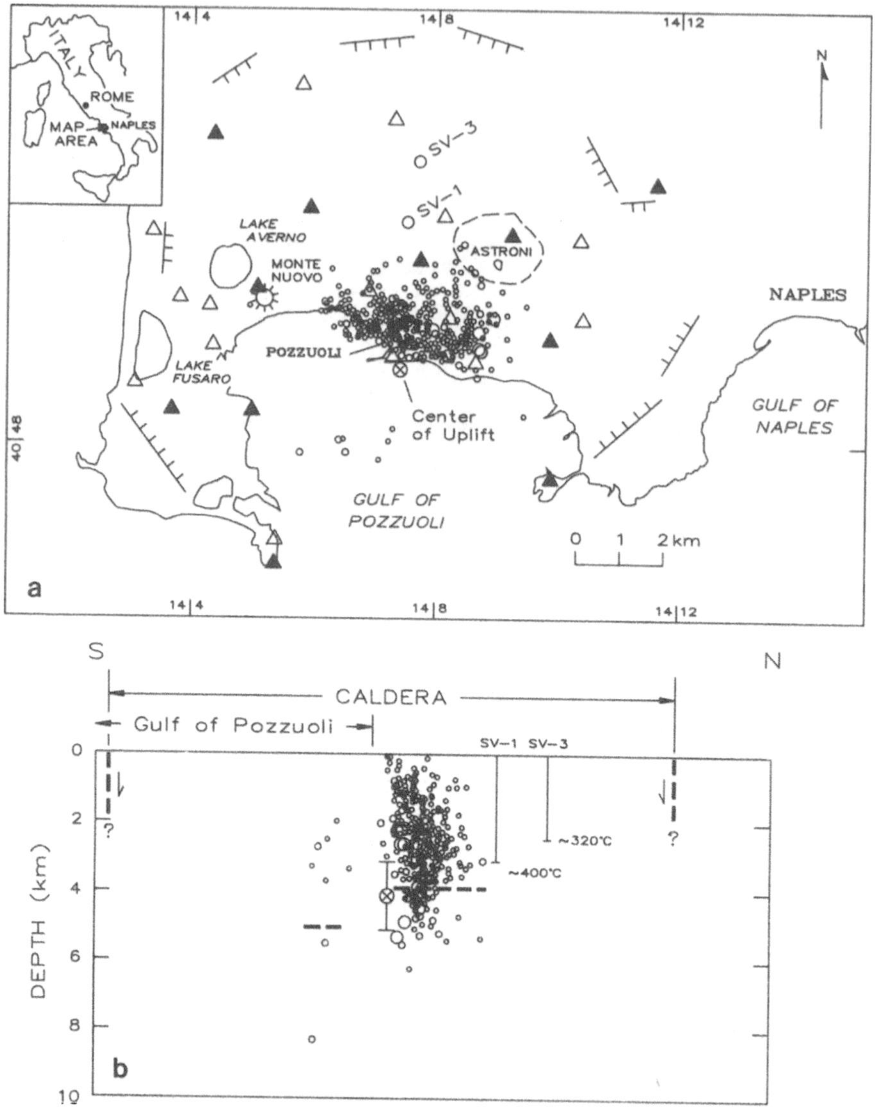

Fig. 6a,b. Seismicity for the Phlegrean Fields caldera adopted from De Natale and Zollo (1986) and Del Pezzo et al. (1987). **a** Map with epicenters indicated by *small open circles*, seismic stations indicated by *open* and *closed triangles*, and caldera-bounding faults indicated by *hachured lines*. *X* is center of uplift. **b** North-south cross-section showing depth distribution for earthquakes in **a**. Drill holes SV-1 and SV-3 marked by *solid vertical lines*. Bottom-hole temperatures indicated for both drill holes. *X with bar* indicates range of depths for the inflation source. *Dashed line* indicates base of the seismogenic zone based on Aster and Meyer's (1988) locations using a three-dimensional velocity model (see De Natale and Pingue this Vol.)

1987). The first episode produced a 2 m uplift at Pozzuoli between mid-1969 and 1972 followed by ten years of quiescence and partial subsidence. The second episode resulted in an additional 1.8 m of uplift between mid-1982 and late-1984 followed by another period of quiescence and slow subsidence. The cumulative uplift at Pozzuoli for the period 1969 through 1988 is about 2.4 m, and the associated deformation field is consistent with a volume increase of approximately 0.15 km^3 centered 3 to 5 km beneath the coast just off Pozzuoli (Berrino et al. 1984; Bianchi et al. 1987; Bonafede et al. 1986; De Natale et al. this Vol.). Few earthquakes accompanied the initial uplift episode, but in mid-1983, nearly a year after the onset of the second uplift episode, intense earthquake swarm activity began occurring beneath Pozzuoli. The strongest of these swarms included an M = 4.2 event on October 4, 1983. This earthquake activity died out in December 1984 with the end of the second uplift episode. Strain rates associated with the uplift during this swarm activity approached 150 μstrain/y (5 $\times 10^{-12} s^{-1}$, or nearly two orders of magnitude greater than the Long Valley strain rate of $3 \times 10^{-14} s^{-1}$ for 1982–87).

Figure 6 shows the epicenters for 450 of the best-recorded earthquakes that occurred in the caldera between August 1983 and May 1984 as located by De Natale and Zollo (1986). These earthquakes were recorded by the 22-station network of vertical- component seismometers operated jointly by the Vesuvius Observatory and Agip (Italian Petrol Agency) and by an additional set of 12 three-component, digital stations operated by the University of Wisconsin from January through March, 1984. De Natale and Zollo (1986) used a homogeneous half-space with P- and S-wave velocities of 3 km/s and 1.7 km/s, respectively, to locate these earthquakes, and they estimate that the resulting hypocentral locations are accurate to within 0.5 km. Aster and Meyer (1988) applied a simultaneous inversion for three-dimensional velocity structure and hypocentral locations using 228 well-recorded earthquakes on the University of Wisconsin three-component stations. Their inversion indicates average P- and S-wave velocities of 2.9 and 1.6 km/s, respectively, for the upper 4 to 5 km of the crust (V_P/V_S = 1.76) but with evidence for anomalously low velocities (roughly 10% in V_P and 14% in V_S) extending to depths of at least 3 km beneath the center of uplift. Their results also suggest high values for the V_P/V_S ratio within this anomalous volume, which has lateral dimensions of 2 to 3 km, with maximum V_P/V_S values approaching 2.2 near the center of uplift (see De Natale and Pingue, this Vol.).

The epicenters located by both De Natale and Zollo (1986) and Aster and Meyer (1988) form an oblate spheroidal cluster, the elongate core of which follows the slightly north of west trend of the coast line and passes 0.5 km north of the center of uplift (Fig. 6a). Focal mechanisms for these earthquakes show dominantly normal solutions with nodal planes generally parallel with the long axis of the seismicity cluster (Del Pezzo et al. 1987). In north-south cross-section (Fig. 6b; De Natale and Zollo 1986; Del Pezzo et al. 1987) the hypocenters define a vertically elongate cluster in the upper 4 to 5 km of the crust. Aster and Meyer (1988), however, find a rather abrupt cut-off in focal depths at 4 km using their three-dimensional velocity structure (dashed line in Fig. 6b; also see De Natale and Pingue this Vol.). Within available resolution, the base of this seismicity cluster coincides with 3-to 5-km depth range for the inflation center driving the uplift (Berrino et al. 1984; Bianchi et al. 1984), although the inflation center appears to be offset by roughly 0.5 km to the south of the seismogenic zone. Note that the core of the seismicity cluster passes within 2 km of the 400 °C bottom-hole temperature measured in the 3-km-deep well, SV-1 (Rosi and Sbrana 1987).

4 Discussion

4.1 Temperatures Beneath the Phlegrean Fields Caldera

The Phlegrean Fields caldera appears at first glance to offer sufficient data for an in situ test for the brittle-plastic transition model illustrated in Fig. 1. Extensive studies summarized by Lirer et al. (1987) and Rosi and Sbrana (1987) provide information on strain rates ($\dot{e} \approx 5 \times 10^{-12} s^{-1}$), depth to the base of the seismogenic zone (4 to 5 km), rock type (trachyte), and geothermal gradients (approximately 100 °C/km with a bottom-hole temperature in excess of 400 °C at a depth of 3 km and within 2 km of the seismogenic volume).

If we simply assume that: (1) the geothermal gradient within the seismogenic volume is close to the 100 °C/km measured in the nearby SV-1 hole (De Vivo et al. 1989; also see Fig. 6); (2) this gradient persists to depths of at least 6 km; and (3) the flow-law parameters for Phlegrean Fields trachytes, which have no modal quartz (Appendix 3 in Rosi and Sbrana 1987), are similar to those for other feldspathic rocks such as diabase or anorthosite (Table 1), we find that the model predicts that the base of the seismogenic zone should occur between 4.5 and 5 km, in close agreement with the maximum focal depths from De Natale and Zollo (1986) and 1 km deeper than the Aster and Meyer (1988) solutions (Fig. 6b). Temperatures predicted for these depths are 600 to 700 °C (Fig. 7).

While the agreement between this straightforward application of the model and the seismicity data is encouraging, it may be misleading. Consider the following uncertainties:

1. The base of the seismogenic zone may correspond to a change in composition. Whether the Mesozoic carbonate rocks, exposed around the periphery of the Phlegrean Fields-Vesuvius volcanic depression, underlie the caldera is not clear (Rosi and Sbrana 1987). If they do, they must be deeper than the 3 km penetrated by existing drill holes, and they would be metamorphosed to a calc-silicate assemblege by the intense magmatism that has persisted in the crust beneath the caldera for more than 34 000 years. Anorthite is the only calc-silicate minerals for which flow-law parameters are available, and its ductile strength is slightly lower than diabase (see anorthosite in Fig. 7). Accordingly, the possibility remains that the base of the seismogenic zone beneath the Phlegrean Fields caldera corresponds to the boundary between a 4-km-thick sequence of trachitic rocks and an underlying sequence of (perhaps) weaker calc-silicate rocks.

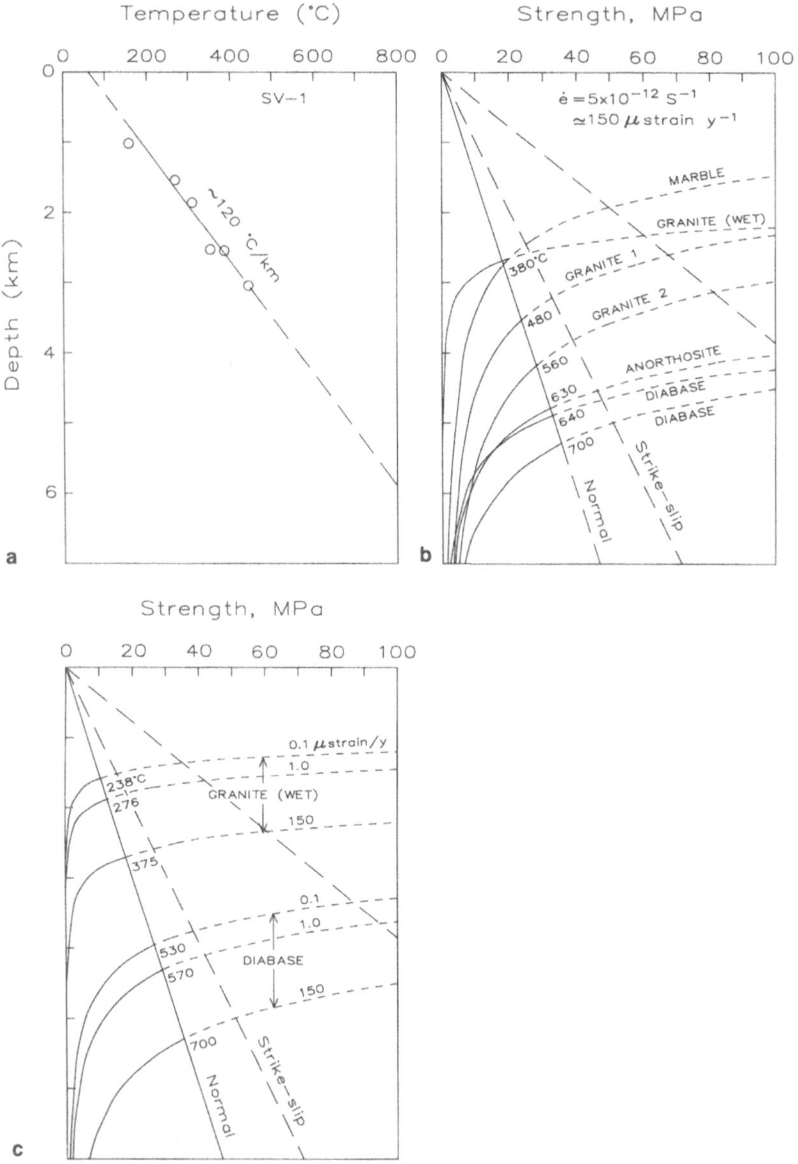

Fig. 7a–c. Variation in strength with composition **b** and strain rate **c** for average temperature gradient in Phlegrean Fields drill hole SV-1 **a**. Temperature data for SV-1 in **a** from De Vivo et al. (1989). Frictional strength curves in **b** are for hydrostatic pore pressure. Strain rate in **b** corresponds to 150 μstrain/y in **c**. Solid lines mark strength profiles defined by Eqs. (1) and (2) (see Fig. 1). *Numbers* at cusps correspond to temperatures T_b at the base of the seismogenic zone for normal faulting typical of the Phlegrean Fields caldera. Flow-law parameters for diabase assumed to approximate behavior of Phlegrean Fields trachytes (see text)

2. The strain rate near the base of the seismogenic zone may be one to two orders of magnitude greater than that measured at the surface. To first order, the deformation measured in both the Phlegrean Fields and Long Valley calderas matches the deformation predicted for a dilitational point source beneath the free surface of an elastic half-space, and the strain in such a model varies inversely with the cube of the distance (R^{-3}) from the point source. In fact, the strain rate probably increases more slowly with depth than R^{-3} to the extent that the crust deviates from a homogeneous, perfectly elastic half-space and the inflation sources involve laterally extensive volumes (i.e., see De Natali et al. this Vol.). As indicated in Fig. 7c, an increase in strain rate by a factor of 150 increases the temperature for the onset of quasi-plastic deformation in a given rock type by roughly 100 °C. Accordingly the temperature at the base of the seismogenic crust in the Phlegrean Fields could approach 800 °C, which would require an increase in the geothermal gradient somewhere below 3 km.
3. Brittle failure associated with earthquakes tends to maintain fracture permeability within seismogenic volumes of the crust. Accordingly, hydrothermal convection may be more important within the seismogenic zone beneath Pozzuoli than in adjacent, relatively aseismic sections of the crust penetrated by the drill holes (Fig. 6). Spatial temperature variations associated with hydrothermal convection within the seismogenic volume would violate assumption 1 above. Furthermore, variations in pore pressure suggested by lateral variations in V_P/V_S (Aster and Meyer 1988) may lead to substantial variations in the depth to the base of the seismogenic zone beneath the central section of the caldera.

4.2 Temperatures Beneath Long Valley Caldera

In Long Valley caldera, granitic (quartzo-feldspathic) rocks dominate the upper 15 to 20 km of crust, with metamorphic roof pendants (dominantly quartz-rich hornfels) forming the upper kilometer or so of the basement beneath the western half of the caldera (Reinhart and Ross 1964). Extensive monitoring data for the mid-1982 to mid-1987 period clearly define both the strain rate ($3 \times 10^{-14} s^{-1}$ throughout the period) and relief on the base of the seismogenic zone (Fig. 4). Sorey et al. (1978) estimate that the total hydrothermal discharge from the caldera is about 2.9×10^8 W. Temperatures measured in the relatively shallow drill holes (<2.1 km), however, reflect convection in the hydrothermal system, and we must await results from the deep hole (target depth 6 km) planned for the southern section of the resurgent dome (Rundle et al. 1986) for information on deeper, conductive geothermal gradients.

Results from a number of independent geophysical studies suggest that magma underlies the western half of the caldera at depths somewhere between 4 to 10 km beneath the southern section of the resurgent dome and between 7 to 12 km beneath the northern section of the resurgent dome (Rundle et al. 1986; Rundle and Hill 1988; Figs. 2 and 3). The caldera-forming eruption 730 000 years ago apparently exhausted the magma chamber underlying the eastern half of the caldera (Lachenbruch et al. 1976). Figures 8, 9, and 10 illustrate some hypothetical one-dimensional thermal models for magma at these depths together with the corresponding strength profiles based on the flow-law parameters for wet and dry granite in a strike-slip stress regime. Two factors are worth keeping in mind regarding these figures: (1) temperatures a given distance from a one-dimensional model for a magma body (infinite lateral extent) will generally be higher than those the same distance from a three-dimensional model (finite lateral extent), and (2) increasing strain rates with depth toward the inflation source beneath the resurgent dome will push the base of the seismogenic zone to greater depths and higher temperatures; as a rule of thumb, recall that a 150-fold increase in strain rate corresponds to a temperature increase of roughly 100 °C with corresponding depth increase governed by the appropriate geotherm.

4.2.1 A Steady-State Model

Lachenbruch and Sass (1977) present a steady-state, one-dimensional thermal model for hydrothermal convection in a surfical layer of thickness h when the heat is supplied by an upper crustal magma chamber at depth H. Their model can be expessed in the following form:

$$Q_c = K(T_m - T_1)/(H - h) = NKT_1/h \tag{3a}$$

$$N = Q_c/Q^*, \tag{3b}$$

where Q_c is the mean heat flow measured at the surface; K is the mean thermal conductivity of the overburden ($z < H$); NK is the effective thermal conductivity through the hydrothermal system ($z < h$) with N corresponding to the Nusselt number for thermal convection; T_m is the magma temperature at $z = H$; and T_1 is the temperature at the base of the hydrothermal system at $z = h$. In Eq. (3b), $Q^* = KT_1/h$, or the conductive heat flow that would be realized through $z < h$ in the absence of hydrothermal convection for a temperature T_1 at $z = h$.

For Long Valley caldera, $T_m \approx 800\,°C$ based on Hildreth's (1979) analysis of temperatures for the Bishop tuff magmas, and $Q_c \approx J/A$, where $J = 2.9 \times 10^8$ W is the heat flux associated with the total hydrothermal discharge from the caldera (Sorey et al. 1978) and A is the area underlain by magma. Thus $Q_c \approx 630\,\mathrm{mW\,m^{-2}}$ if magma underlies the entire caldera ($A \approx 460\,\mathrm{km^2}$), or $Q_c \approx 1260\,\mathrm{mW\,m^{-2}}$ if magma underlies only the western half of the caldera. (Clearly, the average heat flow, Q_c over an even smaller magma body would be quite large; but if $\sqrt{A} < 2H$, this one-dimensional approximation will begin to fail.) If we assume, in addition, that hydrothermal convection occurs throughout the seismogenic zone (persistent earthquake activity maintains fracture permeability in a rock volume), we can identify h with the base of the seismogenic zone and take $T_1 = T_b$ (the temperature for the onset of plastic deformation in Fig. 1.) Under this assumption, the plastic layer above the magma coincides with the conductive thermal layer $h \leqslant z \leqslant H$, and their common thickness is $\Delta h = H - h = K(T_m - T_b)/Q_c$.

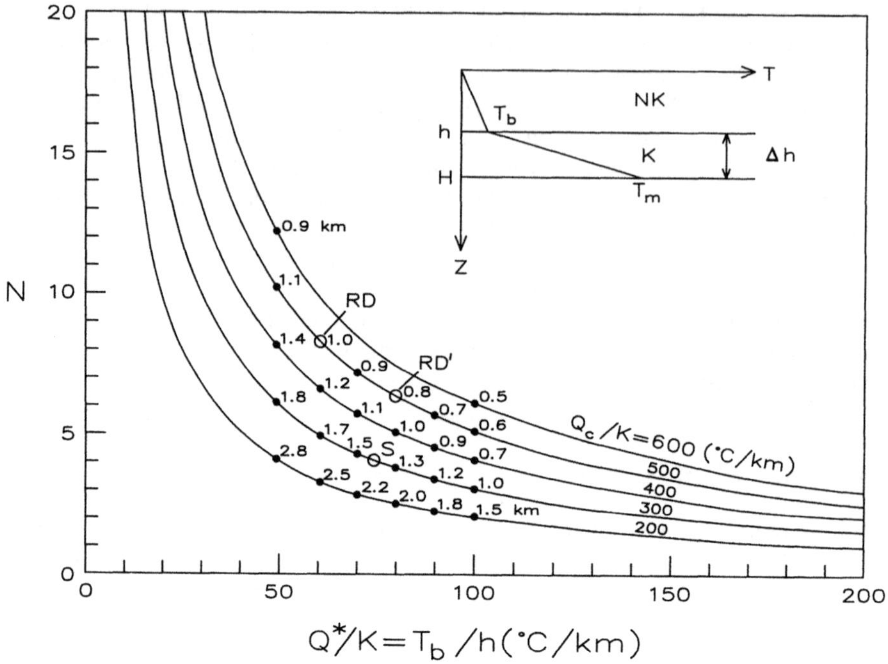

Fig. 8. Variation in N with $T_b/h = Q^*/K$ and a range of values for Q_c/K for the one-dimensional, steady-state model of a hydrothermal system over a magma chamber proposed by Lachenbruch and Sass (1977). The curves are based on Eqs. 3a,b. *Points S* (Sorey 1985) and *RD* (Resurgent Dome) correspond to the 630 and 1260 $mW\,m^{-2}$ geotherms in Fig. 9a,b, respectively. *RD'* corresponds to the 1260 $mW\,m^{-2}$ geotherm for 150 μstrain/y at depth (Fig. 9c). *Numbers* along the curves of constant Q_c/K correspond to the thickness, Δh, of the plastic layer between the magma chamber and the base of the hydrothermal (seismogenic) layer for h = 5 km. For the point S, h = 3 km, and Δh = 2 km

This model is illustrated in Fig. 8 with a plot showing the variation of N with Q^*/K (the effective gradient in the hydrothermal layer) for various values of Q_c/K (the gradient in the conductive layer). Two properties of this model act to keep the base of the seismogenic zone more or less coincident with the base of the hydrothermal system provided N is significatly large and T_1 does not differ too greatly from T_b. Specifically, (1) steep temperature gradients in the conductive layer ($h > z > H$) will confine the onset of plastic deformation to $z \approx h$ because large temperature differences correspond to small increments in depth, and (2) as N increases (hydrothermal convection becomes more vigorous), temperatures within the hydrothermal layer ($z < h$) will deviate from the effective conductive gradient $Q^*/K = T_b/h$ (under the constraint that $T \leqslant T_b$) permitting, at least locally, relatively low temperatures just above $z = h$.

Figure 9a illustrates hypothetical steady-state geotherms for three heat flow values: Q_c = 160, 630, and 1260 $mW\,m^{-2}$. Sorey (1985) proposed the 160 and 630 $mW\,m^{-2}$ geotherms as end-member models for the geothermal regime beneath the caldera. His 630 $mW\,m^{-2}$ model puts the base of the hydrothermal system at 3 km (225 °C) and magma at 5 km corresponding to point S in Fig. 8

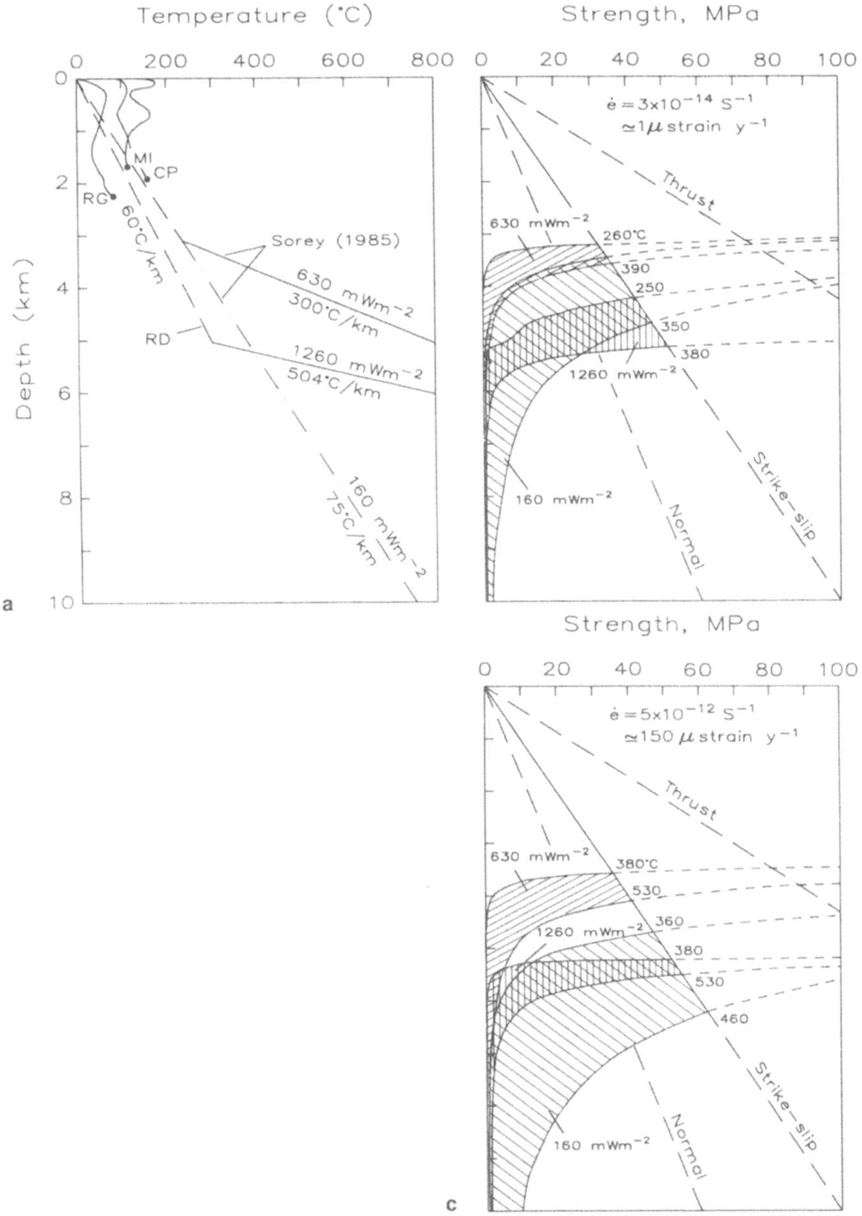

Fig. 9a–c. Steady-state, one-dimensional models for Long Valley caldera thermal and strength profiles. **a** Geotherms for average heat flow values Q_c = 630, 1260, and 160 mW m^{-2} and magma at depths of 5, 6, and 10.5 km, respectively. K = 2.1 W °C^{-1} m^{-1} for the 160 and 630 mW m^{-2} models (Sorey 1985), and K = 2.5 W °C^{-1} m^{-1} for the 1260 mW m^{-2} model. *M1*, *CP*, and *RG* are bottom-hole temperatures for drill holes in the central part of the caldera (see Sorey 1985). **b** Strength profiles for magma at 5, 6, and 10.5 km give range for wet and dry granite (dry granite 2 in Table 1) at 1 μstrain/y. *Numbers at cusps* in the strength profiles (*heavy lines*) correspond to temperatures t_b at the base of the seismogenic zone for strike-slip faulting typical of Long Valley caldera. **c** Same as **b** but for 150 μstrain/y

(N = 4 in this case). The associated strength profile for a 1 μstrain/y strain rate (Fig. 9b) indicates that the base of the seismogenic zone should occur near the base of the hydrothermal layer at depths between 3 and 3.5 km and at temperatures between 250 and 390 °C for wet and dry granite, respectively. (For 150 μstrain/y, the corresponding depths and temperatures are 3.5 to 4 km and 380 to 530 °C; see Fig. 9c.) These depths are too shallow for the seismogenic sections of south moat and the resurgent dome, although they may be appropriate for the west moat of the caldera and the northern section of Mammoth Mountain (see Fig. 5). The average heat flow through west moat, however, must be more than double the 630 mW m^{-2} value if, as seems likely, magma confined to the western half of the caldera supplies the heat carried in the total hydrothermal flux from the caldera.

Sorey (1985) suggested the 160 mW m^{-2} geotherm as a possible background thermal regime where the hydrothermal system has been inactive for the past 0.3 m.y. or more. This geotherm represents a purely conductive gradient (N = 1) for magma at a depth of 10.5 km. Note that although a conductive heat flow of 160 mW m^{-2} would contribute little to the total hydrothermal heat flux from the caldera, the associated geothermal gradient of 75 °C/km is sufficiently steep to confine the base of the seismogenic zone to depths of 3.5 to 4.5 km for wet granite and 5 to 6 km for dry granite (Fig. 9b,c).

The 1260 mW m^{-2} geotherm (Fig. 9a) represents a model in which the heat flux associated with the total hydrothermal discharge from the caldera is supplied by magma confined to the western half of the caldera, and the depth to the base of the hydrothermal system is 5 km, or the average depth to the base of the seismogenic zone beneath the resurgent dome (Fig. 5). The temperature at this depth is taken as the nominal 300 °C value for the onset of plastic deformation in quartzo-feldspathic rocks (Scholz 1988), an assumption consistent with geochemical evidence for maximum temperatures of 280 °C in the Long Valley hydrothermal reservoir (Fournier et al. 1979). This model corresponds to the point RD (resurgent dome) in Fig. 8, for which N = 8.4. The strength profiles for the 1260 mW m^{-2} geotherm and a $3 \times 10^{-14}\,s^{-1}$ strain rate (Fig. 9b) are consistent with the onset of plastic deformation near the base of the hydrothermal system at 5 ± 0.4 km and at temperatures between 250 and 380 °C. The steep conductive gradient (Q_c/K = 504 °C/km) associated with this model requires magma at a depth, H = 6 km, or only 1 km beneath the base of the hydrothermal system (Δh = 1 km).

Figure 8 provides a simple graphical means for evaluating the effect of varying parameters on this model. Consider, for example, increasing the depth and temperature at the base of the hydrothermal system to 5.2 km and 400 °C consistent with the strength profile for $\dot{e} = 3 \times 10^{-12}\,s^{-1}$ in Fig. 9c (i.e., an elevated strain rate at depth). This moves the point RD down the curve of constant Q_c/K to RD′ thereby decreasing both N and the thickness of the plastic layer to N = 6.5 and Δh = 0.8 km, respectively. The depth to magma remains at approximately 6 km because the increase in h is offset by the decrease in Δh. Alternatively, consider increasing the average heat flow, Q_c, which moves RD vertically upward thereby increasing N and decreasing both Δh and the depth to magma. This latter example may be relevant, if the magma beneath the caldera underlies only a section of the resurgent dome and west moat, as suggested by some recent geophysical studies (see Goldstein and Stein 1988). Recall, however, that this one-dimensional model will begin to fail (over-estimate temperatures) unless the smallest horizontal dimension of the magma body is at least a factor of two greater than the depth to the body.

4.2.2 A Transient Model

Figure 10a shows transient conductive temperatures above a cooling, 5-km-thick magma body emplaced 10000 years ago with its upper surface at depths of 4, 6, and 8 km [these transient geotherms were generated using Eq. (5) in Lachenbruch et al. 1976]. The thermal front from these 10000-year intrusions would have propagated only about 2 km above the respective intrusion depths. Note that the geotherm for magma intruded at 6 km is grossly similar in form to the steady-state case for Q_c = 1260 mW m^{-2} and a 5-km-deep hydrothermal system in Fig. 9a. Frequent intrusions (intervals of 5000 years or less) at 6 km would be required, however, before an overlying hydrothermal system would approach a steady state condition because a convecting hydrothermal system would rapidly deplete the heat from a single intrusion of finite thickness (see Lachenbruch et al. 1976; Sorey et al. 1978).

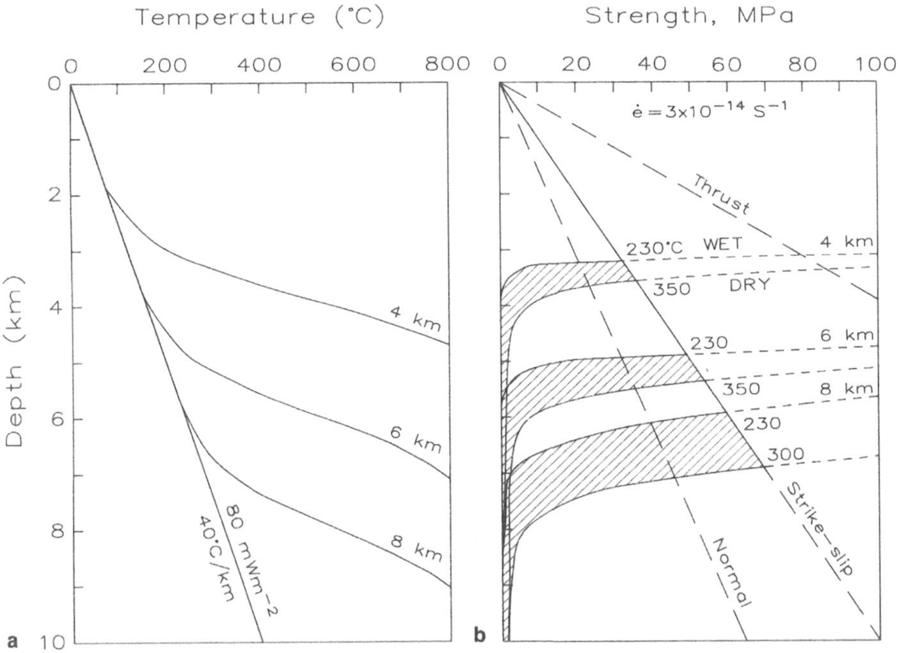

Fig. 10a,b. Transient model for a cooling magma body injected 10000 years ago at depths of 4, 6, and 8 km for K = 2.5 W °C^{-1} m^{-1}. **a** Temperature profiles. **b** Corresponding strength profiles for wet granite and granite 2 (dry). *Numbers at cusps* in strength profiles show temperatures T_b at base of seismogenic zone for strike-slip faulting typical of Long Valley

The strength profiles in Fig. 10b suggest that the base of the seismogenic zone should be little more than 0.5 km above the top of the 4-km-deep body and as much as 1.5 km above the 8-km-deep body. Note the similarity between the strength profiles for the 4- and 6-km-deep bodies and those for the 630 and 1260 $mW\,m^{-2}$ geotherm in Fig. 9. Indeed, the outstanding difference between the models in Figs. 9 and 10 is that the average heat flow for the transient models will be much lower than for the comparable steady-state models and the hydrothermal systems in the transient models will be inactive or in the early stages of development.

4.2.3 Regional Temperatures

To place the above results in Long Valley caldera in a regional context, consider the strength profiles for granite in Figure 11b corresponding to background Basin and Range and Sierra Nevada geotherms (Fig. 12a after Lachenbruch and Sass 1977). Strain rates measured in the regions adjacent to Long Valley caldera are roughly 0.1 μstrain/y, or $3 \times 10^{-15}\,s^{-1}$ (Savage et al. 1987). Earthquake focal depths approach 18 to 20 km beneath the Sierra Nevada south of the caldera (Hill et al. 1989) and 10 to 12 km beneath the Basin and Range province east of the caldera (Cockerham and Corbett 1987) in rough agreement with depths for the base of the seismogenic crust predicted in Fig. 11.

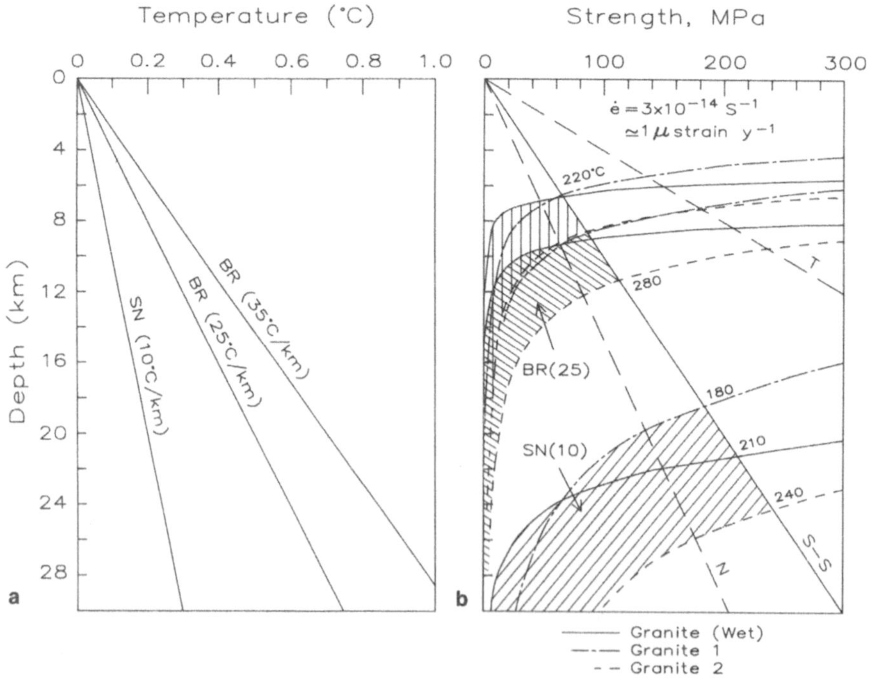

Fig. 11a,b

c

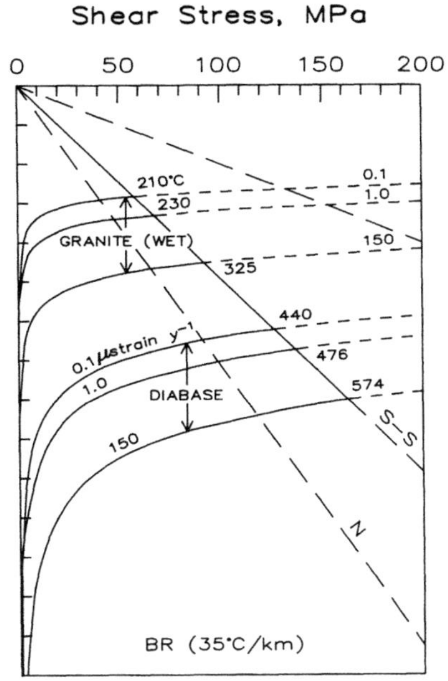

Fig. 11. Regional geotherms for **a** the Basin and Range (*BR*) province and the Sierra Nevada (*SN*) based on Lachenbruch and Sass (1977). **b** Corresponding strength profiles for the three granite flow-law parameters in Table 1. **c** Strength profiles for wet granite and diabase for the BR 35 °C/km geotherm and variable strain rates. *Numbers at cusps* in strength profiles (*heavy lines*) show temperatures at base of seismogenic zone for strike-slip faulting

5 Conclusions

Strength profiles based on laboratory measurements of rock friction and flow-laws parameters for common crustal rock types provide a means of estimating temperatures near the base of seismogenic zones when information is available on local crustal composition and strain rates. Implicit with this approach is the assumption that the base of a seismogenic zone corresponds closely with the temperature at which the first mineral species in the crustal rock enters the quasi-plastic domain, which in turn marks the onset of the temperature-dependent, brittle-plastic transition zone. Temperature estimates based on this assumption, however, still have large uncertainties (the weakening effect of water, for example, is still not quantifed for most rock types), and they are not unique (cool sections of the brittle crust may be aseismic simply because local shear stresses are below the frictional strength). Accordingly, temperatures estimated in this manner must be viewed with a critical eye, and, where possible, they must be calibrated with in situ bore-hole measurements and reliable information on conductive geothermal gradients in the crust.

Application of this approach to the crust beneath the Phlegrean Fields caldera in southern Italy, where the rock type is dominantly feldspathic (trachitic), suggests temperatures ranging from 600 to perhaps as high as 800 °C for the base of the seismogenic zone at depths of 4 to 5 km. The higher value is appropriate for dry conditions in the plastic domain and for a strain rate that increases significantly with depth toward the inflation center beneath Pozzuoli. A temperature near 800 °C at the base of the seismogenic zone is only about 50 °C below the wet solidus for trachitic rocks at these depths (Armienti et al. 1984) and requires that temperature must increase at depths greater than 3 km with respect to the geothermal gradient (≈100 °C/km) measured in the adjacent 3-km-deep well. These results remain blurred, however, by uncertainties in the variation in strain rate with depth, rock type below 3 km, and the role of hydrothermal convection within the seismogenic volume.

In Long Valley caldera in eastern California, where the rock type is dominantly quartzofeldspathic (granitic), this approach suggests temperatures for the base of the seismogenic zone of 250 to 350 °C and perhaps as high as 500 °C beneath the resurgent dome. Depths to the base of the seismogenic zone vary from 8 to 9 km beneath the south moat to less than 5 km beneath the resurgent dome and less than 4 km beneath Mammoth Mountain. Temperatures of 250 to 350 °C at these depths are consistent with a variety of plausible but hypothetical geothermal profiles associated with magma beneath the caldera at depths from 5 to 10 km. The lower of these temperatures (associated with the quasi-plastic strength of wet granite) is comparable with maximum temperatures estimated for the Long Valley geothermal reservoir (280 °C). The higher temperature (500 °C) applies to dry conditions within the plastic domain and an increasing strain-rate with depth toward an inflation center that may be as shallow as 5 km beneath the southern section of the resurgent dome. The one-dimensional thermal models in Figs. 8, 9, and 10 suggest the plastic layer between the base of the brittle (seismogenic) zone and magma is a few hundred meters to 1 km or so thick. This plastic layer may be much thinner, however, if the magma bodies have a limited lateral extent such that the one-dimensional approximations seriously overestimate temperatures a given distance from magma, and/or if the magma bodies were emplaced considerably more recently than 10 000 years ago. In any case, the 6-km-deep drill hole in the southern section of the resurgent dome planned to start in mid-1989 (Rundle et al. 1986) should help define the conductive geothermal regime beneath the caldera and thus, provide an important test for the extrapolation of rheological parameters, determined in the laboratory, to geologic conditions in an active magmatic environment.

Acknowledgements. I've benefited from discussions with many Italian colleagues on the Phlegrean Fields activity, and I'm particularly grateful to R. Scandone, R. Scarpa, and G. De Natale for their guidance on matters relating to this work. The hypocentral locations for earthquakes in the Long Valley caldera region used in this work are due to the diligent work of Robert S. Cockerham. I wish to thank N. Deichmann and L. Rybach for stimulating discussions on the problem of the brittle-plastic transition during my stay at the Swiss Federal Institute of Technology (ETH), Zürich, and Peter Hill for guidance on solidus temperatures. I'm also grateful to Bernard Chouet and Robert Fournier for their constructive reviews of an early version of this manuscript.

References

Armienti P, Barberi F, Innocenti F (1984) A model of the Phlegraean Fields magma chamber in the last 10500 years. Bull Volcanol 47:349–358

Aster RC, Meyer RP (1988) Three-dimensional velocity and hypocenter distribution in the Campi Flegrei Caldera, Italy. Tectonophysics 149:195–218

Bailey RA (1989) Geologic map of Long Valley caldera, Mono-Inyo craters volcanic chain and vicinity, Mono County, California. US Geol Surv Misc Invest Map I-1933

Bailey RA, Dalrymple GB, Lanphere MA (1976) Volcanism, structure, and geochronology of Long Valley caldera, Mono County, California. J Geophys Res 81:725–744

Barberi F, Corrado G, Innocenti F, Luongo G (1984) Phlegraean Fields 1982–1984: brief chronicle of a volcano emergency in a densely populated area. Bull Volcanol 47:175–185

Berrino G, Corrado G, Luongo G, Toro B (1984) Ground deformation and gravity changes accompanying the 1982 Pozzuoli uplift. Bull Volcanol 47:187–200

Bianchi R, Coradini C, Federico C et al. (1987) Modelling of surface deformation in volcanic areas: the 1970–1972 and 1982–1984 crises of Campi Flegrei, Italy. J Geophys Res 92:14,139–14,150

Blackwell DD (1985) A transient model of the geothermal system of the Long Valley caldera, California. J Geophys Res 90:11229–11242

Bonafede M, Dragoni M, Quareni F (1986) Displacement and stress fields produced by a center of dilation and by a pressure source in a viscoelastic half-space: application to the study of ground deformation and seismic activity at Campi Flegri, Italy. Geophys JR Astr Soc 87:455–485

Brace WF, Kohlstedt DL (1980) Limits on lithospehric stress imposed by laboratory experiments. J Geophys Res 85:6248–6252

Byrlee JD (1978) Friction of rocks. Pure Appl Geophys 116:615–626

Caristan Y (1982) The transition from high-temperature creep to fracture in Maryland diabase. J Geophys Res 87:6781–6798

Carle SF (1988) Three-dimensional gravity modeling of the geologic structure of Long Valley caldera. J Geophys Res 93:13237–13250

Carter NL, Anderson DA, Hansen FD, Kranz RL (1981) Creep and creep rupture of granitic rocks. Am Geophys Union Geophys Monogr 24:61–82

Chen WP, Molnar P (1983) Focal depths of intracontinental and intraplate earthquakes and their implications for the thermal and mechanical properties of the lithosphere. J Geophys Res 88:4183–4214

Cockerham RS, Corbett EJ (1987) The July 1986 Chalfant Valley, California, earthquake sequence: preliminary results. Bull Seismol Soc Am 77:280–289

De Natale G, Zollo A (1986) Statistical analysis and clustering features of the Phlegraean Fields earthquake sequence. Bull Seismol Soc Am 76:801–814

De Vivo B, Belkin HE, Barbieri M et al. (1989) The Campi Flegrei (Italy) geothermal system: a fluid inclusion study of the Mofete and San Vito fields. J Volcanol Geotherm Res 36:303–326

Del Pezzo E, De Natale G, Martini M, Zollo A (1987) Source parameters of microearthquakes at Phlegraean Fields (southern Italy) volcanic area. Phys Earth Planet Int 47:25–42

Eaton JP (1966) Crustal structure in northern and central California from seismic evidence. In: Bally EH (ed) Geology of northern California. Bull Calif Div Min Geol 190:419–426

Eichelberger JC, Vogel TA, Younker LW, Miller CD, Heikin GH, Wohletz KH (1988) Structure ans stratigraphy beneath a young phreatic vent: south Inyo crater, Long Valley caldera, California. J Geophys Res 93:13208–13220

Foulger GR (1988) Hengill triple junction, SW Iceland 2. Anomalous earthquake focal mechanisms and implications for processes within the geothermal reservoir and at accretionary plate boundaries. J Geophys Res 93:13507–13523

Fournier RO (1989) Maximim depths of earthquakes as an aid to evaluating covective and conductive heat flux from the Cascade province and adjacent regions. In: Muffler LJP, Weover CS, Blackwell DB (eds) Geological geophysical and tectonic setting of the Cascade Range. US Geol Surv Open File Rep 89–178, 177–198

Fournier RO, Sorey ML, Mariner RH, Truesdell AH (1979) Chemical and isotopic prediction of aquifer temperatures in the geothermal system at Long Valley, California. J Volcanol Geotherm Res 5:17–34

Goldstein NE, Stein RS (1988) What's new at Long Valley? J Geophys Res 93:13 187–13 190

Hansen FD, Carter NL (1982) Creep of selected crustal rocks at 1000 MPa. EOS 63:437

Heard HC (1976) Comparison of the flow properties of rocks at crustal conditions. Phil Trans R Soc Lond A 283:173–186

Hildreth W (1979) Evidence for the origin of compositional zoning in silicic magma chambers. Geol Soc Am Spec Pap 180:43–75

Hill DP (1977) A model for earthquake swarms. J Geophys Res 82:1347–1352

Hill DP (1984) Monitoring unrest in a large silicic caldera, the Long Valley-Inyo craters volcanic complex in east-central California. Bull Volcanol 47:371–395

Hill DP, Bailey RA, Ryall AS (1985a) Active tectonic and magmatic processes beneath Long Valley caldera, eastern California: a summary. J Geophys Res 90:11 111–11 120

Hill DP, Kissling E, Luetgert JH, Kradolfer U (1985b) Constraints on the upper crustal structure of the Long Valley-Mono Craters volcanic complex, eastern California, from seismic refraction measurements. J Geophys Res 90:11 135–11 150

Hill DP, Eaton JP, Ellsworth WL, Cockerham RS, Lester FW, Corbett EJ (1989) The seismotectonic fabric of central California. In: Slemmons DB, Engdahl ER, Blackwell D, Schwartz D, Zoback M (eds) Neotectonics of North America. Geol Soc Am GSMV-1 (in press)

Hobbs BE, Ord A, Teyssier C (1986) Earthquakes in the ductile regime? Pageoph 124:309–336

Johnson MC, Rutherford MJ (1989) Experimental calibration of the aluminum-in-hornblende geobarometer with application to Long Valley caldera volcanic rocks. Geology 17:837–841

Kirby SH (1983) Rheology of the lithosphere. Rev Geophys Space Phys 21:1458–1487

Kirby SH, Kronenberg AK (1987) Rheology of the lithosphere: selected topics. Rev Geophys 25:1219–1244

Lachenbruch AH, Sass JH (1977) Heat flow in the United States and the thermal regime of the crust. In: Heacock JG (ed) The nature and physical properties of the earth's crust. Am Geophys Union Geophys Monogr 20:626–675

Lachenbruch AH, Sass JH, Munroe RJ, Moses TH (1976) Geothermal setting and simple heat conduction models for the Long Valley caldera. J Geophys Res 81:769–784

Langbein J (1989) The deformation of the Long Valley caldera, eastern California, from mid-1983 to mid-1988; measurements using a two-color geodimeter. J Geophys Res (in press)

Lirer L, Luongo G, Scandone R (1987) On the volcanological evolution of Campi Flegri. EOS 68:226–234

Meisner R, Strelau J (1982) Limits of stresses in continental crust and their relationship to the depth-frequency distribution of shallow earthquakes. Tectonics 1:73–89

Miller CD (1985) Holocene eruptions at the Inyo volcanic chain, California – implications for possible eruptions in Long Valley caldera. Geology 13:14–17

Reinhart CD, Ross DC (1964) Geology and mineral deposits of the Mount Morrison quadrangle Sierra Nevada, California. US Geol Surv Prof Pap 385:106

Rosi M, Sbrana A (1987) Phlegrean fields, Quaderni de "La Richerca Scientifica" 114. Consiglio Nazionale Delle Ricerche, Rome, 9:175 pp, 4 plates

Rundle JB, Hill DP (1988) The geophysics of a restless caldera – Long Valley, California. Ann Rev Earth Planet Sci 16:251–271

Rundle JB, Carrigan CR, Hardee H, Luth W (1986) Deep drilling to the magmatic environment in Long Valley caldera. EOS 67:490–491

Savage JC (1983) Strain accumulation in western United States. Ann Rev Earth Planet Sci 11:11–43

Savage JC (1988) Principal component analysis of geodetically measured deformation in Long Valley caldera, eastern California, 1983–1987. J Geophys Res 93:13297–13305

Savage JC, Clark MM (1980) Magmatic resurgence in Long Valley caldera, California: possible cause of the 1980 Mammoth Lakes earthquakes. Science 217:531–533

Savage JC, Cockerham RS (1984) Earthquake swarm in Long Valley, California, January 1983: evidence for dike injection. J Geophys Res 89:8315–8324

Savage JC, Cockerham RS, Estrem JE, Moore LR (1987) Deformation near the Long Valley caldera, eastern California, 1982–1986. J Geophys Res 92:2721–2746
Scholz CH (1988) The brittle-plastic transition and the depth of seismic faulting. Geol Rundsch 77:319–328
Shelton G, Tullis JA (1981) Experimental flow laws for crustal rocks. EOS 62:396
Sibson RH (1982) Fault zone models, heat flow, and the depth distribution of earthquakes in the continental crust of the United States. Bull Seismol Soc Am 72:151–163
Sibson RH (1983) Continental fault structure and the shallow earthquake source. J Geol Soc Lond 140:747–767
Sibson RH (1984) Roughness at the base of the seismogenic zone: contributing factors. J Geophys Res 89:5791–5799
Smith RB, Bruhn RL (1984) Interplate extensional tectonics of the eastern Basin-Range: inferences on structural style from siesmic reflection data, regional tectonics, and termal-mechanical models of brittle-ductile deformation. J Geophys Res 89:5733–5762
Sorey ML (1985) Evolution and present state of the hydrothermal system in Long Valley caldera. J Geophys Res 90:11 219–11 228
Sorey ML, Lewis RE, Olmsted FH (1978) The hydrothermal system of Long Valley caldera, California. US Geol Surv Prof Pap 1044-A:60

Seismic Investigation of the Campi Flegrei: A Summary and Synthesis of Results

R.C. Aster, R.P. Meyer, G. DeNatale, A. Zollo, M. Martini, E. Del Pezzo, R. Scarpa, and G. Iannaccone

Abstract

Campi Flegrei, Italy is a 10-km-diameter Quaternary explosive caldera near Naples and Mount Vesuvius. Numerous eruptive vents lie within it, the last of which formed in 1538 A.D. An elevation change of ≈10 m has occurred within the caldera during the past 2000 years. Recently (mostly 1982–1985), following more than 400 years of subsidence, there has been approximately 2 m of uplift in a region 2.5 km in diameter in the central portion of the caldera. Local earthquakes in the magnitude range $0.6 \leqslant M_D \leqslant 4.2$ and depth range $1.5 < z < 5$ km associated with this uplift were recorded by a portable 3-component digital network and by a telemetered network of single-component stations. The results of seismological studies arising from this data set are considered in consort with geodetic measurements, geothermal drilling, and gravity data to constrain the structure and dynamics of the shallow caldera. The largely aseismic central region, where the greatest uplift occurred is spatially correlated with a 10 mgal gravity low and with a zone of low seismic velocities with a high v_P/v_s ratio. This area is interpreted as an incompetent, highly fractured, water saturated zone of low-density material which extends to at least 3 km. The shear-wave Q of the central caldera measured from direct arrivals is 110 ± 50, somewhat lower than the range of coda-Q estimates for the region (120–250). Hypocenter locations indicate a minimum magma chamber depth of 3.5 km and suggest the presence of an inward-dipping, elliptical ring fault. Fault plane mechanisms are highly heterogeneous, even for highly restricted regions, and thus do not indicate coherent reverse faulting at the perimeter of the ring. Moment tensor inversion indicates that seismic sources are well-modeled by low stress drop brittle fracture due to uplift; no clear indications of non-double couple focal mechanisms are seen which might indicate a significant seismogenic role for intrusion.

1 Introduction

The Campi Flegrei (Phlegraean Fields) are a complex volcanic landform located approximately 15 km west of Naples, Italy (Fig. 1). The area contains numerous explosive eruptive vents that are framed by the rim of a Quaternary (age approximately 40 000 years) explosive caldera with a diameter of approximately 12 km. The caldera is flanked on the east by Mt. Vesuvius and on the west by the volcanic islands of Ischia and Procida. All of these eruptive centers except Procida have been active in historic times.

Bonito (1691) reported a small, apparently phreatic eruption of Solfatara crater in 1198, which did little damage. The most recent Campi Flegrei eruption, in 1538, formed the 100-m-high cone of Monte Nuovo, obliterating the nearby village of Tre Pergole with pyroclastic flows, and inflicting heavy damage in

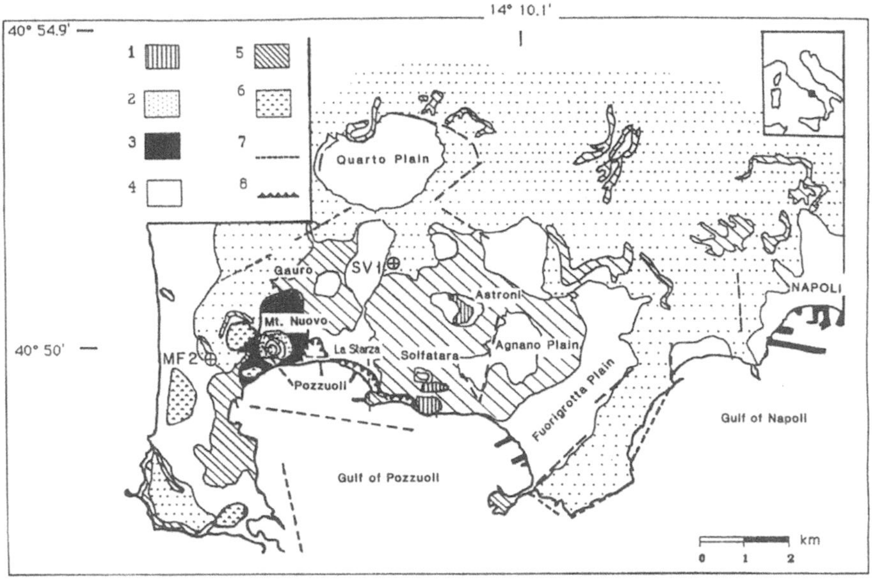

Fig. 1. Volcanological map of the Campi Flegrei. *1* Domes and lava flows; *2* outlying pyroclastic fall deposits; *3* pyroclastic flow deposits from the Monte Nuovo eruption of 1538 A.D.; *4* areas of soil cover and urban development; *5* base surge deposits; *6* lakes; *7* faults associated with caldera formation; *8* limit of the La Starza (≈4500 y b.p.) uplift terrace. *SV1* and *MF2* show locations of deep geothermal wells (after Di Vito et al. 1985)

Pozzuoli, 3 km to the east. Several inches of ash fell in Naples from this eruption. This activity was preceded by several years of perceived seismicity and by an abrupt uplift of approximately 5 m 2 days before (Rosi and Santacroce 1984). Solfatara crater currently displays fumerole (solfataric) activity.

Approximately 10 m of coastal elevation change has occurred in historic times (e.g., Lyell 1873; Dvorak and Mastrolorenzo 1991), and features such as the La Starza marine terrace confirm that large inflationary and deflationary episodes extend back to at least 4500 y B.P. The most famous evidence for this pronounced and highly localized (bradyseismic) ground motion is seen at the Roman temple of Serapeo, constructed in the first century A.D. along the Pozzuoli waterfront. The 8-m-high columns show indisputable evidence of oceanic submersion in the form of marine borings. Parascandola (1947) estimated the elevation of the Serapeo temple in Pozzuoli over the last 2000 years from the density of marine borings on the columns and historical records. Bradyseismic motion has not occurred along the coast outside the caldera. Furthermore, it is clear that the land elevation in the vicinity of Monte Nuovo was approximately 5 m higher in Roman times than at present; a currently submerged first century A.D. Roman port (Port Julius) exists there. Such observations were used by Lyell (1873) and Charles Babbage in the nineteenth century to conclusively usurp the old dogma of the immutability of land elevations.

Following the 1538 Monte Nuovo eruption, the central caldera subsided for over 400 years. About 1968 this trend reversed and a resurgent episode began, resulting in approximately 70 cm of uplift at Pozzuoli by 1972. After February, 1972 a subsidence of 20 cm ensued for the next 8 years. Seismic activity associated with this bradyseismic episode was measurable (Corrado et al. 1976), but caused only minor property damage. Inflation resumed at a higher rate in 1982. The elevation change at the Pozzuoli waterfront from this second episode reached 1.7 m before leveling off in January 1985. The maximum daily uplift rate was substantially higher (approximately 2.5 mm/day) than during the 1970 through 1972 episode (approximately 0.7 mm/day). Swarms of shallow (depth approximately 3 km) earthquakes of duration magnitude $\leqslant 4.0$ occurred in 1983 and 1984, resulting in widespread civic concern and in the eventual evacuation of approximately 30000 people from the Pozzuoli area (more than 300000 people live inside the topographic caldera rim). By January 1985, the latest seismic activity ceased, and only a few events per month are currently being recorded by the local array. The caldera is presently in a period of subsidence which has amounted to approximately 50 cm (5/90).

In August 1983, the University of Wisconsin Geophysical and Polar Research Center (UW) began a cooperative field experiment with triggered, wide-dynamic range, 3-component digital instruments, supplementing the 21-station analog vertical seismic array maintained by the Vesuvius Observatory (VO) and AGIP (Fig. 2). A maximum of 12 UW sites were occupied simultaneously by the digital seismographs. This network was maintained by UW and VO personnel until mid-June 1984. Data were recorded at 100 or 200 Hz sampling rates, with four-pole Butterworth antialiasing filters at 24 and 48 Hz, respectively. 1-Hz resonant geophones were used (Hall-Sears HS10-1 and Geotech S-13). The absolute timing of all digital stations was maintained with a 1σ accuracy of better than 5 ms using broadcasts of the eight-station worldwide Omega navigational network (Schneider et al. 1987).

The generally high level of cultural and oceanic noise at the Campi Flegrei above 1 Hz makes seismic observation difficult, particularly during the daylight hours. Equipment security is also a concern, due to the urban character of much of the area. Despite these difficulties, the UW network was successful, particularly during the night hours, in recording thousands of high-quality digital seismograms with well-defined P- and S-wave arrivals.

Numerous models have been proposed for the uplift and seismicity of the Campi Flegrei. An effective point source in a homogeneous elastic medium (Mogi 1958) was applied by Corrado et al. (1976) and Berrino et al. (1984). Casertano et al. (1976) suggested that the bradyseismic uplift was mainly due to intense thermofluidodynamic processes acting inside a porous medium. They hypothesized that the ground deformation and related seismicity was due to an increase in fluid pore pressure as a result of a high degree of hydrothermal activity and fluid circulation in the shallow crust (0–3 km). Bonasia et al. (1984) proposed a magma-filled dike. Bianchi et al. (1984) proposed a more general finite-element model of finite sources embedded in a heterogeneous elastic half-space. Bonafede et al. (1986), proposed a finite spherical source in a viscoelastic medium. Martini (1986) suggested a mechanism of increased heat flow and

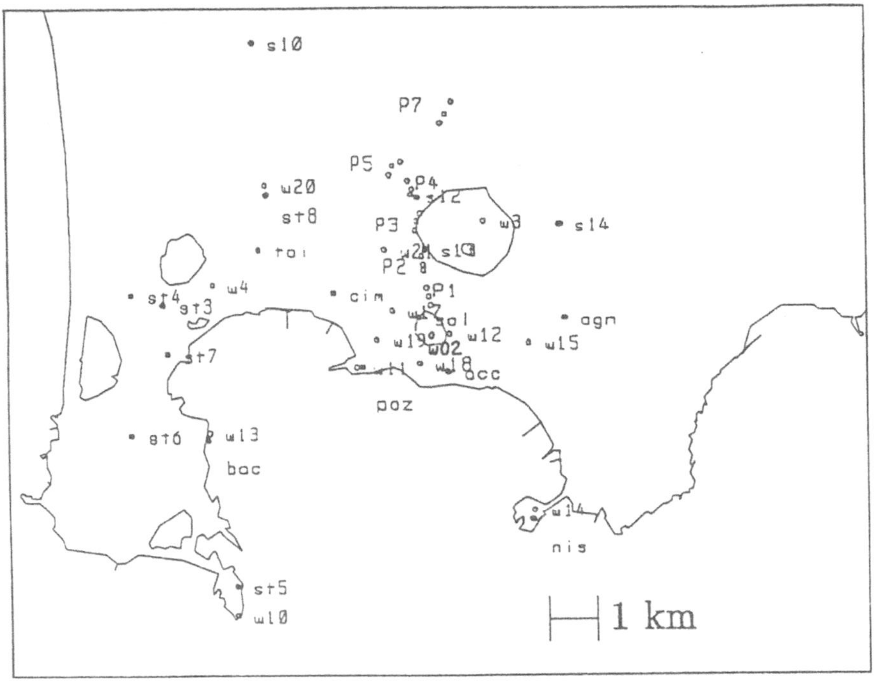

Fig. 2. Locations of seismographs used in this study. UW three-component triggered digital stations are indicated by *open circles*, VO vertical-component telemetered stations by *filled circles*. Stations *p1–p5* and *p7* are UW instruments, each connected to an array of three vertical seismometers with a spacing of ≈0.3 Km

resulting volatile migration into the geothermal system. Cortini et al. (1987) proposed that the pressure source may not be generated by a magma supply, but by the non-equilibrated thermodynamic adjustment of a complex system involving temperature, pressure and chemical composition of a shallow magma chamber undergoing progressive cooling. Recently, De Natale and Pingue (this Vol.) concluded that any pressure source must be located shallower than about 3 km to provide a satisfactory fit to the deformational data. All of these studies were somewhat successful in modeling the deformation (although end-point disagreement is a general difficulty; see Discussion), demonstrating that ground measurements of the type obtained in the Campi Flegrei (tidal gauges and leveling lines) cannot by themselves provide adequate constraints on the detailed character of either source or medium (e.g., Walsh and Decker 1971; Dvorak and Berrino 1991).

Seismic observations can address many important questions regarding the gross features of the Campi Flegrei phenomenon. Source mechanism studies can shed light on the fundamental question of whether active intrusion is primarily or partially responsible for the seismicity. Accurate earthquake locations are capable of providing information on the limits of brittle and/or intruded zones. Path property studies provide information on the elastic and anelastic properties of the medium in and surrounding the seismogenic zone. Shear-wave splitting induced by shear-wave anisotropy (Savage et al. 1989) may provide information on the stress field in the central caldera which is relevant to both seismic source and inflationary studies.

In this work, we summarize the results of seismological studies that primarily used digital microearthquake data from the UW network. Although artificial source studies and/or teleseismic observations would provide key supplements to our knowledge of the caldera, particularly at greater depths, well-constrained conclusions on the distribution of hypocenters, velocity structure, local attenuation of seismic waves and source properties are obtainable solely from local data. We consider the results of such studies along with other geophysical and geological data to examine the general physical processes occurring in the Campi Flegrei caldera.

2 Velocity Structure and Hypocenter Distribution

The shallow velocity structure of the Campi Flegrei was inferred using P- and S-wave arrival times from the UW and VO-AGIP arrays. To obtain a best-fit half-space v_P and v_S velocity model, Aster and Meyer (1988) used P and S arrival times from 197 events recorded at six or more UW stations during 15 March–24 April 1984. Arrival times were used to infer least-square-error hypocenters using a range of half-space models. The location program used was Hypoinverse (Klein 1978), which applies a generalized inverse implemented via singular value decomposition. 147 homogeneous half-space models were tried, independently varying v_p from 2 to 4 km/s in steps of 0.1 km/s and v_P/v_S from 1.5 to 2.1 in steps of 0.1. Each model was evaluated in terms of weighted mean, $\bar{r}$, and the quadrature sum of weighted standard deviations, σ_Q. The weighted means for P and S residuals are:

$$\bar{r}_{P,S} = \frac{\sum_{i=1}^{n_{P,S}} w_{Pi,Si} \cdot r_{Pi,Si}}{\sum_{i=1}^{n_{P,S}} w_{Pi,Si}} \tag{1}$$

where w_{Pi} and w_{Si} are weighting factors derived from P- and S-pick qualities, q_P and q_S, given as:

$$w_{Pi,Si} = (4 - q_{Pi,Si})/4 \tag{2}$$

and n_P and n_S are the number of P and S picks, respectively. The quadrature sum of residuals is:

$$\sigma_Q = [\sigma_P^2 + \sigma_S^2]^{1/2} \tag{3}$$

where the weighted standard devitions of the P and S residuals about the weighted means are:

$$\sigma_{P,S} = \left[\frac{\sum_{i=1}^{n_{P,S}} w_{Pi,Si}^2 (r_{PiSi} - \bar{r}_{P,S})^2}{\sum_{i=1}^{n_{P,S}} w_{Pi,Si}^2} \right]^{1/2} \tag{4}$$

$(v_P, v_P/v_S) = (2.86\,\text{km/s}, 1.76)$ were chosen as parameters for the best-fit homogeneous half-space velocity model and as initial parameters for subsequent iteratively-determined models, based on minimum $\bar{r}_P$, $\bar{r}_S$ and σ_Q values. The weighted rms travel-time residual for all arrivals (P and S) was 0.116 s.

P- and S-wave travel-time residuals were used to study the v_P/v_S variation within the caldera. This work was initiated because clear variations in S-P versus P arrival times were seen in the three-component seismograms recorded by the U.W. array. The simple property that the normalization of each residual, r_i, by the length of its associated straight line ray path, P_i, is an estimate of the differential slowness, Δs_i, along that path with respect to the model slowness, $\bar{s}$,

$$\Delta S_{Pi,Si} = r_{Pi,Si}/\rho_i \tag{5}$$

was used to initially identify correlations between travel-time residuals and ray geometry. An estimate of v_P/v_S based on the ith set of P and S readings at a station is thus:

$$\{v_P/v_S\}_i = \frac{\bar{s}_S + \Delta s_{Si}}{\bar{s}_P + \Delta s_{Pi}} \tag{6}$$

v_P/v_S ratios were evaluated according to equation 6 at station w_{11}, located near the point of highest uplift, for 134 P and S arrival pairs with quality factors of 0 or 1. Plotting these v_P/v_S estimates with respect to azimuth showed a broad range of high-v_P/v_S values, with the highest values in the direction of Solfatara Crater and the center of uplift. The mean v_P/v_S ratio for all arrivals at station w_{11} was significantly higher (1.98) than the mean ratio calculated for all stations in the array (1.78). v_P/v_S ratios as a function of azimuth for stations w_{17} and w_{21} were also high for ray paths in the vicinity of Solfatara Crater and Pozzuoli. Recently, Pujol and Aster (1990) have shown that these velocity variations are also readily resolvable using joint hypocentral determination.

To further localize the velocity anomalies, Aster and Meyer (1988) applied the method of Thurber (1981, 1983), as expanded for v_P and v_S inversion by Eberhart-Phillips (1986). 1842 P and 1006 S weighted pick times from 228 well-located earthquakes recorded at six or more UW stations from 14 February to 6 June 1984 were used in the inversion. Details of the algorithm are discussed extensively in Thurber (1981).

One-dimensional inversions for velocity versus depth alone did not result in significant reductions in data variance. In a previous study, Scarcella (1984) also found no evidence of strong velocity variation with depth in the upper 3 km of the central caldera.

A three-dimensional model with refined hypocenter estimates was obtained in two steps. First, a lateral 9 × 7 km, two-dimensional grid of 63 velocity nodes at 1-km intervals was used, with the half-space velocity model and locations as initial parameters. The output parameters of the two-dimensional inversion were then used as a starting point for a $9 \times 7 \times 3$ km, three-dimensional inversion with 189 velocity nodes spaced at 1-km intervals. The improvement in weighted rms residuals was 13% (0.116 s to 0.101 s, three iterations) in going from the homogeneous half-space to the two-dimensional model and 16% (0.101 s to 0.085 s, four iterations) in going from the two-dimensional model to the three-dimensional model, resulting in a net reduction in weighted rms residuals of 27% (a variance reduction of 47%). v_P and v_S values for the three-dimensional model and v_P/v_S ratios obtained by dividing corresponding P and S velocity node values are contoured in Fig. 3.

Ninety-five percent hypocenter error ellipsoids were estimated from the principal axes of the spatial partition of the covariance matrix (Klein 1978), using an estimated 1σ picking error of 0.03 s, and projected using the algorithm of Aster (1988). Because hypocenter error ellipse dimensions are strongly driven by the weighted rms residual, there was a general decrease in error ellipse dimensions in going from the homogeneous half-space to the three-dimensional velocity models, particularly for events in the well-sampled central caldera. Hypocenter error ellipses obtained using the three-dimensional velocity model are shown in plan and section view in Fig. 4.

About 10% of the locations are beneath the Gulf of Pozzuoli, at depths between 3.5 and 5 km, along a steeply-dipping sheet. The remaining (northern seismic distribution) earthquakes are onshore, with a centroid located about 1 km northeast of the maximum measured uplift and near Solfatara Crater. This is also the region where the largest shocks (all events will $M_D > 3$) and greatest seismic strain release occurred in 1983 and 1984 (Bradisismo Flegreo: Rapporto Sorveglianza 1985). The shallowest seismicity occurs as a distinct, upward-reaching tongue of hypocenters, with a minimum depth of about 2 km, beneath the southwest rim of Solfatara Crater.

The pattern of epicenters vaguely suggests an elliptical ring fault, with an eccentricity of approximately 0.9. The southern edge of the northern seismic zone is close to the La Starza uplift terrace (Fig. 1). No evidence is found to support an appreciably outward-dipping ring structure, such as that proposed for the caldera of Rabaul, New Guinea, by Mori and McKee (1987); rather, a slightly inward-dipping trend is suggested when the epicenters are plotted in section (Fig. 4).

The greatest seismic swarm associated with the recent uplift activity occurred on April 1, 1984. This swarm (De Natale and Zollo 1986) was remarkable in that it was not only highly clustered in both space and time, but occurred in a zone that was devoid of hypocenters for several weeks both preceding and following the episode. The hypocenters of the April 1 swarm are well contained by a 1 km^3 volume. Epicenters for the 1984 time periods 15 March–30 March, 1 April, and 2 April–26 April are shown in Fig. 5a–c.

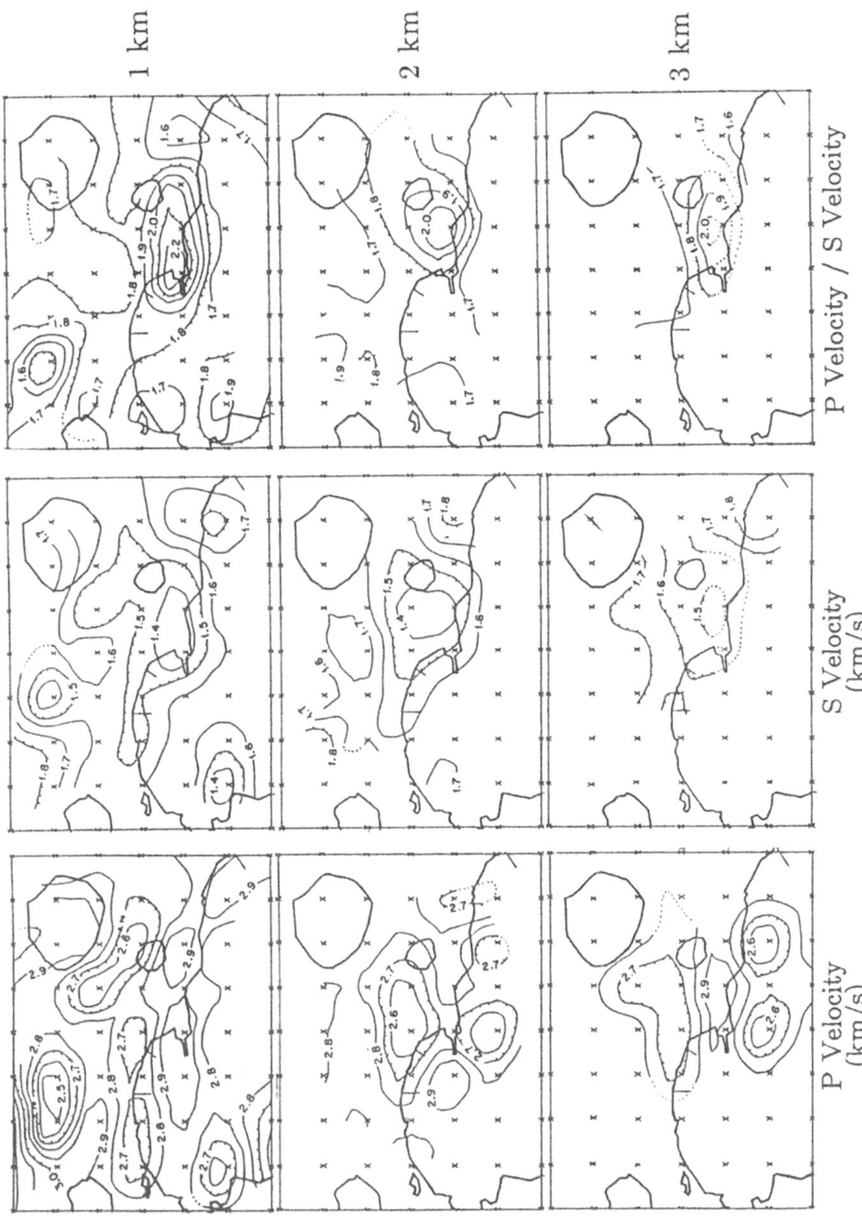

Fig. 3. Contoured v_P, v_S, and v_P/v_S determined for depths of 1, 2, and 3 km. Only those contours that are relatively well-resolved (within the 0.1 diagonal resolution element contours of both P and S) are shown. The contour interval is 0.1. The grid spacing is 1 km (After Aster and Meyer 1988)

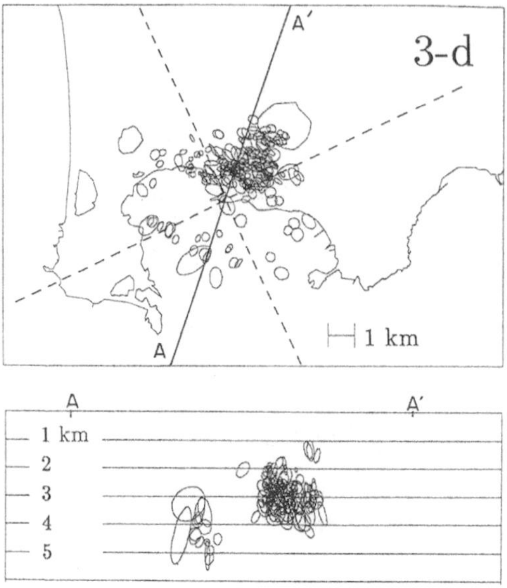

Fig. 4. 2.4σ (90%) hypocenter error ellipses for 228 events using the three-dimensional velocity model. Those hypocenters that fall between the *dotted lines* oriented at ±45° are shown in vertical section along *A-A′* (After Aster and Meyer 1988)

3 Q and Source Parameters

The quality factor, Q, for a general mechanical system, is expressable as

$$Q = 2\pi \frac{E}{\Delta E}, \tag{7}$$

where ΔE is the amount of energy dissipated per cycle of harmonic excitation and E is the peak energy of the system. Shear wave Q, Q_S, was evaluated for the Campi Flegrei from both coda waves and direct arrivals. In addition, we measured the high frequency spectral decay of the acceleration spectrum.

The coda wave quality factor, Q_{Scoda}, was estimated assuming the single-scattering model of Aki and Chouet (1975), where the coda is assumed to be made up primarily of singly-scattered S-to-S waves from randomly-distributed scatterers of uniform density. The resulting formulation for displacement spectral amplitude as a function of time is t, is

$$A(f, t - t_0) = \frac{A_0(f)e^{-\pi f(t-t_0)/Q_{Scoda}(f)}}{t - t_0}, \tag{8}$$

where t_0 is the source origin time.

We suggest that the hypothesis of single-scatter dominance is valid for our data set based on the inequality due to Gao et al. (1983)

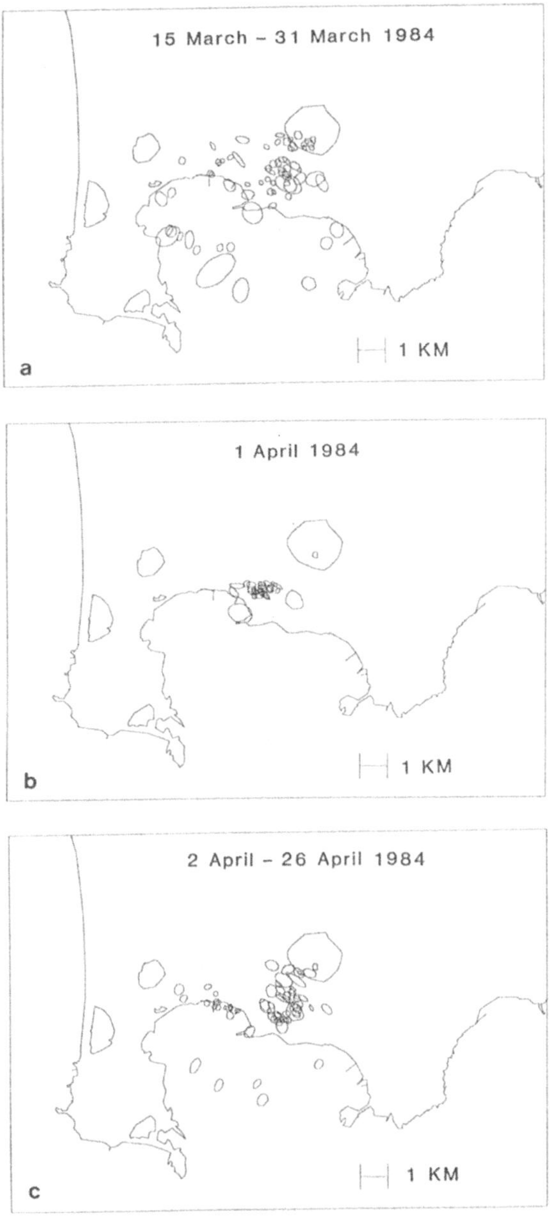

Fig. 5a–c. Epicenters of periods before, during, and after the April 1, 1984 earthquake swarm. Note the strong clustering in space and time and the quiescence of the swarm zone both prior to and following April 1

$$t - t_0 < \frac{l}{0.8\beta} \tag{9}$$

where β is the shear wave velocity and l is the mean free path. For the Campi Flegrei, using $l = 50$ km (Castellano et al. 1984) and $\beta = 2.5$ km/s, we have equation 9 satisfied for $t - t_0 < 36$ s. The majority of observed codas in the Campi Flegrei satisfy this criterion; the mean unfiltered coda length (where the amplitude envelope intersects the noise) was approximately 25 s, implying a sampled hemispherical volume of approximately 30-km radius.

Equation (8) can be linearized to obtain:

$$\ln[A(f, t - t_0) \cdot (t - t_0)] \equiv Y = \ln[A_0(f)] - \pi f(t - t_0)/Q_{Scoda}(f). \tag{10}$$

Thus, for a particular frequency, f_1, $1/Q_{Scoda}$ (f) is:

$$1/Q_{Scoda}(f_1) = -\frac{1}{\pi f}\left[\frac{\partial Y}{\partial t}\right]_{f=f_1}. \tag{11}$$

A smoothed estimate of $A(f, t - t_0)$ was computed via a discrete Fourier transform in a sliding window 3 s wide, moving in 1.5 s steps. The beginning of the coda was taken as the maximum of the envelope of Y, and end of the coda was taken as the time when Y begins to increase because of noise dominance.

117 earthquakes were selected on the basis of the best signal to noise ratio for the analysis of 3-component seismograms at 5 digital stations (Del Pezzo et al. 1987). As a previous analysis (Del Pezzo et al. 1985) showed no significant bias in Q_{Scoda} for different components, Q_{Scoda} was averaged over the three components and events for each station. Q_{Scoda} values in each frequency band exhibit good agreement from station to station, being almost independent of frequency. Q_{Scoda} ranges from 120 to 250 in the frequency band 3–24 Hz. Average station values, are shown in Fig. 6 as a function of frequency along with results from several other seismically active areas where similar estimates have been made. Note that the frequency dependence of Q_{Scoda} is relatively weak for volcanic regions. If one assumes the attenuation formulation of Dainty (1981)

$$1/Q(f) = 1/Q_i + c/(2\pi f l), \tag{12}$$

where $1/Q_i$ represents the intrinsic attenuation due to internal frictional mechanisms and the second term represents frequency-dependent attenuation due to scattering, where l is the mean free path and c is the seismic velocity, the relative independence of Q on frequency in volcanic regions apparent in Fig. 6 may indicate that the predominant mechanism in the Campi Flegrei at these frequencies is intrinsic. This is consistent with the high temperatures and the presence of magmatic bodies which characterize volcanic environments (Kampfmann and Berckhemer 1985)

Q_{Scoda} represents an average value over an approximately 30-km radius half-sphere centered on the Campi Flegrei. Thus, estimates of $Q_{Sdirect}$, which sample only the uppermost few kilometers at the crust could be markedly different. To estimate the attenuation of shallow layers, we computed $Q_{Sdirect}$ from the spectral analysis of the first 2.56 s of the S_H signal.

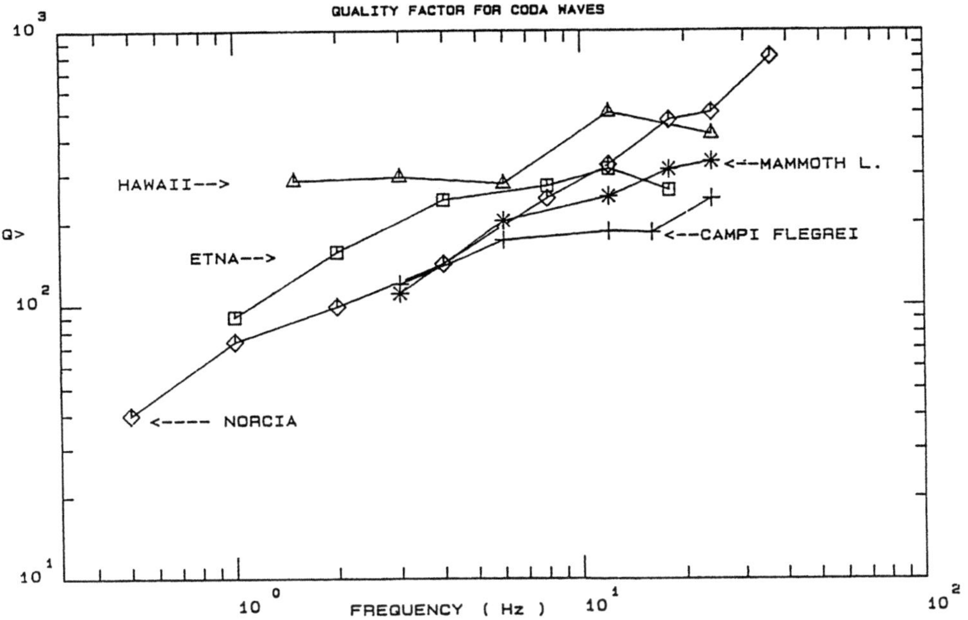

Fig. 6. Q_{Scoda} vs frequency for the Campi Flegrei (station average) and for other seismic areas: Hawaii (Chouet 1976); Mammoth Lakes, California (Del Pezzo pers. comm.); Etna, Sicily (Del Pezzo pers. comm.); Norcia, northern Apennines (Del Pezzo and Zollo 1984)

The S_H wave displacement spectrum may be expressed as (Aki and Richards 1980):

$$a(f) = A_0(f)e^{-\pi f(r/\beta)/Q_{Sdirect}} \tag{13}$$

where r is the hypocenter-station distance and $Q_{Sdirect}$ is assumed to be constant with respect to frequency. At frequencies less than the spectral corner frequency and above the low frequency instrument corner (1 Hz), $A_0(f)$ is constant and we can take the natural logarithm to obtain:

$$Q_{Sdirect} = \left[\frac{-1}{\pi(r/\beta)}\left(\frac{d\ln A(f)}{df}\right)\right]^{-1}. \tag{14}$$

S_H displacement spectra for the same set of data used to estimate Q_{Scoda} were used to compute $Q_{Sdirect}$ from Eq. 14 by Del Pezzo et al. (1985). Averaging over all earthquakes and stations gives $Q_{Sdirect} = 110 \pm 50$.

De Natale et al. (1988) computed the spectral decay parameter, κ, for P and S_H acceleration spectra from 32 Campi Flegrei earthquakes using the acceleration spectra formulation of Anderson and Hough (1984)

$$a(f) = a_0 e^{-\pi\kappa f} \tag{15}$$

where a_0 is constant. The mean κ values were $\bar{\kappa}_P = 0.008 \pm 0.027$ and $\bar{\kappa}_{SH} = 0.015 \pm 0.02$. Negative κ observations are inconsistent with a hypothesis of high frequency spectral amplitudes being dominated by attenuation and reflect the large variation in the spectral estimates and perhaps the nearness of the source. The average value of κ at the Campi Flegrei was much smaller than that found at Friuli, northern Italy by De Natale et al. (1987a) or in California by Anderson and Hough (1984). For an average ray-path length of 4.5 km, the mean value of κ suggests a Q_S value of approximately 190, using

$$\kappa(r) = \frac{r}{Q\beta}. \tag{16}$$

Del Pezzo et al. (1987) used displacement spectra from 181 seismograms to obtain seismic moments, source radii, and stress drops for 32 Campi Flegrei earthquakes. Seismic moments computed from low-frequency spectral levels ranged from $5 \times 10^{10} - 1.3 \times 10^{13}$ N-m. Stress drops were nearly constant with respect to moment, ranging from approximately 2–10 bar using a Brune (1970) model and 10–30 bar using a Madariaga (1976) model.

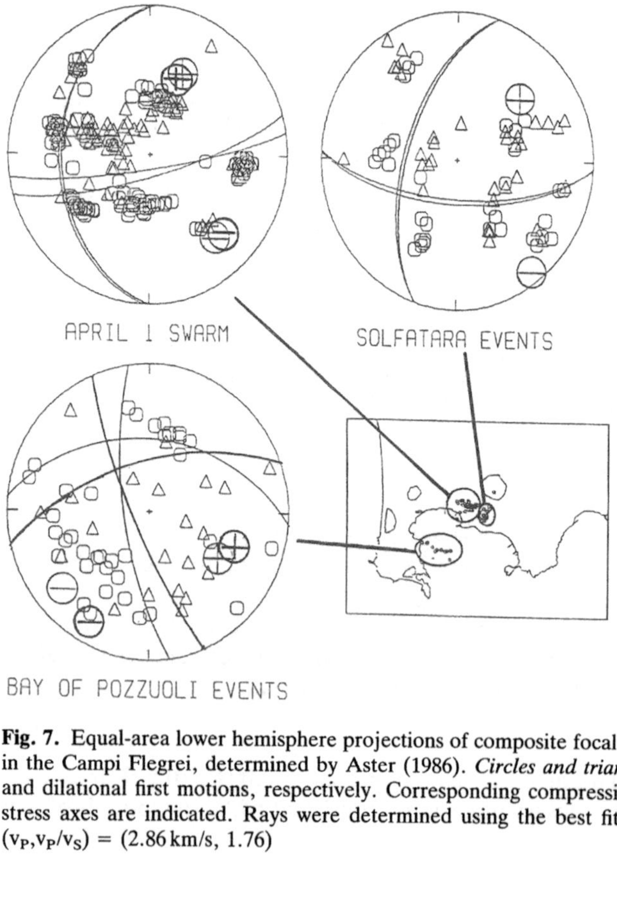

Fig. 7. Equal-area lower hemisphere projections of composite focal mechanisms for three areas in the Campi Flegrei, determined by Aster (1986). *Circles and triangles indicate* compressional and dilational first motions, respectively. Corresponding compressional (+) and tensional (−) stress axes are indicated. Rays were determined using the best fit half-space velocity model, $(v_P, v_P/v_S) = (2.86\,\text{km/s}, 1.76)$

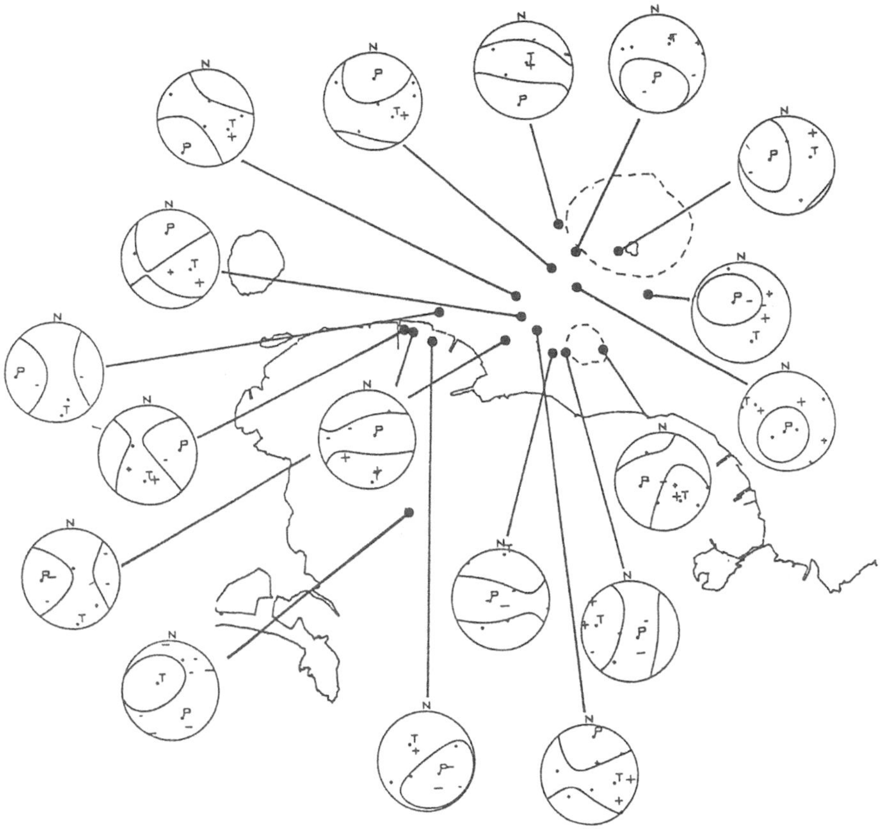

Fig. 8. Lower hemisphere projections of focal mechanism radiation patterns determined De Natale et al. (1987b) from moment tensor inversion (zero trace constraint)

3.1 Fault-Plane Solutions and Moment Tensor Analysis

Gaudiosi and Iannaccone (1984) estimated 15 double-couple fault-plane solutions for Campi Flegrei earthquakes using P-wave first motions from the VO-AGIP network and the method of Wickens and Hodgson (1967). They found most of the faulting in the Solfatara and Pozzuoli areas to be nearly normal, with a highly variable strike. Aster (1986) applied the exhaustive search algorithm of Snoke (1986) using only the clearest first P motions to obtain best-fit composite double-couple solutions for events in three zones: Solfatara crater (13 events, 67 motions), the 1 April, 1984 swarm zone (39 events, 184 motions) and the western Gulf of Pozzuoli (11 events, 57 motions). These composites were largely incoherent (Fig. 7), indicating a wide range of mechanisms even for these highly restricted zones.

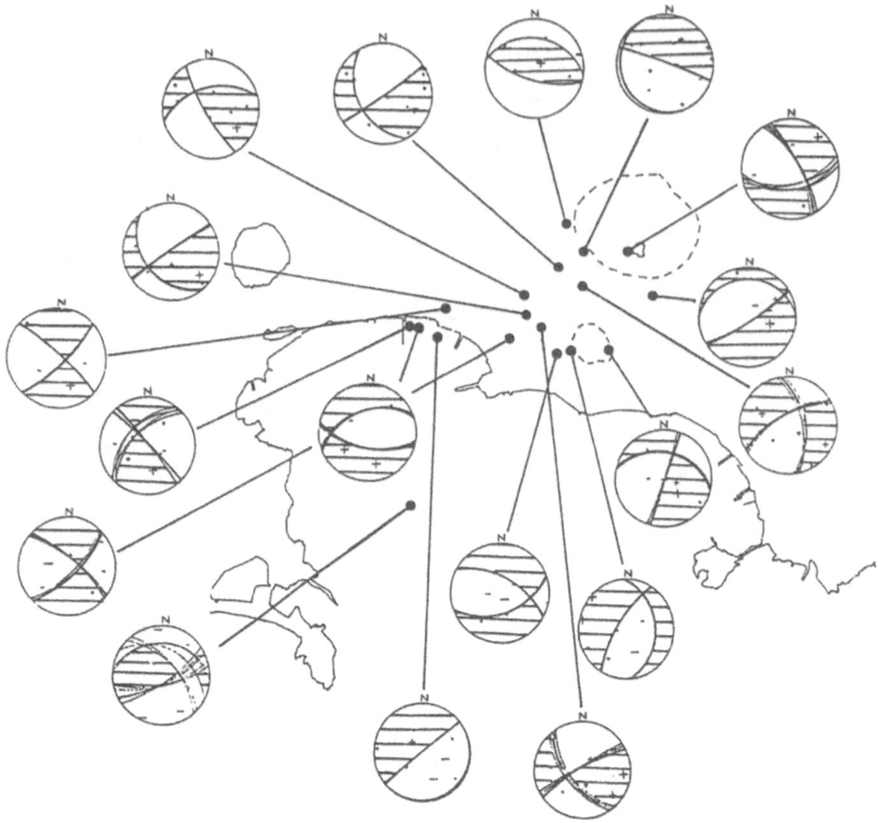

Fig. 9. Lower hemisphere projections of focal mechanism radiation patterns determined De Natale et al. (1987b) from moment tensor inversion (double-couple constraint)

To perform a general investigation of source radiation patterns in the Campi Flegrei and elsewhere, De Natale et al. (1987b) determined focal mechanisms via moment tensor inversion using P- and S_H-wave displacement first pulse amplitudes corrected for attenuation. Resulting moment tensor solutions for 18 events with zero isotropic component and double couple constraints are shown in Figs. 8 and 9, respectively. These earthquakes are well interpreted by double-couple mechanisms. An F-test indicated that the double-couple, zero-trace, and free solutions were not statistically different at the 95% confidence level. Furthermore, zero-trace solutions and double couple solutions were quite similar in the majority of cases. De Natale et al. (1987b) found that solutions which allow isotropic components showed a smaller number of mechanisms which were similar to double couples. However, except for a single ill-constrained event, the variance reductions were not large enough to justify the inclusion of the additional parameter needed to characterize the free solution.

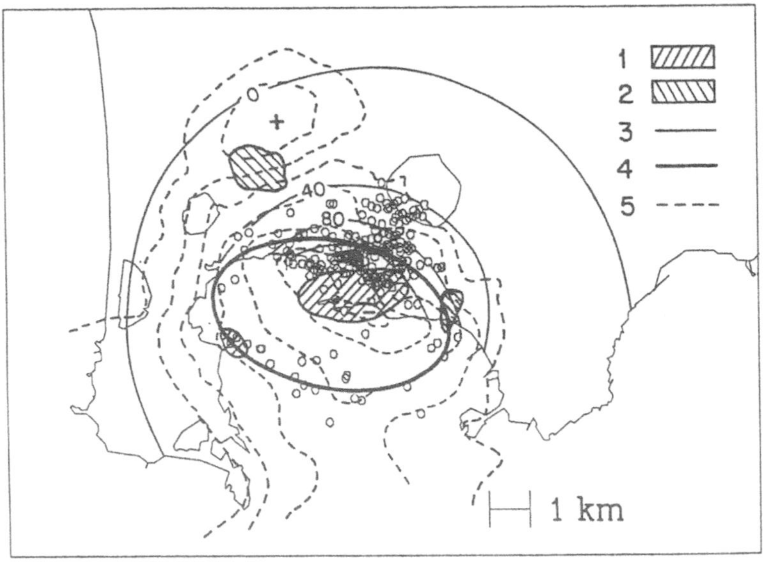

Fig. 10. Earthquake epicenters (February–June, 1984) with other geophysical results superimposed: *1* areas of $v_P/v_S > 1.9$ at 1 km depth (Fig. 3); *2* areas of $v_P/v_S < 1.6$ at 1 km depth (Fig. 3); *3* 40-cm elevation contours for the 1982–1985 uplift (Bradisismo Flegreo: Rapporto Sorveglianza 1985); *4* elliptical trace of the hypothesized ring fault of Aster and Meyer (1988); *5* 1 mgal Bouguer gravity contours (La Torre and Nannini 1980)

4 Discussion

Epicenter locations, the elliptical hypocenter trend, regions of anomalous v_P/v_S, gravity contours, and uplift contours are shown together in Fig. 10. To summarize the geographic correlations:

1. There is excellent agreement between the principal high-v_P/v_S anomaly, the gravity minimum, and the uplift; all are approximately centered on Pozzuoli, near the middle of the caldera.
2. Most seismicity occurs north of the centrally correlated features of 1, with only a few hypocenters occurring in the central anomalous velocity zone.
3. The roughly elliptical hypocenter trend is relatively ill-constrained in the northern seismic zone compared to the east, west and south. The ellipse is approximately centered on Pozzuoli.
4. There is spatial correlation between the northwest gravity high and the principal low-v_P/v_S anomaly, but the three-dimensional inversion did not encompass the entire northwest gravity high due to limited ray-path coverage. This zone is occupied by the prominent yellow tuff cone of Gauro Volcano (Fig. 1).

Additionally, to summarize the results of attenuation and source studies:

5. Q_{Scoda}, sensitive out to a radius of approximately 30 km, ranges from 120 to 250 in the 3–24 Hz frequency band; this degree of variation is weak relative to non-volcanic regions and suggests high intrinsic attenuation.
6. Q_S estimates from S_H-wave displacement spectra, for signals that sample only the caldera, indicate a somewhat lower range than Q_{Scoda} of 110 ± 50.
7. Estimates of the high-frequency decay of acceleration spectra show much scatter, but suggests a Q_S of about 190 for caldera material.
8. Seismic source parameters indicate relatively low (2–10 bar Brune) stress drops, with little or no dependence on moments from $10^{10}-10^{13}$ N-m.
9. Composite focal mechanisms indicate wide variation in faulting, even for very restricted regions (e.g. the region of the April 1, 1984 swarm).
10. Moment tensor analysis yields no clear evidence for non-double couple source mechanisms.

To account for the high-v_P/v_S central velocity anomaly, we propose a low-density, highly fractured zone of liquid water-filled cracks or pores. The observed temperature profile in the SV1 well, approximately 2 km to the north of Pozzuoli (Fig. 1), $dT/dz \approx 140\,°C/km$, $T_{max} = 420\,°C$ at 3040 m (Giberti et al. 1984), is insufficient (by about $dT/dz \approx 260\,°C/km$) to produce steam under lithostatic pressure, using a mean density $\bar{\rho} = 2.0\,g/cm^3$, and the boiling-point curve for pure water below the critical point (21.8 MPa, 374 °C; Fisher 1976). Furthermore, unless the (undrilled) Pozzuoli region has a significantly higher temperature gradient than this, partial melting cannot contribute to the velocity anomaly.

In areas where dT/dz is significantly higher than the values observed in geothermal drilling, and/or where the confining pressure is less than lithostatic (due to a surface-connected crack system) regions of steam will exist. Both of the these situations occur in the fumerole fields at Solfatara Crater and offshore just east of Pozzuoli. Bolognesi et al. (1986) found that fluids from these fumeroles had chemical and isotopic properties that were consistent with a meteoritic origin, implying shallow circulation. At greater depths, drilling revealed evidence of abundant sub-boiling-point water in closed reservoirs at depths up to 3 km.

Liquid water- and steam-filled cracks will have quite different effects on the bulk elastic moduli of rocks, because normal stress-coupling between the opposing faces of cracks will be much greater in the liquid-filled case than in the gas-filled case. Aster and Meyer (1988) estimated the degree of cracking necessary in a liquid-saturated system necessary to account for the velocity anomaly using theory of O'Connell and Budiansky (1976) for the elastic behavior of solids with pervasive ellipsoidal cracks (semiaxes $a > b \gg c$). Assuming the unmodified velocity parameters for the central caldera to be those of the best-fit homogeneous half-space velocity model, where $(v_S, v_P/v_S) = (1.7\,km/s, 1.8)$, then the high v_P/v_S (2.20) and low v_S (1.35 km/s) values obtained in the velocity inversion for material 1 km beneath station w_{11} can be modeled with a zone of liquid-saturated cracks having a crack density parameter, ε, of about 0.5, where:

$$\varepsilon = \frac{2\eta A^2}{\pi p} \tag{17}$$

and A is the elliptical crack area, p is the elliptical crack perimeter, and η is the number of cracks per unit volume.

As the central velocity anomaly was essentially aseismic during the 1982–1985 uplift episode, we hypothesize that the observed seismic properties are primarily due to pre-existing factors. The cracking and fluid migration presumedly responsible for the central velocity anomaly could have been seismically initiated during some past bradyseismic episode, or may alternately represent an inherently highly porous and saturated zone (Tatham 1982). Regardless of the nature of the material, it is clear that the area of the central velocity anomaly incurred considerable deformation during the most recent bradyseismic inflation, when the elevation change reached 1.7 m at the Port of Pozzuoli, and yet remained nearly aseismic. Because none of the geothermal wells intersect this region (Giberti et al. 1984; De Vivo et al. 1987), there are no direct sampling data to suggest why the seismic properties of this zone are so different from those of the area to the north, where more normal v_P/v_S ratios are observed, and where most seismicity occurred. Gravity contours (La Torre and Nannini 1980) indicate that it is a low-density zone. Velocity studies (Aster and Meyer 1988; Scarcella 1984) have found evidence for only a weak dependence of velocity with depth. It thus appears that this central zone represents a region of relatively homogeneous, low-density, high v_P/v_S, low velocity material that has been so highly fractured by past activity or is so inherently weak that it now deforms aseismically. Future drilling is therefore strongly indicated to provide more data on this unusual zone. In the context of the crack model, we must suppose that the shallow, low-v_P/v_S region beneath Gauro Volcano is due to dry cracks (the absence of fumeroles in this area makes it unlikely that there is near-surface steam present).

Kampfmann and Berckhemer (1985) studied the elastic and anelastic response of high-temperature rocks in the laboratory. In particular, they measured the complex shear modulus of various ultrabasic and basic rocks under forced torsional oscillations in the frequency range 0.003 to 30 Hz. Their results show that at 600 °C, Q is only weakly dependent on frequency from 1–30 Hz, with values in the range 100–500. These observations suggest that the weak dependence of Q observed in the Campi Flegrei and in other volcanic areas may be predominantly an effect of high intrinsic attenuation due to high temperatures. The observed Q values from direct and coda wave studies are consistent with this model.

Seismic activity from the most recent uplift episode is characterized by spatial-temporal swarm activity (Del Pezzo et al. 1984; De Natale and Zollo 1986). The climatic swarm during the most recent uplift occurred on 1 April 1984 (Fig. 6), when the VO network recorded approximately 500 earthquakes in 12 h. An important question regarding this swarm is whether it is more appropriately interpreted as a localized intrusive event (i.e. as a dike swarm) or as the rupture of an asperity due to stress accumulation. Widely varying fault plane solutions for the 1 April swarm (Fig. 9) are consistent with fracturing in a region of heterogeneous material properties and/or widely varying stress field, but do not suggest a distinct preference for a particular model. We do, however, favor the asperity hypothesis for several reasons:

1. The hypothesis of a tectonically stressed, relatively weak, and intensively fractured seismogenic zone is in agreement with the high degree of temporal earthquake clustering described by De Natale and Zollo (1986) for earthquakes in the northern seismic zone, and with the constant, low Brune stress drop values of approximately 2–10 bar (Del Pezzo et al. 1987).
2. Low-frequency events which might be generated by intrusion (e.g., Koyanagi et al. 1987) have never been observed in the Campi Flegrei.
3. All of the Campi Flegrei moment tensors, except for one marginal case, are well approximated by double-couple source mechanisms. This is inconsistent with magmatic transport, which might be expected to generate detectable volume moment tensor components.
4. The April 1 swarm fills a seismic gap near the outskirts, rather than in the interior of the seismic and inflationary zone. This favors the idea of stress accumulation at an outlying asperity, although marginal intrusion along a pre-existing zone of weakness remains a possibility.

Because the best evidence is negative (lack of low-frequency events and nearly double-couple moment tensors), further study could still conceivably produce evidence for seismogenic intrusion. It does seem clear, however, that the majority of the observed seismicity is well modeled by brittle faulting in the shallow crust in response to the stress field changes induced by the ground deformation and does require intrusion.

The spatial resolution of a caldera ring feature via seismicity at the Campi Flegrei is intermediate between the cases of Rabaul (very well resolved; Mori and McKee 1987), and Mammoth Lakes, California (very poorly or non-resolved; Cockerham and Pitt 1984). As at Rabaul, there is a fairly abrupt cutoff of hypocenters at the lower terminator of the distribution, establishing an approximately 3.5 km upper limit for the roof of any underlying magma body.

Ninety percent of the earthquakes are located in the northern seismic zone, with a centroid approximately 1 km north of Pozzuoli (Fig. 5). A significant change has not been observed in the extent of this zone by the VO-AGIP network for the period 1 January 1983 through 31 April 1984 (during which the maximum elevation of the uplift changed by more than +1 m). Many attempts to model the ground deformation (De Natale and Pingue) resulted in significant disagreement near the end points; the observed zero-inflation contour is relatively constant, where as the models generally predict a measurable increase in the extent of the uplifted area with time. Furthermore, the theoretical slope of the deformation approaches zero more gradually than the observations indicate (Berrino et al. 1984).

These results suggest that both the uplift and principal seismic areas are geologically confined. In particular, a distinct geological feature, the La Starza terrace (Fig. 1), may act as a southerly décollement for the northern seismic zone. The greatest deformation affects the less dense materials of the central caldera, and the northern seismic zone may represent a relatively brittle region, that is weak or fractured enough to produce low stress-drop events.

The model of the shallow ($z < 3$ km) caldera that emerges from local seismic studies and other available geophysical data is thus one of a heterogeneous lid overlying a pressure source. Indicative of the degree of heterogeneity is the observation that despite a highly coherent uplift, seismic sources have a wide range of focal mechanisms and the seismicity is asymmetric with respect to the center of uplift. Q studies suggest that attenuation is dominated by intrinsic mechanisms, perhaps as a result of locally elevated temperatures and is somewhat lower in the caldera relative to the surrounding region. The low density central caldera is predominantly aseismic with a velocity structure consistent with cracked and/or highly porous, water saturated and relatively incompetent material. There is no evidence for the seismogenic intrusion of magma; the data is consistent with low-stress, brittle fracture of material without dilatational strain. An elliptical trend of epicenters is suggestive of an inward-dipping ring fault, but the wide range of focal mechanisms and the pattern of seismicity are not indicative of coherent motion on such a fault during the recent uplift.

Acknowledgements. The authors gratefully acknowledge the field support of Bruno Tramma and Hans Meyer, to whom this paper is dedicated. This work was funded under NSF grants EAR 85-17919 and INT 84-00920.

References

Aki K, Chouet B (1975) Origin of coda waves: source, attenuation and scattering effects. J Geophys Res 80:3322–3342

Aki K, Richards P (1980) Quantitative seismology, theory and methods, vol I. Freeman, San Francisco, CA, 932 pp

Anderson JG, Hough S (1984) A model for the shape of the Fourier amplitude spectrum of acceleration at high frequencies. Bull Seismol Soc Am 74:1969–1994

Aster R (1986) Hypocenter locations and velocity structure in the Phlegraean Fields, Italy. MS Thesis, Univ Wisconsin-Madison, Madison, Wisconsin, USA, 128 pp

Aster R (1988) On projecting error ellipses. Bull Seismol Soc Am 78:1373–1374

Aster R, Meyer R (1988) Three-dimensional velocity structure and hypocenter distribution in the Phlegraean Fields, Italy. Tectonophysics 149:195–218

Berrino G, Corrado G, Luongo G, Toro B (1984) Changes accompanying the 1982 Pozzuoli uplift. Bull Volcanol 47:187–200

Bianchi R, Coradini A, Federico C, Giberti G, Sartoris G, Scandone R (1984) Modeling of surface ground deformation in the Phlegraean Fields volcanic area, Italy. Bull Volcanol 47:321–330

Bolognesi L, Noto P, Nuti S (1986) Fluid geochemistry in Phlegraean fields area: its contribution to a geochemical model (Abstr). Int Worksh Resumed activity in areas of quiescent volcanism, Firenze, Italy, May 15–17

Bonafede M, Dragoni M, Quareni F (1986) Displacement and stress fields produced by a centre of dilation and by a pressure source in a viscoelastic half space: application to the study of ground deformation and seismic activity at Campi Flegrei, Italy. JR Astr Soc 87:455–485

Bonasia V, Pingue F, Scarpa R (1984) A fluid-filled fracture as possible methanism of ground deformation at Phlegraean Fields, Italy. Bull Volcanol 47:313–320

Bonito M (1691) Terra Tremante o Vero Continuatione de'Treuoti. Library of the Vesuvius Observatory, Napoli, 822 pp

Bradisismo Flegreo: Rapporto Sorveglianza (Phlegraean bradyseism: research report) (1985) Publ Vesuvius Observatory, Naples, Italy, 93 pp

Brune J (1970) Tectonic stress and the spectra of seismic shear waves from earthquakes. Bull Seismol Soc Am 60:4997–5009

Casertano L, Oliveri del Castillo A, Quagliariello M (1976) Hydrodynamics and geodynamics in Phlegraean Fields area of Italy. Nature (Lond) 264:5582

Castellano M, Del Pezzo E, De Natale G, Zollo A (1984) Seismic coda Q and turbidity coefficient at the Phlegraean Fields volcanic area: preliminary results. Bull Volcanol 47:219–224

Chouet B (1976) Source, scattering and attenuation effects on high frequency seismic waves. PhD Thesis, Massachusetts Inst Tech, Cambridge, Mass

Cockerham R, Pitt A (1984) Seismic activity in Long Valley caldera, eastern California, U.S.A.: June 1982 through July 1984, Proc Worksh XIX Active Tectonic and magmatic processes beneath Long Valley caldera, eastern California, vol I and II. USGS Open File Rep 84-939:942

Corrado G, Guerra I, Lo Bascio A, Luongo G, Rampoldi R (1976) Inflation and microearthquake activity of Phlegraean Fields, Italy. Bull Volcanol 40:1–20

Cortini M, Liguori A, Del Pezzo E, Nassisi A (1987) The space-temporal evolution of a bradyseismic crisis in a large volcanic area: a nonequilibrium thermodynamics approach. Earth Planet Sci Lett (submitted)

Dainty A (1981) A scattering model to explain seismic Q observations in the lithosphere between 1 and 30 Hz. Geophys Res Lett 8:1126–1128

D'Amore, F Panichi C (1985) Geochemistry in geothermal exploration. Energy Res 9:277–298

Del Pezzo E, Zollo A (1984) Attenuation of coda waves and turbidity coefficient in central Italy. Bull Seismol Soc Am 74:2655–2659

Del Pezzo E, De Natale G, Zollo A (1984) Space-time distribution of small earthquakes at Phlegraean Fields, south-central Italy. Bull Volcanol 47:201–208

Del Pezzo E, De Natale G, Scarcella G, Zollo A (1985) Q_c of three component seismograms of volcanic microearthquakes at Campi Flegrei volcanic area (southern Italy). Pageoph 123:685–696

Del Pezzo E, De Natale G, Martini M, Zollo A (1987) Source parameters of microearthquakes at Phlegraean Fields (southern Italy) volcanic area. Phys Earth Planet Int 47:25–42

De Natale G, Zollo A (1986) Statistical analysis and clustering features of the Phlegraean Fields earthquake sequence (May 1983–May 1984). Bull Seismol Soc Am 73:801–814

De Natale G, Maderiaga R, Scarpa R, Zollo A (1987a) Source parameter analysis from strong motion records of the Friuli (Italy) earthquake sequence (1976–1977). Bull Seism Soc Am 77:1127–1146

De Natale G, Iannaccone G, Martini M, Zollo A (1987b) Seismic sources and attenuation properties at the Campi Flegrei volcanic area. Pageoph 125:883–917

De Natale G, Faccioli E, Zollo A (1988) Scaling of peak ground motions from digital recordings of small earthquakes at Campi Flegrei, southern Italy. Pageoph 126:37–53

Di Vito M, Lirer L, Mastrolorenzo G, Rolandi G, Scandone R (1985) Volcanological Map of Campi Flegrei (preprint). Universita' degli Studi di Napoli, Dipartimento di Geofisica e Vulcanologia, Napoli, Italy

De Vivo B, Barbieri M, Belkiu H et al. (1987) The Phlegraean Fields (Naples, Italy) geothermal system: fluid inclusions and Sr isotopes from deep geothermal wells. (unpubl)

Dvorak J, Berrino G (1991) Recent ground movement and seismic activity in Campi Flegrei, southern Italy: episodic growth of a resurgent dome. J Geophys Res 96:2309–2324

Dvorak J, Mastrolorenzo G (1991) History of vertical movement in Pozzuoli Bay, southern Italy: the result of regional extension related to evolution of the Tyrrhenian Sea and of local volcanic activity. GSA Spec Pap 263 (in press)

Eberhart-Phillips D (1986) Simulps5. Fortran program, USGS Menlo Park, (unpubl)

Fisher J (1976) The volumetric properties of H_2O – a graphical portrayal. J Res US Geol Surv 4, 2:189–193

Gao L, Lee L, Biswas N, Aki K (1983) Effects of multiple scattering on coda waves in a three-dimensional medium. Pageoph 121:1–16

Gaudiosi G, Iannaccone G (1984) A preliminary study of stress pattern at Phlegraean Fields as inferred from focal mechanisms. Bull Volcanol 47:225–232

Giberti G, Moreno S, Sartoris G (1984) Thermal history of the Phlegraean Fields in the last 50000 years. Bull Volcanol 47:331–341

Kampfmann W, Berckhemer H (1985) High-termperature experiments on the elastic and anelastic behavior of magmatic rocks. Phys Earth Planet Int 40:223–247

Klein F (1978) Hypocenter location program – hypoinverse, part 1: users' guide to versions 1, 2, 3, and 4. USGS Open File Rep 78-694:113

Koyanagi R, Chouet B, Aki K (1987) Origin of volcanic tremor in Hawaii, part I, data from the Hawaiian Volcano Observatory 1969–1985. Volcanism in Hawaii, US Geol Surv Prof Pap 1350:1221–1257

La Torre P, Nannini R (1980) Geothermal well location in southern Italy: the contribution of geophysical methods. Boll Geofis Teor Appl XXII, 87:201–209

Lyell C (1873) Principles of geology, vols I & II. Appleton, New York, 652 pp

Madariaga R (1976) Dynamics of an expanding circular fault. Bull Seismol Soc Am 66:639–666

Martini M (1986) Thermal activity and ground deformation at Phlegraean Fields, Italy, precursor of eruptions or fluctuations of quiescent vulcanism, a contribution of geochemical studies. J Geophys Res 91:12255–12260

Mogi K (1958) Relations between the eruptions of various volcanos and the deformations of the ground surfaces around them. Bull Earthq Res Inst Tokyo Univ 36:99–134

Mori J, McKee C (1987) Outward-dipping ring-fault structure at Rabaul caldera as shown by earthquake locations. Science 235:193–195

O'Connell R, Budiansky B (1976) Seismic velocities in dry and saturated cracked solids. J Geophys Res 79:5413–5426

Parascandola A (1947) I fenomeni bradisismici del Serapeo, Pozzuoli. Genovese, Napoli

Pujol J, Aster R (1990) Joint hypocentral determination and the detection of low-velocity anomalies, an example from the Phlegraean Field earthquakes. Bull Seismol Soc Am 80:129–139

Rosi M, Santacroce R (1984) Volcanic hazard assessment in the Phlegraean Fields: a contribution based on stratigraphic and historical data. Bull Volcanol 47:359–370

Savage M, Shih X, Meyer R, Aster R (1989) Shear-wave anisotropy of active tectonic regions via automated S-wave polarization analysis. Tectonophysics 165:279–292

Scarcella G (1984) Methodi di Inversione di Dati Sismici: applicazioni nel Basso Tirreno e nei Campi Flegrei. Thesis, Univ Naples, Naples, Italy

Schneider J, Aster R, Powell L, Meyer R (1987) Timing of portable seismographs from Omega navigational signals. Bull Seismol Soc Am 77:1457–1478

Snoke A (1986) Focmec. Fortran program, Virginia Polytech Inst State Univ (unpubl)

Tatham R (1982) V_P/V_S and lithology. Geophysics 47:336–344

Thurber C (1981) Earth structure and earthquake locations in the Coyote Lake area, central California. PhD Thesis, Massachusetts Inst Tech, Cambridge, Mass, 332 pp

Thurber C (1983) Earth structure and earthquake locations in the Coyote Lake area, central California. J Geophys Res 88:8226–8236

Walsh J, Decker R (1971) Surface deformation associated with volcanism. J Geophys Res 76:3291–3302

Wickens A, Hodgson J (1967) Computer reevaluation of earthquake mechanism solutions. Publ Dominion Observatory, Ottawa

Seismological and Geodetic Data at Campi Flegrei (Southern Italy): Constraints on Volcanological Models

G. De Natale and F. Pingue

Abstract

Seismological data from local earthquakes and ground deformation measurements carried out during the last ground uplift episode (1982–1984) at Campi Flegrei place considerable constraints on the possible volcanological models for seismo-volcanic crises. The presence of a high level seismicity down to 4 km, tectonic in character, indicates such depth as the shallowest limit for the location of a massive magma chamber, where a ductile behaviour of the rocks is expected. Inversion of vertical and horizontal ground deformations has been performed in terms of extended sources of volume expansion in a half-space, as a superposition of point sources. A gradient method with positive constraints has been used to test the possibility, for a generic source with constraints on location and source strenght, of fitting in the deformation data. The results of such analysis show that the source volume expansion (or, equivalently, overpressure) must be located at depths shallower than about 3 km. This result indicates that the source cannot be a magma chamber, since no data support the presence of large magma bodies at such shallow depths. Deeper sources cannot satisfactorily fit the data. Moreover, the source must be large enough to allow for the total amount of deformation being generated by reasonably small values of overpressure. This last observation precludes the source as a shallow intrusive magma body, since it overlaps with most of the seismicity. The introduction of elastic or viscoelastic heterogeneity of the medium does not seem allow a good fit to the data with sources deeper than 4 km, nor with reasonable overpressures (less than a few tens of MPa), if realistic values of elastic and viscoelastic parameters are considered. In view of these results we believe that an interpretation of brady-seismic crises such as those in 1970–1972 and 1982–1984 cannot be interpreted as due, exclusively, to increased pressure within a magma chamber. Other effects, connected to the heating of shallow rocks and equifers, should also be considered.

1 Introduction

Campi Flegrei is a volcanic area located north of Naples, probably formed by a caldera collapse that occurred about 35 000 years ago. Recent volcanic activity has been confined to the central part of the caldera, including only two eruptions in historical times, the last one in 1538 forming a volcanic cone 130 m high, called Monte Nuovo. The episodes of slow ground movements in this area are named "bradyseism". The last two episodes of strong ground uplift and seismicity occurred in the periods 1970–1972 and 1982–1984. They represent volcanic episodes sharing some similarities with those which recently occurred at Long Valley, USA (Rundle et al. 1985) and at Raboul, New Guinea (McKee et al. 1984). The most striking similarity lies, obviously, in the absence of eruptive activity accompayning seismicity and ground uplift. During the 1982–1984 crisis the total ground uplift amounted to about 1.8 m; more than 10000 micro-earthquakes were recorded, ranging in magnitude from about zero to 4.2 (De Natale and Zollo 1986).

Several models have been formulated in order to explain the bradyseism at Campi Flegrei. Most of them are based on the hypothesis of an increased pressure within a shallow magma chamber and/or within intruded magma bodies (Corrado et al. 1977; Berrino et al. 1984; Bonafede et al. 1986; Bianchi et al. 1984, 1987). Some other researchers tried to explain the crises as due to increased water circulation within the shallowest porous layers of the crust (Casertano et al. 1976; Oliveri del Castillo and Montagna 1985) or as due to the heating of the shallow rocks (Caputo 1979). Ground deformation data collected during the last two episodes and seismic and geochemical data collected between 1982–1985 represent a good data base to constrain the possible volcanological models for the Campi Flegrei bradyseism.

In this paper we will show that elastic models based only on the overpressure in a confined magma chamber are hardly defendable and probably to reject. We will also describe and apply an inversion technique of ground deformation data for the pressure variation (or volume expansion) in a confined volume, with inequality constraints. The results of these analyses indicate that the source of overpressure must be very shallow and laterally extended in order to generate the observed ground deformations with realistic values of overpressure. Such a source could be represented by the heating of well-confined aquifers, through energy transfer from a deeper magma chamber.

2 Main Features of Seismicity and Ground Deformation

Seismic activity occurring during the bradyseismic crisis of 1982–1984 was continuosly monitored by the local network owned by Osservatorio Vesuviano and AGIP (Fig. 1). During the period January–May 1984, an additional 13-station three-component digital seismic network, owned by the University of Wisconsin (Madison) was installed in the area (Fig. 1). Digital three-component recordings from this network allowed us to obtain non-saturated seismograms over a wide range of magnitudes ($0 < M_l < 3.5$) with a high signal/noise ratio and a very high resolution on timing of both P and S waves (of the order of 10 to 50 ms). Several studies about source mechanisms of micro-earthquakes and elastic/anelastic behaviour of the surrounding medium have been performed using these data (De Natale et al. 1987; Del Pezzo et al. 1987; Aster et al. this volume; Del Pezzo et al. 1985). For the purposes of this presentation, the striking features we want to point out are the precise hypocentral locations of the Campi Flegrei micro-earthquakes and the seismic wave velocities in the area. For an extensive discussion about these topics see Aster et al. (this Vol.) where the 3-D velocity structure is shown; Fig. 2 reports earthquake locations computed in the heterogeneous model for the 360 events used in the inversion. As is clear, a high level of seismicity occurred down to 4–4.5 km. Furthermore, earthquake epicentres appear to be concentrated in the central part of the

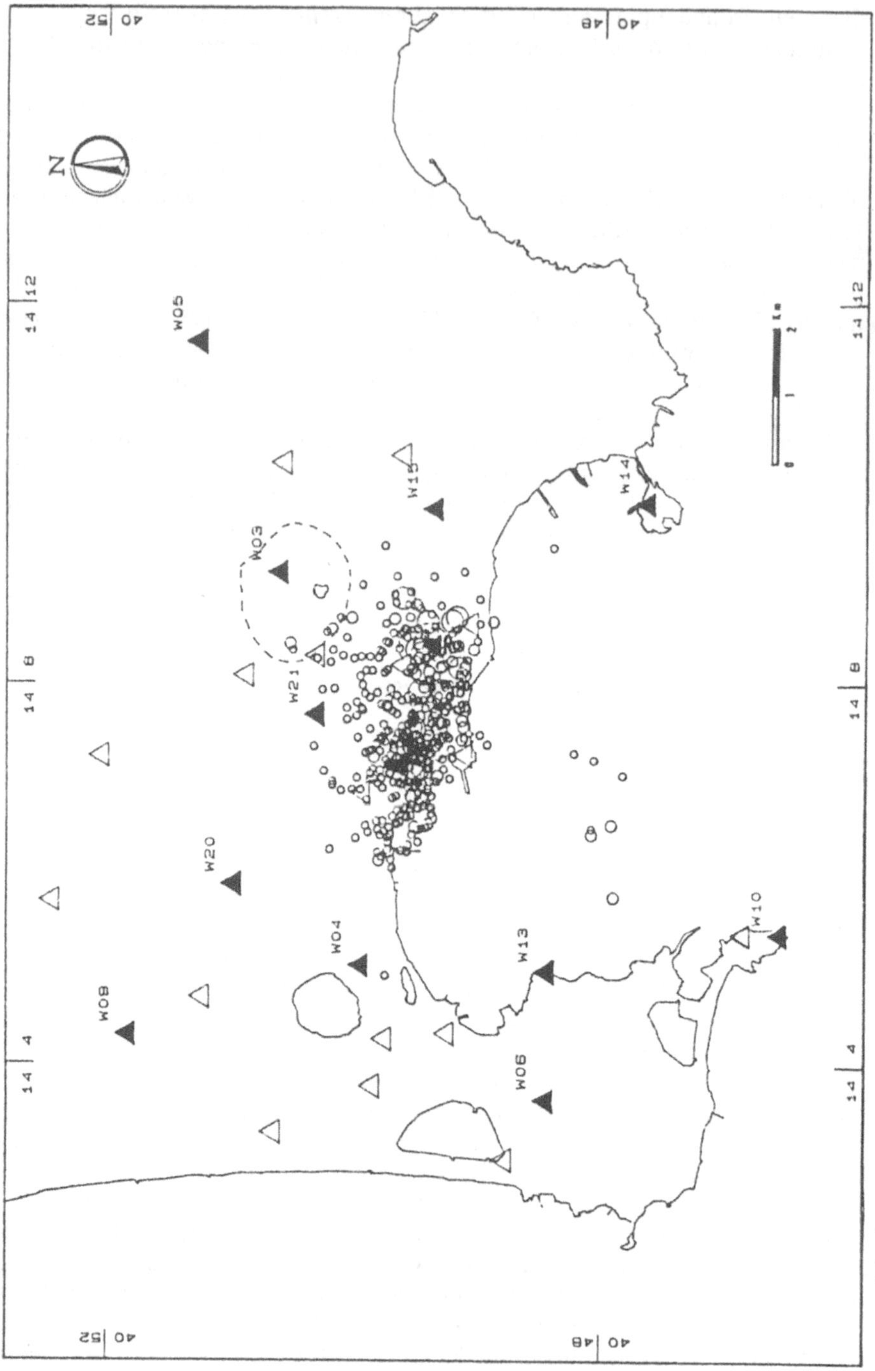

Fig. 1. Seismic networks operating in the area: *open triangles* are OV-AGIP monitoring network and *closed triangles* are UW three-component network

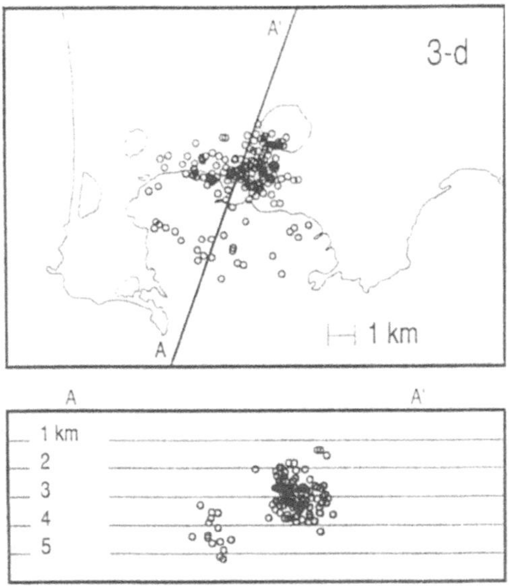

Fig. 2. Hypocentre locations of 360 earthquakes obtained using the 3-D velocity model (after Aster et al. this Vol.)

caldera, far from the faults defining the caldera rim. No time variation of the epicentral area was observed (De Natale et al. 1984). The velocity structure shows an average V_p = 2.8 km/s, an average V_p/V_s = 1.8 and a narrow zone with a markedly higher V_p/V_s ratio (2–2.2) centred on Pozzuoli, closely corresponding to the centre of uplift. This anomalously high V_p/V_s ratio, extending from the surface down to not less than 3 km, indicates the presence of a highly fractured, fluid saturated medium. In particular, it is inconsistent with dry fracturation, which results in a decrease of V_p/V_s ratio (O'Connell and Budianski 1976).

After the M. Nuovo eruption in 1538, the secular trend of vertical deformations at Campi Flegrei shows (Fig. 3) a decrease of the ground level, as inferred by Parascandola (1947). In March 1970, however, measurements performed by Genio Civile of Napoli and IGMI (Italian Military Geographic Institute) on the levelling line Napoli-Pozzuoli-Baia-Miseno (Fig. 4a) indicated that an uplift episode was occurring. Other periodic measurements were then performed, indicating a maximum uplift of about 1.5 m with respect to 1953 ground level, with an uplift rate up to 2 mm/day. The maximum uplift occurred at the bench mark of Pozzuoli.

Since the summer of 1972 ground level began to subside recovering about 20 cm at the end of 1974.

A new episode of ground uplift began in 1982 with a maximum elevation of about 1.8 m measured at Pozzuoli at the end of 1984 (Osservatorio Vesuviano 1984). The shape of vertical ground deformation (Fig. 5b) was similar to that of the previous episode. Since then, ground level has been subsiding and presently

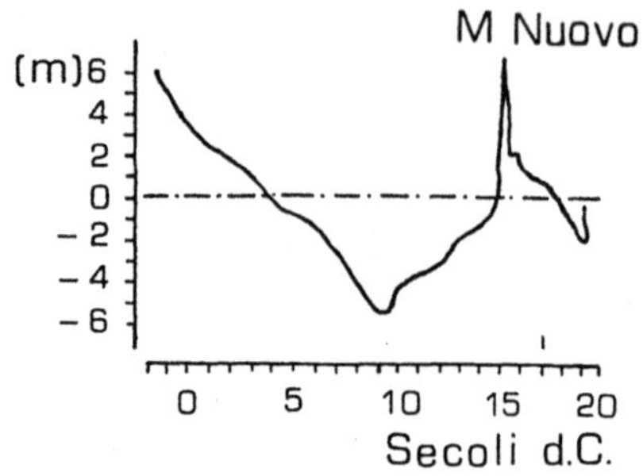

Fig. 3. Height of the floor of Serapeo at Pozzuoli, referred to the sea level (shown as a *broken line*) (After Parascandola 1947)

about 45% of vertical deformation has been recovered (Osservatorio Vesuviano 1988).

During the 1970–1972 crisis, a planimetric network (Fig. 4b) was estabilished (Dequal 1972). Measurements were performed by *Istituto di Topografia del Politecnico di Torino* (De Michelis et al. 1978) and by *Istituto di Topografia, Geodesia e Geofisica Mineraria di Bologna* in 1977, 1979, 1980 (Barbarella et al. 1980, 1983). Data were compensated by keeping the bench mark of Accademia and the direction Accademia-Baia fixed (Barbarella et al. 1983). During June 1983 geodimetric observations were repeated on the same network. Horizontal deformation in the periods 1970–1972 and 1980–1983 (Fig. 6a,b) were obtained after compensations performed by Corrado et al. (1977) and Berrino et al. (1984) by keeping the bench mark in Pozzuoli fixed. With this kind of compensation (altough not rigorous) horizontal displacements are someway minimized [it would be rigorously so if the centre of mass of the network was kept fixed, as in the inner solution (Brunner 1979)]. As is clear from the figures the ratio of h_{max} (maximum horizontal displacement) to v_{max} (maximum vertical displacement) was not less than about 0.5–0.6 for 1970–1972 and about 0.5 for 1980–1983.

2.1 Critical Review of Volcanological Models

Several models for Campi Flegrei bradyseism have been proposed, based on the hypothesis of increased pressure within a magma chamber. Corrado et al. (1977) and Berrino et al. (1984) found the best point source models for the two last episodes of bradyseism. They used Mogi's model (Mogi 1958) for a point source of overpressure embedded in a homogeneous elastic half-space, in order to infer depth, location and enthalpy variation (VΔP) of the source. The best fitting models were located under the benchmark of Pozzuoli at depths of about 2.5–3 km. They interpreted this source as a shallow magma chamber having a radius of 500 m, subject to magma injection from deeper sources and causing considerable overpressure. In the framework of the point source model, the maximum vertical displacement is:

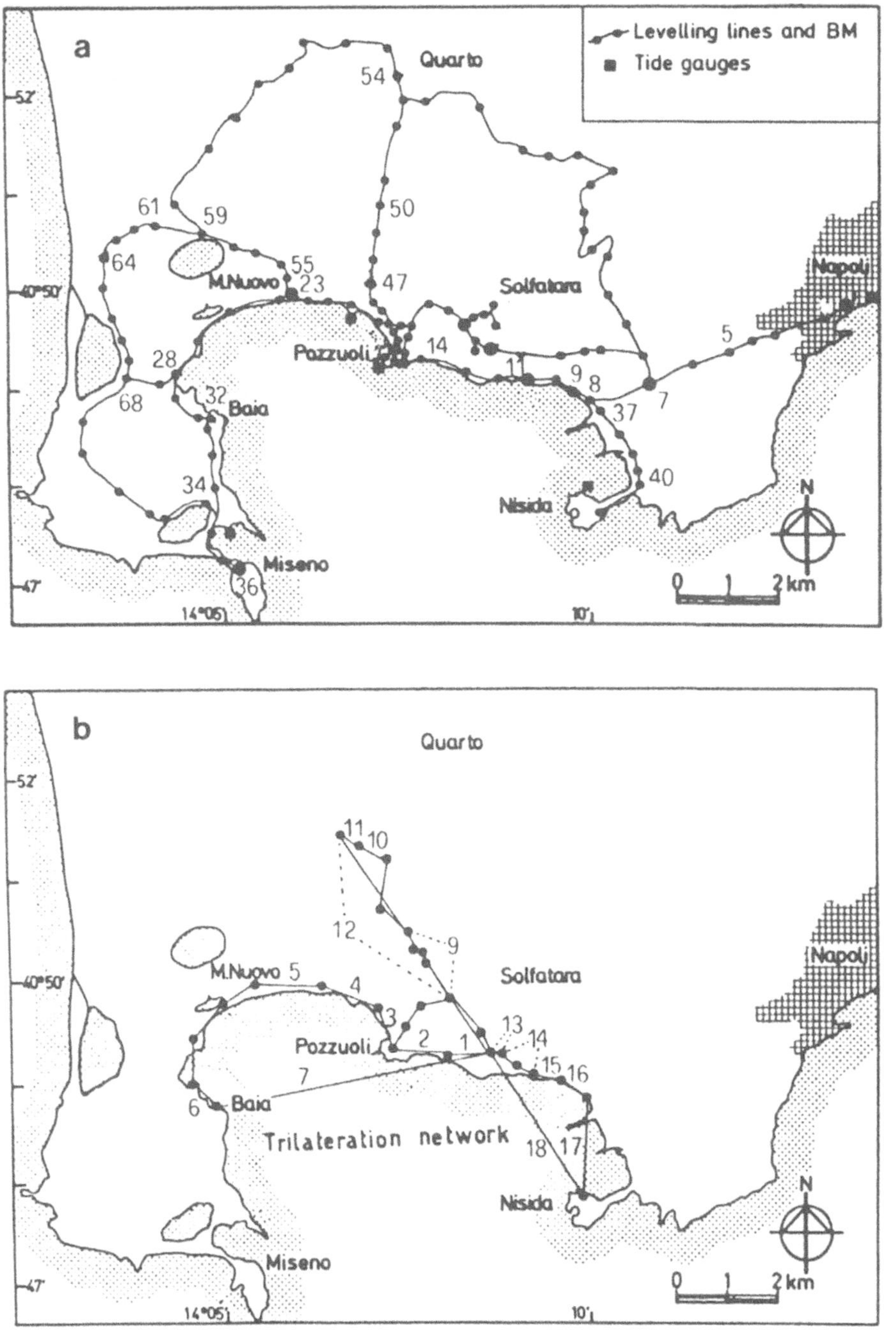

Fig. 4. a Levelling network; **b** Planimetric network (after Berrino et al. 1984, modified). *Dashed lines* show the ends of some lines, when not evident

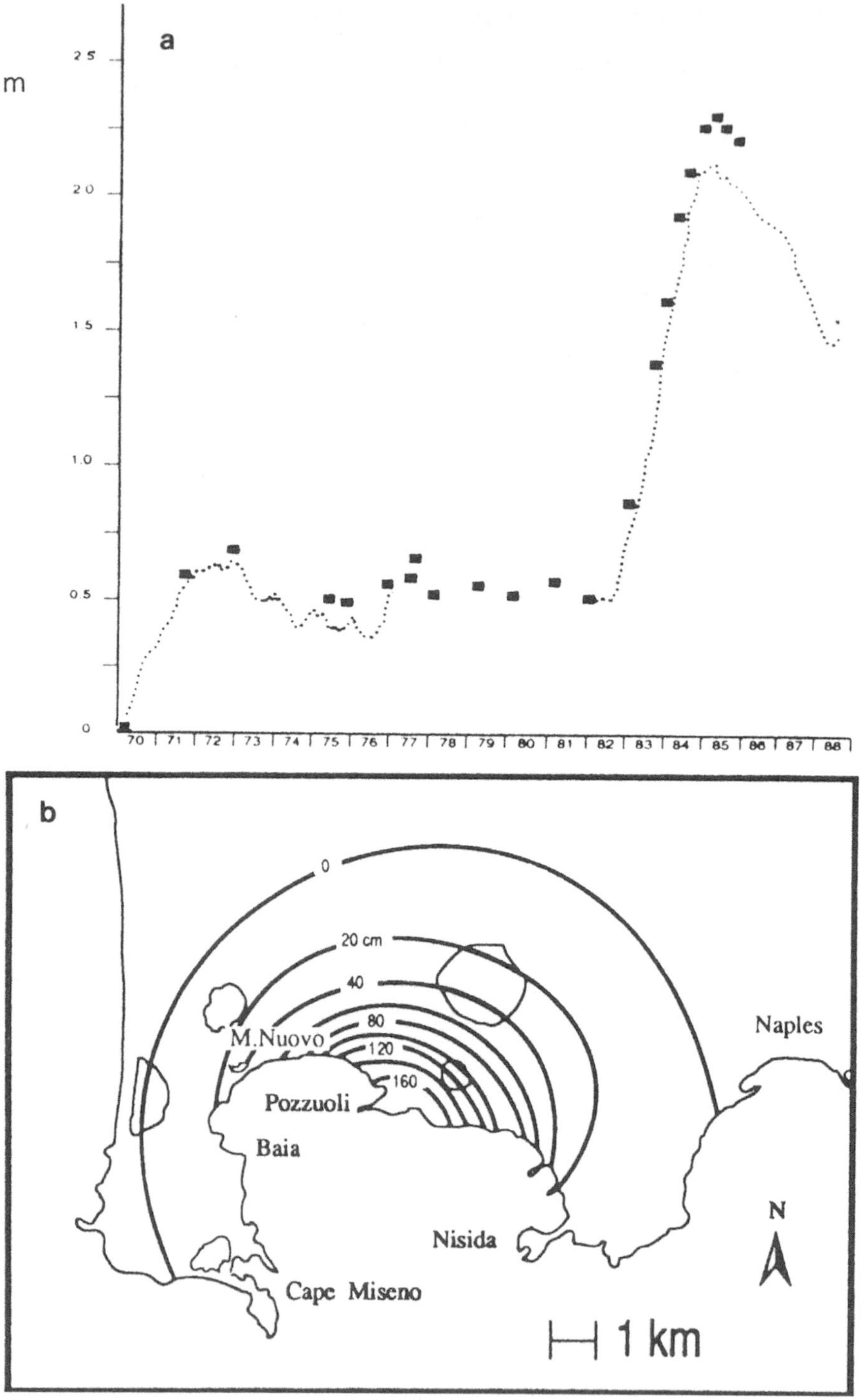

Fig. 5. **a** Temporal variation of the maximum elevation at Pozzuoli, as measured by tide gauge (*dots*) and levelling data (*squares*); **b** schematic contouring of maximum heights reached at the end of the crisis. Data from Osservatorio Vesuviano

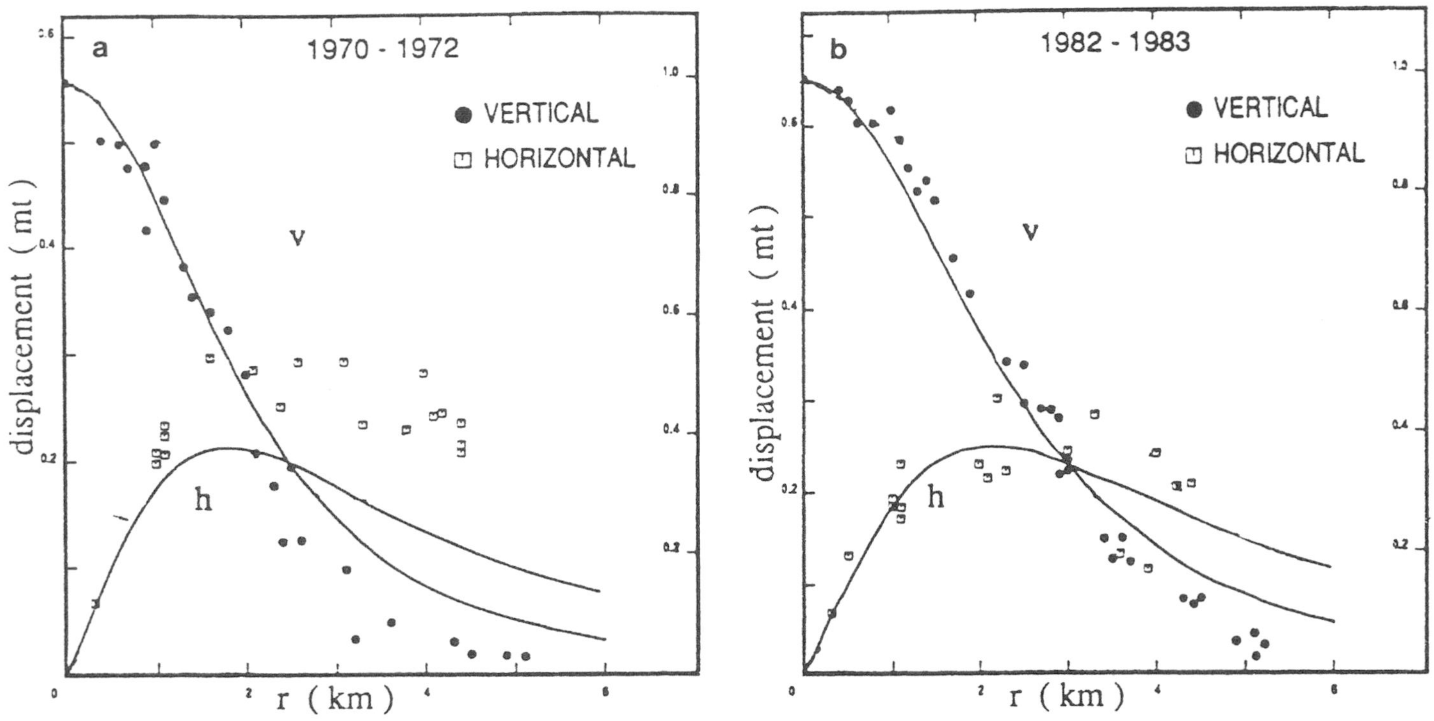

Fig. 6. **a** Ground deformation data as a function of the distance from Pozzuoli, measured in the period 1970–1972. The fit with the best Mogi model is also shown (depth = 2.5 km); **b** the same as a, for the period 1980–1983; the depth of the best Mogi model is 3 km

$$v_{max} = \frac{(1 - \nu)R^3}{\mu d^2}\Delta P,$$

where ν is the Poisson coefficient, μ is the rigidity, d is the depth of the source and R its radius; ΔP is the overpressure. The average rigidity of the area, within the first 3 km of depth, is $\mu = 5$ GPa; for $\nu = 0.25$ (which is an underestimation) and $d = 3$ km, an overpressure $\Delta P = 864$ Mpa for $R = 0.5$ km or $\Delta P = 108$ MPa for $R = 1$ km is required, in order to obtain $v_{max} = 1.8$ m, as observed during the last crisis. These values of overpressure are obviously unrealistic, because they far exceed rock strength (see, for example, Griggs et al. 1960). At the same time, the hypothesis of a magma chamber shallower than 3 km is not realistic, because all the geophysical data [temperature measurements in drilled holes (CNR 1987); active seismic soundings (Ferrucci et al. 1988); earthquake locations (Aster et al. this Vol.)] point out that the top of the magma chamber must be deeper than 4–5 km.

Bonafede et al. (1986), using a mathematical formulation for a point source of pressure embedded in a viscoelastic half-space with Maxwell's rheology, found that only 10 MPa would be sufficient to explain the temporal pattern of vertical deformation during the uplift phase, hypothesizing a source radius of 1 km at a depth of 3 km and a viscosity $\eta = 10^{16}$ Pa $\times$ s. This results because, for a Maxwell's rheology, the ground uplift is not bounded at a constant pressure for $t \rightarrow \infty$. In the framework of their model, the rate of uplift depends on both the viscoelastic parameters and on the time history of the pressure at the source. They used a trapezoidal time function, beginning at $t = 0$ and characterized by three parameters (a, b, c) corresponding to the changes in slope of the pressure rate (see Fig. 7). Considering only the rate of uplift, there is a strong trade off between source and medium parameters, which is difficult to eliminate because of the large uncertainty on the viscosity values (several orders of magnitude). The lower the viscosity, the lower the overpressure required for a given uplift rate. However, a much stronger constraint is given by the subsidence phase of the vertical ground deformation, which began at the end of 1984 and is still in progress. The amount of permanent deformation (for $t \rightarrow \infty$) predicted by this model is given by:

$$v(t \rightarrow \infty) = \frac{3V\Delta P}{16\pi d^2 \eta}(c + b - a).$$

Using low values for η, produces high values of permanent deformation and, consequently, a small amount of subsidence after the pressure has decreased to zero. Figure 7 shows the theoretical ground elevation, as a function of time, for several values of ΔP, and the trapezoidal time functions, for a source with a radius of 1 km. It is not possible to justify the pronounced ground subsidence (about 45% of the maximum uplift) with viscosities lower than about 10^{18} Pa $\times$ s. For higher viscosities, however, the rheology during the uplift phase is practically elastic, so that very high overpressures are required to fit the data (ΔP about 100 MPa). This is because effective rigidity decreases with a characteristic time $\tau = (3K + \mu)\eta/3K\mu$. For $\eta = 10^{16}$ Pa $\times$ s, $\mu = 5$ GPa and $K = 5\mu/3$, τ is about 28 days, so that after two years the effective rigidity is

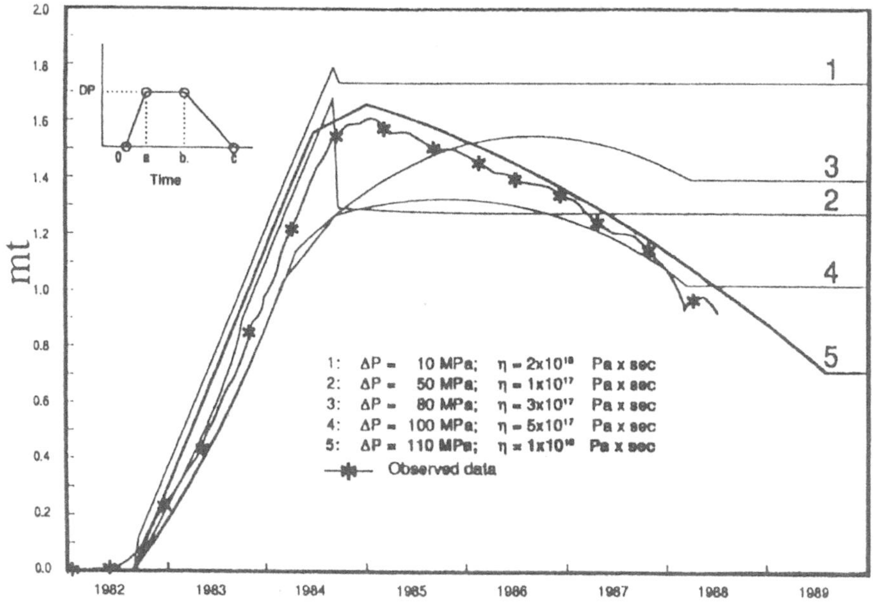

Fig. 7. Theoretical and experimental curves of maximum vertical ground displacement as a function of time. Theoretical curves are for *a* point source embedded in a viscoelastic homogeneous half-space. *Upper left corner*: general shape of time function for pressure generated at source. Times a, b, c used to generate theoretical curves are given as follows: *1* a = 010 days; b = 730 days; c = 0740 days; *2* a = 400 days; b = 730 days; c = 0740 days; *3* a = 550 days; b = 730 days; c = 2000 days; *4* a = 600 days; b = 730 days; c = 2000 days; *5* a = 650 days; b = 830 days; c = 2500 days

almost zero and no recovery of deformation is possible. For $\eta = 10^{18}$ Pa $\times$ sec, τ is about 8 years and the rheology for a stress applied over 2 years is practically elastic. Bianchi et al. (1987) performed a detailed analysis of possible models of pressure sources in heterogeneous elastic media, using a finite element tecnique. They tried to interprete the ground deformation field as due to overpressure within a magma chamber located at 4–5 km depth, embedded in a heterogeneous elastic medium. Their idea was that the lower rigidity in the central caldera with respect to the surrounding rocks could be interpreted in terms of a shallower source, when using a homogeneous model. However, their results showed that the shape of vertical deformation, as due to an oblate spheroidal magma chamber with a centre at 5.5 km of depth, could be justified only by extremely strong rigidity contrasts, namely by a factor 25–250 lower in the central basin with respect to the surrounding rocks. Rigidity values required in the central basin ranged from 1 GPa to 0.1 GPa below about 2 km depth, well below the measured values (about 5 GPa) with seismic methods (Aster et al. this Vol.). In principle, as already discussed, rigidity inferred by seismic methods is not comparable with the effective rigidity seen by ground deformation data, owing to viscous behaviour of the rocks which can be dominant for a characteristic time of about 1 year. However, altough lateral heterogeneities in viscosity could produce, at long term, higher apparent rigidity contrasts than the real ones, a very low viscosity value in the central caldera would be needed in order to reproduce the observed shape for a vertical deformation, which is not hypothesizable because inconsistent with the strong amount of subsidence occurred since 1985.

3 Testing Possible Models for Ground Deformation

Generally, ground deformation data are fitted to an expansion source of well-defined geometry (spheric, ellipsoidal, etc.) and uniform pressure. However, there is a strong trade off between depth and shape of the source (see for example Dieterich and Decker 1975), particularly for shallow sources. On the other hand, constraints are often available on the location of magma chambers, coming from other kinds of data (natural seismicity, seismic soundings, drill holes, temperature field measurements, etc.). Our approach is to find the feasibility of a generic model, satisfying particular constraints on location and source "intensity", rather than fitting the data to a well-specified source model.

3.1 Inverse Method

The method is based on the inversion of ground deformation data for the distribution of source "intensity" (adiabatic volume expansion or, equivalently, overpressure) within a volume of arbitrary shape and location. A representation in terms of a grid of N point sources is used for the expanding volume. For a homogeneous half-space, the direct problem linking ground deformation at $\mathbf{x_i}$ on the surface to source expansion at ζ_j, in the framework of this approximation is (Mogi 1958):

$$u_k(\mathbf{x_i}) = \sum_{j=1}^{N} G_k(\mathbf{x_i},\zeta_j)q(\zeta_j),$$

where:

$$G_k(\mathbf{x_i},\zeta_j) = \frac{3(1-\nu)d_k}{4\pi\mu r_{ij}^3}$$

$q(\zeta_j) = V_j \, \Delta P_j$ = enthalpy variation associated to the j-th source

$r_{ij}^2 = \| \mathbf{x_i} - \zeta_j \|$

$d_k = x_{ik} - \zeta_{jk}$.

When using line elongations as data, the appropriate Green function H for the elongation between the bench marks located at $\mathbf{x_i}$ and $\mathbf{x_h}$ is:

$$H(\mathbf{x_i},\mathbf{x_h},\zeta_j) = (x_{il} - x_{hl})\,(G_l(\mathbf{x_i},\zeta_j) - G_l(\mathbf{x_h},\zeta_j)) + (x_{i2} - x_{h2})\,(G_2(\mathbf{x_i},\zeta_j) - G_2(\mathbf{x_h},\zeta_j)).$$

These equations can be inverted for the pressure variation of the N point sources by ordinary linear inverse methods. When a dense distribution of point sources is used to approximate the expanding volume, the inverse problem is effectively underdetermined and an infinite number of solutions exist. However, when dealing with volcanic sources of uplift, the only meaningful solutions are those with positive values of pressure variation ($\Delta P_i > 0$ for all i). Adding to the previous system the inequality constraint $\Delta P_i > 0$ for $i = 1,N$ the set of solutions drops from infinite to possibly null and least squares solutions must be searched for. A further constraint $\Delta P_i < \Delta P_{max}$ (for $i = 1,N$) can be added in order to select solutions with physically realistic values of overpressure.

Least squares solutions to the system, satisfying the two kinds of inequality constraints, can be found by an iterative gradient tecnique based on the minimization of the total variance. The gradient of the total squared error $E(\mathbf{q}) = \| \mathbf{u} - G\mathbf{q} \|$ can be written as $\nabla E(\mathbf{q}) = -2G^T(\mathbf{u} - G\mathbf{q})$; starting from an initial solution $\mathbf{q_0}$, a correction $\boldsymbol{\delta}\mathbf{q}$ can be found as:

$$\boldsymbol{\delta}\mathbf{q} = -\nabla E(\mathbf{q_0})\,\boldsymbol{\Delta}\mathbf{q},$$

where $\boldsymbol{\Delta}\mathbf{q}$ is an approriate positive multiplicative factor; the improved model is:

$$\mathbf{q} = \mathbf{q_0} + \boldsymbol{\delta}\mathbf{q}$$

with the additional constraints $0 \leqslant \mathbf{q_i} \leqslant V_i \Delta P\mathrm{max}$ (for $i = 1,N$).

The process is iterated until the total squared error becomes lower than a specified maximum value. The previous constraints are imposed at each step by forcing the values of q^i out of the allowed range to the respective limiting values. It is possible to demonstrate (Ward and Valensise 1989) that, without the inequality constraints, the gradient solution tends toward the minimum norm solution for a large number of iterations. This is because the correction to the starting value is proportional to the gradient $\nabla E(\mathbf{q})$, so that point sources that give a negligible contribution to the deformation field are largely ignored and no significant corrections are applied. A noteworthy feature of such a method is that the shape of the source and the enthalpy distribution are totally free within the constraints imposed by the allowed ranges for location and intensity. Several methods could be used to get estimates on the reliability of the solution, such as errors and resolution (see, for example, Ward and Barrientos 1986). However, we are interested here only in the possibility of a model with constraints to give an acceptable fit to the data and not into the details of the model itself. For this reason, we do not need error and resolution estimates.

3.2 Data Analysis

In order to test the appropriateness of various models for the expanding source, 69 vertical displacement data and 19 line elongation data, collected in 1980–1983, have been used. Line elongation data have been used instead of horizontal ground displacements because they are independent from the survey data compensation, whose reference points are somewhat arbitrary, as already mentioned.

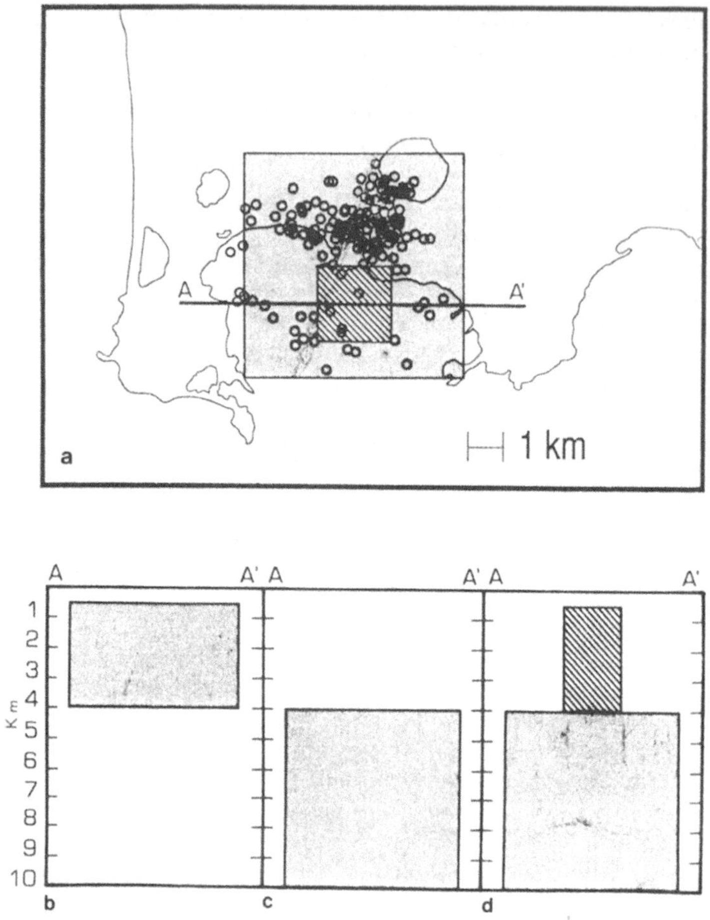

Fig. 8. a Surface projection of the expanding volume for models 1, 2 and 3 and for the shallow intrusion of model 3 (*shaded area*); **b** E-W section of the expanding volume of model 1; **c** the same as b for model 2; **d** the same as b for model 3

Models with three source distributions have been tested. The source volume in model 1 is represented by a three-dimensional grid with sides 6 km long, whose top is located at 0.5 km depth and grid spacing of 300 m on horizontal plane and 250 m on depth (Fig. 8a,b). The maximum allowed overpressure is $\Delta P = 8$ MPa for data in the period 1980–1983, corresponding to $\Delta P = 25$ MPa at the end of 1984, when the elevation reached the maximum of about 1.8 m. We implicitly assume that the shape of ground deformation did not change during the whole 1982–1984 crisis, so that data in the considered periods are representative of the whole period, apart from a constant multiplicative factor. This is true for the vertical data, as revealed by the repeated levelling surveys performed at about 1-month intervals by Osservatorio Vesuviano (Osservatorio Vesuviano 1984,

1988), and we have no reason to doubt that this was the case also for horizontal data, although no further measurements were performed after 1983.

The source volume for model 2 is built from a three-dimensional grid with sides 6 km long, located between 4 and 10 km depth and the same grid spacing of model 1; it represents the depth range in which a magma chamber can be located (Fig. 8a,c).

The source volume for model 3 is the same as model 2 with an additional smaller grid reaching 0.5 km depth (Fig. 8a,d). The latter takes into account the possibility of a shallow intrusion from the main magma body, whose extent is laterally constrained to match the aseismic central zone. The same inequality constraints on ΔP of model 1 apply to models 2 and 3 also.

Results of the inversions as applied on the basis of the above models are summarized in the following:

With model 1 a good fit is obtained, for both vertical and horizontal data. In order to satisfy the constraint $\Delta P < 25$ MPa, overpressure, as shown in Fig. 9, is concentrated at shallow depth, in the range 0.5–2 km; the RMS residual is 7.7 cm. Figure 10 shows the fit to vertical and line elongation data of this model.

The fit to vertical data is very good, whereas line elongation data are satisfactorily fitted except for two lines (Nos. 11, 12), which presents the maximum

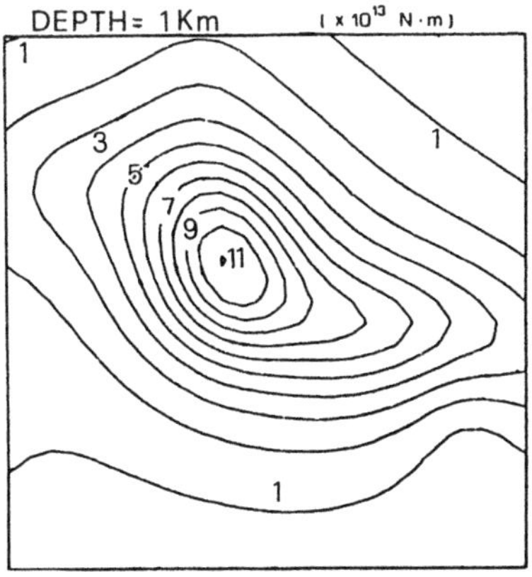

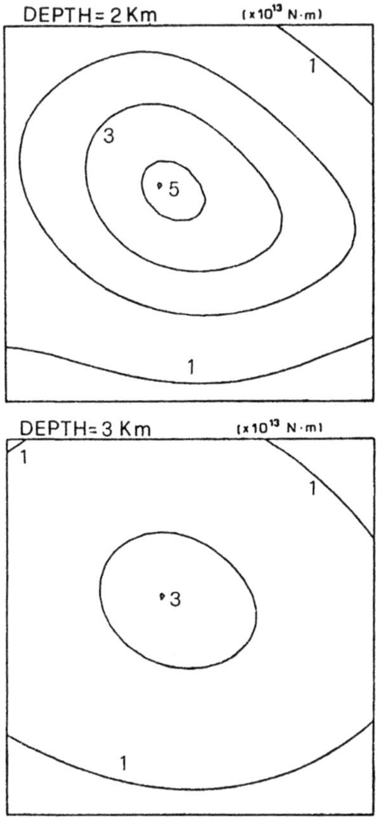

Fig. 9. Enthalpy distribution for model 1, at three different depths

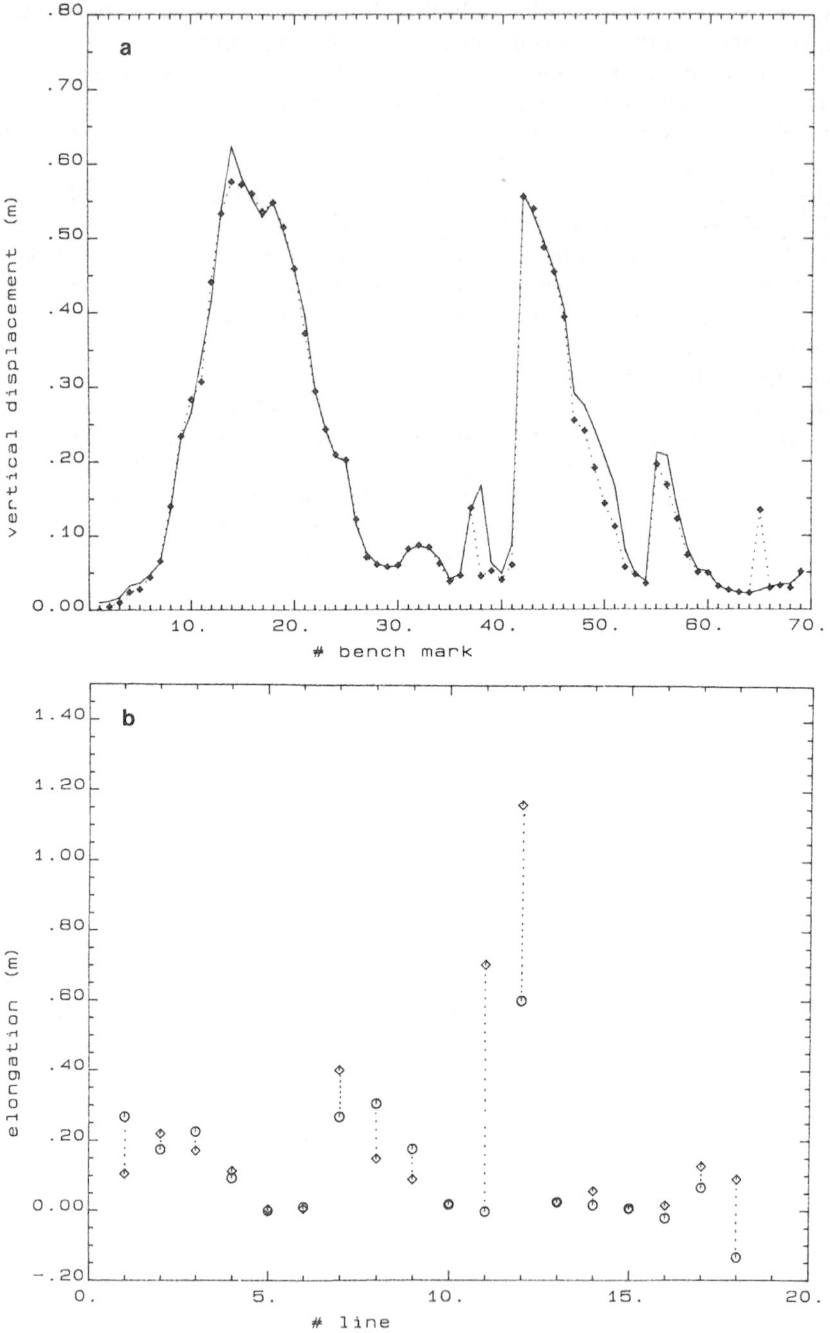

Fig. 10. **a** Fit of model 1 to vertical ground deformations; *diamonds* represent observed data; **b** Fit of model 1 to line elongation data; *diamonds* represent observed data and *circles* theoretical ones. Refer to Fig. 4 for location of bench marks and line numbers

elongations. Line No. 11 is very short (about 300 m) and shows about 70 cm of positive elongation in the farthest zone from the centre of caldera. The observation of a consistent dilatation in this zone is also supported by the independent finding of the high positive elongation (about 1.15 m) measured on line No. 12, which is almost parallel to line No. 11 in the northernmost part. These findings are probably associated to the aseismic detachment of preexisting structures.

With model 2, no satisfactory solutions can be found, the main discrepancy coming from theoretical vertical displacements, always much less confined and consistently lower than the experimental ones. Also releasing the superior constraint for ΔP, no reliable solutions can be found, giving the correct radial decay from the centre of the vertical displacements. The RMS residual for the model 2 is 16.8 cm.

In the framework of model 3, no satisfactory solutions exist, when the upper limit for ΔP is applied; the corresponding RMS is 13.6 cm. When this constraint is released, solutions exist giving a good fit to the data, but with overpressure values higher than about 100 MPa. The much higher overpressure values required for model 3, with respect to model 1, are obviously due to the smaller volume allowed for the source, constrained by the seismic area, consistently with the hypothesis that it represents a massive magma intrusion. An additional observation is that the lower part of the grid, in the range 4–10 km, simulating the main magma body, is largely ignored by the solution, which tends to assign the most overpressure to the shallowest volumes, representing the intrusion. This is partly due to the great sensitivity of the data to these shallower sources, but, also in view of the results obtained with model 2, it clearly indicates that the contribution of the deeper magma body would be negligible with respect to that of a massive shallow intrusion.

4 Discussion and Conclusion

Most of models proposed to explain the bradyseism at Campi Flegrei are based on the hypothesis of an increased pressure within a shallow magma chamber. The main argument against such a hypothesis, as shown in this paper, is that the source of expansion responsible for the ground uplift must be shallower than 2–3 km, a depth where no data support the presence of large magma bodies. If the source were so small, as to go undetected (as, for example, the top of a thin magmatic intrusion), the overpressure required to generate the total amount of ground uplift (about 1.8 m) would be extremely high (over 100 MPa). To analyze the ground deformation data at Campi Flegrei, we have followed an approach different from what is commonly used. In fact, using specific models of pressure source, such as the sphere or the dyke and so on, shape, intensity and depth are not univocally determined and models with different shapes should give equivalent fit to the data with different depths and intensity. For this reason, we believe it is more meaningful to use the constraints coming from other kind of data to fit the data with a general kind of expanding source, without any specific shape, but with other physical constraints, like the depth range and the maximum overpressure. Another important advantage of using our inverse method is that it naturally takes into account the ratio between horizontal and vertical data, which is strongly influenced by the source shape. At Campi Flegrei, in particular, many researchers have found inconsistencies between source models inferred separatedly from vertical and horizontal data (Corrado et al. 1977). As emphasized by many kinds of data (e.g., temperature measurements (CNR 1987), earthquake locations (Aster et al. this Vol.), active seismic surveys (Ferrucci et al. 1991) the main magma chamber should be located deeper than 4–5 km. Our results indicate that increase of pressure within a depth range 4–10 km, in a homogeneous elastic half-space, cannot produce the observed shape for the ground deformation data. Hence, in the framework of the above hypotheses, overpressure within a confined magma chamber cannot be the only source for the bradyseism. If, in addition, the effect of a magmatic intrustion, confined to lie within the central aseismic zone, is allowed, ground deformations can be satisfactorily fit, but only with very high overpressures (over 100 MPa).

The hypothesis of a consistent magmatic intrusion at depth of about 1–2 km requires some additional discussion. It is in sharp contrast with several kinds of data, including geochemical indicators and seismological observations. Geochemical observations point out the absence of direct magmatic contributions in the gases sampled at fumaroles during the whole crisis (Martini 1986; Tedesco et al. 1988). Seismological observations indicate the occurrence of considerable seismicity, tectonic in character, spread over a large area, except for an aseismic central zone at sea; no volcanic tremor or low frequency earthquakes have been observed (Aster et al. this Vol.). The hypothesis of an intrusion confined in the aseismic zone, the only one where a ductile rheology due to the presence of molten rocks can hold, is, however, in contrast with geochemical data and with the lack of volcanic tremor and low frequency earthquakes. On the basis of our results, the only way to fit ground deformation data with reasonable values of overpressure is for a source located within 1–3 km of depth, with a considerable lateral extent (about 2 km radius), as shown by the results for model 1. The lateral extension of the source is needed in order to give the correct shape to the vertical displacements with low overpressure. Such a shallow source cannot be thought, obviously, to represent a magma chamber. It can be rather interpreted as a thermal expansion effect of the shallowest rocks, probably aquifers. This effect would be caused by energy transfer from the deep magma chamber, for example, in the form of vapour migration toward the surface. This hypothesis is supported by the observation that changes in geochemical indicators, in particular changes of about 30 °C of temperature in fumaroles, preceeded the changes in geophysical observables (ground movements and seismicity) both at the beginning and at the end of the 1982–1984 crisis (Tedesco et al. 1988), indicating a high efficiency in the heat transfer, at least in the area around Pozzuoli. Such a mechanism could be also related to the high v_p/v_s ratio zone detected at the centre of caldera by travel times data, indicating a strong fracturing and fluid saturation down to about 3 km, in a narrow zone around Pozzuoli.

As a final remark, we should consider in more detail the possibility that a substantial inhomogeneity of the rheological parameters in the area can produce the observed ground displacements with a considerably deeper source. As already discussed, very high rigidity contrasts, as required in order to produce significant changes in the shape of the deformations, are not realistic and the only possibility could be strong variations in viscosity between the caldera basin and the external rocks. Very low viscosity values in the central basin would be able to produce, over long times, low values of the apparent rigidity with respect to the external rocks, but in this case the shape of ground deformations would exibit a notable time dependence, which is not observed. In addition, viscosity of the calderic rocks is constrained by the large amount of subsidence occurring since 1985, to values for which the apparent rigidity after 2 years of stress application is only slightly lower than the elastic values.

References

Barbarella M, Folloni G, Gubellini A, Russo P (1980) Risultati dei recenti rilievi delle reti di controllo planimetrico nell'area flegrea. Pub Ist Topografia, Geodesia Geof Min, Bologna

Barbarella M, Gubellini A, Russo P (1983) Rilievo ed analisi dei recenti movimenti orizzontali del suolo nell'area flegrea. Atti II Conv Gr Naz Geof Terra Solida 659–669

Berrino G, Corrado G, Luongo G, Toro B (1984) Ground deformation and gravity changes accompanying the 1982 Pozzuoli uplift. Bull Volcanol 47-2:187–200

Bianchi R, Coradini A, Federico C, Giberti JP, Sartoris G, Scandone R (1984) Modelling of surface ground deformation in the Phlegraean Fields volcanic area, Italy. Bull Volcanol 47-2:321–330

Bianchi R, Coradini A, Federico C et al. (1987) Modeling of surface ground deformation in volcanic areas: the 1970–1972 and 1982–1984 crises of Campi Flegrei, Italy. J Geophys Res 92:14,139–14,150

Bonafede M, Dragoni M, Quareni F (1986) Displacement and stress fields produced by a centre of dilation and by a pressure source in a viscoelastic half-space: application to the study of ground deformation and seismic activity at Campi Flegrei. Geophys J R Astr Soc 87:455–485

Brunner FK (1979) On the analysis of geodetic network for determination of the incremental strain tensor. Surv Rev XXV, 192:56–67

Caputo M (1979) Two-thousand years of geodetic and geophysical observation in the Phlegraean Fields near Naples. Geophys J R Astr Soc 56:319–328

Carapezza M, Nuccio PM, Valenza M (1984) Geochemical surveillance of the Solfatara of Pozzuoli (Phlegraean Fields) during 1983. Bull Volcanol 47-2:303–311

Casertano L, Oliveri Del Castillo A, Quagliariello MT (1976) Hydrodinamics and geodynamics in Phlegraean Fields area of Italy. Nature (Lond) 264:5582

Cioni R, Corazza E, Marini L (1984) The gas/steam ratio as indicator of heat transfer at the Solfatara fumaroles, Phlegraean Fields (Italy). Bull Volcanol 47-2:295–302

CNR (1987) Rasi M, Sbrana A (eds) Quaderni de "La Ricerca Scientifica". Prog Fin Geodin 114-9: 175 pp

Corrado G, Guerra I, Lo Bascio A, Luongo G, Rampoldi F (1977) Inflation and microearthquake activity of Phlegraean Fields, Italy. Bull Volcanol 40-3:169–188

Dall'Aglio M, Martini M, Tonani F (1972) Rilevamento geochimico delle emanazioni vulcaniche dei campi Flegrei. Quad Ric Sci 53 CNR:113–136

Del Pezzo E, De Natale G, Scarcella G, Zollo A (1985) Qc of three component seismogram of vulcanic microearthquake at Campi Flegrei volcanic area (Southern Italy). Pageoph 123:685–696

Del Pezzo E, De Natale G, Martini M, Zollo A (1987) Source parameters of microearthquake at Phlegraean Fields (southern Italy) volcanic area. Phys Earth Plan Int 47:25–42

De Michelis AM, Dequal S, Folloni G (1978) Geodetic surveying of soil movements in the Phlegraean Fields area. Riv Cat Serv Tech Erar XXXIII, 1:11–26

De Natale G, Zollo A (1986) Statistical analysis and clustering features of the Plegraean Fields earthquake sequence (May 1983–May 1984). Bull Seismol Soc Am 73:801–814

De Natale G, Zollo A, Del Gaudio C, Ricciardi GP, Martini M (1984) Error analysis in hypocentral locations at Phlegraean Fields. Bull Volcanol 47-2:209–218

De Natale G, Iannaccone G, Martini M, Zollo A (1987) Seismic sources and attenuation properties at Campi Flegrei volcanic area. Pageoph 125, 6:883–917

Dequal S (1972) Rilievo degli spostamenti planimetrici di punti nella zona di Pozzuoli. Quad Ric Sci 53, CNR:231–262

Dieterich JH, Decker RW (1975) Finite element modeling of surface defounation associated with volcauism. J Geophys Res 80, 29:4094–4102

Ferrucci F, De Natale G, Hirn A, Mirabile I, Virieux J (1991) P-SV conversions at a shallow boundary beneath Campi Flegrei caldera (Naples, Italy): possible evidence for the magma chamber. J Geophys Res (in press)

Griggs DT, Turner FJ, Heard HC (1960) Deformations of rocks at 500 to 800 °C. Mem Geol Soc Am 79:39–104

Martini M, Cellini Legittimo P, Piccardi G, Giannini L (1984) Composition of hydrothermal fluids during the bradyseismic crisis wich commenced at Phlegraean Fields in 1982. Bull Volcanol 47-2:267–273

McKee CO, Lowenstein PL, De Saint Ours P, Talai B, Itikarai I, Mori JJ (1984) Seismic and ground deformation crises at Rabaul Caldera: Prelude to an eruption? Bull Volcanol 47, 2:397–411

Martini M (1986) Thermal activity and ground deformation at Phlegraean Fields, Italy: precursors of eruptions or fluctuations of quiescent volcanism? A contribution to geochemical studies. J Geophys Res 91:12255–12260

Mogi K (1958) Relations of the eruptions of various volcanoes and the deformations of the ground surface around them. Bull Earth Res Inst 36:99–134

O'Connel R, Budiansky B (1976) Seismic velocities in dry and saturated cracked solids. J Geophys Res 79:5413–5426

Oliveri del Castillo A, Montagna S (1985) Nuovi elementi sulla connessione tra termofluido-dinamica e geodinamica ai Campi Flegrei (Napoli) – Bradisismo 1982–1984. Arte Tipog SAS, Napoli

Osservatorio Vesuviano (1988) Rapporto sulla sorveglianza del bradisismo flegreo (Aprile, 1988). Int Rep, 6 pp

Osservatorio Vesuviano (1984) Bradisismo Flegreo, Rapporto Sorveglianza Gennaio 1982-Gennaio 1984. Int Rep, 93 pp

Parascandola A (1947) I fenomeni bradisismici del Serapeo di Pozzuoli. Genovese, Napoli

Rundle JB (1978) Viscoelastic crustal deformation by finite quasi-static sources. J Geophys Res 83:5937–5945

Rundle JB et al. (1985) Seismic imaging in Long Valley, California, by surface and borehole techniques: an investigation of active tectonics. EOS Trans 66:194–201

Tedesco D, Pece R, Sabroux JC (1988) No evidence of a new magmatic gas contribution to the Solfatara volcanic gas, during the bradyseismic crisis at Campi Flegrei (Italy). Geophys Res Let 15, 12:1441–1444

Ward SN, Barrientos SE (1986) An inversion for slip distribution and faul shape from geodetic observations of the 1983, Borah Peak, Idaho, earthquake. J Geophys Res 91, B5:4909–4919

Ward SN, Valensise GR (1989) Fault parameters and slip distribution of the 1915 Avezzano, Italy, earthquake derived by geodetic observations. Bull Seismol Soc Am 79, 3:690–710

Possibilities of Detecting Volcanic Seismic Sources at Campi Flegrei Caldera

G. Del Re, G. Iannaccone, G. La Volpe, and M. Martini

Abstract

Moment tensor solutions of ten earthquakes, recoraed at Campi Flegrei volcanic area, are obtained. Amplitudes of P and SH direct waves, recorded by a digital three-component network, were employed as imput data for an inversion technique used to obtain the moment tensor components. Since seismic sources differing from the classical double couple can be expected in this area, a detailed analysis of all the factors that can influence moment tensor solutions was performed. In particular, local site effect was estimated by coda method to correct amplitudes of body waves used as imput data in the moment tensor inversion. Uncertainty of take-off angles due to poor knowledge of a detailed velocity model of the area and hypocentre location errors were carefully analyzed. Taking into account these factors, we show that moment tensor solutions can be interpreted in terms of double-couple source mechanisms.

1 Introduction

The focal mechanism is a basic parameter of seismic source studies. The technique using the polarities of P-wave first motions is widely used both for tectonic and for volcanic earthquakes. The dynamic model is a fault-slip due to a double couple system of forces, whereas the separation of polarities on the focal sphere by a quadrant-type pattern allows inference of azimuth, dip and slip of the nodal planes and direction of stress field. The application of focal mechanism studies is performed with an a priori model of a shear dislocation.

Recently, the use of dense seismic networks and the progress in analytical and numerical modelling of the seismic source mechanism revealed that some volcanic earthquakes can be explained by a non-double couple force model.

However, the polarities method is routinely applied to study source parameters of earthquakes recorded in volcanic areas (Scarpa et al. 1983; Karpin and Thurber 1987).

Julian and Sipkin (1985) showed that a large number of first motions of two events that occurred in May 1980 at Long Valley caldera were inconsistent with the radiation pattern expected from a fault-slip source model and the discrepancies might be minimized by hypothesizing as a source mechanism a compensated linear vector dipole (CLVD). The CLVD is a mechanism theoretically proposed by Knopoff and Randall (1970) but never observed. Julian and Sipkin explained the CLVD source as due to the injection of magma fluid in a preexisting fracture which subsequently expanded rapidly.

Shimizu et al. (1987) investigated volcanic earthquakes that occurred at the Izu-Bonin Arc after an eruption. They considered that the events might be generated by a sudden opening or closing of tensile-shear cracks due to rapid excess or decrease in pressure of intrusive magma.

Ukawa and Ohtake (1987) studied a small earthquake (M = 2.7) which occurred 14 months before the Izu-Ooshima volcanic eruption. Analyzing the amplitude ratio of S to P waves, the frequency content of recorded signals and the polarization pattern of shear waves, they suggested that a probable source mechanism was a traction force caused by magma flow through a conduit producing unidirectional viscous force on the conduit wall.

Although evidence for non-double couple may be highly interesting, we think that is very important to analyze carefully all the possibilities that can influence the data analysis.

Trehu et al. (1981) extensively studied two moderately sized Atlantic ridge earthquakes which occurred in 1970 and 1972. Since fault plane solutions of ridge earthquakes frequently indicate non-orthogonal nodal planes, the authors performed a complete study of source mechanisms using both surface and body waves. Without invoking departures from a double-couple, source mechanism, they showed that the data were compatible with a pure shear fault. The apparent non-orthogonality of nodal planes could be attributed to structural effects. The shallowness of the sources produced an interference in the direct and reflected phases (pP and sP) masking the first motions recorded by seismic stations producing an erroneous distribution of polarities on the focal sphere.

The aim of this paper is to perform an analysis of focal mechanisms of a set of earthquakes recorded at Campi Flegrei volcanic area using amplitudes of P and S direct waves. We evaluated, by an inverse method, the moment-tensor components for a point source. This representation of the source is useful because it is not restricted to a double couple mechanism and gives information on the possible departures from this classical source model. The geological and geophysical characteristics of the area suggest that earthquakes with "volcanic source" could occur. We show that taking into account uncertainty on knowledge of velocity structure, errors on hypocentre location and site amplification effects, we are forced to confirm the double couple source mechanism for the analyzed data.

2 Brief Volcanological and Geophysical Features of Campi Flegrei

Campi Flegrei is a volcanic area located west of the city of Naples, southern Italy. The volcanic history is characterized by a massive eruption of alkali trachytic ignimbrites followed by a caldera collapse that occurred 35 000 years ago. The following volcanic activity took place in the depressed part of the caldera with a migration of eruptions near and around a local geomorphological discontinuity (Starza marine terrace). This discontinuity delimits a sector of

Campi Flegrei probably uplifted by a magma injection at shallow depth (Rosi et al. 1983). The last historical eruption of Monte Nuovo (1538) also occurred close to the Starza terrace.

In the last two decades, two marked ground uplift episodes characterized the activity of the area. The former started in early 1970 producing a maximum vertical displacement of 170 cm (Corrado et al. 1976). The latter began during the summer of 1982 producing a comparable uplift (Berrino et al. 1984). The two episodes had common geophysical features:

- very shallow seismicity (depths of hypocentres less than 5 km);
- low energy of earthquakes (maximum observed magnitude $M_L = 4.2$);
- similar shape of uplift on land (approximately circular distribution of vertical ground deformation centred in the town of Pozzuoli).

Ground deformation has been modelled by an increase in pressure in a spherical magma chamber embedded in a viscoelastic half-space (Bonafede et al. 1986). Seismic activity is located in the central part of Campi Flegrei and is interpreted as related to fracturing processes along the Starza structure although intrusive events are also hypothesized (De Natale et al. 1987; Aster et al. this Vol.).

2.1 Data

During 1984 a digital seismic network equipped with 12 three-component short-period stations operated at Campi Flegrei. Digital stations were characterized by a wide dynamic range (106 dB) and by a flat velocity response curve in the frequency range 2–25 Hz.

A seismic monitoring network consisting of 22 analog stations equipped with vertical component seismometers operated in the area in the same period. During the bradyseismic crisis more than 10 000 events were recorded by the seismic network. Figure 1 reports a set of well-located micro-earthquakes showing the distribution of the seismicity. Starting from this set of data we selected 233 records according to a visual inspection of seismograms to obtain data suitable for the following analysis. In particular, only records with a clear P and S first arrival were chosen (Fig. 2).

To determine focal mechanisms we evaluated moment-tensor components using a method based on the inversion technique of P and SH direct wave amplitudes. The main advantage, with respect to the standard techniques based only on the first P pulse polarities, is the possibility of considering mechanisms different from those consisting of double couple forces. A detailed description of the method is reported in Aster et al. (this Vol.).

Input data for moment tensor analysis are amplitudes of direct P and S waves and location of seismic stations on the focal sphere, defined by the azimuth from the north of the source-receiver direction and the take-off angle.

We rotate three-component seismograms on the ray coordinate system. By an automated method, searching for the maximum value on the first half-wave following the P-wave arrival, we read the amplitudes of P radial waves. All the

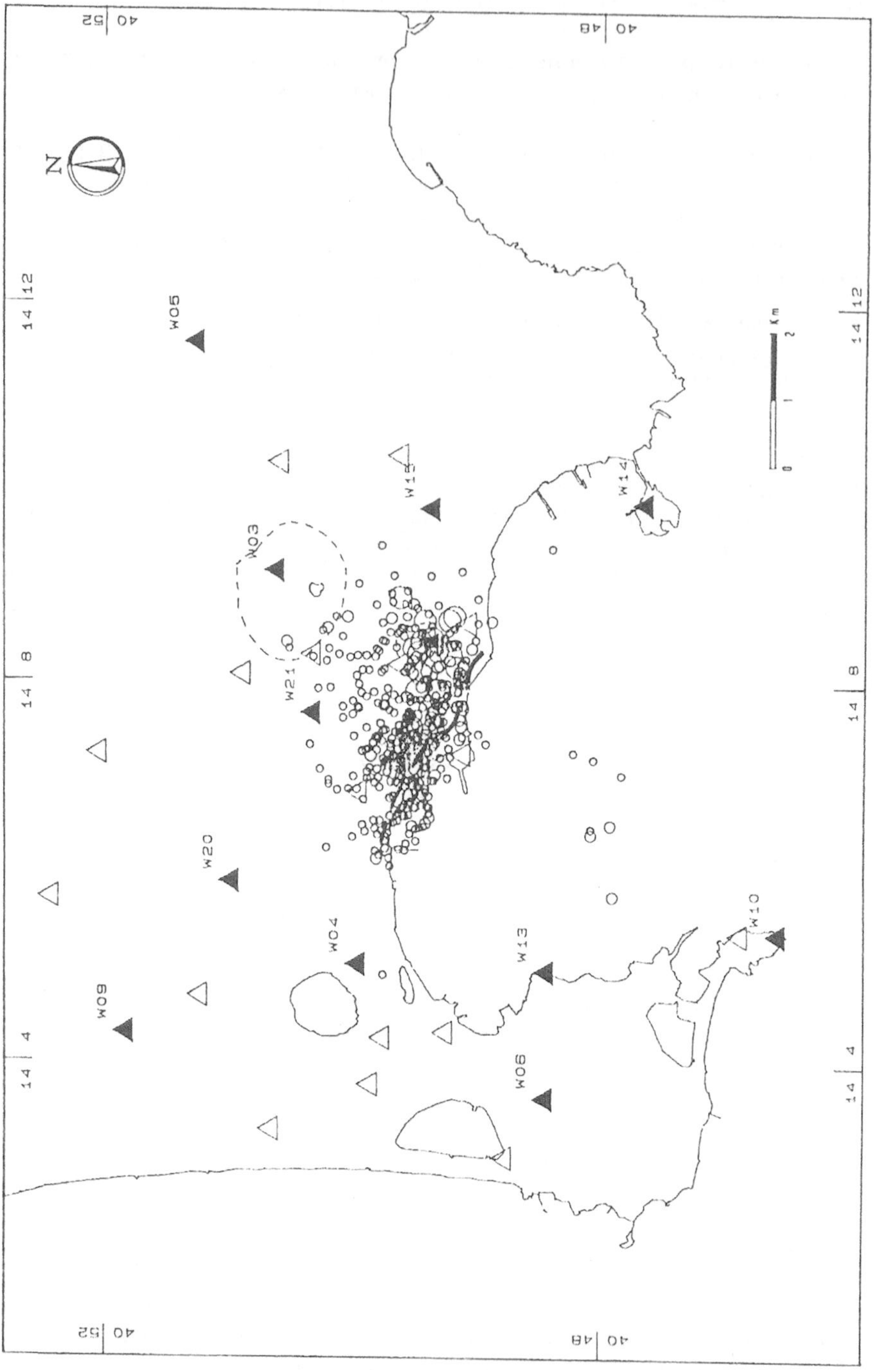

Fig. 1. Epicentral map of a set of earthquakes representative of overall seismicity of the area. *Solid triangles* are three-component digital stations, *Open triangles* represent the seismic monitoring network (one-component analog stations). The *broken line* represents the limits of the Starza marine terrace

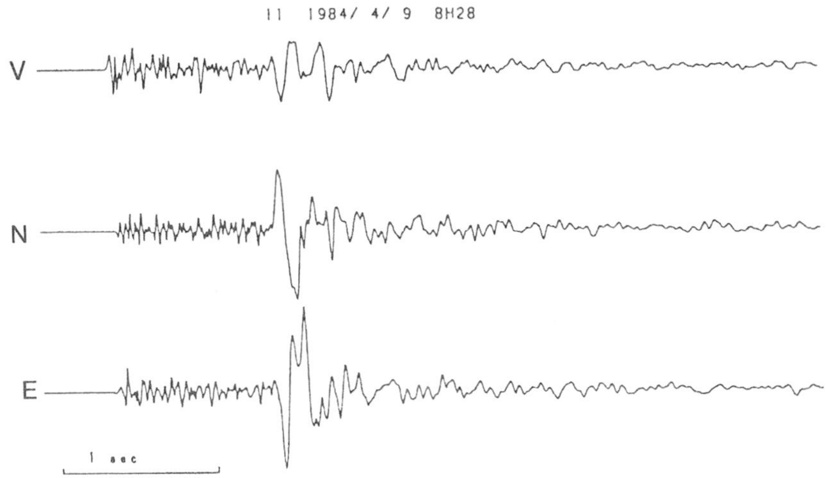

Fig. 2. Example of a three-component seismogram selected for moment-tensor analysis

readings were verified by a visual inspection of the plotted data. Amplitudes of S waves were manually read from the seismograms.

Ten events with a minimum number of eight amplitudes were considered for our analysis. Before computing the moment tensor components with the previously described method, we evaluated the possible sources of errors that may lead to spurious solutions. They can be summarized by amplitude errors and take-off uncertainty. We neglect errors on epicentral location because according to the network configuration and the quality of selected data, these errors do not significantly influence the analysis. The next paragraphs show how these errors have been taken into account.

2.2 Evaluation of Site Amplification Effect by Coda Waves

The main sources of errors using amplitude data are seismic noise and site amplification effects.

Local seismic noise is very small for our data set. In fact, we always select data with a signal to noise ratio greater than 20 dB. Site amplification effect was the main source of error for amplitude data recorded at Campi Flegrei area. De Natale et al. (1987), analyzing source spectra parameters at Campi Flegrei, observed a systematic amplification or a filtering effect at some seismic stations. We evaluated the site effect by a method developed by Aki and Phillips (1986) based on the scattered wave field generated by inhomogeneities of the medium.

On the basis of the hypothesis developed by Aki and Chouet (1975), we consider coda waves from small local earthquakes as waves scattered by randomly distributed heterogeneities in the lithosphere. We used the single backscattering model in which scattering is considered to be a weak process where loss of seismic energy by scattering is neglected. The single scattering hypothesis is justified if the source-receiver distances are much smaller than the mean distances among scatterers (mean free path). This is true for Campi Flegrei, because we observe coda wave durations of less than 30 s, corresponding to travel distances of the order of a few kilometers, whereas the measured mean free path is about 50 km (Castellano et al. 1984). But, even if the model is not exactly correct, this does not influence the estimation of the site effect, because we evaluate the site effect by coda-wave spectral-ratios between the data of the two stations.

Backscattered waves recorded at two (close) stations A and B sample different volumes during the last portion of their travel paths. Therefore, if any site effect exists, the amplitudes of waves travelling toward A and B will be differently modified by propagation through the two volumes just below the stations. These volumes act like filters on the signals whose response may be determined by coda-wave spectral ratios. For a given station, the estimated site effect will concern a volume whose radius is about half way between the two stations.

The moving Fourier spectrum of scattered waves is expressed by:

$$A(f, t) = Ao(f) * t \quad * \exp(-\pi f t/Q), \tag{1}$$

where Ao(f) is the source term, t is the origin time of the earthquake and Q is the quality factor. If the spectral amplitude is modified around the station by site effects, then (1) becomes:

$$A(f, t) = Ao(f) * Si(f) * t \quad * \exp(-\pi f t/Q), \tag{2}$$

in which Si(f) represents the site response.

It is possible to calculate the site effect by evaluating the ratios:

$$\ln \frac{[A(f, t) * t]}{[Ao(f, t) * t]} \tag{3}$$

which expresses the amplification of each station with respect to a reference.

The method we used involves the following steps:

1. Computation of A(f, t) by fast Fourier transform algorithm in a 3-s sliding window, moving along the coda by 1.5-s steps. For each window centred at time t*, the term A(f, t*) is averaged in six spectral bands with central frequency fc equal to 1.7, 3, 6, 12, 16, 24 Hz and with a width of 0.6 fc. This procedure gives a smoothed estimate of the spectrum as a function of t.
2. Selection of the onset and the end of the coda. This is automatically obtained by looking at the changes in the slope of quantity A(f, t) * t, plotted as a function of the time.

Equation (2) can be linearized performing the logarithm, so a good estimate of ratios (3) is obtained from the values calculated by the least squares fit at the time tc, central in the coda of the reference station. The choice of the time tc, for all the stations, cancels any effect of distance related to geometrical spreading and moreover, the term A(f, tc) is better estimated at the centre of the data set, at least for the reference station.

Our reference station is W21, chosen for its spectrum regularity.

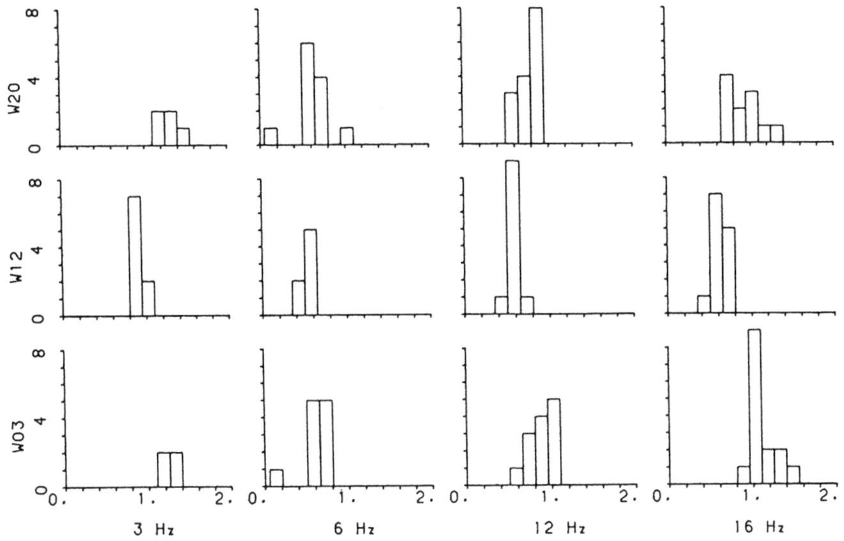

Fig. 3. Example of the site amplification distribution for the frequency bands analyzed (stations W03, W12 and W20, with reference station W21)

To reduce computation time we do not remove the instrumental response curve from the recording file of each earthquake. Then our site effect includes not only geological amplification but also an instrument curve.

Figure 3 gives results obtained from site effect analysis. Assuming a Gaussian distribution, the uncertainties on the obtained values are equal to the standard deviation. The overall results of the attenuation versus frequency are shown in Fig. 4. Station W17 has a similar behaviour to the reference station, the others are all attenuated. In particular, stations W11 and W13 are almost four times smaller than the reference one.

2.3 Influence of Velocity Model and Hypocentral Error on Take-Off Angles

The further key parameter that appears in the equations used for the moment-tensor inversion is the take-off angle. The uncertainties in the take-off angles are due both to hypocentral errors and to poor knowledge of the velocity model.

Aster and Meyer (1988) using the digital network shown in Fig. 1, obtained a three-dimensional velocity model of the central part of the Campi Flegrei volcanic area. Figure 5 shows the contoured Vp values determined. Starting from this figure we selected two possible velocity models: a homogeneous halfspace ($Vp = 2.8\,km/s$) for stations W11, W12 and W17 and two different linear velocity models for the other stations ($Vp = 2.2 + 0.3 * h$ and $Vp = 2 + 0.5 * h$). These are two ring-shaped models that reproduce the caldera shape.

We prefer a linear model because using a layered velocity model the uncertainty on the take-off angles becomes critical when the events are located

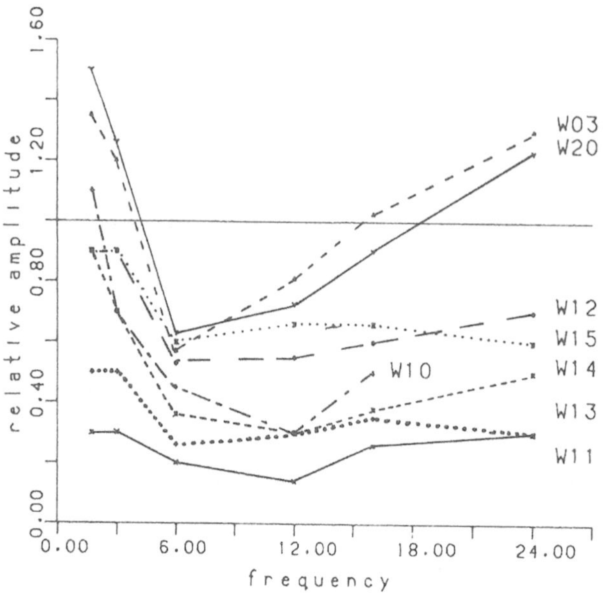

Fig. 4. Site response effect, as average of the site amplification in each frequency band with respect to the reference station W21, is plotted versus frequency. Site effect includes the different gain levels and response curve of the stations

near a boundary and because there is no evidence of strong discontinuities for the first 3 km of the crust at Campi Flegrei (Aster and Meyer 1988).

To analyze the variability of the take-off angle from errors in the depth, we evaluated, for all the seismic stations networked, the dependence on the take-off angle from hypocentral depth for the selected velocity models. Figure 6 shows this dependence for two stations and for two earthquakes. We show data only for a station close to the epicentre (station W12 $\Delta = 0.8$ km) and for a more distant station (W20 $\Delta = 4.4$ km) relative to a very shallow event ($h = 0.8$ km) and a deeper one ($h = 2.2$ km). The differences between the take-off angles computed from the different velocity models are very small, amounting to about 1° for the closest stations and a maximum of 10° for more distant ones.

The variability of the take-off angle to error in the depth is more evident for stations close to the epicentre of very shallow earthquakes. In fact, for the station W12 located at 0.8 km from the epicentre of the event at 0.8 km in depth, the error produces a change of about 25° in the take-off angle. This change is less sensitive for the distant station W20 where it amounts to about 10°.

2.4 Moment-Tensor Solutions

Amplitude data of P and SH direct waves, selected as previously described, were corrected for site amplification effect deduced by the coda method and

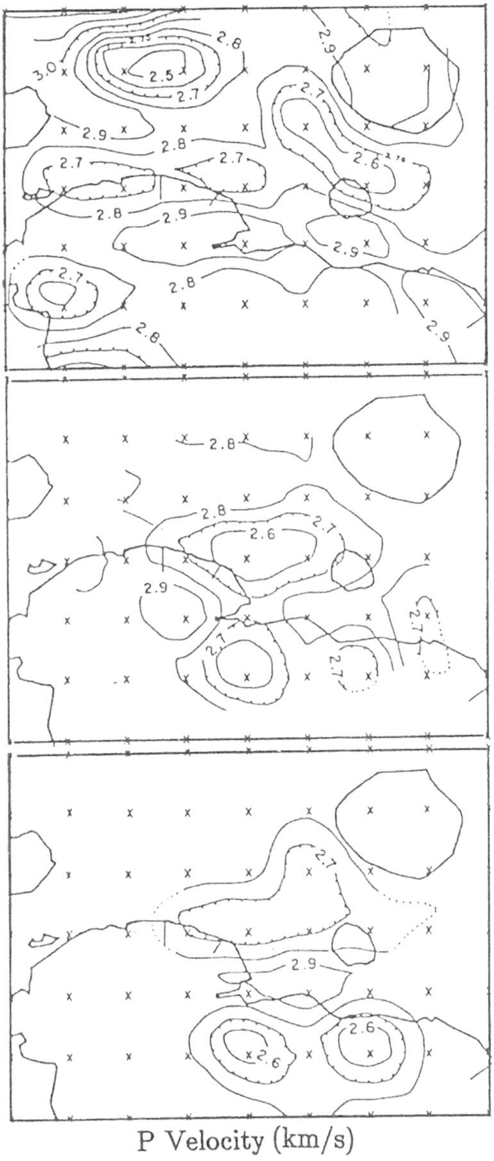

Fig. 5. Contoured three-dimensional Vp model as determined for depths of 1, 2 and 3 km (After Aster and Meyer 1988)

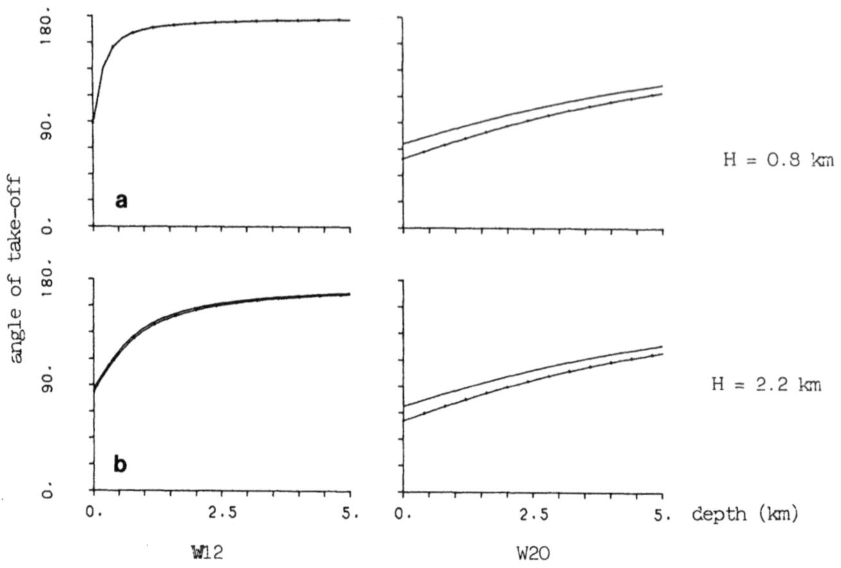

Fig. 6a,b. Dependence of the uncertainty on take-off angles on the errors in depth and on poor knowledge of the velocity model for two events: **a** shallow (0.8 ± 0.3 km), and **b** deeper (2.2 ± 0.4 km), using two realistic velocity models for the area (a common homogeneous velocity model for the stations W11, W12, and W17 and two linear velocity models for the others: Vp = 2.2 + H∗0.3 and Vp = 2 + H∗0.5 (*solid line* and *dotted continuous* line, respectively). The figure shows the different behaviours of the nearby station W12 and of the more distant station W20

used as input data for moment tensor inversion. For each earthquake two inversions were performed: with zero-trace constraint and double couple constraint.

The results of the inversions are given in Fig. 7.

The main features are that most zero trace solutions approach a double couple mechanism and the orientation of principal axes is similar between the two sets of solutions. Some solutions from zero trace inversion are practically the same as the double-couple constraint (Nos. 4, 5, 6, 8). Only one event, No. 9, shows a significant departure from a pure double-couple source mechanism.

To test the variability of moment-tensor solutions according to the errors of take-off angles, we performed the inversion using the extreme values of take-off angles deduced from Fig. 5.

Since we are interested in checking the stability of solutions obtained, we selected two events: a clear double-couple solution (No. 6) and an event that shows a zero trace solution different from a classical double couple (No. 9). Figure 8 shows, for the same two earthquakes, the Wulff projection of nodal planes for double-couple and zero trace solutions using the extreme values of take-off angles for all stations.

The double-couple solution of event No. 6 is very stable, changing only a few degrees in dip and strike of the planes. The zero trace solution is more sensitive

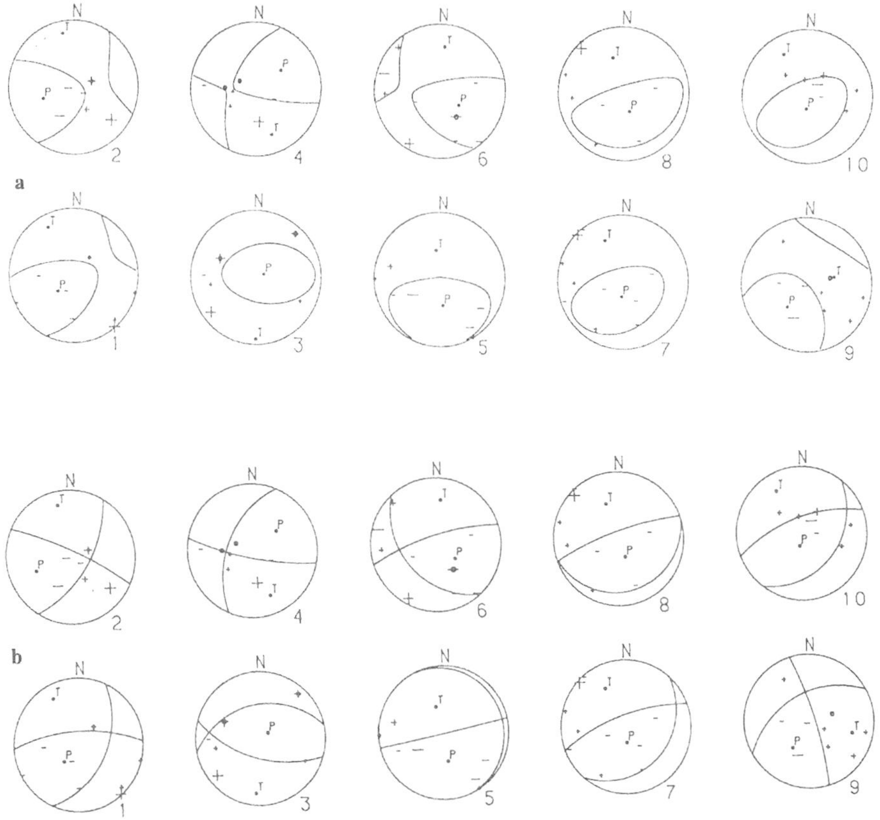

Fig. 7a,b. Wulff projection, lower hemisphere of nodal lines deduced from moment-tensor inversion. The size of symbol is proportional to the amplitude. **a** Solutions constrained to zero-trace moment tensor corresponding to a source with no volume change. **b** Solutions constrained to double-couple sources, zero-trace, and null-determinant of moment tensor

to the variation of take-off angle. In the first case, it approaches a double couple very well, then a departure from a double-couple mechanism becomes evident. The solutions of event No. 9 behave similarly. The variation of strike and dip of double couple amounts to about 15°. The zero trace solution shows, as seen in Fig. 7, a departure from the double couple, but with the variation of the take-off angle this difference becomes non-significant and the solution is almost a shear fault. Furthermore, event No. 9, taking into account the possible variation of input parameters of moment tensor, approaches a double-couple mechanism.

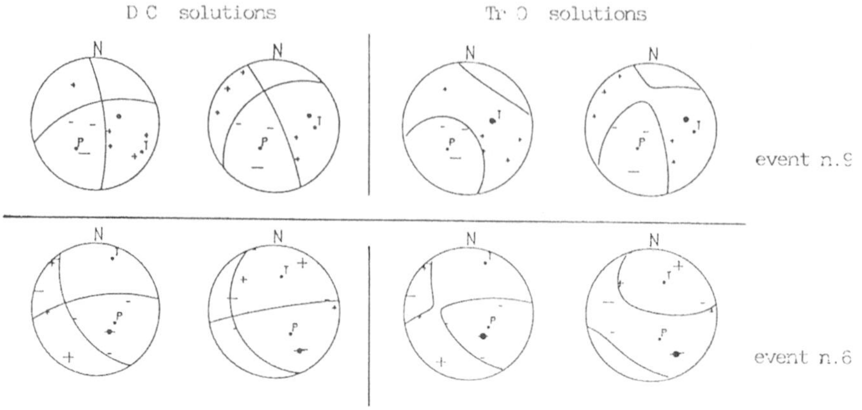

Fig. 8. Dependence of moment tensor solutions on the variation of take-off angles due to uncertainty of the velocity model and hypocentral errors for two selected events (No. 6 and No. 9)

3 Discussion and Conclusion

Ten earthquakes which occurred at Campi Flegrei volcanic area during the 1983–84 bradyseismic crisis are investigated. Moment-tensor components have been obtained by an inverse method using the amplitude of P and SH direct waves. Two inversions with an increasingly more stringent constraint were performed. The no-volume change constraint was first applied and the double-couple solution was therefore imposed. Since, in this area, source mechanisms different from the classical double couple can be expected, a detailed analysis of input parameters has been performed in order to evaluate the reliability of deduced moment-tensor solutions.

Errors affecting amplitude data were examined and local site response evaluated using the coda method. Then amplitudes of body waves were corrected for the site effect.

Uncertainty of emergent angles at source were estimated, both depending on lack of knowledge of a detailed velocity model of Campi Flegrei area and error in depth of earthquakes. Two realistic velocity models were deduced and according to the hypocentral error, the range of variations of take-off angle was computed for each seismic station. With these values, inversions to obtain the moment-tensor components were performed. Results show that most of the solutions are consistent with a double-couple source mechanism. Taking into account the possible variations of input parameters used for moment-tensor inversion, also the few solutions that apparently show a departure from a double-couple solution can be interpreted in terms of shear faults.

In conclusion, we do not think any evidence of "volcanic" seismic source (for example, tensile cracks due to fluid injection) can be deduced from earthquakes recorded at Campi Flegrei volcanic area and that the seismic activity can be related to shear fracture processes induced by the ground deformation.

References

Aki K, Chouet B (1975) Origin of coda waves: source, attenuation and scattering effects. J Geophys Res 80:3332–3342

Aki K, Phillips WS (1986) Site amplification of coda waves from local earthquakes in central California. Bull Seismol Soc Am 76:627–648

Aster RC, Meyer RP (1988) 3-Dimensional velocity structure and hypocenter distribution in the Campi Flegrei caldera, Italy. Tectonophysics 3:195–218

Berrino G, Corrado G, Luongo G, Toro B (1984) Ground deformation and gravity changes accompanying the 1982 Pozzuoli upflit. Bull Volcanol 47(2):187–200

Bonafede M, Dragoni M, Quareni F (1986) Displacement and stress fields produced by a centre of dilatation and by a pressure source in a viscoelastic half-space: application to the study of ground deformation and seismic activity at Campi Flegrei. Geophys J R Astr Soc 87:455–485

Castellano M, Del Pezzo E, De Natale G, Zollo A (1984) Seismic coda Q and turbidity coefficient at Phlegraean Fields volcanic area: preliminary results. Bull Volcanol 47:218–224

Corrado G, Guerra I, Lo Bascio A, Luongo G, Rampoldi R (1976) Inflation and microearthquake activity at Phlegraean Fields, Italy. Bull Volcanol 40:1–20

De Natale G, Iannaccone G, Martini M, Zollo A (1987) Seismic sources and attenuation properties at the Campi Flegrei volcanic area. Pageoph 125:883–917

Julian BR, Sipkin S (1985) Earthquake processes in the Long Valley Caldera area, California. J Geophys Res 90:11155–11169

Karpin TL, Thurber CH (1987) The relationship between swarms and magma transport: Kilauea volcano, Hawaii, Pageoph 125, 6:971–991

Knopoff L, Randall J (1970) The compensated linear vector dipole: a possible mechanism for deep earthquakes. J Geophys Res 75:4957–4963

Rosi M, Sbrana A, Principe C (1983) The Phlegraean Fields: structural evolution, volcanic history and eruptive mechanisms. J Volcanol Geotherm Res 17:237–288

Scarpa R, Patane' G, Lombardo G (1983) Space-time evolution of seismic activity at Mount Etna during 1974–1982. Ann Geophys 1, 6:451–462

Shimizu H, Ueki S, Koyama J (1987) A tensile-shear crack model for the mechanism of volcanic earthquakes. Tectonophysics 144:287–300

Trehu AM, Nabelek JL, Solomon SC (1981) Source characterization of two Reykjanes Ridge earthquakes: surface waves and moment tensor; P waveforms and nonorthogonal nodal planes. J Geophys Res 86, B3:1701–1724

Ukawa M, Ohtake M (1987) A monochromatic earthquake suggesting deep-seated magmatic activity beneath the Izu-Ooshima volcano, Japan. J Geophys Res 92:12649–12663

Apparent Variation of Coda Q in Phlegraean Fields During the Bradyseismic Crisis of 1982–1984

R. Ortiz, A.M. Correig, J.L. Diez, and M.L. Muñoz

Abstract

Coda Q^{-1} has been computed at Plegraean Fields as a function of frequency for very close events, ranging from 0.7 to 3 km, from which coda waves have sampled the region containing the magmatic chamber as well as the region of maximum uplift. Values of coda Q^{-1} are very high, one order of magnitude higher than those obtained by other authors for a wider zone, and show statistically significative temporal and spatial variations. These variations persist upon interpretation of Q^{-1} data through a continuous relaxation model. Spatial variation of Q^{-1} is interpreted in terms of the spatial distribution of temperature due to a magmatic intrusion generated 10500 years ago. Temporal variation of Q^{-1} is qualitatively interpreted in terms of variations of pressure as well as variation in the degree of saturation of pore fluids.

1 Introduction

In the period 1982–1984 a bradyseismic crisis occurred at Plegraean Fields (Italy). As a part of an agreement for Scientific Cooperation between the CSIC (Spain) and the CNR (Italy) for volcanologic problems, a prototype of a digital seismic recorder system was installed by the CSIC at Phlegraean Fields for a period of 9 days in October 1983, and for a period of 11 days in September 1984. The fieldwork was carried out in collaboration with the Gruppo Nazionale per la Vulcanologia (Italy). The aim of this study is to present the analysis and results of the data recorded during these two periods.

Coda Q has been determined from the digital records. Prior to our study, Castellano et al. (1984) and Del Pezzo et al. (1985) made similar studies of the same zone. The differences between their study and ours lie in the fact that we carried out the analysis for a shorter lapse time of coda waves, thus sampling a much reduced volume which contained the magmatic chamber along with the uplifted zone. We found evidence of temporal and spatial variations of Q^{-1} (f) that could be related to the thermal processes of the magmatic chamber and surrounding medium.

2 Instrumentation

The digital seismic recording system is a four-channel event detector with an LTA/STA algorithm (Lee and Stewart 1981) working with digital implementation and supported by two low power 8-bit CMOS microprocessors. The system uses a 12-bit high-speed CMOS A-D converter, where the signal is digitized at 64 samples/s with a 16-s time recording length. An auxiliary channel is available for time signal or event marks. A chart recorder is used as a high-speed monitor. The seismic events are digitally stored in $5\frac{1}{4}''$ floppy disks (Muñoz 1988). It is a light portable piece of equipment, specifically designed to operate in microseismicity on hardly accessible areas (Ortiz et al. 1987).

In our fieldwork in the Phlegraean Fields, this seismic system was directly connected to the demodulator's output of the local FM telemetric network of the Vesuvius Observatory Network (Guerra et al. 1974) at the recording center of Arco Felice. Only the vertical component was used for the digital recording system. In 1983, only one station placed in Darsena (DAR-1983) was connected to the digital recording system for the period 20 to 28 October. In 1984, seismic stations located at Darsena (DAR-1984), Solfatara (SOL-1984), Baia (BAI-1984) and Accademia (ACA-1984) were connected to the digital recording equipment for the period 11 to 21 September. For this latter period, only events detected by two or more stations were recorded. The locations of the seismic stations are shown in Fig. 1, along with the uplift contour lines.

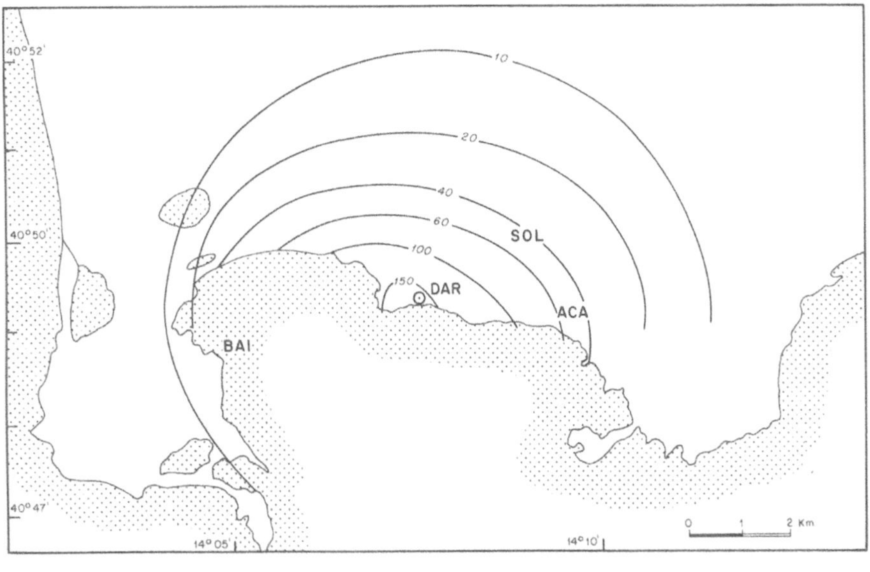

Fig. 1. Location of the stations of the seismic network (*SOL*, *DAR*, *BAI*, *ACA*). The *continuous line* represents the isolines of the uplift (in cm) measured during the period January 1982 to September 1983

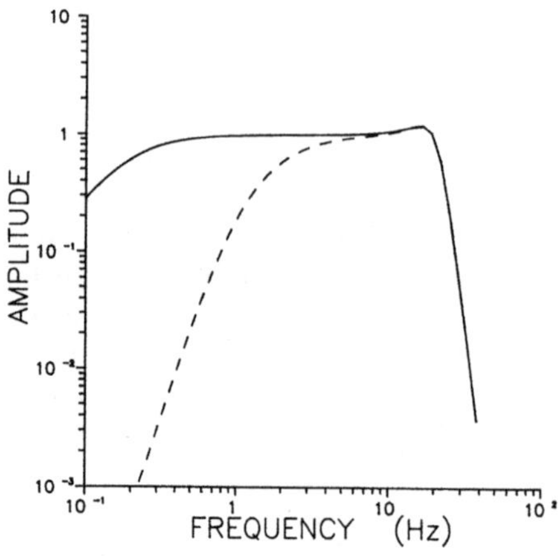

Fig. 2. Instrumental response of the seismic station

Figure 2 shows the instrumental response of the system. It is mainly dominated by the response of the geophones and the telemetry of the Vesuvian Observatory Network.

3 Earthquake Location

In September 1984 four stations (vertical component) were in operation. The data are generally very noisy because the network is located near the sea and near a densely inhabited area. Nineteen events with a reasonably good signal-to-noise ratio were recorded by the four stations. Hypocenters and origin time were calculated in two steps. First, epicenters were located by computer-assisted graphical methods, looking, through trial and error, for the best fit of the intersection of the circles defined by the P-wave travel time and, whenever possible, S-wave travel time (Aguacil 1987). Once the convergence was achieved, the second step was to use the graphically determined hypo-center with its corresponding origin time as a trial input to the computer program BASIC-HYPO (Mendoza and Morgan 1985). The program was modified in order to work with very short distances (of the order of a few kilometers) by changing from polar to rectangular coordinates. Model A of De Natalle et al. (1984) was used as a velocity model. Epicenters are distributed on a "ring" around station DAR, with hypocentral distances to station DAR ranging from 0.7 to 3 km. Taking as a reference station DAR, station SOL is located at 1.6 km, station ACA at 2.2 km and station BAI at 3.3 km. Epicentral locations are shown in Fig. 3, and are in good agreement with those obtained by Del Pezzo et al. (1985) and De Natale et al. (1984, 1987). Unfortunatelly, no absolute time was available, only a relative one common to the four stations. Therefore, the located events cannot be identified with those located by the seismic network of Osservatorio Vesuviano.

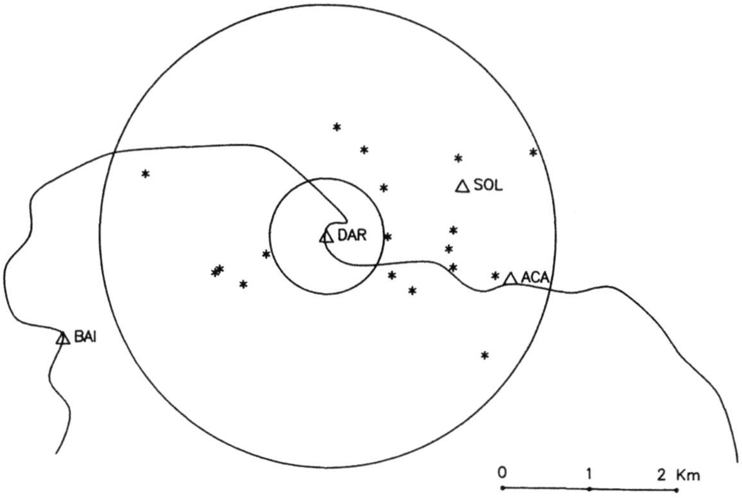

Fig. 3. Epicentral locations of the events used in the study

In October 1983 only station DAR (vertical component) was in operation and 29 events were selected on the basis of the signal-to-noise ratio. Obviously, no epicentral location could be deduced. However, an estimation of the hypocentral distance was possible from S-P delays, and the travel times were computed by assuming a mean velocity of 2.9 km/s (Castellano et al. 1984; Del Pezzo et al. 1985). Epicenters were selected to range between 1 and 3 km to station DAR, to agree with the distance range of the 1984 selected events. To obtain a rough estimation of the committed errors, this method was applied to the events recorded in 1984, with known hypocentral distances. It was found that the distances were overestimated by 15–20%. It is assumed that the epicenter location follows the same pattern as in 1984.

4 Evaluation of Coda Q

The method to evaluate coda-Q is based on the assumption that coda waves are single S to S-scattered waves from randomly distributed heterogeneities. Del Pezzo et al. (1985) discuss the validity of this assumption. Aki and Chouet (1975) derived a model for the amplitude of coda waves by considering, among other hypotheses, the source and the receiver located at the same point. This assumption is valid for lapse times greater than twice the travel time of S-waves. In the present study, however, because of the short recording length (16 s, including an average pre-event length of 4 s), the earlier part of the coda had to be used, and consequently the source-receiver separation must be considered. This separation is included in the single scattering model developed by Sato (1977). For a lapse time greater than twice the travel time of S-waves, Sato and Aki and Chouet models are equivalent.

According to Sato's model, and following Correig et al. (1988), the reduced amplitude of coda waves as a function of time can be written:

$$\ln[A(r, t, \omega)\, r\, K(a)^{-1/2}] = \ln C(\omega) - bt/2, \quad (1)$$

where

$$b = \omega/Q_c. \quad (2)$$

$A(r, t, \omega)$ is the rms amplitude observed at a distance r and frequency ω, t is the lapse time from the origin, $C(\omega)$ is the source factor, Q_c the coda-Q factor and $K(a)$ is defined by:

$$K(a) = \frac{1}{a} \ln \frac{a + 1}{a - 1}, \quad (3)$$

where $a = t/t_0$ and t_0 is the arrival time of the S-wave.

4.1 Data Analysis

Data analysis follows the classical method of Aki and Chouet (1975) applied to Sato's model (Sato 1977). For details, see Roeckher et al. (1982) and Correig et al. (1990).

The seismograms of the selected events (vertical component) were band-pass filtered using six-pole Butterworth filters with center frequencies at 1.5, 3, 6, 12 and 24 Hz and respective band widths of 1, 2, 4, 8 and 16 Hz. Figure 4 shows an example of a seismogram before and after the process of filtering. Amplitudes were obtained for each filtered seismogram using an overlapping moving window of 1 s. Once the average peak-to-peak rms amplitude was determined for the first 1-s window, the window was shifted 0.25 s and the average amplitude was again determined. This procedure was continued until the signal was lost in noise or until the end of the record. This process was carried out for each filtered seismogram. The rms time series was fitted to Eq. (1) by linear regression, and by means of Eq. (2) coda-Q values were obtained. Q-values were accepted when the correlation coefficient was equal or greater than 0.85. Figure 5 shows an example of the exponential fit of a seismogram filtered at 12 Hz to Eq. (1). As a test, coda-Q has been computed for a given distance r as well as for the same station and event by randomly varying the hypocentral distance r up to maximum of 25%. As a result, the variation of the new obtained coda-Q is no larger than 20%. Thus, exact knowledge of the hypocentral distance r, although important, is not critical.

Following Jackson and Anderson (1970), once coda-Q has been obtained use of Q^{-1} instead of Q was made, because: (1) if a seismic wave is traveling over regions of different anelastic properties, it is attenuated according to an average of Q^{-1} and (2) when attenuation is carried by more than one mechanism, the total attenuation will be described by the sum of Q^{-1} from the separate mechanisms. Q^{-1} has been average for each station and for each frequency band. Obtained results are presented in Table 1.

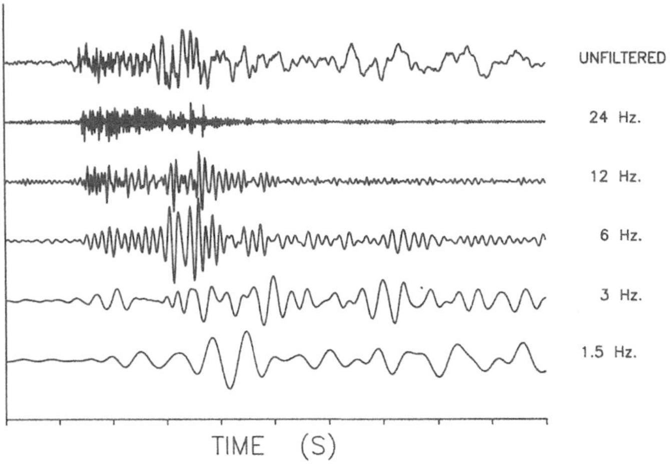

Fig. 4. Example of an seismogram filtering: original seismogram (*above*) and seismogram filtered at 1.5, 3, 6, 12 and 24 Hz

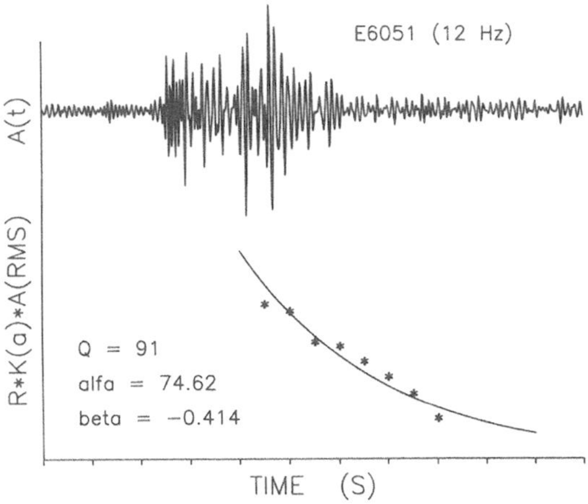

Fig. 5. Example of the exponential fit of the recorded amplitude to Eq. (1) for the retrieval of Q, along with the filtered seismogram at 12 Hz. *Alpha* denotes ln[C(ω)] and *beta* b

Del Pezzo et al. (1985) report coda-Q from the Phlegraean Fields for the period August–September 1983. Their results are summarized in Table 2. Comparing their data with those from station DAR – 83 (see Table 1A), it appears that our Q^{-1} data are higher than those from Del Pezzo et al. (1985) by a factor ranging from 10 at 1.5 Hz to 2.5 at 12 Hz, and still higher than those obtained by Castellano et al. (1984). As a general characteristic, results obtained in the present study and those reported by Castellano et al. (1984) show that $Q^{-1}(f)$ descreases with frequency, whereas results reported by Del Pezzo et al. (1985) show that $Q^{-1}(f)$ decreases from 1.7 to 6 Hz and $Q^{-1}(f)$ is constant from 6 to 18 Hz. At least qualitatively, this difference can be explained by the short coda

Table 1. Mean code Q^{-1}, number of samples (NS) and standard error of the mean (SEM) as a function of frequency

A Station DAR-1983

Frequency Hz	$Q^{-1} * 1000$	NS	SEM
1.5	120.83	4	26.68
3	67.40	8	18.54
6	27.50	29	5.73
12	8.64	29	0.75

B Station DAR-1984

Frequency Hz	$Q^{-1} * 1000$	NS	SEM
1.5	73.87	14	8.69
3	43.94	17	4.31
6	19.24	17	2.08
12	7.52	17	0.88
24	6.32	9	0.99

C Station SOL-1984

Frequency Hz	$Q^{-1} * 1000$	NS	SEM
3	41.98	6	4.42
6	12.38	19	1.31
12	6.66	19	0.72

D Station BAI-1984

Frequency Hz	$Q^{-1} * 1000$	NS	SEM
3	41.29	6	5.77
6	12.89	14	1.51
12	5.33	14	0.70

E Station ACA-1984

Frequency Hz	$Q^{-1} * 1000$	NS	SEM
3	58.37	5	13.06
6	15.52	18	2.52
12	6.91	18	1.39

duration of the studied events, 6 to 10 s; compared to that of Del Pezzo et al. (1985), who obtained a coda duration of 15 to 40 s, resulting in a sampling of the backscattered waves of a hemisphere with radious of about 8 – 10 km in our case, and a radious of about 25–40 km in the Del Pezzo's study (Pulli 1984). Thus, the coda waves of the presently studied events sample a more reduced and surficial (and thus inhomogeneous) zone, roughly corresponding to the uplifted zone.

Table 2A. Mean coda Q^{-1} as deduced from the mean coda-Q, standard deviation (SD) and standard error of the mean (SEM) of the vertical component of coda-Q values as a function of frequency reported by Del Pezzo et al. (1985) for the Phlegraean Fields

Frequency Hz	$Q^{-1} * 1000$	Q	SD of Q	SEM of Q
1.7	10.31	97	47	16
3	6.41	156	67	22
6	3.34	299	102	24
12	3.47	288	68	14
18	3.34	299	106	15

B Mean coda Q^{-1} as deduced from the mean coda Q, standard deviation (SD) and sample variance (SV) of the vertical component of coda-Q values as a function of frequency reported by Castellano et al. (1984) for the Phlegraean Fields

Frequency Hz	$Q^{-1} * 1000$	Q	SD of Q	SEM of Q
1.7	7.41	135	27	10
3	4.44	225	53	20
6	3.77	265	93	21
12	3.07	326	112	23
18	2.31	433	131	36

4.2 Coda Q^{-1} Results

Table 1 clearly shows the decrease in Q^{-1} for increasing frequencies for each station. Following Aki (1980), coda Q^{-1} has been fitted to the power law function:

$$Q(f)^{-1} = Q_0{}^{-1} * f^n. \tag{4}$$

The fit has been applied to stations DAR-1983 and DAR-1984 and for all available data (ALL). Figure 6 shows the variation of Q^{-1} as a function of frequency for station DAR-1984. As a reference, the fit has also been applied to the data reported by Castellano et al. (1984). It should be noted that the differences, surprisingly showing the results reported by Castellano et al. (1984) with respect to those reported by Del Pezzo et al. (1985), and reproduced by De Natale et al. (1987), were apparently deduced from the same set of data or, at least, for the same period of time and from the same instrument. Results presented in Table 3 clearly show the differences already noted in Tables 1 and 2. As expected, due to of the difference in the sampled volume, values of $Q_0{}^{-1}$ and n from Castellano et al. (1984) are quite different from those of the present study, showing a much lower attenuation. Although not so striking, values of $Q_0{}^{-1}$ and n for DAR-1983 are apparently different than those for DAR-1984 and ALL.

In order to test this apparent difference in Q^{-1} values, following Peng et al. (1987), a statistical test of significance of the differences between coda Q^{-1} for

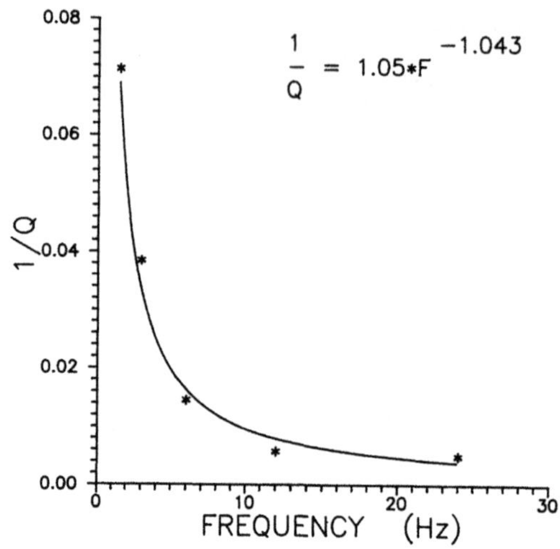

Fig. 6. Variation of Q^{-1} as a function of frequency for station DAR-1984

Table 3. Values of the parameters Q_0^{-1} and n obtained by fitting the power law $Q(f)^{-1} = Q_0^{-1} * f^n$ to data recorded at station DAR for the years 1983 (DAR-1983) and 1984 (DAR-1984)[a]

Station	Q_0^{-1}	SD	n	SD
DAR-1983	0.23	1.24	−1.27	0.13
DAR-1984	0.11	1.23	−0.96	0.10
ALL	0.11	1.20	−1.04	0.09
CAS	0.0084	1.13	−0.44	0.62

[a] For comparison, the values of these parameters, obtained by fitting all available data, are given (ALL) along with the values of these parameters obtained by fitting the coda-Q values reported by Castellano et al. (1984) (CAS). SD = standard deviation, r = correlation coefficient.

the different groups has been applied. To carry out the statistical test of significance, the following statistics are defined (Fisher and Yates 1970):

$$d = \frac{\bar{x}_1 - \bar{x}_2}{\sqrt{s_1^2 - s_2^2}} \tag{5}$$

$$s_i = \frac{\sigma_i}{\sqrt{N_i - 1}}, \tag{6}$$

where $\bar{x}_i$ is the mean of coda Q^{-1} for group i, s_i is the standard error of the mean for group i, σ_i is the standard error of all samples in group i, and N_i is the number of samples in group i.

This method is appropiate for the difference of two means for which the errors are due to different causes. The difference between two means is considered to be significant when d in Eq. (5) exceeds the critical values for different sets of sampling numbers given in Table VI of Fisher and Yates (1970) at the 95% confidence level. Otherwise, the difference is not considered to be significant. The critical values of d depend not only on the sampled numbers of coda Q^{-1} for the different groups, but they also vary with the scale factor (θ), where:

$$\tan\theta = \frac{s_1}{s_2}. \tag{7}$$

The test of significance has been applied to data presented in Table 1 for frequencies of 6 and 12 Hz, for which the larger number of observations was obtained. It appears that all groups have passed the test, so that all groups are significatively different. This means that coda Q^{-1} for DAR-1983 is statistically different from that of DAR-1984 (temporal variation), and that for the year 1984 coda Q^{-1} for station DAR is statistically different than that for the other stations (spatial variation).

In order to explain the variation of coda Q^{-1} the data are phenomenologically interpreted in terms of a continuous relaxation model. For the continuous relaxation model (CRM) the specific dissipation function satisfies the relation (Liu et al. 1976):

$$Q^{-1}(\omega) = \frac{2}{\pi Q_m}\tan^{-1}\left[\frac{\omega(s_2 - s_1)}{\omega^2 + s_1 s_2}\right], \tag{8}$$

where ω is the frequency, s_1 and s_2 are the low- and high-frequency cutoffs, respectively, and Q_m is the flat part of the spectrum. The relaxation times are defined as $t_1 = s_1{}^{-1}$ and $t_2 = s_2{}^{-1}$. On the other hand, it is well known that in an attenuating medium body waves must be dispersive in order to satisfy causality. Attenuation and dispersion are related through the Kramers-Kroning relations. It has been shown (Correig et al. 1989) that in a high attenuative medium body wave dispersion can be measured through the arrival time of filtered seismograms, once corrected by instrument and filter group delay. The variation in arrival time with frequency is measured with respect to a given frequency taken as a reference. The filtered seismograms for coda-Q determination have been used, and as a reference we have taken the 6 Hz frequency. The arrival time as a function of frequency can be written as (Correig et al. 1989):

$$t(\omega) = t_u\left[\frac{1}{2\pi Q_m}\ln\frac{\omega^2 + s_2{}^2}{\omega^2 + s_1{}^2} + 1\right], \tag{9}$$

where t_u is the travel time of the unfiltered signal. From Eq. (9) the predicted dispersion $t(f) - t(6)$ can be computed and compared with observations. Table 4 shows the measured body wave dispersion for every station and for all stations. There is no apparent spatial or temporal variation of the dispersion.

The retrieval of $Q_m{}^{-1}$, s_1 and s_2 was carried out in two steps. First, $Q_m{}^{-1}$ and $t_1 = s_1{}^{-1}$ were obtained, assuming an arbitrary value for s_2 of 10000, through a

Table 4. Observed body wave dispersion[a]

A Station DAR-1983

Frequency Hz	D(f) (unc)	NS	SEM	D(f) (corr)
1.5	0.172	2	–	0.002
3	0.070	4	0.031	0.029
6	0.000	24	–	0.000
12	−0.094	24	0.011	−0.105

B Station DAR-1984

Frequency Hz	D(f) (unc)	NS	SEM	D(f) (corr)
1.5	0.396	6	0.096	0.226
3	0.118	10	0.022	0.077
6	0.000	11	–	0.000
12	−0.069	11	0.014	−0.080
24	−0.084	3	0.010	−0.100

C Station SOL-1984

Frequency Hz	D(f) (unc)	NS	SEM	D(f) (corr)
3	0.031	2	–	0.000
6	0.000	12	–	0.000
12	−0.042	12	0.089	−0.053

D Station BAI-1984

Frequency Hz	D(f) (unc)	NS	SEM	D(f) (corr)
3	0.063	2	–	0.022
6	0.000	12	–	0.000
12	−0.042	12	0.011	−0.053

E Station ACA-1984

Frequency Hz	D(f) (unc)	NS	SEM	D(f) (corr)
3	0.177	3	0.038	0.136
6	0.000	15	–	0.000
12	−0.057	15	0.014	−0.068

F All stations

Frequency Hz	D(f) (unc)	NS	SEM	D(f) (corr)
1.5	0.339	8	0.086	0.169
3	0.103	20	0.017	0.062
6	0.000	85	–	0.000
12	−0.058	85	0.057	−0.069
24	−0.125	4	0.042	−0.141

[a] The dispersion is measured as the difference in arrival time for a seismogram filtered at a given frequency f minus the arrival time for the seismogram filtered at 6 Hz. D(f) (unc) denotes the dispersion without correction for instrument and filter group delay, and D(f) (corr) denotes the true dispersion, that is, the difference in arrival time corrected for group delay. NS = number of samples and SEM = standard error of the mean.

Table 5. Fit of the coda-Q values obtained for each station to a continuous relaxation model[a]

Station	Q_n^{-1}	t_1	$\chi^2(Q_m^{-1}, t_1)$	t_2	$\chi^2(t_2)$
DAR-1983	0.750	0.390	0.458E-04	15	0.458E-04
DAR-1984	0.172	0.136	0.778E-05	6	0.764E-05
SOL-1984	0.193	0.191	0.759E-04	4	0.741E-04
BAI-1984	0.190	0.185	0.731E-04	4	0.717E-04
ACA-1983	0.266	0.195	0.206E-03	4	0.201E-04
ALL	0.309	0.167	0.910E-05	3	0.154E-04
PEZ-1983	0.015	0.057	0.223E-05	–	–
CAS-1983	0.008	0.023	0.103E-05	–	–

[a] Q_n^{-1} and t_1 have been obtained from coda-Q values by assuming $t_2 = 10000$ and, in a second step, t_2 was obtained through body wave dispersion. The χ^2 criterion is a measure of the goodness-of-fit. As a reference, the fits to the data reported by Del Pezzo et al. (1985) (PEZ-1983) and Castellano et al. (1984) (CAS-1984) are given. For more details, see text.

nonlinear fit to Eq. (8) by means of the procedure of the grid search procedure (Bevington 1969). $Q^{-1}(f)$ values presented in Tables 1 and 2 were used and the iterative process was repeated until the change in χ^2 of a given iteration was less than 1/1000 of the previous value of χ^2. Results are presented in Table 5. Values of Q_m^{-1} are very large (high attenuation), an order of magnitude larger than that reported by Del Pezzo et al. (1985) and Castellano et al. (1984), and it appears that there is a clear difference in the values of Q_m^{-1} for 1983 and 1984, and practically no spatial differences for 1984. The lower values of Castellano and Del Pezzo could be due to their larger lapse times for coda-Q determination. From now on, the term spatial variation of Q_m^{-1} will refer to the variation of Q^{m-1} values of our determinations with respect to those of Castellano et al. (1984) and Del Pezzo et al. (1985). Values of t_1 show a smooth temporal and spatial variation. The higher values correspond to station DAR and show a decrease from 1983 to 1984. For the year 1984, stations SOL, BAI and ACA show a constant t_1. Again, values of t_1 deduced by Castellano and Del Pezzo are an order of magnitude lower. An estimation of $t_2 = s_2^{-1}$ was made through Eq. (9) by substituting Q_m^{-1} and t_1 by the values previously found. Results are presented in Table 5. Surprisingly, values of t_2 are very close to values of t_1, thus defining a very narrow absorption band.

5 Discussion

To explain the temporal and spatial variation of coda Q^{-1} we must bear in mind, first of all, that the sampled region is very small, a hemispherical shell with a radius of about 10 km, and secondly, the location of the stations: station DAR (1983 and 1984) is located at the point of maximum uplift, station SOL at a point of moderate uplift and stations BAI and ACA at points of low uplift.

Thus, the region that the coda waves have sampled corresponds, in its largest part, to the deformed region. On the other hand, results obtained in the present study show a very high ${Q_m}^{-1}$ (an order of magnitude higher than that obtained by Castellano et al. (1984) and Del Pezzo et al. (1985).

As previously noted, values of t_1 and t_2 are very close, so that in a first approximation this narrow absorption band can be approximated by a Debye absorption peak. For a Debye absorption peak, $Q^{-1}(\omega)$ is expressed as (Liu et al. 1976):

$$Q^{-1}(\omega) = M\frac{\omega t}{1 + \omega^2 t^2}, \tag{10}$$

where ω is the frequency, t the relaxation time and, for the purpose of this presentation, M is a constant.

By assuming that $Q^{-1}(\omega)$ can be explained as an activated process (Jackson and Anderson 1970) and making the approximation suggested by Weertman (1970),

$$t = t_0 \exp(-gT_m/T), \tag{11}$$

where g is a constant, T_m the melting temperature of the solidus, T the absolute temperature, and t_0 the atomic jump time, of the order of 10^{-13} s (Anderson et al. 1977). A suggested value for g is 7 (Meissner 1986). By considering $T_m/T = 2$, $\omega^2 t^2 \ll 1$, so that Eq. (10) can be rewritten as:

$$Q^{-1} = C \exp(-gT_m/T), \tag{12}$$

where $C = M\omega t_0$. In order to compare Q^{-1} from distinct determinations, we consider the value of Q^{-1} at the Debye peak, i.e. at $\omega = t^{-1}$. In this way C is a constant ($C = Mt^{-1}t_0$) and Q^{-1} becomes a function of only the temperature T of the medium.

The actual distribution of temperatures at Phlegraean Fields can be approximated due to the cooling of a magmatic chamber that was developed 10500 years ago (Armienti et al. 1984). The temperature at a distance r from the center of the intrusion and at time t is given by (Carlslaw and Jaeger 1980):

$$T(r,t) = \frac{T_0 - T_a}{2}\left\{\operatorname{erf}\frac{r+a}{2\sqrt{kt}} - \operatorname{erf}\frac{r-a}{2\sqrt{kt}} - 2\frac{\sqrt{kt}}{r\sqrt{\pi}}\left[\exp\left(\frac{-(r-a)^2}{4kt}\right) - \exp\left(\frac{-(r+a)^2}{4kt}\right)\right]\right\} + Ta, \tag{13}$$

where T_0 is the temperature of the intrusion, T_a the ambient temperature, a the radius of the intrusion, k the thermal diffusivity of the medium and t the time since the formation of the intrusion.

If the constant C is known, Eq. (12) provides us with a means to calculate Q^{-1} in the space surrounding the magmatic chamber as the Q^{-1} averaged over the sampled volume, that is

$$Q_s^{-1} = \frac{1}{V_s}\int_{vs} Q^{-1}T(r)\,dV_s, \tag{14}$$

where V_s stands for the sampled volume. By introducing Eq. (12) into Eq. (14) we obtain

$$Q_s^{-1} = \frac{C}{\frac{4}{3}\pi r_s^3} \int_0^{r_s} \exp\left[-g\frac{T_m}{T(r)}\right] 4\pi r^2 dr. \tag{15}$$

Equation (15) explicitly gives us the average Q^{-1} corresponding to a spherical volume of radius r_s as a function of the distribution of temperatures $T(r)$ generated by the magmatic intrusion.

Because of the difficulty of computing the constant M that appears in Eq. (12) (related to the relaxed and unrelaxed modules), by computing Eq. (15) at two different distances, r_i and r_j, and dividing, the relative variation of Q^{-1} is obtained:

$$\frac{Q_i^{-1}}{Q_j^{-1}} = \frac{r_j^3}{r_i^3} \frac{\int_0^{r_i} \exp[-g\,T_m/T(r)] r^2 dr}{\int_0^{r_j} \exp[-g\,T_m/T(r)] r^2 dr} \tag{16}$$

This equation allows us to compare the results obtained in this work with those obtained by Castellano et al. (1984) and Del Pezzo et al. (1985). In doing this comparison, it is assumed that coda Q^{-1} is representative of intrinsic Q^{-1}. Because of the small sampled volume, this may indeed be the case (Correig et al. 1989; Aki pers. comm.).

For the comparison we consider as representative of our results the average Q^{-1} of all stations (ALL). From Table 5 we obtain:

$$Q^{-1}(\text{CAS-1983})/Q^{-1}(\text{ALL}) = 0.026$$

$$Q^{-1}(\text{PEZ-1983})/Q^{-1}(\text{ALL}) = 0.045.$$

In applying Eqs. (13) and (16) the following values are used: $t = 10\,500$ years, $T_0 = 1273\,K$, $T_m = 1123\,K$, $a = 1.5\,km$, $k = 1.2$ E-06 $m^2 s^{-1}$ (Armienti et al. 1984) and $g = 7$ (Meissner 1986). From Eq. (16) the following ratios $Q^{-1}(r_i)/Q^{-1}(r_j)$ have been computed, where r_k stands for the radius of the spherical shell containing the region sampled by coda waves:

$$Q^{-1}(30)/Q^{-1}(10) = 0.409, \quad Q^{-1}(30)/Q^{1}.002^{1}(7) = 0.017;$$

$$Q^{-1}(35)/Q^{-1}(10) = 0.036, \quad Q^{-1}(35)/Q^{1}.002^{1}(7) = 0.012;$$

$$Q^{-1}(40)/Q^{-1}(10) = 0.029, \quad Q^{-1}(40)/Q^{1}.002^{1}(7) = 0.001.$$

By changing the values of temperatures and g, the above ratios are not substantially changed. Thus, we can conclude that the differences in Q^{-1} found between the results obtained by Castellano et al. (1984) and Del Pezzo et al. (1985) with respect to the ones actually presented may be satisfactorily explained by the distribution of temperatures originated by the magmatic intrusion and its further evolution.

The other point to discuss is the temporal variation of coda Q^{-1} at station DAR. As already stated, the statistical test applied at 6 and 12 Hz showed a significative difference between 1983 and 1984. Interpreted in terms of a relaxation model, Q_m^{-1} for 1983 is four times larger than that for 1984, and t_1 for 1983 is three times larger than that for 1984. Such a variation cannot be due to temperature variations of the medium because for only 1 year it would be unnoticed. One physical parameter that has suffered a significant change during this year is the ground deformation, with the maximum uplift centered at station DAR. Bonafede et al. (1986) studied this ground deformation in terms of a pressure source in a viscoelastic half-space and found that the measured uplift could be explained by means of a variation in pressure of 20–100 bar. Apparently, Eq. (11) provides a model for the variation of the relaxation time t in terms of the variation in pressure, noting that the product gT_m is related to E + pV, where E is the activation energy, V the activation volume and p the pressure. However, this analysis is beyond the scope of this study.

Another possible mechanism to explain the temporal variation of coda Q^{-1} is the variation in the degree of saturation of the pore fluids (Winkler and Nur 1979). This mechanism predicts that the attenuation should increase with the degree of saturation and then rapidly decrease at total saturation. Near the boiling temperature of pore fluid, attenuation may be strongly affected (Johnston et al. 1979). In 1983, the system of geothermal aquifers in the area experienced a very large variation. On the basis of the systematic geochemical surveillance on the fumaroles located in the area, they appeared to present maximum values in the convective input of magmatic origin (Martini et al. 1984). It is estimated that they occurred in September–October 1983 just during the period of operation of seismic station DAR-1983, producing an increase in ebullition of shallow aquifers and thus modifying the state of pore fluid saturation. This process explains a variation in Q^{-1} of one or two orders of magnitude.

6 Conclusions

Coda Q^{-1} has been retrieved from the analysis of coda waves. Statistically significative temporal and spatial variations of Q^{-1} have been found.

A physical interpretation is proposed for the spatial variation of Q^{-1} in terms of the spatial distribution of temperature due to the generation of a magmatic intrusion. The underlying mechanism consists in activation processes, as described by Jackson and Anderson (1970), which predicts the variation of $Q^{-1}(\omega)$ as a Debye absorption peak. The interpretation of $Q^{-1}(\omega)$ observations along with body wave dispersion in terms of a continuous relaxation model confirms the Debye shape.

The temporal variation of Q^{-1} has been qualitatively interpreted in terms of variation of pressure and/or the variation in the degree of saturation of pore fluids. However, more theoretical and observational work is needed for a quantitative interpretation.

Acknowledgements. This research was supported in part by the Comision Asesora para la Investigación Científica y Técnica (CICYT) under grants 449 and PB86-0431-C05-03.

References

Aguacil G (1987) PEDES. Programa para la localización de sismos próximos. Observatorio de Cartuja, Granada

Aki K (1980) Scattering and attenuation of shear waves in the lithosphere. J Geophys Res 85:6496–6504

Aki K, Chouet B (1975) Origin of coda waves: source, attenuation and scattering effects. J Geophys Res 80:3322–3342

Anderson DL, Kanamori H, Hart RS, Liu HP (1977) The Earth as a seismic absorption band. Science 196:1104–1106

Armienti P, Barberi F, Innocenti F (1984) A model of the Phlegraean Fields magma chamber in the last 10500 years. Bull Volcanol 47:349–358

Bevington, PR (1969) *Data reduction and error analysis for the physical sciences* McGraw-Hill, New York, 336 pp

Bonafede, M, Dragoni M, Quareni F (1986) Displacement and stress fields produced by a center of dilatation and by a pressure source in a viscoelastic half-space: application to the study of ground deformation and seismic activity at Campi Flegrei, Italy. Geophys JR Astr Soc 87:455–485

Carslaw, HS, Jaeger JC (1980) Conduction of heat in solids, 2nd edn. Oxford Univ Press, Oxford

Castellano, M, Del Pezzo E, De Natale G, Zollo A (1984) Seismic coda Q and turbidity coefficient at the Phlegraeán Fields volcanic area: preliminary results. Bull Volcanol 47:219–224

Correig AM, Mitchell BJ, Ortiz R (1990) Seismicity and coda Q values in the eastern Pyrenees: first results from the La Cerdanya seismic network. Pageoph 132:311–329

Correig AM, Mitchell BJ (1989) Attenuative body wave dispersion at La Cerdanya (Eastern Pyrenees). Phys Earth Planet Int 57:304–310

Del Pezzo E, De Natale G, Scarcella G, Zollo A (1985). Q_c of three component seismograms of volcanic microearthquakes at Campi Flegrei volcanic area – southern Italy. Pageoph 123:683–696

De Natale G, Zollo A, Del Gaudio C, Ricciardi GP, Martini M (1984) Error analysis in hypocentral locations at Phlegraean Fields. Bull Volcanol 47:209–218

De Natale G, Iannacconde G, Martini M, Zollo A (1987) Seismic sources and attenuation properties at the Campi Flegrei volcanic area. Pageoph 125:883–917

Fisher RA, Yates F (1970) Statistical tables for biological, agricultural and medical research. Hafner, Darien, Conn, pp 1–146

Guerra I, Lo Bascio A, Luongo G, Nazzaro A (1974) La sorveglianza delle aree vulcaniche e la rete sismica dei Campi Flegrei. Boll Geofis 26:238–252

Jackson DD, Anderson DL (1970) Physical mechanisms of seismic-wave attenuation. Rev Geophys Space Phys 8:1–63

Johnston DH, Toksoz MN, Timur A (1979) Attenuation of seismic waves in dry and saturated rocks: II. Mechanisms. Geophysics 44:691–711

Lee WHK, Stewart SW (1981) Principles and applications of microearthquake networks. Adv Geophys, Academic Press, Lond New York, 293 pp

Liu HP, Anderson DL, Kanamori H (1976) Velocity dispersion due to anelasticity: implications for seismology and mantle composition. Geophys J R Astr Soc 47:41–58

Martini M, Cellini P, Piccardi G, Giannini L (1984) Composition of hydrothermal fluids during the bradyseismic crisis which commenced at Phlegraean Fields in 1982. Bull Volcanol 47:267–273

Meissner R (1986) The continental crust. Int Geophys Ser 34, Academic Press, Lond New York, 426 pp

Mendoza J, Morgan D (1985) BASIC-HYPO: a basic language hypocenter location program user's guide. Standford Univ, Standford, 32 pp
Muñoz ML (1988) Metodologia para el analisis de la sismicidad volcánica en fases eruptivas. Desarrollo de un procesador específico. PhD Thesis. Univ Complutense Madrid, 343 pp
Ortiz R, Garcia A, Gimenez E (1987) Desarrollo de instrumentación geofísica para su operación en condiciones antarticas. Act II Symp Español Estud Antarticos, CSIC, Madrid, pp 197–202
Peng JY, Aki K, Chouet B et al. (1987) Temporal change in coda Q associated with the Round Valley, California, earthquake of November 23, 1984. J Geophys Res 92:3507–3526
Pulli J (1984) Attenuation of coda waves in New England. Bull Seism Soc Am 74:1149–1166
Roeckner SV, Tucker B, King, Hatzfeld D (1982) Estimates of Q in central Asia as a Function of frequency and depth using the coda of locally recorded earthquakes. Bull Seismol Soc Am 72:129–149
Sato H (1977) Energy propagation including scattering effects: single isotropic scattering approximation. J Phys Earth 25:27–41
Weertman J (1970) The creep strength of the earth's mantle. Rev Geophys Space Phys 8:145–168
Winkler K, Nur A (1979) Pore fluids and seismic attenuation in rocks. Geophys Res Lett 6:1–4

Dynamic Parameters of Volcaniclastic Soils and Rocks of Campi Flegrei (Naples, Italy)

F.M. GUADAGNO, C. NUNZIATA, and A. RAPOLLA

Abstract

The Campi Flegrei caldera (Italy), has been the site of intense seismic activity over the past decade. This area is densely populated and includes several industries. For proper use and management of the territory, evaluation of the dynamic properties of near-surface rocks, is necessary.

The most widespread materials are tuffs (volcaniclastic rocks) and pozzolana (volcaniclastic soils). Their dynamic properties were evaluated in the laboratory using the torsional resonant column and ultrasonic pulse methods. They are expressed in terms of compression, shear wave velocities, and material damping. Tests were conducted to evaluate the influence on the dynamic parameters of shearing strain amplitude, confining pressure and, as regards pozzolanas, also of confinement time, stress history and initial unit weight.

The hardening due to hydrothermal processes, very common in this area, was observed to play an important role in the dynamic behaviour of the volcaniclastic soils. In fact, an altered pozzolana behaves like a tuff, although it has a lower shear modulus and failure pressure. Pozzolana soils, on the other hand, behave like sands.

1 Introduction

Campi Flegrei is a highly urbanized volcanic and geothermal area located west of the city of Napoli (Fig. 1). It consists of a 35 000 year old caldera about 12 km in diameter hosting monogenetic pyroclastic volcanoes (Rosi et al. 1983).

Campi Flegrei has periodically been the site of upward and downward ground movements, at least since the Roman period.

Over the last few decades, the area has been affected by well detected and very intense uplift. From 1969 through 1972, a total vertical uplift of 170 cm occurred (Corrado et al. 1977). This movement was accompanied by an increase in fumarolic activity and very low seismic activity. From July 1982 until the end of 1984, Campi Flegrei was subjected to another series of movements with maximum uplift of 160 cm (Berrino et al. 1984). During this period, intensive seismic activity was recorded. This activity was characterized by hundreds of earthquakes with a maximum magnitude of 4 and with focal depths shallower than about 4 km. These earthquakes generally had a rather short duration and were characterized by a 10–20 Hz dominant frequency (De Natale et al. 1988).

As far as engineering and applied geology and geophysics are concerned, interest in this area is directed towards repairing old buildings and planning new urban developments. For urban planning, it is, therefore, crucial to inves-

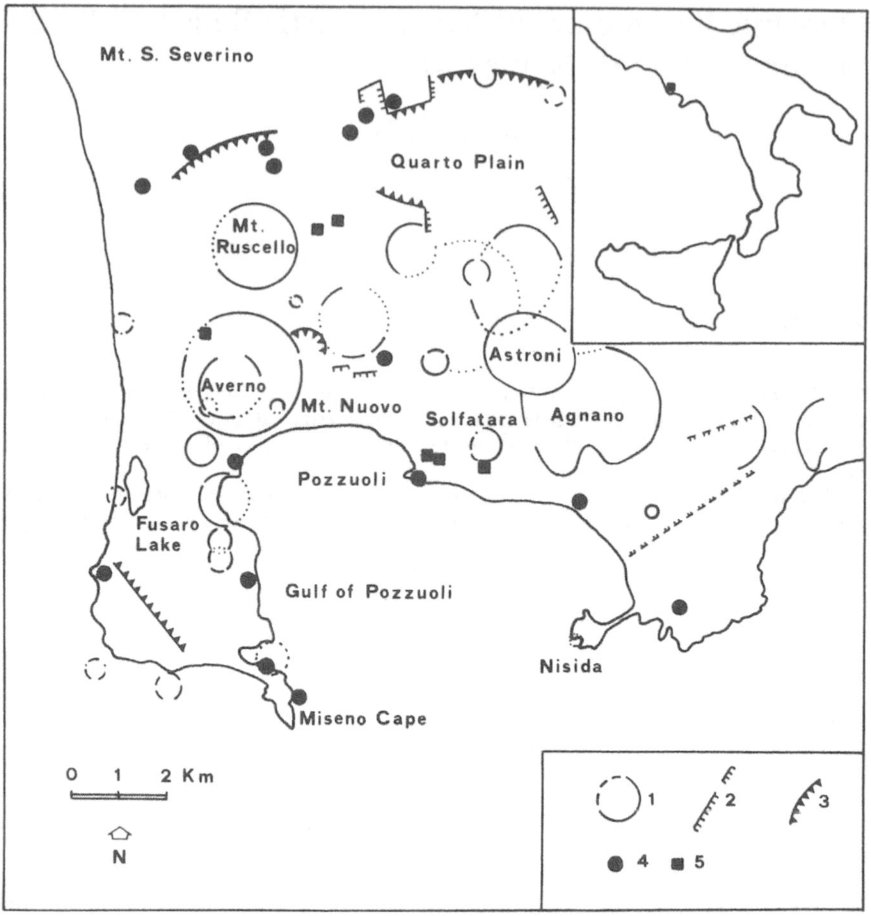

Fig. 1. Volcano tectonic sketch map of Campi Flegrei. *1* Crater; *2* fault; *3* caldera rim; *4* tuff sampling site; *5* volcaniclastic soil sampling site

tigate the area to gather information about the seismic site response, the lithostratigraphic situation near the surface and the dynamic properties of the local soils. This area is, geologically, most complex because of the interrelations between volcanic activity and geomorphological variation.

As a preliminary approach aimed at characterizing the response of these soils to earthquake shaking, laboratory measurements of dynamic properties are reported herein. Pyroclastic materials, and in particular ash (locally named pozzolana), which are the most widespread, are particularly interesting because they are primarily composed of volcanic glass and because diagenetic phenomena can have a large influence on their behaviour.

Hereinafter, we present results relative to samples typical of local volcaniclastic material.

2 Methodologies

The dynamic properties of the volcaniclastic soils and rocks were evaluated principally using resonant column (RC) and ultrasonic pulse (UP) laboratory methods.

The resonant column was a "fixed free" type with which solid cylindrical specimens were forced to vibrate in torsional motion to determine shear modulus G, and material damping ratio D (Isenhower 1979). Testing was conducted to evaluate the influence on G and D of isotropic confining pressure, shearing strain amplitude, confinement time, stress history and initial unit weight (void ratio) (Guadagno et al. 1988).

Ultrasonic pulse testing was carried out with 54 kHz transducers in tight contact with rock materials. As regards soil materials, measurements were performed on specimens confined in a one-dimensional consolidation device by using transducers in the top and base caps (Carrara et al. 1986). Measurements were performed at various stages during a normal consolidation test. By measuring wave travel times and knowing sample geometrical parameters, compression and shear wave velocities and thus shear moduli were determined. UP measurements were conducted also to investigate the influence of high vertical pressure on the shear modulus of volcaniclastic materials.

2.1 Testing Materials

The most interesting and widespread materials in Campi Flegrei are the tuffs (volcaniclastic rocks) and the pozzolana (volcaniclastic soils) that generally cover the tuffs. The thickness of the soil deposits vary from a few meters to more than 60 m. The original material forming the tuffs and the volcanic soils is generally the same, with the volcaniclastic rocks being subjected to hardening by hydrothermal post-depositional phenomena.

As regards tuffs, the Yellow Neapolitan Tuff (YNT) and the Yellow Tuff (YT) (Di Girolamo et al. 1984) were both tested. Block samples of tuff were taken from quarries and drillings (Fig. 1). The dry unit weights of YNT and YT samples varied from 11 and 13.7 kN/m^3 and from 10.2 to 12.8 kN/m^3, respectively. The water content was generally about 8%.

Undisturbed block samples of pozzolana were taken from hand-excavated pits. Although they resulted from different volcanic formations, the lithological and the grain-size characteristics are practically the same. These samples were not altered and were in primary sedimentation condition. The natural unit weights vary from 11.4 to 13.5 kN/m^3, the water contents being between 23 and 45%. Based on grain size, all samples classify as sandy silts (ML). Generally, the specific gravity of these soils varies from 2.2 to 2.5. Undisturbed specimens, carved from the block samples, were tested in their natural condition.

Disturbed samples of volcaniclastic soil material, obtained by drilling in various outcrops, were tested too. Testing was conducted on reconstituted samples, by the Ladd method (1978), with maximum and minimum dry unit weights. The petrographic characteristics are very close to those determined for the undisturbed samples, but the grain size can be somewhat larger, with two samples classifying as well-graded sand. Disturbed samples of volcaniclastic material, which had undergone weathering and rill-wash processes, were also tested. Such material is mainly composed of scoriaceus and pumiceous rounded elements and generally classified as poorly graded sand.

Finally, an undisturbed sample, weakly welded as a result of hydrothermal processes, was tested to evaluate the influence of hardening on the dynamic parameters of the volcaniclastic materials.

3 Results

Dynamic parameters of soils depend upon several factors which include: mean principal effective pressure (σ_o), duration of confinement, void ratio (e), stress history, shearing strain amplitude (γ) and strain rate. However, the most important factor to consider for earthquake excitation is the shearing strain amplitude. Below a low-strain amplitude (generally a threshold of about 0.001% is assumed), soils exhibit linear behaviour. At these small strains, the shear modulus and material damping assume the maximum (Gmax) and minimum value (Dmin), respectively. At strains greater than the threshold, dynamic soil properties exhibit non-linear characteristics.

Dynamic parameters of rocks, on the other hand, are not generally influenced, or scarcely influenced, by the above mentioned factors.

The results of laboratory measurements on volcaniclastic rock and soil samples are reported, and a comparison with in situ measurements is also presented.

3.1 Volcaniclastic Rocks

Resonant column tests on three YNT samples show that maximum shear modulus (Gmax) and minimum damping ratio (Dmin) are scarcely influenced by mean confining pressure, at least in the considered range (Fig. 2). In particular, Gmax and Dmin range between 764 and 875 MPa and 0.81 and 1.11%, respectively. UP measurements on all tuff samples showed similar behaviour as regards the vertical pressure. Compression wave velocities of YNT and YT samples varied between 1200–1750 m/s and 1000–1600 m/s, respectively. Shear wave velocities from UP and RC tests varied between 650–1000 m/s for YNT samples and 550–800 m/s for YT samples. A plot (Fig. 3) of all P- and S-wave velocities versus dry unit weights does not show any significant variation. In particular, Vp values seem to be in very good agreement with the Nafe-Drake relationship (1963) for sedimentary rocks in the considered range.

RC tests were conducted to evaluate the influence of shearing strain amplitude, too. Shear modulus and damping ratio are practically constant, changing slightly at the maximum strain amplitudes reached during the tests (Fig. 4).

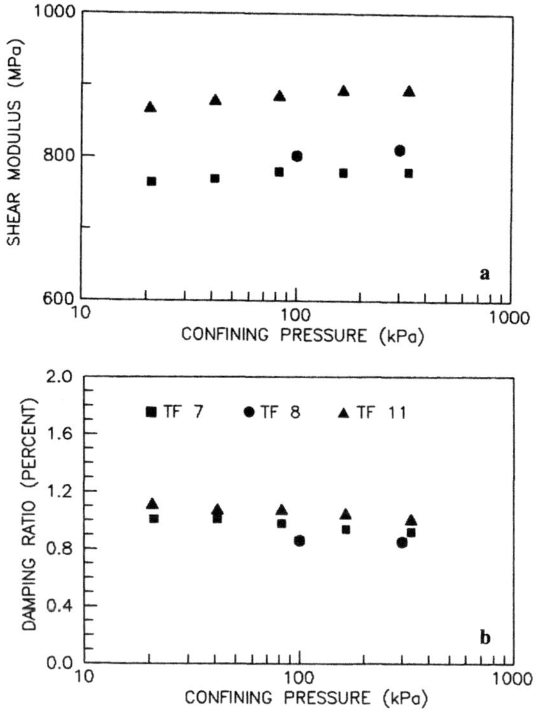

Fig. 2a,b. Variation in maximum shear modulus and minimum damping ratio with mean effective confining pressure for tuff, from RC tests

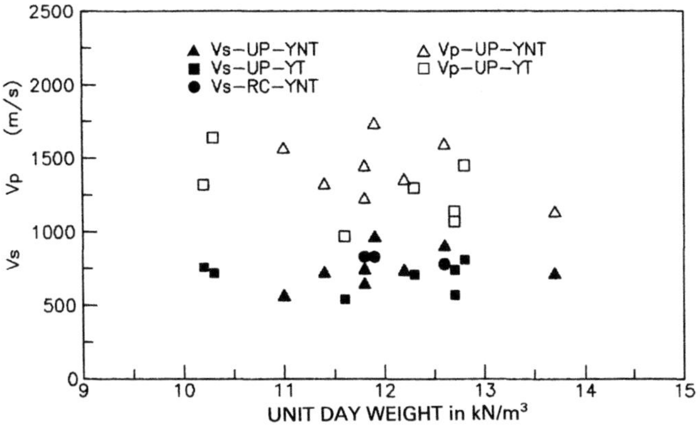

Fig. 3. Compression and shear wave velocities for tuffs from RC and UP tests versus unit dry weight

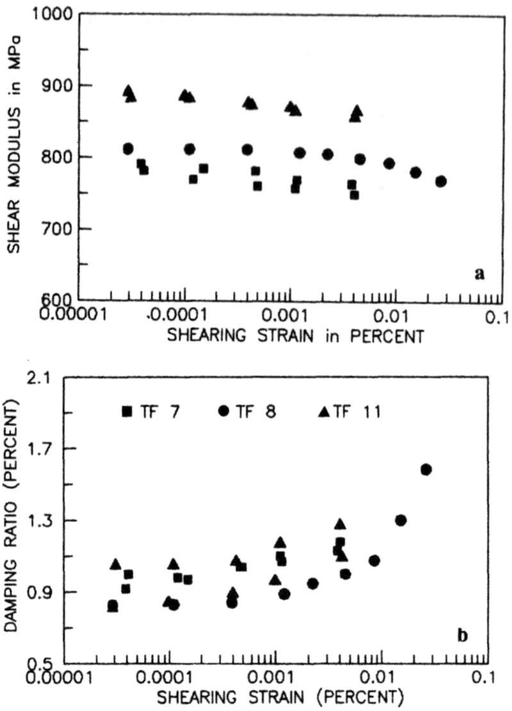

Fig. 4a,b. Variation in shear modulus and damping ratio with the shearing strain amplitude, at different confining pressures, for tuff from resonant column tests

3.2 Volcaniclastic Soils

RC and UP testings were conducted to evaluate the influence on G and D of isotropic confining pressure, initial unit weight (void ratio), confinement time, stress history and shearing strain amplitude. Since the shearing strain amplitude is the most important factor in the evaluation of dynamic parameters of soils, the results are presented separately for low and large strains.

3.3 Dynamic Properties at Low Strain Amplitudes

The results of RC tests performed on undisturbed pozzolana specimens are presented in Fig. 5. The influence of mean effective stress on Gmax and Dmin is clearly shown. It is evident that Gmax increases and Dmin decreases with increasing confining pressure. These relationships fall in a narrow range, even though the materials came from different geological sites. Also, the behaviour exhibited in Fig. 5 is typical of sands (Hardin and Drnevich 1972; Hardin 1978). It is interesting to note that reconstituted specimens (sample 4) with maximum

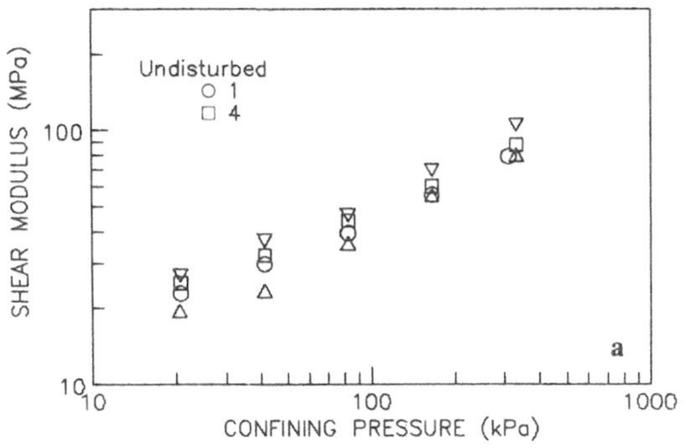

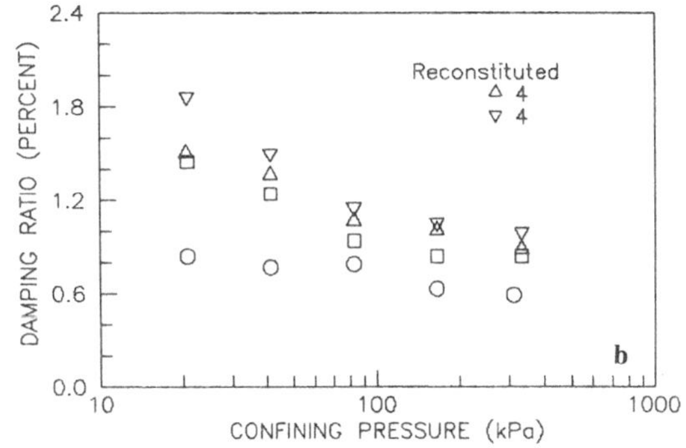

Fig. 5a,b. Variation in maximum shear modulus and minimum damping ratio with mean effective confining pressure for undisturbed and reconstituted pozzolana samples, from RC tests

(e = 1.071) and minimum (e = 1.201) dry unit weights nicely bracket the moduli for the undisturbed specimens, but exhibit slightly large damping ratios.

The influence of mean effective stress on the dynamic properties of reconstituted specimens is presented in Fig. 6. Two general types of materials were studied: unweathered pozzolana and weathered scoria soil material. As shown in Fig. 6, Gmax values are higher and Dmin values are lower for the weathered pyroclastics than for the unweathered ones. This seems to be due to the different petrographic characteristics, the weathered material being in fact richer in scoriaceous elements.

The empirical relationship recommended by Hardin (1978) relating shear modulus and mean effective stress for cohesionless soils, for the studied undisturbed specimens results as:

$$\mathrm{Gmax} = 143/F(e)(Pa)^{1-0.4}(\sigma_0)^{0.4}$$

The factor 143 is the material stiffness coefficient (S), F(e) is a function of void ratio and is equal to $0.3 + 0.7\,e^2$, Pa is the atmospheric pressure (kPa) and 0.4 is the slope of the $\log \mathrm{Gmax}$-$\log \sigma_0$ relationship (n). These values are within the range of values typical of sands, but are lower than the average. For the

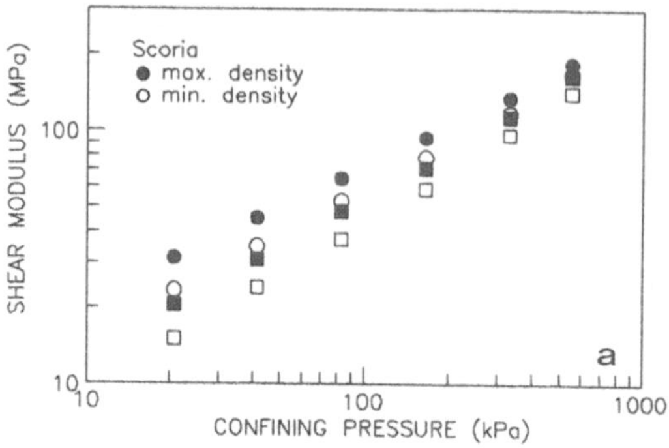

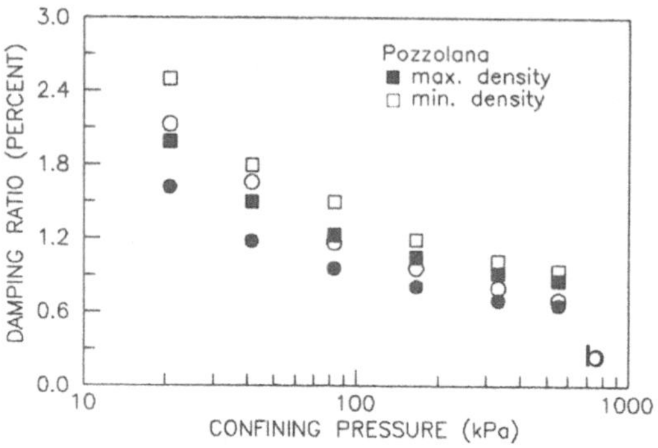

Fig. 6a,b. Variation in maximum shear modulus and minimum damping ratio with mean effective confining pressure for reconstituted scoria and pozzolana samples, from RC tests

reconstituted specimens, S = 130 and n = 0.67. The relative change in values of the coefficients follows the pattern expected for disturbance.

The results from UP tests carried out in the consolidation tests are presented in Fig. 7. In these tests, the mean effective stress has been evaluated from the vertical effective stress and a coefficient of earth pressure at rest, Ko (equal to 1 − sin φ′ with φ′ between 34 and 38°) (Jaky 1948).

The results agree reasonably well with those obtained from resonant column tests in the overlapping pressure range. However, at higher pressures, the positive gradient of Gmax with σ_0 is reduced, especially above 1000 kPa. This effect can be attributed not only to the variation in particle contacts but also to rupture of the fragile pumiceous particles.

From RC testings performed on reconstituted pozzolana samples with different grain-size characteristics, the slope of log Gmax-log σ_0 relationship turns out to be practically the same for the samples classified as well graded sand and poorly graded sand.

To evaluate the influence of the stress history on the dynamic parameters, a test was carried out on pozzolana specimens reconstituted to the minimum unit weight. In particular, after every load level (20.7, 41.4, 82.7, 16.7, 331, 551.6 kPa), the specimens were unloaded to the initial pressure (20.7 kPa). The

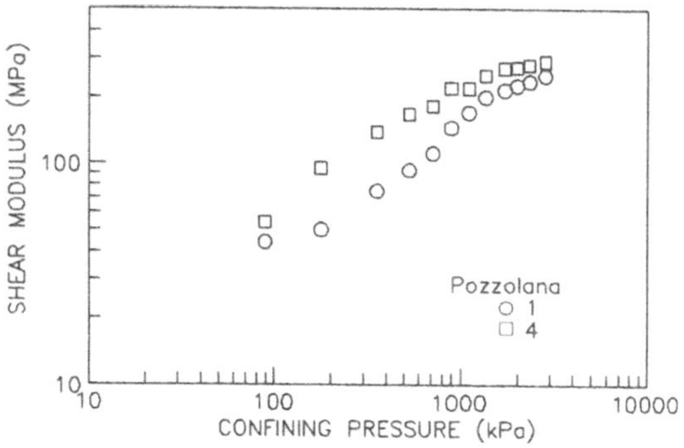

Fig. 7. Variation in maximum shear modulus with mean effective confining pressure for pozzolana samples, from UP tests

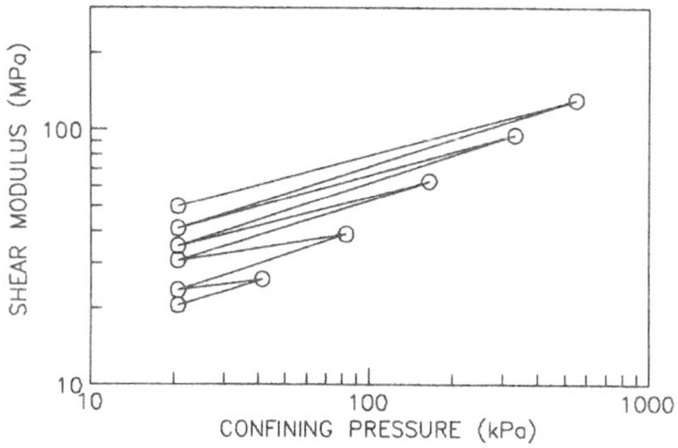

Fig. 8. Influence of stress history on maximum shear modulus for reconstituted pozzolana samples from resonant column tests

large effect of the overconsolidation, that induced a 500% Gmax increase, in connection with increased values of the OCR is shown in Fig. 8.

The influence of the duration of confinement time on the dynamic parameters has also been studied. Gmax is observed to increase at a slow rate. The Ng and Nd parameters (Anderson and Stokoe 1978) are generally less than 5%. These values are typical of sandy materials.

A preliminary study of the effect of welding by alteration on the dynamic moduli was carried out. The relationship between Gmax and σ_0, from resonant column tests, for a partially welded pozzolana sample is shown in Fig. 9. The influence of particle welding seems to be clear: the dynamic behaviour of welded pozzolana is like that of tuffs, being quite independent of pressure, except that Gmax values are lower. Moreover, the unconfined failure pressure of the welded pozzolana was less than 500 kPa, whereas typical values for tuffs are generally higher than 3000 kPa.

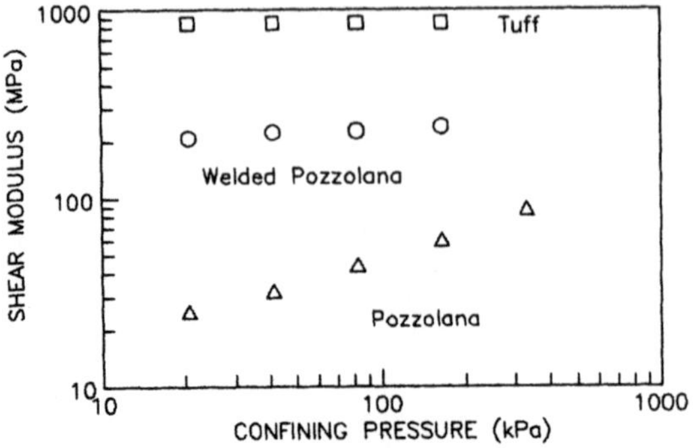

Fig. 9. Variations in maximum shear modulus with confining pressure of various pyroclastic materials

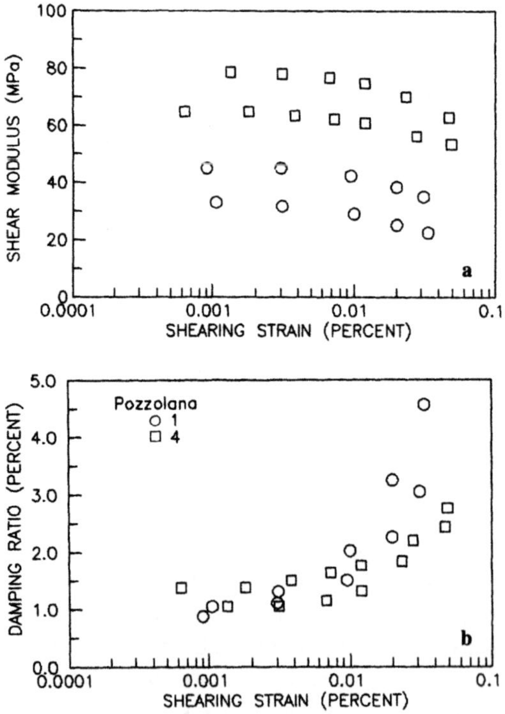

Fig. 10a,b. Variation in shear modulus and damping ratio with shearing strain for undisturbed pozzolana samples

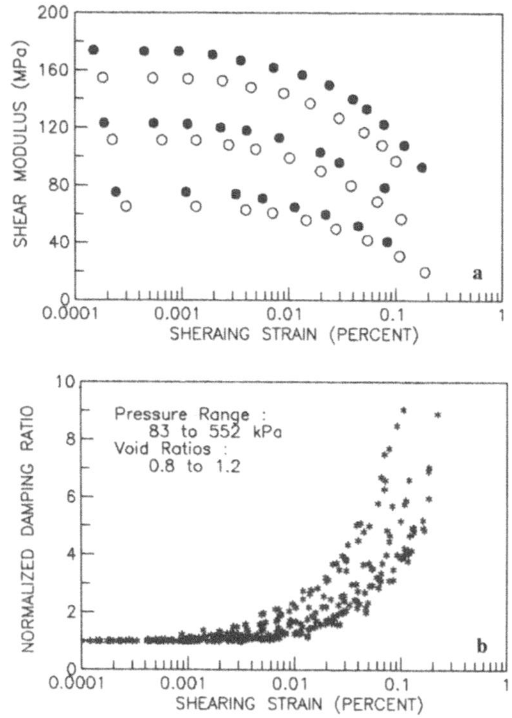

Fig. 11a,b. Variation in shear modulus and damping ratio with shearing strain for minimum and maximum density pozzolana samples

3.4 Dynamic Moduli at High Shearing Strain Amplitudes

The RC method was used to evaluate G and D of the pyroclastic soils at strains in the range where material behaviour becomes quite non-linear. The variation in these properties with strain for undisturbed pozzolana specimens is shown in Fig. 10, for confining pressures ranging from 42 to 330 kPa. The decrease in G and the increase in D at strains above about 0.002% is clearly shown. This influence of strain on dynamic properties is typical of soils in general. The 0.002% threshold can be considered as typical of sands.

The influence of strain amplitude on the dynamic behaviour of reconstituted specimens of pozzolana is shown in Fig. 11 for confining pressures ranging from 83 to 330 kPa. This behaviour is essentially the same as that of the undisturbed specimens. This comparison is more easily made by plotting the normalized shear modulus, G/Gmax, and the normalized damping ratio, D/Dmin, versus strain as in Fig. 12. The variations in normalized moduli and damping ratio with strain, fall into reasonably narrow ranges. The primary factor causing scattering, especially at strains above 0.01%, is the large variation in confining pressure. These results are typical of sands.

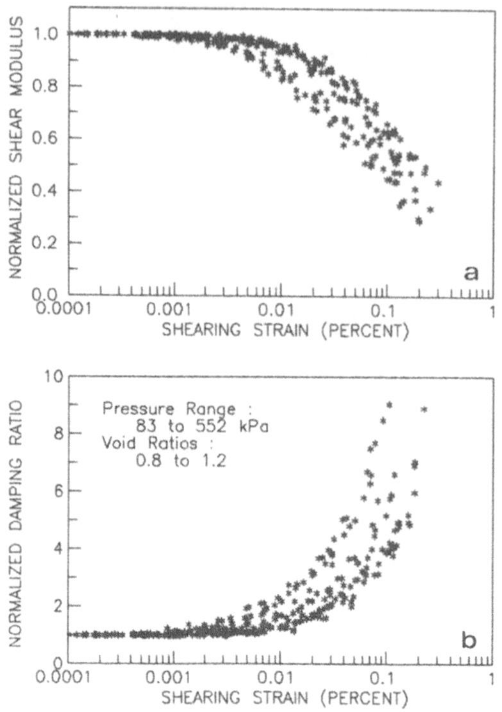

Fig. 12a,b. Variation in G/Gmax and D/Dmin ratios with shearing strain for all the examined pozzolana samples

4 Conclusions

Measurements of the dynamic parameters of Campi Flegrei volcaniclastic materials have been conducted. This study is of importance for the evaluation of the site seismic response, and for quantitative modelling of ground deformation and seismic activity.

The original material forming the tuffs and the volcanic soils are the same, with the volcaniclastic rocks being later subjected to hardening by hydrothermal post-depositional phenomena.

Laboratory (resonant column and ultrasonic pulse) tests show that the dynamic behaviour of the tuffs is typical of lithoid rocks. Moreover, compression and shear wave velocities do not seem to change with dry unit weight; this is in agreement with the Nafe and Drake (1963) relationship for sedimentary rocks. The hardening due to hydrothermal phenomena plays an important role in the dynamic behaviour of Campi Flegrei materials. In fact, an RC test shows that a partially welded pozzolana behaves like a tuff, even though shear modulus and failure pressure are lower.

The shear modulus and material damping of undisturbed and reconstituted volcaniclastic soils from Campi Flegrei result in being dependent upon mean

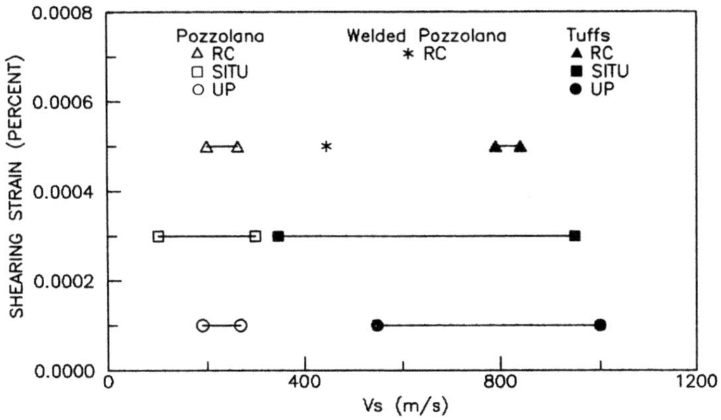

Fig. 13. Shear wave velocities of tuffs and volcanoclastic soils from laboratory tests (RC and UP) and in situ measurements

effective stress, void ratio and shearing strain amplitude in a manner typical of sands. At small strains (below 0.002%) Gmax increases and Dmin decreases with increasing σ_o and the log Gmax-logσ_o relationship is linear. However, at higher confining stresses (more than several hundred kPa), this relationship becomes non-linear because of particle crushing. Confinement time has little effect on dynamic properties, while stress history can have important effects.

In situ measurements, both by seismic refraction surveys and down-hole tests, have been carried out in the same sampled outcroppings (Carrara et al. 1987). All the measured S-wave velocities relative to the examined volcaniclastic materials, referred at the same 150 kPa pressure, are shown in Fig. 13 at the shearing strain amplitudes corresponding to the different tests and falling in the range of elastic behaviour. In our opinion, the spreading of in situ measurements may be due to the different physical and chemical conditions of the materials and, in particular, to the hardening of the soils and to the presence of fractures in the rock bodies. This results clearly in the overlapping field belonging to the welded pozzolana.

The results of this study suggest the need for a detailed study of the dynamic parameters of the local materials when a seismic response analysis is requested. This conclusion is valid in general, and in particular in a volcanic and geothermal area such as Campi Flegrei.

References

Anderson DG, Stokoe KH II (1978) Shear modulus: a time-dependent soil property. Dyn Geotech Testing, ASTM, SPT 654:66–90

Berrino G, Corrado G, Luongo G, Toro B (1984) Ground deformation and gravity changes accompanying the 1982 Pozzuoli uplift. Bull Volcanol 47-2:187–200

Carrara E, Guadagno FM, Rapolla A et al. (1986) Determinazione dei parametri elastici dinamici in laboratorio su campioni di pozzolane e tufi flegrei. Boll Soc Nat Napoli XCV:211–227

Carrara E, Guadagno FM, Nunziata C, Rapolla A, Roberti N (1987) Dynamic properties of pyroclastic soils and rocks of the Phlegraean Fields. In: The volcaniclastic rocks of Campania (southern Italy). Spec Issue Soc Naz Sc Lett Art Napoli, pp 41–68

Corrado G, Guerra I, Lo Bascio A, Luongo G, Rampoldi R (1977) Inflation and microearthquake activity of Phlegraean Fields, Italy. Bull Volcanol 40-3:169–188

De Natale G, Faccioli E, Zollo A (1988) Scaling of peak ground motion from digital recordings of small earthquakes at Campi Flegrei, southern Italy. Pageoph 126:1

Di Girolamo P, Ghiara MR, Lirer L, Munno R, Rolandi G, Stanzione D. (1984) Vulcanologia e petrologia dei Campi Flegrei, Boll Soc Geol It 103:349–413

Guadagno FM, Rapolla A, Ni SH, Stokoe KH II (1988) Dynamic properties of pyroclastic soils of the Phlegraean Fields, Naples, Italy. Proc 9th World Conf Earth Eng Tokyo, 3, pp 35–40

Hardin BO (1978) The nature of stress-strain behaviour for soils. Proc Geotherm Eng Div Spec Conf Earthq Eng Soil Dyn I ASCE Pasadena, CA, pp 3–90

Hardin BO, Drnevich VP (1972) Shear modulus and damping in soils: design equation and curves. J Soil Mech Foun Div, ASCE, 98:667–692

Isenhower WM (1979) Torsional Simple Shear/resonant Column Properties of San Francisco Bay mud. Geotech Eng Thesis GT80-1, Civil Eng Dpt, Univ Texas at Austin, Tx

Jaky J (1948) Pressure in soils. 2nd ICSMFE 1:103–107

Ladd RS (1978) Preparing test specimens using undercompaction. Geotech Test J GTJODJ 1-1:16–23

Nafe JE, Drake CL (1963) Physical properties of marine sediments. In: Hill MN (ed) The sea. 3 Intersci New York, pp 794–85

Rosi M, Sbrana A, Principe C (1983) The Phlegraean Fields: structural evolution, volcanic history and eruptive mechanism. J Volcanol Geotherm Res 17:273–288

Seed HB, Wong RT, Idriss IM, Tokimatsu K (1984) Moduli and damping factor for dynamic analyses of cohesionless soils. Rep EERC 84–14, Earthq Eng Res Cent, Univ California, Berkeley CA

A Thermo-Mechanical Model of the Inflation and Seismicity of Volcanic Calderas: An Application to the Campi Flegrei System

M. Como and M. Lembo

Abstract

A thermo-mechanical model of calderas, based on the assumption that the rocks behave as no-tension materials, is proposed. The medium surrounding the magma chamber is considered, in compression, as a non-homogeneous elastic material – with elastic moduli depending on the temperature – and incapable of sustaining any tensile stress. This assumption allows us to follow the development of fractures at the ground surface and in the subsoil. The latter can be correlated with the localization of earthquake hypocentres and with the occurrence of eruptions. A finite element analysis of the model is performed. Comparisons between the ground surface deformations, observed at Campi Flegrei during the 1982–1984 crisis, and the deformations predicted by the model are developed. The proposed model seems to give a reasonable picture of many of the complex phenomena – ground uplift, fractures of the earth surface and in the subsoil, seismic activity – that occur in the volcanic area of Campi Flegrei.

1 Introduction

Earth uplifts from great depths in volcanic areas, accompanied by local intensive seismic activity and eventual eruption crises, are usually explained to reflect the seesaw of pressure in shallow magma reservoirs or the occurrence of an intrusive event (Mogi 1958; Walsh and Decker 1971). As magma moves below, it forces the land surface up and down in an earthquake-triggering action. The complex phenomena occurring in the volcanic area of Campi Flegrei (Italy) belong to this context (Barberi et al. 1978; Nunziata and Rapolla 1981; Armienti et al. 1983; Lirer et al. 1987). A still unanswered question is their modelling, many studies having been made to give a thermo-mechanical explanation of the inflation and of the seismicity that occurred, particularly after the last crisis in the years 1982–84 (Armienti et al. 1984; Bianchi et al. 1984; Bianchi et al. 1987; Aster et al. 1988; Como and Lembo 1988).

In this framework, the aim of the present work is to develop a new suitable thermo-mechanical model of calderas, able to represent the gradual fracturing of the rocks under increasing overpressures in the chamber. In the following pages, the main aspects of the proposed model are outlined and, referring to the Campi Flegrei Caldera, we examine the response of the model to changes in some of the parameters involved.

2 The New Thermo-Mechanical Model of the Campi Flegrei Caldera

2.1 Geometry

The system is represented by a cylindrical body with an internal cavity under pressure, fed from below by a vertical conduit. The 7 km radius of the cylinder is equal to the radius of the caldera; the height of the cylinder is equal to its diameter, i.e., 14 km. In the finite element analysis of the model, the nodes on the lateral surface of the cylinder have been assumed to be fixed and the vertical displacements of the nodes on the bottom of the cylinder have been assumed to be zero. These simplifying assumptions take into account the great difference in stiffness between the rocks inside and around the caldera whose rim is in contact with much stiffer materials. Numerical investigations on a more complex model, in which the cylinder is surrounded by stiffer rocks, gave results equivalent to those corresponding to the model with fixed lateral surface.

In a first numerical exploration, following studies of many Authors on the subject (Armienti et al. 1983; Armienti et al. 1984; Bianchi et al. 1984; Bianchi et al. 1987), the cavity is assumed to be a flat ellipsoid, whose major (horizontal) and minor (vertical) semi-axes have lengths of 2 and 1 km, respectively. Two different locations of the chamber have been considered, choosing the depth of the cavity center of 6 km in model 1, while it is 5 km in model 2 and model 3. Correspondingly, the top of the chamber, in the three analyzed models, is at a depth of 5 and 4 km (Fig. 1).

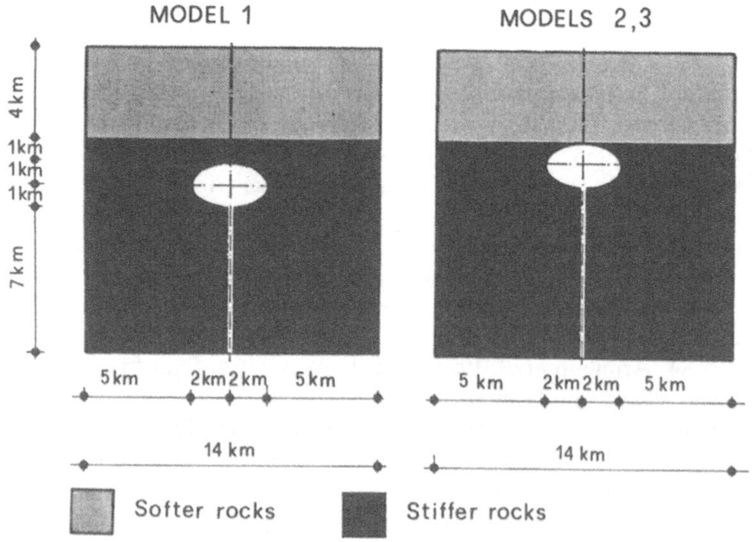

Fig. 1. The assumed geometrical models of the caldera

The whole system is axi-symmetric with a vertical axis of symmetry passing through the caldera center and intersecting the land surface at a point corresponding to the Pozzuoli harbour.

2.2 Geology

Two different rock materials are assumed to represent the subsoil inside the caldera for all three models (Fig. 1). A first stratum, from the Earth's surface to a depth of 4 km, is composed of volcanic rocks (as a pyroclastic cover and tuffs followed by intercalations of lavas and stony tuffites) more fractured and softer in the central part of the caldera. From 4 km depth downwards, the materials are considered to be stiffer and composed of the same tuffites, but compacted by a strong hydrothermal action, or of methamorphic and carbonate rocks (Rosi et al. 1983; Bruni et al. 1985).

2.3 Temperature Field

A stationary thermic field is assumed in the cylindrical body. It represents, with some simplifying approximations, the stationary temperature field due to the heat conduction from the magma chamber and from its feeding conduit to the Earth's surface (Fig. 2). The thermal field is defined by the following values of the temperatures T:

- at the boundary of the magma chamber and of its feeding conduit, T = 1000 °C;
- at the earth surface, T = 0 °C;
- at the deep base of the cylinder, T varies linearly from 1000 °C at the center to 500 °C at the boundary;
- on the lateral surface of the cylinder, along the vertical direction, T varies linearly from 500 °C at the bottom to 0 °C at the surface;
- in the interior of the cylinder, along any line starting from the center of the chamber, T is assumed to vary, according to a decay law similar to the one considered in Bianchi et al. (1984), from 1000 °C, at the boundary of the chamber, to the value present at the intersection of the considered line with the cylinder boundary.

In the assumed thermal field we can recognize the presence, in the central part of the caldera, of a thermal gradient of about 150 °C/km, of the same order as the gradient detected by AGIP (Bruni et al. 1985) in the first 3 km of subsoil in the deep drilling of S. Vito 1, the nearest to the caldera axis.

2.4 Modelling the Mechanical Properties of Rocks

The rock is considered to be a "no-tension" material, i.e., incapable of sustaining tractions. In compression, the material is assumed elastic with varying

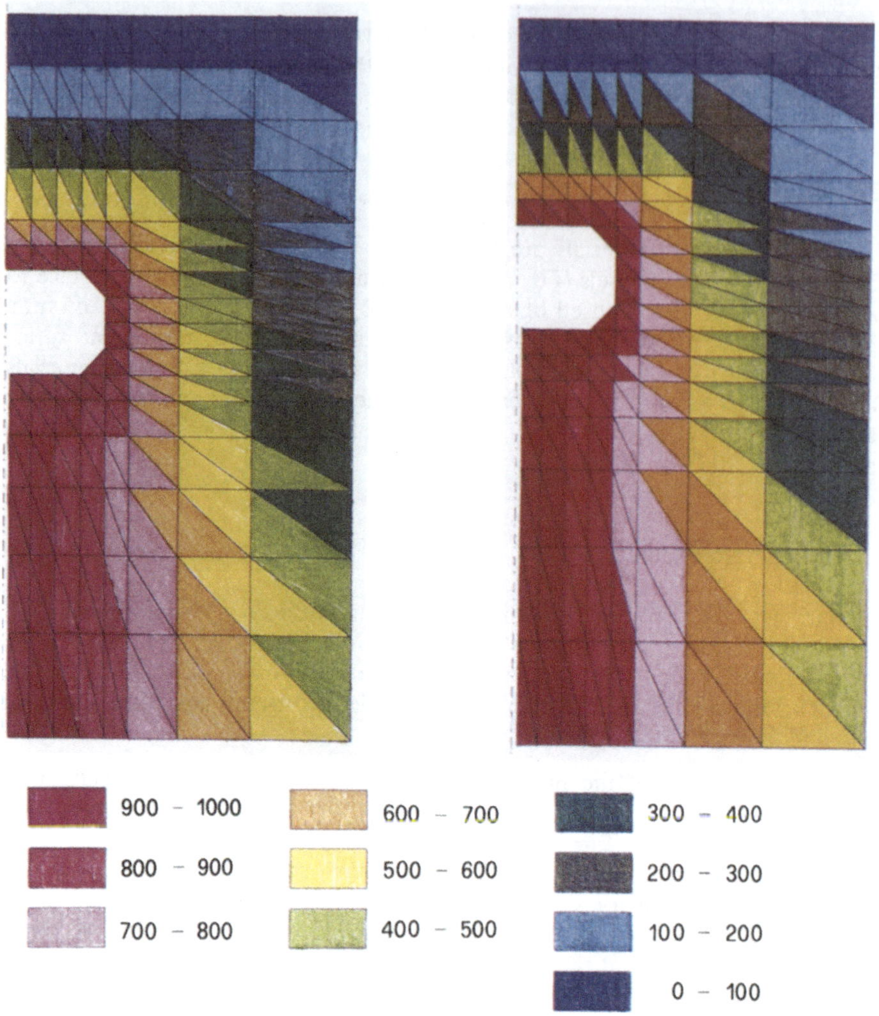

Fig. 2. The assumed temperature (°C) field in the caldera

moduli from point to point because of the variations in the vertical and confining pressures and in temperature. On the other hand, in order to consider the visco-elasticity and the temperature-creep, the values of the elastic moduli have been suitably reduced to respect their "instantaneous" values, which can be determined by means of dynamical tests in situ.

2.5 Assumption of Fracturing Material

The assumption of zero strength in traction is justified by the fact that all the rock materials exhibit a tensile strength considerably lower than the compressional one. This assumption, more particularly satisfied by volcanic rocks as tuffs and tuffites, implies that, at any point of the body, stress fields are only compressive, i.e.:

$$\sigma \leqslant 0, \tag{1}$$

in the sense that the maximum eigenvalue of the stress tensor cannot be positive. According to this assumption, the strain tensor is composed of an elastic (ε_e) and a "fractured" (ε_c) part:

$$\varepsilon = \varepsilon_e + \varepsilon_c. \tag{2}$$

The elastic component ε_e is connected to the stress σ by:

$$\sigma = \mathbb{C}\varepsilon_e, \tag{3}$$

where $\mathbb{C}$ is the elastic tensor, depending on the local value of the temperature. The "fracture" strain ε_c is connected to the stress σ by means of the orthogonality condition (Romano and Romano 1985)

$$\sigma \cdot \varepsilon_c = 0. \tag{4}$$

The no-tension elastic model seems suitable for studying the behaviour of the Campi Flegrei caldera. The initial geostatic stress field, except a thin layer near the surface, is in fact characterized by highly compressive values of the maximum principal stress, of an order of magnitude much larger than the corresponding values of the tensile strengths. Tensile ruptures and fractures, due to the decompression effects in the subsoil as consequence of the chamber overpressure, are thus barely influenced by the fact that average tensile strengths are zero or some hundreds of t/m^2. On the other hand, near the surface the pyroclastic cover is characterized by a very small horizontal confining pressure and vanishing tensile strength.

When the tensile strength of the rock is attained, a local fracture occurs with consequent exchange of energy, from elastic to kinetic. An earthquake is thus produced with the hypocenter located where the fracture occurred. Also in this case, the elastic no-tension model can give useful information on the hypocenter locations if we correlate the fractures produced in the ideal no-tension model with the ruptures that occur in the real rock mass. Then, the assumption of no-tension material allows the evaluation of the gradual fracturing of the rocks under increasing overpressure in the magma chamber.

The model neglects compressive fracturing. In fact: (1) the considered strong overpressures, up to 2×10^4 t/m^2 ($\approx 2 \times 10^2$ MPa) (to which wide tensile fractures, spread all over the subsoil, correspond) produce increments of compressive stresses which, in a small region only around the chamber, are of a magnitude order up to two to three times the initial geostatic stress: (2) all the rocks, and particularly carbonates and metamorphic rocks (Baidyuk 1969) acquire extensive plastic properties and increase their strength at pressure measured in thousands or tens of thousands of t/m^2.

With increasing temperature, as has been shown by tests up to 800 °C (Baidyuk 1969), rocks reduce in strength but further increase in ductility. Consequently, in the neighbourhood of the magma chamber, where the pressure is of an order of at least 10000 t/m^2 and the temperature ranges from 600 °C to

800 °C, the rocks exhibit a marked plastic behaviour with considerable capacity to suffer large deformations without rupture.

In conclusion, one may obtain:

1. The final stress resulting from the superposition of the stress, due to the overpressure in the chamber, to the lithostatic initial stress.
2. The evolution of strains and fractures, at the surface and at depth, and their connection with the global state of the stress in the subsoil.
3. The correlation between the global fracturing of the rocks and the hypocenter locations of the associated earthquakes.
4. The correlation of the evolution of the fracturing with the possibility of an eruption, which can be in fact attained when, under increasing overpressure in the magma chamber, a continuous system of fractures, connecting the chamber with the surface, is formed.

2.6 Rigidity of the Rocks

According to the results obtained by deep drillings, geophysical inspections and tests (Nunziata and Rapolla 1981; Bruni et al. 1985; Aster et al. 1988), the rocks of the Campi Flegrei Caldera, at least for the first 3 km in depth, exhibit uniform rigidity in the vertical direction and increasing rigidity in the radial direction, from the center of the caldera towards the rim. The average values of the S and P wave velocities in fact are $V_s = 1.6\,km/s$, $V_p = 2.9\,km/s$, while at the center and on the rim we have $V_s = 1.4\,km/s$ and $V_s = 1.8\,km/s$, with a value of the Poisson ratio equal to 0.26. The corresponding average "dynamical" values of the Young and rigidity moduli are $E_{dyn} \approx 1.31 \times 10^6\,t/m^2$ and $G_{dyn} \approx 5.2 \times 10^5\,t/m^2$.

The temperatures recorded in the first 3 km of depth – higher at the center than around the boundary of the caldera – together with their influence on the rock stiffness (Baidyuk 1969), cannot explain the detected radial variation of the rock stiffness. Therefore, the radially increasing rigidity of the rock has to be attributed to a different consistency due to the state of fracturing, history, etc. Consequently, in our models, for the first 4 km of depth, a linear variation of the basic value of the rigidity modulus, i.e., the value of the rigidity modulus at 0 °C, has been assumed.

Inflation at Campi Flegrei, during the last crisis, occurred over a period of about 2 years. Accordingly, in the thermo-mechanical model of the Campi Flegrei caldera, creep effects of the rocks have to be considered. Creep in most rocks is of the general visco-elastic kind. Unfortunately, little experimental data are available and, for lack of a better approximation, linear visco-elasticity can be considered. This last assumption can, in fact, be roughly accepted because increases in temperature and hydrostatic pressure, respectively, increase and decrease creep effect (Baidyuk 1969). As a consequence of the assumed linear behaviour, no stress redistribution occurs due to creep under sustained loading. The stress distribution in the subsoil can, thus, be simply computed on the basis of elastic analysis by using long-term moduli.

The long-term moduli of the rocks have been obtained by a suitable reduction of the corresponding "dynamical" values E_{dyn}, G_{dyn}, detected from dynamic tests in situ. According to many results obtained by tests on the visco-elastic behaviour of rocks (Lama and Vutukuri 1978), we assume, for a loading condition acting for a period of about 2 years

$$E_{lt} = 0.5\,E_{dyn}; \quad G_{lt} = 0.5\,G_{dyn}. \tag{5}$$

Thus, we obtain an average value of $G_{lt} \approx 2.6 \times 10^5\,t/m^2$ for the long-term rigidity modulus relative to the first 3 km in depth. In these calculations a suitable linear variation of the elastic moduli with the temperature (Nadai 1950) has been assumed.

No direct information is available for deeper rocks. It seems justified, from a geological point of view, to assume, for great depths, the presence of different and more consistent rocks, such as tuffites compacted by a strong hydrothermal action, or methamorphic and carbonate rocks. At the same time, while the increase in the vertical and confining pressure determines an increase in the rock rigidity, the higher temperature produces a softening action. The actual lack of experimental results and knowledge on the matter cannot allow us to consider this effect with sufficient accuracy. In our models, from 4 km downward, the basic elastic moduli at 0 °C have been assumed to be equal to 2.5 times the corresponding value for the more superficial rocks. The influence of the temperature at depth, on the other hand, strongly affects the distribution of rigidity and produces a considerable softening of the rocks under the magma chamber and, at deeper levels, near the feeding conduit. This heterogeneity of the rock under the magma chamber, however, does not produce relevant effects on the ground deformations at the earth surface as it has been possible to analyze.

In conclusion, the moduli of the rock at a point p, located at radial distance r from the axis of the caldera and at a depth z from the surface, where the temperature is T_p °C, have been determined by:

$$\begin{aligned} E_p &= k_{ET} \cdot k_r \cdot k_z \cdot \bar{E}_{lt} \\ G_p &= k_{GT} \cdot k_r \cdot k_z \cdot \bar{G}_{lt}, \end{aligned} \tag{6}$$

where:

1. $\bar{E}_{lt}$, $\bar{G}_{lt}$ are the average values, in the first 3 km in depth, of long-term Young and rigidity moduli;
2. k_{ET} and k_{GT} are coefficients which reflect the variation of the moduli with the temperature and are defined by:

   $$k_{ET} = (2.68 - 1.68 \times T_p \times 10^{-3})/(2.68 - 0.84 \times T_3 \times 10^{-3})$$
   $$k_{GT} = (2.84 - 1.84 \times T_p \times 10^{-3})/(2.84 - 0.92 \times T_3 \times 10^{-3})$$

   (T_3 is the temperature at the depth of 3 km on the symmetry axis);
3. k_r and k_z are coefficients which reflect the variations of the moduli along the radial and vertical directions and are defined by:

   a. for $0 < z \leqslant 4$ km
      $k_r = 0.75 + 0.5 \times r/3.5$, if $0 < r \leqslant 3.5$ km

$k_r = 1.25$, if $r > 3.5\,\text{km}$
$k_z = 1$

b. for $z > 4\,\text{km}$
$k_r = 1$
$k_z = 2.5$.

The corresponding calculated dynamic moduli G of the models are in agreement with the values resulting from the measured S-wave velocities in the first 3 km in depth.

2.7 Initial Geostatic Stress Field

The "initial" state of stress is represented by the lithostatic stress distribution depending only on the weight of the rock. At the "initial" state the following stress field is assumed:

1. A vertical compressive stress σ_z due to the weight of the rock:

$$\sigma_z = \gamma z; \tag{7}$$

2. Horizontal compressive stresses σ_h given by:

$$\sigma_h = \sigma_z \nu/(1 - \nu) = k_o \sigma_z, \tag{8}$$

where γ is the weight per unit volume and ν is the Poisson ratio.

In the rocks around the chamber or at deeper levels, because of the presence of high temperature, it is assumed $\nu = 0.5$ which implies $\sigma_h = \sigma_z$. At higher levels, the horizontal/vertical stress ratio k_o gradually decreases. To be precise, it has been assumed $k_o = 0.3$ from the Earth's surface down to 4 km in model 1 and down to 3 km in model 2. In model 3 it has been assumed $k_o = 0.5$ for the first 3 km. For greater depths the ratio k_o linearly increases so as to reach the unit value at the top of the magma chamber. This value is then taken to be fixed at deeper levels (Fig. 3). In brief, from the top of the magma chamber to deeper levels an initial hydrostatic state of stress is assumed. In the magma chamber a similar initial state of internal pressure is considered. Subsequent overpressures in the chamber due, for instance, to new magma injections, will produce additional stress and strain and new ground inflation.

2.8 Numerical Solution

From a mechanical point of view, according to all assumptions previously examined, the caldera cylindrical body represents a non-homogeneous and non-linear elastic system that, from an initial geostatic stress state, is subject to gradually increasing overpressures in an ellipsoidal-shaped hole. The assumption of no-tension rock material requires the use of an iterative numerical process. The approach followed, in the framework of finite elements method, was first proposed by Zienkiewicz et al. (1968) and can be described as follows.

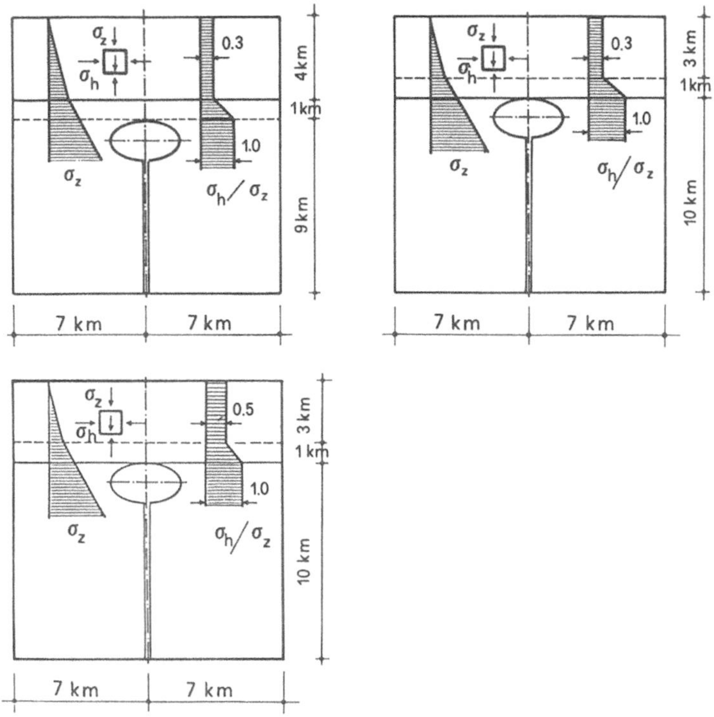

Fig. 3. The assumed variation law for the initial geostatic stresses in the caldera

a. For an assigned value of the overpressure in the chamber, analyze the problem as linear elastic and compute the principal stresses in each toroidal finite element adding the initial stresses.
b. At the end of stage a., we find that certain tensile principal stresses have developed. As the material is assumed to be incapable of sustaining tension, they are eliminated without permitting any joint displacement. In order to maintain equilibrium, "restraining" forces have to be temporarily applied to the body at this stage. In terms of finite element analysis, these forces are evaluated element by element.
c. As the "restraining" forces do not exist, their effect has to be removed from the body by superposition of "equal" but "opposite" nodal forces. The body is now reanalysed for the effect of such forces and the computed stresses are added to those at the end of stage b., when tension has been eliminated. During the application of the "de-restraining" forces the material behaviour is again assumed to be linearly elastic and, when the principal stresses are computed, one may find that tension may still develop. These tensions will, however, be much reduced as compared with their values at the previous stage.
d. If at the end of stage c. principal tensions are still in existence, steps b. and c. are repeated until all tensile stresses are reduced to a negligible figure.

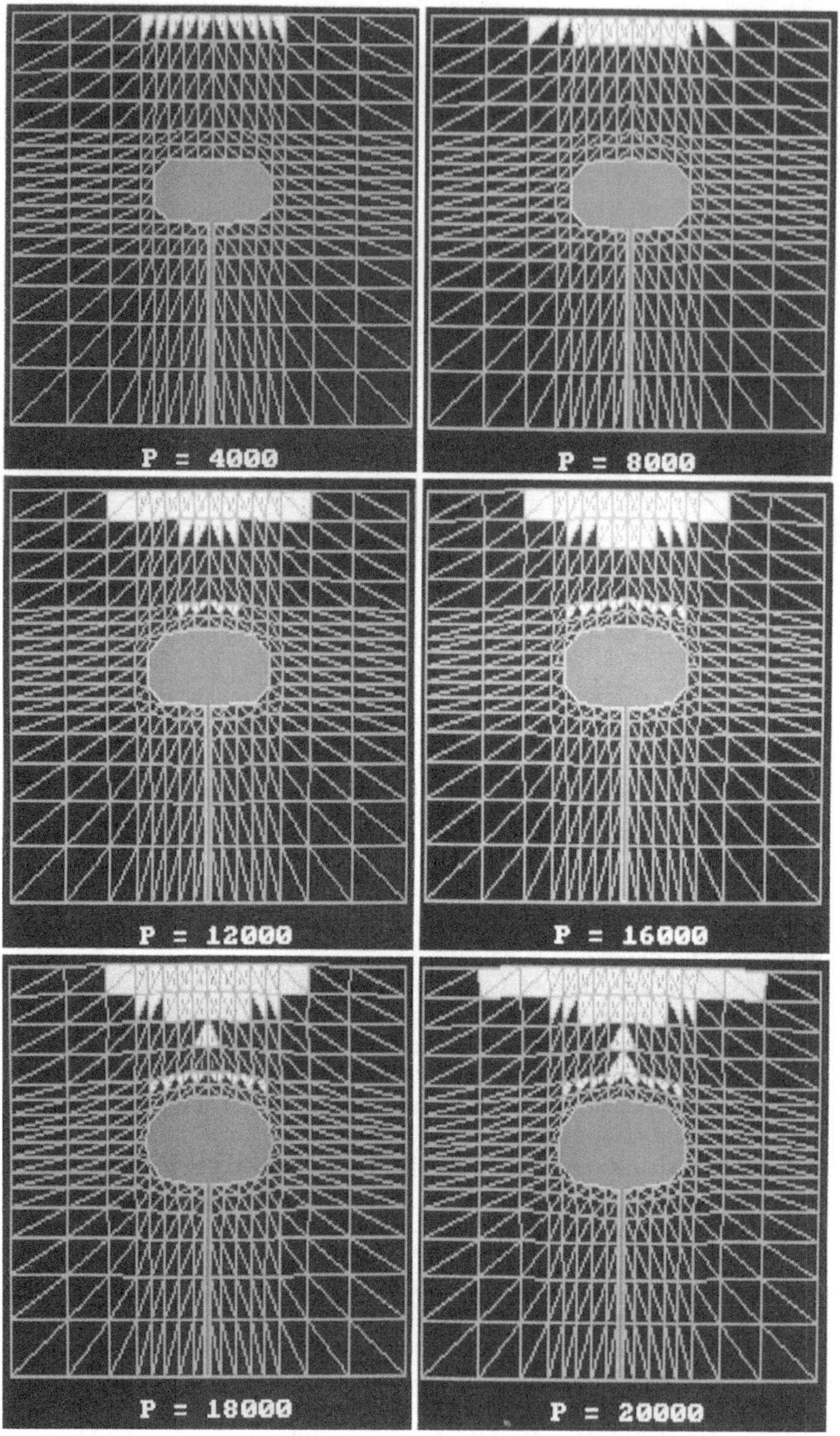

Fig. 4. Fracturing in model 1

3 Response of the Model

First numerical results have been obtained by using an IBM personal computer. Stress and strain evolution in the three assumed models of the caldera has been analyzed by applying subsequent overpressures P in the ellipsoidic cavity, starting from the initial lithostatic state, defined by the initial pressure P_{il}. Then the total pressure in the magma chamber, at the generic stage of the loading, is:

$$P_{tot} = P_{il} + P. \tag{9}$$

Figures 4, 5 and 6 represent the gradual propagation of the fracturing at the Earth's surface and underground for the three considered models of the caldera. The white colouring of an element points out that in this element a circumferential fracture strain is developed, in consequence of an enlargement of the parallel circles. The corresponding fractures will be characterized by diffused or discontinuous cracks along radial direction on vertical planes. The red dash indicates the presence of radial fracture strains and therefore of circular cracks.

Figures 7, 8 and 9 show, for the three models, the incremental evolution of the uplift and of the horizontal displacements at the earth surface. To be precise, they represent the differences w and u between the uplifts W and the horizontal displacements U corresponding to subsequent values of the overpressure in the chamber. Hence we get:

$$w = W(P_{il} + P + \Delta P) - W(P_{il} + P), \tag{10}$$

$$u = U(P_{il} + P + \Delta P) - U(P_{il} + P) \tag{11}$$

where the pressure increments ΔP are $2000\,t/m^2$ for model 1 and $1000\,t/m^2$ for models 2 and 3.

The evaluation of the response of the models, corresponding to two subsequent values of the total pressure in the magma chamber, is useful for comparing the observed strain at the surface (Berrino et al. 1984; Ortolani and Pagliuca 1984). At ground level, in fact, during an inflation, only the strain changes corresponding to an increment of the total pressure in the chamber can be detected.

3.1 Uplifts and Horizontal Displacements at the Surface

Figures 7, 8 and 9 show that the system gradually increases its non-linearity, because of the gradual spreading of the fractures at the surface as well as underground.

Figures 10 and 11 represent the evolution of the maximum uplift and of the maximum horizontal displacement at the surface by increasing overpressures in the chamber. The horizontal displacements are more influenced by the gradual fracturing of the rocks than the vertical ones. Because of non-linear behaviour of the models, the circle at the surface, where the maximum horizontal displacement is attained, gradually shrinks under increasing overpressures.

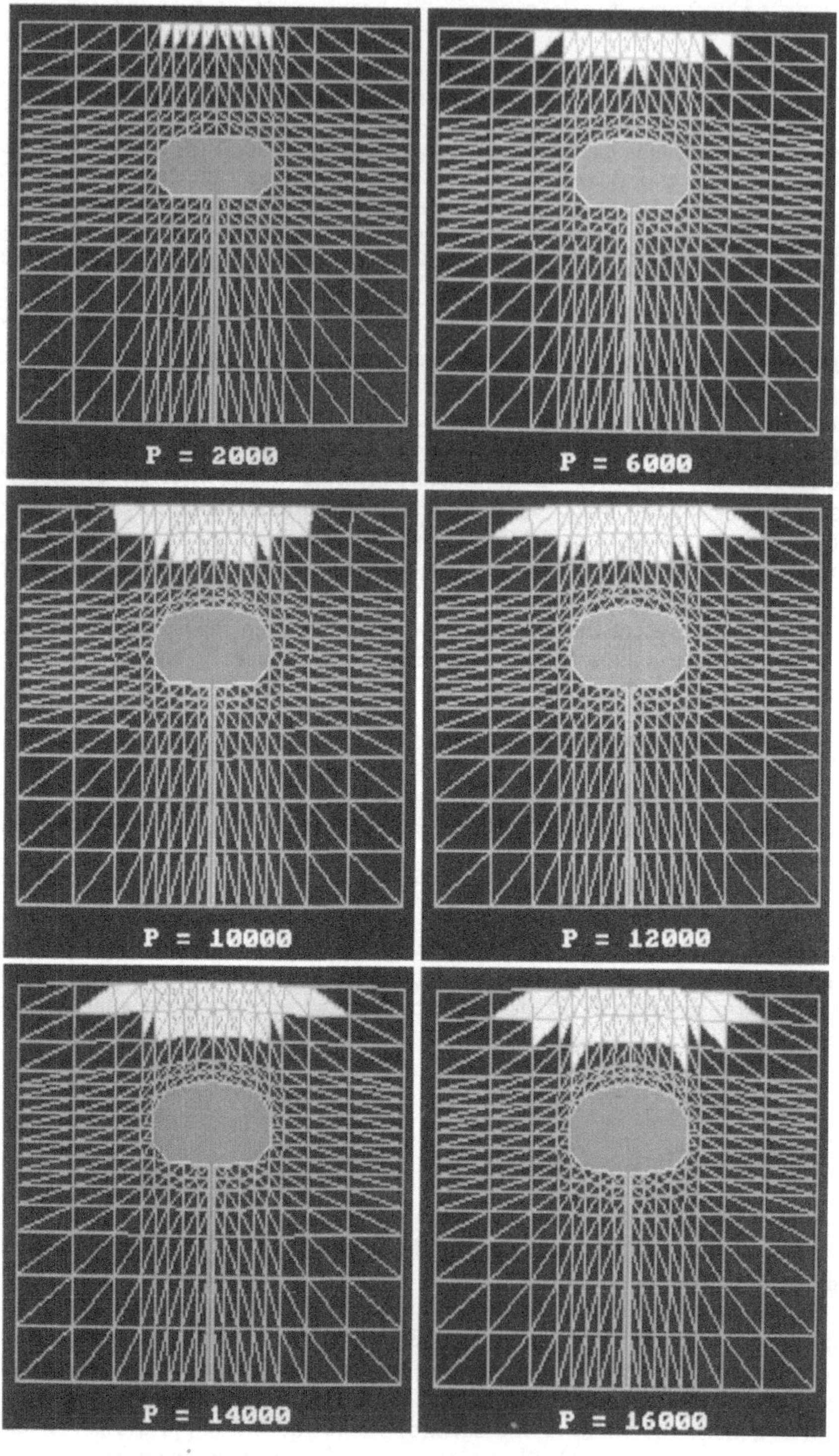

Fig. 5. Fracturing in model 2

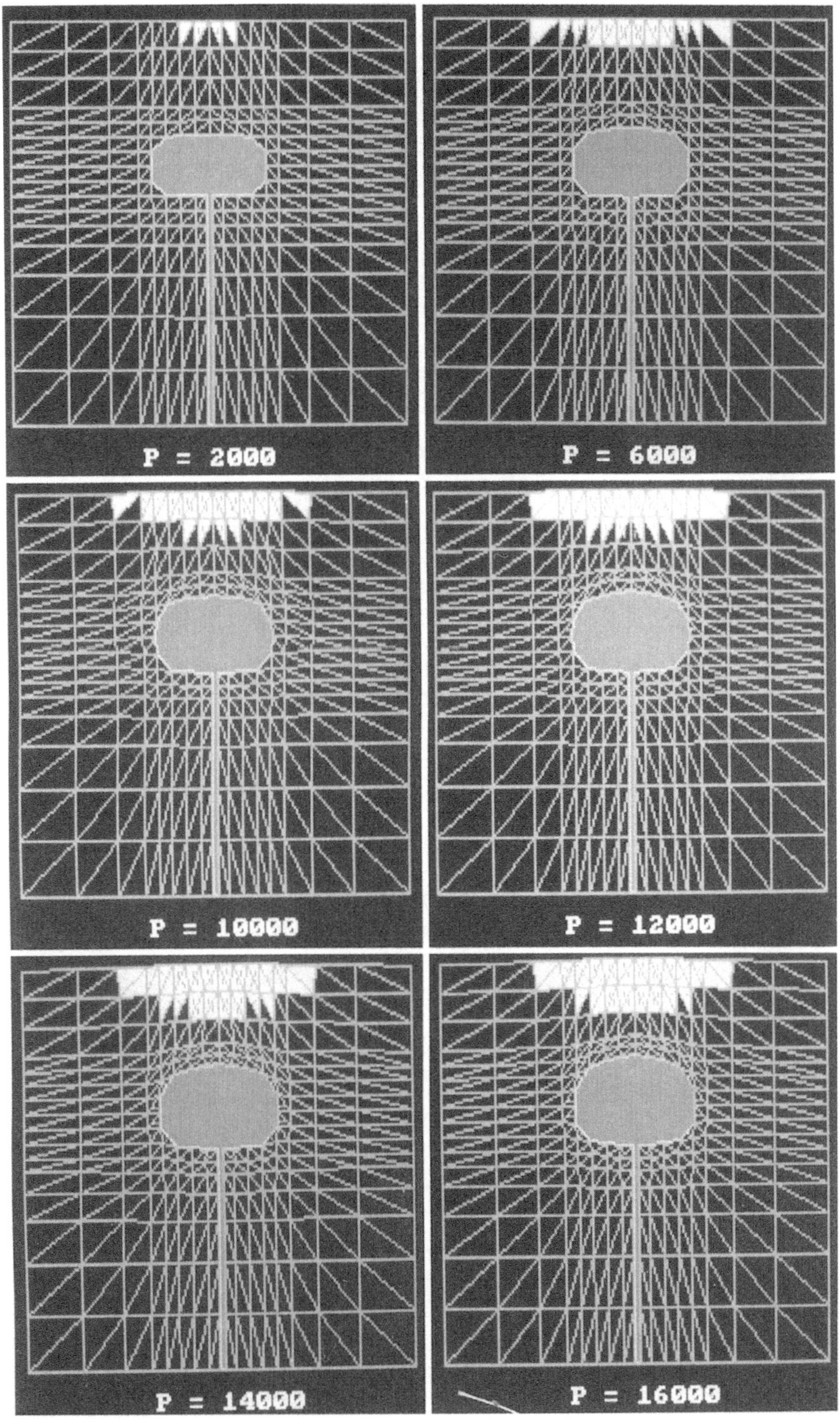

Fig. 6. Fracturing in model 3

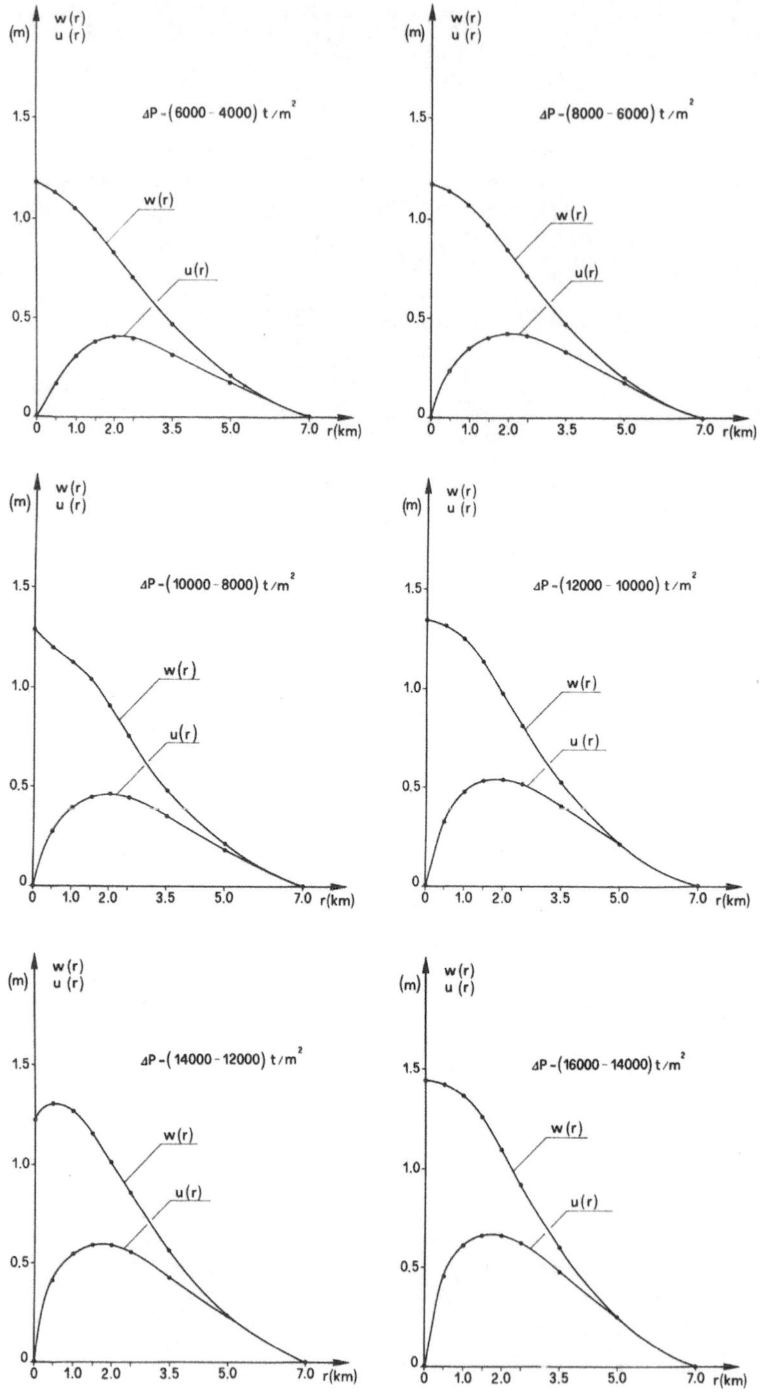

Fig. 7. Incremental displacements in model 1

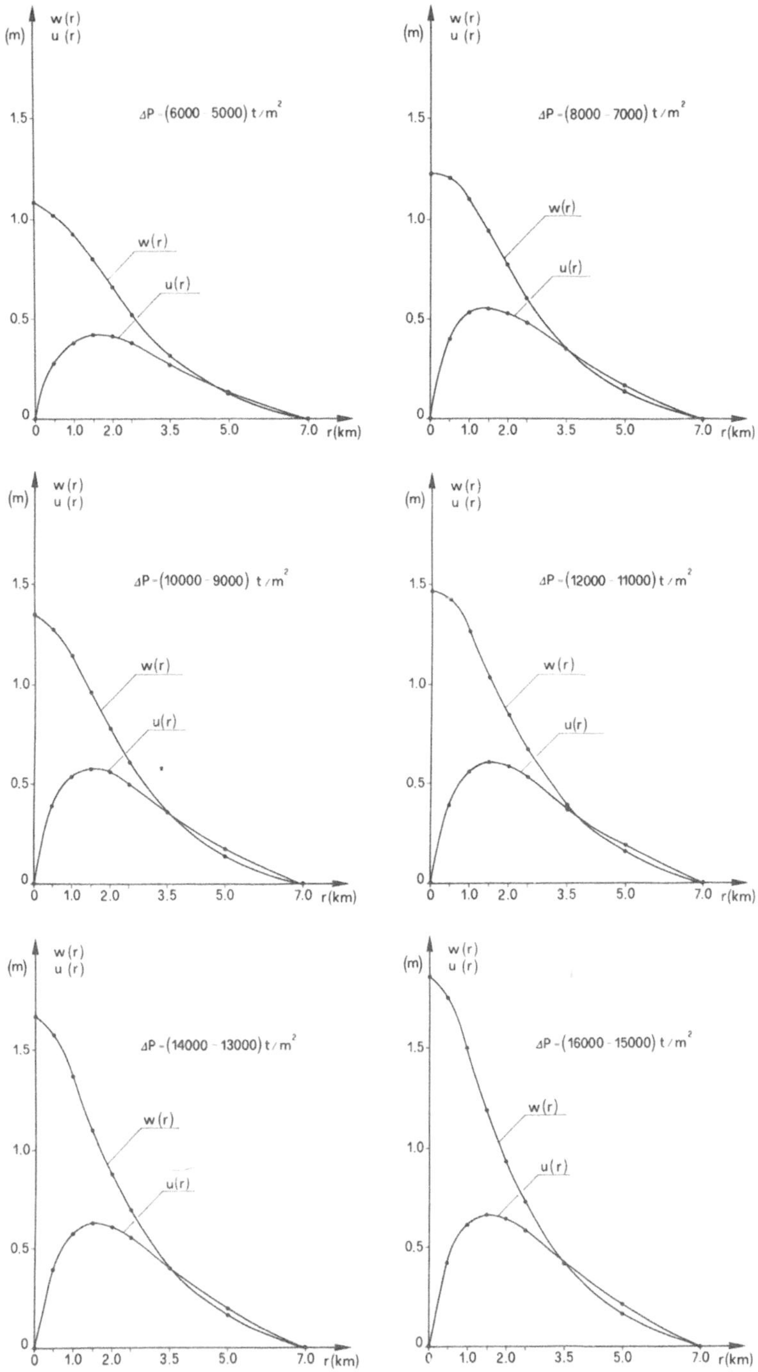

Fig. 8. Incremental displacements in model 2

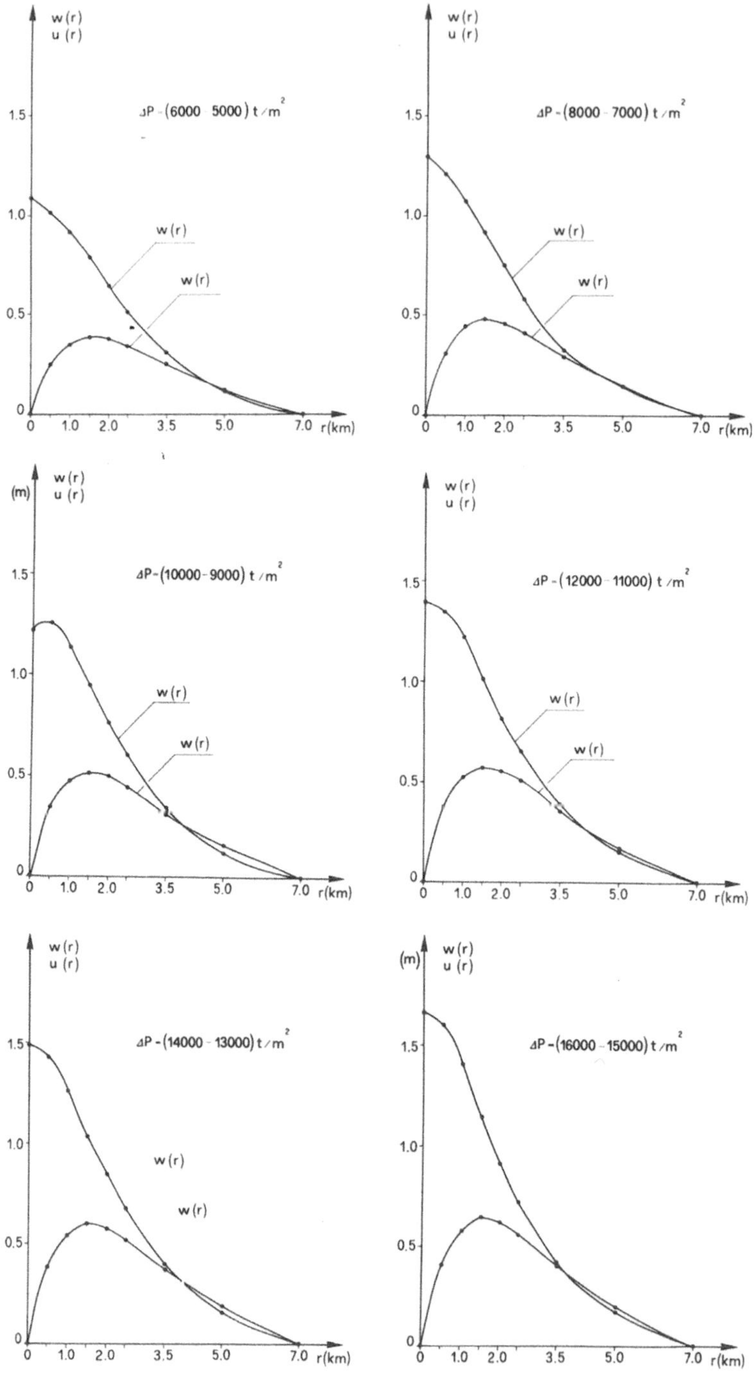

Fig. 9. Incremental displacements in model 3

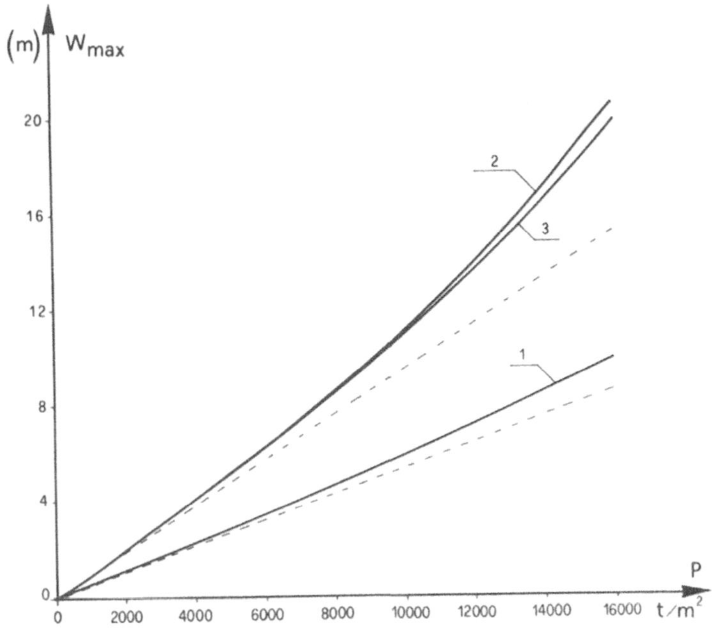

Fig. 10. Maximum uplift for the three models

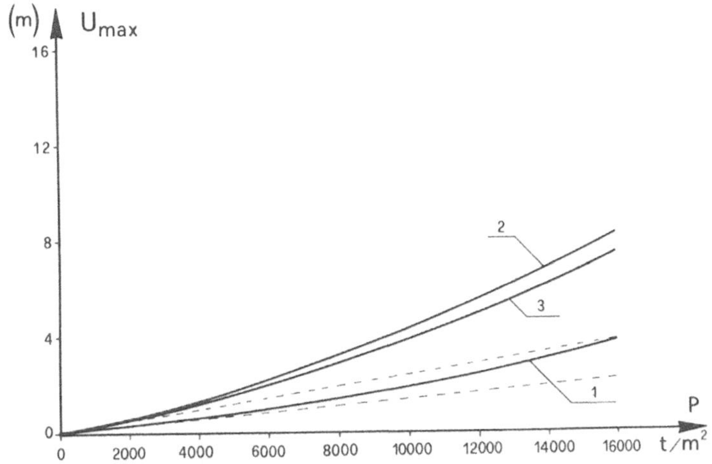

Fig. 11. Maximum horizontal displacements for the three models

3.2 Strains at the Surface

The incremental evolution of the ground deformations at earth surface can be detected by comparing Figs. 4, 5 and 6, where the fractures corresponding to an assigned value of the pressure in the chamber are represented. The comparison of two subsequent pictures furnishes a first estimate of the strain changes at ground level. A more refined evaluation of the strain increments, on the other hand, can be obtained from the values of the horizontal displacements at the surface.

Model 2 and model 1 exhibit, respectively, under an assigned value of the overpressure, the most and the least relevant fracturing at the surface. Models 2 and 3 present the same kind of ground fracturing, except that for model 3 higher values of the overpressures have to be applied.

3.3 Deep Fracturing of the Rocks

The evolution of the fractures at depth – that can compared with the locations of hypocenters of the associated earthquakes – is very different in model 1 and in the two other models.

3.3.1 Model 1

The underground fracturing is characterized by a gradual deepening of the fractures from shallower levels and by the formation of fractures at depth, clustering at the top of the stiffer rocks. Other deep fractures gradually appear just around the chamber (Fig. 4).

3.3.2 Models 2 and 3

These models also exhibit a gradual deepening of the fractures, whereas the formation of fractures at depth, over the chamber, is completely different. They cluster around a cylindrical surface, with a vertical axis, tangent to the chamber. Other deep fractures, as in model 1, gradually appear just around the chamber (Figs. 5, 6).

3.4 Occurrence of Eruptions

In a first approximation, it can be assumed that an eruption state is attained when the development of a continuous system of fractures connects the magma chamber with the surface. Figure 4 shows that model 1 reaches this state by means of a system of fractures occurring in the axial zone of the caldera, whereas in model 2 and model 3 (Figs. 5, 6) the connection occurs over the lateral boundary of the chamber.

3.5 Comparison Between the Detected Uplifts During the Last Inflation at Campi Flegrei and the Predicted Uplifts of the Three Models

In Fig. 12 a comparison is made between the detected uplifts during the last crisis, i.e., in the years 1982–84, at Campi Flegrei, and the predicted ones. The uplift curve predicted by model 1 has been obtained by applying an increment of $2000\,t/m^2$ to the overpressure of $14\,000\,t/m^2$. The curves relative to the models 2 and 3 correspond to an increment of $1000\,t/m^2$ subsequent to overpressures of $12\,000\,t/m^2$ and of $14\,000\,t/m^2$, respectively.

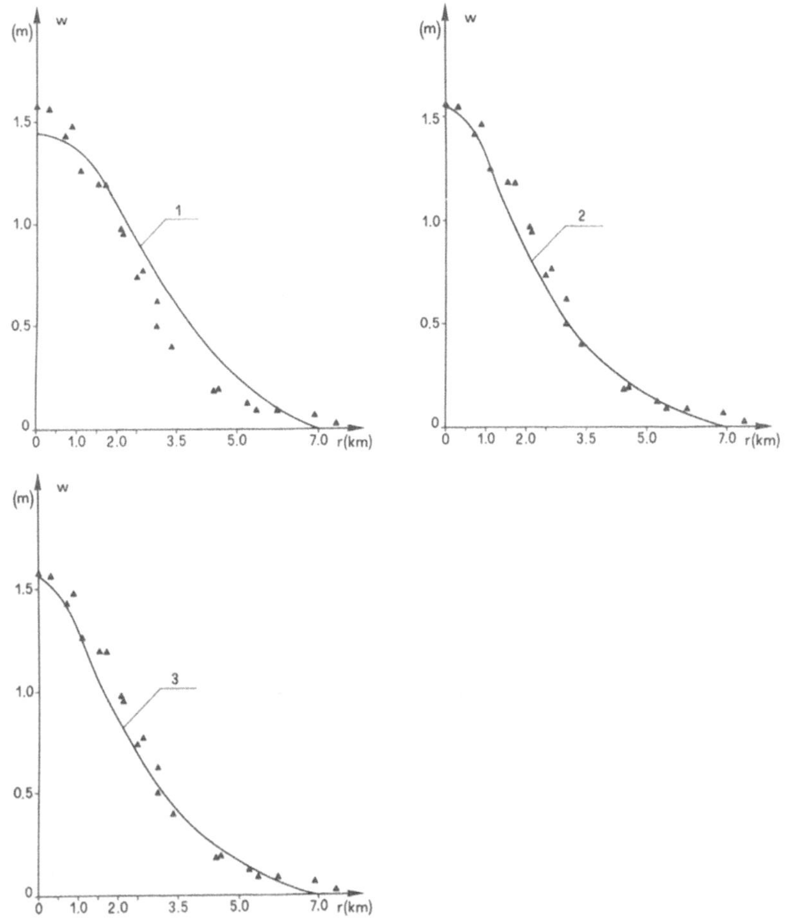

Fig. 12. Comparison between the uplifts measured from 1982 to 1984 and those predicted by the models

4 Conclusions

The proposed thermo-mechanical model of the inflation of volcanic calderas seems to give a reasonable picture of many of the complex phenomena; ground uplift, fractures on the earth surface and at depth, seismic activity; those occurring in the volcanic area of Campi Flegrei.

Increases of pressure in the chamber, together with the spread of fractures, result in inflation increments that well match the axi-symmetric bulge-shaped distribution of detected vertical displacements. In particular, when an increment $\Delta P = 1 \times 10^3\,t/m^2$ is applied to an existing overpressure P, equal to $1.2 \times 10^4\,t/m^2$ in model 2 and to $1.4 \times 10^4\,t/m^2$ in model 3, the vertical inflation increments predicted by the models are in good agreement with the measured data (Fig. 12). At the same time, the magnitude of strain increments at the surface is of the order of 10^{-4}, equal to the that of the detected strains. Moreover, the assumed fracturing behaviour of the rock allows us to obtain a description of the gradual deep fracturing development, which can be related, on the one hand, to the hypocenter localizations and, on the other, to the risk of the occurrence of an eruption. For instance, according to model 2, for the pressure increments ranging from $1.4 \times 10^4\,t/m^2$ to $1.6 \times 10^4\,t/m^2$ (Fig. 5), the seismic activity should be more pronounced along the vertical cylindrical surface tangent to the magma chamber than along its axis, with hypocenters at 3–4 km depth. This result is consistent with the essentially aseismic behaviour of the central part of the caldera (Aster et al., this vol.). The deep fracturing development predicted by this model could suggest, for further pressure increments, the occurrence of an eruption from a vent located on the vertical projection, on the ground surface, of the lateral boundary of the chamber.

References

Armienti P, Barberi F, Bizouard H (1983) The Phlegrean Fields magma evolution within a shallow chamber. J Volcanol Geotherm Res 17:289–311

Armienti P, Barberi F, Innocenti F (1984) A model of the Phlegraean Fields magma chamber in the last 10500 years. Bull Volcanol 44-2:349–358

Aster RC, Meyer RP, De Natale G, Zollo A, Martini M, Del Pezzo E, Scarpa R, Iannuccone G (1992) Seismic Investigation of the Campi Flegrei (this vol.)

Baidyuk BV (1969) Mechanical properties of rocks at high temperatures and pressures. Transl from Russian, Consultants Bureau, New York

Barberi F, Innocenti F, Lirer L, Munno R, Pescatore T (1978) The Campania Igninbrite: a major prehistoric eruption in the Neapolitan area (Italy). Bull Volcanol 41-1:1–22

Berrino G, Corrado G, Luongo G, Toro B (1984) Ground deformation and gravity changes accompanying the 1982 Pozzuoli uplift. Bull Volcanol 44-2:187–199

Bianchi R, Coradini A, Federico C, Giberti G, Sartoris G, Scandone R (1984) Modelling of surface ground deformation in the Phlegrean Fields volcanic area. Bull Volcanol 44-2:321–330

Bianchi R, Coradini A, Federico C (1987) Modelling of surface deformation in volcanic areas: the 1970–72 and 1982–84 crises of Campi Flegrei, Italy. J Geophys Res 92:14139–14150

Bruni P., Chelini W, Sbrana A, Verdiani G (1985) Deep exploration of the S Vito area, Pozzuoli, Naples: Well S. Vitol. Proc 3rd Int Sem Eur Geotherm Update, Munich 29 Nov–1 Dec 1983, EUR 8853 EN, pp 390–406

Como M, Lembo M (1988) Sull'istituzione di un modello termomeccanico del Bradisismo Flegreo. In: Omaggio a Giulio Ceradini. Dipart Ing Strutt Geotec, Univ Roma La Sapienza, Esagrafica, Roma, pp 249–263

Lama RD, Vutukuri VS (1978) Handbook on mechanical properties of the rock. Trans Tech Publ, Clausthal

Lirer L, Luongo G, Scandone R (1987) On the volcanological evolution of Campi Flegrei. Eos Trans AGU 68:226–234

Mogi K (1958) Relations between eruptions of various volcanoes and the deformations of the ground surface around them. Bull Earthq Res Inst Univ Tokio 36:99–134

Nadai A (1950) Theory of flow and fracture of solids. McGraw Hill, New York

Nunziata O, Rapolla A (1981) Interpretation of gravity and magnetic data in the Phlegrean Fields geothermal area, Naples, Italy. J Volcanol Geotherm Res 9:209–230

Ortolani F, Pagliuca S (1984) Le lesioni nel suolo del centro storico di Pozzuoli. . . . Dipart Scie Terr, Univ Napoli

Parascandola A (1983) I fenomeni bradisismici del Serapeo di Pozzuoli, 1947. New printing, Genovese, Napoli.

Romano G, Romano M (1985) Elastostatics of structures with unilateral conditions on strain and displacements. Int Conf Unilateral Probl Struct Anal, Ravello, 1983, CISM, Springer Berlin Heidelserg New York, pp 315–338

Rosi M, Sbrana S, Principe V (1983) The Phlegrean Fields: structural evolution, volcanic history and eruptive mechanism. Volcanol Geotherm Res 17:273–288

Walsh JB, Decker W (1971) Surface deformation associated with volcanism. J Geophys Res 76:3291–3302

Zienkiewicz C, Vallapian S, King IP (1968) Stress analysis of rock as a no-tension material. Geotechnique 18:56–66

Subject Index

GPSR Compliance

The European Union's (EU) General Product Safety Regulation (GPSR) is a set of rules that requires consumer products to be safe and our obligations to ensure this.

If you have any concerns about our products, you can contact us on ProductSafety@springernature.com

In case Publisher is established outside the EU, the EU authorized representative is:

Springer Nature Customer Service Center GmbH
Europaplatz 3
69115 Heidelberg, Germany

Batch number: 10704711

Printed by Printforce, the Netherlands